Lecture Notes in Computer Science 11536

Commenced Publication in 1973
Founding and Former Series Editors:
Gerhard Goos, Juris Hartmanis, and Jan van Leeuwen

More information about this series at http://www.springer.com/series/7407

João M. F. Rodrigues · Pedro J. S. Cardoso ·
Jânio Monteiro · Roberto Lam ·
Valeria V. Krzhizhanovskaya ·
Michael H. Lees · Jack J. Dongarra ·
Peter M. A. Sloot (Eds.)

Computational Science – ICCS 2019

19th International Conference
Faro, Portugal, June 12–14, 2019
Proceedings, Part I

Editors
João M. F. Rodrigues
University of Algarve
Faro, Portugal

Pedro J. S. Cardoso
University of Algarve
Faro, Portugal

Jânio Monteiro
University of Algarve
Faro, Portugal

Roberto Lam
University of Algarve
Faro, Portugal

Valeria V. Krzhizhanovskaya
University of Amsterdam
Amsterdam, The Netherlands

Michael H. Lees
University of Amsterdam
Amsterdam, The Netherlands

Jack J. Dongarra
University of Tennessee at Knoxville
Knoxville, TN, USA

Peter M. A. Sloot
University of Amsterdam
Amsterdam, The Netherlands

ISSN 0302-9743 ISSN 1611-3349 (electronic)
Lecture Notes in Computer Science
ISBN 978-3-030-22733-3 ISBN 978-3-030-22734-0 (eBook)
https://doi.org/10.1007/978-3-030-22734-0

LNCS Sublibrary: SL1 – Theoretical Computer Science and General Issues

This Springer imprint is published by the registered company Springer Nature Switzerland AG
The registered company address is: Gewerbestrasse 11, 6330 Cham, Switzerland

Preface

Welcome to the 19th Annual International Conference on Computational Science (ICCS - https://www.iccs-meeting.org/iccs2019/), held during June 12–14, 2019, in Faro, Algarve, Portugal. Located at the southern end of Portugal, Algarve is a well-known touristic haven. Besides some of the best and most beautiful beaches in the entire world, with fine sand and crystal-clear water, Algarve also offers amazing natural landscapes, a rich folk heritage, and a healthy gastronomy that can be enjoyed throughout the whole year, attracting millions of foreign and national tourists. ICCS 2019 was jointly organized by the University of Algarve, the University of Amsterdam, NTU Singapore, and the University of Tennessee.

The International Conference on Computational Science is an annual conference that brings together researchers and scientists from mathematics and computer science as basic computing disciplines, as well as researchers from various application areas who are pioneering computational methods in sciences such as physics, chemistry, life sciences, engineering, arts, and humanitarian fields, to discuss problems and solutions in the area, to identify new issues, and to shape future directions for research.

Since its inception in 2001, ICCS has attracted an increasingly higher quality and numbers of attendees and papers, and this year was no exception, with over 350 participants. The proceedings series have become a major intellectual resource for computational science researchers, defining and advancing the state of the art in this field.

ICCS 2019 in Faro was the 19th in this series of highly successful conferences. For the previous 18 meetings, see: http://www.iccs-meeting.org/iccs2019/previous-iccs/.

The theme for ICCS 2019 was "Computational Science in the Interconnected World," to highlight the role of computational science in an increasingly interconnected world. This conference was a unique event focusing on recent developments in: scalable scientific algorithms; advanced software tools; computational grids; advanced numerical methods; and novel application areas. These innovative novel models, algorithms, and tools drive new science through efficient application in areas such as physical systems, computational and systems biology, environmental systems, finance, and others.

ICCS is well known for its excellent line-up of keynote speakers. The keynotes for 2019 were:

- Tiziana Di Matteo, King's College London, UK
- Teresa Galvão, University of Porto/INESC TEC, Portugal
- Douglas Kothe, Exascale Computing Project, USA
- James Moore, Imperial College London, UK
- Robert Panoff, The Shodor Education Foundation, USA
- Xiaoxiang Zhu, Technical University of Munich, Germany

This year we had 573 submissions (228 submissions to the main track and 345 to the workshops). In the main track, 65 full papers were accepted (28%); in the workshops, 168 full papers (49%). The high acceptance rate in the workshops is explained by the nature of these thematic sessions, where many experts in a particular field are personally invited by workshop organizers to participate in their sessions.

ICCS relies strongly on the vital contributions of our workshop organizers to attract high-quality papers in many subject areas. We would like to thank all committee members for the main track and workshops for their contribution to ensure a high standard for the accepted papers. We would also like to thank Springer, Elsevier, and Intellegibilis for their support. Finally, we very much appreciate all the local Organizing Committee members for their hard work to prepare this conference.

We are proud to note that ICCS is an A-rank conference in the CORE classification.

June 2019

João M. F. Rodrigues
Pedro J. S. Cardoso
Jânio Monteiro
Roberto Lam
Valeria V. Krzhizhanovskaya
Michael Lees
Jack J. Dongarra
Peter M. A. Sloot

Organization

Workshops and Organizers

Advanced Modelling Techniques for Environmental Sciences – AMES

Jens Weismüller
Dieter Kranzlmüller
Maximilian Hoeb
Jan Schmidt

Advances in High-Performance Computational Earth Sciences: Applications and Frameworks – IHPCES

Takashi Shimokawabe
Kohei Fujita
Dominik Bartuschat

Agent-Based Simulations, Adaptive Algorithms, and Solvers – ABS-AAS

Maciej Paszynski
Quanling Deng
David Pardo
Robert Schaefer
Victor Calo

Applications of Matrix Methods in Artificial Intelligence and Machine Learning – AMAIML

Kourosh Modarresi

Architecture, Languages, Compilation, and Hardware Support for Emerging and Heterogeneous Systems – ALCHEMY

Stéphane Louise
Löic Cudennec
Camille Coti
Vianney Lapotre
José Flich Cardo
Henri-Pierre Charles

Biomedical and Bioinformatics Challenges for Computer Science – BBC

Mario Cannataro
Giuseppe Agapito
Mauro Castelli

Riccardo Dondi
Rodrigo Weber dos Santos
Italo Zoppis

Classifier Learning from Difficult Data – CLD^2

Michał Woźniak
Bartosz Krawczyk
Paweł Ksieniewicz

Computational Finance and Business Intelligence – CFBI

Yong Shi
Yingjie Tian

Computational Methods in Smart Agriculture – CMSA

Andrew Lewis

Computational Optimization, Modelling, and Simulation – COMS

Xin-She Yang
Slawomir Koziel
Leifur Leifsson

Computational Science in IoT and Smart Systems – IoTSS

Vaidy Sunderam

Data-Driven Computational Sciences – DDCS

Craig Douglas

Machine Learning and Data Assimilation for Dynamical Systems – MLDADS

Rossella Arcucci
Boumediene Hamzi
Yi-Ke Guo

Marine Computing in the Interconnected World for the Benefit of Society – MarineComp

Flávio Martins
Ioana Popescu
João Janeiro
Ramiro Neves
Marcos Mateus

Multiscale Modelling and Simulation - MMS

Derek Groen
Lin Gan
Stefano Casarin
Alfons Hoekstra
Bartosz Bosak

Simulations of Flow and Transport: Modeling, Algorithms, and Computation - SOFTMAC

Shuyu Sun
Jingfa Li
James Liu

Smart Systems: Bringing Together Computer Vision, Sensor Networks, and Machine Learning - SmartSys

João M. F. Rodrigues
Pedro J. S. Cardoso
Jânio Monteiro
Roberto Lam

Solving Problems with Uncertainties - SPU

Vassil Alexandrov

Teaching Computational Science - WTCS

Angela Shiflet
Evguenia Alexandrova
Alfredo Tirado-Ramos

Tools for Program Development and Analysis in Computational Science - TOOLS

Andreas Knüpfer
Karl Fürlinger

Programme Committee and Reviewers

Ahmad Abdelfattah
Eyad Abed
Markus Abel
Laith Abualigah
Giuseppe Agapito
Giovanni Agosta
Ram Akella
Elisabete Alberdi
Marco Aldinucci
Luis Alexandre
Vassil Alexandrov
Evguenia Alexandrova
Victor Allombert
Saad Alowayyed
Stanislaw Ambroszkiewicz
Ioannis Anagnostou
Philipp Andelfinger
Michael Antolovich
Hartwig Anzt
Hideo Aochi

Rossella Arcucci
Tomasz Arodz
Kamesh Arumugam
Luiz Assad
Victor Azizi Tarksalooyeh
Bartosz Balis
Krzysztof Banas
João Barroso
Dominik Bartuschat
Daniel Becker
Jörn Behrens
Adrian Bekasiewicz
Gebrail Bekdas
Stefano Beretta
Daniel Berrar
John Betts
Sanjukta Bhowmick
Bartosz Bosak
Isabel Sofia Brito
Kris Bubendorfer
Jérémy Buisson
Aleksander Byrski
Cristiano Cabrita
Xing Cai
Barbara Calabrese
Carlos Calafate
Carlos Cambra
Mario Cannataro
Alberto Cano
Paul M. Carpenter
Stefano Casarin
Manuel Castañón-Puga
Mauro Castelli
Jeronimo Castrillon
Eduardo Cesar
Patrikakis Charalampos
Henri-Pierre Charles
Zhensong Chen
Siew Ann Cheong
Andrei Chernykh
Lock-Yue Chew
Su Fong Chien
Sung-Bae Cho
Bastien Chopard
Stephane Chretien
Svetlana Chuprina
Florina M. Ciorba
Noelia Correia
Adriano Cortes
Ana Cortes
Jose Alfredo F. Costa
Enrique Costa-Montenegro
David Coster
Camille Coti
Carlos Cotta
Helene Coullon
Daan Crommelin
Attila Csikasz-Nagy
Loïc Cudennec
Javier Cuenca
Yifeng Cui
António Cunha
Ben Czaja
Pawel Czarnul
Bhaskar Dasgupta
Susumu Date
Quanling Deng
Nilanjan Dey
Ergin Dinc
Minh Ngoc Dinh
Sam Dobbs
Riccardo Dondi
Ruggero Donida Labati
Goncalo dos-Reis
Craig Douglas
Aleksandar Dragojevic
Rafal Drezewski
Niels Drost
Hans du Buf
Vitor Duarte
Richard Duro
Pritha Dutta
Sean Elliot
Nahid Emad
Christian Engelmann
Qinwei Fan
Fangxin Fang
Antonino Fiannaca
Christos Filelis-Papadopoulos
José Flich Cardo
Yves Fomekong Nanfack
Vincent Fortuin
Ruy Freitas Reis
Karl Frinkle
Karl Fuerlinger
Kohei Fujita
Wlodzimierz Funika
Takashi Furumura
Mohamed Medhat Gaber
Jan Gairing
David Gal
Marco Gallieri
Teresa Galvão
Lin Gan
Luis Garcia-Castillo
Delia Garijo
Frédéric Gava
Don Gaydon
Zong-Woo Geem
Alex Gerbessiotis
Konstantinos Giannoutakis
Judit Gimenez
Domingo Gimenez
Guy Gogniat
Ivo Gonçalves
Yuriy Gorbachev
Pawel Gorecki
Michael Gowanlock
Manuel Graña
George Gravvanis
Marilaure Gregoire
Derek Groen
Lutz Gross
Sophia Grundner-Culemann
Pedro Guerreiro
Kun Guo
Xiaohu Guo
Piotr Gurgul
Pietro Hiram Guzzi
Panagiotis Hadjidoukas
Mohamed Hamada
Boumediene Hamzi
Masatoshi Hanai
Quillon Harpham

William Haslett
Yiwei He
Alexander Heinecke
Jurjen Rienk Helmus
Alvaro Herrero
Bogumila Hnatkowska
Maximilian Hoeb
Paul Hofmann
Sascha Hunold
Juan Carlos Infante
Hideya Iwasaki
Takeshi Iwashita
Alfredo Izquierdo
Heike Jagode
Vytautas Jancauskas
Joao Janeiro
Jiří Jaroš
Shantenu Jha
Shalu Jhanwar
Chao Jin
Hai Jin
Zhong Jin
David Johnson
Anshul Joshi
Manuela Juliano
George Kallos
George Kampis
Drona Kandhai
Aneta Karaivanova
Takahiro Katagiri
Ergina Kavallieratou
Wayne Kelly
Christoph Kessler
Dhou Khaldoon
Andreas Knuepfer
Harald Koestler
Dimitrios Kogias
Ivana Kolingerova
Vladimir Korkhov
Ilias Kotsireas
Ioannis Koutis
Sergey Kovalchuk
Michał Koziarski
Slawomir Koziel
Jarosław Koźlak
Dieter Kranzlmüller
Bartosz Krawczyk
Valeria Krzhizhanovskaya
Paweł Ksieniewicz
Michael Kuhn
Jaeyoung Kwak
Massimo La Rosa
Roberto Lam
Anna-Lena Lamprecht
Johannes Langguth
Vianney Lapotre
Jysoo Lee
Michael Lees
Leifur Leifsson
Kenneth Leiter
Roy Lettieri
Andrew Lewis
Jingfa Li
Yanfang Li
James Liu
Hong Liu
Hui Liu
Zhao Liu
Weiguo Liu
Weifeng Liu
Marcelo Lobosco
Veronika Locherer
Robert Lodder
Stephane Louise
Frederic Loulergue
Huimin Lu
Paul Lu
Stefan Luding
Scott MacLachlan
Luca Magri
Maciej Malawski
Livia Marcellino
Tomas Margalef
Tiziana Margaria
Svetozar Margenov
Osni Marques
Alberto Marquez
Paula Martins
Flavio Martins
Jaime A. Martins
Marcos Mateus
Marco Mattavelli
Pawel Matuszyk
Valerie Maxville
Roderick Melnik
Valentin Melnikov
Ivan Merelli
Jianyu Miao
Kourosh Modarresi
Miguel Molina-Solana
Fernando Monteiro
Jânio Monteiro
Pedro Montero
James Montgomery
Andrew Moore
Irene Moser
Paulo Moura Oliveira
Ignacio Muga
Philip Nadler
Hiromichi Nagao
Kengo Nakajima
Raymond Namyst
Philippe Navaux
Michael Navon
Philipp Neumann
Ramiro Neves
Mai Nguyen
Hoang Nguyen
Nancy Nichols
Sinan Melih Nigdeli
Anna Nikishova
Kenji Ono
Juan-Pablo Ortega
Raymond Padmos
J. P. Papa
Marcin Paprzycki
David Pardo
Héctor Quintián Pardo
Panos Parpas
Anna Paszynska
Maciej Paszynski
Jaideep Pathak
Abani Patra
Pedro J. S. Cardoso
Dana Petcu
Eric Petit
Serge Petiton
Bernhard Pfahringer

Daniela Piccioni
Juan C. Pichel
Anna Pietrenko-Dabrowska
Laércio L. Pilla
Armando Pinto
Tomasz Piontek
Erwan Piriou
Yuri Pirola
Nadia Pisanti
Antoniu Pop
Ioana Popescu
Mario Porrmann
Cristina Portales
Roland Potthast
Ela Pustulka-Hunt
Vladimir Puzyrev
Alexander Pyayt
Zhiquan Qi
Rick Quax
Barbara Quintela
Waldemar Rachowicz
Célia Ramos
Marcus Randall
Lukasz Rauch
Andrea Reimuth
Alistair Rendell
Pedro Ribeiro
Bernardete Ribeiro
Robin Richardson
Jason Riedy
Celine Robardet
Sophie Robert
João M. F. Rodrigues
Daniel Rodriguez
Albert Romkes
Debraj Roy
Philip Rutten
Katarzyna Rycerz
Augusto S. Neves
Apaar Sadhwani
Alberto Sanchez
Gabriele Santin
Robert Schaefer
Olaf Schenk
Ulf Schiller
Bertil Schmidt
Jan Schmidt
Martin Schreiber
Martin Schulz
Marinella Sciortino
Johanna Sepulveda
Ovidiu Serban
Vivek Sheraton
Yong Shi
Angela Shiflet
Takashi Shimokawabe
Tan Singyee
Robert Sinkovits
Vishnu Sivadasan
Peter Sloot
Renata Slota
Grażyna Ślusarczyk
Sucha Smanchat
Maciej Smołka
Bartlomiej Sniezynski
Sumit Sourabh
Hoda Soussa
Steve Stevenson
Achim Streit
Barbara Strug
E. Dante Suarez
Bongwon Suh
Shuyu Sun
Vaidy Sunderam
James Suter
Martin Swain
Grzegorz Swisrcz
Ryszard Tadeusiewicz
Lotfi Tadj
Daniele Tafani
Daisuke Takahashi
Jingjing Tang
Osamu Tatebe
Cedric Tedeschi
Kasim Tersic
Yonatan Afework Tesfahunegn
Jannis Teunissen
Andrew Thelen
Yingjie Tian
Nestor Tiglao
Francis Ting
Alfredo Tirado-Ramos
Arkadiusz Tomczyk
Stanimire Tomov
Marko Tosic
Jan Treibig
Leonardo Trujillo
Benjamin Uekermann
Pierangelo Veltri
Raja Velu
Alexander von Ramm
David Walker
Peng Wang
Lizhe Wang
Jianwu Wang
Gregory Watson
Rodrigo Weber dos Santos
Kevin Webster
Josef Weidendorfer
Josef Weinbub
Tobias Weinzierl
Jens Weismüller
Lars Wienbrandt
Mark Wijzenbroek
Roland Wismüller
Eric Wolanski
Michał Woźniak
Maciej Woźniak
Qing Wu
Bo Wu
Guoqiang Wu
Dunhui Xiao
Huilin Xing
Miguel Xochicale
Wei Xue
Xin-She Yang

Dongwei Ye
Jon Yosi
Ce Yu
Xiaodan Yu
Reza Zafarani
Gábor Závodszky
H. Zhang
Zepu Zhang
Jingqing Zhang
Yi-Fan Zhang
Yao Zhang
Wenlai Zhao
Jinghui Zhong
Xiaofei Zhou
Sotirios Ziavras
Peter Zinterhof
Italo Zoppis
Chiara Zucco

Contents - Part I

ICCS Main Track

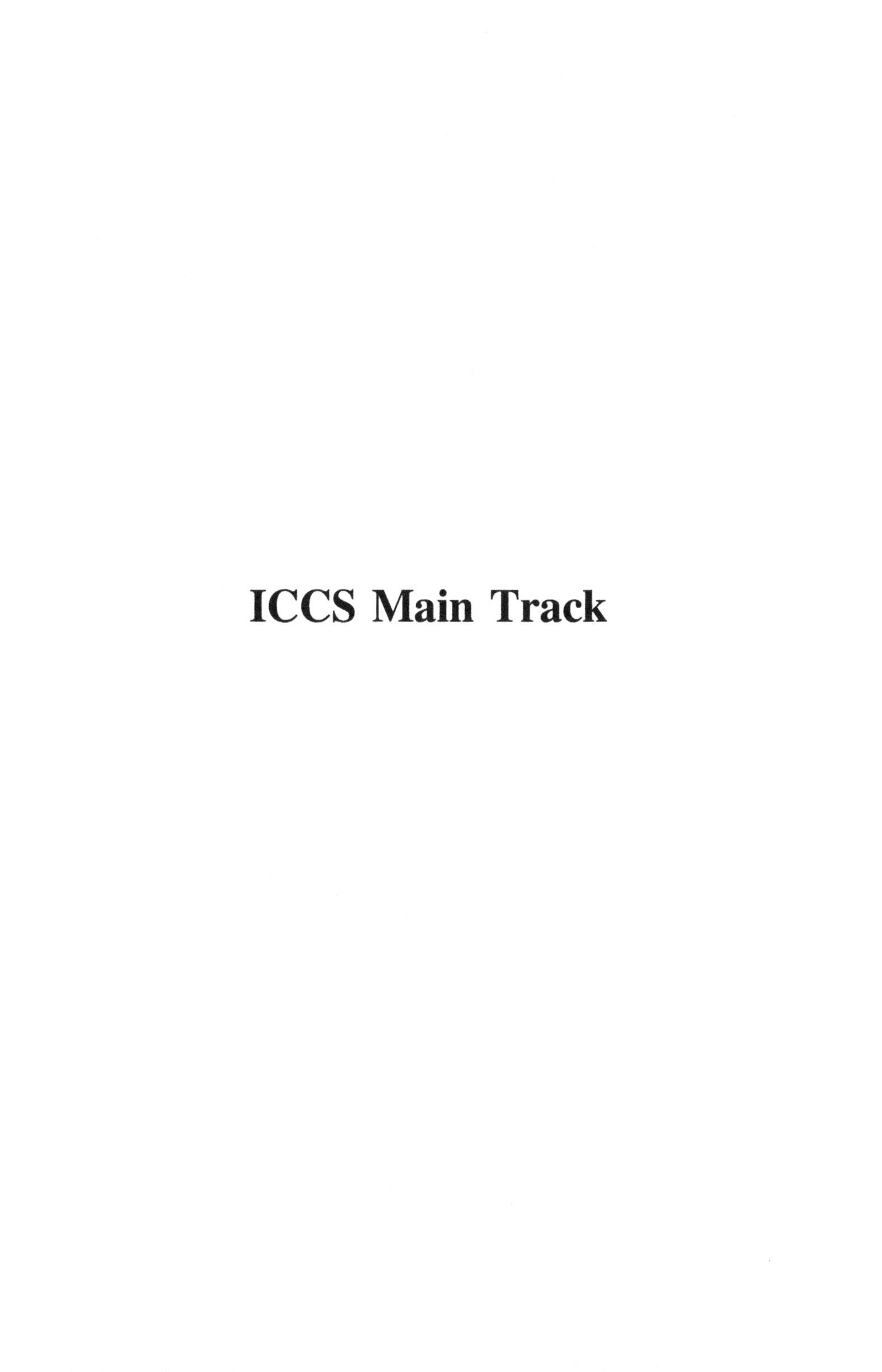

ICCS Main Track

Efficient Computation of Sparse Higher Derivative Tensors

Jens Deussen(✉) and Uwe Naumann

Software and Tools for Computational Engineering,
RWTH Aachen University, Aachen, Germany
{deussen,naumann}@stce.rwth-aachen.de

Abstract. The computation of higher derivatives tensors is expensive even for adjoint algorithmic differentiation methods. In this work we introduce methods to exploit the symmetry and the sparsity structure of higher derivatives to considerably improve the efficiency of their computation. The proposed methods apply coloring algorithms to two-dimensional compressed slices of the derivative tensors. The presented work is a step towards feasibility of higher-order methods which might benefit numerical simulations in numerous applications of computational science and engineering.

Keywords: Adjoints · Algorithmic differentiation · Coloring algorithm · Higher derivative tensors · Recursive coloring · Sparsity

1 Introduction

The preferred numerical method to compute derivatives of a computer program is Algorithmic Differentiation (AD) [1,2]. This technique produces exact derivatives with machine accuracy up to an arbitrary order by exploiting elemental symbolic differentiation rules and the chain rule. AD distinguishes between two basic modes: the forward mode and the reverse mode. The forward mode builds a directional derivative (also: tangent) code for the computation of Jacobians at a cost proportional to the number of input parameters. Similarly, the reverse mode builds an adjoint version of the program but it can compute the Jacobian at a cost that is proportional to the number of outputs. For that reason, the adjoint method is advantageous for problems with a small set of output parameters, as it occurs frequently in many fields of computational science and engineering.

For some methods second or even higher derivatives are required. In [3] it was already shown that Halley-methods using third derivatives are competitive to the Newton's method. The use of higher-order Greeks to evaluate financial options is discussed in [4]. The moments method approximates moments of a function by

Supported by the DFG project "Aachen dynamic optimization environment" funded by the German Research Foundation.

J. M. F. Rodrigues et al. (Eds.): ICCS 2019, LNCS 11536, pp. 3–17, 2019.
https://doi.org/10.1007/978-3-030-22734-0_1

propagating uncertainties [5,6]. To improve the accuracy of the approximated moments higher-order terms in the Taylor series that require higher derivatives need to be involved. The method has been applied to robust design optimization using third derivatives [7].

One possibility to obtain higher derivatives is to propagate Taylor series [8]. Another is to reapply the AD modes to an already differentiated program [2]. In [9] it is shown that computing a projection of the third derivative is proportional to the cost of calculating the whole Hessian. Nevertheless, the computation of higher derivatives is expensive even for adjoint algorithmic differentiation methods.

In this paper, we focus on the exploitation of symmetry and investigate the case where the d-th derivative tensor is considered sparse. Besides the application of coloring techniques for solving linear systems [10] these can be used to exploit the sparsity of Jacobian and Hessian matrices. The general idea is to assign the same color to those columns (or rows) of the corresponding matrix that are structurally orthogonal, thus they can be computed at the same time. In [11] a comprehensive summary of coloring techniques is given. We designed procedures to make solutions of the coloring problem applicable to the computation of third and higher derivative tensors. Due to symmetries of these tensors the colors obtained by a Hessian coloring can be applied multiple times. Furthermore, these colors can be used for a compression of higher derivative tensors to perform a coloring on the compressed two-dimensional slices of the tensor. We call this approach recursive coloring.

For the computation of sparse fourth derivatives we introduced three different approaches: The first is to use the colors of the Hessian three times. The second is to use the colors of the Hessian once to compress the third derivative, reapply the coloring algorithms to the two-dimensional slices of the compressed three-tensor and use the colors of the third derivative twice. The third is to use the colors of the Hessian and the third derivative slice to compress the fourth derivative and again color each two-dimensional slice. We generated random sparse polynomials for matrices from the Matrix Market collection [12] for the computation of sparse fourth derivative tensors to evaluate the efficiency of the three approaches.

The paper is organized as follows: In Sect. 2 there is a brief introduction to AD. Section 3 gives an overview of coloring techniques to exploit the sparsity of Hessian matrices. Subsequently in Sect. 4 we describe the procedures to exploit sparsity for third and higher derivatives. A description of the test cases and their implementation can be found in Sect. 5. Furthermore, this section provides numerical results to compare the introduced procedures. The last section gives a conclusion and an outlook.

2 Algorithmic Differentiation in a Nutshell

AD is a technique that transforms a *primal function* or *primal code* by using the chain rule to compute additionally to the function value the derivative of that function with respect to a given set of input (and intermediate) variables.

For simplicity and w.l.o.g. we will only consider multivariate scalar functions. Thus the given primal code has a set of n inputs $\mathbf{x}$ and a single output y.

$$f : \mathbb{R}^n \rightarrow \mathbb{R}, \quad y = f(\mathbf{x})$$

Furthermore, we assume that we are interested in the derivatives of the output variable with respect to all input variables. The first derivative of these functions is the gradient $\nabla f(\mathbf{x}) \in \mathbb{R}^n$, the second derivative is the Hessian $\nabla^2 f(\mathbf{x}) \in \mathbb{R}^{n \times n}$, and the third derivative is the three-tensor $\nabla^3 f(\mathbf{x}) \in \mathbb{R}^{n \times n \times n}$. In this paper a two-dimensional subspace of a tensor will be referred as a *slice*.

Schwarz's theorem says that the Hessian is symmetric if f has continuous second partial derivatives

$$\frac{\partial^2 y}{\partial x_j \partial x_k} = \frac{\partial^2 y}{\partial x_k \partial x_j}. \tag{1}$$

This generalizes to third and higher derivatives. Thus, a derivative tensor of order d has only $\binom{n+d-1}{d}$ structurally distinct elements.

AD is applicable if f and its corresponding implementation are locally differentiable up to the required order. The two basic modes are introduced in the following sections.

2.1 Tangent Mode AD

The tangent model can be simply derived by differentiating the function dependence. In Einstein notation this yields

$$y^{(1)} = \frac{\partial y}{\partial x_j} x_j^{(1)}.$$

The notation implies summation over all the values of $j = 0, \ldots, n-1$. The superscript $^{(1)}$ stands for the tangent of the variable [2]. This approach can be interpreted as an inner product of the gradient $\nabla f(\mathbf{x}) \in \mathbb{R}^n$ and the tangent $\mathbf{x}^{(1)}$ as

$$y^{(1)} = f^{(1)}(\mathbf{x}, \mathbf{x}^{(1)}) = \nabla f(\mathbf{x}) \cdot \mathbf{x}^{(1)}. \tag{2}$$

For each evaluation with $\mathbf{x}^{(1)}$ set to the i-th Cartesian basis vector $\mathbf{e}_i$ in $\mathbb{R}^n$ (also called seeding), an entry of the gradient can be extracted from $y^{(1)}$ (also called harvesting). To get all entries of the gradient by using this model, n evaluations are required which is proportional to the number of input variables. The costs of this method are similar to the costs of a finite difference approximation but AD methods are accurate up to machine precision.

2.2 Adjoint Mode AD

The adjoint mode is also called reverse mode, due to the reverse computation of the adjoints compared to the computation of the values. Therefore, a data-flow reversal of the program is required, to store additional information on the computation (e.g. partial derivatives) [13], which potentially leads to high memory requirements.

Again following [2], first-order adjoints are denoted with a subscript $_{(1)}$

$$x_{(1),j} = \frac{\partial y}{\partial x_j} y_{(1)}.$$

This equation is computed for each $j = 0, \ldots, n-1$.

Reverse mode yields a product of the gradient with the adjoint $y_{(1)}$

$$\mathbf{x}_{(1)} = f_{(1)}(\mathbf{x}, y_{(1)}) = y_{(1)} \cdot \nabla f(\mathbf{x}). \tag{3}$$

By setting $y_{(1)} = 1$ the resulting $\mathbf{x}_{(1)}$ contains all entries of the gradient. A single adjoint computation is required.

2.3 Higher Derivatives

One possibility to obtain second derivatives is to nest the proposed AD modes. Hence there are four combinations to compute second derivatives. The tangent-over-tangent model result from the application of tangent mode to the first-order tangent model in (2):

$$y^{(1,2)} = f^{(1,2)}(\mathbf{x}, \mathbf{x}^{(1)}, \mathbf{x}^{(2)}) = \frac{\partial^2 y}{\partial x_j \partial x_k} x_j^{(1)} x_k^{(2)}.$$

By seeding the Cartesian basis of $\mathbb{R}^n$ for the tangents $\mathbf{x}^{(1)}$ and $\mathbf{x}^{(2)}$ independently the entries of the Hessian can be computed with n^2 evaluations of $f^{(1,2)}$. The other three methods apply the adjoint mode at least once and thus their computational complexity is different to the pure tangent model. To obtain the whole Hessian a second-order adjoint model needs to be evaluated n times.

Applying the tangent mode to (3) yields the second-order adjoint model

$$x_{(1),k}^{(2)} = f_{(1),k}^{(2)}(\mathbf{x}, \mathbf{x}^{(2)}, y_{(1)}) = \frac{\partial^2 y}{\partial x_k \partial x_j} x_j^{(2)} y_{(1)} \ . \tag{4}$$

These models are simplified versions of the complete second-order models by setting mixed terms to zero. Detailed derivations can be found in [1] and [2].

Recursively applying the two basic modes yields even higher derivative models. As an example the simplified third-order adjoint model via tangent-over-tangent-over-adjoint can be written as

$$x_{(1),l}^{(2,3)} = f_{(1),l}^{(2,3)}(\mathbf{x}, \mathbf{x}^{(2)}, \mathbf{x}^{(3)}, y_{(1)}) = \frac{\partial^3 y}{\partial x_l \partial x_k \partial x_j} x_j^{(2)} x_k^{(3)} y_{(1)}. \tag{5}$$

Instead of seeding all combinations of Cartesian basis vector for both tangents which requires n^2 evaluations of $f_{(1)}^{(2,3)}$ we could exploit symmetry and evaluate the adjoint model only for those combinations of directional derivatives that fulfill $n \geq i_2 \geq \ldots \geq i_{d-1} \geq i_d \geq 1$. Thus, the tangents $\mathbf{x}^{(j)}$ are set to $\mathbf{e}_{i_j}$ for $2 \leq j \leq d$. which results in $\binom{n+d-2}{d-1}$ evaluations of the d-th order adjoint model.

3 Coloring Techniques

For the computation of higher derivatives it is indispensable to exploit the sparsity and symmetry of the particular matrices or tensors due to the expenses that come along with their computation. We aim for a reduction of the number of projections that are needed for the computation of the sparse derivative. In [14,15] it is shown that the symmetric graph coloring problems considered in this section are equivalent to a matrix partitioning problem and can be used to enable the determination of the corresponding derivative matrices.

Coloring techniques can be used to exploit sparsity of Jacobian and Hessian matrices [11]. The general idea is to identify columns (or rows) of the corresponding matrix that can be computed at the same time and assign the same *color* to these columns.

Definition 1. *We define $\mathcal{C}_j$ to denote the set that contains the indices of all columns belonging to color j. These sets are grouped in a set of sets $\mathcal{C}^d$ for the d-th derivative. $\mathcal{C}^d$ can be used to compute a compressed derivative by seeding $\sum_{j \in \mathcal{C}_k} \mathbf{e}_j$ for each color k instead of $\mathbf{e}_j$ for each column.*

In the following we outline the difference between direct and indirect coloring heuristics for Hessian matrices. For the proposed heuristics we assume multivariate scalar functions $f : \mathbb{R}^n \rightarrow \mathbb{R}$. Furthermore, we assume their sparsity patterns of all derivatives to be known.

Definition 2. *The sparsity pattern of the d-th derivative is a set that contains the (multi-)indices of those elements that are non-zero. It is denoted by $\mathcal{P}^d$.*

The detection of sparsity patterns is out of the scope of this paper. For further reading about sparsity pattern detection see for example [16,17].

3.1 Hessian Coloring with Direct Recovery

A direct approach to find a suitable Hessian coloring considering symmetry of the Hessian (1) is the so called *star coloring*. This heuristic ensures that each element of the Hessian is computed at least once and can be directly extracted from the computed derivative projections. Star coloring uses an adjacency graph where each node belongs to a column (or row). Each non-zero element $\frac{\partial^2 y}{\partial x_j \partial x_k}$ with $(j, k) \in \mathcal{P}^2$ in the matrix is an edge connecting node j and k in the graph. Then, a distance-1 coloring is applied to the graph in that the same color cannot be assigned to adjacent nodes. Furthermore, there is an additional condition that every path of length four uses at least three colors. The nodes and thus the columns of the Hessian with the same color can be computed simultaneously.

3.2 Hessian Coloring with Indirect Recovery

Another heuristic described in [11] is *acyclic coloring* that also applies distance-1 coloring to the adjacency graph. This coloring technique has the additional condition that each cycle in the graph needs at least three distinct colors. Since it might assign the same color to columns that are not structurally orthogonal, the derivative matrix can be compressed even more. Nevertheless, a recovery of these non-zero elements implies the solution of a system of linear equations.

Both methods for Hessian coloring reduce the number of required derivative projections from n to $|\mathcal{C}^2|$.

4 Computation of Sparse Derivative Tensors

In this section, we show how to use coloring heuristics introduced in Sect. 3 to obtain sparse higher derivative tensors efficiently. Section 4.1 focuses on the reapplication of colors for higher derivatives. In Sect. 4.2 a recursive coloring approach for the computation of higher derivatives is described. After that, we propose the incomplete recursive coloring in Sect. 4.3. We will use the higher-order adjoint models from Sect. 2.3, that is (5) for third derivatives.

The proposed algorithms assume multivariate scalar functions $f : \mathbb{R}^n \rightarrow \mathbb{R}$ for simplicity. For the more general case of multivariate vector functions the algorithms can be applied to each output variable individually. It is also possible to combine the proposed approaches with Jacobian coloring for vector functions to make their derivative computation more efficient, but this is future work.

We will illustrate the algorithms in this section for

$$f(\mathbf{x}) = \sum_{i=0}^{n-1} x_i^3 + \sum_{i=1}^{n-2} x_{i-1} \cdot x_i \cdot x_{i+1}, \tag{6}$$

with $n = 6$. The sparsity pattern of the Hessian and the corresponding adjacency graph are visualized in Fig. 1. The colors in this figure are a solution for the symmetric graph coloring problem obtained by acyclic coloring.

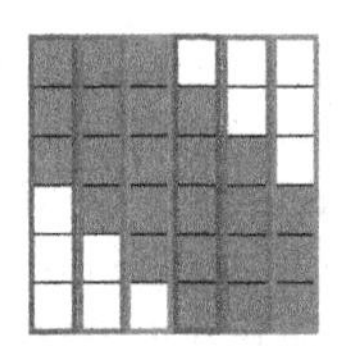

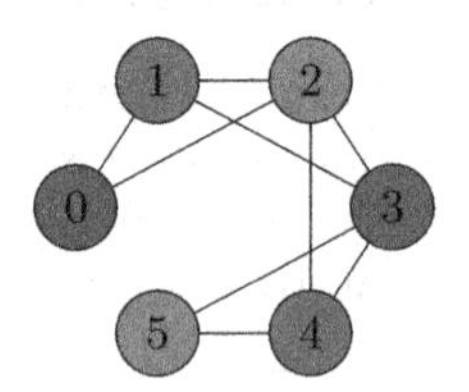

Fig. 1. Sparsity pattern of the Hessian (left) and the corresponding adjacency graph (right) with colors obtained by acyclic coloring for the function from (6) (Color figure online)

4.1 Reapplication of Colors

In [3] a so called *induced* sparsity of the third derivative was introduced where

$$\left(\nabla^2 f_{jk} = 0 \ \vee \ \nabla^2 f_{jl} = 0 \ \vee \ \nabla^2 f_{kl} = 0\right) \quad \Rightarrow \quad \nabla^3 f_{jkl} = 0. \tag{7}$$

Definition 3. *We call a third-order tensor* induced sparse *if only those elements that fulfill* (7) *have zero entries. The same can be generalized to any order $d \geq 2$: a d-th-order tensor is called* induced sparse *if only those elements of the tensor are zero that are induced by a zero in the $(d-1)$-th derivative. The induced sparsity pattern of the d-th derivative is denoted as $\tilde{\mathcal{P}}^d$.*

We extended the concept of induced sparsity to show that the colors obtained by a Hessian coloring can be used for the higher derivative computation.

Lemma 1. *Every solution $\mathcal{C}$ of the coloring problem for a symmetric matrix $A \in \mathbb{R}^{n \times n}$ with sparsity pattern $\mathcal{P}_A$ is also valid for a symmetric matrix $B \in \mathbb{R}^{n \times n}$ with sparsity pattern $\mathcal{P}_B$ if $\mathcal{P}_B \subseteq \mathcal{P}_A$.*

Proof. In case of $\mathcal{P}_B \subseteq \mathcal{P}_A$ we can consider a matrix A' with the same sparsity pattern as A as a composition of matrix B with its values at $(j,k) \in \mathcal{P}_B$, numerical zeros at $(j,k) \in \mathcal{P}_A \backslash \mathcal{P}_B$ and structural zeros at $(j,k) \notin \mathcal{P}_A$. The coloring result $\mathcal{C}$ obtained for A can be used to determine matrix A' and thus B beside the numerical zeros. ■

Theorem 1. *A solution $\mathcal{C}^2$ of the coloring problem for a Hessian matrix can be used for seeding the $(d-1)$ tangents of a d-th-order adjoint model to compute the d-th derivative tensors for $d \geq 2$.*

Proof. Mathematical induction can be used to prove that Theorem 1 holds for all natural numbers $d \geq 2$.

Basis: Since [14,15] showed that the symmetric graph coloring problems can be transformed to matrix partitioning problems, colors $\mathcal{C}^2$ obtained for the coloring problem can be used as seed vectors for the tangent of the second-order adjoint model (4). Thus, Theorem 1 holds for $d = 2$.

Inductive Step: Given that Theorem 1 holds for the $(d-1)$-th derivative, we can apply the Hessian colors to compress this derivative at least $d-2$ times. We need to show that the Hessian colors $\mathcal{C}^2$ can be used $d-1$ times for the d-th derivative.

The induced sparsity from (7) holds for any order $d \geq 2$ and the sparsity pattern of each slice $\tilde{\mathcal{P}}^d_{i_j}$ is a subset of the sparsity pattern of the $(d-1)$-th derivative $\mathcal{P}^{d-1}$, where i_j is the index for the direction of tangent $\mathbf{x}^{(j)}$. Furthermore, the induced pattern is an overestimation of the actual sparsity pattern $\mathcal{P}^d_{i_j}$ which yields

$$\mathcal{P}^d_{i_j} \subseteq \tilde{\mathcal{P}}^d_{i_j} \subseteq \mathcal{P}^{d-1} \qquad \text{for} \qquad 0 \leq i_j \leq n-1 \,, \tag{8}$$

with $2 \leq j \leq d$.

Lemma 1 is applicable since (8) holds such that the coloring results used to compute the $(d-1)$-th derivative can also be used for the computation of the d-th derivative. This includes that the Hessian colors can be applied for the first $d-2$ tangents due to the assumption that the theorem holds for $d-1$. Seeding these tangents yields $\mathbf{x}^{(j)} = \sum_{k \in \mathcal{C}^2_{i_j}} \mathbf{e}_k$ for $2 \leq j \leq d-1$. Thus, it remains to show that $\mathcal{C}^2$ can also be applied to the last tangent $\mathbf{x}^{(d)}$.

Using the Hessian colors for the $d-2$ tangents leads to a compression of the d-th-order tensor which is a projection of the derivative tensor in direction of the colors

$$\sum_{k_2 \in \mathcal{C}^2_{i_2}} \cdots \sum_{k_{d-1} \in \mathcal{C}^2_{i_{d-1}}} \nabla^d f_{k_2,\ldots k_{d-1}} \tag{9}$$

and reduces the size of the corresponding tangent directions i_j with $2 \leq j \leq d-1$ from n to $|\mathcal{C}^2|$. For each combination of the directions we will receive a two-dimensional slice of the tensor of size $n \times n$.

Again (8) shows that the sparsity pattern of each of these slices is still a subset of the sparsity pattern of the $(d-1)$-th derivative

$$\bigcup_{k \in \mathcal{C}_{i_d}} \mathcal{P}^d_k \subseteq \mathcal{P}^{d-1} \quad \text{for} \quad 0 \leq i_d \leq |\mathcal{C}| - 1. \tag{10}$$

Recursively applying (10) for $(d-2)$ tangents yields

$$\bigcup_{k_2 \in \mathcal{C}^2_{i_2}} \cdots \bigcup_{k_{d-1} \in \mathcal{C}^2_{i_{d-1}}} \mathcal{P}^d_{k_2,\ldots k_{d-1}} \subseteq \mathcal{P}^2 \quad \text{for} \;\; 0 \leq i_j \leq |\mathcal{C}^2| - 1, \;\; 2 \leq j \leq d-1.$$

Hence, due to Lemma 1 $\mathcal{C}^2$ can also be used to seed tangent $\mathbf{x}^{(d)}$. Thus, with the inductive step, we have shown that the Hessian colors $\mathcal{C}^2$ are $(d-1)$ times applicable for the computation of the d-th derivative. ■

Applying the Hessian colors for all tangents requires $|\mathcal{C}^2|^{d-1}$ model evaluations of the d-th-order adjoint model to compute the whole d-th derivative tensor.

Example 1. The induced sparsity pattern of the third derivative of (6) is shown in Fig. 2. It can be seen, that the actual sparsity pattern (gray) is a subset of the induced sparsity pattern (striped).

Using the Hessian colors from Fig. 1 for the first tangent results in the compressed derivative tensor visualized in Fig. 3 (left). The sparsity pattern of each slice of the compressed tensor is (a subset of) the Hessian sparsity pattern. Thus, the Hessian colors are applied again.

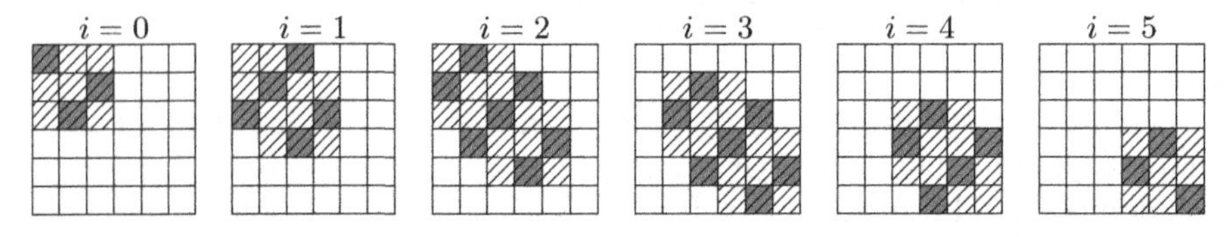

Fig. 2. Induced (striped) and actual (gray) sparsity pattern of the third derivative

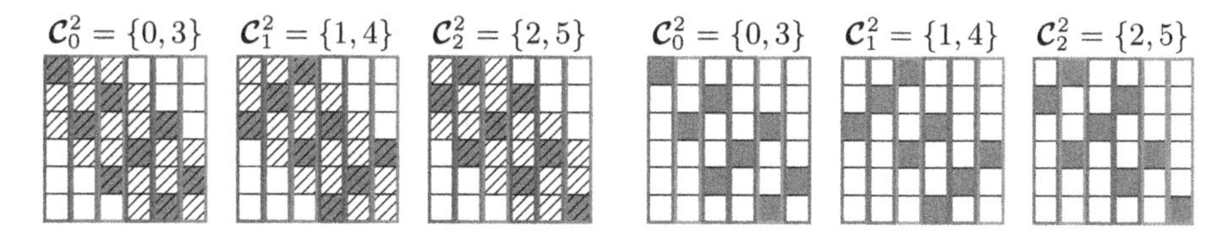

Fig. 3. Sparsity pattern of the third derivative projections by using Hessian colors of function (6) with resulting colorings for reapplication of Hessian colors (left) and recursive coloring results (right) (Color figure online)

4.2 Recursive Coloring

Since the actual sparsity patterns of the higher derivatives are assumed to be known, this information should be used to design more efficient algorithms.

Algorithm 1.
Full Recursive Coloring

1: **procedure** RecCol(d)
2:   **if** $d > 2$ **then**
3:     $\mathbf{C}^{d-1} \leftarrow$ RecCol($d-1$)
4:     **for all** $\mathcal{C}_L^{d-1} \in \mathbf{C}^{d-1}$ **do**
5:       **for** $k \leftarrow 0, |\mathcal{C}_L^{d-1}| - 1$ **do**
6:         $l \leftarrow \begin{pmatrix} L & \mathcal{C}_{L,k}^{d-1} \end{pmatrix}$
7:         $\mathcal{C}_l^d \leftarrow$ Col(proj$\left(\mathcal{P}^d, l\right)$)
8:         $\mathbf{C}^d \leftarrow \begin{pmatrix} \mathbf{C}^d & \mathcal{C}_l^d \end{pmatrix}$
9:       **end for**
10:     **end for**
11:     **return** $\mathbf{C}^d$
12:   **else**
13:     $\mathcal{C}_\emptyset^2 \leftarrow$ Col($\mathcal{P}^2$)
14:     **return** $\mathcal{C}_\emptyset^2$
15:   **end if**
16: **end procedure**

Algorithm 2.
Incomplete Recursive Coloring

1: **procedure** IncRecCol(d, o)
2:   **if** $d > 2$ **then**
3:     $\mathbf{C}^{d-1} \leftarrow$ IncRecCol($d-1$)
4:     **for all** $\mathcal{C}_L^{d-1} \in \mathbf{C}^{d-1}$ **do**
5:       **for** $k \leftarrow 0, |\mathcal{C}_L^{d-1}| - 1$ **do**
6:         $l \leftarrow \begin{pmatrix} L & \mathcal{C}_{L,k}^{d-1} \end{pmatrix}$
7:         **if** $d \leq o$ **then**
8:           $\mathcal{C}_l^d \leftarrow$ Col(proj$\left(\mathcal{P}^d, l\right)$)
9:         **else**
10:           $\mathcal{C}_l^d \leftarrow \mathcal{C}_l^{d-1}$
11:         **end if**
12:         $\mathbf{C}^d \leftarrow \begin{pmatrix} \mathbf{C}^d & \mathcal{C}_l^d \end{pmatrix}$
13:       **end for**
14:     **end for**
15:     **return** $\mathbf{C}^d$
16:   **else**
17:     $\mathcal{C}_\emptyset^2 \leftarrow$ Col($\mathcal{P}^2$)
18:     **return** $\mathcal{C}_\emptyset^2$
19:   **end if**
20: **end procedure**

Definition 4. *Recursive coloring is defined as a technique that recursively solves the symmetric graph coloring problem for two-dimensional slices (fixing the compressed directions) of a compressed d-th derivative tensor* (9) *and to use the solution of this problem for seeding tangent* $\mathbf{x}^{(d)}$*. The other tangents need to be set previously for the compression.*

Algorithm 1 describes the procedure for a d-th derivatives. To compress the tensor of order d the coloring results of all lower derivatives are required. These colors are computed by the recursive call in line 3. Note that the variable $\mathbf{C}^{d-1}$ is a list that contains all coloring results $\mathcal{C}_L^{d-1}$ as a set of sets. The subscript L of $\mathcal{C}_L^d$ is the list of the $d-2$ tangent directions for the compression of the

d-th derivative tensor that led to the coloring. In line 6 the k-th color of $\mathcal{C}_L^{d-1}$ is attached to L. This new list is used in line 7 to compress the tensor and to obtain the coloring for one slice of the compressed d-th-order tensor by solving the coloring problem for a symmetric matrix. After that, the new colors are attached to $\mathbf{C}^d$.

The following lemma will be used to show that the recursive coloring is at least as good as the reapplication of colors.

Lemma 2. *The number of colors of the optimal solution $\mathcal{C}^A$ of the coloring problem for a symmetric matrix $A \in \mathbb{R}^{n \times n}$ with sparsity pattern $\mathcal{P}_A$ is larger than that for a symmetric matrix $B \in \mathbb{R}^{n \times n}$ with sparsity pattern $\mathcal{P}_B$ if $\mathcal{P}_B \subseteq \mathcal{P}_A$, such that $|\mathcal{C}^B| \leq |\mathcal{C}^A|$.*

Proof. In a proof by contradiction we assume the optimal coloring results $|\mathcal{C}^B| > |\mathcal{C}^A|$. Since $\mathcal{P}_B \subseteq \mathcal{P}_A$ we can apply Lemma 1 such that $\mathcal{C}^A$ is also a coloring for B, which implies that $\mathcal{C}^B$ was not optimal. Thus, the assumption was wrong. ■

Theorem 2. *The optimal number of adjoint model evaluations obtained with a recursive coloring is less than or equal to the optimal number of adjoint model evaluations resulting from the reapplication of colors.*

Proof. Assuming that both approaches have the same compression for the first $d-2$ tangents, it is necessary to show that the number of adjoint model evaluations for recursive coloring is less than those of the reapplication of colors

$$\sum_{\mathcal{C}_L^{d-1} \in \mathbf{C}^{d-1}} \sum_{k \in \mathcal{C}_L^{d-1}} |\mathcal{C}_{\{L,k\}}^d| \leq \sum_{\mathcal{C}_L^{d-1} \in \mathbf{C}^{d-1}} \sum_{k \in \mathcal{C}_L^{d-1}} |\mathcal{C}_L^{d-1}| , \tag{11}$$

where $\mathbf{C}^{d-1}$ is the list of coloring results for the $(d-1)$-th derivative and L contains those directions that are used for the compression of this derivative.

Since (10) holds for any coloring $|\mathcal{C}|$ we can apply Lemma 2 to show

$$|\mathcal{C}_{\{L,k\}}^d| \leq |\mathcal{C}_L^{d-1}| \quad \text{for} \quad k \in \mathcal{C}_L^{d-1} , \tag{12}$$

such that the minimal number of colors obtained for two-dimensional slices in $\mathbb{R}^{n \times n}$ of the compressed d-th derivative tensor is less than or equal to those used for the compression. Thus, (12) directly yields (11). ■

By using heuristics to obtain a solution of the coloring problems, it might be that the solution $\mathcal{C}_{\{L,k\}}^d$ is worse than $\mathcal{C}_L^{d-1}$ for some k. In that case the algorithm should select the colors used for tangent $\mathbf{x}^{(d-1)}$ by setting $\mathcal{C}_{\{L,k\}}^d = \mathcal{C}_L^{d-1}$.

The resultant seeding can be stored in a tree structure. The seed of the adjoint is stored in the root of the tree, which has $|\mathcal{C}^2|$ children. These nodes with depth 1 contain the first tangent directions that correspond to the Hessian colors. Analogous, the nodes with depth 2 store the coloring results of the compressed third derivative. In general, the compression is done by the information stored in all nodes on the path connecting that node with the root. In case of a node with depth 2, this will be only the direction that is stored in its parent node. To obtain the complete higher derivative each path from a leaf to the root needs to be used as a seeding for the adjoint model.

The list L in Algorithm 1 does not need to be stored explicitly, because this information is already available in the tree structure. Depending on the coloring results the tree does not need to be balanced. Furthermore, due to (10) we know that each node has no more children than its parent node.

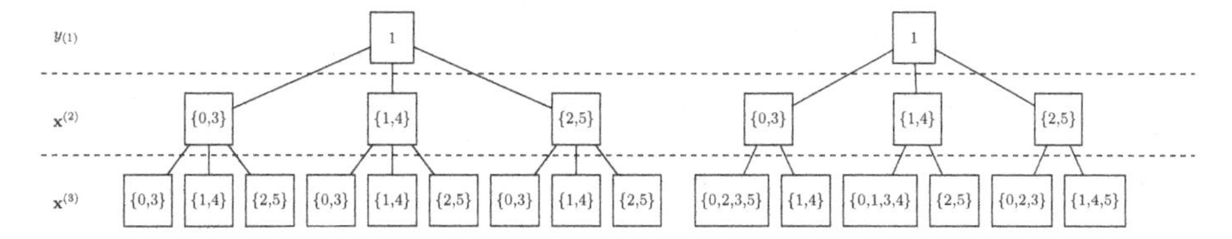

Fig. 4. Seeding trees for computing the third derivative tensor of the example for the reapplication of Hessian colors (left) and the recursive coloring (right)

Example 2. The results of the recursive coloring are illustrated in Fig. 3 (right) for the example function. After the compression with the first tangent $x^{(2)} = \boldsymbol{\mathcal{C}}^2_{i_2}$ the resulting slices are colored again. In case of $i_j = 0$ the resulting coloring is $\{\{0, 2, 3, 5\}, \{1, 4\}\}$. The seeding trees for both approaches, the reapplication of colors and the recursive coloring are shown in Fig. 4.

4.3 Incomplete Recursive Coloring

The incomplete recursive coloring is an approach that combines the reapplication of colors from Sect. 4.1 and the recursive coloring from Sect. 4.2. Instead of applying a full recursive coloring, the recursion can be stopped on any level and the computed colors can be reapplied for all tangents that belong to higher derivatives. This is possible due to Lemma 1 and (8). The additional check if the recursive coloring should be stopped in line 7 of Algorithm 2 is the main modification. In that case the previously computed colors are reused (line 10) and the list in the subscript needs to be updated.

Definition 5. *We call an incomplete recursive coloring a d-th-order approach if it uses colorings of derivatives up to d-th-order. The approach of using the Hessian colors for all tangents from Sect. 4.1 is an incomplete recursive coloring of second order.*

5 Experiments and Results

In this section we compare the different coloring approaches presented in Sect. 4 for a computation of a fourth derivative tensor. For evaluating the approaches we use a tool that creates sparse fourth-order polynomials from sparsity patterns. It is stored in a data structure that is an extension of the compressed tube storage [3] for super-symmetric induced four-tensors. The tool takes the sparsity ratio of the third ρ_3 and the fourth ρ_4 derivative.

Definition 6. *We define the sparsity ratio ρ_d to describe the ratio of non-zero elements in the d-th derivative compared to the number of elements that are induced by the $(d-1)$-th derivative. Given that the gradient is dense, ρ_2 is the ratio of non-zero elements to distinct elements in the Hessian.*

Due to the fact that the proposed approaches only apply colorings to two-dimensional (sub-)spaces of the derivative tensors, coloring algorithms implemented in *ColPack* [18] can be used. Particularly, we use the acyclic coloring algorithm with indirect recovery to obtain the expected colors.

In a first set of tests we want to investigate the efficiency of the second-order approach compared to an approach disregarding the sparsity (see Sect. 2.3).

After that, we vary the parameters for the sparsity of the higher derivative tensors ρ_3 and ρ_4. In case of $0 < \rho_3 < 1$ or $0 < \rho_4 < 1$ the sparsity patterns of the higher derivatives are generated randomly such that we perform 1000 computations for each of the parameter combinations and average the results. As a test case we selected the `nos4` matrix from [12].

In a third test, we compute the derivatives of polynomials generated from several matrices from the database. A selection of these matrices is listed in Table 1, in which the first column is the name in the database, the second is the size and the third is the sparsity ratio ρ_2 of the matrix. For this test we assume that the higher derivatives are induced sparse ($\rho_3 = 1$ and $\rho_4 = 1$) which

Table 1. Selected matrices from [12] with the ratio of projections required for the second-order approach to a computation disregarding sparsity

Label	n	ρ_2	Proj. ratio	Label	n	ρ_2	Proj. ratio
`nos4`	100	0.069	$1.258 \cdot 10^{-2}$	`can_1072`	1072	0.012	$4.857 \cdot 10^{-6}$
`bcsstk04`	132	0.215	$6.220 \cdot 10^{-2}$	`dwt_1242`	1242	0.008	$1.072 \cdot 10^{-6}$
`bcsstk05`	153	0.109	$1.314 \cdot 10^{-2}$	`bcsstk09`	1083	0.017	$1.929 \cdot 10^{-5}$
`bcsstm07`	420	0.043	$8.562 \cdot 10^{-4}$	`bcsstm10`	1086	0.020	$2.725 \cdot 10^{-5}$
`nos3`	960	0.018	$3.322 \cdot 10^{-5}$	`bcsstm12`	1473	0.010	$7.674 \cdot 10^{-6}$
`dwt__992`	992	0.018	$1.346 \cdot 10^{-5}$	`bcsstk11`	1473	0.016	$3.293 \cdot 10^{-5}$

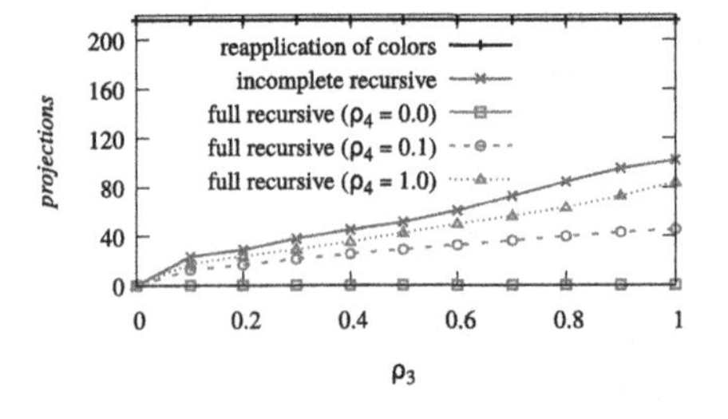

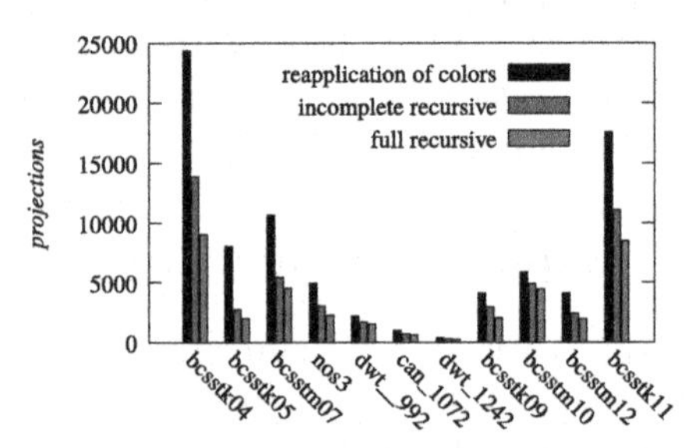

Fig. 5. Number of projections of the three coloring approaches for the `nos4` example (left) and for a set of test cases from [12] and induced sparsity (right) (Color figure online)

will result in a somehow worst-case estimation of the efficiency of the third- and fourth-order approaches. This test only requires the Hessian sparsity pattern.

5.1 Numerical Results

The results for the efficiency of the second-order approach is given in the last column of Table 1. It can be seen that this approach becomes more efficient if the matrices become sparser. The number of derivative projections required for the second-order approach are less than 7% of those required with the symmetric approach disregarding sparsity.

Figure 5 (left) visualizes the results for the second tests. The black line shows the number of seeds that are required for the Hessian coloring approach. It is constant because the Hessian coloring is independent on the sparsity of the higher derivatives. The red line shows the incomplete recursive coloring up to third order, which means that the colors of the compressed third derivative slices are computed. The blue lines stand for the full recursive coloring. These lines are isolines with a constant parameter for the sparsity of the fourth derivative.

It can be seen that the recursive colorings up to third and fourth order become more efficient the sparser the corresponding derivatives are. In the special case of $\rho_3 = 0$ or $\rho_4 = 0$ the number of adjoint model evaluations is zero for the approaches that use these parameters. In the induced sparse case, the incomplete recursive coloring up to third order still reduces the number of required projections to 47.2% of the Hessian coloring approach, while the fully recursive approach is even better with 38.9%. So even in the case where the higher derivatives are as dense as possible dependent on the sparsity of the Hessian, the recursive approaches yield good savings in terms of adjoint model evaluations.

Similar results can be observed for the other test matrices. Figure 5 (right) shows the results for the matrices from Table 1. Again the black bar denotes the second-order approach, the red bar stands for the incomplete third-order recursive coloring and the blue bar shows the full recursive coloring approach. The average savings compared to the second-order approach are 75.9% for the incomplete recursive coloring and 68.5% for the full recursive coloring only considering test cases with more than a single color for the Hessian matrix. In the case of third and fourth derivative that are sparser than the induced sparsity the recursive coloring approaches become more efficient.

6 Conclusion and Outlook

In this paper we have introduced procedures to make coloring techniques applicable for the computation of higher derivative tensors. We proposed two basic concepts to achieve this: application of previous computed colors and recursive coloring. By combining both concepts we came up with the incomplete recursive coloring. Depending on how much sparsity information is available the recursion depth of this approach can be adjusted.

The results show that even in the case where only the Hessian sparsity is known the savings compared to an approach disregarding sparsity are significant. Including higher sparsity information increases these savings further. Assuming that the costs of the coloring is considerably lower than an adjoint evaluation, additional colorings of the compressed slices can be accepted. Furthermore, the (incomplete) recursive coloring of the higher derivative tensors can be considered to be done in compile-time to build up the seeding tree. This seeding tree can be used multiple times in the execution to obtain derivatives at various points.

In the future, we intend to provide a software package making the (incomplete) recursive coloring setup available to efficiently generate seeding trees for the computation of higher derivatives with the most common AD tools. For the parallelization of the proposed algorithms it is necessary to find a suitable load balancing. Another interesting future study should consider direct coloring methods for higher derivative tensors. We expect that direct coloring further decreases the number of required projections and thus the costs of the higher derivative computation.

References

1. Griewank, A., Walther, A.: Evaluating Derivatives: Principles and Techniques of Algorithmic Differentiation. SIAM (2008)
2. Naumann, U.: The Art of Differentiating Computer Programs: An Introduction to Algorithmic Differentiation, SE24 in Software, Environments and Tools. SIAM (2012)
3. Gundersen, G., Steihaug, T.: Sparsity in higher order methods for unconstrained optimization. Optim. Methods Softw. **27**, 275–294 (2012)
4. Ederington, L.H., Guan, W.: Higher order greeks. J. Deriv. **14**, 7–34 (2007)
5. Smith, R.C.: Uncertainty Quantification: Theory, Implementation, and Applications. SIAM (2013)
6. Christianson, B., Cox, M.: Automatic propagation of uncertainties. In: Bücker, M., Corliss, G., Naumann, U., Hovland, P., Norris, B. (eds.) Automatic Differentiation: Applications, Theory, and Implementations. LNCSE, vol. 50, pp. 47–58. Springer, Heidelberg (2006). https://doi.org/10.1007/3-540-28438-9_4
7. Putko, M.M., Taylor, A.C., Newman, P.A., Green, L.L.: Approach for input uncertainty propagation and robust design in CFD using sensitivity derivatives. J. Fluids Eng. **124**, 60–69 (2002)
8. Griewank, A., Utke, J., Walther, A.: Evaluating higher derivative tensors by forward propagation of univariate Taylor series. Math. Comput. **69**, 1117–1130 (2000)
9. Gower, R.M., Gower, A.L.: Higher-order reverse automatic differentiation with emphasis on the third-order. Math. Program. **155**, 81–103 (2016)
10. Jones, M.T., Plassmann, P.E.: Scalable iterative solution of sparse linear systems. Parallel Comput. **20**, 753–773 (1994)
11. Gebremedhin, A.H., Manne, F., Pothen, A.: What color is your Jacobian? Graph coloring for computing derivatives. SIAM Rev. **47**, 629–705 (2005)
12. Boisvert, R.F., Pozo, R., Remington, K., Barrett, R.F., Dongarra, J.J.: Matrix Market: a web resource for test matrix collections. In: Boisvert, R.F. (ed.) Quality of Numerical Software. IFIPAICT, pp. 125–137. Springer, Boston (1997). https://doi.org/10.1007/978-1-5041-2940-4

13. Hascoët, L., Naumann, U., Pascual, V.: "To be recorded" analysis in reverse-mode automatic differentiation. FGCS **21**, 1401–1417 (2005)
14. Coleman, T.F., Moré, J.J.: Estimation of sparse Hessian matrices and graph coloring problems. Math. Program. **28**, 243–270 (1984)
15. Coleman, T.F., Cai, J.-Y.: The cyclic coloring problem and estimation of sparse Hessian matrices. SIAM J. Algebraic Discrete Methods **7**, 221–235 (1986)
16. Gay, D.M.: More AD of nonlinear AMPL models: computing Hessian information and exploiting partial separability. In: Computational Differentiation: Applications, Techniques, and Tools, pp. 173–184. SIAM (1996)
17. Gower, R.M., Mello, M.P.: Computing the sparsity pattern of Hessians using automatic differentiation. ACM TOMS **40**, 10:1–10:15 (2014)
18. Gebremedhin, A.H., Nguyen, D., Patwary, M.M.A., Pothen, A.: ColPack: software for graph coloring and related problems in scientific computing. ACM TOMS **40**, 1:1–1:31 (2013)

Rational Approximation of Scientific Data

Youssef S. G. Nashed(✉), Tom Peterka, Vijay Mahadevan, and Iulian Grindeanu

Mathematics and Computer Science Division, Argonne National Laboratory, Lemont, IL 60439, USA
{ynashed,tpeterka,mahadevan,iulian}@anl.gov

Abstract. Scientific datasets are becoming increasingly challenging to transfer, analyze, and store. There is a need for methods to transform these datasets into compact representations that facilitate their downstream management and analysis, and ideally model the underlying scientific phenomena with defined numerical fidelity. To address this need, we propose nonuniform rational B-splines (NURBS) for modeling discrete scientific datasets; not only to compress input data points, but also to enable further analysis directly on the continuous fitted model, without the need for decompression. First, we evaluate three different methods for NURBS fitting, and compare their performance relative to unweighted least squares approximation (B-splines). We then extend current state-of-the-art B-spline adaptive approximation to NURBS; that is, adaptively determining optimal rational basis functions and weighted control point locations that approximate given input data points to prespecified accuracy. Additionally, we present a novel local adaptive algorithm to iteratively approximate large data input domains. This method takes advantage of NURBS local support to refine regions of the approximated model, acting locally on both input and model subdomains, without affecting other regions of the global approximation. We evaluate our methods in terms of approximated model compactness, achieved accuracy, and computational cost on both synthetic smooth functions and real-world scientific data.

Keywords: Piecewise approximation · Adaptive methods · Domain partitioning · Parallel algorithms

1 Introduction

Advancing science through high-performance computing (HPC) depends on managing, analyzing, and visualizing data generated from large-scale simulations or experiments. Much attention has been paid to how best to scale compute capabilities in terms of extreme concurrency and high numbers of operations per

This work is supported by Advanced Scientific Computing Research, Office of Science, U.S. Department of Energy, under Contract DE-AC02-06CH11357, program manager Laura Biven.

J. M. F. Rodrigues et al. (Eds.): ICCS 2019, LNCS 11536, pp. 18–31, 2019.
https://doi.org/10.1007/978-3-030-22734-0_2

second. That, in addition to the prevalence of IoT and scientific observational devices, led to unprecedented data volumes and rates. However, there is currently a gap between our increased ability to generate raw data and our ability to store, analyze, and produce scientific results based on these data. This paper seeks to bridge this gap by building upon a fundamentally different kind of data model, termed ***Multivariate Functional Approximation*** (MFA) [19], that conserves resources while improving data understanding and sharing. The new model, which can accommodate many types of scientific datasets on HPC architectures, provides compression and facilitates analytical reasoning not possible before. Moreover, the accuracy of the model is known and guaranteed to user prescription.

The MFA model relies on fitting piecewise smooth functionals to multi-dimensional discrete input data. NURBS basis functions are chosen in MFA because they allow the model to be directly usable in downstream analytics and visualization without the need to evaluate (***decode***) the entire model. This is due to well established NURBS features: NURBS models are continuous across all the input domain, differentiable up to the degree of the used basis functions, and preserve geometric and statistical properties of the input discrete data points. Model continuity provides implicit and inexpensive point evaluation anywhere in the input domain, meaning the model can be sampled at a different mesh resolution than the original discrete data. Differentiability is of particular interest to applications relying on gradient fields, useful for feature detection and tracking, and high-order derivatives, to guide mathematical optimization algorithms for example. Those features are the main reason for NURBS being the *de facto* standard for modeling three-dimensional shapes in Computer Aided Design (CAD) applications.

In this paper, we extend NURBS models to high dimensional scientific data, evaluate different methods for fitting a NURBS model in contrast to a nonrational B-spline model, present adaptive refinement methods to guarantee the accuracy of the model in representing input data, and develop an algorithm to refine locally without resolving the model over the whole input domain or compromising the continuity of the model. The motivation behind this local refinement algorithm is to fit large input domains for which a global solve is expensive, more so if the solve needs to be repeated multiple times to reach a fitting error bound.

2 Background

Parametric representations of curves, surfaces, volumes, and hyper-volumes traditionally involve fitting polynomials to known input points. However, the main drawback of polynomial fitting is that its basis function is 'global'; i.e., the fitted value at a given domain location depends on values from all input points across the whole domain. A solution to this problem is to employ additional basis functions that define a local support of the polynomial fit, such as spline bases [9], and radial basis functions [17]. Here, we choose spline bases for their simplicity and desired features highlighted in the previous section, making them suitable as a new data representation for scientific applications.

2.1 B-spline Approximation

A basis spline (B-spline) function is a piecewise polynomial, where each polynomial piece is defined over a subdomain or partition of the whole input domain. The locations where these partitions meet are termed ***knots***. Knot values are usually stored in D knot vectors, each of the form $T_d = \{t_1, ..., t_{k_d}\}$, where k_d is the number of knots for the d_{th} dimension of domain dimensionality D. Each T_d vector is sorted in nondecreasing order, and defined in parameter space $U \in [0, 1]^D$. The domain parameterization function, $M(\bar{x})$, is a mapping function $\mathbb{R}^D \rightarrow [0, 1]^D$, from a point in input coordinates $\bar{x}$ to a point in parameter space $\bar{u} \in U$. This mapping function is usually determined beforehand, along with knot spacing (uniform vs non-uniform), spline basis degree p, and number of ***control points*** $n = \{n_1, ..., n_D\}$. Control points define the shape of the approximated model, so the main objective of B-spline fitting is to find control point locations, or ***encoding***, that, when decoded, closely represent variations in the input variable field (see Fig. 1). In practice, n is usually prespecified by the user, and the number of knots for the d_{th} dimension is directly calculated as $k_d = n_d + p + 1$. Automatically deciding these parameters given input data points is an active research area, and beyond the scope of this paper. The adaptive refinement methods presented here start from an initial number of knots, and control points, then iteratively increase knot resolution where the fitting error is above a certain error metric bound.

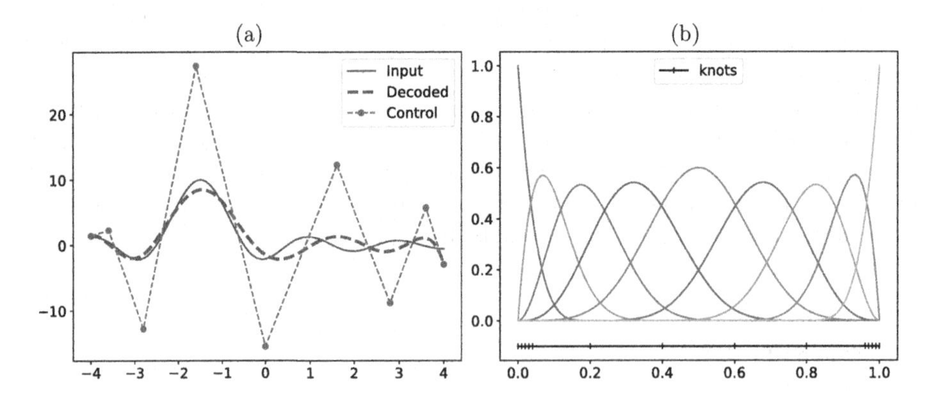

Fig. 1. (a) A 1D curve (blue) and its quartic B-spline approximation (green) using 9 control points (red). (b) A plot showing the 9 4-th degree basis functions (arbitrary colors) used by the B-spline in (a), along with knot locations in parameter space. (Color figure online)

A decoded value from a B-spline model is computed as

$$V(\bar{u}) = \sum_{i=1}^{n} P_i \prod_{d=1}^{D} N_{i,p}(u_d), \quad (1)$$

where $V(\bar{u})$ is the decoded hypervolume value at $\bar{u} = \{u_1, ..., u_D\}$, and $N_{i,p}(u_d)$ is the p-degree B-spline basis function for control point P_i at parameter location u_d. $N_{i,p}(\bar{u})$ can be precomputed and stored in memory, or computed as needed on the fly, by the Cox-de Boor recursion formula [7,8].

As seen from Eq. 1, decoding one point from the MFA model involves a tensor product of the basis functions with the d-dimensional mesh of control points. Encoding an MFA model for input point field $I = \{y_1, ..., y_m\}$ of size m is traditionally achieved by solving a least squares problem derived from the L_2 norm defining the accuracy of the encoded model by the sum of squared errors (SSE) metric [20],

$$SSE = \sum_{i=1}^{m} \|V(M(x_i)) - I(x_i)\|^2 . \quad (2)$$

2.2 NURBS

NURBS, as the name suggests, is the rational extension of B-splines. It was first introduced and used within CAD software tools because the B-spline formulation is incapable of accurately representing conics [22]. For our specific use case here, using NURBS results in a more compact model than using B-splines.

The difference between the rational equations of NURBS and B-splines is associating a weight variable w_i with each control point P_i. Decoding a NURBS MFA model follows a similar approach to Eq. 1 for B-splines, with the exception of substituting $N_{i,p(\bar{u})}$ with rational basis functions $R_{i,p}(\bar{u})$, defined as

$$R_{i,p}(\bar{u}) = \frac{N_{i,p}(\bar{u}) w_i}{\sum_{j=1}^{n} N_{j,p}(\bar{u}) w_j}. \quad (3)$$

NURBS encoding, however, is not as straightforward as the decoding modification. The division in Eq. 3 results in a nonlinear problem. That is why most NURBS implementations resort to using uniform weights (all set to 1) that are solved like B-spline models, and then manually tweaked by users for additional model shape control.

3 Adding Weights

There has been some work on approaches for solving the nonlinear problem of finding NURBS control point locations and weights simultaneously. Laurent-Gengoux and Mekhilef [15] manually derive analytical gradients of the nonlinear problem w.r.t. the control point locations and weights, and also the knot locations. The gradients are then employed by a numerical optimization method to minimize a cost function with appropriate geometric continuity constraints. The method relies on a specific complicated and time-consuming optimization algorithm, and the derivation is only valid for cubic (degree 3) NURBS. In order to overcome these shortcomings, Xie et al. [28] propose an alternating linear projection approach, where the control point locations are first determined using least squares, then their corresponding weights are found using conjugate gradient descent optimization methods. These two steps are then repeated until convergence. An alternative approach is an explicit two-step linear scheme proposed by Ma and Kruth [16]. In this explicit method, control point weights are first identified from a homogeneous system using symmetric eigenvalue decomposition and linear programming, and their locations are subsequently found using least squares. A third method involves the use of automatic differentiation to directly calculate the analytical derivative of the error function in Eq. 2 w.r.t both P_i and w_i in one step. Other notable approaches that also attempt to directly find control point locations and weights at the same time include the use of metaheuristic techniques, such as evolutionary and swarm intelligence search algorithms [23,26].

3.1 Automatic Differentiation

Automatic differentiation (AD) [21], also known as Algorithmic Differentiation, solves many problems with symbolic and numerical differentiation. It can automatically provide derivatives, high-order derivatives, and partial derivatives with respect to many input parameter functions defined in computer source code [12]. AD approaches typically use a computational graph that is traversed to compute the root node derivative by aggregating the partial derivatives along all paths to leaf nodes, applying the chain rule for every edge weight [4].

The past few years have seen a growing interest in AD from both academia and industry, fueled by a need for generic, user-friendly deep learning toolkits. The backpropagation learning algorithm used to train deep neural networks is a special case of reverse mode AD [24]. Consequently, several libraries available for deep learning also include high performance routines for AD [3,27]. TensorFlow [1] is a Python based deep learning API provided by Google implemented with a GPU backend. In this paper, we extend our previous use of Tensorflow for solving inverse problems [18] to automatically calculate gradients for the NURBS fitting forward model and error metric previously defined in Eq. 2.

3.2 Evaluation

In this section we present a comparison of three different methods for fitting an MFA NURBS model: Xie'12 [28], Ma&Kruth'95 [16], and AD [21]. B-spline (unweighted) fitting is also included in the evaluation to assess the benefit of solving for control point weights. Throughout the experiments presented within this paper, we use one synthetic dataset, and one scientific dataset generated from a production HPC simulation code.

Datasets. For the synthetic data we use the *sinc* cardinal sine function of the form $y = sin(x)/x$, that we can generate in any dimensionality and resolution. In order to increase the range and slope of the data, we scaled the sinc function by a factor of 10. The 1D sinc function is $f(x) = 10sin(x)/x$. In 2D, $f(x,y) = 10sinc(x)sinc(y)$, and so forth for higher dimensions. The sinc function was chosen for its smoothness; because of its high degree of continuity, the MFA is able to model such data efficiently. In contrast to the sinc data, *S3D* is a turbulent combustion data set generated by an S3D simulation [6] of fuel jet combustion in the presence of an external cross-flow [13]. This dataset is non-smooth, with sharp edges and high-frequency details, and is representative of actual data one would encounter in scientific experiments or simulations. The domain is 3D (x, y, z) $(704 \times 540 \times 550)$, and the range variable $f(x, y, z)$ is the magnitude of the 3D velocity vector at each domain point. We slice this dataset to produce 1D and 2D cross-sections.

Experiment and Results. We ran the four fitting algorithms with sinc, generated for 1D input curve ($m = 1000$), and S3D 1D curve sliced in the X-axis ($m = 704$). The experiments were performed by varying the number of control points from the minimum number ($n = p+1$) to half the input points ($n = m/2$). For all runs, we use quartic curve fitting ($p = 4$), and the L-BFGS-B optimizer [5] for the gradients provided by the Xie'12 and AD methods, for which the weights bounds were set within the range $[10^{-4}, 1]$ to keep the weights positive. The plots in Fig. 2 report the accuracy of the fitting as defined by the SSE metric in log-log scale.

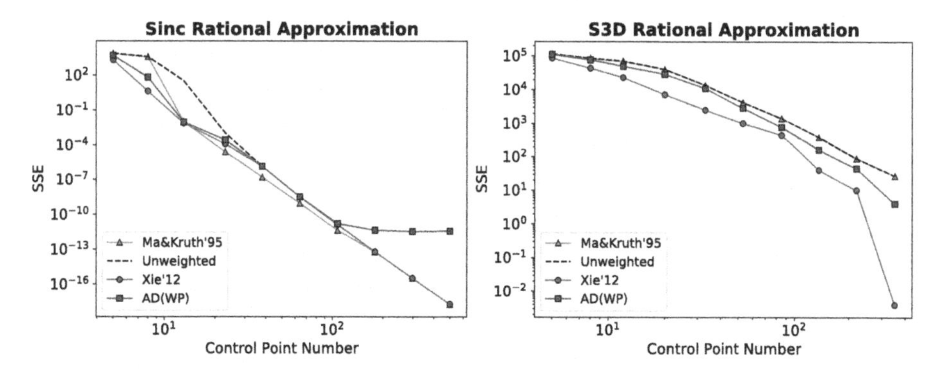

Fig. 2. Rational approximation accuracy plots with increasing control point number. Left plot, synthetic data (sinc). Right plot, combustion simulation data (S3D).

The left plot shows that all methods perform well on the smooth synthetic input, with rational algorithms generally providing superior approximations, when the ratio of input points to control points, (m/n), is higher; this directly translates to better compression factors. The reason AD accuracy stagnates around $SSE = 10^{-13}$ is because it is running on the GPU and using single-precision (32-bit format), while the other methods are using double-precision (64-bit). S3D results, shown in the right plot of Fig. 2, favor iterative gradient-based methods (Xie'12 and AD) over explicit methods (Ma&Kruth'95 and Unweighted) when dealing with nonsmooth real input data.

4 Rational Adaptivity

Next we turn to extending our B-spline adaptive algorithm [19] to NURBS; solving for control point weights in addition to their locations. Adaptivity is achieved through increasing knot resolution where an error metric is above a given error bound ϵ, which, in turn, leads to additional control points in this underrepresented region. Here we use Normalized Squared Error (NSE) as the error metric used for adaptivity, but other metrics can similarly be used depending on the scientific application. We implemented two variants of this approach: one that splits all the knot spans where the user-set error bound is violated, termed ***wga*** for weighted global all; the other approach splits one knot span at each adaptive iteration, termed ***wg1*** for weighted global one. The two variants of our rational adaptive method are presented in Algorithm 1.

Algorithm 1. Global adaptive encoding algorithm

```
1: function GLOBALADAPTIVE(I, T, ε, splitAll)
2:   do
3:     R ← basis(T)
4:     P, w ← RationalEncode(I, R)
5:     E ← Decode(P, w, R) − I
6:     NSE ← E²/(max(I) − min(I))
7:     toSplit ← findKnots(NSE, T, ε, splitAll)
8:     T_new ← ∅
9:     for all t_i ∈ T do
10:      T_new = T_new ⋃ t_i
11:      if i ∈ toSplit then
12:        T_new = T_new ⋃{t_i + ((t_{i+1} − t_i)/2)}
13:    T = T_new
14:  while toSplit ≠ ∅
15:  return P, w, T
16: function FINDKNOTS(NSE, T, ε, splitAll)
17:  if splitAll then
18:    return {where(NSE > ε, T)}          ▷ returns indices in T where NSE > ε
19:  else
20:    maxNSE ← max(NSE)
21:    if maxNSE > ε then
22:      return {where(NSE = maxNSE, T)}   ▷ returns the index of maxNSE in T
23:    else
24:      return ∅
```

The algorithm starts with an initial knot distribution stored in T, then, using one of the methods presented in Sect. 3, computes a rational encoding, for input field I, in terms of control point locations and weights, P and w. To compute the encoded model accuracy we perform a global decode operation, as specified by Eq. 1, and compare the decoded model to the input field. From the calculated error field E, the locations of errors above the error bound ϵ

are mapped to knot values in T, and their associated knot spans are split in the middle. This process is repeated until there are no more knot spans to split, which is either due to E being below ϵ everywhere in the input domain, or not enough input points within the spans to be split. The latter scenario means a rational polynomial of degree p is not sufficient to model the discrete input points within the given knot spacing, with the desired error bound ϵ. Remedies to this problem include modifying the overall NURBS degree used for fitting, the use of domain decomposition techniques to assign different degrees to different subdomains, or a hierarchy of different subdomains with varying spline properties at each level of the hierarchy [11]. In this paper, we leave the investigation into NURBS degree and initial conditions for future work. Therefore, the rational adaptive methods, as presented here, might not converge for certain inputs.

4.1 Local Rational Adaptivity

The global adaptive algorithm presented in the previous subsection is computationally inefficient for two main reasons: (1) it includes a global encode and decode at the beginning of every adaptive iteration, and (2) the global rational encode procedure, in Algorithm 1 line 3, ignores the results of the previous iteration, starting from scratch over the whole domain of input points. To address these shortcomings, we develop a local adaptive algorithm that is able to incrementally refine a rational model. The algorithm steps are listed in Algorithm 2, and highlighted in Fig. 5.

Algorithm 2. Local adaptive encoding algorithm

```
1: function LocalAdaptive(I, T, ε, p, n, m)
2:    R ← basis(T)
3:    P, w ← RationalEncode(I, R)
4:    E ← Decode(P, w, R) − I
5:    NSE ← E^2/(max(I) − min(I))
6:    while mean(NSE) > ε do
7:       toSplit ← findKnots(NSE, T, ε, False)
8:       T_new, P, w, pStart, pEnd ← knotInsertion(toSplit, T, P, w)
9:       R ← basis(T_new)
10:      localStart ← max(pStart − p, 1)      ▷ expand the local domain to include constraints
11:      localEnd ← min(pEnd + p, n)
12:      R_local ← R[localStart : localEnd]             ▷ extract local basis from R
13:      I_local ← I[where(R_local > 0, I)]             ▷ extract local input points from I
14:      P_local, w_local ← LocalRationalEncode(I_local, R_local, p)   ▷ local solve with p constraints
15:      E_local ← Decode(P_local, w_local, R_local) − I_local
16:      P[localStart : localEnd] ← P_local             ▷ update control point locations
17:      w[localStart : localEnd] ← w_local             ▷ update control point weights
18:      NSE[where(R_local > 0, NSE)] ← E_local^2/(max(I) − min(I))      ▷ update error
19:      T = T_new
20:   return P, w, T
```

Our local adaptive method, ***wl1***, relies on two important modifications to Algorithm 1 to address its shortcomings. The global rational encode and decode procedures are replaced with their local counterparts; acting on subdomains of I, R, P, and w. In particular, the rational encode optimization problem is amended to include boundary equality constraints, in order to preserve p-continuity at the local subdomain edges [11] (see Fig. 5, bottom row). The other modification incorporates the standard knot insertion algorithm [20] instead of a global encode after splitting a knot span. Standard knot insertion is an algorithm similar to DeCasteljau's method for subdividing Bézier curves [10] in that it adds one knot, and in turn one control point, to a spline model without changing the shape of the decoded curve. For a 1D curve, this is achieved by removing $p-1$ control points, belonging to the knot to be split, and calculating positions for p new control points explicitly using a triangular recursion scheme. In the case of 2D surfaces, $p-1$ control point rows and columns are removed, and p control point rows and columns are added as depicted in Fig. 5.

4.2 Evaluation

We compare the three variants of our rational adaptive encoding algorithms (wga, wg1, wl1) with their unweighted counterparts (ga, g1, l1), with varying error bounds, in terms of compression factors (m/n), and total runtime. In order to provide a fair walltime comparison, we unify the encoding procedures to use AD for all the six methods, coupled with a Sequential Least Squares Programming (SLSQP) optimizer [14] that is able to handle constrained nonlinear optimization problems. This approach highlights the flexibility of our AD method for solving either nonlinear problems with constraints, without constraints, or the linear problem of determining just control point locations given their weights, with minimal changes to the underlying forward model, and without needing to change the derivative calculations.

Experiments and Results. First, we provide an evaluation for the 1D sinc, and S3D datasets with the same parameters as presented in Sect. 3. For initial conditions for the adaptive algorithms, we use $p = 4$, and $n = p+1$. The plots in Fig. 3 report the different methods compression factors and run times in log-log scale. The results in Fig. 3 further reinforce our finding of rational approximations generally outperforming their equivalent unweighted models, i.e. NURBS are better than B-splines. It is also not surprising that the variants of our adaptive algorithm that split one knot span at a time (wg1, wl1) mostly provide better compression factors (fewer control points), than the variants that split all knot spans that violate the error bound. In fact, the flat portions of wga and ga lines signify these methods 'overshooting,' or splitting more knots than needed. In terms of run time, the rational local method is the worst performing among all the tested methods. This is attributed to the overhead needed for solving the constrained optimization problem which seems to outweigh the benefits gained from a local solve. Moreover, since all methods start from the minimum number of control points, the local method, which refines the decoded curve from previous iterations, does not fare well compared to methods that recompute the whole model at every iterations. In order to support these claims, we conduct an experiment where we compare the rational adaptive methods with increasing number of input points. The plots in Fig. 4 show the performance of our rational adaptive algorithms on the S3D 1D curve resampled at different domain

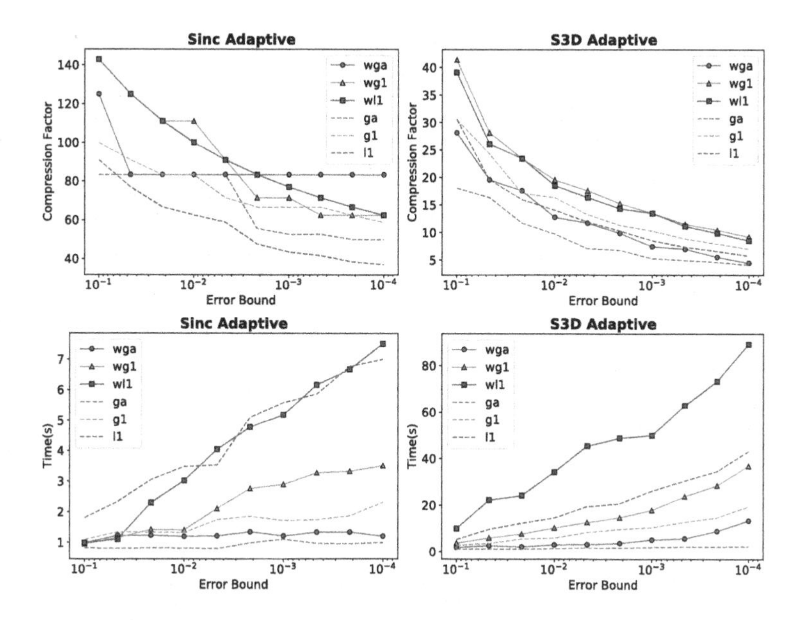

Fig. 3. Adaptive algorithms comparison with varying error bounds. Solid lines are used for the rational methods, while unweighted methods are depicted with dashed lines.

spacings, employing cubic spline interpolation, $\epsilon = 10^{-4}$, and $n = 50$ initial control points distributed uniformly along the resampled input domain. It is worth noting, this initial number of control points was chosen arbitrarily without any manual tuning. From the plots in Fig. 4, we can see that wg1 run time starts scaling exponentially at 20k points. As the number of input points increases, while the number of initial control points is kept constant, wg1 requires more iterations to reach the desired fitting accuracy. Since each iteration contains global rational encode and decode operations, wg1 iterations eventually become more expensive than wl1 iterations, which are composed of constrained local rational encode and decode operations.

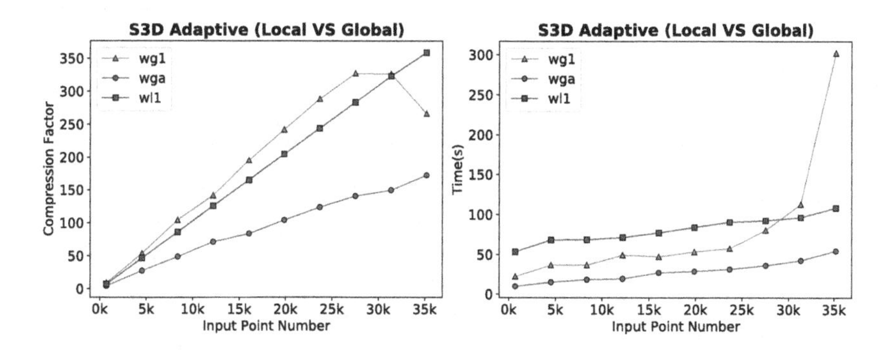

Fig. 4. Rational adaptive approximation of resampled S3D curve.

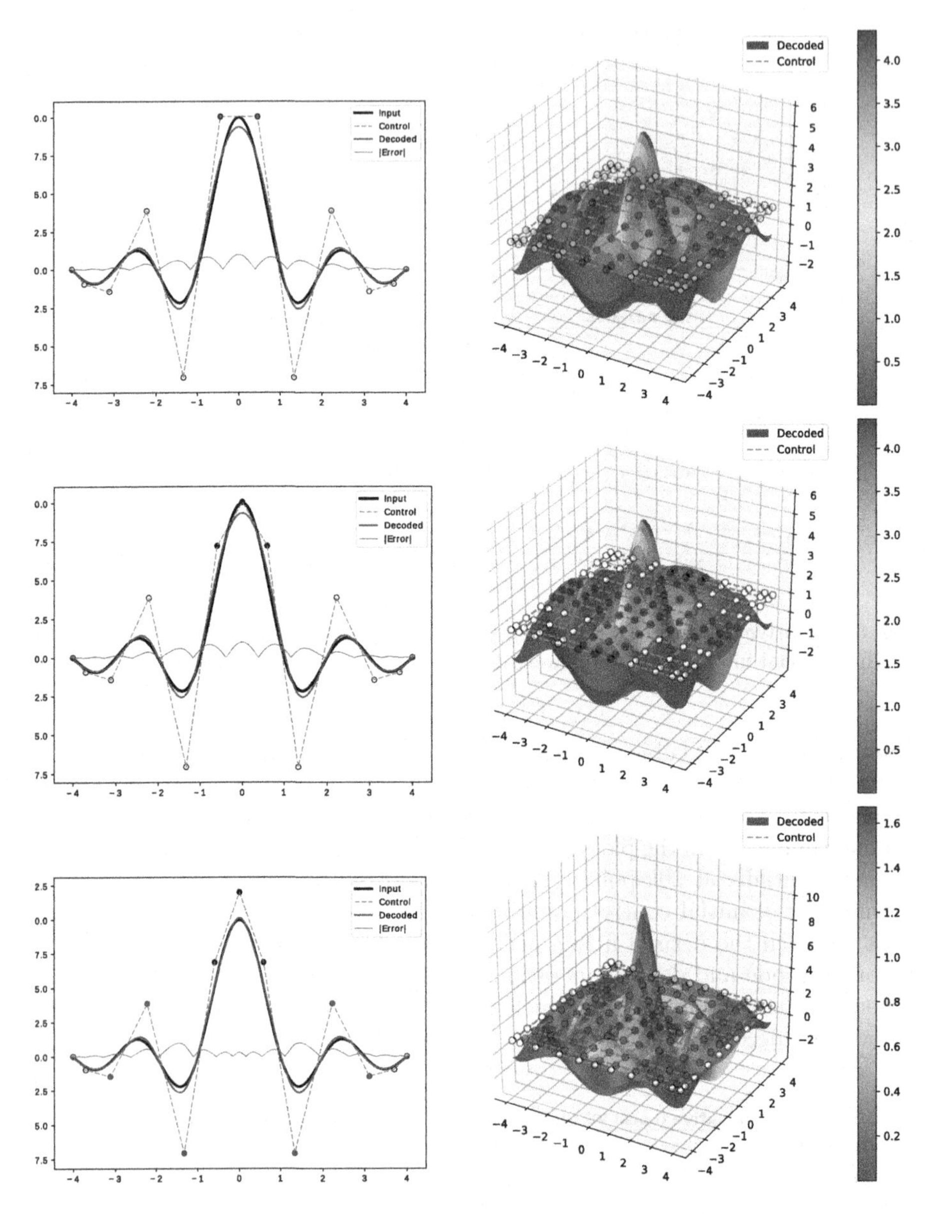

Fig. 5. Local adaptive algorithm steps on 1D (left) and 2D (right) sinc function using a cubic ($p = 3$) NURBS model. The 3D plots on the right show the decoded surface colored by the error magnitude $|E|$, while the plots on the left overlay the error curve (red) with the input (black) and decoded (green) curves. First step: insert a knot where the error is highest by removing $(p-1)$ control points (top row, red circles) and adding p new control points (middle row, black circles), without changing the decoded curve/surface. Second step: (bottom row) perform a local rational encode for the new control points (black circles) with equality constraints for p boundary control points (red circles). Repeat for next highest error location. (Color figure online)

Second, we compare the six adaptive encoding methods on a 2D slice of the S3D dataset, using AD for encoding, SLSQP as the optimization algorithm, $p = 4$, and $n = \{10, 10\}$. These tests show all the methods presented in this paper are directly generalizable to high dimensions, and are able to model real scientific inputs (see Fig. 6). The results are presented in Table 1, and are consistent with the results reported for 1D datasets, with the exception of high run time of the wga and ga methods. This is due to the global encoding complexity scaling with the number of control points.

Table 1. 2D S3D Dataset w/2.3×10^4 Input Points, Desired $\epsilon = 0.1$

Methods	Output Ctrl Pts	Cmpr Fctr	Actual SSE	Actual $NMSE$	Actual $max(E)$	Time (s)
wga	2.1×10^3	10.8	4.1×10^5	$5.2 \times 10 - 2$	3.0×10^{-1}	31899
ga	2.1×10^3	10.8	7.2×10^5	9.1×10^{-2}	4.0×10^{-1}	700
wgl	3.6×10^2	65.8	9.4×10^5	1.1×10^{-1}	3.1×10^{-1}	233
gl	4.0×10^2	59.4	1.1×10^6	1.4×10^{-1}	3.2×10^{-1}	34
wll	6.2×10^2	38.0	7.5×10^5	9.4×10^{-2}	2.9×10^{-1}	559
ll	6.2×10^2	38.0	9.5×10^5	1.2×10^{-1}	3.2×10^{-1}	129

The results of a sample run of our local adaptive algorithm on S3D is shown in Fig. 6. The encoded model is able to capture rapid changes in the input field by increasing knot and control point resolution in high turbulence regions.

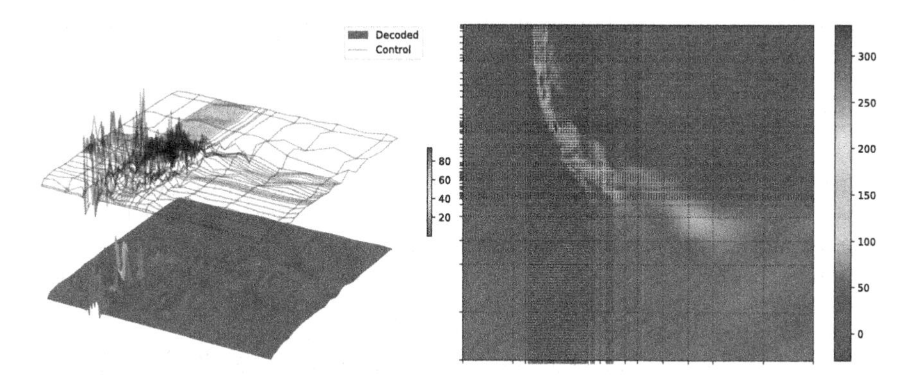

Fig. 6. A visualization of a decoded model for S3D 2D slice. Left, the decoded surface colored by error magnitude, with the control mesh (black) offset upwards for clarity. Right, the decoded surface projected on a 2D image, with the knot locations grid depicted as dotted lines.

The main limitation of the methods presented here, which stems from adopting the standard NURBS model, is their reliance on tensor products of basis function for each domain dimension. While this leads to added cost in terms of extra memory/computation, generally, it results in calculations that can be very easily vectorized. Furthermore, the approximation accuracy should benefit from the overall increased degrees of freedom. The implication of the tensor product formulation for adaptive algorithms is that it requires additional hyper planes of control points in every domain dimension when splitting just one knot span. We are currently working on a T-Splines [2] extension to our work that is compatible with the local adaptive algorithm presented here. The T-Spline formulation will restrict the added control points to the edges of the local region being refined. We are also investigating hybrid local refinement techniques, in which multiple subdomains can be solved in parallel as long as they do not overlap in the unconstrained control points. Additionally, since the quality of the model depends on the nature of the input dataset, we are researching methods for determining initial conditions (parameterization function, knot spacing, number of control points, fitting degree) based on properties of the input data.

5 Conclusion

This paper is primarily concerned with rational approximations, which we show to fit input data more precisely and compactly compared to unweighted ones. Here, we solve a nonlinear problem of finding NURBS optimal control point locations and their associated weights that accurately approximate given input points to user-set error bounds. Additionally, by taking advantage of the local support property of NURBS, we developed an algorithm that is able to locally refine a given approximation on a subset of the input domain. This effectively reduces the computational burden by restricting the iterative gradient-based optimization locally in subdomains of both the approximation and the input domains, and naturally lends the algorithm to a parallel implementation [25].

References

1. Abadi, M., Agarwal, A., Barham, P., Brevdo, E., Chen, Z., Citro, C., Corrado, G.S., Davis, A., Dean, J., Devin, M., et al.: TensorFlow: large-scale machine learning on heterogeneous distributed systems. arXiv preprint arXiv:1603.04467 (2016)
2. Bazilevs, Y., Calo, V.M., Cottrell, J.A., Evans, J.A., Hughes, T.J.R., Lipton, S., Scott, M.A., Sederberg, T.W.: Isogeometric analysis using T-splines. Comput. Methods Appl. Mech. Eng. **199**(5–8), 229–263 (2010)
3. Bergstra, J., Breuleux, O., Bastien, F., Lamblin, P., Pascanu, R., Desjardins, G., Turian, J., Warde-Farley, D., Bengio, Y.: Theano: a CPU and GPU math compiler in Python. In: Proceedings of 9th Python in Science Conference, pp. 1–7 (2010)
4. Bischof, C.H., Hovland, P.D., Norris, B.: On the implementation of automatic differentiation tools. Higher-Order Symbolic Comput. **21**(3), 311–331 (2008)
5. Byrd, R.H., Lu, P., Nocedal, J., Zhu, C.: A limited memory algorithm for bound constrained optimization. SIAM J. Sci. Comput. **16**(5), 1190–1208 (1995)
6. Chen, J.H., Choudhary, A., De Supinski, B., DeVries, M., Hawkes, E.R., Klasky, S., Liao, W.K., Ma, K.L., Mellor-Crummey, J., Podhorszki, N., et al.: Terascale direct numerical simulations of turbulent combustion using S3D. Comput. Sci. Discov. **2**(1), 015001 (2009)

7. Cox, M.G.: The numerical evaluation of B-splines. IMA J. Appl. Math. **10**(2), 134–149 (1972)
8. De Boor, C.: On calculating with B-splines. J. Approximation Theory **6**(1), 50–62 (1972)
9. De Boor, C., De Boor, C., Mathématicien, E.U., De Boor, C., De Boor, C.: A Practical Guide to Splines, vol. 27. Springer, New York (1978)
10. De Casteljau, P.d.F.: Shape Mathematics and CAD, vol. 2. Kogan Page (1986)
11. Forsey, D.R., Bartels, R.H.: Hierarchical b-spline refinement. ACM Siggraph Comput. Graph. **22**(4), 205–212 (1988)
12. Griewank, A., Juedes, D., Utke, J.: Algorithm 755: ADOL-C: a package for the automatic differentiation of algorithms written in C/C++. ACM Trans. Math. Softw. (TOMS) **22**(2), 131–167 (1996)
13. Grout, R., Gruber, A., Yoo, C.S., Chen, J.: Direct numerical simulation of flame stabilization downstream of a transverse fuel jet in cross-flow. Proc. Combust. Inst. **33**(1), 1629–1637 (2011)
14. Kraft, D.: A software package for sequential quadratic programming. Forschungsbericht- Deutsche Forschungs- und Versuchsanstalt fur Luft- und Raumfahrt (1988)
15. Laurent-Gengoux, P., Mekhilef, M.: Optimization of a NURBS representation. Comput. Aided Design **25**(11), 699–710 (1993)
16. Ma, W., Kruth, J.P.: Nurbs curve and surface fitting and interpolation (1995)
17. Mai-Duy, N., Tran-Cong, T.: Approximation of function and its derivatives using radial basis function networks. Appl. Math. Model. **27**(3), 197–220 (2003)
18. Nashed, Y.S., Peterka, T., Deng, J., Jacobsen, C.: Distributed automatic differentiation for ptychography. Procedia Comput. Sci. **108**, 404–414 (2017)
19. Peterka, T., Nashed, Y., Grindeanu, I., Mahadevan, V., Yeh, R., Trixoche, X.: Foundations of multivariate functional approximation for scientific data. In: Proceedings of 2018 IEEE Symposium on Large Data Analysis and Visualization (2018)
20. Piegl, L., Tiller, W.: The NURBS Book. Springer Science & Business Media, Heidelberg (2012). https://doi.org/10.1007/978-3-642-97385-7
21. Rall, L.B.: Automatic Differentiation: Techniques and Applications. Springer, Heidelberg (1981)
22. Rogers, D.F.: An Introduction to NURBS: With Historical Perspective. Elsevier, Oxford (2000)
23. Sarfraz, M.: Computer-aided reverse engineering using simulated evolution on NURBS. Virt. Phys. Prototyping **1**(4), 243–257 (2006)
24. Schmidhuber, J.: Deep learning in neural networks: an overview. Neural Netw. **61**, 85–117 (2015)
25. Toselli, A., Widlund, O.: Domain Decomposition Methods-Algorithms and Theory, vol. 34. Springer Science & Business Media, Heidelberg (2006)
26. Ulker, E.: Nurbs curve fitting using artificial immune system. Int. J. Innov. Comput. Inf. Control **8**, 2875–87 (2012)
27. Walter, S.F., Lehmann, L.: Algorithmic differentiation in python with algopy. J. Comput. Sci. **4**(5), 334–344 (2013)
28. Xie, W.C., Zou, X.F., Yang, J.D., Yang, J.B.: Iteration and optimization scheme for the reconstruction of 3D surfaces based on non-uniform rational B-splines. Comput. Aided Design **44**(11), 1127–1140 (2012)

Design of a High-Performance Tensor-Vector Multiplication with BLAS

Cem Bassoy(✉)

Fraunhofer IOSB, 76275 Ettlingen, Germany
cem.bassoy@iosb.fraunhofer.de

Abstract. Tensor contraction is an important mathematical operation for many scientific computing applications that use tensors to store massive multidimensional data. Based on the Loops-over-GEMMs (LOG) approach, this paper discusses the design of high-performance algorithms for the mode-q tensor-vector multiplication using efficient implementations of the matrix-vector multiplication (GEMV). Given dense tensors with any non-hierarchical storage format, tensor order and dimensions, the proposed algorithms either directly call GEMV with tensors or recursively apply GEMV on higher-order tensor slices multiple times. We analyze strategies for loop-fusion and parallel execution of slice-vector multiplications with higher-order tensor slices. Using OpenBLAS, our parallel implementation attains 34.8 Gflops/s in single precision on a Core i9-7900X Intel Xeon processor. Our parallel version of the tensor-vector multiplication is on average 6.1x and up to 12.6x faster than state-of-the-art approaches.

1 Introduction

Numerical multilinear algebra has become ubiquitous in many scientific domains such as computational neuroscience, pattern recognition, signal processing and data mining [4,11]. Tensors representing large amount of multidimensional data are decomposed and analyzed with the help of basic tensor operations where the contraction of tensors plays a central role [5,6]. To support numeric computations, the development and analysis of high-performance kernels for the tensor contraction have gained greater attention. Based on the Transpose-Transpose-GEMM-Transpose (TGGT) approach, [2,13] reorganize tensors in order to perform a tensor contraction with an optimized matrix-matrix multiplication (GEMM) implementation. A more recent method, GEMM-like Tensor-Tensor Multiplication (GETT), is to design algorithms according to high-performance GEMM [1,9,14]. Other methods are based on the LOG approach in which algorithms utilize GEMM with multiple tensor slices [7,10,12]. Focusing on class 3 compute-bound tensor contractions with free tensor indices, most implementations of the above mentioned approaches reach near peak performance of the computing machine [9,12,14].

In this work, we design and analyze high-performance algorithms for the tensor-vector multiplication that is used in many numerical algorithms, e.g. the

J. M. F. Rodrigues et al. (Eds.): ICCS 2019, LNCS 11536, pp. 32–45, 2019.
https://doi.org/10.1007/978-3-030-22734-0_3

higher-order power method [2,5,6]. Our analysis is motivated by the observation that implementations for class 3 tensor contractions do not perform equally well for tensor-vector multiplications. Our approach is akin to the one proposed in [7,12] but targets the utilization of general matrix-vector multiplication routines (`GEMV`) using `OpenBLAS` [15] without code generation. We present new recursive in-place algorithms that compute the tensor-vector multiplication by executing `GEMV` with slices and fibers of tensors. Moreover, except for few corner cases, we demonstrate that in-place tensor-vector multiplications with any contraction mode can be implemented with one recursive algorithm using multiple slice-vector multiplications and only one `GEMV` parameter configuration. For parallel execution, we propose a variable loop fusion method with respect to the slice order of slice-vector multiplications. Our algorithms support dense tensors with any order, dimensions and any non-hierarchical layouts including the first- and the last-order storage formats for any contraction mode. We have quantified the impact of the tensor layout, tensor slice order and parallel execution of slice-vector multiplications with varying contraction modes. The runtime measurements of our implementations are compared with those presented in [1,9,14]. In summary, the main findings of our work are:

- A tensor-vector multiplication is implementable by an in-place algorithm with 1 `DOT` and 7 `GEMV` parameter configurations supporting all combinations of contraction mode, tensor order, dimensions and non-hierarchical storage format validating the second recipe in [10] with a precise description.
- Algorithms with variable loop fusion and parallel slice-vector multiplications can achieve the peak performance of a `GEMV` with large slice dimensions. The use of order-2 tensor slices helps to retain the performance at a peak level.
- A `LOG`-based implementation is able to compute a tensor-vector product faster than `TTGT`- and `GETT`-based implementations that have been described in [1,9, 14]. Using symmetrically shaped tensors, an average speedup of 3 to 6x for single and double precision floating point computations can be achieved.

The remainder of the paper is organized as follows. Section 2 presents related work. Section 3 introduces the terminology used in this paper and defines the tensor-vector multiplication. Algorithm design and methods for parallel execution is discussed in Sect. 4. Section 5 describes the test setup and discusses the benchmark results in Sect. 6. Conclusions are drawn in Sect. 7.

2 Related Work

The authors in [10] discuss the efficient tensor contractions with highly optimized `BLAS`. Based on the `LOG` approach, they define requirements for the use of `GEMM` for class 3 tensor contractions and provide slicing techniques for tensors. The slicing recipe for the class 2 categorized tensor contractions contains a short description with a rule of thumb for maximizing performance. Runtime measurements cover class 3 tensor contractions.

The work in [7] presents a framework that generates in-place tensor-matrix multiplication according to the `LOG` approach. The authors present two strategies for efficiently computing the tensor contraction applying `GEMM`s with tensors. They report a speedup of up to 4x over the `TTGT`-based `MATLAB` tensor toolbox library discussed in [2]. Although many aspects are similar to our work, the authors emphasize the code generation of tensor-matrix multiplications using high-performance `GEMM`'s.

The authors of [14] present a tensor-contraction generator `TCCG` and the `GETT` approach for dense tensor contractions that is inspired from the design of a high-performance `GEMM`. Their unified code generator selects implementations from generated `GETT`, `LoG` and `TTGT` candidates. Their findings show that among 48 different contractions 15% of `LoG` based implementations are the fastest. However, their tests do not include the tensor-vector multiplication where the contraction exhibits at least one free tensor index.

Using also the `GETT` approach, the author presents in [9] a runtime flexible tensor contraction library. He describes block-scatter-matrix algorithm which uses a special layout for the tensor contraction. The proposed algorithm yields results that feature a similar runtime behavior to those presented in [14].

3 Background

Notation. An order-p *tensor* is a p-dimensional array or hypermatrix with p modes [8]. For instance, scalars, vectors and matrices are order-0, order-1 and order-2 tensors. We write a, $\mathbf{a}$, $\mathbf{A}$ and $\underline{\mathbf{A}}$ in order to denote scalars, vectors, matrices and tensors. In general we will assume the order of a tensor to be p and explicitly mention it otherwise. Each dimension n_r of the r-th mode shall satisfy $n_r > 1$. The p-tuple $\mathbf{n}$ with $\mathbf{n} = (n_1, n_2, \ldots, n_p)$ will be referred to as a *dimension tuple*. We will use round brackets $\underline{\mathbf{A}}(i_1, i_2, \ldots, i_p)$ or $\underline{\mathbf{A}}(\mathbf{i})$ together with a multi-index $\mathbf{i} = (i_1, i_2, \ldots, i_p)$ to identify tensor elements. The set of all multi-indices of a tensor is denoted by $\mathcal{I}$ which is defined as the Cartesian product of all index sets $I_r = \{1, \ldots, n_r\}$. The set $\mathcal{J} = \{0, \ldots, \bar{n} - 1\}$ contains (relative) memory positions of an order-p tensor with $\bar{n} = n_1 \cdot n_2 \cdots n_p$ contiguously stored elements with $|\mathcal{I}| = |\mathcal{J}|$. A *subtensor* denoted by $\underline{\mathbf{A}}'$ shall reference or view a subset of tensor elements where the references are specified in terms of p index ranges. The r-th *index range* shall be given by an index pair denoted by $f_r : l_r$ with $1 \leq f_r \leq l_r \leq n_r$. We will use $:_r$ to specify a range with all elements of the r-th index set. A subtensor is an order-p' *slice* if all modes of the corresponding order-p tensor are selected either with a full index range or a single index where p' with $p' \leq p$ is the number of all non-singleton dimensions. A *fiber* is a tensor slice with only one dimension greater than 1.

Non-Hierarchical Storage Formats and Memory Access. Given a dense order-p tensor, we use a *layout tuple* $\boldsymbol{\pi} \in \mathbb{N}^p$ to encode non-hierarchical storage formats such as the well known first-order or last-order layout. They contain permuted tensor modes whose priority is given by their index. For $1 \leq k \leq p$, an element π_r of *k-order layout* tuple is defined as $k-r+1$ if $1 < r \leq k$ and r in any other case. The well-known first- and last-order storage formats are then given by $\boldsymbol{\pi}_F = (1, 2, \ldots, p)$ and $\boldsymbol{\pi}_L = (p, p-1, \ldots, 1)$. Given a layout tuple $\boldsymbol{\pi}$ with p modes, the π_r-th element of a *stride tuple* is given by

$$w_{\pi_r} = n_{\pi_1} \cdot n_{\pi_2} \cdots n_{\pi_{r-1}} \quad \text{for } 1 < r \leq p. \tag{1}$$

With $w_{\pi_1} = 1$, tensor elements of the π_1-th mode are contiguously stored in memory. In contrast to hierarchical storage formats, all tensor elements with one differing multi-index element exhibit the same stride.

The location of tensor elements within the allocated memory space is determined by the storage format of a tensor and the corresponding *layout function*. For a given layout and stride tuple, a layout function $\lambda_{\mathbf{w}} : \mathcal{I} \rightarrow \mathcal{J}$ maps a multi-index to a scalar index according to

$$\lambda_{\mathbf{w}}(\mathbf{i}) = \sum_{r=1}^{p} w_r (i_r - 1) \tag{2}$$

With $j = \lambda_{\mathbf{w}}(\mathbf{i})$ being the relative memory position of an element with a multi-index $\mathbf{i}$, reading from and writing to memory is accomplished with j and the first element's address of $\underline{\mathbf{A}}$.

Tensor-Vector Multiplication. Let $\underline{\mathbf{A}}$ be an order-p input tensor with a dimension tuple $\mathbf{n} = (n_1, \ldots, n_q, \ldots, n_p)$ and let $\mathbf{b}$ be a vector of length n_q with $p > 1$. Let $\underline{\mathbf{C}}$ be a tensor with $p-1$ modes with a dimension tuple $\mathbf{m} = (n_1, \ldots, n_{q-1}, n_{q+1}, \ldots, n_p)$. A mode-$q$ tensor-vector multiplication is denoted by $\underline{\mathbf{C}} = \underline{\mathbf{A}} \times_q \mathbf{b}$ where

$$c_{i_1, \ldots, i_{q-1}, i_{q+1}, \ldots, i_p} = \sum_{i_q=1}^{n_q} a_{i_1, \ldots, i_q, \ldots, i_p} \cdot b_{i_q} \tag{3}$$

is an element of $\underline{\mathbf{C}}$. Equation (3) is an inner product of a fiber of $\underline{\mathbf{A}}$ and $\mathbf{b}$. The mode q is its *contraction mode*. We additionally term $\boldsymbol{\pi}$ as the layout tuple of the input tensor $\underline{\mathbf{A}}$ with a stride tuple $\mathbf{w}$ that is given by Eq. (1). With no transposition of $\underline{\mathbf{A}}$ or $\underline{\mathbf{C}}$, elements of the layout tuple $\boldsymbol{\varphi}$ of the mode-q tensor-vector product $\underline{\mathbf{C}}$ are given by

$$\varphi_j = \begin{cases} \pi_k & \text{if } \pi_k < \pi_q \\ \pi_k - 1 & \text{if } \pi_k > \pi_q \end{cases} \quad \text{and} \quad j = \begin{cases} k & \text{if } k < q \\ k-1 & \text{if } k > q \end{cases} \tag{4}$$

for $k = 1, \ldots, p$. The stride tuple $\mathbf{v}$ is given by Eq. (1) using the shape $\mathbf{m}$ and permutation tuple $\boldsymbol{\varphi}$ of $\underline{\mathbf{C}}$.

4 Algorithm Design

4.1 Standard Algorithms with Contiguous Memory Access

The control and data flow of the basic tensor-vector multiplication algorithm implements Eq. (3) with a single function. It uses tree recursion with a control

```
1  tensor_times_vector_recursive(A, b, C, n, i, q, q̂, r)
2    if r = q̂ then
3      tensor_times_vector_recursive(A, b, C, n, i, q, q̂, r − 1)
4    else if r > 1 then
5      for i_{π_r} ← 1 to n_{π_r} do
6        tensor_times_vector_recursive(A, b, C, n, i, q, q̂, r − 1)
7    else
8      for i_q ← 1 to n_q do
9        for i_{π_1} ← 1 to n_{π_1} do
10         C(i_1, ..., i_{q−1}, i_{q+1}, ..., i_p) += A(i_1, ..., i_q, ..., i_p) · b(i_q)
```

Algorithm 1. Recursive implementation of the tensor-vector multiplication starting with $r = p$ for $p \geq 2$ and $1 \leq \pi_1 \neq q \leq p$ with better data locality for large dimensions. Iteration along mode $\hat{q}$ with $\hat{q} = (\pi^{-1})_q$ is moved into the inner-most recursion level.

flow akin to one of Algorithm 1 in [3]. Instead of combining two scalars elementwise in the inner-most loop, the tensor-vector multiplication algorithm computes an inner product and skips the iteration over the q-th index set, i.e. the q-th loop. The algorithm supports tensors with arbitrary order, dimensions and any non-hierarchical storage format. However, it accesses memory non-contiguously if the storage format does not prioritize the q-th mode with $\pi_1 \neq q$ and $w_q > 1$, see Eq. (1). The access pattern can be enhanced by modifying the tensor layout, i.e reordering tensor elements according to the storage format. A reordering however, limits its overall performance of the contraction operation [12].

As proposed in [3], elements can be accessed according to the storage format using a permutation tuple. In this way, the desired index set for a given recursion level can be selected. By inserting the q-th (contraction) loop into an already existing branch for $r > 1$ additionally simplifies the algorithm's control-flow. Yet the loop-reordering forces the first $\bar{n}_{k-1} = \prod_{r=1}^{k-1} n_{\pi_r}$ elements of $\underline{\mathbf{C}}$ to be accessed n_q-times with $\pi_k = q$. If the number of reaccessed elements exceeds the last-level cache size, cache missus occur resulting in a poor performance of the algorithm with longer execution times.

Algorithm 1 mproves the data locality if the number of elements $\bar{n}_{k-1}$ exceeds the cache size. By nesting the π_1-th loop inside the i_q-th loop, the function only reuses n_{π_1} elements. This is done by inserting an `if`-statement at the very beginning of the function which skips the q-th loop when $r = \hat{q}$ with $\hat{q} = (\pi^{-1})_q$ where $\hat{q}$ is the index position of q within $\boldsymbol{\pi}$. The proposed algorithm constitutes the starting point for `BLAS` utilization.

4.2 Extended Algorithms Utilizing BLAS

The number of reused elements in Algorithm 1 can be further minimized by tiling the inner-most loops. Instead of applying loop transformations as proposed

Table 1.

Case	Order p	Layout $\boldsymbol{\pi}$	Mode q	Routine	FORMAT	M	N	LDA
1	1	-	1	DOT	-	n_1	-	-
2	2	$(1,2)$	1	GEMV	ROW	n_2	n_1	n_1
3	2	$(1,2)$	2	GEMV	COL	n_1	n_2	n_1
4	2	$(2,1)$	1	GEMV	COL	n_2	n_1	n_2
5	2	$(2,1)$	2	GEMV	ROW	n_1	n_2	n_2
6	>2	any	π_1	GEMV	ROW	$\bar{n}_q$	n_q	n_q
7	>2	any	π_p	GEMV	COL	$\bar{n}_q$	n_q	$\bar{n}_q$
8	>2	any	$\pi_2, \pi_3, \ldots, \pi_{p-1}$	GEMV*	COL	$\hat{n}_q$	n_q	$\hat{n}_q$

Parameter configuration of the DOT- and GEMV with eight cases executing a tensor-vector multiplication with respect to the order p, layout $\boldsymbol{\pi}$ and contraction mode q. All three parameters determine the values of FORMAT, M, N and LDA. GEMV* denotes a multiple execution of GEMV with different tensor slices. In case of order-2 and order-$\hat{q}$ slices, the number of rows must be equal to $\hat{n}_q = n_{\pi_1}$ and $\hat{n}_q = w_q$, respectively. The number of rows for case 6 and 7 is given by $\bar{n}_q = \prod_{r=1}^{p} n_r / n_q$.

in [9,14], we apply highly optimized routines to fully or partly execute tensor contractions as it is done in [7,12] for class 3 tensor operations. The function and parameter configurations for the tensor multiplication can be divided into eight cases.

Case 1 ($p = 1$)*:* The tensor-vector product $\underline{\mathbf{A}} \times_1 \mathbf{b}$ can be computed with a DOT operation $\mathbf{a}^T\mathbf{b}$ where $\underline{\mathbf{A}}$ is an order-1 tensor, i.e. a vector $\mathbf{a}$ of length n_1.

Case 2–5 ($p = 2$)*:* Let $\mathbf{A}$ be an order-2 tensor, i.e. matrix with dimensions n_1 and n_2. If $m = 2$ and if $\mathbf{A}$ is stored according to the column-major $\boldsymbol{\pi} = (1,2)$ or row-major format $\boldsymbol{\pi} = (2,1)$, the tensor-vector multiplication can be trivially executed by a GEMV routine using the tensor's storage format. The two remaining cases for $m = 1$ require an interpretation of the order-2 tensor. In case of the column-major format $\boldsymbol{\pi} = (1,2)$, the tensor-vector product can be computed with a GEMV routine, interpreting the columns of the matrix as rows with permuted dimensions. Analogously, a GEMV routine executes a tensor-vector multiplication with $\boldsymbol{\pi} = (2,1)$.

Case 6–7 ($p > 2$)*:* General tensor-vector multiplications with higher-order tensors execute the GEMV routine multiple times over different slices of the tensor. There are two exceptions to the general case. If $\pi_1 = q$, a single GEMV routine is sufficient for any storage layout. The tensor can be interpreted as a matrix with $\bar{n}_q = \prod_{r=1}^{p} n_r / n_q$ rows and n_q columns. The leading dimension LDA for $\pi_1 = q$ is n_q. Tensor fibers with contiguously stored elements are therefore interpreted as matrix rows. In case of $\pi_p = q$, the leading dimension LDA is given by $\bar{n}_q$ where all fibers with the exception of the dimension π_p are interpreted as matrix columns. The interpretation of tensor objects does not copy data elements.

Case 8 ($p > 2$): For the last case with $\pi_1 \neq q$ and $\pi_p \neq q$, we provide two methods that loop over tensor slices. Lines 8 to 10 of Algorithm 1 perform a slice-vector multiplication of the form $\mathbf{c}' = \mathbf{A}' \cdot \mathbf{b}$. It is executed with a `GEMV` with no further adjustment of the algorithm. The vector $\mathbf{c}'$ denotes a fiber of $\underline{\mathbf{C}}$ with n_u elements and $\mathbf{A}'$ denotes an order-2 slice of $\underline{\mathbf{A}}$ with dimensions n_u and n_v such that

$$\mathbf{A}' = \underline{\mathbf{A}}(i_1, \ldots, :_u, \ldots, :_v, \ldots, i_p) \quad \text{and} \quad \mathbf{c}' = \underline{\mathbf{C}}(i_1, \ldots, :_u, \ldots, i_p) \tag{5}$$

where $u = \pi_1$ and $v = q$ or vice versa. Algorithm 1 needs a minor modification in order to loop over order-$\hat{q}$ slices. With $\hat{q} = (\pi^{-1})_q$, the conditions in line 2 and 4 are changed to $1 < r \leq \hat{q}$ and $\hat{q} < r$, respectively. The modified algorithms therefore omits the first $\hat{q}$ modes $\pi_1, \ldots, \pi_{\hat{q}}$ including $\pi_{\hat{q}} = q$ where all elements of an order-$\hat{q}$ slice are contiguously stored. Choosing the first-order storage format for convenience, the order-$\hat{q}$ and order-$(\hat{q}-1)$ slices of both tensors are given by

$$\underline{\mathbf{A}}' = \underline{\mathbf{A}}(:_1, \ldots, :_q, i_{q+1}, \ldots, i_p) \text{ and } \underline{\mathbf{C}}' = \underline{\mathbf{C}}(:_1, \ldots, :_{q-1}, i_{q+1}, \ldots, i_p). \tag{6}$$

The fiber $\mathbf{c}'$ of length $w_q = n_1 \cdot n_2 \cdots n_{q-1}$ is the one-dimensional interpretation of $\underline{\mathbf{C}}'$ and the order-2 slice $\mathbf{A}'$ with dimensions w_q and n_q the two-dimensional interpretation of $\underline{\mathbf{A}}'$. The slice-vector multiplication in this case can be performed with a `GEMV` that interprets the order-$\hat{q}$ slices as order-2 according to the description. Table 1 summarizes the call parameters of the `DOT` or `GEMV` for all order, storage format and contraction mode combinations.

4.3 Parallel Algorithms with Slice-Vector Multiplications

A straight-forward approach for generating a parallel version of Algorithm 1 is to divide the outer-most π_p-th loop into equally sized iterations and execute them in parallel using the `OpenMP parallel for` directive [3]. With no critical sections and synchronization points, all threads within the parallel region execute their own sequential slice-vector multiplications. The outer-most dimension n_{π_p} determines the degree of parallelism, i.e. the number of parallel threads executing their own instruction stream.

Fusing additional loops into a single one improves the degree of parallelism. The number of fusible loops depends on the tensor order p and contraction mode q of the tensor-vector multiplication with $\hat{q} = (\pi^{-1})_q$. In case of mode-q slice-vector multiplications, loops $\pi_{\hat{q}+1}, \ldots, \pi_p$ are not involved in the multiplications and can be transformed into one single loop. For mode-2 slice-vector multiplications all loops except π_1 and $\pi_{\hat{q}}$ can be fused. When all fusible loops are lexically present and both parameters are known before compile time, loop fusion and parallel execution can be easily accomplished with the `OpenMP collapse` directive. The authors of [7] use this approach to generate parallel tensor-matrix functions.

With variable number of dimensions and a variable contraction mode, the iteration count of slice-vector multiplications and the slice selection needs to be determined at compile or run time. If $\bar{n}$ is the number of tensor elements of $\underline{\mathbf{A}}$, the total number of slice-vector multiplications with mode-$\hat{q}$ slices is given by $\bar{n}' = \bar{n}/w_q$. Using Eq. (1), the strides for the iteration are given by $w_{\pi_{\hat{q}+1}}$ for $\underline{\mathbf{A}}$ and $v_{\pi_{\hat{q}}}$ for $\underline{\mathbf{C}}$. In summary, one single parallel outer loop with an iteration count $\bar{n}'$ and an increment variable j iteratively calls mode-$\hat{q}$ slice-vector multiplications with adjusted memory location $j \cdot w_{\pi_{\hat{q}+1}}$ and $j \cdot v_{\pi_{\hat{q}}}$ for $\underline{\mathbf{A}}$ and $\underline{\mathbf{C}}$, respectively. The degree of parallelism $\prod_{r=\hat{q}+1}^{p} n_r$ decreases with increasing $\hat{q}$ and corresponds for $\hat{q} = p-1$ to the first parallel version. Tensor-vector multiplications with mode-2 slice-vector multiplications are further optimized by fusing additional $\hat{q}-2$ loops.

5 Experimental Setup

Computing System. The experiments were carried out on a Core i9-7900X Intel Xeon processor with 10 cores and 20 hardware threads running at 3.3 GHz. It has a theoretical peak memory bandwidth of 85.312 GB/s resulting from four 64-bit wide channels with a data rate of 2666MT/s. The sizes of the L3-cache and each L2-cache are 14 MB and 1024 KB. The source code has been compiled with `GCC v7.3` using the highest optimization level `-Ofast` and `-march=native`, `-pthread` and `-fopenmp`. Parallel execution for the general case (8) has been accomplished using `GCC`'s implementation of the `OpenMP v4.5` specification. We have used the `DOT` and `GEMV` implementation of the `OpenBLAS` library `v0.2.20`. The benchmark results of each function are the average of 10 runs.

Tensor Shapes. We have used *asymmetrically-shaped* and *symmetrically-shaped* tensors in order to provide a comprehensive test coverage. *Setup 1* performs runtime measurements with *asymmetrically-shaped* tensors. Their dimension tuples are organized in 10 two-dimensional arrays $\mathbf{N}_q$ with 9 rows and 32 columns where the dimension tuple $\mathbf{n}_{r,c}$ of length $r+1$ denotes an element $\mathbf{N}_q(r,c)$ of $\mathbf{N}_q$ with $1 \leq q \leq 10$. The dimension $\mathbf{n}_{r,c}(i)$ of $\mathbf{N}_q$ is 1024 if $i = 1$, $c \cdot 2^{15-r}$ if $i = \min(r+1, q)$ and 2 for any other index i with $1 < q \leq 10$. The dimension $\mathbf{n}_{r,c}(i)$ of $\mathbf{N}_1$ is given by $c \cdot 2^{15-r}$ if $i = 1$, 1024 if $i = 2$ and 2 for any other index i. Dimension tuples of the same array column have the same number of tensor elements. Please note that with increasing tensor order (and row-number), the contraction mode is halved and with increasing tensor size, the contraction mode is multiplied by the column number. Such a setup enables an orthogonal test-set in terms of tensor elements ranging from 2^{25} to 2^{29} and tensor order ranging from 2 to 10. *Setup 2* performs runtime measurements with *symmetrically-shaped* tensors. Their dimension tuples are organized in one two-dimensional array $\mathbf{M}$ with 6 rows and 8 columns where the dimension tuple $\mathbf{m}_{r,c}$ of length $r+1$ denotes an element $\mathbf{M}(r,c)$ of $\mathbf{M}$. For $c = 1$, the dimensions of $\mathbf{m}_{r,c}$ are given by 2^{12}, 2^8, 2^6, 2^5, 2^4 and 2^3 with descending row number r from 6 to 1. For $c > 1$, the remaining dimensions are given by $\mathbf{m}_{r,c} = \mathbf{m}_{r,c} + k \cdot (c-1)$ where k is 2^9, 2^5, 2^3, 2^2, 2, 1 with descending row number r from 6 to 1. In this setup, shape tuples of a column do not yield the same number of subtensor elements.

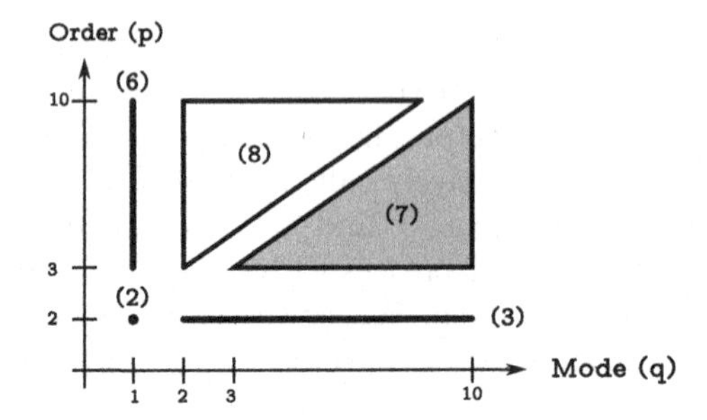

Fig. 1. Schematic contour view of the following average performance maps for the tensor-vector multiplication with tensors that are stored according to the first-order storage format. Each case x in Table 1 affects a different region x within the performance map. Performance values are the arithmetic mean over the set of tensor sizes with 32 and 8 elements in case of the first and second test setup, respectively. Contraction mode $q = p$ for $q > p$ where p is the tensor order.

Performance Maps. Measuring a single tensor-vector multiplication with the first setup produces $2880 = 9 \times 32 \times 10$ runtime data points where the tensor order ranges from 2 to 10, with 32 shapes for each order and 10 contraction modes. The second setup produces $336 = 6 \times 8 \times 7$ data points with 6 tensor orders ranging from 2 to 7, 8 shapes for each order and 7 contraction modes. Similar to the findings in [3], we have observed a performance loss for small dimensions of the mode with the highest priority. The presented performance values are the arithmetic mean over the set of tensor sizes that vary with the tensor order and contraction mode resulting in a three dimensional performance plot. A schematic countour view of the plots is given in Fig. 1 which is divided into 5 regions. The cases 2, 3, 6 and 7 generate performance values within the regions 2, 3, 6 and 7 where only a single parallel `GEMV` is executed, see Table 1. Please note that the contraction mode q is set to the tensor order p if $q > p$. Performance values within region 8 result from case 8 which executes `GEMV`'s with tensor slices in parallel.

The following analysis considers four parallel versions `SB-P1`, `LB-P1`, `SB-PN` and `LB-PN`. `SB` (small-block) and `LB` (large-block) denote parallel slice-vector multiplications where each thread recursively calls a single-threaded `GEMV` with mode-2 and mode-$\hat{q}$ slices, respectively. `P1` uses the outer-most dimension n_p for parallel execution whereas `PN` applies loop fusion and considers all fusible dimensions for parallel execution.

6 Results and Discussion

Matrix-Vector Multiplication. Figure 2 shows average performance values of the four versions `SB-P1`, `LB-P1`, `SB-PN` and `LB-PN` with asymmetrically-shaped tensors. In case 2 (region 2), the shape tuple of the two-order tensor is equal to (n_2, n_1) where n_2 is set to 1024 and n_1 is $c \cdot 2^{14}$ for $1 \leq c \leq 32$. In case 6 (region

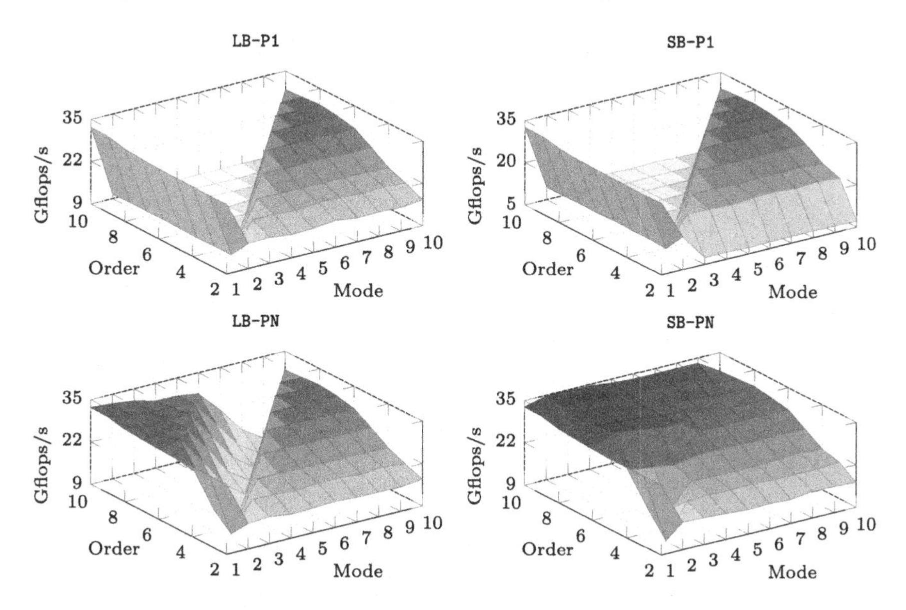

Fig. 2. Average performance maps of four tensor-vector multiplications with varying tensor orders p and contraction modes q. Tensor elements are encoded in single-precision and stored contiguously in memory according to the first-order storage format. Tensors are *asymmetrically-shaped* with dimensions.

6), the p-order tensor is interpreted as a matrix with a shape tuple $(\bar{n}_1, n_1)$ where n_1 is $c \cdot 2^{15-r}$ for $1 \leq c \leq 32$ and $2 < r < 10$. The mean performance averaged over the matrix sizes is around 30 Gflops/s in single-precision for both cases. When $p = 2$ and $q > 1$, all functions execute case 3 with a single parallel `GEMV` where the 2-order tensor is interpreted as a matrix in column-major format with a shape tuple (n_1, n_2). In this case, the performance is 16 Gflops/s in region 3 where the first dimension of the 2-order tensor is equal to 1024 for all tensor sizes. The performance of `GEMV` increases in region 7 with increasing tensor order and increasing number of rows $\bar{n}_q$ of the interpreted p-order tensor. In general, `OpenBLAS`'s `GEMV` provides a sustained performance around 31 Gflops/s in single precision for column- and row-major matrices. However, the performance drops with decreasing number of rows and columns for the column-major and row-major format. The performance of case 8 within region 8 is analyzed in the next paragraph.

Slicing and Parallelism. Functions with `P1` run with 10 Gflops/s in region 8 when the contraction mode q is chosen smaller than or equal to the tensor order p. The degree of parallelism diminishes for $n_p = 2$ as only 2 threads sequentially execute a `GEMV`. The second method `PN` fuses additional loops and is able to generate a higher degree of parallelism. Using the first-order storage format, the outer dimensions $n_{q+1}, \ldots, n_p$ are executed in parallel. The `PN` version speeds up the computation by almost a factor of 4x except for $q = p - 1$. This explains the notch in the left-bottom plot when $q = p - 1$ and $n_p = 2$.

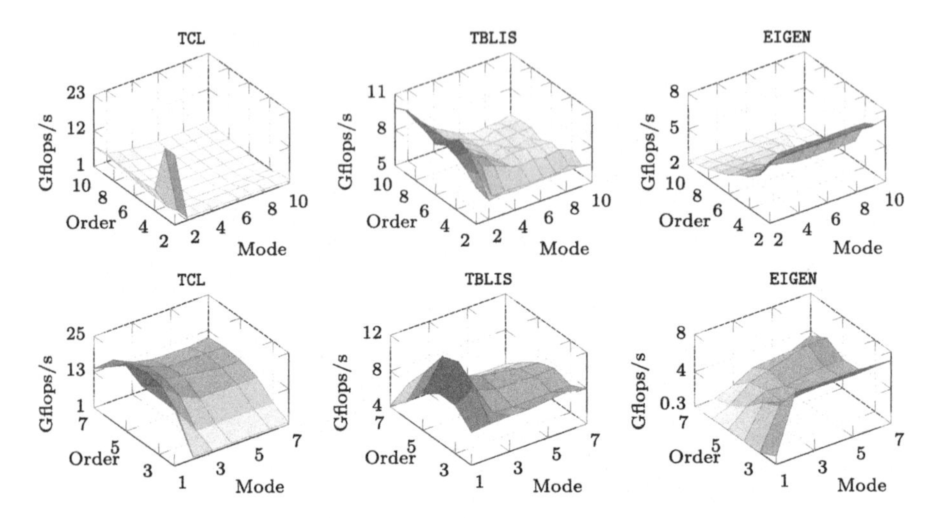

Fig. 3. Average performance maps of tensor-vector multiplication implementations using *asymmetrically-shaped* (top) and *symmetrically-shaped* (bottom) tensors with varying contraction modes and tensor order. Tensor elements are encoded in single-precision and stored contiguously in memory according to the first-order storage format.

In contrast to the LB slicing method, SB is able to additionally fuse the inner dimensions with their respective indices $2, 3, \ldots, p-2$ for $q = p-1$. The performance drop of the LB version can be avoided, resulting in a degree of parallelism of $\prod_{r=2}^{p} n_r / n_q$. Executing that many small slice-vector multiplications with a GEMV in parallel yields a mean peak performance of up to 34.8 (15.5) Gflops/s in single (double) precision. Around 60% of all 2880 measurements exhibit at least 32 Gflops/s that is GEMV's peak performance in single precision. In case of symmetrically-shaped tensors, both approaches achieve similar results with almost no variation of the performance achieving up on average 26 (14) Gflops/s in single (double) precision.

Tensor Layouts. Applying the first setup configuration with asymmetrically-shaped tensors, we have analyzed the effects of the blocking and parallelization strategy. The LB-PN version processes tensors with different storage formats, namely the 1-, 2-, 9- and 10-order layout. The performance behavior is almost the same for all storage formats except for the corner cases $q = \pi_1$ and $q = \pi_p$. Even the performance drop for $q = p - 1$ is almost unchanged. The standard deviation from the mean value is less than 10% for all storage formats. Given a contraction mode $q = \pi_k$ with $1 < k < p$, a permutation of the inner and outer tensor dimensions with their respective indices $\pi_1, \ldots, \pi_{k-1}$ and $\pi_{k+1}, \ldots, \pi_p$ does influence the runtime where the LB-PN version calls GEMV with the values w_m and n_m. The same holds true for the outer layout tuple.

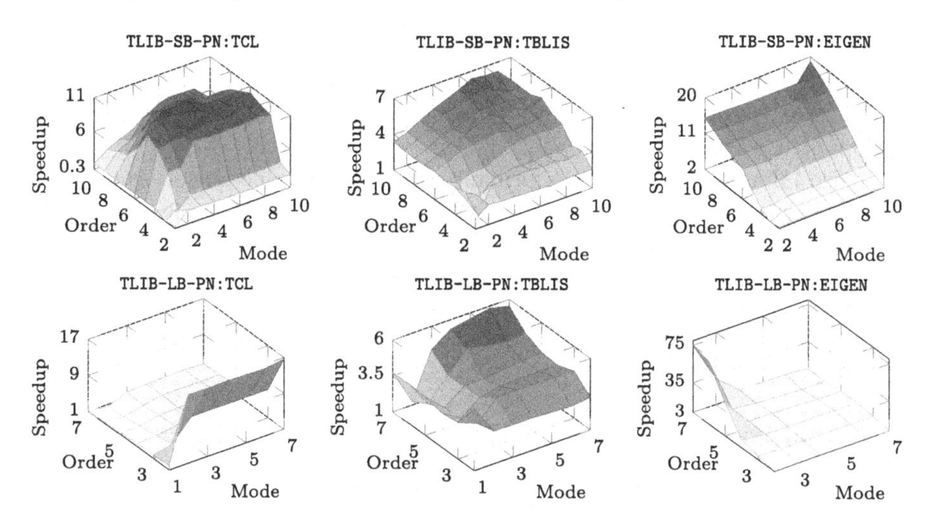

Fig. 4. Relative average performance maps of tensor-vector multiplication implementations using *asymmetrically* (top) and *symmetrically* (bottom) shaped tensors with varying contraction modes and tensor order. Relative performance (speedup) is the performance ratio of TLIB-SB-PN (top) and TLIB-LB-PN (bottom) to TBLIS, TCL and EIGEN, respectively. Tensor elements are encoded in single-precision and stored contiguously in memory according to the first-order storage format.

Comparison with Other Approaches. The following comparison includes three state-of-the-art libraries that implement three different approaches. The library TCL (v0.1.1) implements the (TTGT) approach with a high-perform tensor-transpose library HPTT which is discussed in [14]. TBLIS (v1.0.0) implements the GETT approach that is akin to BLIS's algorithm design for matrix computations [9]. The tensor extension of EIGEN (v3.3.90) is used by the Tensorflow framework and performs the tensor-vector multiplication in-place and in parallel with contiguous memory access [1]. TLIB denotes our library that consists of sequential and parallel versions of the tensor-vector multiplication. Numerical results of TLIB have been verified with the ones of TCL, TBLIS and EIGEN.

Figure 3 illustrates the average single-precision Gflops/s with asymmetrically- and symmetrically-shaped tensors in the first-order storage format. The runtime behavior of TBLIS and EIGEN with asymmetrically-shaped tensors is almost constant for varying tensor sizes with a standard deviation ranging between 2% and 13%. TCL shows a different behavior with 2 and 4 Gflops/s for any order $p \geq 2$ peaking at $p = 10$ and $q = 2$. The performance values however deviate from the mean value up to 60%. Computing the arithmetic mean over the set of contraction modes yields a standard deviation of less than 10% where the performance increases with increasing order peaking at $p = 10$. TBLIS performs best for larger contraction dimensions achieving up to 7 Gflops/s and slower runtimes with decreasing contraction dimensions. In case of symmetrically-shaped tensors, TBLIS and TCL achieve up to 12 and 25 Gflops/s in single precision with a standard deviation between 6% and 20%, respectively. TCL and TBLIS behave similarly and perform better with increasing contraction dimensions. EIGEN executes faster with decreasing order and increasing contraction mode with at most 8 Gflops/s at $p = 2$ and $q \geq 2$.

Figure 4 illustrates relative performance maps of the same tensor-vector multiplication implementations. Comparing TCL performance, TLIB-SB-PN achieves an average speedup of 6x and more than 8x for 42% of the test cases with asymmetrically shaped tensors and executes on average 5x faster with symmetrically shaped tensors. In comparison with TBLIS, TLIB-SB-PN computes the tensor-vector product on average 4x and 3.5x faster for asymmetrically and symmetrically shaped tensors, respectively.

7 Conclusion and Future Work

Based on the LOG approach, we have presented in-place and parallel tensor-vector multiplication algorithms of TLIB. Using highly-optimized DOT and GEMV routines of OpenBLAS, our proposed algorithms is designed for dense tensors with arbitrary order, dimensions and any non-hierarchical storage format. TLIB's algorithms either directly call DOT, GEMV or recursively perform parallel slice-vector multiplications using GEMV with tensor slices and fibers.

Our findings show that loop-fusion improves the performance of TLIB's parallel version on average by a factor of 5x achieving up to 34.8/15.5 Gflops/s in single/double precision for asymmetrically shaped tensors. With symmetrically shaped tensors resulting in small contraction dimensions, the results suggest that higher-order slices with larger dimensions should be used. We have demonstrated that the proposed algorithms compute the tensor-vector product on average 6.1x and up to 12.6x faster than the TTGT-based implementation provided by TCL. In comparison with TBLIS, TLIB achieves speedups on average of 4.0x and at most 10.4x. In summary, we have shown that a LOG-based tensor-vector multiplication implementation can outperform current implementations that use a TTGT and GETT approaches.

In the future, we intend to design and implement the tensor-matrix multiplication with the same requirements also supporting tensor transposition and subtensors. Moreover, we would like to provide an in-depth analysis of LOG-based implementations of tensor contractions with higher arithmetic intensity.

Project and Source Code Availability. TLIB has evolved from the Google Summer of Code 2018 project for extending Boost's uBLAS library with tensors. The sequential tensor-vector multiplication of TLIB is part of Boost uBLAS v1.70.0. Parallel versions with results can be found at https://github.com/bassoy/ttv.

Acknowledgements. The author would like to thank Volker Schatz and Banu Sözüar for proofreading. He also thanks Stefan Seefeld for helpful discussions about TLIB and Thomas Perschke and Michael Arens for their support. He gratefully acknowledges stipends from Google.

References

1. Abadi, M., et al.: TensorFlow: a system for large-scale machine learning. In: Proceedings of the 12th USENIX Conference on Operating Systems Design and Implementation, OSDI 2016, pp. 265–283. USENIX Association, Berkeley (2016)
2. Bader, B.W., Kolda, T.G.: Algorithm 862: MATLAB tensor classes for fast algorithm prototyping. ACM Trans. Math. Softw. **32**, 635–653 (2006)
3. Bassoy, C., Schatz, V.: Fast higher-order functions for tensor calculus with tensors and subtensors. In: Shi, Y., Fu, H., Tian, Y., Krzhizhanovskaya, V.V., Lees, M.H., Dongarra, J., Sloot, P.M.A. (eds.) ICCS 2018. LNCS, vol. 10860, pp. 639–652. Springer, Cham (2018). https://doi.org/10.1007/978-3-319-93698-7_49
4. Karahan, E., Rojas-López, P.A., Bringas-Vega, M.L., Valdés-Hernández, P.A., Valdes-Sosa, P.A.: Tensor analysis and fusion of multimodal brain images. Proc. IEEE **103**(9), 1531–1559 (2015)
5. Kolda, T.G., Bader, B.W.: Tensor decompositions and applications. SIAM Rev. **51**(3), 455–500 (2009)
6. Lee, N., Cichocki, A.: Fundamental tensor operations for large-scale data analysis using tensor network formats. Multidimens. Syst. Signal Process. **29**(3), 921–960 (2018)
7. Li, J., Battaglino, C., Perros, I., Sun, J., Vuduc, R.: An input-adaptive and in-place approach to dense tensor-times-matrix multiply. In: 2015 SC-International Conference for High Performance Computing, Networking, Storage and Analysis, pp. 1–12. IEEE (2015)
8. Lim, L.H.: Tensors and hypermatrices. In: Hogben, L. (ed.) Handbook of Linear Algebra, 2nd edn. Chapman and Hall, Boca Raton (2017)
9. Matthews, D.A.: High-performance tensor contraction without transposition. SIAM J. Sci. Comput. **40**(1), C1–C24 (2018)
10. Napoli, E.D., Fabregat-Traver, D., Quintana-Ortí, G., Bientinesi, P.: Towards an efficient use of the BLAS library for multilinear tensor contractions. Appl. Math. Comput. **235**, 454–468 (2014)
11. Papalexakis, E.E., Faloutsos, C., Sidiropoulos, N.D.: Tensors for data mining and data fusion: models, applications, and scalable algorithms. ACM Trans. Intell. Syst. Technol. (TIST) **8**(2), 16 (2017)
12. Shi, Y., Niranjan, U.N., Anandkumar, A., Cecka, C.: Tensor contractions with extended BLAS kernels on CPU and GPU. In: 2016 IEEE 23rd International Conference on High Performance Computing (HiPC), pp. 193–202, December 2016
13. Solomonik, E., Matthews, D., Hammond, J., Demmel, J.: Cyclops tensor framework: reducing communication and eliminating load imbalance in massively parallel contractions. In: 2013 IEEE 27th International Symposium on Parallel & Distributed Processing (IPDPS), pp. 813–824. IEEE (2013)
14. Springer, P., Bientinesi, P.: Design of a high-performance gemm-like tensor-tensor multiplication. ACM Trans. Math. Softw. (TOMS) **44**(3), 28 (2018)
15. Wang, Q., Zhang, X., Zhang, Y., Yi, Q.: AUGEM: automatically generate high performance dense linear algebra kernels on x86 CPUs. In: 2013 International Conference for High Performance Computing, Networking, Storage and Analysis (SC), pp. 1–12. IEEE (2013)

High Performance Partial Coherent X-Ray Ptychography

Pablo Enfedaque[1(✉)], Huibin Chang[1,2], Bjoern Enders[3], David Shapiro[4], and Stefano Marchesini[1]

[1] Computational Research Division, Lawrence Berkeley National Laboratory, Berkeley, USA
pablo.enfedaque@gmail.com
[2] School of Mathematical Sciences, Tianjin Normal University, Tianjin, China
[3] National Energy Research Scientific Computing Center, Lawrence Berkeley National Laboratory, Berkeley, USA
[4] Advanced Light Source, Lawrence Berkeley National Laboratory, Berkeley, USA

Abstract. During the last century, X-ray science has enabled breakthrough discoveries in fields as diverse as medicine, material science or electronics, and recently, ptychography has risen as a reference imaging technique in the field. It provides resolutions of a billionth of a meter, macroscopic field of view, or the capability to retrieve chemical or magnetic contrast, among other features. The goal of ptychography is to reconstruct a 2D visualization of a sample from a collection of diffraction patterns generated from the interaction of a light source with the sample. Reconstruction involves solving a nonlinear optimization problem employing a large amount of measured data—typically two orders of magnitude bigger than the reconstructed sample—so high performance solutions are normally required. A common problem in ptychography is that the majority of the flux from the light sources is often discarded to define the coherence of an illumination. Gradient Decomposition of the Probe (GDP) is a novel method devised to address this issue. It provides the capability to significantly improve the quality of the image when partial coherence effects take place, at the expense of a three-fold increase of the memory requirements and computation. This downside, along with the fine-grained degree of parallelism of the operations involved in GDP, makes it an ideal target for GPU acceleration. In this paper we propose the first high performance implementation of GDP for partial coherence X-ray ptychography. The proposed solution exploits an efficient data layout and multi-gpu parallelism to achieve massive acceleration and efficient scaling. The experimental results demonstrate the enhanced reconstruction quality and performance of our solution, able process up to 4 million input samples per second on a single high-end workstation, and compare its performance with a reference HPC ptychography pipeline.

1 Introduction

Ptychography [1] permits imaging macroscopic specimens at nanometer wavelength resolutions while retrieving chemical, magnetic or atomic information

J. M. F. Rodrigues et al. (Eds.): ICCS 2019, LNCS 11536, pp. 46–59, 2019.
https://doi.org/10.1007/978-3-030-22734-0_4

about the sample. It has proven to be a remarkably robust technique for the characterization of nano materials, and it is currently used in scientific fields as diverse as condensed matter physics [2], cell biology [3], materials science [4] and electronics [5], among others. Ptychography is based on recording the distribution of the diffraction patterns produced by the interaction of an X-ray beam (illumination) with a sample. The diffracted signal contains information about features much smaller than the size of the beam, making it possible to achieve higher resolutions than with standard scanning transmission techniques. Only the intensities of the diffracted illumination are measured, and one has to retrieve the corresponding phases to be able to reconstruct an image of the sample. To solve this problem, diffraction patterns are obtained from overlapping regions of the sample, producing a redundancy that can be used to recover the original phases of the signal.

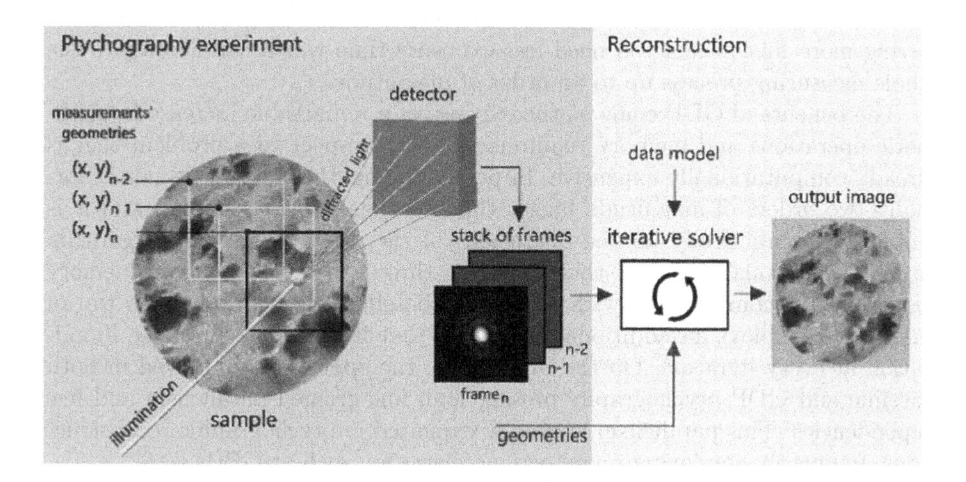

Fig. 1. Overview of a ptychography experiment and reconstruction. An illumination source (X-ray beam) consecutively scans regions of the sample to produce a stack of phase-less intensities. The stack and the geometry of the measurements are fed to an iterative solver that retrieves the phases and reconstructs an image of the original sample.

An overview of ptychography is depicted in Fig. 1. First, a sample is repetitively scanned with an X-ray beam, producing diffraction patterns that are recorded on a 2D detector. Each measurement is stored as a frame, and its exact location in the sample is also registered. Secondly, the stack of frames and the measurements' geometries are fed to a non-linear iterative solver that recovers the phases of the measurements. The solver optimizes based on two main constraints: (1) the match between overlapping regions of the frames, and (2) the match with a given model for the data. After the solver reaches an exit condition, the output is the overlap of the stack of frames (now with phases) in their corresponding geometries. This overlap corresponds to the 2D reconstructed image of the sample.

Normally, the ptychography reconstruction problem can only be solved if the illumination employed is coherent. The (spatial) coherence of an illumination defines how correlated are different points of its wavefront. To achieve higher coherence, X-ray microscopes employ apertures to filter the illumination, producing an homogeneous wavefront where different points are virtually identical in phase and amplitude. This solution wastes the majority of the X-ray flux, which is left behind the aperture. Overall, research institutions employ considerable resources to produce brighter X-ray sources, while over 90% of the photons are discarded to produce a coherent illumination.

A recent study from the CAMERA team at the Lawrence Berkeley National Laboratory (LBNL) proposed a novel algorithm that allows a ptychographic reconstruction with an incoherent source of illumination. The new algorithm, named Gradient Decomposition of the Probe (GDP) [6], has been proven to achieve successful reconstructions with significantly incoherent illuminations. The GDP algorithm also allows for a faster experimental data acquisition time: having more flux means you need less exposure time which can accelerate the whole measuring process up to an order of magnitude.

The benefits of GDP come at the expense of a remarkable increase in arithmetic operations and memory requirements with respect to a problem that is already computationally expensive. In ptychography, the stack of frames is normally two orders of magnitude bigger than the image reconstructed, and it is employed in practically all the operations of the solver. The GDP algorithm employs additional variables that require a three-fold increase of the memory footprint and computation with respect to baseline ptychography. On top of that, GDP employs an additional sub-solver that iteratively refines the illumination at every iteration. On the bright side, the operations employed in both baseline and GDP ptychography present high fine-grained parallelism and few dependencies. This parallelism is usually exploited in ptychographic reconstructions, frequently employing many-core accelerators, such as GPUs [7–9].

In this paper we propose the first high performance implementation of a partial coherent ptychography solution using GDP. We design an implementation that exploits the GDP parallelism and data requirements, making use of multiple GPU devices to achieve state-of-the-art reconstruction times. We compare the performance of the proposed implementation with that of baseline SHARP [7], a reference HPC ptychography solution, heavily optimized and also multi-GPU accelerated. The experimental results demonstrate how our implementation achieves only 2.5 times slower reconstruction times, on average, compared with standard coherent ptychography, while handling with 3 times more data and performing 4 to 5 times more arithmetic operations. The proposed solution has the key benefit of being able to process non-coherent illumination measurements, potentially leading to more flux utilization, increased robustness to non-stable sample exposures, and the capability to use less measurements when employing partially coherent illumination sources. Experimental results also assess the increased quality of the proposed method and implementation when handling partially coherent data, as compared with that of baseline coherent ptychography.

The paper is structured as follows. Section 2 overviews the main concepts regarding ptychography reconstruction, and introduces the GDP model. Section 3 presents the proposed algorithm and implementation with a detailed description of the challenges behind its design and the techniques employed, and Sect. 4 assesses its performance through experimental results. The last section summarizes this work.

2 Overview of Ptychography and GDP

A coherent ptychography problem can be defined as follows. A X-ray illumination (or probe) ω scans through a sample u, while a 2D detector collects a sequence J of phaseless intensities f. The goal is to retrieve a reconstruction of the sample u from the sequence of intensity measurements f. In a discrete setting, $u \in \mathbb{C}^n$ is a 2D image with $\sqrt{n} \times \sqrt{n}$ pixels, $\omega \in \mathbb{C}^{\bar{m}}$ is a localized 2D probe with $\sqrt{\bar{m}} \times \sqrt{\bar{m}}$ pixels, and $f_j = |\mathcal{F}(\omega \circ \mathcal{S}_j u)|^2$ is a stack of phaseless measurements, with $f_j \in \mathbb{R}_+^{\bar{m}}$ $\forall 0 \leq j \leq J-1$. The operation $|\cdot|$ represents an element-wise absolute value of a vector, whereas $\circ$ denotes an element-wise multiplication, and $\mathcal{F}$ represents a normalized 2-dimensional discrete Fourier transform. Each $\mathcal{S}_j \in \mathbb{R}^{\bar{m} \times n}$ corresponds to a binary matrix that selects a region j of size $\bar{m}$ from the sample u.

Besides recovering the sample u, in a ptychographic experiment the illumination is rarely perfectly known, and thus both sample and illumination need to be retrieved jointly. This is commonly referred to as blind ptychographic phase retrieval [10]. The joint problem can be formulated as:

$$\text{To find } \omega \in \mathbb{C}^{\bar{m}} \text{ and } u \in \mathbb{C}^n, \quad s.t. \quad |\mathcal{A}(\omega, u)|^2 = f, \tag{1}$$

where bilinear operators $\mathcal{A} : \mathbb{C}^{\bar{m}} \times \mathbb{C}^n \to \mathbb{C}^m$ and $\mathcal{A}_j : \mathbb{C}^{\bar{m}} \times \mathbb{C}^n \to \mathbb{C}^{\bar{m}}$ $\forall 0 \leq j \leq J-1$, are denoted as:

$$\begin{aligned} \mathcal{A}(\omega, u) &:= (\mathcal{A}_0^T(\omega, u), \mathcal{A}_1^T(\omega, u), \cdots, \mathcal{A}_{J-1}^T(\omega, u))^T, \\ \mathcal{A}_j(\omega, u) &:= \mathcal{F}(\omega \circ \mathcal{S}_j u), \end{aligned}$$

and $f := (f_0^T, f_1^T, \cdots, f_{J-1}^T)^T \in \mathbb{R}_+^m$.

There are multiple algorithms designed to solve the ptychography problem. The most popular ones are the extended Ptychographic Iterative Engine (ePIE) [11], Difference Map [10,12], Maximum Likelihood (ML) method [13], Proximal Splitting algorithm [14], Relaxed Averaged Alternating Reflections (RAAR) [15] based algorithms [7], and generalized Alternating Direction Method of Multipliers (ADMM) [16,17] for blind ptychography [18,19].

When using a partial coherent illumination, modeling the ptychohraphy problem is more challenging. GDP proposes a model based on describing the illumination as the superposition of a single coherent illumination convolved with a separable translational kernel. This way, the partial coherence effect can be handled using this single illumination, its gradient, and the variance of the convolution kernel. Following this idea, GDP is based on the following model:

$$\sum_{\xi} |\mathcal{F}_{x\to q}(\omega(x-\xi)\mathcal{S}_j u(x))|^2 \kappa(\xi) = f_{pc,j}(q), \tag{2}$$

where f_{pc} represents a sequence of partially coherent intensity measurements, $\kappa(\xi)$ is a 2D kernel with variance (second order moments) σ_1^2, σ_1^2 and σ_{12} ($\sigma_1^2\sigma_2^2 - \sigma_{12}^2 \geq 0$). Then, the Taylor expansion of ω can be derived and simplified as:

$$f_{pc} \simeq |\mathcal{A}(\tilde{\omega}, u)|^2 + \sigma_1^2 |\mathcal{A}(\nabla_1\tilde{\omega}, u)|^2 + \sigma_2^2 |\mathcal{A}(\nabla_2\tilde{\omega}, u)|^2, \tag{3}$$

with:

$$\tilde{\omega} := \omega + \frac{1}{2}(\sigma_1^2 \nabla_{11}\omega + \sigma_1^2 \nabla_{22}\omega + 2\sigma_{12}\nabla_{12}\omega),$$

and ∇_1, ∇_2, ∇_{11}, ∇_{22}, ∇_{12} corresponding to the forward first and second order finite difference operators (gradients) with respect to x, y, xx, yy and xy directions. Considering the sequence of measurements j, we can define the nonlinear operator $\mathcal{G}_j : \mathbb{C}^{\bar{m}} \times \mathbb{C}^n \times \mathbb{R}^2 \to \mathbb{R}_+^{\bar{m}}$ as:

$$\mathcal{G}_j(\tilde{\omega}, u, \sigma) := |\mathcal{A}_j(\tilde{\omega}, u)|^2 + \sigma_1^2 |\mathcal{A}_j(\nabla_1\tilde{\omega}, u)|^2 + \sigma_2^2 |\mathcal{A}_j(\nabla_2\tilde{\omega}, u)|^2,$$

with $\sigma := (\sigma_1, \sigma_2)$, and finally, we can establish the GDP nonlinear optimization model as:

$$\min_{\tilde{\omega}, u, \sigma} \frac{1}{2} \sum ||\sqrt{f_{pc,j}} - \mathcal{G}_j(\tilde{\omega}, u, \sigma)||^2, \tag{4}$$

where $||\cdot||$ represents the L^2 norm in Euclidean space.

3 High Performance GDP Solution

The GPD model is proposed in [6] together with an algorithm employing the ADMM framework to efficiently solve the derived subproblems (GDP-ADMM). In this work we design an implementation of GDP-ADMM and also propose a novel one employing the RAAR algorithm (GDP-RAAR). In the following section we focus on GDP-RAAR to describe the implementation, although the main insights and operations are common to both. The implementations and algorithms of this work are developed inside the SHARP framework, and some of the technologies and operations are common to the baseline coherent solutions. In this section, we focus on the main key operations unique to the GDP method; please refer to [7,9] for a detailed description of other aspects of the end-to-end solution not described in here.

The challenge deriving from the GDP model is threefold. First, the algorithm requires to maintain in memory additional high dimensional variables. Second, the main ptychography operations need to be reformulated to handle the new problem. Third, to solve the illumination refinement, an additional inner solver needs to be considered at every ptychography iteration. The standard memory footprint of the ptychography problem involves the following structures.

There are two main inputs: (1) a stack of 2D frames ($frames_m[x, y, z]$) containing the floating point values from the original measured intensities, and (2) a vector containing the coordinates of each one of the frames in the sample 2D image ($int2\ coord[z]$). Then, at least three additional structures are required in the iteration process: $sample[i, j]$, $illum[x, y]$, and $frames_s[x, y, z]$, containing a 2D image of the object, the refined illumination and the stack of solution frames (with phases), respectively. Each one of these contains phase and amplitude information, and thus they are stored as complex values ($float2$).

The main idea behind the GPD model is to fit the stack of frames with constraints implying the original illumination and also its gradient on the x and y directions. Because of this, we need to consider three different variables for the stack of solution frames: $frames_{s1}$, $frames_{s2}$, and $frames_{s3}$, each one $float2$ with size $[x, y, z]$. This increase in memory requirements is very relevant performance-wise. A real case example: to generate an image with size 1024×1024, a stack of measured frames of size $1500 \times 256 \times 256$ is collected, which represents a ratio of 1:94 output/input. When using GDP, every pixel in the output ($float2$) is iteratively produced from 94×1 $float$ $frames_m$ $\times$ 3 $float2$ $frames_s$ elements, which constitutes a ratio of 2:658 in floating point values. On top of it, practically all the operations involved in a ptychography reconstruction are memory bounded, so proper memory managing and locality becomes a key factor to achieve performance.

3.1 The Implementation

Algorithm 1 describes the high level outline of the proposed implementation using the new GDP-RAAR algorithm[1]. Note that all the operations are performed in GPU, using either custom CUDA kernels or Thrust operands. Most of the operations are implemented in a fused fashion in order to minimize GPU global memory transfers.

In this work we propose an scheme where all three $frames_{s1,2,3}$ variables are stored as a single interleaved memory structure to maximize locality and performance. In Algortihm 1, $frames_s$ stores the three $frames_{s1,2,3}$ variables, with a total size $[x \times y \times z \times n\_interleaved\,]$ and $n\_interleaved = 3$ being the stride. The motivation behind this design is related to the topology of the operations performed (how inputs contribute to the outputs). Baseline ptychography involves four kind of core operations: (1) *Split*, (2) *Overlap*, (3) 2D Fast Fourier Transforms (FFTs), and (4) and point-wise additions, multiplications, divisions, etc.

[1] Algorithm 1 presents a simplified outline of the method. Multiple operations and memory structures regarding regularization terms, stabilizers, background removal optimization, etc. have been omitted for simplicity.

Algorithm 1. ***GDP-RAAR***

Parameters: $frames_m[x \times y \times z]$, $coord[z]$, $iter_{max}$, $n\_interleaved = 3$, $tolerance$

1: **allocate** $sample[i \times j]$, $illum[x \times y \times n\_interleaved]$,
   $frames_s[x \times y \times z \times n\_interleaved]$
2: $illum = \boldsymbol{InitializeIllum}(frames_m)$
3: $sample = 1$, $frames_s = illum \times \boldsymbol{Split_{int}}(sample)$
4: **for** $k = 0$ **to** $iter_{max} - 1$ **do**
5: $\quad frames_s = \boldsymbol{ForwardFT}(frames_s,\ stride = n\_interleaved)$
6: $\quad frames_s = \boldsymbol{UpdateFrames}(frames_s, frames_m)$
7: $\quad frames_s = \boldsymbol{InverseFT}(frames_s,\ stride = n\_interleaved)$
8: $\quad illum = \boldsymbol{GDescent}(min\{||\boldsymbol{Split_{int}}(sample) \times (I, \nabla_1, \nabla_2)illum_1 - frames_s||\})$
9: $\quad sample = \dfrac{\boldsymbol{Overlap_{int}}(frames_s \times illum^*)}{\boldsymbol{Overlap_{int}}(|illum|^2)}$
10: $\quad residual = \boldsymbol{ComputeResidual}(frames_s, frames_m)$
11: $\quad$ **if** $residual < tolerance$ **then break**
12: $\quad frames_s = \boldsymbol{RAAR\_Update}(illum \times \boldsymbol{Split_{int}}(sample), frames_s)$
13: **end for**
14: **return** $sample$, $illum$

The standard *Overlap* operation takes as inputs $frames[x, y, z]$ and $coord[z]$, and adds each frame into a 2D image[2], on its respective coordinate, as follows:

$$\begin{aligned}&sample[\,:\,,\,:\,] = 0\\&for(\ i = 0;\ i < z;\ i++)\{\\&\qquad sample[\ coord[\ i\ ]\ ] += frames[\ :,\ :,\ i\ ]\ \},\end{aligned}$$

with the index":" referring to the full slice in a dimension. The *Split* operation does the opposite: for each coordinate, a frame is extracted from an input 2D image, constructing an output 3D stack of frames. In GDP, the *Overlap* and *Split* operations are performed considering the three stack of frames variables. Each $frames_{s1,2,3}$ is added into a single image for the Overlap, and a single image is split into three stack of frames. The interleaved strategy mentioned above permits to maximize data locality in these operations.

Algorithm 2 presents the interleaved Overlap CUDA kernel ($\boldsymbol{Overlap_{int}}$) implemented for GDP. The thread to data mapping is as follows: each CUDA thread block processes a single frame from $frames_s$, iterating over the frame with a stride of samples equal to the thread block size (*block_dim*) (line 12). For each pixel in the original frame size $[x, y]$, each thread accumulates in a local register the contribution from all three $frames_{s1,2,3}$ variables, iterating over the interleaved stride (line 7). Then, the accumulated value can be written into the

[2] Note that normalization may be required afterwards.

Algorithm 2. $\boldsymbol{Overlap_{int}}$

Parameters: $sample[i \times j]$, $frames_s[x \times y \times z \times n\_interleaved]$, $coord[z]$, $n\_interleaved = 3$

1: $frame\_size = x \times y$
2: $n\_frame = block\_id$
3: **for** $f = thread\_id$ **to** $frame\_size$ **do**
4: $\quad accum\_output = 0$
5: $\quad out\_index = \boldsymbol{ComputeSampleCoord}(f, coord\,[n\_frame], frame\_size)$
6: $\quad frames\_index = (frame\_size \times n\_frame + f) \times n\_interleaved$
7: $\quad$ **for** $p = 0$ **to** $n\_interleaved$ **do**
8: $\quad\quad accum\_output + = frames_s[frames\_index + p\,]$
9: $\quad\quad p = p + 1$
10: $\quad$ **end for**
11: $\quad sample[out\_index] = \boldsymbol{AtomicAddition}(sample[out\_index], accum\_output)$
12: $\quad f = f + block\_dim$
13: **end for**

output image employing a single atomic addition operation (line 11). The atomic operation is employed as an efficient way to handle the collision caused by having coordinates from different frames overlapping into the sample.

The interleaved Split operation ($\boldsymbol{Split_{int}}$) is handled similarly as in the Overlap case. The main difference is that no atomic operation is required in it. When splitting the sample into frames, each frame is normally multiplied by the illumination. In GDP, each $frames_{s1,2,3}$ variable needs to be multiplied by the illumination, its horizontal gradient and its vertical gradient, respectively. To reduce computation, the gradients of the illumination are computed once per iteration and stored as an interleaved variable with size $[x \times y \times n\_interleaved\,]$. This strategy permits performing efficient straightforward point-wise operations between the interleaved frames and the interleaved illumination variables. In Algorithm 1, *illum* stores the interleaved illumination structures; it is initialized in $\boldsymbol{InitializeIllum}$ (line 2) employing the information from the measured frames ($frames_m$) to generate an initial guess. Then, the same function computes and stores the x and y gradients of the produced illumination.

The interleaved strategy is also beneficial when computing the $L2$ norm of the $frames_{s1,2,3}$ variables. In $\boldsymbol{UpdateFrames}$ and in the background noise modeling (not shown in Algortihm 1 for simplicity) the sum of the square root of the $L2$ norm needs to be computed, benefiting again from the enhanced locality of the interleaved structures. In the case of the 2D FTT operations, the interleaved layout actually reduces memory locality. To handle this issue with minimum performance impact we employ the in-build strided FFT feature implemented in $cuFFT$. This allows to transparently process our data through FFTs and back, without having to handle any reorganization of it or additional computation.

The GDP model requires to solve an additional subproblem for the illumination refinement step (Algorithm 1 line 8). The standard refinement is performed by fixing the current estimate of the *sample* and minimizing the difference with the $frames_s$, solving a problem in the form $||\boldsymbol{Split}(sample) \times illum - frames_s||$, which is a linear problem with diagonal matrix that can be solved in a single step. In GDP, the illumination refinement step couples the GDP expansion of the illumination $(I, \nabla_1, \nabla_2)illum_1$, with the $\boldsymbol{Split}_{int}(sample)$, and the interleaved $frames_s$ in the form:

$$||\boldsymbol{Split}_{int}(sample) \times (I, \nabla_1, \nabla_2)illum_1 - frames_s||,$$

with I and $illum_1$ referring to the identity matrix and the original illumination variable, respectively. This poses a linear problem with a sparse band-diagonal matrix, which we solve using the gradient descent algorithm with a fixed step size, referred in Algorithm 1 as ***GDescent***. Instead of using the conjugate gradient described in the original GDP paper, we choose the gradient descent algorithm because it avoids the reduction operations used to compute the step size and conjugate directions scaling factors, thus offering an increased performance. The algorithm is implemented using custom CUDA kernels that allow pre-computing in place multiple factors, a custom manipulation of the interleaved structures, and permits fusing the iterating process with pre- and post-process operations, like the x, y gradient computation of the new illumination as a last step of the refinement.

The GDP-RAAR implementation proposed in this paper is also accelerated using multi-GPU over MPI/NCCL. The partition employed is similar to the one used in [7,9]. The main idea is to divide both $frames_m$ and $frames_s$ variables across different GPUs so that each independent device process only a subset of frames. Then, communication is required every time the sample and the illumination are updated. The communication operation is essentially an *AllReduce* directive (*Reduce* and *Broadcast*) that performs a summation of the partial results of each independent GPU. The communication directives are performed in-place, using NCCL if it is installed on the system, or over standard MPI otherwise. Given the significant increase of the problem size of GDP, the proposed implementation greatly benefits from multi GPU execution when executed on high-end workstations. An additional feature is that the communication frequency can be adapted to occur every N iterations, employing previous iteration data for the non-local areas of the image. For the proposed GDP implementation, this features enhances performance only with small problem sizes per GPU (<30 millions measured samples).

4 Experimental Results

The results presented below have been executed in a dual socket workstation with two Intel Xeon E5-2683 v4, with a clock frequency of 2.10 GHz and 16 cores each. The machine is equipped with 4 dual-slot Tesla K80 GPUs, for a total of 8 GK210B devices, each device with 2496 CUDA cores. The implementations reported here have been compiled using gcc 5.4.0 and nvcc 8.0. The profiling results have been obtained with both Nvidia visual and inline profilers, nvvp and nvprof, respectively. All execution times and performance results consider the full pipeline execution time, including loading the data from memory, GPU runtime initialization, memory allocation and transfers, and writing back the reconstructed image and illumination. The experiments below all employ the GDP-RAAR algorithm described in the previous section but the reconstruction results and performance are also comparable when using GDP-ADMM. All experiments are measured using 100 solver iterations, which is enough to achieve convergence using standard tolerance thresholds for the datasets presented in here. The performance analysis and results below can also be extrapolated when running more iterations.

The first experiment, reported in Fig. 2, evalutes the reconstruction quality of the proposed GDP-RAAR algorithm when retrieving a partial coherent illumination and sample, as compared with the baseline RAAR method from SHARP.

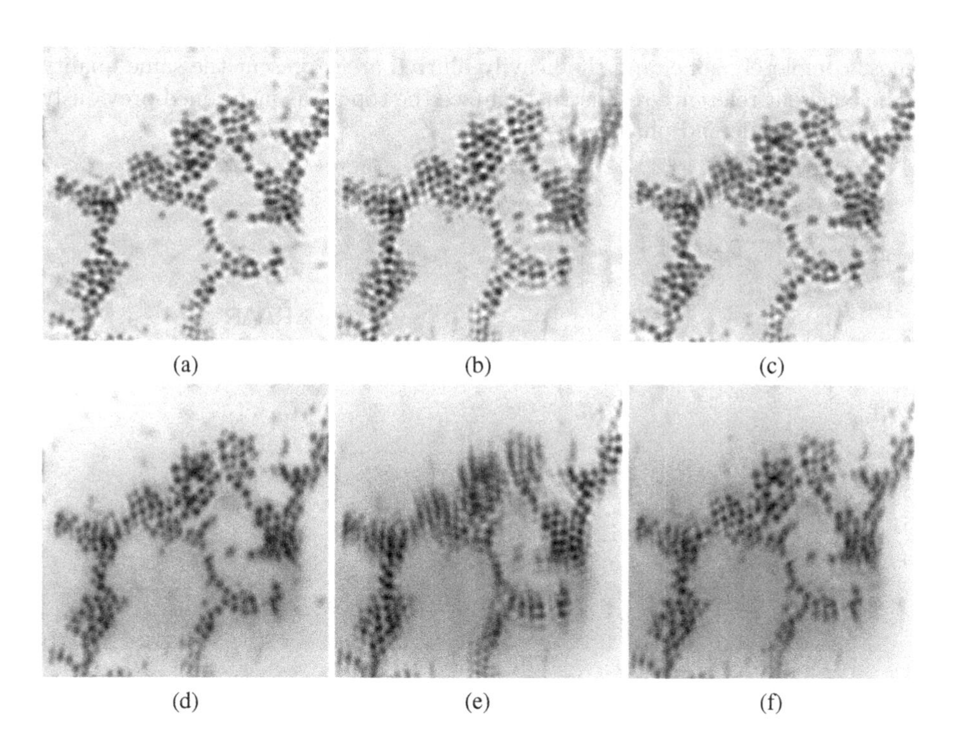

Fig. 2. First column: baseline reconstruction using a standard coherent illumination using RAAR. Second column: reconstruction using a partial coherent illumination using RAAR. Third column: reconstruction using a partial coherent illumination using the proposed GDP-RAAR algorithm and implementation. Top row (a, b, c) corresponds to the amplitude images retrieved, whereas the bottom row (d, e, f) depicts the phase images from the same reconstructions.

In order to perform this test, a sample was measured with a standard coherent illumination first (Fig. 2 first column) and then it was measured again using a partial coherent illumination (Fig. 2 second and third columns). The first column represents our reference reconstruction and the second and third columns report an actual partial coherence experiment, reconstructed using RAAR and GDP-RAAR, respectively. Top and bottom rows correspond to the amplitude and phase contrast, respectively, from the same reconstruction. This experiment was conducted at the Advanced Light Source (ALS) in 2018, at the COSMIC beamline, and the sample corresponds to a conglomerate of nanometer-sized gold particles of uniform shape and size. Both coherent and partial coherent experiments have virtually the same configuration, with both datasets containing 1600 frames of size 256×256 each. We can clearly see in Fig. 2 second column how the RAAR algorithm introduces severe ghosting artifacts, specially around the contour of the sample. Some areas of this reconstruction become significantly blurry, specially on the top-right features of the amplitude image and top-center and top-left areas of the phase image. The results reported in the third column of Fig. 2 show how the main artifacts introduced by RAAR are removed by the proposed GDP-RAAR method. When using GDP, the ghosting artifacts are almost completely gone, and the heavily blurred areas present the same quality as the coherent reference reconstruction (see the top areas mentioned previously in both amplitude and phase).

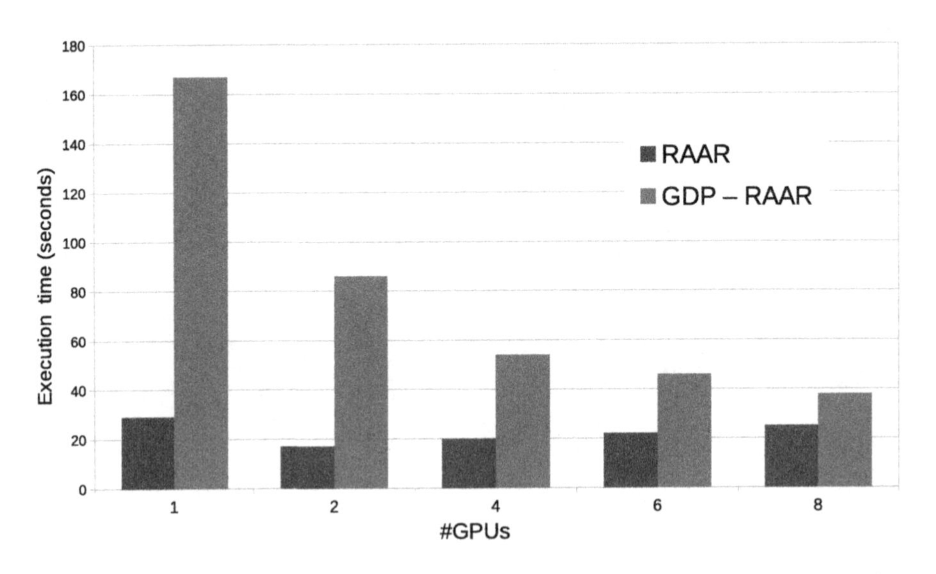

Fig. 3. Execution times of the proposed GDP-RAAR implementation, compared to those of baseline RAAR, when running on 1 to 8 GK210B GPUs. The dataset and configuration are the same presented in the experiment reported in Fig. 2.

The following test evaluates the execution time of the reconstruction results presented in the previous experiment. Results are reported in Fig. 3, and show the time in seconds of RAAR and GDP-RAAR, when being executed on a single GPU and on different multi-GPU settings. First, we can see the significant increase in execution time required by GDP-RAAR, presenting execution times ranging from 5.5 to 1.5 times slower than RAAR. This is consistent with the increase in arithmetic operations required in GDP-RAAR: almost all arithmetic involve 3 times more data, whereas the additional illumination solver performs 20 inner iterations per outer iteration, each iterations requiring multiple point-wise multiplications, divisions, and gradients with high dimensional data. The proposed implementation scales with the number of GPUs, achieving speedups of 1.94, 3.09 and 4.39 when using 2, 4, and 8 GPUs, respectively. The reported scaling is remarkably good, specially considering the fact that communication across GPUs is performed three times per outer iteration, in order to share the sample and illumination structures. This communication can significantly slow down execution, as seen in the time results reported by RAAR. The amount of computation and problem size that baseline RAAR handles is much less than GDP-RAAR, and that is why the speedup gain with the increase of GPUs is lower, as independent devices are not close to reach resource saturation. The communication overhead on its turn becomes higher with the number of independent executions, effectively reducing the performance of RAAR when running on 4, 6 and 8 GPUs.

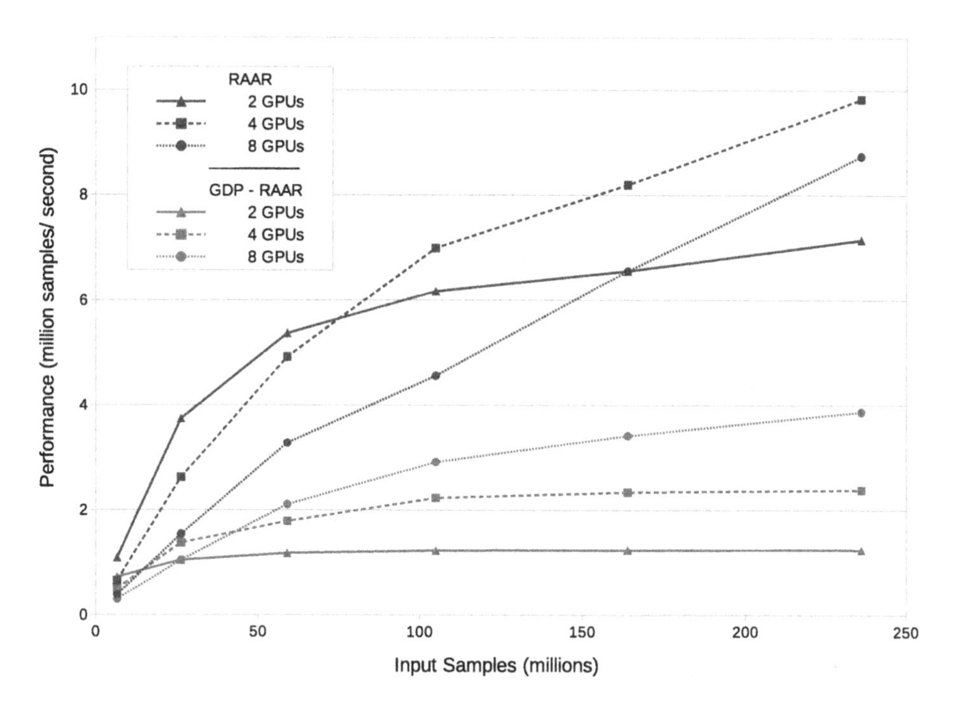

Fig. 4. Performance and scalability of the proposed GDP-RAAR implementation compared to that of baseline RAAR, both executed on 2, 4 and 8 GK210B GPUs. The size of the datasets employed range from $100 \times 256 \times 256$ to $2500 \times 256 \times 256$ measured frames.

The final experiment, reported in Fig. 4, analyzes the scalability of the proposed method with respect to the problem size. In this case we employ a dataset from an experiment performed in the ALS on 2015 that measured a cluster of iron catalyst particles. We have selected different size slices of said experiment to asses the performance of the proposed implementation with different input sizes. The performance metric is given in samples/second (the higher the better). The horizontal axis presents the different size datasets for their number of measured intensity samples (in millions). As a reference, we report the performance of baseline RAAR, together with the proposed method, and each algorithm is executed on 2, 4 and 8 GPUs. The experiment reveals how the proposed implementation achieves an almost perfect linear scaling when running on 2 GPUs. When running on 4 and 8 GPUs, the scaling achieved is better than linear due to the devices not being saturated at first by the smaller individual problem sizes. This effect is very noticeable with the RAAR results, where an (almost) saturation point is only reached with 2 GPUs and the biggest problem sizes. We can also see how the speedup achieved by the GDP-RAAR multi gpu execution effectively scales with the problem size: the biggest dataset (240 million samples) achieves an speedup of 1.92 and 3.13 when running on 4 and 8 GPUs, respectively, with respect to a dual-GPU execution.

5 Conclusions

This paper presents the first GPU-accelerated implementation of GDP for high performance partial coherent ptychography. We tackle the significant increase of computational costs of GDP to produce a solution with a minimum performance loss, while maintaining all the features offered by the method. We design our implementation using an efficient interleaved data layout strategy that enhances the memory locality and overall performance of the core operations of the solver. Multi-GPU parallelism is exploited, achieving linear scaling and capability to process up to 4 million measured samples per second, on a single high-end workstation. We also demonstrate how our implementation achieves a drastic increase of reconstruction quality when dealing with partially coherent light sources, with respect to standard ptychography. The proposed solution has the increased benefit of being able to employ more flux, potentially reducing the acquisition time up to an order of magnitude, while being more robust to non-stable sample exposures. It also offers the capability to use less measurements when employing partially coherent sources. The proposed implementation is currently installed and being used at the ptychography COSMIC beamline at the Advanced Light Source at LBNL, and the binaries and source code are also open to other DOE light sources.

Acknowledgments. This work was partially funded by the Center for Applied Mathematics for Energy Research Applications, a joint ASCR-BES funded project within the Office of Science, US Department of Energy, and the Advanced Light Source under contract number DOE-DE-AC03-76SF00098. This work was also partially supported by the National Natural Science Foundation of China (11871372).

References

1. Rodenburg, J.M.: Ptychography and related diffractive imaging methods. Adv. Imaging Electron Phys. **150**, 87–184 (2008)
2. Shi, X., et al.: Soft x-ray ptychography studies of nanoscale magnetic and structural correlations in thin $SmCo_5$ films. Appl. Phys. Lett. **108**(9), 094103 (2016)
3. Giewekemeyer, K., et al.: Quantitative biological imaging by ptychographic x-ray diffraction microscopy. Proc. Nat. Acad. Sci. **107**(2), 529–534 (2010)
4. Shapiro, D.A., et al.: Chemical composition mapping with nanometre resolution by soft x-ray microscopy. Nat. Photonics **8**(10), 765–769 (2014)
5. Holler, M., et al.: High-resolution non-destructive three-dimensional imaging of integrated circuits. Nature **543**(7645), 402–406 (2017)
6. Chang, H., Enfedaque, P., Lou, Y., Marchesini, S.: Partially coherent ptychography by gradient decomposition of the probe. Acta Crystallogr. Sect. A: Found. Adv. **74**(3), 157–169 (2018)
7. Marchesini, S., et al.: SHARP: a distributed, GPU-based ptychographic solver. J. Appl. Crystallogr. **49**(4), 1245–1252 (2016)
8. Nashed, Y.S., Vine, D.J., Peterka, T., Deng, J., Ross, R., Jacobsen, C.: Parallel ptychographic reconstruction. Opt. Express **22**(26), 32 082–32 097 (2014)
9. Enfedaque, P., Chang, H., Krishnan, H., Marchesini, S.: GPU-based implementation of ptycho-ADMM for high performance X-ray imaging. In: Shi, Y., et al. (eds.) ICCS 2018. LNCS, vol. 10860, pp. 540–553. Springer, Cham (2018). https://doi.org/10.1007/978-3-319-93698-7_41
10. Thibault, P., Dierolf, M., Bunk, O., Menzel, A., Pfeiffer, F.: Probe retrieval in ptychographic coherent diffractive imaging. Ultramicroscopy **109**(4), 338–343 (2009)
11. Maiden, A.M., Rodenburg, J.M.: An improved ptychographical phase retrieval algorithm for diffractive imaging. Ultramicroscopy **109**(10), 1256–1262 (2009)
12. Elser, V.: Phase retrieval by iterated projections. J. Opt. Soc. Am. A **20**(1), 40–55 (2003)
13. Thibault, P., Guizar-Sicairos, M.: Maximum-likelihood refinement for coherent diffractive imaging. New J. Phys. **14**(6), 063004 (2012)
14. Hesse, R., Luke, D.R., Sabach, S., Tam, M.K.: Proximal heterogeneous block implicit-explicit method and application to blind ptychographic diffraction imaging. SIAM J. Imaging Sci. **8**(1), 426–457 (2015)
15. Luke, D.R.: Relaxed averaged alternating reflections for diffraction imaging. Inverse Prob. **21**(1), 37–50 (2005)
16. Glowinski, R., Le Tallec, P.: Augmented Lagrangian and Operator-Splitting Methods in Nonlinear Mechanics. SIAM, Philadelphia (1989)
17. Wu, C., Tai, X.-C.: Augmented Lagrangian method, dual methods and split-Bregman iterations for ROF, vectorial TV and higher order models. SIAM J. Imaging Sci. **3**(3), 300–339 (2010)
18. Chang, H., Enfedaque, P., Marchesini, S.: Blind ptychographic phase retrieval via convergent alternating direction method of multipliers. SIAM J. Imaging Sci. **12**(1), 153–185 (2019). https://doi.org/10.1137/18M1188446
19. Chang, H., et al.: Advanced denoising for x-ray ptychography. Opt. Express **27**(8), 10395–10418 (2019). http://www.opticsexpress.org/abstract.cfm?URI=oe-27-8-10395

Monte Carlo Analysis of Local Cross–Correlation ST–TBD Algorithm

Przemyslaw Mazurek(✉) and Robert Krupinski

Department of Signal Processing and Multimedia Engineering,
West Pomeranian University of Technology Szczecin, Szczecin, Poland
{przemyslaw.mazurek,robert.krupinski}@zut.edu.pl

Abstract. The Track–Before–Detect (TBD) algorithms allow the estimation of the state of an object, even if the signal is hidden in the background noise. The application of local cross–correlation for the modified Information Update formula improves this estimation for extended objects (tens of cells in the measurement space) compared to the direct application of the Spatio–Temporal TBD (ST–TBD) algorithm. The Monte Carlo test was applied to evaluate algorithms by using a variable standard deviation of the additive Gaussian noise. The proposed solution does not require prior knowledge of the size or measured values of the object. Mean Absolute Error for the proposed algorithm is much lower, close to zero to about 0.8 standard deviation, which is not achieved for the ST–TBD.

Keywords: Track–Before–Detect · Tracking · Algorithm analysis · Monte Carlo · Cross–correlation

1 Introduction

Tracking algorithms are applied in numerous civil and military applications [6]. Detection algorithms could be used to estimate basic object parameters, such as only position. Tracking allows you to combine subsequent measurements into paths. Tracking allows the filtering of incoming signals to reduce noise and predict the state of the objects. Such filtering is very important because false measurements lead to distortions in tracking. The signal strength associated with several close observations is not an appropriate criterion for choosing a particular observation as an object signal [6].

Different types of measurement data can be processed: radar or video signals are typical. The complexity of tracking systems depends on SNR (Signal–to–Noise Ratio) and the number of tracked objects [7], and real–time tracking of a single object in low SNR scenarios is very sophisticated. It is well known that the Detection and Tracking (classical) approach can only be used in cases of high and medium SNR [6]. Detection based on the threshold algorithm leads to a large number of false detections, so further processing of the tracking part is not possible. The tracking algorithm (e.g. Benedict–Bordner, Kalman, and EKF) enables

J. M. F. Rodrigues et al. (Eds.): ICCS 2019, LNCS 11536, pp. 60–70, 2019.
https://doi.org/10.1007/978-3-030-22734-0_5

suppression of false detection only to a certain level using a motion model [5]. The selection of a fixed or adaptive threshold algorithm is not a solution to the problem. A too low threshold level leads to an increase in the number of observations that are treated as potentially possible. A too high threshold level leads to the omission of observations that may be associated with the object.

An alternative approach is based on Tracking and Detection (Track–Before–Detect). It uses tracking of all possible trajectories without first detecting objects, even if no object is in the range. This is possible due to the processing of raw measurements, without thresholding, as in classical systems [24]. The measured values are accumulated on trajectories, so the signal is filtered (improved SNR), and then detection is possible after tracking. The computational cost of such a solution is extremely high, but in many applications it is acceptable due to safety and expected reliability.

1.1 Related Work

There are many TBD algorithms (e.g. Velocity Filters [10], Viterbi TBD [22], ST–TBD [17], SLRT [24], Directional Filters [26]) and all of them support tracking of many objects. This is an internal feature of TBD algorithms. The cost of calculations for most TBD algorithms is constant, regardless of the number of objects and requires high computing power. The reduction of the number of calculations by reducing the number of analyzed possible paths is available in the Particle Filters TBD algorithms [25]. Unfortunately, the main problem with the Particle Filters TBD method is the difficult initialization of the algorithm, especially when the object can appear in range at any time.

A typical tracking system assumes zero mean noise and a positive value for a large point object, so a single pixel (or cell) is excited by the tracked object [24]. Otherwise, additional preprocessing is required for conditioning the input signal [16,18,24]. Tracking a larger object that occupies dozens of pixels, sometimes with values below the zero value, requires a dedicated preprocessing algorithm. Large extended objects are considered in Sect. 2. Objects with known signal characteristics (profile) can be improved by applying matching filters as shown in [16]. Very specific objects are noise objects (only noise is observed from the object), therefore local signal distributions are used [18,19].

1.2 Content and Contribution of the Paper

The proposed solution for tracking extended objects uses the modified ST–TBD with local analysis of cross–correlation between neighboring measurements.

The ST–TBD algorithm is oriented to the processing of individual cells (pixels) of the input signal (image), and the use of cross-correlation allows better use of information about neighbor values in determining the potential position for the next measurement. This solution can be applied to extended objects without significantly increasing the computing cost and losing information about the similarity of the pixels representing the object in motion.

Using the local cross–correlation and by comparing the current and previous image frame, it is possible to sharpen the measurement values, which is necessary to improve the SNR, as discussed in Sect. 3. The analysis of the algorithm's performance is possible only through the Monte Carlo approach, as presented in Sect. 4. The discussion is presented in Sect. 5. The final conclusions and further work are presented in Sect. 6.

2 Data

There are several types of objects being tracked in image processing applications and a typical area of objects with a single pixel. Sometimes the neighborhood pixels are excited because of the imperfections of the sensor. An extended type object includes tens or more pixels, so the direct use of TBD algorithms is not possible, because object features can be treated as separate objects. Classic algorithms use the conversion of extended objects to a single pixel or the estimation and processing of positions. An example of an extended object is shown in Fig. 1 that may have low contrast in poor weather or lighting conditions. The obtained image may also be deteriorated due to the measurement noise caused by poor lighting, long distance and sensor noise.

Fig. 1. Example of good quality measurement (left), low contrast measurement (center), and measurement of the noisy low contrast (right) for the aircraft.

An additional problem with TBD is blurring caused by the Motion Update formula (explicitly or not). Sharpening is a good solution, but in the case of a low SNR it leads to the emphasis of noise. The solution proposed in this article uses local correlation to sharpen. The example of a cross–correlation that shows the use of local cross–correlation is shown in Figs. 2 and 3. The highest peak value is for $c(v = 2)$ in the 25 position, which results in the appropriate spatial offset $(v = 2)$ (the object's velocity is 2) and the position of the object. The test case with no background noise is shown in Fig. 2, so detection and tracking is possible using very simple algorithms. Detection and tracking is possible due to the contrast between the object signal and the background. Tracking an object when the signal values for the object are close to the background noise are shown in Fig. 3.

In this article, the Monte Carlo method [20] was used to compare the basic ST–TBD algorithm with the local cross–correlation ST–TBD for 1D signals. The extension to 2D cases (images) is possible, but is not taken into account due to the high cost of simulation.

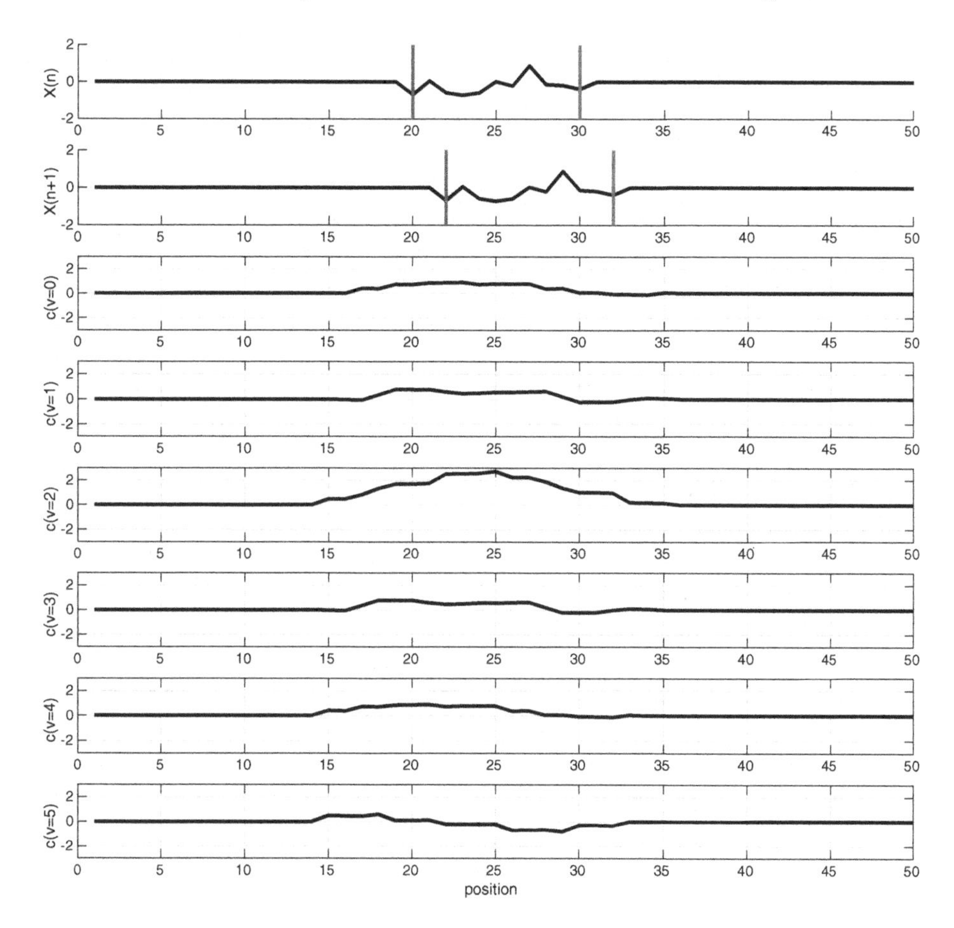

Fig. 2. Example of local cross–correlations $c(.)$ for two 1D $X(n)$ and $X(n+1)$ signals (two time moments: n and $n+1$) with an object (between vertical lines) for several possible shifts of objects in space. No background noise.

It is assumed that the extended object has a size of 11 pixels (cells). The pixel values of the object are randomly selected at the beginning of the test (obtained from a uniform random number generator). The object velocity is an integer value in the range 0–10 and the velocity value is chosen randomly. A velocity value of 0 corresponds to the static position of the object (no movement).

The measurement is disturbed by the additive Gaussian noise. The standard deviation of this noise is controlled, configured in a random number generator, which allowed testing of various SNR cases.

There are 1500 cells associated with the position, so the input image has a resolution 1×1500 for the assumed 1D case. This allows you to test the object for 100 frames, because 100 frames for a maximum velocity of 10, gives a maximum of 1000 pixels of movement. Estimated and known positions are compared after 100 frames to determine the position error.

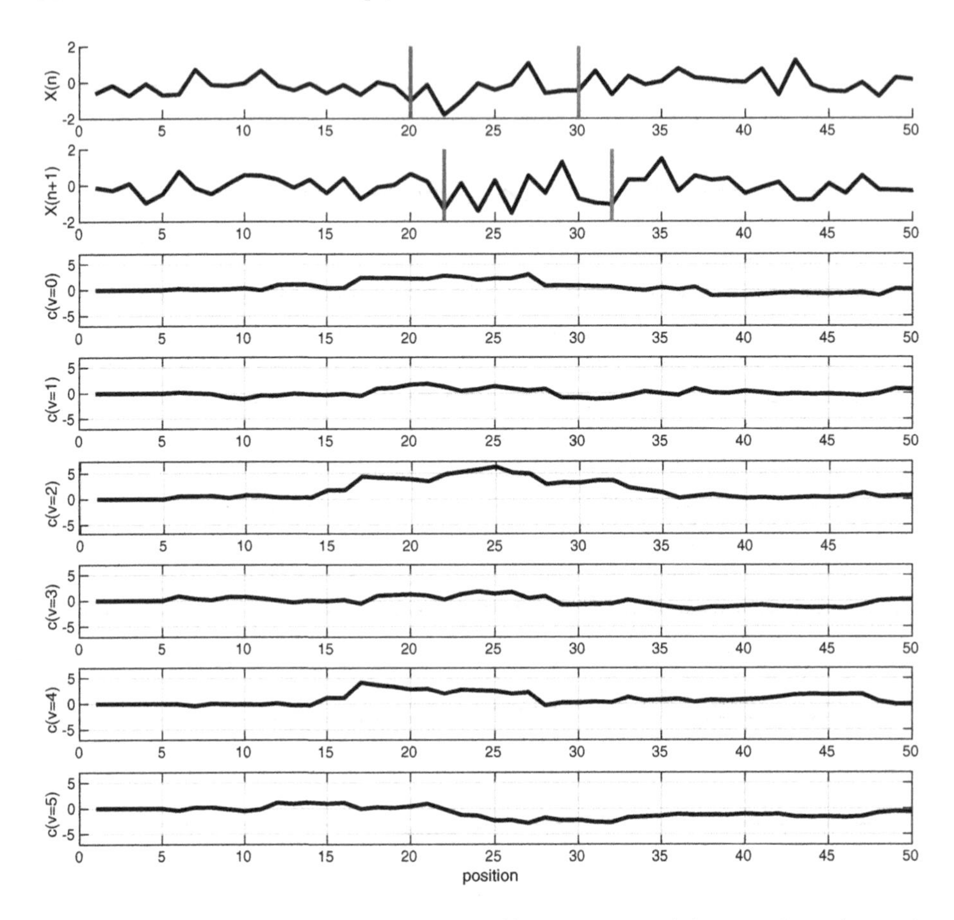

Fig. 3. Example of local cross–correlations $c(.)$ for two 1D $X(n)$ and $X(n+1)$ signals (two time moments: n and $n+1$) with a low contrast object (between vertical lines) for several possible shifts of objects in space.

3 Method

The proposed tracking solution is based on the ST–TBD algorithm that uses the Motion Update formula (2) to predict the state and Information Update formula (3) to process a new input signal:

Start

$$P(k = 0, s) = 0 \tag{1}$$

For $k \geq 1$ and $s \in S$

$$P^{-}(k, s) = \int_{S} q_k(s|s_{k-1}) P(k-1, s_{k-1}) ds_{k-1} \tag{2}$$

$$P(k,s) = \alpha P^{-}(k,s) + (1-\alpha)X(k) \tag{3}$$

EndFor

End

where: S is a state-space, s is a state (spatial and velocity components), k is a time step, α is a smoothing coefficient $\alpha \in (0,1)$, $X(k)$ is an observed value, $P(k,s)$ is the estimated value of state, $P^{-}(k,s)$ is the predicted value of state, $q_k(s|s_{k-1})$ denotes the transition between states (a Markov matrix [24]).

The Markov matrix determines the state changes and can generally describe the probability of changing position and velocity to any other. This matrix can be additionally variable over time. The Markov matrix does not support the transition between velocities in this test, because the assumed velocity of the object is constant. The transition model for a single cell is shown in Fig. 4. The variable velocity of the object causes that the transition leads to several neighboring targets and therefore the blur of the state space occurs. The knowledge of the object's movement enables optimization of this matrix and reduction of blur.

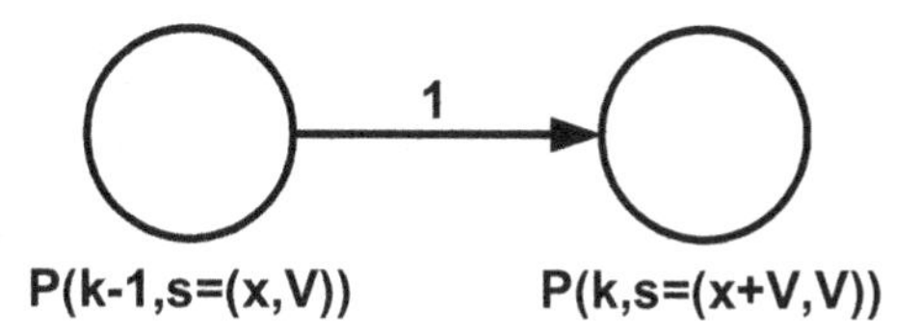

Fig. 4. The simplest transition model.

The local cross-correlation ST-TBD uses a different Information Update formula (3):

$$P(k,s) = \alpha P^{-}(k,s) + (1-\alpha)C^{N}\left[X(k,s), X(k-1,s)\right], \tag{4}$$

where: $C^{N}[.,.]$ is a cross-correlation function for a local window with a length of N.

The state-space can be defined in various ways. In the considered approach, the direct mapping of the input signal to the state-space is selected. One-dimensional measurement uses the following formula:

$$\begin{aligned} &C^{N}\left[X(k, s=(x,V)), X(k-1, s=(x,V))\right] = \\ &= \sum_{i=-\frac{N-1}{2}}^{\frac{N-1}{2}} X(k, x+i)X(k-1, x+i-V), \end{aligned} \tag{5}$$

where x is a position, which leads to a 2D state-space. The local window length is N. Velocity is the number of cells for the movement of local window between the time moments k and $k+1$ and is denoted by V. A similar formula can be obtained for 2D inputs (images) and this leads to a 4D state-space.

4 Results

The Monte Carlo test can be used to analyze the performance of the algorithm. This method is based on calculating the results for different sets of values (vectors) related to the object signal as well as various random vectors associated with the background signal. A single test cannot be used to evaluate the algorithm, even if the algorithm is deterministic. The problem is the influence of random input data on the output results, so the algorithm evaluation requires many tests for the same statistical parameters, such as standard deviation of the background input noise. The vector of the signal associated with the object is also generated using a random number generator. The object velocity is another parameter that is randomly selected. The evaluation of the algorithm requires many repetitions and the average results related to the position error could be presented for the comparison of different algorithms or the configurations of a single algorithm. The smooth curves of this error are observed if an acceptable convergence is achieved. The noisy curves are obtained if convergence is not achieved and such results cannot be used to formulate conclusions. The Monte Carlo algorithm is a simple approach and more advanced sampling algorithms are also available, such as MCMC (Markov Chain Monte Carlo) [23].

Standard ST–TBD and local cross–correlations are compared (Fig. 5). In each variant there are 10,000 test repetitions and two smoothing factors are tested: 0.95 or 0.98. The local cross–correlation algorithm is tested with the local window $N = 11$ and $N = 21$.

The fixed velocity and one–dimensional tracking case were adopted with a non–negative motion vector to simplify the calculation. The maximum value of the space–state is assumed as the detection criterion, therefore the location and velocity of the extended object are estimated. The estimation error is a function of additive Gaussian noise, which is the main disruptive factor. The obtained results are shown in Fig. 5. The corrected ST–TBD algorithm assumes detection of any pixel of the extended object instead of the center, which is important due to the random value of the signal object. The correction effect on the result is noticeable as a vertical shift in the result graph, however it is very small (Fig. 5 top–left).

5 Discussion

The number of test cases has been selected to obtain smooth curves (Fig. 5). The convergence was analyzed using a variable number of tests, and 10,000 tests were sufficient to determine the properties. The advantage of the Monte Carlo test is the possibility of comparative testing of various algorithms, the influence of parameters and responses for different classes of tracked objects. The obtained results indicate a significant improvement of the proposed method in relation to the standard algorithm (ST–TBD only).

The standard algorithm is single–pixel oriented, and the denoise of measurements is not as effective as the local cross–correlation between neighboring

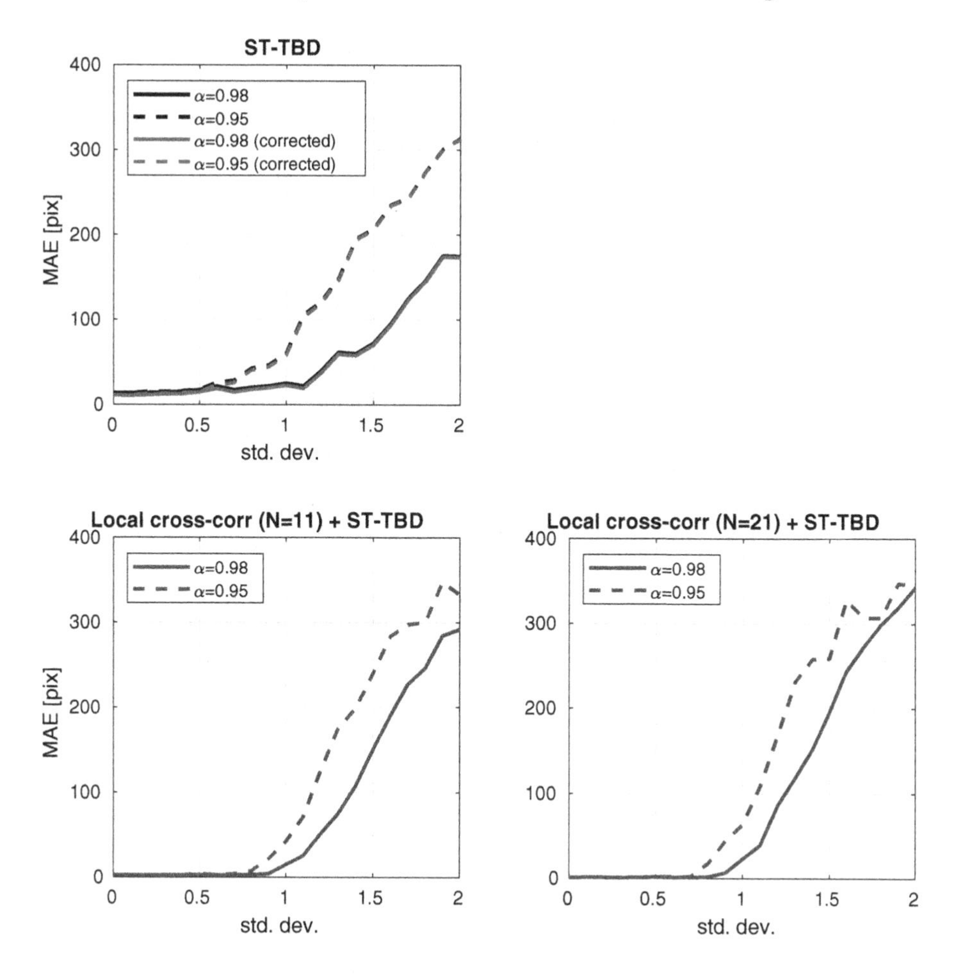

Fig. 5. Mean absolute error (MAE) for the Monte Carlo comparison test: the standard algorithm (ST-TBD only) and local cross-correlation with ST-TBD one.

measurement frames. MAE for the proposed algorithm is much lower, close to zero to about 0.8 standard deviation (Fig. 3), which is not achieved for the standard algorithm, even if a correction is applied. The increase in MAE for standard deviation around 1 is expected behavior. This is a region in which a significant influence of the smoothing factor is observed.

A high value of $\alpha = 0.98$ allows filtering the noise, but in real scenarios this value can not be very high. The smoothing factor reduces the impact of trajectory changes, so it should be estimated for a specific application. Higher errors are observed in the case of large standard deviations (> 1.2). Interestingly, MAE is lower for the standard algorithm, but high MAE values show a general problem for all algorithms.

The influence of the local window size is not relevant to the considered case. This is important, because the size of the object is in many cases an unknown parameter. The cost of calculations is not significant compared to the overall algorithm. The main cost is the calculation of the Motion Update formula (2). The local cross–correlation is calculated for known fixed velocities V, so it is fast with a cost similar to the 1D FIR filter for the 1D input data.

The real–time implementation is not considered in this document, but efficient parallel or non–parallel implementations are possible if the Markov matrix is not considered as a typical matrix. It is a sparse matrix and numerous implementation optimization techniques are possible. The proposed Information Update formula (4) only requires the calculation of local cross–correlations. It is possible to use parallel processing including SIMD (Single Instruction, Multiple Data) instructions with MAC (Multiply and Accumulate) instructions.

Processing using OpenMP [8], MPI [12] and CUDA [11,21] is possible for the considered algorithm. Evaluation of algorithms using the Monte Carlo test is very important, because many computers could be used independently for computations.

Tracking systems should have high noise immunity. Parameters of algorithms should be chosen so that the trajectories considered correspond to the real behavior of objects. This means that it is necessary to examine the application of optimization methods [2,4,9,14], thus increasing the credibility of the system and reducing the already large calculation budget.

The article assumes the comparison of the potential location of the object using a local cross–correlation, but the use of other measures may allow a potential improvement in tracking quality [1] as well as clustering [3,13,15].

6 Conclusions and Further Work

The proposed algorithm can be extended for 2D measurement spaces (such as video or radar images). ST–TBD, like many other TBD algorithms, enables combining data from many of the same or different types of sensors, which is important for improving the quality of tracking. The local cross–correlation assumes the preservation of the signal from the extended object in the neighborhood measurement frames, even if the values associated with the object are unknown.

The achieved result improves the tracking of the hidden signal in the background noise, which is clearly visible in Fig. 3. Detection of the object's position is not possible directly, but the TBD processing used together with the local cross–correlation allows the detection and estimation of position and velocity.

Further work will be related to the application to 2D tracking scenarios and other TBD algorithms.

Acknowledgment. This work is supported by the UE EFRR ZPORR project Z/2.32/I/1.3.1/267/05 "Szczecin University of Technology - Research and Education Center of Modern Multimedia Technologies" (Poland).

References

1. Abualigah, L.: Feature Selection and Enhanced Krill Herd Algorithm for Text Document Clustering. Studies in Computational Intelligence. Springer, Cham (2018). https://doi.org/10.1007/978-3-030-10674-4
2. Abualigah, L., Khader, A.T., Said Hanandeh, E.: A combination of objective functions and hybrid krill herd algorithm for text document clustering analysis. Eng. Appl. Artif. Intell. **73**, 111–125 (2018)
3. Abualigah, L., Khader, A.T., Said Hanandeh, E.: Hybrid clustering analysis using improved krill herd algorithm. Appl. Intell. **48**(11), 4047–4071 (2018)
4. Abualigah, L.M., Khader, A.T.: Unsupervised text feature selection technique based on hybrid particle swarm optimization algorithm with genetic operators for the text clustering. J. Supercomput. **73**(11), 4773–4795 (2017)
5. Blackman, S.: Multiple-Target Tracking with Radar Applications. Artech House, Dedham (1986)
6. Blackman, S., Popoli, R.: Design and Analysis of Modern Tracking Systems. Artech House, Dedham (1999)
7. Boers, Y., Ehlers, F., Koch, W., Luginbuhl, T., Stone, L.D., Streit, R.L.: Track before detect algorithms. J. Adv. Signal Process. **2008**, 2 (2008). https://doi.org/10.1155/2008/413932. Article ID 413932, Hindawi Publishing Corporation EURASIP
8. Chapman, B., Jost, G., Pas, R.V.D.: Using OpenMP: Portable Shared Memory Parallel Programming (Scientific and Engineering Computation). The MIT Press, Cambridge (2007)
9. Clerc, M.: From theory to practice in particle swarm optimization. In: Panigrahi, B.K., Shi, Y., Lim, H.M. (eds.) Handbook of Swarm Intelligence. Adaptation, Learning, and Optimization, pp. 3–36. Springer, Heidelberg (2011). https://doi.org/10.1007/978-3-642-17390-5_1
10. Dragovic, M.: Velocity Filtering for Target Detection and Track Initiation, vol. DSTO-TR-1406. Weapons Systems Division, Systems Sciences Laboratory (2003)
11. Farber, R.: CUDA Application Design and Development. Morgan Kaufmann, San Francisco (2011)
12. Karniadakis, G., Kirby, R.: Parallel Scientific Computing in C++ and MPI. Cambridge University Press, New York (2003)
13. Kaufman, L., Rousseeuw, P.: Finding Groups in Data: An Introduction to Cluster Analysis. Wiley, Hoboken (1990)
14. Kennedy, J., Eberhart, R.: Particle swarm optimization. In: Proceedings of ICNN 1995 - International Conference on Neural Networks, vol. 4, pp. 1942–1948, November 1995
15. King, R.S.: Cluster Analysis and Data Mining: An Introduction. Mercury Learning & Information, Dulles (2014)
16. Mazurek, P.: Comparison of different measurement spaces for spatio–temporal recurrent track–before–detect algorithm. In: Choraś, R.S. (ed.) Image Processing and Communications Challenges 3. AISC, vol. 102, pp. 157–164. Springer, Heidelberg (2011). https://doi.org/10.1007/978-3-642-23154-4_18
17. Mazurek, P.: Parallel distributed downsampled spatio-temporal track-before-detect algorithm. In: 2014 19th International Conference on Methods and Models in Automation and Robotics (MMAR), pp. 119–124, September 2014

18. Mazurek, P.: Preprocessing using maximal autocovariance for spatio-temporal track-before-detect algorithm. In: Choras, R.S. (ed.) Image Processing and Communications Challenges 5. AISC, vol. 233, pp. 45–54. Springer, Heidelberg (2014). https://doi.org/10.1007/978-3-319-01622-1_6
19. Mazurek, P.: Noise objects tracking using multiple order statistics and spatio-temporal track-before-detect algorithm. In: Choraś, R. (ed.) IP&C 2016. AISC, vol. 525, pp. 112–119. Springer, Cham (2017). https://doi.org/10.1007/978-3-319-47274-4_13
20. Metropolis, N.: The Beginning of the Monte Carlo Method. Los Alamos Science (1987). http://library.lanl.gov/la-pubs/00326866.pdf
21. Sanders, J., Kandrot, E.: CUDA by Example: An Introduction to General-Purpose GPU Programming. Addison-Wesley, Upper Saddle River (2010)
22. Scott, T.A., Nilanjan, R.: Biomedical Image Analysis: Tracking. Morgan & Claypool, San Rafael (2005)
23. Spall, J.C.: Introduction to Stochastic Search and Optimization: Estimation, Simulation, and Control. Willey, Hoboken (2003)
24. Stone, L., Barlow, C., Corwin, T.: Bayesian Multiple Target Tracking. Artech House, Norwood (1999)
25. Torstensson, J., Trieb, M.: Particle Filtering for Track Before Detect Applications. Master's thesis, Division of Automatic Control, Department of Electrical Engineering, Linköping University (2005)
26. Zhang, T., Li, M., Zuo, Z., Yang, W., Sun, X.: Moving dim point target detection with three-dimensional wide-to-exact search directional filtering. Pattern Recogn. Lett. **28**, 246–253 (2007)

Optimization of Demodulation for Air–Gap Data Transmission Based on Backlight Modulation of Screen

Dawid Bak[1], Przemyslaw Mazurek[1(✉)], and Dorota Oszutowska–Mazurek[2]

[1] Department of Signal Processing and Multimedia Engineering, West Pomeranian University of Technology Szczecin, Szczecin, Poland
{dawid.bak,przemyslaw.mazurek}@zut.edu.pl
[2] Department of Epidemiology and Management, Pomeranian Medical University, Szczecin, Poland
adorotta@pum.edu.pl

Abstract. Air–gap is an efficient technique for the improving of computer security. Proposed technique uses backlight modulation of monitor screen for data transmission from infected computer. The optimization algorithm for the segmentation of video stream is proposed for the improving of data transmission robustness. This algorithm is tested using Monte Carlo approach with full frame analysis for different values of standard deviations of additive Gaussian noise. Achieved results show improvements for proposed selective image processing for low values of standard deviation about ten times.

Keywords: Air–gap transmission · Digital demodulation · Image processing · Network security · Monte carlo simulations

1 Introduction

Problems of computer security is currently one of the most important problems of technical civilization. There are many methods of attacking computers or computer networks, in particular remote methods that do not require direct access.

The emergence of this type of problem resulted in the emergence of a number of defense methods. Some of these methods use technical means, such as firewalls or antivirus programs. Users' awareness of possible threats is also important for the security of computers or computer networks. Organizational security methods are, for example, software or hardware audits.

Air–gap is a very efficient method of improving the security of individual computers or computer systems [6]. Isolation through the lack of communication interfaces reduces the possibility of attacks. Air–gap breaking by the establishing unconventional communication interfaces is possible by the infection of secured

J. M. F. Rodrigues et al. (Eds.): ICCS 2019, LNCS 11536, pp. 71–80, 2019.
https://doi.org/10.1007/978-3-030-22734-0_6

computer or computer system [7]. The methods used for infection are not taken into account in this work.

Establishing one–way or two–way communication is important for both attackers and for prevention that is important for security engineers [8]. The data rate achieved is usually very low: a single bit is sent per second. Such transmission allows attack on PIN codes, passwords and very short confidential data. A typical air–gap attack is a one-way attack where data is sent outside of a protected computer or computer network. This type of attack is a long–term process that requires the attacker to listen in close proximity to the protected area.

1.1 Content and Contribution of the Paper

Air–gap transmission techniques are very active research area and some recent development are related to the application of different physical mediums. Example optical channels are: QR code embedding in image [13] (VisiSploit), network LEDs modulation in routers [18] (xLED), infrared transmission [16] (aIR–Jumper). Radio based transmission attacks are for example: USBee [15], AirHopper [14] and GSMem [17]. These types of transmission channels are not covered by typical protection, in particular audits, and therefore are very dangerous. Intentional data transmission hiding could be used for monitoring devices also [21].

In previous works, we proposed using a computer screen flashing for data transmission [4,5]. In the absence or very low user activity, very small changes in screen brightness are used that are not noticeable to humans.

It is possible to receive data correctly from a large distance using image observation with a camera and a telescope, as well as digital image processing and demodulation algorithms A typical modulation applied in this type of solutions is BFSK (Binary Frequency–Shift Keying) [10,19], which allows estimation of transmission settings like symbol and keyed frequencies [5].

Previous work assumed the processing of full–frame video sequences for estimation the transmitted signal [5]. Each frame of the image was transformed by averaging to a single value, i.e. the video sequence was converted into a one-dimensional signal The signal analysis for a single image pixel is not effective due to the low SNR (Signal–to–Noise Ratio) value. Averaging the whole frame of the image allows improved SNR, and the best solution is to average only the selected image area, such as the computer screen image. The problem dealt with in this article is the automatic determination of the area subjected to averaging. Determining the image area where brightness modulation occurs allows the SNR to be improved, because areas where there is no modulation are just a source of undesirable noise. This is particularly important in the case of an indirect attack, where the computer screen is not visible directly, but only the glow in the room is visible (Fig. 1). Very often screens of computers and keyboards, for security reasons, are set so that they are not visible from the outside.

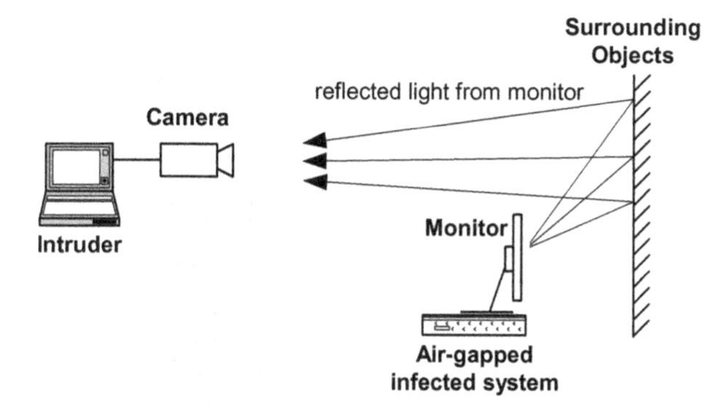

Fig. 1. Scheme of measurement of monitor brightness changes.

What's more, diffusion foils are used to block the possibility of direct observation of the room through the windows, but diffusion foils does not eliminate the glow of monitors.

The application of the optimization algorithm allows searching in the frame for the image of areas for which the brightness changes are the greatest to reduce the impact of noise. Areas for which the brightness changes are small bring the most noise to the signal and should be omitted in the analysis. Areas that should be omitted are those that are illuminated by other sources (other monitors, sources of light in the room).

The proposed algorithm is an offline algorithm, and the data is processed after recording the video image.

Example scenarios and acquisition configurations are presented in Sect. 2. Details of algorithm and processing method are considered in Sect. 3. Result are presented in Sect. 4 and discussion is provided in Sect. 5. Final conclusions and further work are considered in Sect. 6.

2 Data

The evaluation of the algorithm is based on an empirical test as proof of concept.

The source of the data is the laptop of which the screen is not visible directly. The laptop has software that modulates the brightness of the monitor. Brightness changes are not visible to a human directly, which means that the signal has a relatively small SNR.

The second laptop is positioned so that its screen is visible in the frame (Fig. 2). This laptop is a source of local interference, due to strong changes in screen brightness. Additional sources of interference are the lamps that illuminate the corridor. The purpose of the optimization algorithm is to detect areas from which transmitted data can be obtained, while avoiding interference areas.

Data is recorded using the ZWO ASI1600MM–COOL camera with a monochromatic sensor [26] and a resolution of 16 Mpix, and Canon EF 100 mm

Fig. 2. Light disturbance computer from the left and back of observed computer screen from the right.

1:2.8L IS USM lenses. The camera was placed on a tripod to ensure video recording without shifting the essential elements of the scene (Fig. 3).

ASICAP v.1.3 software for MS Windows and USB 3.0 interface has been used for video sequence acquisition. The camera sensor supports ROI (Region of Interest), that is used to reduce USB bandwidth and storage requirements, so the recorded picture frames have a resolution of 1024×768. Measured HFOV (Horizontal Field of View) is about 3° for the selected ROI, and the distance between the camera and the computer screen is about 29 m. This camera has advanced cooling (thermoelectric cooler and fan) for sensor noise reduction, but this option was not used. Recording speed is 100 fps and 10 ms exposure time is selected for the flickering reduction due to fluorescent lamps sources in corridor. Recorded video sequences are of high quality because they are raw frames, which enables to control the influence of noise (Fig. 4).

3 Method

Direct observation of the computer screen is sometimes not possible due to spatial relations. The light emitted by the computer screen can be observed indirectly by reflections from the surrounding objects. Glass, plastic, metal materials are particularly interesting because they can reflect light strongly in a specific direction. The Lambertian surfaces are also interesting because the light is scattered in all directions.

Fig. 3. Acquisition system.

Fig. 4. Light disturbance computer from the left and back of observed computer screen from the right (camera view).

An optimization algorithm, based on local full–frame search, is required to process as many picture frames as possible using the demodulation algorithm. Signal processing part is described in [4] and [5]. The detection using two band–pass filters and rectifiers was proposed in [19] and is not considered in this paper. There are two sources of interfering signals: light sources and camera noise. The input image is reduced to 16×12 pixels for reducing picture noise and processing complexity.

The aim of proposed algorithm is the calculation of segmented image S with positive pixel values $S(x, y)$ that corresponds to segmented area number. Pixel position (x, y) corresponds to the signal from video sequence $V(x, y, k)$, where x

and y are image coordinates and k is the frame number. Two pixels S_i and S_j belong to the same segmented area $S_i(.,.) = s$ and $S_j(.,.) = s$ if the similarity between corresponding video sequences is found using Euclidean metric:

$$d(i,j) = \left(\sum_k (V_i(x_i, y_i, k) - V_j(x_j, y_j, k))^2 \right) < T, \tag{1}$$

where T is threshold and $d(i,j)$ binary value (1 - similar, 0 - not similar).

The problem is the number of calculations required for $d(i,j)$, large image resolutions and long sequences. Local comparisons can be used to reduce the calculation due to spatial similarity and this variant is considered.

Algorithm randomly selects starting position (x_i, y_i) uses spatial neighborhoods:

$$(x_i - 1, y_i), \tag{2}$$

$$(x_i + 1, y_i), \tag{3}$$

$$(x_i, y_i - 1), \tag{4}$$

$$(x_i, y_i + 1) \tag{5}$$

as a second position (x_j, y_j) if they are not assigned to any s. New positions are marked as possible points for starting new comparisons with own neighborhoods. This algorithm behaves like local fill algorithm [25]. This process is repeated until the assignment to the same s region is possible. New position is randomly selected from not assigned yet positions to any s with new s value and repeated until all positions are not used.

Modified distance criteria (1) should be used due to the problem of direct comparison of values. Adjacent pixels can be illuminated in the same way, but with different average values, so the basic algorithm (1) discards pixels that, for example, belong to different scene objects, even if the lighting changes are similar. Proposed algorithm uses the removal of mean values using the following formulas:

$$d^*(i,j) = \left(\sum_k \left(V_i^*\left(x_i, y_i, k\right) - V_j^*\left(x_j, y_j, k\right)\right)^2 \right) < T, \tag{6}$$

$$V^*(x, y, k) = V(x, y, k) - mean\left(V(x, y, .)\right), \tag{7}$$

where V^* is corrected V sequence and $d^*(i,j)$ is a new similarity metric.

The threshold value is adaptively selected by testing the number of areas achieved, and the number of areas should typically be 5–20, so multiple passes of the segmentation algorithm are necessary to select the threshold value. Demodulation is processed individually for each A_s area with a common value of s. This area uses the following signal fusion formula:

$$A_s(k) = \frac{1}{N_s} \sum_s \left(V(x, y, k) - mean\left(V(x, y, .)\right)\right), \tag{8}$$

where N_s is the number of pixels with the same $S(x, y) = s$ assignment.

Each demodulated sequence should be checked using an additional CRC (Cyclic Redundancy Check) code for the final selection of the sequence of bits and this topic is out of the scope of the paper.

4 Results

Basic method of air gap transmission using the computer screen brightness changes assumes BFSK modulation and 0.18 bit/s transmission is used. The speed results from assumed slow changes in the brightness of the monitor screen. The Hamming distance between known and received binary sequences is used as the quality criterion. Binary keying signals that have higher resolution are used instead of comparison of demodulated bits. This approach reduces final result influences in demodulation algorithm.

Additive Gaussian noise images with known and controlled standard deviation are applied to all video frames in order to analysis of the sensitivity. Particular video sequence is tested 100 times for the selected standard deviation value and number of segmented region in 5–20 range.

Box and whisker plots for full frame processing and selective processing are shown in Fig. 5. These Monte Carlo test [22] allows the comparison of algorithm and the determination of properties basing on intensive numerical tests. Example segmentation results are shown in Fig. 6.

5 Discussion

Obtained results (Fig. 5) show importance of proposed solution. Achieved segmentation results influences Hamming errors. Proposed solution gives about ten times less errors comparing to entire image frame processing. Noised video sequences introduces Hamming errors, but proposed algorithm is still superior (Fig. 5). The results presented using the Monte Carlo analysis show how the signal degradation curve behaves for many tests for different standard deviation values of the interfering video noise.

The computing cost is very large for the proposed algorithm, because the entire video sequence is analyzed. It is possible to reduce it by the application only to a short sequence. Obtained area of interest can be used for the selection of a pixel for the entire video sequence. The largest regions in Fig. 6 are exemplary results of the proposed algorithm (marked using dark red color). This area includes both the background behind the monitor and the monitor housing, so it is not an optimal area, which requires the selection of a different threshold value. The obtained results, however, show how strong the level of interference from the monitor with the directly visible screen (Fig. 5).

The backlight modulation can be used for transmission of data from a room with diffusers in the windows The diffusers are used for protection against direct observation of the computer screen using a telescope. This method also accepts blurred images.

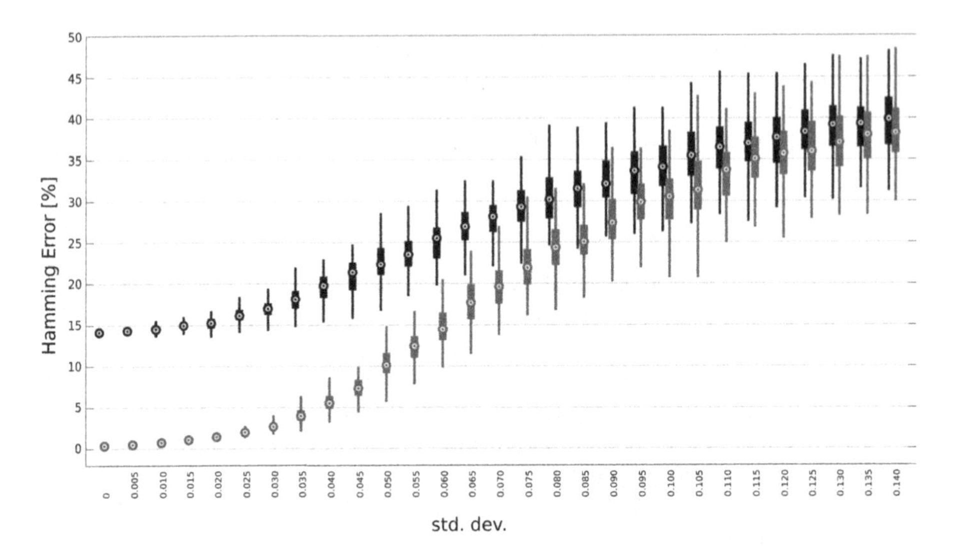

Fig. 5. Monte Carlo analysis of Hamming errors for full frame (black) and selective processing (red) algorithms. (Color figure online)

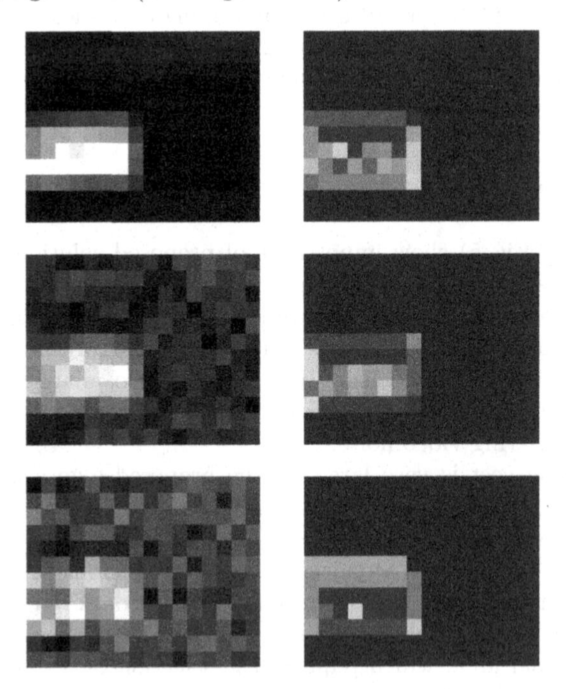

Fig. 6. Examples of processed image (top) and segmented regions (bottom) after rescalling with additive Gaussian noise with std.dev.: 0 (top), 0.055 (middle) and 0.120 (bottom). (Color figure online)

The cost of the calculations depends on the image resolution, so quite small resolution has been used. A serious problem is the necessity of processing quite long video sequences (Eqs. (7) and (8)). Processing using OpenMP [9], MPI [20, 24] and CUDA [12,23] is possible for the considered algorithm.

6 Conclusions and Further Work

Backlight modulation of computer screen is very import technique for air–gap data transmission and proposed segmentation extends possibilities by the automatic selection of video region. Light reflection from walls and surrounding objects could be used, if no direct visibility of screen is possible.

Optical techniques for the data transmission requires direct visibility of source, but this method shows the possibility of indirect data transmission. Very important property of the optical techniques is the possible large distance observation using telescopes and high sensitivity cameras.

The search task for the best area having the largest SNR can be performed using the optimization algorithm [1–3,11]. In addition, the selection of the threshold T value can also be performed automatically using the optimization algorithm. These tasks will be the subject of further work.

Additional further work will be related to the improving of bit rate also, because it is important limitation for numerous air–gap communication methods. Such improvement gives better utilization of short time slots without user activities that could disturb data transmission.

Acknowledgment. This work is supported by the UE EFRR ZPORR project Z/2.32/I/1.3.1/267/05 "Szczecin University of Technology – Research and Education Center of Modern Multimedia Technologies" (Poland).

References

1. Abualigah, L.M.Q.: Feature Selection and Enhanced Krill Herd Algorithm for Text Document Clustering. SCI, vol. 816. Springer, Cham (2019). https://doi.org/10.1007/978-3-030-10674-4
2. Abualigah, L., Khader, A.T., Said Hanandeh, E.: A combination of objective functions and hybrid krill herd algorithm for text document clustering analysis. Eng. Appl. Artif. Intell. **73**, 111–125 (2018)
3. Abualigah, L.M., Khader, A.T.: Unsupervised text feature selection technique based on hybrid particle swarm optimization algorithm with genetic operators for the text clustering. J. Supercomput. **73**(11), 4773–4795 (2017)
4. Bak, D., Mazurek, P.: Air-gap data transmission using screen brightness modulation. In: 2018 International Interdisciplinary Ph.D. Workshop (IIPhDW), pp. 147–150, May 2018
5. Bak, D., Mazurek, P.: Air-gap data transmission using backlight modulation of screen. In: Choraś, M., Choraś, R.S. (eds.) Image Processing and Communications Challenges 10, vol. 892, pp. 96–103. Springer, Cham (2019). https://doi.org/10.1007/978-3-030-03658-4_12

6. Bryant, W.: International conflict and cyberspace superiority: theory and practice. In: International Conflict and Cyberspace Superiority: Theory and Practice, pp. 1–239, 01 2015
7. Carrara, B.: Air-Gap Covert Channels. Ph.D. thesis, School of Electrical Engineering and Computer Science, Faculty of Engineering, University of Ottawa (2016)
8. Carrara, B., Adams, C.: Out-of-band covert channels—a survey. ACM Comput. Surv. **49**(2), 23:1–23:36 (2016)
9. Chapman, B., Jost, G., Pas, R.V.D.: Using OpenMP: Portable Shared Memory Parallel Programming (Scientific and Engineering Computation). The MIT Press, Cambridge (2007)
10. Chitode, J.: Digital Communication. Technical Publications, Pune (2010)
11. Engelbrecht, A.P.: Fundamentals of Computational Swarm Intelligence. Wiley, New York (2006)
12. Farber, R.: CUDA Application Design and Development. Morgan Kaufmann, San Francisco (2011)
13. Guri, M., Hasson, O., Kedma, G., Elovici, Y.: An optical covert-channel to leak data through an air-gap. In: 2016 14th Annual Conference on Privacy, Security and Trust (PST), pp. 642–649, December 2016
14. Guri, M., Solewicz, Y., Daidakulov, A., Elovici, Y.: Acoustic data exfiltration from speakerless air-gapped computers via covert hard-drive noise ('DiskFiltration'). In: Foley, S.N., Gollmann, D., Snekkenes, E. (eds.) ESORICS 2017. LNCS, vol. 10493, pp. 98–115. Springer, Cham (2017). https://doi.org/10.1007/978-3-319-66399-9_6
15. Guri, M., Monitz, M., Elovici, Y.: USBEE: air-gap covert-channel via electromagnetic emission from USB. In: 2016 14th Annual Conference on Privacy, Security and Trust (PST), pp. 264–268, December 2016
16. Guri, M., Bykhovsky, D., Elovici, Y.: Air-jumper: Covert air-gap exfiltration/infiltration via security cameras & infrared (IR). CoRR abs/1709.05742 (2017)
17. Guri, M., Kachlon, A., Hasson, O., Kedma, G., Mirsky, Y., Elovici, Y.: Gsmem: Data exfiltration from air-gapped computers over GSM frequencies. In: 24th USENIX Security Symposium (USENIX Security 15), pp. 849–864. USENIX Association, Washington, D.C. (2015)
18. Guri, M., Zadov, B., Daidakulov, A., Elovici, Y.: XLED: covert data exfiltration from air-gapped networks via router leds. CoRR abs/1706.01140 (2017)
19. Haykin, S.S.: Digital Communications. Wiley, New York (1988)
20. Karniadakis, G., Kirby, R.: Parallel Scientific Computing in C++ and MPI. Cambridge University Press, New York (2003)
21. Mazurek, P., Bak, D.: Embedded software monitoring using pulse width modulation as a communication channel for low pin count microcontroller applications. In: Silhavy, R. (ed.) CSOC2018 2018. AISC, vol. 763, pp. 319–330. Springer, Cham (2019). https://doi.org/10.1007/978-3-319-91186-1_33
22. Metropolis, N.: The Beginning of the Monte Carlo Method. Los Alamos Science (1987). http://library.lanl.gov/la-pubs/00326866.pdf
23. Sanders, J., Kandrot, E.: CUDA by Example: An Introduction to General-Purpose GPU Programming. Addison-Wesley, Upper Saddle River (2010)
24. Snir, M., Otto, S., Huss-Lederman, S., Walker, D., Dongarra, J.: MPI: The Complete Reference. MIT Press, Cambridge (1996)
25. Torbert, S.: Applied Computer Science. Springer, Cham (2016). https://doi.org/10.1007/978-3-319-30866-1
26. ZWO: ASI1600 Manual (2016). revision 1.1

Reinsertion Algorithm Based on Destroy and Repair Operators for Dynamic Dial a Ride Problems

Sven Vallée[1,2](✉), Ammar Oulamara[1], and Wahiba Ramdane Cherif-Khettaf[1]

[1] University of Lorraine LORIA (UMR 7503) laboratory, Campus Scientifique, 615 Rue du Jardin botanique, 54506 Vandœuvre-les-Nancy, France
{sven.vallee,Oulamara,Ramndanec}@loria.fr
[2] Padam, 19 rue des feuillantines, 75005 Paris, France
sven@padam.io

Abstract. The Dial-a-Ride Problem (DARP) consists in serving a set of customers who specify their pickup and drop-off locations using a fleet of vehicles. The aim of DARP is designing vehicle routes satisfying requests of customers and minimizing the total traveled distance. In this paper, we consider a real case of dynamic DARP service operated by *Padam* (www.padam.io.) in which customers ask for a transportation service either in advance or in real time and get an immediate answer about whether their requests are accepted or rejected. A fleet of fixed number of vehicles is available during a working period of time to provide a transportation service. The goal is to maximize the number of accepted requests during the service. In this paper, we propose an original and novel online *Reinsertion Algorithm* based on destroy/repair operators to reinsert requests rejected by the online algorithm used by *Padam*. The proposed algorithm was implemented in the optimization engine of *Padam* and extensively tested on real hard instances up to 1011 requests and 14 vehicles. The results show that our method succeeds in improving the number of accepted requests.

Keywords: Dynamic DARP · Insertion and reinsertion heuristics · Computational experiments

1 Introduction

Road transport is still responsible for the bulk of transport emissions in terms of greenhouse gases and air pollutants. Every day, congested roads are a huge cost to the large cities in the world. However, profound change lies ahead for the transport sector in the world. A series of technological innovation and disruptive business models has led to a growing demand for new mobility services. At the same time, the sector is responding to the pressing need to make transport more efficient and sustainable. Digital technologies are a driving force of this process of innovation in transport sector. These technologies create a truly multimodal

J. M. F. Rodrigues et al. (Eds.): ICCS 2019, LNCS 11536, pp. 81–95, 2019.
https://doi.org/10.1007/978-3-030-22734-0_7

transport system integrating all modes of transport into one mobility service, allowing people and cargo to travel smoothly from door to door.

In this paper, we focus on the transportation service developed by Padam. The service consists in creating dynamic bus lines according to the customer demands. In Padam's system, customers submit demands of transportation (origin and destination locations) via a mobile application, either in advance, i.e. few days before the service, or in real-time for an immediate service. Customers specify either when they wish to be picked up or when they have to be at their destination. The transportation service is operated by mini-bus with restricted number of places. All potential stop locations of buses are predefined and correspond to POI in cities where a bus can stop without affecting traffic, such as train or metro station, administrative buildings and so on. Thus, pickup and destination locations of customers are then associated to their nearest predefined locations and the customer will be serviced at these predefined locations instead of original pickup and destination locations. Once a customer submits its request, the Padam's optimization service decides whether the request can be accepted or not, i.e. whether the request can be inserted in the existing rides or not. When solutions exist, several offers are then proposed to the customer around its requested time-window, among which he will choose the most convenient for himself. The transport operation is outsourced in *Padam*'s service, and contract is negotiated with third parties transport companies. In such contract, the number of mini-bus as well as the shifts of working hours of drivers is specified for each weekday. The number of mini-bus for each day is determined by *Padam* based on historical data of transportation demands. The starting locations of rides are decision variables fixed by the optimization engine depending on the number and the localization of requests of customers. Furthermore, as the transport cost is a fixed cost (i.e. outsourcing cost), the main objective of *Padam*'s service is servicing as many demands as possible during the ride shifts.

In this work, we improve optimization algorithm implemented in *Padam's* service. More precisely, we consider a dynamic dial-a-ride problem with online requests of transportation. Since a solution must be proposed to a customer in real time, i.e. in a few seconds for each request, a heuristic approach is proposed. The proposed method is based on a neighborhood search algorithm for reinsertion of requests rejected by the online insertion algorithm. The *Reinsertion Algorithm* uses construction and destruction operators such as those used in an ALNS meta-heuristic [11]. It should be noted that the reinsertion techniques for the dynamic DARP are not widely used in the literature. To the best of our knowledge, only the paper [10] has proposed a reinsertion algorithm for dynamic DARP in which the objective is to reduce the number of vehicles while in our case the number of vehicles is imposed. The remainder of this paper is structured as follows. In Sect. 2 we provide a selective review on papers related to DARP problems. In Sect. 3 we give more details on the constraints and characteristics of our problem. In Sect. 4 we describe the proposed approach. Experimental results are presented in Sect. 5. The paper concludes with a short summary and an outlook on future research in Sect. 6.

2 Related Work

The Dial-a-Ride Problems (DARPs) have been investigated in the literature for over 30 years. The basic version of DARP consists of serving a set of users who specify their departure and arrival locations using a fleet of vehicles. The user specifies either desired pickup time or desired drop-off time. The basic objective function used in the literature is the total distance traveled by the vehicles [3]. There are two main versions of the problem: static and dynamic DARP. In the first case, all requests are known in advance before computing a solution while in the second case a part of the user requests arrives in real time, vehicle routes being adjusted in real-time to meet new demand. Dynamic DARP has received much less attention than its static counterpart. See [6] for a recent survey on both static and dynamic cases.

To solve dynamic DARP, literature studies have been focused on fast heuristics to insert new requests [7] and on meta-heuristic methods to optimize the system between the appearance of two consecutive requests [1,12]. Most studies have proposed to combine the online insertion and the optimization system between requests [2,4,8,14]. In [8], the authors proposed an online-regret based algorithm for a 2-phase optimization procedure by using available idle time to continuously optimize the solution. [2] presented hybrid method proceeding in two phases. In the first phase a simple insertion scheme is used to generate a feasible solution, which is improved in the second phase with a tabu search algorithm. The tabu search algorithm is stopped each time a new request appears. In [14], a similar approach to the one proposed by [2] has been developed to solve a real problem but the improvement phase uses an Adaptive Large Neighborhood Search metaheuristic. The study in [4] presented a two-phase insertion algorithm based on route perturbations. Every time a new request appears, the insertion is evaluated for each route within an appropriate neighborhood of the current one.

The aim of this paper is to design an efficient online reinsertion heuristic subject to a fixed and limited number of vehicles. By reinsertion heuristic we mean that whenever the online system cannot insert a new customer, we try to rearrange the current solution to try to insert him anyway. This procedure must be fast (the computation time must not exceed few seconds) To the best of our knowledge, only two papers have developed a reinsertion procedure for the DARP [9,10]. In these two studies, the number of vehicles is unlimited, which is not the case in our problem. In [9], the authors developed a heuristic for the static multi-vehicles DARP in order to reduce the number of used vehicles. Their heuristic improves the parallel insertion heuristic of [7]. [10] adapts the latter algorithm to the dynamic case, where the main objective is to reduce the number of used vehicles.

In this paper, we present a new *Reinsertion Algorithm* for dynamic DARP based on destroy/repair operators. This algorithm allows us to test more reinsertion possibilities than the approach proposed in [9] and [10], while remaining fast and simple. We study the performance of our reinsertion heuristic both in terms of the vehicle duration and of the number of accepted requests. Extensive tests on real hard instances provided by Padam are performed.

3 Problem Description

When a user wants to book a ride, he specifies (via a mobile application) temporal and geographical informations: pickup and drop-off locations, number of passengers and requested hour. The requested hour concerns either the desired pickup or drop-off time, in which case the customer's request is said to be pickup-oriented (PO) or delivery-oriented (DO), respectively. This kind of choice is frequently used in Dial-A-Ride systems [3]. The road network is modeled as weighted graph with a set V of nodes and a set E of edges. Nodes represent predefined pickup and drop-off locations and each user's request is associated to the nearest predefined location. Nodes are determined by a statistical study on user travel patterns by combining several data sources, which is not presented in this paper. The set of edges depicts paths between nodes, and each edge (i, j) has weights d_{ij} and t_{ij} which correspond to the shortest distance and the shortest travel time between nodes i and j, respectively.

A homogeneous fleet of vehicles fixed by the transporter is available to serve requests. Vehicles have their own time window (beginning and end of service) and start location. When the system receives a request, the optimization engine determines one or more proposals using a fast heuristic (described in Sect. 4.1), and the proposals are sent to the customer. When the customer confirms one of the proposals, the request is inserted in the appropriate ride and a time window constraint is added on the request (see Sect. 4.1). Furthermore, the system imposes a maximum ride time M_{ij} between pick-up i and drop-off j of each inserted request in a ride where M_{ij} is the maximum detour that a customer can accept and is proportional to the shortest travel time t_{ij} between i and j, i.e. $M_{ij} \leq \gamma_k \times t_{ij}$, with $\gamma_k \geq 0$. The value of γ_k is selected from a set Γ of coefficients already predefined in the system and depends on the value of t_{ij}. For example, for any travel time t_{ij} in the interval $[10, 20]$ of minutes, the value of γ is 1.3. The set Γ helps us to models the fact that the acceptable deviation is not the same for short and long travel time. Thus, for each insertion of a new request, the optimization engine must respect time-window and gamma constraints of each already inserted request in addition to vehicle constraints, namely maximum capacity and service time window. The main goal is then to serve a maximum number of requests under described constraints while minimizing the service time duration of the vehicles. Given the dynamic nature of the problem, it is clear that it is not possible to directly maximize the number of served requests. Instead, our approach will use the total duration of the rides as an objective to be optimized, the duration of a ride being the sum of the travel time between it's successive visited nodes. The idea is to create rides with fewer useless detour and with 'straight' travels so that they serve more requests at the end of the service.

4 Solving Approaches

In this section, we first present the algorithm currently used by the company. We call it the *online insertion algorithm*. We then present our new *reinsertion heuristic* algorithm designed to improve the responses of the system.

4.1 Online Insertion Algorithm

Each time a new request appears, a fast insertion heuristic is launched to get a quick proposal list for the customer (generally in less than 1 s time). This *Online Insertion Algorithm* is a greedy heuristic which tests each possible insertion position (i.e pickup/drop-off position) in each ride. If feasible insertions are found, the heuristic output several proposals that are submitted to the customer, each one at a different time like timetables in public transportation system. The idea is to take into account the wishes of the customer while keeping in the foreground the concept of shared transportation. The customer is free to choose one of the proposals or to refuse them. The proposals are differentiated by their pickup and drop-off hours. If h is the requested hour of the customer, we assume that the customer will accept a proposal if the pickup (drop-off) hour of a PO (DO) request is within $TW_r = [h - W, h + W]$, where the value of W is often around 20 min (see [14] for more details).

Let's assume that the customer chooses a proposal with h_p and h_d as pickup and drop-off time. To ensure that subsequent insertions will not disturb the initial commitment toward the customer, we impose a time-window around the pickup (TW_p) and drop-off (TW_d) hours as follows:

$$TW_p = [h_p - PWB, h_p + PWA]$$

$$TW_d = [h_d - DWB, h_d + DWA]$$

where PWB, PWA, DWB, DWA are parameters fixed by the company. These time-windows (which can as tight as 10 min wide) ensure that subsequent clients can be inserted in the same ride while maintaining a high-quality service for the new customer.

4.2 Reinsertion Algorithm

When dealing with real situations, it can happen that a customer gets no satisfactory answer or even no answer at all. It means that the current arrangement of the rides doesn't allow us to serve him. In this case, rather than simply letting the customer refuse proposals, we try to move other already inserted and not yet served requests to see whether we can find an arrangement allowing us to insert the new customer while respecting all the other constraints. The rearrangement must be done in a few seconds to keep a low response time to the customer. We call the corresponding heuristic the *Reinsertion Algorithm*. The algorithm is based on destroy/repair neighborhoods search [13]. The idea is to appropriately destroy part of the solution by removing existing requests and then reinserting them together with the new request. Thus, each iteration of the algorithm consists in a destroy/repair steps. Iterations are performed until the calculation time limit is reached and the best feasible solution found (if any) is chosen. The maximum operating time of the reinsertion heuristic is noted as MT. The purpose is to find a proposal close to the request hour of the customer (Sect. 4.1).

Algorithm 1 describes our reinsertion heuristic. It is different from a pure Large neighborhood Search (LNS) framework [13] in that once the search finds a new solution, it is no longer improved, but rather returns to the original solution. This is necessary for practical issues. Indeed, at this stage, we are not sure that the customer will validate our proposal. Even if he validates it, he can decide to do it a little time later, for example one minute later. It is possible that other requests may be accepted during this period, which could make the proposed insertion impossible. In order to design an effective validation procedure, we must keep the reinsertion process as simple as possible and disrupt the current solution as little as possible.

Algorithm 1. Reinsertion Heuristic

Input: Current solution s, new request r, list of remove operators LDO, list of repair operators LRO

Output: Best feasible solution s^b if it exists otherwise Empty

1: $L \leftarrow EmptyList()$;
2: **while** time is not over **do**
3: $op \leftarrow SelectRemoveOperators(LDO)$;
4: $n \leftarrow ChooseNumberRequest()$
5: $LR \leftarrow RemoveRequests(op, n)$;
6: $LR \leftarrow LR \cup \{r\}$;
7: $s_{new} \leftarrow ReinsertRequests(LR, LRO, s)$; ▷ See Algorithm 2
8: **if** $feasible(s_{new})$ **then** $L \leftarrow L \cup s_{new}$;
9: **end while**
10: **if** L not empty **then** return $GetBestSolution(L)$;

We now present the components of the Algorithm 1.

4.2.1 Destroy Step

Given the current arrangement of the rides, the first step of the algorithm is to choose a set of requests to be removed from their current places (destroy step). To achieve it, we use three different destroy operators. Each operator selects the requests to be removed among a list LI of already inserted requests in rides. Note that LI does not necessarily contain all requests. Indeed, it is not relevant to remove requests which are distant (in time) from the requested hour h of the new customer. We then define a time window $[h - W - T, h + W + T]$ with T a free parameter and restrict LI to requests whose pickup hour or drop-off hour is in the time window $[h - W - T, h + W + T]$. If T is too small, the set of candidate requests will be small and will not provide enough opportunity to insert the new customer. However, if it is too high, the search will spend a lot of time deleting requests that are not relevant to insert the new request. Section 5.2.1 studies the impact of several values of T. Beside the list of candidates for removal, destroy operators also receive the number k of requests to select. k is randomly chosen (at each iteration) between k_{min} and k_{max}, which are two parameters. Furthermore, at each iteration, the destruction operator used is randomly selected. We now describe our three removal operators:

Random operator: Select k requests randomly.

Worst operator: Select the k requests with the largest savings, i.e. the difference between the objective value of the current solution and the objective value of solution once the requests are removed. In order to increase diversification, this operator is randomized as follow: all requests of LI are sorted in decreasing order of saving values in a list L. A random number y is sampled between 0 and 1 and the request at the position $\lfloor y^{p_r}|L|\rfloor$ where $|L|$ is the size of L and p_r a chosen parameter. This is repeated until k requests have been chosen.

Relatedness operator: Choose a request randomly and select $k - 1$ related requests. The relatedness measure between request i and j is defined as follow:

$$\frac{1}{2}\left(t_{p_i,p_j} + t_{d_i,d_j}\right) + \frac{1}{2}\left(|u_{p_i} - u_{p_j}| + |u_{d_i} - u_{d_j}|\right)$$

where p_i and d_i are respectively the pickup and drop-off nodes, u_{p_i} and u_{d_i} the service time of pickup and drop-off of request i and t_{n_1,n_2} the travel time between nodes n_1 and n_2. This operator is also randomized as for the worst operator. The parameter controlling the randomness is called p_w. In this case $p_w = p_r$.

4.2.2 Repair Step

Once the requests to be removed have been selected, they are removed from their current rides. We then try to reinsert them in the best way, including the request of the new customer. If all requests can be inserted while respecting constraints of the already existing requests, we obtain a feasible solution. To reinsert requests we use three repair operators. Each operator is always called during the repair process, in a sequential way. If several operators find a feasible solution, the best one is selected and used as the outcome of the current iteration (see Algorithm 2).

Deep Greedy operator: Perform the best insertion among all feasible insertions of all remaining requests to be reinserted. The best insertion is defined as the insertion with the minimal increase in the objective function.

Algorithm 2. ReinsertRequests Procedure

Input: List of requests to reinsert LR, list of repair operators LRO, current solution s

Output: best feasible solution s^b if it exists otherwise Empty

1: $L \leftarrow EmptyList()$;
2: **for** each operator o in LRO **do**
3: $s_{new} \leftarrow InsertRequests(o, LR, s)$;
4: **if** $feasible(s_{new})$ **then** $L \leftarrow L \cup s_{new}$;
5: **end for**
6: **if** L not empty **then** return $GetBestSolution(L)$;

Regret operator: Insert the request with the largest regret. The regret measure is defined in [5] and evaluates the difficulty to insert the request later. The idea is to find for each request i its best insertion in each vehicle k, with insertion cost c_{ik}. We then construct a matrix in which each row represents a request and each column a vehicle. If no feasible insertion exists, an arbitrary large value is used instead. A regret of a request i is then computed as:

$$\sum_k \left(c_{ik} - \min_j c_{ij} \right)$$

Request with largest regret is then chosen and inserted in its best position.

Priority Operator: The idea is to insert first requests at positions which will have small impact on others request insertion. To do this, we first select a request, then a ride to insert this request. The following steps are repeated until the number of requests to be inserted is reached: (1) for each request, compute (by testing all possible insertions) the number of rides in which it is possible to insert it, (2) take the request with the smallest number, (3) compute for each ride (where insertion of the selected request is possible) the number of requests that can be inserted into it, (4) take the ride with the smallest number.

5 Experiments

The purpose of this section is to assess the benefits of the *Reinsertion Algorithm* (Sect. 4.2) by running computer simulations on realistic data. The simulations were executed on a server with 16 Intel Processor Core cadenced to 3 GHZ. For each instance, we run a simulation with *online algorithm* as described in Sect. 4.1 to get reference results called RV. The *Reinsertion Algorithm* being stochastic, we run 10 simulations on each instance and use the average results called V for comparison for each experiment. When comparing V against a reference value RV, we compute the relative percentage improvement $\frac{V-RV}{RV} * 100$.

5.1 Instances

We have 19 instances divided in two groups A and B. The groups A and B contain 10 and 9 instances respectively. Each group models a different real transport context, with different service and geographical features. Instances of the same group differ by the number of requests. Each instance is named G_N, where G is the group name and N the number of requests. Table 1 presents fleet information and geographical data of each Group. Column *'Nodes'* indicates the number of nodes of the underlying graph (i.e all pickup/drop-off nodes), *'Service Duration'* is the time span where vehicle serve customer, *'Vehicle Capacity'* is the maximum capacity of a vehicle, *'Area'* is the number of km^2 covered by the set of nodes and *'Fleet Size'* is the number of vehicles *'Requests'* gives the minimum and maximum number of requests among the instances of the group. The two groups are very different: A is a large territory with short service span representing commuting transport (morning or evening) with mini-bus whereas B is far much smaller with larger bus running throughout the day, representing public transport in a neighborhood area.

Table 1. Characteristics of groups A and B

Group	Nodes	Service Duration	Vehicle Capacity	Area (km^2)	Fleet Size	Requests
A	473	3 h30	8	25	13	120–360
B	90	12 h	30	5	6	200–1011

Table 2 presents customer constraints of each Group. Columns PWA/PWB and DWA/DWB concern Time-Window range of requests as explained in Sect. 4.1, *'Gamma Levels'* and *'Gamma Values'* the gamma parameters (see Sect. 3), W is defined in Sect. 4.2 and $DwellingTime$ represents the service time at each pickup or drop-off node, i.e the time requested for people to get in and out of the bus. All temporal parameters are expressed in minutes. Based on the knowledge of the various instances and the operational experience of the company, we can say that group A is harder than group B, mostly because the associated territory is more extended and the associated requests don't follow easy geographical patterns.

Table 2. Service quality for each instance group

Group	PWB/PWA	DWB/DWA	Gamma Levels	Gamma Values	W	Dwelling Time
A	0/10	10/13	[10, 20]	[2, 1.8, 1.7]	20	1
B	5/8	10/10	[5, 10]	[2.5, 2, 1.8]	20	0.5

5.2 Impact of Parameters

Before testing our *Reinsertion Algorithm*, we want to study the impact of its important parameters T and MT defined in Sect. 4.2. Concerning the values of free parameters, preliminary experiment led us to set $k_{min} = 3$, $k_{max} = 10$ and $p_r = 4$.

5.2.1 Impact of T

Our first objective is to evaluate the impact of the value of T (defined in Sect. 4.2) on the final percentage of served requests. In this evaluation, the maximum allowed time MT for the experiment is set to 1 s. The Fig. 2 shows the results of the experiments plotted separately for each group. The behavior on the two groups is pretty clear: the performance increases up to 15 min and then decreases when compared to $T = 15$. Thus, $T = 15$ has a better performance than $T = 0$ since it provides a large search space of solutions. The performance associated with values beyond 15 min decreases because more time is expended in removing non-relevant requests. The decrease in performance is different between the two groups. We observe that in group B, this performance deteriorates when T increases and falls below $T = 0$ for values greater than 60 min. However, we observe that in group A this performance tends to stabilize around an average value when $T > 15$ and always performs better than $T = 0$. Figure 1 gives us insights to explain this behavior. *Neighborhood* represents the average number of requests that are candidates for removal (size of the list LI see Sect. 4.2.1) and *Feasible* computes the proportion of the number of feasible moves in relation to all the moves tested by the algorithm. These values are obtained by running one simulation on each instance. We observe that the size of the neighborhoods is on average similar for both groups. However, it appears that finding feasible moves is more harder in group A, which confirms the hardness of these instances (cf. Sect. 5.1).

T	0	60	120
Neighborhood	23.88	55.18	62.98
Feasible	1.91	1.51	1.29

(a) Group A

T	0	60	120
Neighborhood	13.67	47.10	66.78
Feasible	14.77	8.85	6.14

(b) Group B

Fig. 1. Mean number of booking and feasible moves in Neighborhood according to the value of T for each Group

We keep the value $T = 15$ for future experiments. The fact that the same value gives best results on both groups is probably related to the fact that the W values are similar (see Table 2).

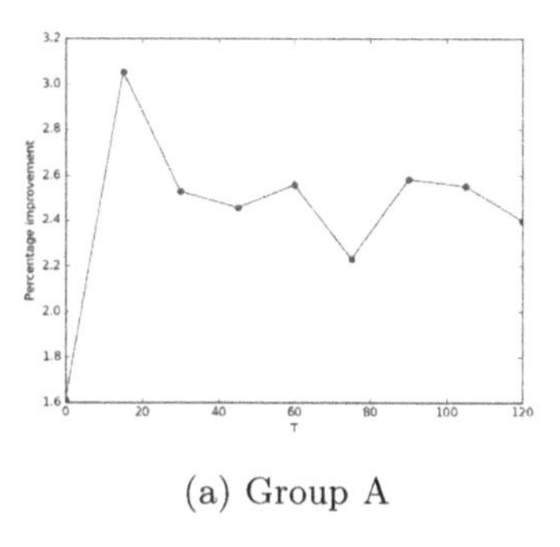

(a) Group A

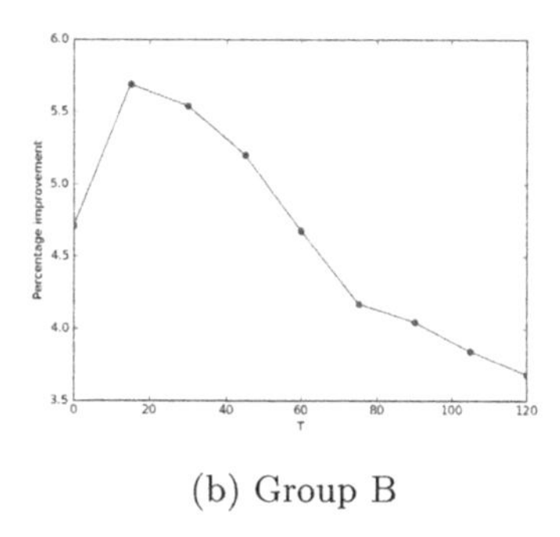

(b) Group B

Fig. 2. Relative improvement according to the value of T for each group of instances.

5.2.2 Impact of Maximum Allowed Time

In this section, we are interested in varying the maximum time MT allowed for the *Reinsertion Algorithm*. We perform independent tests for each value of MT. Figure 3 shows the impact of the maximum allowed time on the two groups of instances. In group B, we observe significant improvement between 1 and 3 s. The gain continues to increase up to 5 s in a less significant way. The behavior is different in group A. We observe a clear improvement when passing from 1 to 2 s, but no clear improvement up to 5 s. We even observe that $MT = 3$ and $MT = 4$ perform slightly worse than $MT = 2$. These variations are mostly due to the stochastic nature of the algorithm and the natural variance it produces. The group B takes more advantage of the increase of maximum time than group A, probably for the same reasons as those given in Sect. 5.2.1.

Based on these observations, we choose to limit the maximum allowed time for the *Reinsertion Algorithm* to 3 s, considering that it allows to keep fast answer time and that it doesn't degrade the performances. Note that 2 s could also be an acceptable value.

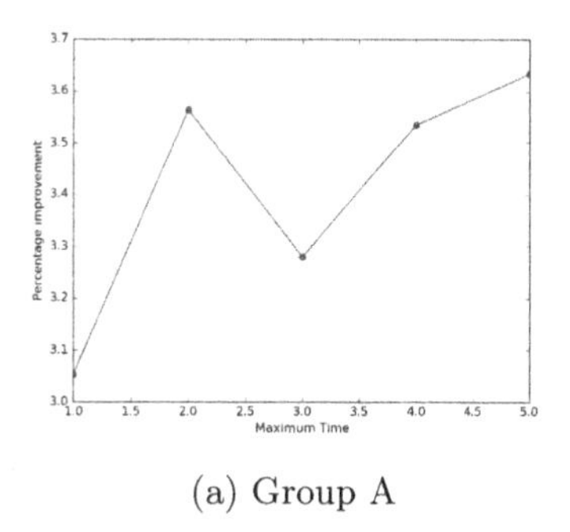

(a) Group A

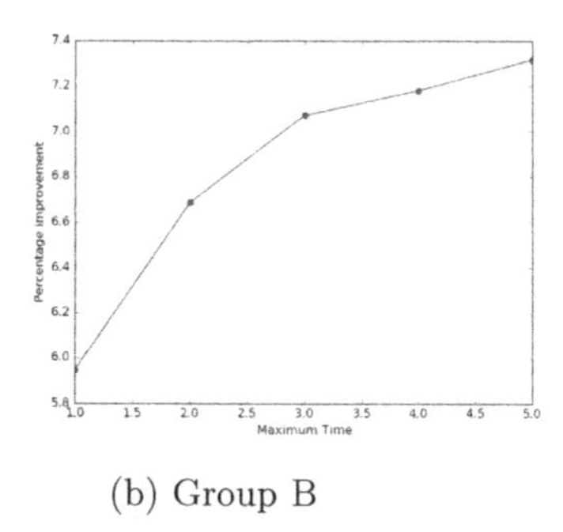

(b) Group B

Fig. 3. Relative Improvement according to the maximum allowed time (seconds) for each group of instances.

5.3 Benefits of Reinsertion

The purpose of this section is to evaluate the benefits of running the proposed *Reinsertion Algorithm* on dynamic Dial-a-Ride problems. We use the percentage of served requests as a metric to evaluate the performance of the *Reinsertion Algorithm* but we also show its impact on the total duration of rides (defined in Sect. 3), which is the objective minimized by algorithms implemented in Padam's service.

Table 3 presents results for group A. It exposes results with 13 vehicles (number of vehicles used in practice for this group of instances) and also for 12 and 14 vehicles for each instance. The meaning of the columns is as follows: *O* is for the percentage of served requests with online algorithm (algorithm implemented in Padam's service), *O+R* is the percentage of served requests with online and reinsertion algorithms, *Imp* is the relative improvement of reinsertion over online in terms of served requests and *ImpD* is the relative improvement of reinsertion over online in terms of total duration of the rides.

We first observe that instances of group A are very hard, because we are often far from satisfying all requests with the online algorithm. However, the *Reinsertion Algorithm* provides an improvement on almost all instances, with an average improvement of 3.64% (respectively 4.56% and 5.95%) with 13 vehicles (respectively with 12 and 14 vehicles). This is done with approximately the same rides duration, which shows that the reinsertion can satisfy more requests with similar vehicle costs. By comparing the results obtained with 12 and 13 vehicles, we observe that the online algorithm serves on average 45.82% of requests with 12 vehicles compared to 49.68% with 13 vehicles, while the reinsertion serves on average 48.17% with 12 vehicles. It means that despite the hardness of the instances, *Reinsertion Algorithm* fills more than half of the gap caused by the removal of a vehicle and even eliminates the need to add a vehicle in the case of the instance A_177 and A_120. The results also show that with 13 vehicles, the *Reinsertion Algorithm* succeeds in inserting on average as many requests as with the online algorithm with 14 vehicles, improving the result of some instances.

An analysis of the results obtained by varying the number of vehicles shows that the relative improvement of the online algorithm with 14 vehicles compared to the same algorithm with 13 vehicles is 3.8%, and the relative improvement of the *Reinsertion Algorithm* is 63% when comparing 14 vehicles against 13 vehicles. Then, we can conclude that the relative improvement of the online algorithm is lower when passing from 13 to 14 vehicles, while the relative improvement of the *Reinsertion Algorithm* is significantly larger.

In rare cases, the *Reinsertion Algorithm* implies a deterioration in the percentage of served requests. We observe on most of theses instances that the final duration of the ride is nevertheless higher. This may be explained by the fact that the reinsertion process can sometimes accept hard requests that would otherwise have been rejected, degrading the rides too much and making it more difficult to insert some subsequent requests. This seems however to be quite rare.

Table 4 presents results for group B. These results are more consistent than for group A with an improvement in all instances. With 6 vehicles, the

Table 3. Performance of Reinsertion Algorithm on instances of group A.

Ve	nb of vehicles = 12				nb of vehicles = 13				nb of vehicles = 14			
INS	O	O+R	Imp	ImpD	O	O+R	Imp	ImpD	O	O+R	Imp	ImpD
A_120	67.5	76.67	12.1	−0.92	74.17	82.5	11.24	0.45	75.83	83.08	9.56	−0.25
A_145	53.10	53.52	0.78	0.13	57.24	60.62	5.90	−2.58	62.06	62.55	0.78	−2.65
A_159	60.38	64.40	6.67	1.38	66.04	68.49	3.71	2.76	65.41	70.75	8.17	4.19
A_177	42.37	45.14	6.53	6.55	44.63	46.89	5.06	−1.49	46.33	48.64	5.00	−2.67
A_216	43.52	41.99	−3.51	1.27	44.44	44.03	−0.94	−0.37	45.37	47.45	4.59	−0.27
A_233	51.07	53.22	4.20	1.03	55.79	58.88	5.54	3.43	60.94	62.70	2.88	0.64
A_271	34.32	35.43	3.23	−0.15	36.53	37.49	2.62	−2.79	38.75	42.43	9.52	0.09
A_288	34.03	33.40	−1.84	2.51	34.72	34.375	−0.99	0.77	35.07	37.26	6.24	0.65
A_288b	45.49	51.32	12.82	−1.9	54.86	54.03	−1.52	0.32	55.55	57.88	4.19	−0.20
A_360	26.39	27.61	4.63	−0.28	28.33	29.97	5.78	−1.51	30.27	32.86	8.53	−0.29
Avg.	45.82	48.17	4.56	1.10	49.68	51.73	3.64	−0.19	51.56	54.56	5.95	−0.07

Table 4. Performance of Reinsertion Algorithm on instance of group B.

Ve	nb of vehicles = 5				nb of vehicles = 6				nb of vehicles = 7			
INS	O	O+R	Imp	ImpD	O	O+R	Imp	ImpD	O	O+R	Imp	ImpD
B_200	84.5	88.95	5.27	5.27	90.5	92.75	2.49	−2.42	89.5	92.0	2.79	1.93
B_301	82.06	91.06	10.97	3.95	89.04	92.29	3.66	−1.61	91.03	94.15	3.43	−0.80
B_398	86.93	90.70	4.34	1.29	88.94	91.21	2.54	0.39	87.44	91.88	5.08	1.99
B_498	76.51	85.94	12.34	2.42	85.94	90.84	5.70	2.40	87.35	93.88	7.47	0.97
B_607	72.32	82.19	13.64	0.85	80.72	88.37	9.47	1.88	84.51	91.57	8.34	−0.38
B_709	69.25	78.22	12.95	0.87	80.82	84.79	4.92	0.96	84.34	89.01	5.53	1.32
B_802	67.83	75.01	10.58	−0.29	72.57	82.77	14.05	2.08	78.05	86.66	11.02	0.06
B_909	61.45	75.01	17.29	0.48	73.68	81.38	10.45	1.17	76.98	85.45	11.00	2.90
B_1011	61.03	69.00	13.06	−0.02	70.23	77.51	10.37	0.44	75.57	82.11	8.65	0.38
Avg.	73.54	81.46	11.16	1.65	81.38	86.88	7.07	0.59	83.86	89.64	7.04	0.93

reinsertion gives an average improvement of 7.07% with approximately similar rides duration, which is a significant improvement. With 5 vehicles, the reinsertion achieves an average improvement of 11.16% resulting on an average of 81.48% of served requests. This average percentage is larger than the results achieved with the online algorithm with 6 vehicles. This is especially true on 6 instances (two-thirds of the group instances). It means that the *Reinsertion Algorithm* allows on average to save 1 vehicle out of 6, which corresponds to a considerable benefit for the transporter. This is also true when comparing results with 6 and 7 vehicles: reinsertion with 6 vehicles performs better than the online algorithm with 7 vehicles on all instances, meaning that the reinsertion prevents the transporter to add a vehicle while having better performances and similar costs. This is another huge benefit.

We also observe that with a fixed number of vehicles, the *Reinsertion Algorithm* tends to perform better when the number of requests increases. This could potentially be explained by the fact that the higher the number of requests, the more the online algorithm is far from optimal arrangement for these requests.

6 Conclusion

In this paper, we have studied a real problem of dynamic DARP raised by the company *Padam*. We have proposed an original and novel re-insertion algorithm for rejected requests based on repair and destroy operators. Our approach explores many reinsertion possibilities since it allows to cover a wide neighbourhood. We conducted extensive experiments on hard and realistic instances up to 1011 requests and 14 vehicles to evaluate the benefits of our proposed method. We have shown that it allows us to respond to a greater number of requests and that it saves an average of one bus on almost half of the cases, which is a significant advantage for *Padam*.

References

1. Attanasio, A., Cordeau, J.-F., Ghiani, G., Laporte, G.: Parallel tabu search heuristics for the dynamic multi-vehicle dial-a-ride problem. Parallel Comput. **30**(3), 8–15 (2004)
2. Beaudry, A., Laporte, G., Melo, T., Nickel, S.: Dynamic transportation of patients in hospitals. OR Spectr. **32**(1), 77–107 (2010)
3. Cordeau, J.-F., Laporte, G.: The dial-a-ride problem: models and algorithms. Ann. Oper. Res. **153**(1), 29–46 (2007)
4. Coslovich, L., Pesenti, R., Ukovich, W.: A two-phase insertion technique of unexpected customers for a dynamic dial-a-ride problem. Eur. J. Oper. Res. **175**(3), 1605–1615 (2006)
5. Diana, M., Dessouky, M.M.: A new regret insertion heuristic for solving large-scale dial-a-ride problems with time windows. Transp. Res. Part B Methodol. **38**(6), 539–557 (2004)
6. Ho, S.C., Szeto, W.Y., Kuo, Y.-H., Leung, J.M.Y., Petering, M., Tou, T.W.H.: A survey of dial-a-ride problems: literature review and recent developments. Transp. Res. Part B Methodol. **111**, 395–421 (2018)
7. Jaw, J.-J., Psaraftis, H.N., Odone, A.R., Wilson, N.H.M.: A heuristic algorithm for the multi-vehicle advance request dial-a-ride problem with time windows. Transp. Res. Part B **20**(3), 243–257 (1986)
8. Lois, A., Ziliaskopoulos, A.: Online algorithm for dynamic dial a ride problem and its metrics. Transp. Res. Procedia **24**, 377–384 (2017)
9. Luo, Y., Schonfeld, P.: A rejected-reinsertion heuristic for the static dial-a-ride problem. Trans. Res. Part B **41**, 736–755 (2007)
10. Luo, Y., Schonfeld, P.: Online rejected-reinsertion heuristics for dynamic multivehicle dial-a-ride problem. J. Transp. Res. Board **2218**, 59–67 (2011)
11. Ropke, S., Pisinger, D.: An adaptive large neighborhood search heuristic for the pickup and delivery problem with time windows. Transp. Sci. **40**(4), 455–472 (2006)

12. Santos, D.O., Xavier, E.C.: Taxi and ride sharing: a dynamic dial-a-ride problem with money as an incentive. Expert Syst. Appl. **42**(19), 6728–6737 (2015)
13. Shaw, P.: Using constraint programming and local search methods to solve vehicle routing problems. In: Maher, M., Puget, J.-F. (eds.) CP 1998. LNCS, vol. 1520, pp. 417–431. Springer, Heidelberg (1998). https://doi.org/10.1007/3-540-49481-2_30
14. Vallée, S., Oulamara, A., Cherif-Khettaf, W.R.: Maximizing the number of served requests in an online shared transport system by solving a dynamic DARP. Computational Logistics. LNCS, vol. 10572, pp. 64–78. Springer, Cham (2017). https://doi.org/10.1007/978-3-319-68496-3_5

Optimization Heuristics for Computing the Voronoi Skeleton

Dmytro Kotsur[1(✉)] and Vasyl Tereshchenko[2]

[1] Software Competence Center Hagenberg (SCCH), Softwarepark 21, 4232 Hagenberg, Austria
dkotsur@gmail.com
[2] Taras Shevchenko National University of Kyiv, Volodymyrska Str. 60, Kiev 01033, Ukraine
vtereshch@gmail.com

Abstract. A skeletal representation of geometrical objects is widely used in computer graphics, computer vision, image processing, and pattern recognition. Therefore, efficient algorithms for computing planar skeletons are of high relevance. In this paper, we focus on the algorithm for computing the Voronoi skeleton of a planar object represented by a set of polygons. The complexity of the considered algorithm is $O(N \log N)$, where N is the total number of polygon's vertices. In order to improve the performance of the skeletonization algorithm, we proposed theoretically justified shape optimization heuristics, which are based on polygon simplification algorithms. We evaluated the efficiency of such heuristics using polygons extracted from MPEG 7 CE-Shape-1 dataset and measured the execution time of the skeletonization algorithm, computational overheads related to the introduced heuristics and the influence of the heuristic onto the accuracy of the resulting skeleton. As a result, we established the criteria allowing us to choose the optimal heuristics for Voronoi skeleton construction algorithm depending on the critical system's requirements.

Keywords: Voronoi diagram · Voronoi graph · Skeleton · Optimization · Heuristics

1 Introduction

The skeletal representation of the planar object is essential for many problems of computer vision and pattern recognition, image processing, computer graphics and visualization [1]. Skeletons are widely used for shape matching [2, 3], optical character recognition [4] and image retrieval [2, 5]. In the area of biomedical image processing, skeletonization methods are extensively applied to compute the central line of thin objects. For example, one can extract the skeletal graph representing the retinal blood vessels topology [6, 7]. A similar technique can be applied to segment biological neural networks [8]. One can also use skeletonization methods to segment cellular filamentous structures using microscopy images [9, 10]. Thus, fast and accurate algorithms for computing the skeleton of the geometrical objects are of high relevance.

J. M. F. Rodrigues et al. (Eds.): ICCS 2019, LNCS 11536, pp. 96–111, 2019.
https://doi.org/10.1007/978-3-030-22734-0_8

Related Work. Existing algorithms for computing the skeleton can be classified based on the type of processed data. For example, morphological thinning techniques are extensively used for computing the skeleton of a binary image [11–13]. They allow us to obtain a pixel-level representation of the thin skeleton. On the next step, such representation can be converted into a graph using the vectorization methods described [14, 15]. However, the accuracy of the skeleton is bounded by the resolution of the pixel grid. Moreover, many of these methods are not rotation-invariant [11, 12].

Other techniques are based on central line tracing. They are commonly used to segment thin-line structures on an image (e.g., axons, dendrites of neurons [16], blood vessels [17], filamentous structures [17, 18]). These methods can directly represent the skeleton as a connected graph. However, due to the iterative nature of these methods, the execution time may vary significantly.

Another class of methods allows us to compute a skeleton of an object, whose shape is represented by simple polygons. Such polygons can be either sampled directly from a vector graphics data or can be extracted from a binary image using the tracing techniques (e.g., Marching squares [19]). Methods to construct the straight skeleton using the polygon shrinking technique with $O(N \log N)$ complexity are described in papers [20, 21]. A linear complexity method for a simple polygon without holes was introduced in [22]. A more general approach for constructing the skeleton of an arbitrary object with holes employs the Voronoi diagram [23, 24], which has computational complexity $O(N \log N)$, N is a number of primitives. In comparison to the techniques above, this approach allows us to directly compute a rotation-invariant thin skeleton of an object as a graph. Moreover, one can also employ the properties of the Voronoi diagram to solve the related geometrical problems [25] (e.g., finding fast a convex hull, nearest neighbor, maximal inscribed disk). However, due to a large number of the processed simple primitives, such method can become computationally costly. Therefore, we focus on the Voronoi-based skeletonization methods and on heuristic techniques allowing us to speed up such methods by employing the shape simplification techniques.

2 Problem Statement

We assume that a planar object has G_1-continuous boundaries (except for a finite number of G_0-continuous points – see *critical points* below). The object's boundaries are represented by a set of simple planar polygons $\mathcal{S} := \{\mathcal{P}^0, \mathcal{P}^1, \ldots, \mathcal{P}^m\}$, where polygon $\mathcal{P}^k$ is defined as an ordered set of its vertices $p_1^k, p_2^k, \ldots, p_{M_k}^k$. Polygon $\mathcal{P}_0$ corresponds to the outer contour of the object. $R := \mathcal{P}_0 \setminus \bigcup_{i=1}^{m} \mathcal{P}_i$ defines the object's domain.

Let's denote the set of open line segments (LS') corresponding to the polygon $\mathcal{P}_k$ by $\mathcal{L}_k := \mathcal{L}(\mathcal{P}_k) = \left\{ l_i^k := \left(p_i^k, p_{i+1}^k\right) | i = 1, \ldots, M_k, p_{M_k+1}^k = p_1^k \right\}$ and the set of all vertices (line segment's endpoints) by $\mathcal{Q} = \bigcup_{k=0}^{m} \bigcup_{j=1}^{M_k} \{p_i^k\}$. A set $\mathcal{L} := \bigcup_{k=1}^{m} \mathcal{L}_k$ contains all line segments of $\mathcal{S}$.

Definition 1. The *Voronoi cell* [29] corresponding to an element $u \in \mathcal{L} \cup \mathcal{Q}$ is defined as a locus of points:

$$\mathcal{VC}(u) = \left\{ p \in \mathbb{R}^2 | dist(p, u) \leq dist(p, w), w \neq u, w \in \mathcal{L} \cup \mathcal{Q} \right\} \quad (1)$$

Definition 2. The *Voronoi diagram* [29] of a set of line segments $\mathcal{L}$ (with endpoints $\mathcal{Q}$) is defined as a set of all Voronoi cells:

$$\mathcal{VD}(\mathcal{L}, \mathcal{Q}) = \bigcup_{u \in \mathcal{L} \cup \mathcal{Q}} \{\mathcal{VC}(u)\} \quad (2)$$

Remark 1. The most of the computational algorithms (e.g., "Divide and Conquer" [26], Fortune's algorithm [27]) represent the boundaries between neighboring Voronoi cells in terms of the Voronoi graph [28, 29] $G_{\mathcal{S}} = (V_{\mathcal{S}}, E_{\mathcal{S}})$ with a set of the Voronoi vertices $V_{\mathcal{S}}$ and a set of Voronoi edges $E_{\mathcal{S}} \subseteq V_{\mathcal{S}} \times V_{\mathcal{S}}$.

Definition 3. Let's assume that a polygon $\mathcal{P}$ approximates boundary of geometrical object and a vertex p of $\mathcal{P}$ corresponds to the point of the boundary, where object is G_0-continouous (but not G_1-continouous). Then vertex p is called *critical points* (vertices) of the polygon $\mathcal{P}$.

Remark 2. Vertices of polygon $\mathcal{P}$ corresponding to G_1-continouous part of object's boundary, might induce redundant edges of the Voronoi diagram – the bisectors between consecutive line segments l_i and l_{i+1} sharing common non-critical endpoint p_i. In order to obtain an approximate Voronoi diagram of an object represented by $\mathcal{S}$, such redundant edges corresponding to all non-critical points of $\mathcal{S}$ should be removed [29].

Definition 4. An *approximate Voronoi diagram* $\mathcal{VD}_a(\mathcal{S})$ [29] for a planar object represented by a set of polygons $\mathcal{S}$ is obtained as a subgraph $G_{\mathcal{S}}^a$ of the Voronoi graph $G_{\mathcal{S}}$ by removing the edges of $G_{\mathcal{S}}$ corresponding to the bisectors between two consecutive line segments l_i and l_{i+1} sharing a common non-critical vertex p_i.

Definition 5. The *Voronoi skeleton* [23] of a planar object represented by $\mathcal{S}$ is a subset of the approximate Voronoi diagram $\mathcal{VD}_a(\mathcal{S})$ located inside object's region R.

Remark 3. Thus, the Voronoi skeleton of $\mathcal{S}$ is obtained by removing (or trimming) the edges of $G_{\mathcal{S}}^a$, which do not locate in R.

Problem Statement: Given a set of polygons $\mathcal{S}$, which represent a planar object, construct the Voronoi skeleton of $\mathcal{S}$.

3 Algorithm

In this section, we describe the algorithm for computing the Voronoi skeleton. In Subsect. 3.2 we show an algorithm for transforming the Voronoi graph G_S into the final Voronoi skeleton. The complexity analysis of the algorithm is shown in Subsect. 3.3.

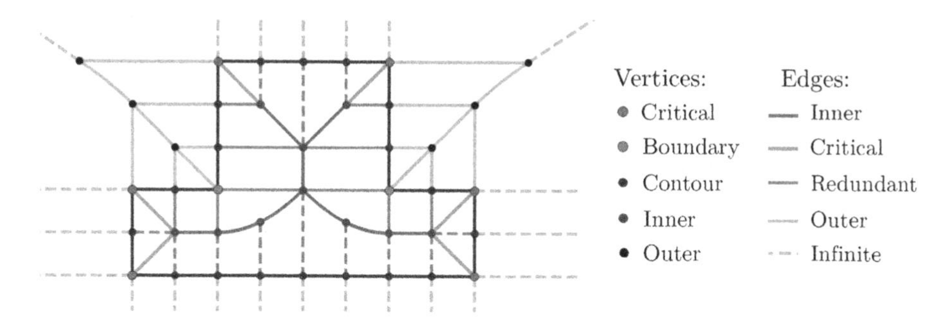

Fig. 1. Examples of the labeled Voronoi vertices and edges.

3.1 Algorithm Description

Input: $\mathcal{S} := \{\mathcal{P}^1, \mathcal{P}^2, \ldots, \mathcal{P}^m\}$ – the set of polygons, each vertex p_i^k of the polygon $\mathcal{P}^k$ has a binary attribute isCritical$[p_i^k] \in \{\text{True}, \text{False}\}$. Polygon $\mathcal{P}^k$ is oriented such that its interior of the object is to the right for any its line segment (LS) $l \in \mathcal{P}^k$.
Algorithm:

1. Compute Voronoi diagram of line segments $\mathcal{L}$ (with endpoints $\mathcal{Q}$) $\Rightarrow$ Obtain Voronoi graph $G_S = (V_S, E_S)$ represented as doubly-connected edge list (DCEL) [28];
2. Using the breadth-first search (BFS) algorithm to traverse the Voronoi graph G_S and label its edges and vertices (see Subsect. 3.2);
3. Remove the edges G_S with labels "R" or "O" and vertices with labels "B" or "O";
4. Remove isolated vertices of G_S if any exist;

3.2 Labeling Voronoi Graph

We traverse the edges and vertices of the Voronoi graph G_S and label them according to their role in a resulting graph of the Voronoi skeleton.

Listing 1. Pseudocode of the Voronoi graph labeling step with auxiliary functions

```
function InitQueue(Q)
begin
  Q := EmptyQueue();
  for edge e
    Label[e] = "None";
    if isInfinite(e)
      EnQueue(e, Q);
      v : = non-null vertex of e;
      Label[v] := "None";
    else
      v1, v2 := non-null vertices of e;
      Label[v1] := Label[v2] := "None";
  return Q;
end;

procedure TraverseBFS(Q)
begin
  while not Empty(Q) do
    e := DeQueue(Q);
    v := Null;
    if (isInfinite(e)) then
      v := non-null vertex of e;
      LabelInfiniteEdge(e);
    else
      v := vertex of e with "None" label;
      LabelFiniteEdge(e);
    for edge e incident to v do
      // Add non-labeled edges to queue
      if (Label[e] = "None") then
        EnQueue(e, Q);
end;

procedure LabelInfiniteEdge(e)
begin
  c1, c2 := cells of e and Twin(e);
  v := non-null vertex of e;
  if Type(c1) = "EP" && Type(c2) = "EP" then
    Label[e] := Label[v] := "O"; // Outer
  else
    p := unique EP of line segment;
  if v coincides with p then
    Label[v] := (isCritical[p]) ? "C" : "B";
    Label[e] := "O"; // Outer
  else
    Label[v] := "I"; // Inner
    if isCritical[p] then
      Label[e] := "C"; // Critical
      Trim e to p;
    else
      Label[e] := "R"; // Redundant
end;

procedure LabelFiniteEdge(e)
begin
  v0 := labeled vertex of e;
  v1 := unlabeled vertex of e;
  c1, c2 := cells of e and Twin(e);
  if Label[v0] = "I" || Label[v0] = "O" then
    if Type(c1) = "LS" && Type(c2) = "LS" then
      if LS' of c1 and c2 share endpoint p then
        if isCritical[p] then
          Label[v1] := "C"; // Critical
          Label[e] := (Label[v0]="I") ? "C" : "O";
        else
          Label[v1] := "B"; // Boundary
          Label[e] := (Label[v0]="I") ? "R" : "O";
      else
        Label[v1] := Label[v0];
        Label[e] := Label[v0];
    else // Edge between LS and EP
      if c1 and c2 belong to the same LS then
        p := line segment's endpoint;
        if p coincides with v1 then
          Label[v1] := (isCritical[p]) ? "C" : "B";
          if Label[v0] = "O" then
            Label[e] := "O"; // Outer
          else
            Label[e] := (isCritical[p]) ? "C" : "R";
        else
          Add new vertex v with position p to Gs
          Replace e by e0 := (v, v0), e1 := (v, v1);
          Label[v] := (isCritical[p]) ? "C" : "B";
          if Label[v0] = "O" then
            Label[v1] := "I";
            if isCritical[p] then
              Label[e1] := "C";
            else
              Label[e1] := "R";
              Label[e0] := "O";
          else
            Label[v1] := Label[e1] := "O";
            Label[e0] := (isCritical[p]) ? "C" : "R";
      else // bisector is a parabolic arc
        Label[v1] := Label[v0];
        Label[e] := (Label[v0]="O")?"O":"I";
  else // Critical or Boundary
    if v1 is located to the right of c1 or c2 then
      Label[v1] := "Inner";
      Label[e] := (Label[v0] = "C") ? "C" : "R";
    else
      Label[v1] := "Outer";
      Label[e] := (Label[v0] = "C") ? "C" : "O";
end;
```

Definition 6. Voronoi vertex (cf., Fig. 1) is called (label abbreviation is in parenthesis):

- *Inner* ("I"), if the vertex is located inside the object's polygon;
- *Outer* ("O"), if the vertex is located outside the object's polygon;
- *Critical* ("C"), if it coincides with one of the critical vertices of the object's polygon;
- *Boundary* ("B"), if it coincides with one of the non-critical vertices of the polygon;

Definition 7. Voronoi edge (cf., Fig. 1) e is called (label abbreviation is in parenthesis):

- *Inner* ("I"), if it locates in R and doesn't touch (intersect) any polygon from $\mathcal{S}$; $\Rightarrow$ both vertices of e are labeled as *Inner*;
- *Critical* ("C"), if it locates in R and adjacent to a critical vertex;
- *Outer* ("O"), if the edge locates outside $R \Rightarrow$ both vertices of e are labeled as *Outer*;
- *Redundant* ("R"), if it locates in R and touches polygon's non-critical vertex;

The pseudocode illustrating the labeling procedure and the related functions is shown in Listing 1. Firstly, we initialize the queue `Q` of the breadth-first search (BFS) algorithm (see function `InitQueue(Q)`) by all infinite edges of the Voronoi graph $G_{\mathcal{S}}$. A common data structure `Label[•]` is used to store labels of Voronoi edges and vertices according to the definitions 6 and 7. Then, starting from infinite edges we label all remaining edges and vertices of $G_{\mathcal{S}}$ in function `TraverseBFS(Q)`. At each iteration of BFS algorithm we label current edge and the adjacent non-labeled vertex. In Listing 1 `(Condition)?Value1:Value2` denotes to the ternary conditional operator.

3.3 Complexity Analysis

Lemma 1. The complexity of Step 1 of the skeletonizing algorithm is $O(N \log N)$, where N is a number of points in a polygon.

Proof. At the Step 1 we construct the Voronoi diagram for polygon's line segments using Fortune's algorithm. According to [27] the complexity of this step is $O(M \log M)$, M - number of line segments. Since $N \sim M$, Step 1 has complexity $O(N \log N)$. ■

Lemma 2. The complexity of Step 2 of the skeletonizing algorithm is O(N), where N is a number of the points in an input polygon.

Proof. Step 2 is about labeling the edges and vertices of the Voronoi graph using BFS traverse algorithm. Note that the Voronoi graph is a planar connected graph. Therefore, Euler's formula $|V| - |E| + f = 2$ take place, where $|V|$, $|E|, f$ is a number of vertices, edges and faces of a graph. If $|V| = N$, then the number of edges $|E| = O(N)$. The BFS algorithm traverses all edges of the Voronoi graph. Since all operations within one BFS iteration can be performed in O(1), the complexity of BFS routine is $O(|E| + |V|) =$ O(N). Thus, the complexity of Step 2 is O(N). ■

Lemma 3. The complexity of Steps 3–4 of the skeletonizing algorithm is O(N), where N is a number of the points in an input polygon.

Proof. One edge can be removed from DCEL in O(1) by reassigning the pointers [25, 28]. According to Lemma 2, the number of edges $|E| = O(N)$. Therefore, the complexity of Step 3 is $O(N)$. A single isolated vertex can be removed from DCEL in O(1). Therefore, the complexity of Step 4 is $O(N)$. ■

Theorem 1. The complexity of the skeletonizing algorithm is O(N log N), where N is a number of the points in an input polygon.

Proof. According to analysis of the complexities of each algorithm's step provided in Lemmas 1–3, the total complexity of skeletonizing algorithm is O(N log N). ■

4 Optimization and Heuristics

We introduce an optimization heuristic allowing us to compute fast the Voronoi skeleton by reducing the number of vertices of input polygons. The main idea behind the optimization procedure is illustrated by the following lemma.

Lemma 4. Let $\mathcal{P} = \{p_1, p_2, \ldots, p_N\}$ be a polygon and l_i denotes the line segment between points p_i and p_{i+1} of a polygon $\mathcal{P}$, $i = 1, \ldots, N$, $(p_{N+1} = p_0)$. The polygon $\mathcal{P}'$ is obtained by subdividing line segments $l_i, i = 1, \ldots N$ of a polygon $\mathcal{P}$ such that line segment l_i is replaced by a polyline $p_{i,1}, p_{i,2} \ldots, p_{i,R_i}$ of points on l_i, $i = 1, 2, \ldots, N$ $(p_{i,1} = p_i,\ p_{i,R_i} = p_{i+1})$. Then the Voronoi skeletons $\mathcal{VS}(\mathcal{P})$ and $\mathcal{VS}(\mathcal{P}')$ constructed using the skeletonizing algorithm above are equal (in terms of the Hausdorff distance between the corresponding Voronoi graphs).

Proof. The Voronoi diagram of line segments of $\mathcal{P}$ and $\mathcal{P}'$ consist of the bisectors of the following types: a bisector between two line segment's interiors, a bisector between a line segment's interior and an endpoint, bisector between two endpoints. Let's consider these cases separately:

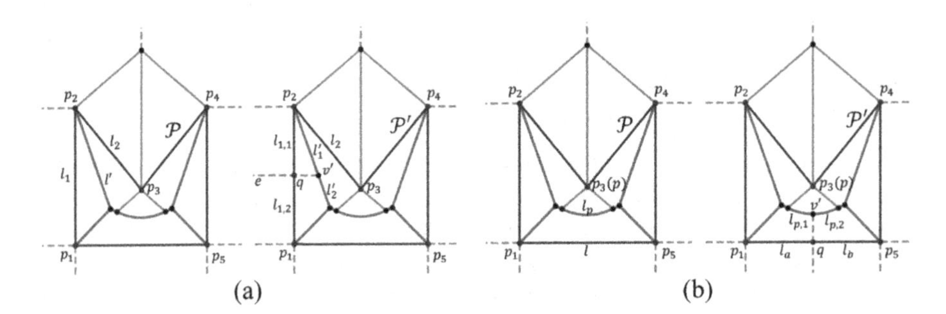

Fig. 2. Illustration of the proof of Lemma 4

Case 1 (see Fig. 2a). The bisector between two line segment's interiors l_1 and l_2 is a line segment l' [27, 28]. Let's suppose that in $\mathcal{P}'$ line segment l_2 remains the same and l_1 is subdivided into two parts $l_{1,1}$ and $l_{1,2}$ connected by a shared endpoint q. Then, the Voronoi cell corresponding to l_1 in $\mathcal{VD}(\mathcal{P})$ will be split into two Voronoi cells (corresponding $l_{1,1}$ and $l_{1,2}$) of $\mathcal{VD}(\mathcal{P}')$ by the Voronoi edge e such that e is a bisector between $l_{1,1}$ and $l_{1,2}$ which passes through q and is perpendicular to l_1 (and therefore, $l_{1,1}$ and $l_{1,2}$). Thus, the Voronoi edge e will divide bisector line segment l' in $\mathcal{VD}(\mathcal{P})$ into two parts l'_1 and l'_2 in $\mathcal{VD}(\mathcal{P}')$ such that l'_1 is a Voronoi edge of the Voronoi cell of

$l_{1,1}$ and l'_2 is a Voronoi edge of the Voronoi cell of $l_{1,2}$. Note that l'_1, l'_2 and edge e are connected together by a newly introduced Voronoi vertex v'. The remaining part of the Voronoi diagrams for $\mathcal{P}'$ and $\mathcal{P}$ stays the same. The BFS labeling procedure (see Step 2 of the algorithm above) for Voronoi edges and vertices of $\mathcal{VD}(\mathcal{P}')$ will split the introduced in $\mathcal{VD}(\mathcal{P}')$ Voronoi edge e into two parts e_1 and e_2: one part will be labeled as "Outer" and the other part will be labeled as "Redundant". Therefore, both parts will be removed at Step 3 of the skeletonizing algorithm and the resulting Voronoi skeleton $\mathcal{VS}(\mathcal{P}')$ will contain the line segment edges l'_1, l'_2 connected by v'.

Case 2. In case of a line segment's interior l and an endpoint p, two possible scenarios take place. First scenario is when p is an endpoint of l. In this case Voronoi diagram contains an edge e' coming through p and perpendicular l. The edge e' can be either removed or not by BFS procedure depending on the type of p. Subdividing l into two parts l_a and l_b which share an endpoint q will introduce a new edge e parallel to e', which will be classifies as "Redundant" and removed from the final skeleton. The second scenario (see Fig. 2b) is when p is not an endpoint of l. Then the bisector between p and l is a parabolic arc l_p, which is subdivided into two parts $l_{p,1}$, $l_{p,2}$ if we split l into l_a and l_b. The analysis in this case is the similar to the Case 1 except that now l'_1 and l'_2 are parabolic arcs $l_{p,1}$ and $l_{p,2}$, respectively.

Case 3. The bisector between two different endpoints of $\mathcal{VD}(\mathcal{P}')$ or $\mathcal{VD}(\mathcal{P})$ is an infinite edge (ray), which is classified at Step 2 of the algorithm above as "Outer" and, therefore, removed from both $\mathcal{VS}(\mathcal{P})$ and $\mathcal{VS}(\mathcal{P}')$ at Step 3.

The case of single subdivision $(L = 1)$ of polygon's line segment for different possible bisectors of the Voronoi diagram is covered above. The general case for several subdivisions L can be proved by induction on L.

Let's assume that for $L = n$ subdivisions of $\mathcal{P}$ holds that $\mathcal{VS}(\mathcal{P})$ and $\mathcal{VS}(\mathcal{P}')$ are equal. The polygon $\mathcal{P}''$ is obtained from $\mathcal{P}'$ by subdividing an arbitrary line segment of $\mathcal{P}'$ into two line segments. Therefore, we can apply one of the proved cases for a single subdivision above and obtain that Voronoi skeletons $\mathcal{VS}(\mathcal{P})$ and $\mathcal{VS}(\mathcal{P}'')$ are equal. Thus, by induction $\mathcal{VS}(\mathcal{P})$ and $\mathcal{VS}(\mathcal{P}')$ are equal for any $L > 0$. ∎

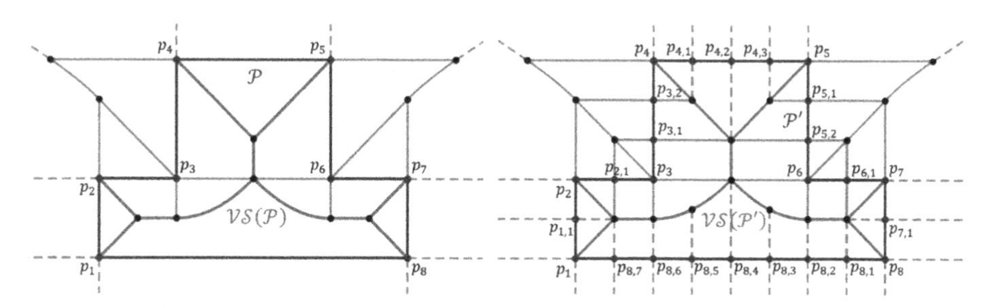

Fig. 3. The Voronoi skeletons (red) for polygon P (blue) and its subdivided version P' (blue) and respective Voronoi diagrams (gray). (Color figure online)

Remark. It follows from Lemma 4 that the Voronoi skeleton $\mathcal{VS}(\mathcal{P}')$ for a subdivided polygon $\mathcal{P}'$ is the same (w.r.t. Hausdorff distance) as the Voronoi skeleton $\mathcal{VS}(\mathcal{P})$ for the original polygon $\mathcal{P}$ (see Fig. 3). However, in comparison to $\mathcal{VS}(\mathcal{P})$, $\mathcal{VS}(\mathcal{P}')$ is represented with a larger number of Voronoi edges and vertices. Therefore, the concept of the Voronoi skeleton with a minimal number of vertices/edges take place. Applying Lemma 4 in the reverse direction allows us to reduce the number of vertices and edges of the Voronoi skeleton. This in turn reduces the execution time of skeletonization algorithm and compresses the resulting graph representation of a skeleton.

Therefore, our aim is to design a heuristic based on simplification operation (reverse to subdivision) and obtain polygon $\mathcal{P}$ from $\mathcal{P}'$. According to Lemma 4 simplification procedure (algorithm) should meet the following requirement:

Simplification Requirement (SR): The polygon simplification heuristic removes the points corresponding to colinear connected line segments of the polygon representing such line segments by a single line segment.

Table 1. The overview of polygon (polyline) simplification algorithms.

Name of algorithm (Abbr.)	Average complexity	Worst-case complexity	SR
Ramer-Douglas-Peucker (DP) [30]	$O(N \log N)$	$O(N^2)$	Yes
Visvalingam-Whyatt (VW) [31]	$O(N \log N)$	$O(N \log N)$	Yes
Reumann-Witkam (RW) [32]	$O(N)$	$O(N)$	Yes
Opheim (OP) [33]	$O(N)$	$O(N)$	Yes
Lang (LA) [34]	$O(NK)$	$O(NK^2)$	Yes
Zhao-Saalfeld (ZS) [35]	$O(N)$	$O(N)$	Yes
Rapso (RA) [36]	$O(N)$	$O(N)$	No
Li-Openshaw (LO) [37]	$O(N)$	$O(N)$	No
N^{th} point (NP) [38]	$O(N)$	$O(N)$	No
Circle (CI) [38]	$O(N)$	$O(N)$	No
Perpendicular distance (PD) [38]	$O(NK)$	$O(N)$	Yes

Thus, we introduce Step 0 of the skeletonizing algorithm: simplify each polygon of a set $\mathcal{S}$ by reducing the points associated with colinear connected line segments (SR). This operation can be performed using one of the existing polygon simplification algorithms satisfying the simplification requirement (SR).

Table 2. Suitable polygon simplification algorithms, their parameter and heuristics.

Algorithm	Parameter(s)	Heuristics for 2nd parameter
DP	$\varepsilon > 0$ – tolerance parameter;	No
VW	$A > 0$ – minimum triangle area;	No
RW	$\varepsilon > 0$ – distance tolerance;	No
OP	$\varepsilon_{min}, \varepsilon_{max} > 0$ – tolerances;	$\varepsilon_{max} = +\infty$ (large number)
LA	$\varepsilon > 0$ – distance tolerance; $R \in \mathbb{N}$ – size of search region;	$R = \theta \cdot N$, N – number of points; $\theta \in \{0.05, 0.1, 0.2, 0.25, 0.5, 1.0\}$.
ZS	$\varepsilon > 0$ – sector bound error;	No
PD	$\varepsilon > 0$ – distance tolerance; $K \in \mathbb{N}$ – number of repetitions;	$R = \theta \cdot N$, N – number of points; $\theta \in \{0.05, 0.1, 0.2, 0.25, 0.5, 1.0\}$.

Analysis of Simplification Algorithms. We have analyzed the most commonly used algorithms for polygon (polyline) simplification and summarized the results in Table 1.

However, certain simplification strategies do not agree with the simplification requirement (SR). For example, a naive N^{th} point [38] method merely removes every N^{th} point from a polygon ignoring its geometry. Circle simplification [38] method groups together points forming spatial clusters based on the distance threshold. Then, a single representative point replaces each such cluster. Li-Openshaw [37] and Rapso [36] algorithms simplify polyline based on spatial pixel (or hexagon-based) grid. These algorithms instead solve the problem of polyline digitization (useful for solving the problem of optimal map rescaling). Therefore, we consider only the algorithms fulfilling SR (see Table 2). Most of the analyzed algorithms have complexity $O(N)$ except DP [30] and VW [31] algorithms with $O(N \log N)$ complexity. In order to select the algorithm, which shows the best performance improvement, has the smallest computational overhead and influences the resulting skeleton the least, we investigated these algorithms empirically as described in the evaluation section.

5 Evaluation

We evaluate the performance of the skeletonization algorithm in terms of the execution time and measure the influence of the introduced heuristics onto the accuracy, execution time of the overall algorithm. We also estimate the computational overheads related to the line simplification algorithms.

Dataset. In order to evaluate the performance of the skeletonization algorithm and individual optimization heuristics, we used polygons obtained from MPEG 7 CE-Shape-1 dataset. These polygons were extracted from binary images using the Marching Squares algorithm [19]. In total the dataset consists of 1282 polygons (see Fig. 4).

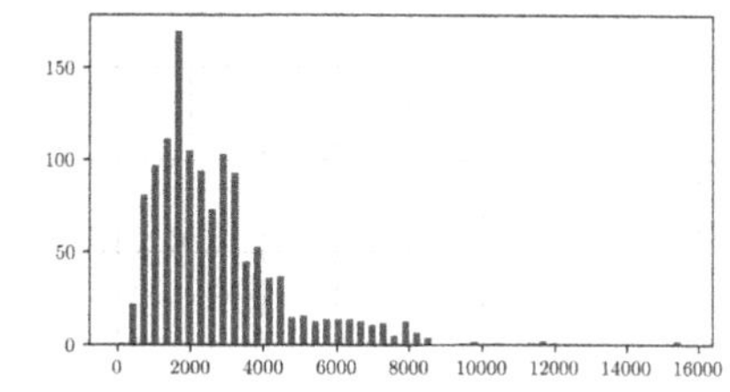

Fig. 4. Distribution of polygon's sizes

Measures. We have measured the following quantities:

1. *Execution time* (ms) of each simplification algorithm, skeletonizing algorithm with (without) the mentioned heuristics and overall execution time. The experiments were carried on Intel Core i7, 2.2 GHz, 16 Gb RAM.
2. *Hausdorff distances* d_H (errors) [39] between the simplified and original polygons and also between the ground truth skeleton and one obtained using the skeletonization with heuristics;
3. *Simplification rate* (%) of the polygon is computed as follows:

$$SR(P, P') = \frac{|P| - |P'|}{|P|} \cdot 100\% \quad (3)$$

where P and P' are original and simplified polygons, respectively. $|P|$ is the number of vertices of P (large values of $SR(P, P')$ correspond to small $|P'|$ w.r.t. $|P|$).

Parameters. The parameters of the simplification algorithms (see Table 2) were chosen using the line search method such that the maximum simplification rate is achieved for a given threshold value of Hausdorff distance d_H between the simplified polygon P' and an original polygon P. This allows us to compare different simplification algorithm with respect to the maximum tolerable error. The established parameters of the simplification algorithms for the respective values of d_H are shown in Table 3.

For the algorithms with two parameters we applied additional heuristics to choose the value of the second parameter (see Table 2). These heuristics were devised to achieve the maximum simplification rate for a given Hausdorff error threshold d_H. It was established that for LA and PD algorithms the optimal value of θ is 0.25 (for $\theta > 0.25$ the simplification rate does not increase, but the execution time of these simplification algorithms rises).

Table 3. Parameters of the simplification algorithms.

Hausdorff distance d_H	Algorithm parameters						
	DP	VW	RW	OP	LA (0.25)	ZS	PD (0.25)
0.001	0.001	0.0007	0.001	0.001	0.001	0.001	0.001
0.005	0.005	0.0025	0.005	0.005	0.005	0.005	0.005
0.01	0.01	0.005	0.009	0.009	0.01	0.01	0.01
0.05	0.05	0.025	0.04	0.04	0.05	0.05	0.05
0.1	0.1	0.05	0.08	0.08	0.1	0.1	0.1
0.5	0.5	0.25	0.4	0.4	0.5	0.5	0.5
1.0	1	0.5	0.8	0.8	1	1	1

Evaluation Results. We have measured the execution time of each suitable simplification algorithm for fixed values of Hausdorff error thresholds d_H (see Fig. 5a). These measurements show the computational overheads related to the optimization step of the skeleton algorithm. In order to compare the quality of the simplification algorithms, we measured the respective simplification rates for given values of d_H.

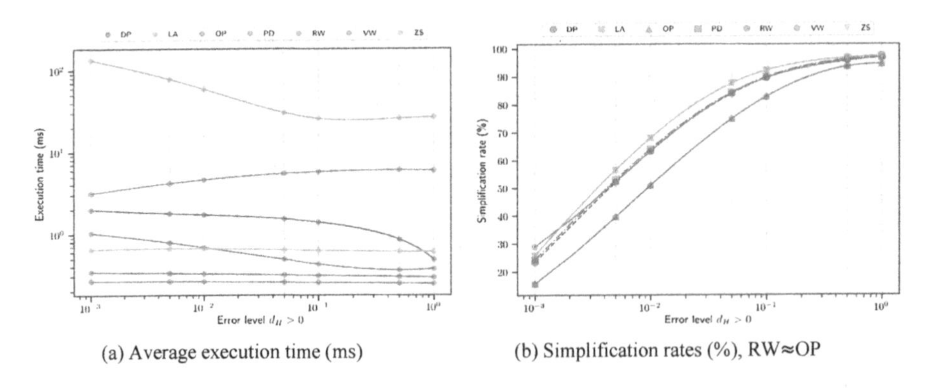

(a) Average execution time (ms)

(b) Simplification rates (%), RW≈OP

Fig. 5. Execution time (ms), simplification rates (%) of optimization heuristics

Figure 5b shows that the algorithms of LA and ZS have the most substantial extent of polygon simplification (compression) for a given d_H having nearly identical dependency curves. PD, VW, and PD algorithms achieve slightly smaller simplification rates showing almost undistinguishable behavior for most of the cases. However, VW algorithm overperforms other algorithms for small values of $d_H < 0.002$. OP and RW algorithms have the lowest simplification rates with nearly identical dependency curves.

In Fig. 5 one notices that despite being the fastest, algorithms of OP and RW have the smallest simplification rate and, therefore, might not guarantee the fastest execution of the skeletonization algorithm. Therefore, we measured the total execution time of the skeletonization algorithm depending on the value of d_H taking into account the overhead time of the simplification heuristics (see Fig. 6a).

Based on Fig. 6a we can choose the fastest optimization heuristics. However, different values of d_H threshold might affect the accuracy of the final skeleton. Therefore, we investigated the influence of d_H on the result of the skeletonization algorithm. We calculated the skeletonization error as Hausdorff distance between the ground truth skeleton and the result of optimized skeletonization algorithm (see Fig. 6b).

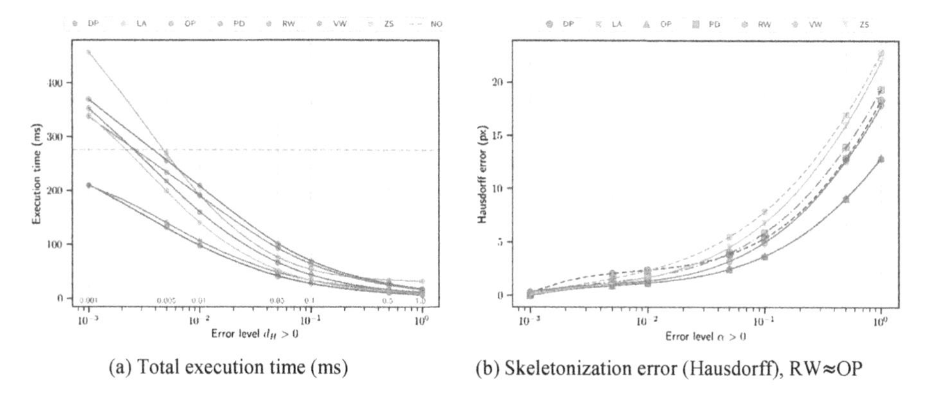

(a) Total execution time (ms)

(b) Skeletonization error (Hausdorff), RW≈OP

Fig. 6. Total average execution time (ms) and skeletonization errors;

6 Discussion

Figure 6a shows that DP- and VW-based heuristics reduce the computational time to the greatest extent. Only these two heuristics overperform the optimization-free approach (NO) for small values of $d_H \leq 0.001$. The optimization based on OP and RW algorithms shows the smallest skeletonization error among the other approaches (see Fig. 6b). However, for $d_H < 0.002$ these algorithms have outsized computational overheads eliminating the whole effect of the optimization. Therefore, it is reasonable to use OP and RW algorithm only for $d_H > 0.002$. Note that the variance of skeletonization error for different heuristics decreases as $d_H \to 0$ (see Fig. 6b).

We computed 2-sample t-test to validate the hypothesis that DP- and VW-based optimizations produce different average skeletonization errors. The test showed that the errors produced by DP and VW optimizations are undistinguishable (p-value ≈ 0.24).

Another hypothesis testing was performed to distinguish the execution time between DP and VW heuristics. It showed that for the most of the cases (except $d_H = 0.001$) DP algorithm overperforms VW (p-value <0.001).

Speed-Accuracy Trade-Off. Figure 6 shows that none of the tested algorithms minimizes the accuracy and execution time of the skeletonizing method at the same time. Therefore, the choice of the heuristics is a trade-off between accuracy and the execution time. Based on the performed computational experiments the following conclusions are drawn:

1. If accuracy of the resulting skeleton is critical, then for $d_H > 0.002$ the optimization can be performed using OP or RW algorithms. However, for $d_H < 0.002$ the only reasonable optimization is using the DP or VW algorithms;
2. If execution time of the algorithm is more critical than the accuracy, then optimization can be performed using DP or VW algorithms, which according to the provided experiments give 1.7 times less accurate result then RW and OP heuristics;

Pruning Effect of Polygon Simplification. It was experimentally discovered, that the introduced optimization heuristics influences the skeleton in a similar way as pruning methods [40]. Figure 7 shows that for large values of d_H (see bottom row) simplification heuristics tends to regularize shape of the object in a way that the branches of the skeleton corresponding to small shape perturbation disappear (cf., Fig. 7, top row). Therefore, such optimization allows us not only to speed-up the execution of the skeletonization, but also to achieve a pruning effect and remove the noisy branches of the skeleton.

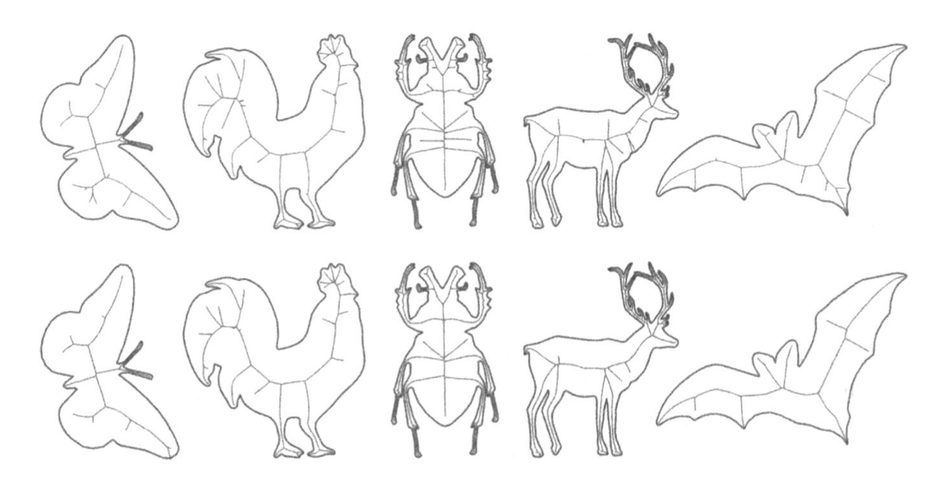

Fig. 7. Examples of the optimized Voronoi skeletons for shapes from MPEG 7 CE-Shape-1 dataset. Optimization heuristics is DP. For $d_H = 0.001$ (top row of images) skeletons contain redundant branches in comparison to $d_H = 1.0$ (the bottom row of images).

7 Conclusion

We proposed optimization heuristics for computing the Voronoi skeleton of the polygonal data. This topic is of relevance due to its direct relation to efficient processing of the vectorized geometrical representations (e.g., for image processing, computer vision, computer graphics). We illustrated in detail the main steps of the Voronoi-based skeletonization algorithm and determined that its complexity is $O(N \log N)$, where N is the number of vertices in a polygon. We also established an optimization criterion (requirement) and proposed theoretically justified optimization heuristics based on the polygon simplification algorithms. In order to evaluate the efficiency of the proposed heuristic, a series of computational experiments were conducted using the polygons from MPEG 7 CE-Shape-1 dataset. Seven state-of-the-art simplification algorithms were evaluated to determine the most suitable optimization heuristic fulfilling the established criterion. We measured the execution time of the skeletonization algorithm with and without the heuristic optimizations and determined the computational overheads related to such heuristics. We also determined the accuracy of the skeleton produced by the optimized algorithm based on the proposed heuristics. As a result, we established the criteria, which allow us to choose the optimal heuristics depending on the system's requirement. For example, DP- and VW-based heuristics allow us to speed up the skeleton computation at least by 30%. It was discovered experimentally, that the optimization heuristics have a pruning effect onto the resulting skeleton.

Acknowledgements. The research reported in this paper has been partly supported by the Austrian Ministry for Transport, Innovation and Technology, the Federal Ministry of Science, Research and Economy, and the Province of Upper Austria in the frame of the COMET center SCCH.

References

1. Saha, P.K., Borgefors, G., Sanniti di Baja, G.: A survey on skeletonization algorithms and their applications. Pattern Recognit. Lett. **76**, 3–12 (2016)
2. Sundar, H., Silver, D., Gagvani, N., Dickinson, S.: Skeleton based shape matching and retrieval. In: 2003 Shape Modeling International, Seoul, pp. 130–139. IEEE (2003)
3. Xie, J., Heng, P.-A., Shah, M.: Shape matching and modeling using skeletal context. Pattern Recognit. **41**, 1756–1767 (2008)
4. Chaudhuri, A., Mandaviya, K., Badelia, P., Ghosh, S.K.: Optical character recognition systems for different languages with soft computing. SFSC, vol. 352. Springer, Cham (2017). https://doi.org/10.1007/978-3-319-50252-6
5. Torres, R.d.S., Falcão, A.X.: Contour salience descriptors for effective image retrieval and analysis. Image Vis. Comput. **25**, 3–13 (2007)
6. Rezaee, K., Haddadnia, J., Tashk, A.: Optimized clinical segmentation of retinal blood vessels by using combination of adaptive filtering, fuzzy entropy and skeletonization. Appl. Soft Comput. **52**, 937–951 (2017)

7. Lasso, W., Morales, Y., Torres, C.: Image segmentation blood vessel of retinal using conventional filters, Gabor transform and skeletonization. In: 2014 XIX Symposium on Image, Signal Processing and Artificial Vision, Columbia. IEEE (2014)
8. Al-Kofahi, Y., Dowell-Mesfin, N., Pace, C., Shain, W., Turner, J.N., Roysam, B.: Improved detection of branching points in algorithms for automated neuron tracing from 3D confocal images. Cytom. Part A. **73A**, 36–43 (2008)
9. Faulkner, C., et al.: An automated quantitative image analysis tool for the identification of microtubule patterns in plants. Traffic **18**, 683–693 (2017)
10. Beil, M., Braxmeier, H., Fleischer, F., Schmidt, V., Walther, P.: Quantitative analysis of keratin filament networks in scanning electron microscopy images of cancer cells. J. Microsc. **220**, 84–95 (2005)
11. Changxian, S., Yulong, M.: Morphological thinning based on image's edges. In: ICCT 1998, 1998 International Conference on Communication Technology. Publishing House of Construction Materials, Beijing (1998)
12. Zhang, T.Y., Suen, C.Y.: A fast parallel algorithm for thinning digital patterns. Commun. ACM **27**, 236–239 (1984)
13. Yan, T.-Q., Zhou, C.-X.: A continuous skeletonization method based on distance transform. In: Huang, D.-S., Gupta, P., Zhang, X., Premaratne, P. (eds.) ICIC 2012. CCIS, vol. 304, pp. 251–258. Springer, Heidelberg (2012). https://doi.org/10.1007/978-3-642-31837-5_37
14. Chen, J., Du, M., Qin, X., Miao, Y.: An improved topology extraction approach for vectorization of sketchy line drawings. Vis. Comput. **34**, 1633–1644 (2018)
15. Hilaire, X., Tombre, K.: Robust and accurate vectorization of line drawings. IEEE Trans. Pattern Anal. Mach. Intell. **28**(6), 890–904 (2006)
16. Acciai, L., Soda, P., Iannello, G.: Automated neuron tracing methods: an updated account. Neuroinformatics **14**, 353–367 (2016)
17. De, J., et al.: A graph-theoretical approach for tracing filamentary structures in neuronal and retinal images. IEEE Trans. Med. Imaging **35**, 257–272 (2016)
18. Stein, A.M., Vader, D.A., Jawerth, L.M., Weitz, D.A., Sander, L.M.: An algorithm for extracting the network geometry of three-dimensional collagen gels. J. Microsc. **232**, 463–475 (2008)
19. Maple, C.: Geometric design and space planning using the marching squares and marching cube algorithms. In: 2003 International Conference on Geometric Modeling and Graphics, 2003, London. IEEE Computer Society (2003)
20. Aichholzer, O., Aurenhammer, F., Alberts, D., Gärtner, B.: A novel type of skeleton for polygons. In: Maurer, H., Calude, C., Salomaa, A. (eds.) J.UCS The Journal of Universal Computer Science, pp. 752–761. Springer, Heidelberg (1996). https://doi.org/10.1007/978-3-642-80350-5_65
21. Eppstein, D., Erickson, J.: Raising Roofs, crashing cycles, and playing pool: applications of a data structure for finding pairwise interactions. Discret. Comput. Geom. **22**, 569–592 (1999)
22. Chin, F., Snoeyink, J., Wang, C.A.: Finding the medial axis of a simple polygon in linear time. Discret. Comput. Geom. **21**, 405–420 (1999)
23. Ogniewicz, R., Ilg, M.: Voronoi skeletons: theory and applications. In: Proceedings 1992 IEEE Computer Society Conference on Computer Vision and Pattern Recognition, Champaign. IEEE Computer Society Press (1992)
24. Siddiqi, K., Pizer, S.M. (eds.): Medial Representations. Springer, Dordrecht (2008). https://doi.org/10.1007/978-1-4020-8658-8
25. Preparata, F.P., Shamos, M.I.: Computational Geometry. Springer, New York (1985). https://doi.org/10.1007/978-1-4612-1098-6

26. Shamos, M.I., Hoey, D.: Closest-point problems. In: 16th Annual Symposium on Foundations of Computer Science, Berkley. IEEE (1975)
27. Fortune, S.: A sweepline algorithm for Voronoi diagrams. Algorithmica **2**, 153–174 (1987)
28. de Berg, M., Cheong, O., van Kreveld, M., Overmars, M.: Computational Geometry. Springer, Heidelberg (2008). https://doi.org/10.1007/978-3-540-77974-2
29. Okabe, A., Boots, B., Sugihara, K., Chiu, S.N., Kendall, D.G. (eds.): Spatial Tessellations. Wiley, Hoboken (2000)
30. Douglas, D.H., Peucker, T.K.: Algorithms for the reduction of the number of points required to represent a digitized line or its caricature. Cartographica **10**, 112–122 (1973)
31. Visvalingam, M., Whyatt, J.D.: Line generalisation by repeated elimination of points. Cartogr. J. **30**, 46–51 (1993)
32. Reumann, K., Witkam, A. P. M.: Optimizing curve segmentation in computer graphics. In: International Computing Symposium, Elsevier, North Holland, pp. 467–472 (1973)
33. Opheim, H.: Fast data reduction of a digitized curve. Geo-Processing **2**, 33–40 (1982)
34. Lang, T.: Rules for robot draughtsman. Geogr. Mag. **42**, 50–51 (1969)
35. Zhao, Z., Saalfeld, A.: Linear-time sleeve-fitting polyline simplification algorithms. In: Proceedings of the Annual Convention and Exposition. Technical Papers, Seattle, USA, pp. 214–223 (1997)
36. Raposo, P.: Scale-specific automated line simplification by vertex clustering on a hexagonal tessellation. Cartogr. Geogr. Inf. Sci. **40**, 427–443 (2013)
37. Nie, H., Huang, Z.: A new method of line feature generalization based on shape characteristic analysis. Metrol. Meas. Syst. **18**, 597–606 (2011)
38. Song, J., Miao, R.: A novel evaluation approach for line simplification algorithms towards vector map visualization. Int. J. Geo-Inf. **5**, 223 (2016)
39. Taha, A.A., Hanbury, A.: An efficient algorithm for calculating the exact hausdorff distance. IEEE Trans. Pattern Anal. Mach. Intell. **37**, 2153–2163 (2015)
40. Beristain, A., Graña, M., Gonzalez, A.I.: A pruning algorithm for stable Voronoi skeletons. J. Math. Imaging Vis. **42**, 225–237 (2011)

Transfer Learning for Leisure Centre Energy Consumption Prediction

Paul Banda[(✉)], Muhammed A. Bhuiyan, Kevin Zhang, and Andy Song

RMIT University, Melbourne 3000, Australia
paul.banda@student.rmit.edu.au

Abstract. Demand for energy is ever growing. Accurate prediction of energy demand of large buildings becomes essential for property managers to operate these facilitates more efficient and greener. Various temporal modelling provides a reliable yet straightforward paradigm for short term building energy prediction. However, newly constructed buildings and recently renovated buildings, or buildings that have energy monitoring systems newly installed, do not have sufficient data to develop accurate energy demand prediction models. In contrast, established buildings often have vast amounts of data collected which may be lying idle. The model learned from these buildings with huge data can be useful if transferred to buildings with little or no data. An ensemble tree-based machine learning algorithm and datasets from two leisure centres and an office building in Melbourne were used in this transfer learning investigation. The results show that transfer learning is a promising technique in predicting accurately under a new scenario as it can achieve similar or even better performance compared to learning on a full dataset. The results also demonstrated the importance of time series adaptation as a method of improving transfer learning.

Keywords: Transfer learning · Energy use prediction · Leisure centre

1 Introduction

Efficient use of energy is undoubtedly a subject of great importance in sustainability as the increase of world population and economic growth keep adding pressure on energy supply. One key area of efficient energy management is in the building sector. According to [23], almost 39% of total energy consumption in the US is from buildings. China's energy usage by buildings is expected to reach as high as 35% by 2030 [5]. In Europe, buildings account for 40% energy usage which is equivalent to 36% CO_2 emissions [18]. Recreational facilities are attracting attention because of the increased awareness of health and fitness in modern lifestyle [6]. Sports facilities account for 8% of the total building energy usage in Europe [8]. Leisure centres, for example in Australia, often offer an array of different activities under one roof, such as swimming pools (indoor and outdoor), physical fitness centres, spas, and children's play park. Such arrangement

J. M. F. Rodrigues et al. (Eds.): ICCS 2019, LNCS 11536, pp. 112–123, 2019.
https://doi.org/10.1007/978-3-030-22734-0_9

exhibit complex and high energy use profiles that present energy management challenges for building managers. Accurate prediction of energy use at these leisure centres is of paramount importance for the building managers or owners as it enables them to make informed decisions to manage better and optimise the operational performances of their buildings.

Widely used techniques in the literature for building energy consumption prediction, are broadly classified under engineering methods (white box methods such as EnergyPlus by [7], and DOE-2 by [3], statistical methods, and artificial intelligence (black box) methods [1]. Comprehensive discussions on the above techniques and their advantages and disadvantages can be found in recent reviews [9,10] and [22]. In this study, we introduce transfer learning to facilitate the prediction process. The main thrust of transfer learning is the notion that it ignores the condition that training and testing data must obey the same distribution. Just like human beings can acquire knowledge while learning tasks and leverage that knowledge to solve related tasks. Transfer learning operates intuitively on the same principles. Thus, by utilising knowledge gained from one task, transfer learning overcomes the isolated learning limitation in the traditional machine and deep learning methods [20,21] and [14].

In another work [16], investigated energy forecasting in the context of cross-building transfer with limited historical data by leveraging data from other buildings. In this work, reinforcement learning algorithms were combined with a deep belief network to improve the former's continuous state estimation capabilities.

In another research, [12] developed a transfer learning methodology for residential buildings climate control. They developed a generalized online transfer learning algorithm which leveraged forecasting knowledge from the source data to enhance the prediction of the target house. Their work utilised simulated residential houses created in EnergyPlus, as a test bed for the developed online transfer methodology and yielded positive results in the period within the first five weeks of the target dataset.

Recently, [19] proposed an inductive transfer learning algorithm that is sensitive to seasonality and trends present in electricity consumption data. The algorithm is applicable in a supervised transfer learning setting, that is, it requires limited data from the target building, and its extent of operation is bound only to similar buildings. A prediction accuracy increase of up to 11.2% using data from additional schools was reported on the target school with only one month of data.

This work is the first to explore transductive transfer learning for building energy studies, using different building types (leisure centres and office buildings) in a supervised learning setting. The most skilful of the five machine learning methods developed and evaluated on the task of building energy consumption prediction at two leisure centres and an office building is selected for transfer learning. The work investigates the application feasibility of transfer learning and how to improve its performances for improved energy consumption prediction using limited measured data. An ensemble tree based algorithm is tested in the transfer learning task to transfer knowledge amongst buildings with different energy consumption distributions. The work presents some of the initial results of the ongoing transfer learning experiment with a much broader scope.

2 Transductive Transfer Learning

Given a source domain, a target domain and a learning task, energy consumption prediction in our case, transfer learning aims to help improve the learning of the target predictive function in a new leisure center (Don Tatnell) using the knowledge in another leisure center (Waves) and an office building. In the traditional machine learning, to ensure the accuracy and high reliability of the model obtained by training, there are two basic assumptions: (1) the training sample used for learning and the new test sample satisfy the condition of independent and identical distribution; (2) There must be enough training samples available to learn a good model. While in transfer learning these assumptions are no longer necessary.

2.1 Predictive Algorithms

This section gives a brief overview of the adopted predictive algorithms for the energy consumption prediction exercise. A total of five predictive algorithms that is, decision trees, random forest(RF), lightGBM, k nearest neighbour(k-NN)and ensemble extra trees(EET) are considered for the input-output mapping task.

Random forest is an ensemble based learning algorithm used for both regression and classification problems [4]. RF is an ensemble of models which uses the decision tree approach to data collection. Initially, an individual tree is trained by taking note of a random subset of observations. A random subset of the variable is then considered to split the decision thus creating a diverse set of trees essential for improving overall prediction performance of the ensemble model.

LightGBM is a recently launched algorithm, which by using histogram-based algorithms buckets continuous feature values into discrete bins. This enhances training and results in reduced memory usage. While most decision tree algorithms grow their trees by level (depth)-wise, LightGBM instead grows trees leaf-wise (best-first). The leaf with max delta loss is chosen to grow. However, when data is small. Leaf-wise often results in over-fitting. More details on the model performance are found in [15].

The k-nearest neighbour algorithm is one of the most straightforward supervised learning regression algorithms to implement but gives highly competitive results. The k-NN algorithm's primary assumption is that similar things exist nearby. The algorithm computes a distance value between the output and each item in the training data-set. The k-NN then picks the items with the k lowest distances and conduct a "majority vote" among those data points. The value of k is determined by either by trial and error or by cross-validation to find an optimal value [2].

Decision tree takes the form of a tree-like structure where numerous aspects and attributes are considered to predict the electrical energy demand. For evaluation, a recursive algorithm is used to identify the attributes with the highest information [13].

Ensemble extra trees (extremely randomised trees) implement a meta estimator that fits randomised decision trees on various dataset sub-samples and then averages them to have improved predictive accuracy and at the same time to control over-fitting. Ensemble extra trees differentiate itself from other tree-based ensembles, in that it splits nodes by selecting cut-points entirely at random and incorporates the whole learning sample as opposed to a bootstrap replica to grow the trees. The individual tree predictions are aggregated to give the final prediction, by arithmetic average in regression type of problems and a majority vote in classification type of problems. Tree complexity and size is controlled by adjusting two parameters namely, $max_{depth}(\mathrm{D}_{\max})$ and $min_{sampleleaf}$ [11].

2.2 Evaluation Metrics

Assessment of the models' performance was done using standard evaluation metrics namely, the mean square error (MSE), mean absolute error (MAE) and R-squared (R^2). The MSE is the mean of the square of the errors. The closer the MSE is to zero the more ideal. The larger the MSE, the larger the error. The MSE's basic value is in selecting one prediction model over another. R^2 describes the proportion of variance of the dependent variable that is explained by the regression model. A low R^2 value shows a low level of correlation, meaning a regression model that is not valid, but not in all cases. MAE gives the mean error (positive) for all test data. Note one cannot look at these metrics in isolation in sizing up the model. These performance evaluation metrics are calculated using Eqs. (2 to 5) as follows:

$$R^2 = 1 - \frac{\sum(y - y')^2}{\sum(y - \bar{y}')^2}, \tag{1}$$

$$RMSE = \sqrt{\Sigma\frac{(y' - y)^2}{N}}, \tag{2}$$

$$\mathrm{MAE} = \frac{1}{n}\sum_{i=1}^{n}|y' - y| \tag{3}$$

$$\mathrm{MSE} = \frac{1}{n}\sum_{i=1}^{n}(y' - y)^2 \tag{4}$$

where y is the measured energy consumption value, y' is the energy consumption predicted value, $\bar{y}$ represents the average values of the corresponding variables, N is the number of data points considered, CV is the coefficient of variation, σ is the standard deviation and μ is the population mean.

3 Development of Prediction Models

3.1 Building Description

The office building and two leisure centres namely, Waves and Don Tatnell leisure centres are in Melbourne and are both run and managed by Kingston municipality. Waves leisure centre generally comprises of a standalone Aquatic and Leisure Centre situated along longitude and latitude 145.0577 °E and 37.9516 °S respectively. The centre comprises an aquatic area, the health and fitness areas and ancillary facilities comprise of male and female toilet and change rooms, staff rooms, school change, family change, creche, retail store, kiosk with associated food and storage areas, Mezzanine floor party hire area, general administration, reception area and foyerentry area. The building area sits on approximately 5500 m^2 of land on a concrete slab base, with rendered masonry walls and covered with a pitched tin roof and fitted with aluminium framed windows and doors.

Don Tatnell leisure centre also houses various indoor leisure activities under one roof including, a fitness centre, a spa, indoor swimming pool, a formal pool and an occasional day-care centre. Don Tatnell leisure centre location is to the northeast corner of the site, with the main entrance facing east. It is constructed from a concrete slab base, with rendered masonry walls and covered with a pitched tin roof and fitted with aluminium framed windows and doors. The longitude and latitude of the site are approximately 145.0924 °E and 37.9911 °S respectively. Both leisure centres are open every day of the week, from 6am–9pm weekdays and 7am–6pm weekends.

The longitude and latitude of the office building site are approximately 145.0 °E and −37.8°S respectively. True north is about −57° from the front elevation (Nepean Highway or North Eastern side) of the building. The building is rectangular and comprises approximately 10,500 m^2 floor area with seven stories of office space. The building facade primarily comprises painted precast concrete panels of 200 mm thickness. Within the panels are vision glass window and spandrel combination sets. The glazing height varies from 2450 mm for the ground level to 1950 mm for levels 1 to 6. The width of glazing varies from 1980 mm to 3280 mm, while the front and rear have feature combinations on the centre of the facade.

3.2 Dataset Description

The electrical energy consumption datasets of the three buildings do not contain missing values. All data points correspond with correct timestamp values. However, the datasets did include some outliers. The system on few instances recorded a series of zero values then suddenly sums up the total power usage for that given period (with zero readings) with a very high value which would not typically be consumed in 15 min. A total of 20 potential input variables were investigated to test their impact on electrical power demand prediction at the two leisure centres and the office building. These inputs are: maximum and minimum temperature (Tmax, Tmin), dry bulb air temperature (T °C), mean temperature (Tmean °C), dew temperature (DewT °C), maximum and minimum wind (Umax, Umin), prevailing wind speed (U), gusty winds (Ug), relative humidity (RH%), wind direction (Ud), average wind speed (Umean), average wind direction (Udmean), year, week of the year, weekday, day of the year, month of the year, hour of the day, and 15 min. The climatic variables of a nearby weather station were obtained from the Bureau of Meteorology with a 15-min resolution. The two leisure centres' training and validation datasets contain data from 01/05/2017 00:00 to 24/09/2018 09:30 while the office building dataset ranges from 01/06/2011-24/03/2018. Melbourne is classified according to the Köppen climate classification as a temperate oceanic climate. The city warms up in summer with mean temperatures between 14–25.3 °C and winters averaging between 6.5–14.2 °C.

3.3 Statistical Description of the Data

This section provides a brief insight into the energy consumption of the three datasets used for experimentation. Table 1 gives the statistical description of the energy consumption profiles at the sites. The office building has the highest number of historical energy consumption observations followed by Waves, and then Don Tatnell has the least observations recorded. On average, Waves leisure has the highest energy consumption rate followed by Don Tatnell and then the office building.

Table 1. Statistical summary of energy consumption profiles at the three cites

Parameter	Waves centre	Don tatnell	Office building
Count	48998	48050	236268
Mean	55.702233	23.84	19.79
Standard deviation	11.71	8.71	16.53
Min	0	6.50	0.00
25%	45	14.03	7.42
50%	60	24.84	9.79
75%	65	30.63	33.64
Max	94	53.97	176.54

The energy consumption distribution patterns for the Don Tatnell leisure centre, Waves leisure centre and the office building are shown in Fig. 1.

The two leisure centres show a seemingly similar electrical consumption distribution shape relative to the office building. Don Tatnell leisure centre has a somewhat almost symmetric distribution while Waves leisure centre shows some skewness to the left and the office building being skewed to the right. The majority of the office building energy consumption readings fall within the

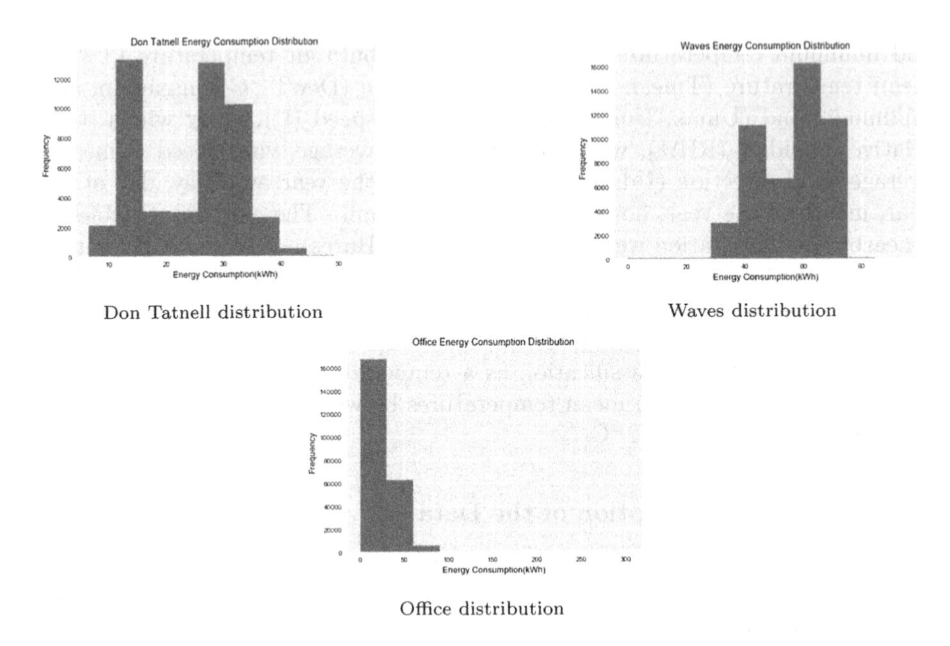

Don Tatnell distribution

Waves distribution

Office distribution

Fig. 1. Energy consumption histograms of the three buildings

0–35 kWh range while Waves leisure centre has a range between 60–70 kWh. It is worth mentioning that leisure centres tend to consume more energy than office buildings; however, little building research exists in building energy performance literature regarding these leisure centres.

3.4 Selection of Candidate Inputs

Among the 20 potential inputs, $Temperature$, $Tmean$, $Tmax$ and $Tmin$ showed high correlation and the same is true with $Umean$, $Umax$, $Ugust$. This means that they have a similar effect on the dependent variables (energy consumption) as such, choosing only one as an input in the model is equally effective than using all. Following this explanation, $Tmean$ and $Umean$ were adopted to represent temperature and wind-related inputs respectively, bringing the number of input variables down from 20 to 13 inputs.

3.5 Data Transformation

Due to the vast differences in numerical ranges between the input and output values standardisation of the inputs values was carried out. Standardisation scales each feature such that the distribution is centred around 0, with a standard deviation of 1. Standardisation allows comparability among inputs and it also enhances the training process since the numerical condition of optimisation is improved as opposed to without standardisation. As such, the mean and standard deviation for each feature is calculated, and then the feature is scaled using Eq. 5:

$$z - score = (x_i - \mu)/\sigma. \tag{5}$$

where x_i is the observed value, μ is the population mean and σ is the population standard deviation.

3.6 Transfer Learning Experiment Set-Up

The most skilful of the five machine learning algorithm described in Sect. 2.1 are selected in the transfer learning investigation. Initially, the models are developed and fine-tuned from scratch on the office building (D_O), Don Tatnell (D_C) and Waves leisure centres (D_W). For comparison purposes models developed using Don Tatnell dataset will act as the baseline models to test transfer learning effect. During the development of all models, the training set sizes are varied gradually from 1% (to simulate the lack of data) to 80% (enough data). The earlier developed office building (D_O)and Waves leisure centre (D_W) models are then retrained using data from Don Tatnel for predicting energy consumption at Don Tatnell centre using a similarly sized training dataset of between 1% to 80%. The results are then compared using the evaluation metrics already described in Sect. 2.2.

All learning algorithms were implemented using Python programming language. The development of machine learning models was done using the scikit-learn (Python programming language library) [17]. All model development and experimental tasks were conducted on a Windows machine (Intel Core i5 2.40 GHz 8GB RAM).

4 Results and Discussion

4.1 Model Selection for Transfer Learning

Four evaluation metrics that are RMSE, MAE, MSE and R^2 are used for evaluation of the skill of prediction models. The performance of the five models in energy consumption prediction for Waves Leisure Centre is summarized in Table 2. All ensemble based tree models had equally good performance with slight variations amongst themselves. The decision tree algorithm has the least performance followed by the k-NN algorithm. The EET model with the least amount of error is adopted for the transfer learning task having MSE, MAE, RMSE and R^2 values of 12.89, 2.52, 3.59 and 0.913 respectively. Following this result, the EET model became the model of choice in the transfer learning experimentation.

The optimum number of trees (M), maximum tree depth (D_{max}), the minimum number of samples needed for splitting a node (n_{min}) and the attribute selection strength parameter (K) are 260, 8, 10 and 9 respectively for the best performing model. The (EET) algorithm was again selected and fine-tuned for

Table 2. Waves leisure centre energy consumption prediction results

Model	MSE	MAE	RMSE	R^2
Decision tree	28.83	3.72	5.37	0.805
K-Nearest Neighbour	22.87	3.52	4.78	0.85
Random forest	14.41	2.67	3.79	0.903
Ensemble Extra trees	12.89	2.52	3.59	0.913
LightGBM	13.58	2.64	3.68	0.908

office building energy consumption prediction. The EET model obtained MSE, MAE, RMSE and R^2 values of 0.48, 0.43, 0.69 and 0.97 respectively. Overall the EET model demonstrated better performance on office building relative to the leisure centre. This result is as expected and demonstrates the complexity of the leisure centre prediction exercise and also the larger dataset available at the office building.

4.2 Transfer Learning Results

Following a series of investigations, this section outlines the finding and gives discussion around the results. The performances of the developed EET models in the transfer learning exercise are summarised in Table 3 and 4. Initially, the EET models are developed using historical data and fine-tuned for the energy consumption prediction of the three buildings before selection for transfer learning. After that, the most skilful models according to the discussed evaluation metrics, are then set aside for the task of transfer learning.

Table 3. Transfer learning results using an undifferenced time-series

	With transfer				Without transfer (Direct don)				With transfer			
	Pre-trained with Office (D_O)				Direct Don (D_C)				Pre-trained with Waves (D_W)			
Training %	MSE	MAE	RMSE	R^2	MSE	MAE	RMSE	R^2	MSE	MAE	RMSE	R^2
80	7.45	1.87	2.73	0.89	8.55	2.0	2.92	0.87	6.85	1.76	2.62	0.9
50	8.62	2.1	2.94	0.88	9.55	2.2	3.1	0.87	7.76	1.92	2.79	0.89
20	9.91	2.25	3.15	0.87	11.19	2.41	3.35	0.85	8.62	2.06	2.94	0.89
10	10.6	2.4	3.26	0.86	11.56	2.5	3.4	0.85	8.96	2.13	2.99	0.88
8	10.38	2.35	3.22	0.86	11.59	2.49	3.41	0.85	9.08	2.14	3.01	0.88
5	10.24	2.27	3.2	0.87	11.67	2.45	3.42	0.84	9.34	2.11	3.06	0.88
3	10.43	2.3	3.23	0.86	11.49	2.45	3.39	0.85	9.68	2.16	3.11	0.87
1	13.13	2.66	3.62	0.83	13.92	2.78	3.73	0.82	11.4	2.42	3.38	0.85
0.8	11.78	2.53	3.43	0.84	12.77	2.6	3.57	0.83	10.65	2.34	3.26	0.86
0.5	12.4	2.59	3.52	0.84	14.01	2.81	3.74	0.82	11.02	2.35	3.32	0.85
0.3	12.28	2.57	3.5	0.84	14.56	2.86	3.82	0.81	11.49	2.38	3.39	0.85
0.1	33.97	4.57	5.82	0.55	31.03	4.44	5.57	0.59	25.6	3.9	5.06	0.66

Table 4. Transfer learning results using a differenced time-series

	With transfer			Without transfer (Direct don)			With transfer		
	Pre-trained with Office (D_O)			Direct Don (D_C)			Pre-trained with Waves (D_W)		
Training %	MSE	MAE	RMSE	MSE	MAE	RMSE	MSE	MAE	RMSE
80	8.75	1.8	2.96	11.03	2.18	3.32	8.73	1.8	2.95
50	8.83	1.81	2.97	11.14	2.19	3.38	8.82	1.81	2.97
20	9.5	1.88	3.08	12.14	2.32	3.48	9.54	1.88	3.09
10	9.56	1.91	3.09	12.25	2.33	3.5	9.6	1.92	3.1
8	9.61	1.92	3.1	12.12	2.32	3.48	9.66	1.92	3.11
5	9.72	1.93	3.12	12.08	2.32	3.48	9.78	1.94	3.13
3	9.75	1.94	3.12	11.8	2.31	3.44	9.81	1.95	3.13
1	10.19	2.01	3.19	12.5	2.39	3.54	10.33	2.04	3.21
0.8	10.3	2.03	3.21	12.95	2.44	3.6	10.38	2.06	3.22
0.5	10.22	2.03	3.2	12.32	2.38	3.51	10.34	2.05	3.22
0.3	10.42	2.05	3.23	12.77	2.41	3.57	10.51	2.08	3.42
0.1	11.15	2.25	3.34	13.37	2.55	3.66	11.15	2.25	3.34

Office building prediction has seven years worth of historical data, while both leisure centres have only sixteen months worth of available historical data. To test the effect of transfer learning on models the training data for all models is varied between 0.1% (simulating data shortage scenario) to 80% (simulating enough data situation) and the performance of the models is monitored consequently by tracking the evaluation metrics adopted. Particular importance is given to the instances were models are trained with few data as that represents the primary motivation for the investigation.

As seen in Table 3, it is evident that both instances of pre-trained models (with transfer learning) do have superior performances relative to the cases where the models are trained from scratch (no transfer learning). This phenomenon is observable at all training data sizes under consideration with the weakest performance being for training data sizes less than 10%. It is also noted that while pre-trained models do perform better than training from scratch, models pre-trained on Waves centre dataset have lower error metrics as opposed to those pre-trained on office buildings. This may be because of the similarities in building operations, location and building form. Overall, on average, models pre-trained on Waves centre relative to those that are trained from scratch, have lower MSE, MAE and RMSE values by 19%, 14% and 10% respectively. On the other hand, those models pre-trained on the office building have lower MSE, MAE and RMSE values by 7%, 5% and 4% respectively compared to Don Tatnell models. This shows that pre-trained models did have better performance than models developed from scratch with the models pre-trained on waves having the most superior performance.

In the last section of the investigation, differenced (lag = 1) time-series data is considered under the same investigation conditions above, and the results do show a similar trend as observed earlier. As expected, pre-trained models show superior performance relative to the one trained from scratch. While in the earlier example models pre-trained on waves leisure centre had superior performance to that training on the office building, results show almost similar performance on training set sizes greater and 10%. On average, both pre-trained models (D_O and D_W) show lower MSE, MAE and RMSE values of 19%, 16% and 10% as compared to those trained from scratch (D_c) models. However, of particular note is the rather seemingly equal performance by pre-trained models from Waves centre and the office building.

Nonetheless models trained on differenced time-series show lower error values in general across all training data sizes and models (pre-trained and direct). Thus, transfer learning with differenced time series recorded lower error values (Table 4) as compared to instances were the time-series is undifferenced (Table 3).

5 Conclusion

The need for fast and accurate models for building energy consumption prediction continues to increase particularly with the rise in the need for renewable energy sources and ever-changing smart grid networks. Data acquisition is a costly exercise concerning both time and financial resources. Due to the general problem of inadequate training data, the authors try to show the benefits of the well-known transfer learning paradigm. The knowledge learnt from buildings with vasts amount of data was leveraged for use in the building with little data. This study investigated the applicability of transfer learning and ways of maximising its benefits in the task of building energy consumption prediction. To exemplify the transfer learning problem case three buildings comprising an office building and two leisure centres, are analysed and machine learning models, based on mainly ensemble decision trees, are developed and tested. The benefits of this approach are evident in our experimental results. Transfer learning demonstrated advantage, in terms of prediction accuracy, even comparing with models with adequate training data sets. The study also concluded that differencing a time series improves transfer learning. This advantage is independent of underlying learning methods. Hence we conclude that transfer learning is valid and a useful method for building energy consumption prediction in complex facilities like leisure centres. With this approach, extensive data collection prior to the learning becomes less essential.

The authors are currently investigating ways of improving transfer learning on these time-series related problems using deep learning models as an extension of this current research work. Unlike classical machine learning models, deep learning techniques have abilities to learn the temporal dependence automatically and naturally handle temporal structures found within time series data.

References

1. Ahmad, M.W., Mourshed, M., Yuce, B., Rezgui, Y.: Computational intelligence techniques for hvac systems: a review. Build. Simul. **9**, 359–398 (2016)
2. Altman, N.S.: An introduction to kernel and nearest-neighbor nonparametric regression. Am. Stat. **46**(3), 175–185 (1992)
3. Birdsall, B., Buhl, W.F., Ellington, K.L., Erdem, A.E., Winkelmann, F.C.: Overview of the doe-2 building energy analysis program, version 2.1 d (1990)

4. Breiman, L.: Random forests. Mach. Learn. **45**(1), 5–32 (2001)
5. Chen, H., Lee, W., Wang, X.: Energy assessment of office buildings in china using china building energy codes and leed 2.2. Energ. Buildings **86**, 514–524 (2015)
6. Costa, A., Keane, M.M., Torrens, J.I., Corry, E.: Building operation and energy performance: monitoring, analysis and optimisation toolkit. Appl. Energy **101**, 310–316 (2013)
7. Crawley, D.B., Lawrie, L.K., Pedersen, C.O., Winkelmann, F.C.: Energy plus: energy simulation program. ASHRAE J. **42**(4), 49–56 (2000)
8. Eurostat, L., et al.: Energy indicators for sustainable development: guidelines and methodologies (2008)
9. Foucquier, A., Robert, S., Suard, F., Stéphan, L., Jay, A.: State of the art in building modelling and energy performances prediction: a review. Renew. Sustain. Energy Rev. **23**, 272–288 (2013)
10. Fumo, N.: A review on the basics of building energy estimation. Renew. Sustain. Energy Rev. **31**, 53–60 (2014)
11. Geurts, P., Ernst, D., Wehenkel, L.: Extremely randomized trees. Mach. Learn. **63**(1), 3–42 (2006)
12. Grubinger, T., Chasparis, G.C., Natschläger, T.: Generalized online transfer learning for climate control in residential buildings. Energ. Buildings **139**, 63–71 (2017)
13. Hastie, T., Tibshirani, R., Friedman, J.: Unsupervised learning. The Elements of Statistical Learning. SSS, pp. 485–585. Springer, New York (2009). https://doi.org/10.1007/978-0-387-84858-7_14
14. Huang, J.T., Li, J., Yu, D., Deng, L., Gong, Y.: Cross-language knowledge transfer using multilingual deep neural network with shared hidden layers. In: 2013 IEEE International Conference on Acoustics, Speech and Signal Processing (ICASSP), pp. 7304–7308. IEEE (2013)
15. Ke, G., et al.: Lightgbm: a highly efficient gradient boosting decision tree. In: Advances in Neural Information Processing Systems, pp. 3146–3154 (2017)
16. Mocanu, E., Nguyen, P.H., Kling, W.L., Gibescu, M.: Unsupervised energy prediction in a smart grid context using reinforcement cross-building transfer learning. Energ. Buildings **116**, 646–655 (2016)
17. Pedregosa, F., et al.: Scikit-learn: machine learning in python. J. Mach. Learn. Res. **12**, 2825–2830 (2011)
18. Recast, E.: Directive 2010/31/eu of the european parliament and of the council of 19 May 2010 on the energy performance of buildings (recast). Off. J. Eur. Union **18**(06), 2010 (2010)
19. Ribeiro, M., Grolinger, K., Elyamany, H.F., Higashino, W.A., Capretz, M.A.: Transfer learning with seasonal and trend adjustment for cross-building energy forecasting. Energ. Buildings **165**, 352–363 (2018)
20. Thomas, S., Ganapathy, S., Hermansky, H.: Multilingual MLP features for low-resource LVCSR systems. In: 2012 IEEE International Conference on Acoustics, Speech and Signal Processing (ICASSP), pp. 4269–4272. IEEE (2012)
21. Vu, N.T., Imseng, D., Povey, D., Motlicek, P., Schultz, T., Bourlard, H.: Multilingual deep neural network based acoustic modeling for rapid language adaptation. In: 2014 IEEE International Conference on Acoustics, Speech and Signal Processing (ICASSP), pp. 7639–7643. IEEE (2014)
22. Wei, Y., et al.: A review of data-driven approaches for prediction and classification of building energy consumption. Renew. Sustain. Energy Rev. **82**, 1027–1047 (2018)
23. Zuo, J., Zhao, Z.Y.: Green building research-current status and future agenda: a review. Renew. Sustain. Energy Rev. **30**, 271–281 (2014)

Forecasting Network Throughput of Remote Data Access in Computing Grids

Volodimir Begy[1,2(✉)], Martin Barisits[1], Mario Lassnig[1], and Erich Schikuta[2]

[1] CERN, Geneva, Switzerland
volodimir.begy@cern.ch
[2] University of Vienna, Vienna, Austria

Abstract. Computing grids are key enablers of computational science. Researchers from many fields (High Energy Physics, Bioinformatics, Climatology, etc.) employ grids for execution of distributed computational jobs. These computing workloads are typically data-intensive. The current state of the art approach for data access in grids is data placement: a job is scheduled to run at a specific data center, and its execution commences only once the complete input data has been transferred there. An alternative approach is remote data access: a job may stream the input data directly from arbitrary storage elements. Remote data access brings two innovative benefits: (1) the jobs can be executed asynchronously with respect to the data transfer; (2) when combined with data placement on the policy level, it can aid in the optimization of the network load, since these two data access methodologies partially exhibit nonoverlapping bottlenecks. However, in order to employ this technique systematically, the properties of its network throughput need to be studied carefully. This paper presents experimentally identified parameters of remote data access throughput, statistically tested formalization of these parameters and a derived throughput forecasting model. The model is applicable to large computing workloads, robust with respect to arbitrary dynamic changes in the grid infrastructure and exhibits a long-term prediction horizon. Its purpose is to assist various stakeholders of the grid in decision-making related to data access patterns. This work is based on measurements taken on the Worldwide LHC Computing Grid at CERN.

Keywords: Grid computing · Remote data access · Network modeling · Applied time series forecasting · Applied machine learning

1 Motivation

Computing grids [5] are collections of geographically sparse data centers. Employing a wide range of heterogenous hardware, these participating facilities work together to reach a common goal in a coordinated manner. Grids store vast amounts of scientific data, and numerous users run computational jobs to analyze these data in a highly distributed and parallel fashion. For example,

J. M. F. Rodrigues et al. (Eds.): ICCS 2019, LNCS 11536, pp. 124–137, 2019.
https://doi.org/10.1007/978-3-030-22734-0_10

within the World-Wide LHC Computing Grid (WLCG) more than 150 computing sites are employed by the ATLAS experiment at CERN. WLCG stores more than 450 petabytes of ATLAS data, which is used for distributed analysis by more than 5000 users.

In order to achieve a modular architecture, the grid resources are typically divided into three major classes: storage elements, worker nodes and network. The function of storage elements is to replicate and persist the data for the long term using various technologies (hard disks, solid state drives or tapes). On the other hand, the worker nodes perform CPU intensive tasks. WAN and LAN networks facilitate the data transfer between storage elements and worker nodes. Such strict division of labour has led to the fact that the current best practice for data access in the grid is data placement [3]. Data placement is performed by dedicated distributed data management systems. Consider a job scheduled to run at the data center DC_1, while the required input replica is located at the data center DC_2. The job may commence its execution only after the completion of the following workflow: (1) the input replica is transferred from the relevant storage element at DC_2 to a storage element at DC_1; (2) the input replica is staged-in from the relevant storage element at DC_1 to the worker node. This simplistic approach has two major disadvantages: (1) the jobs are staying idle while waiting for the input data; (2) due to the limited infrastructure resources the distributed data management system handling the data placement may queue the transfers up to several days. An alternative approach is remote data access, which allows to directly stream the input replica from a local or remote storage element.

To demonstrate the potential of remote data access to optimize the resource utilization when combined on policy level with traditional data placement, consider two following scenarios.

In the first example, a computational workload is submitted to a worker node at the data center DC_1. The jobs require the input replica R, which is persisted in the remote storage element at the data center DC_2. Currently, the storage element in the local data center DC_1 has a high load because of transfers to various other data centers. Assume that the bottleneck is only present at this local storage element. This is recognized by the distributed data management system, and it queues the transfer of the input replica R. Employing remote data access, the bottleneck can be immediately bypassed.

In another example, the input replica is located at the local storage element. However, the worker node does not have enough disk space available for stage-in. The distributed data management system queues the file transfer, until the required disk space becomes available. In contrast, using remote data access to the local storage element, the input data can be streamed in fractions without persisting the complete input file.

Given the magnitude of real-life computing grids, it is obvious that many more potential scenarios will benefit from remote data access in the context of resource utilization. This has motivated us to study the properties of its network throughput. Furthermore, a forecasting model of the network throughput is needed for coordination of scientific workloads.

2 Related Work

A substantial body of related work in the literature focuses on prediction of network throughput [1,2,7,11,13,17,18]. However, many authors present only short-term forecasts. In addition, numerous proposed models operate on low-level metrics, which cannot be obtained globally in a computing grid. Finally, the results from the older contributions are outdated with respect to the contemporary network requirements in computing grids. Below we present some notable contributions in more detail.

The authors in [4] describe a method for derivation of short-term throughput predictions based on regression, whose coefficients are continuously updated based on a running time window. The effect of concurrent traffic is not explicitly investigated in this work. The experiments are performed with data probes of up to 16 megabytes. In contrast, our model delivers long-term predictions, taking into account the concurrent traffic of parallel threads and processes. Our work is based on experiments with file transfers totaling up to tens of gigabytes.

Kim et al. [9] have presented an analytical model for prediction of TCP throughput with varying amounts of parallel streams. The performance of the model is not demonstrated for extended forecasting horizons. The model operates on low-level parameters, such as maximum segment size, round-trip time and packet loss ratio. In a grid computing setting such low-level metrics are typically not available globally, since they are not monitored by the data management systems. On the other hand, our model requires only realistically obtainable metrics on file accesses.

Mirza et al. [12] have applied support vector regression for TCP throughput prediction, considering following features as inputs: estimated available bandwidth, queueing delay and packet loss. During the experimental evaluation of the model in the wide area setting, the authors make certain simplifying assumptions (selection of nodes with no or minimal concurrent third-party workloads and concrete operating system), which are impractical for computing grids. Only transfers of 2 MB files are considered.

A novel approach for prediction of throughput in data centers is proposed in [14]. First, various components of a data center (topology, configuration, etc.) are modeled in a highly abstract fashion. Then, the initial meta-model is transformed into a Queueing Petri Net. In order to derive predictions, the net is executed by a simulator. The approach is evaluated in a minimal isolated LAN environment.

Recent work in [10] demonstrates feature engineering for throughput prediction based on a wide spectrum of Globus logs. File transfers, which are expected to encounter high third-party concurrent loads during the execution are filtered out from the analyzed dataset. Moreover, the authors assume that nodes within a given site may be treated as equivalent. Our detailed investigation reveals that this is not the case.

3 Parameters of Remote Data Access Throughput

The first phase of this study is the experimental identification of the crucial parameters, which impact remote data access throughput. In our experiment the throughput is indirectly represented by transfer time for a given filesize. Note that these quantities can be easily converted to one another. The hardware of the communicating hosts is a black box. The properties of the hosts are represented exclusively by the observed network throughput. A single computational job is equivalent to a process. Within a process, a new file access is executed by an individual thread. The null hypothesis for the experiment states that for a computational job streaming a given remote file, the total transfer time T cannot be linearly regressed onto the following independent variables: (1) S, the size of the original file; (2) *ConTh*, the cumulative concurrent traffic originating from other threads in the same process during the streaming of the original file; (3) *ConPr*, the cumulative concurrent traffic originating from other processes in the given computing workload during the streaming of the original file. The alternative hypothesis states that such a regression represents the true relationship between the mentioned independent and dependent variables.

The independent variable *ConPr* stands for traffic, which is created as part of the investigated computing workloads. Other traffic originating from latent workloads is present on the investigated links. In practice it is not possible to obtain complete metrics about a grid. Nevertheless, our minimal approach (following the Occam's razor principle) of monitoring a single additive workload allows us to construct an accurate forecasting model for regression coefficients. The dynamic aspects of the grid are implicitly learned by the model.

To sample the data, 12 computational jobs are launched on a single worker node at the CERN data center (Switzerland) in the period of 28.04.2018 00:00–30.04.2018 14:30. In 15-min steps the processes initiate remote accesses to the storage element GRIF-LPNHE_SCRATCHDISK at the GRIF-LPNHE data center (France). At each step, various numbers of concurrent processes (up to 12) launch various amounts of concurrent threads (up to 4). The threads stream files of different sizes (300 MB–3 GB). This allows us to sample a wide range of data for the independent variables. Each launched file access is treated as an observation in the final dataset. After such sampling and transformations we derive 1122 observations, each having values for the variables T, S, *ConTh* and *ConPr*. The protocols used for remote data access are WebDAV/HTTP. Note that such an intrusive sampling design was necessary, because currently these metrics are not logged in WLCG. Once proper remote data access monitoring is implemented grid-wide, the observations can be mined from logs unintrusively. Since WLCG employs commodity hardware, the presented findings are universally generalizable.

A multiple linear regression is fit to the sampled data using ordinary least squares, resulting into the following equation:

$$T = 0.023568 * S + 0.043705 * ConTh + 0.001538 * ConPr \tag{1}$$

The fit has an F-statistic of 4.514e+04 on 3 degrees of freedom, 1119 residual degrees of freedom and a p-value of $< 2.2e{-16}$. Thus, the null hypothesis is rejected.

Note that the coefficient of *ConTh* is about 28.4 times greater than the coefficient of *ConPr*. This means that concurrent traffic among threads within one process is penalized to a much higher extent in terms of throughput than concurrent traffic among different processes.

To check for interaction among the variables *ConTh* and *ConPr*, two reduced multiple linear regression models are fit on accordingly adjusted datasets using ordinary least squares. The resulting models are demonstrated below:

$$T = 0.02565 * S + 0.04293 * ConTh \tag{2}$$

which is statistically significant, having an F-statistic of 3.119e+04 on 2 degrees of freedom and 340 residual degrees of freedom and a p-value of $< 2.2e{-16}$, and

$$T = 0.023195 * S + 0.001575 * ConPr \tag{3}$$

which is also statistically significant with an F-statistic of 2.035e+04 on 2 degrees of freedom and 832 residual degrees of freedom and a p-value of $< 2.2e{-16}$. These fits are depicted in Figs. 1 and 2.

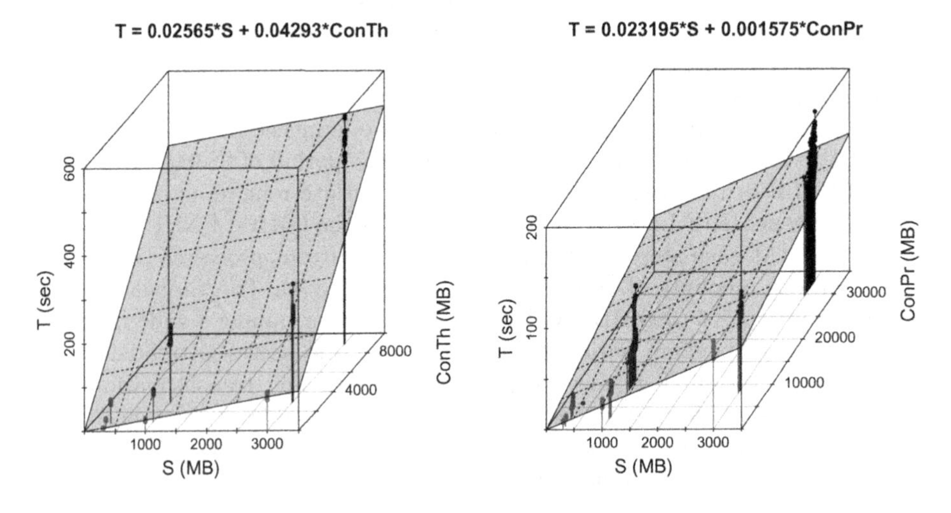

Fig. 1. Regression fit T ∼ 0+S+ConTh

Fig. 2. Regression fit T ∼ 0+S+ConPr

The coefficient of *ConTh* in the reduced model exhibits a log change of only −0.018 in relation to the full model. The coefficient of *ConPr* has a log change of 0.024. This is a hint that in the true relationship, there is no interaction between the variables *ConTh* and *ConPr*.

4 Forecasting Model

The idea behind our novel throughput forecasting model is the following. We have demonstrated that transfer time T can be regressed onto the variables S, $ConTh$ and $ConPr$. To facilitate modeling of transfer time T at various points in time, multivariate time series of regression coefficients are formed. Computing grids are highly complex and dynamic systems. Thus, it is practically impossible to capture all the independent variables, which impact the response T. The uncaptured variables remain latent. The 3 presented independent variables S, $ConTh$ and $ConPr$ explain a significant portion of the variability in the response, but not all of it. The effect of the latent variables manifests itself in the variability of the coefficient values throughout time, as can be seen further in their time series plots. Using time series analysis and forecasting techniques, the effect of the latent variables is implicitly captured by learning the patterns in the time series data. The derived model allows to accurately calculate the transfer time (and thus throughput) given arbitrary amounts for S, $ConTh$ and $ConPr$ at any past, present or forecasted time step. Such calculations can be iteratively applied to every file access in a complex distributed workload.

Consider the time series plot in Fig. 3, which shows transfer times for 2 threads, transferring 2 gigabytes and 300 megabytes respectively, within 1 computing job in the time window 10.03.2018 22:00–12.04.2018 21:30. The observations are sampled once every 30 min. The computing job runs in the worker node pool ANALY_AUSTRALIA and streams the data from a local storage element AUSTRALIA-ATLAS_SCRATCHDISK. Once per day the job gets resubmitted to a new worker node in the pool, which is selected by the workload management system. This is done to emulate the real-life workflow during job submission. Throughout the rest of the paper a sampling day is equivalent to 48 sampling points. It is evident from the daily shifts along the y-axis that various worker nodes exhibit either very different, or very similar throughputs. We assume that these large shifts are caused by major differences in the hardware of various nodes. Given a single worker node - storage element pair, the captured independent variables explain the vast majority of the variability in transfer time T. Thus, the regression model needs to be constructed for each worker node - storage element pair separately. In what follows, we normalize the sampled data with respect to the daily shifts in T to emulate a time series for a single worker node - storage element pair.

Let a, b and c denote the coefficients of the regression

$$T \sim 0 + a * S + b * ConTh + c * ConPr \tag{4}$$

The time series plot in Fig. 3 demonstrate that given a single worker node - storage element pair, the coefficient values slightly vary over time due to latent factors. The sampled time series can be transformed into a multivariate time series of coefficients a and b. The reduced regression is written as

$$T \sim 0 + a * S + b * ConTh \tag{5}$$

Let T_1 and T_2 be the sampled transfer times for the 2 GB and 300 MB files respectively. Since the transfer of the 300 MB file always terminates first in our data, the coefficients a and b can be estimated for each time step using the following system of equations:

$$\begin{cases} T_1 = 2000 * a + 300 * b \\ T_2 = 300 * a + (2000 * (T_2/T_1)) * b \end{cases} \tag{6}$$

after substitution we get:

$$a = -((17 * T_1 * T_2)/(100 * (9 * T_1 - 400 * T_2))) \tag{7}$$

$$b = (T_1 - 2000 * a)/300 \tag{8}$$

Note that we did not sample data, which would allow us to reconstruct the time series for the coefficient c. This is due to the demonstrated fact that a vast amount of concurrent traffic among processes is required, in order to reach a noticeable effect. Such sampling would be highly intrusive during a complete month. The complete multivariate coefficient time series can be unintrusively mined from logs, once proper remote data access monitoring is implemented.

Figure 4 demonstrates the transformation of the sampled time series of transfer times into time series of normalized regression coefficients. The variate for the coefficient c is simulated based on the previous demonstration that in case of remote data access from CERN to GRIF-LPNHE data centers, its value on average was 28.4 times smaller, than that of the coefficient b.

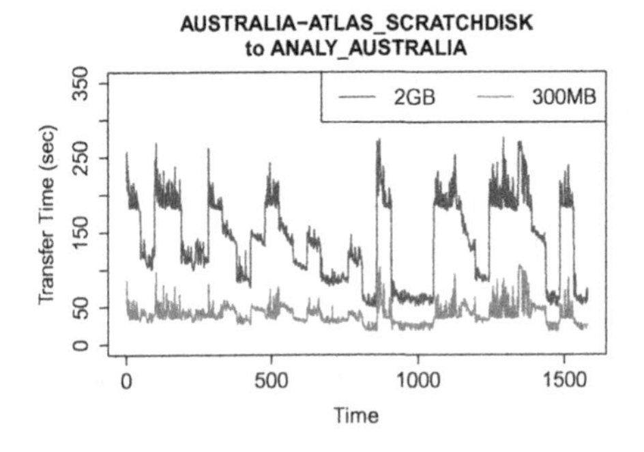

Fig. 3. Australia: transfer times

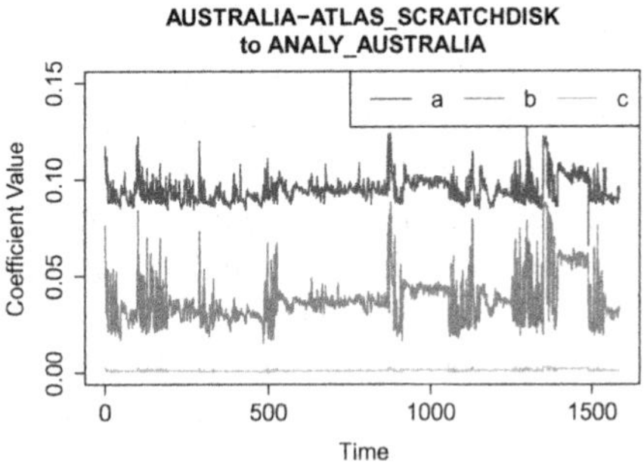

Fig. 4. Australia: normalized coefficients

We have performed an analogous sampling and analysis for the storage element - worker node pool pair BNL-OSG2_SCRATCHDISK - ANALY_BNL_LONG (USA). The transfer time and coefficient time series are shown in Figs. 5 and 6. In infrequent instances in Fig. 6 the values of the coefficients are negative, because the time series is normalized with respect to the first sampling day.

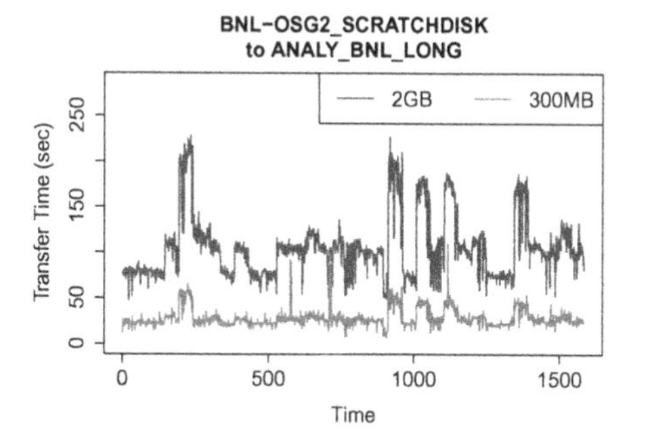

Fig. 5. BNL: transfer times

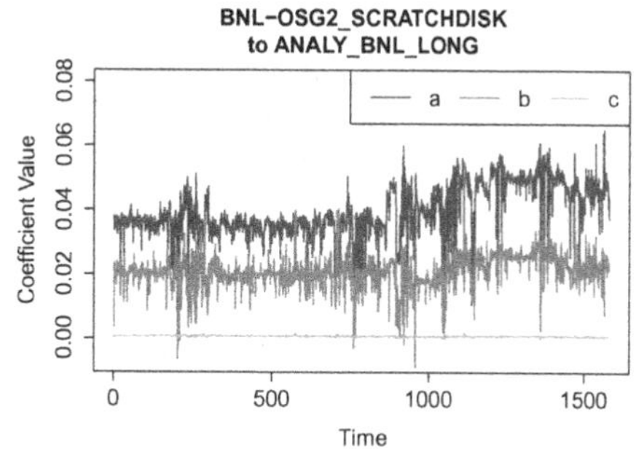

Fig. 6. BNL: normalized coefficients

4.1 Input Features

The forecasting model individually predicts the 3 variates of the multivariate coefficient time series. When forecasting a single variate, following features are considered as inputs to the forecasting model: (1) lagged values of the variate itself; (2) lagged values of 2 other variates; (3) slightly lagged values of the time series representing the overall traffic at the data center of interest.

Consider Fig. 7, which depicts the cross-correlation functions between the normalized coefficients a and b sampled at data centers Australia-ATLAS and BNL-ATLAS.

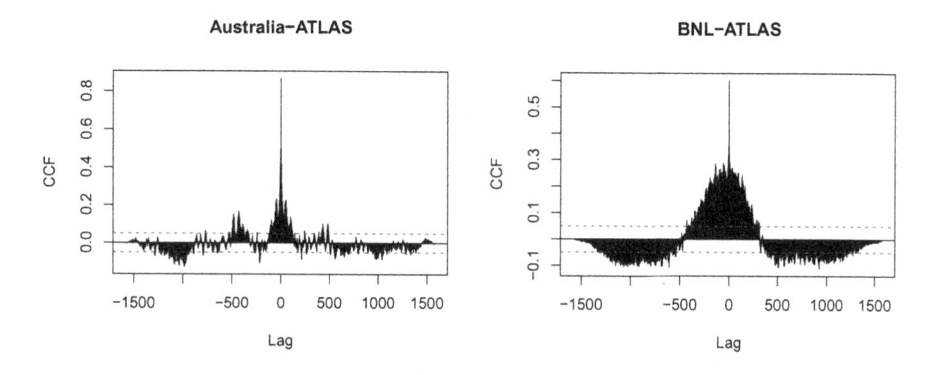

Fig. 7. CCF between normalized coefficients a and b

It is evident from the CCF plots that the values are highly symmetric around the origin. Thus, we do not use other variates as input features to predict a given variate, since they do not posses any additional information for forecasting.

Without a posteriori knowledge, a plausible hypothesis is that the time series of overall network load at the data center of interest leads the amplitude of the coefficient time series with a positive cross-correlation at small negative lags (i.e. if the amount of the overall traffic increases, the time to transfer an additional unit will increase shortly after). We have mined 4 different traffic time series from the logs recorded by the WLCG's distributed data management system Rucio [6] for the investigated period at the data center Australia-ATLAS: (1) intra-data center traffic in bytes; (2) inter-data center traffic in bytes; (3) sum of intra- and inter-data center traffic in bytes; (4) amount of intra-data center file accesses. It can be seen from the cross-correlation plots in Fig. 8 that none of the 4 mentioned traffic time series is significantly leading the normalized coefficient a at small lags with positive cross-correlation. Due to the symmetric CCF between the coefficients a and b, this result is also present in the analysis of CCF between the mentioned traffic categories and the normalized coefficient b. The same analysis delivers insignificant results at the BNL-ATLAS data center. Thus, the overall traffic time series is not used as an input feature for the prediction of the coefficients.

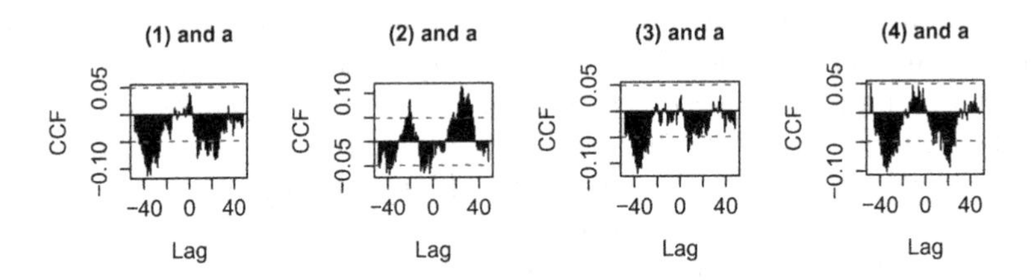

Fig. 8. Australia: CCF between traffic time series and normalized coefficient a

Finally, only the lagged values of the variate of interest will be used as inputs to predict the future values.

We have not sampled authentic data for derivation of coefficient c due to the high associated cost. We assume that in the demonstrated scenarios the CCF between c and other variates exhibits an analogous behavior.

4.2 Benchmark of Forecasting Models

The time series of the normalized coefficient a at the Australia-ATLAS (Fig. 4) and BNL-ATLAS (Fig. 6) data centers are used for demonstrational purposes in this paper. Due to the shown symmetric CCF between coefficients a and b it is known that the coefficient b exhibits an analogous behavior. We assume, that so does the coefficient c.

The aforementioned time series are split into 16 datasets respectively, each spanning a period of 2 sampling days. 80% of each dataset are used for the construction of the model, the remaining 20% are used for its validation. Thus, we are forecasting the series for the next 9.5 h. Following methods are benchmarked in the context of coefficient time series forecasting: (1) Auto-Regressive Integrated Moving Average (ARIMA) [15]; (2) Long Short-Term Memory Artificial Neural Networks (LSTM ANN) [8].

ARMA is a statistical state of the art approach for derivation of a mathematical model describing the time series of interest in terms of its past values (AR part) and forecast errors (MA part). ARMA models work with stationary time series. ARIMA, the integrated version of ARMA, allows to work with time series, which become stationary after a certain degree of differencing. The model parameters are identified based on the analysis of the (partial) auto-correlation function and stationarity tests. Such models are computationally cheap to construct. They remain one of the dominant forecasting methods due to their simplicity and effectiveness.

Consider Fig. 9, which depicts the auto-correlation functions of the normalized coefficients a and b at both the data centers.

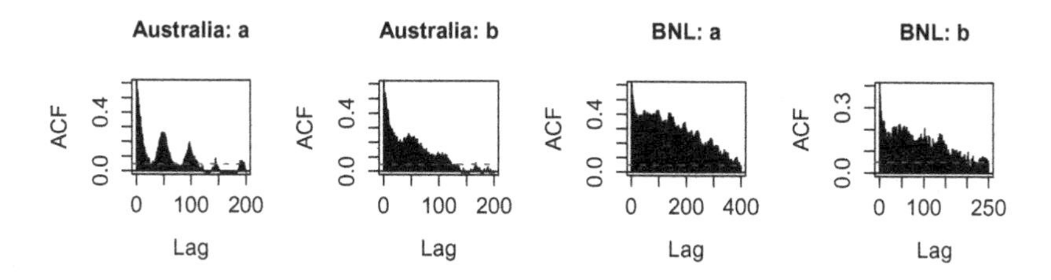

Fig. 9. ACF of normalized coefficients a and b

It is evident from the figure that all the time series have a seasonal component with a period of 48 points, which is equivalent to a day. To fit ARIMA models we use the *auto.arima()* function from R, which will estimate the model parameters based on a number of statistical tests (KPSS, OCSB and others). When passing a time series object to the function, we indicate the discovered seasonality.

Long Short-Term Memory neural networks are a kind of recurrent neural networks. Employing a gating architecture and a recurrent state, LSTM nets are optimized to partially overcome the vanishing gradient problem and learn from arbitrarily extended lags. They have been used for time series forecasting by researchers in different domains, e.g. for traffic speed prediction, detection of faults in industrial data, etc. It has been also demonstrated at the recent time series forecasting competition CIF 2016 [16] that LSTM has delivered outstanding results, outperforming common statistical and soft computing methods. In contrast to ARIMA, LSTM models can learn more families of functions.

We use the PyTorch library to construct LSTM models. At each timestep the input is 1- and the output is 19-dimensional. Prior to the training the data is projected onto the $[-1; 1]$ interval with the goal to achieve stronger gradients. The models are trained using the mean squared error (MSE) loss, since no further normalization is required. We use the Adam algorithm for the optimization of the model parameters. The first 3 datasets from each of the data centers (6 in total) are used to perform a grid search and determine the optimal values for hyperparameters based on the validation set approach. The number of considered hidden layers varied from 1 to 3, the number of training epochs from 1 to 2000 and the learning rate from 1e−4 to 1e−2. A dropout layer with probability 0.5 is introduced between the hidden layers for regularization. Based on the results of the grid search, the final model has one hidden layer, and is trained for 1827 epochs (which corresponds to early stopping regularization) with a learning rate of 1e−2. Figure 10 shows the validation loss for different combinations of epochs and hidden layers, given the optimal learning rate.

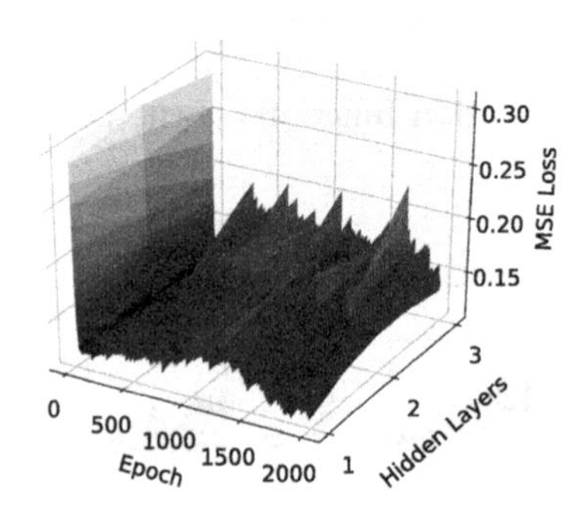

Fig. 10. Grid search validation loss

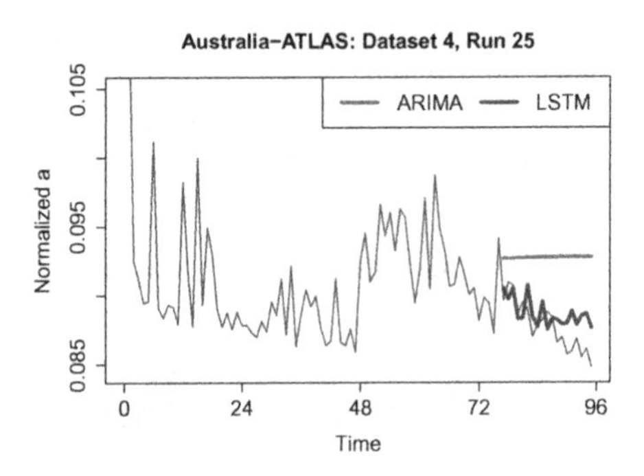

Fig. 11. LSTM outperforming ARIMA

The remaining 13 datasets from each of the data centers (26 in total) are used to compare the performance of ARIMA and LSTM. Given the input time series, the *auto.arima()* function returns a deterministically computed ARIMA model. In contrast, LSTM instances will differ from run to run due to the stochastic aspects of their training. To facilitate an extensive comparison of the performance of both methods, we train 50 LSTM instances on each dataset. The scale of the LSTM forecasts is reversed from [−1; 1] to the original one. Normalized root mean squared error (NRMSE) is used to compare loss among all datasets and forecasting methods. For each of the datasets the corresponding LSTM training runs are considered, and the log changes between ARIMA and LSTM NRMSE (holding ARIMA NRMSE as reference) are computed, resulting into a distribution. Our findings show that given a dataset, one approach will systematically outperform the other. Consider Figs. 11 and 14, which show concrete examples of LSTM outperforming ARIMA(3, 0, 2) with non-zero mean and ARIMA(0, 1, 1)(0, 0, 1) [48] with drift outperforming LSTM.

We have performed the augmented Dickey-Fuller test implying stationarity under the alternative hypothesis on each dataset (on its fraction used for model construction). We have then tested for an association between the ADF test statistic and the type of the forecasting method, which dominates in terms of performance given the dataset. Unfortunately, this analysis did not deliver significant results. Thus, the method, which is more beneficial on average needs to be selected. Figure 12 shows boxplots of all measured log NRMSE changes for the two separate data centers. The median log NRMSE changes are 0.27 and 0.14 for Australia-ATLAS and BNL-ATLAS data centers respectively. According to these results, ARIMA models exhibit better performance. We attribute

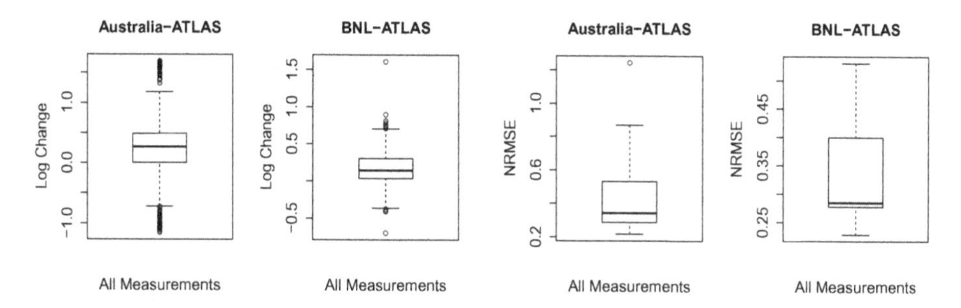

Fig. 12. Log NRMSE changes

Fig. 13. Performance of ARIMA

this to the fact that low-capacity models are able to robustly fit high-variance data and adopt ARIMA for our throughput forecasting model. The boxplots of ARIMA NRMSE across all datasets are shown in Fig. 13.

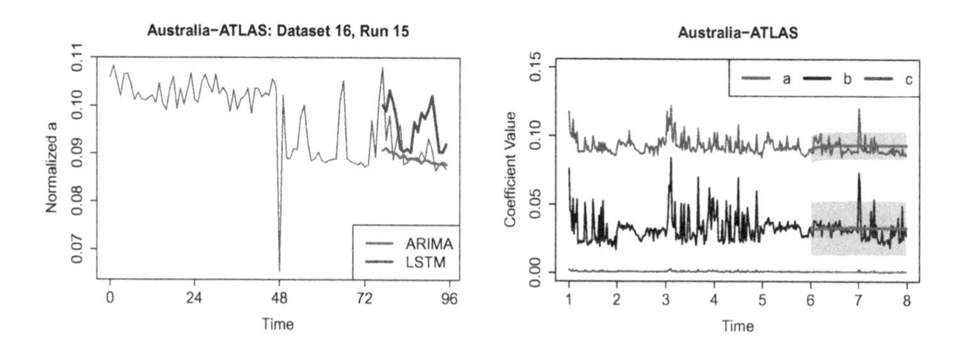

Fig. 14. ARIMA outperforming LSTM

Fig. 15. Long-term forecasts

The forecasting model delivers short- and long-term predictions. Figure 15 demonstrates multivariate 2 days forecasts. The mean forecasts are depicted in red and their 95% confidence intervals in purple. Note that the confidence intervals miss only few outliers. Figure 14 demonstrates that by increasing the time resolution, the forecasts become more accurate in the short-term.

5 Conclusions and Future Work

In this paper we have demonstrated that the network throughput of remote data access in computing grids can be framed as a multiple linear regression. The regression needs to be fit for each worker node - storage element pair. The estimates of the regression coefficients can be mined from logs in form of time series. This requires realistically obtainable logs on file accesses from only 2 concurrent processes, one of them having 2 concurrent threads. Given a proper implementation of remote data access monitoring, these time series can be mined unintrusively. We have identified useful input features for the forecasting model. The ARIMA model learns to explain the residual variance in the response induced by the latent factors, e.g. changes in the grid infrastructure. A benchmark of LSTM ANN has partially shown good results, but did not significantly outperform ARIMA. The model can be constructed with a predefined resolution and deliver short- and long-term forecasts with the state of the art accuracy.

These findings allow to robustly model the remote data access throughput for any virtual links in the grid. The model was constructed and validated using performance metrics from randomly selected data centers in the World-Wide LHC Computing Grid.

Currently we are focusing on the following aspects for the future work:

- Formalization and forecasting of network throughput of other data access techniques in the grid (data placement and stage-in).
- Construction of a grid simulator with a heavy emphasis on various data access techniques.
- Evolutionary optimization of data access patterns within bags of jobs with the objective to minimize the joint data transfer time. This constitutes a constrained optimization problem. The solution fitness evaluation employs the aforementioned simulator.

References

1. Akioka, S., Muraoka, Y.: Extended forecast of CPU and network load on computational grid. In: IEEE International Symposium on Cluster Computing and the Grid. CCGrid 2004, pp. 765–772 (2004). https://doi.org/10.1109/CCGrid.2004.1336711
2. Arslan, E., Guner, K., Kosar, T.: Harp: predictive transfer optimization based on historical analysis and real-time probing. In: Proceedings of the International Conference for High Performance Computing, Networking, Storage and Analysis. SC 2016, pp. 25:1–25:12. IEEE Press, Piscataway (2016). http://dl.acm.org/citation.cfm?id=3014904.3014938
3. Chervenak, A., et al.: Data placement for scientific applications in distributed environments. In: Proceedings of the 8th IEEE/ACM International Conference on Grid Computing. GRID 2007, pp. 267–274. IEEE Computer Society, Washington (2007). https://doi.org/10.1109/GRID.2007.4354142
4. Faerman, M., Su, A., Wolski, R., Berman, F.: Adaptive performance prediction for distributed data-intensive applications. In: SC 1999: Proceedings of the 1999 ACM/IEEE Conference on Supercomputing, pp. 36–36 (1999). https://doi.org/10.1109/SC.1999.10048
5. Foster, I., Kesselman, C.: The Grid 2: Blueprint for a New Computing Infrastructure. The Elsevier Series in Grid Computing. Elsevier, Boston (2004)
6. Garonne, V., et al.: Rucio - the next generation of large scale distributed system for atlas data management. J. Phys. Conf. Ser. **513**(4), 042021 (2014). http://stacks.iop.org/1742-6596/513/i=4/a=042021

7. Hacker, T.J., Athey, B.D., Noble, B.: The end-to-end performance effects of parallel tcp sockets on a lossy wide-area network. In: Proceedings 16th International Parallel and Distributed Processing Symposium, pp. 10 pp- (2002). https://doi.org/10.1109/IPDPS.2002.1015527
8. Hochreiter, S., Schmidhuber, J.: Long short-term memory. Neural Comput. **9**(8), 1735–1780 (1997). https://doi.org/10.1162/neco.1997.9.8.1735
9. Kim, J., Yildirim, E., Kosar, T.: A highly-accurate and low-overhead prediction model for transfer throughput optimization. Cluster Comput. **18**(1), 41–59 (2015). https://doi.org/10.1007/s10586-013-0305-4
10. Liu, Z., Balaprakash, P., Kettimuthu, R., Foster, I.: Explaining wide area data transfer performance. In: Proceedings of the 26th International Symposium on High-Performance Parallel and Distributed Computing. HPDC 2017, pp. 167–178. ACM, New York (2017). https://doi.org/10.1145/3078597.3078605
11. Lu, D., Qiao, Y., Dinda, P.A., Bustamante, F.E.: Characterizing and predicting TCP throughput on the wide area network. In: 25th IEEE International Conference on Distributed Computing Systems (ICDCS 2005), pp. 414–424 (2005). https://doi.org/10.1109/ICDCS.2005.17
12. Mirza, M., Sommers, J., Barford, P., Zhu, X.: A machine learning approach to tcp throughput prediction. IEEE/ACM Trans. Netw. **18**(4), 1026–1039 (2010). https://doi.org/10.1109/TNET.2009.2037812
13. Nine, M.S.Q.Z., Guner, K., Kosar, T.: Hysteresis-based optimization of data transfer throughput. In: Proceedings of the Fifth International Workshop on Network-Aware Data Management. NDM 2015, pp. 5:1–5:9. ACM, New York (2015). https://doi.org/10.1145/2832099.2832104
14. Rygielski, P., Kounev, S.: Data center network throughput analysis using queueing petri nets. In: 2014 IEEE 34th International Conference on Distributed Computing Systems Workshops (ICDCSW), vol. 00, pp. 100–105 (2014). https://doi.org/10.1109/ICDCSW.2014.11
15. Shumway, R., Stoffer, D.: Time Series Analysis and Its Applications: With R Examples. Springer Texts in Statistics. Springer, New York (2010)
16. Štěpnička, M., Burda, M.: On the results and observations of the time series forecasting competition CIF 2016. In: 2017 IEEE International Conference on Fuzzy Systems (FUZZ-IEEE), pp. 1–6 (2017). https://doi.org/10.1109/FUZZ-IEEE.2017.8015455
17. Vazhkudai, S., Schopf, J.M.: Predicting sporadic grid data transfers. In: Proceedings 11th IEEE International Symposium on High Performance Distributed Computing, pp. 188–196 (2002). https://doi.org/10.1109/HPDC.2002.1029918
18. Wolski, R.: Experiences with predicting resource performance on-line in computational grid settings. SIGMETRICS Perform. Eval. Rev. **30**(4), 41–49 (2003). https://doi.org/10.1145/773056.773064

Accurately Simulating Energy Consumption of I/O-Intensive Scientific Workflows

Rafael Ferreira da Silva[1(✉)], Anne-Cécile Orgerie[2], Henri Casanova[3], Ryan Tanaka[3], Ewa Deelman[1], and Frédéric Suter[4]

[1] USC Information Sciences Institute, Marina del Rey, CA, USA
{rafsilva,deelman}@isi.edu
[2] Univ Rennes, Inria, CNRS, IRISA, Rennes, France
anne-cecile.orgerie@irisa.fr
[3] Information and Computer Sciences, University of Hawaii, Honolulu, HI, USA
{henric,ryanyt}@hawaii.edu
[4] IN2P3 Computing Center, CNRS, Villeurbanne, France
frederic.suter@cc.in2p3.fr

Abstract. While distributed computing infrastructures can provide infrastructure-level techniques for managing energy consumption, application-level energy consumption models have also been developed to support energy-efficient scheduling and resource provisioning algorithms. In this work, we analyze the accuracy of a widely-used application-level model that have been developed and used in the context of scientific workflow executions. To this end, we profile two production scientific workflows on a distributed platform instrumented with power meters. We then conduct an analysis of power and energy consumption measurements. This analysis shows that power consumption is not linearly related to CPU utilization and that I/O operations significantly impact power, and thus energy, consumption. We then propose a power consumption model that accounts for I/O operations, including the impact of waiting for these operations to complete, and for concurrent task executions on multi-socket, multi-core compute nodes. We implement our proposed model as part of a simulator that allows us to draw direct comparisons between real-world and modeled power and energy consumption. We find that our model has high accuracy when compared to real-world executions. Furthermore, our model improves accuracy by about two orders of magnitude when compared to the traditional models used in the energy-efficient workflow scheduling literature.

Keywords: Scientific workflows · Energy-aware computing · Workflow profiling · Workflow scheduling

1 Introduction

Computational workloads that require from a few hours to a few months of execution are commonplace in scientific simulations. These simulations often

J. M. F. Rodrigues et al. (Eds.): ICCS 2019, LNCS 11536, pp. 138–152, 2019.
https://doi.org/10.1007/978-3-030-22734-0_11

comprise individual computational (but often I/O-intensive) tasks with some dependency structure, which is why many scientists today formulate their computational problems as scientific workflows [26]. To obtain simulation results within acceptable time-frames, large scientific workloads are executed on distributed computing infrastructures such as grids and clouds [21]. The need to manage energy consumption across the entire suite of information and communication technology has received significant attention in the last few years [1,3]. As a result, large data-centers have developed techniques for managing cooling and energy usage at the infrastructure level. Concurrently, researchers have investigated application-level techniques and algorithms to enable energy-efficient executions [19]. In the context of scientific workflows, researchers have proposed a range of energy-aware workflow task scheduling or resource provisioning algorithms [8,17,20,23,30]. Results therein are obtained based on a model of power consumption that is easy to instantiate but that makes strong assumptions: power consumption is considered to be linearly correlated with CPU utilization, and equally divided among virtual machines or CPU cores within a computational node. An interesting question is whether this model is accurate in practice, and whether it can be applied to I/O-intensive workflow executions.

Our broad objective in this work is to characterize the energy consumption behavior of complex workflow applications that execute on distributed platforms. We profile real scientific workflow applications on a distributed platform that comprises multi-socket, multi-core compute nodes equipped with power meters. We select two widely scientific workflows, each of which has some I/O-intensive tasks, for which we conduct a comprehensive analysis of the power and energy consumption of their execution. Via this analysis, we quantify the accuracy, or lack thereof, of power consumption models commonly used in the energy-efficient workflow scheduling literature. We then propose a more accurate power consumption model. More specifically, this work makes the following contributions:

1. The power and energy consumption profiles of two real I/O-intensive scientific workflow applications;
2. A comprehensive analysis of these profiles with respect to resource utilization and I/O operations;
3. An evaluation of the accuracy of the power model that is widely used in workflow scheduling research;
4. A power consumption model for I/O-intensive workflows that accounts for the allocation of cores to sockets, CPU utilization, and I/O operations;
5. An experimental evaluation of the proposed model that shows that it can produce nearly accurate energy consumption estimates, with improvements over traditional models by almost two orders of magnitude.

2 Workflow Characterization

The analysis presented in this work is based on the execution of two production scientific workflow applications on the Grid'5000 [2] platform. Grid'5000 is a testbed for experiment-driven research, which provides resource isolation and advanced monitoring and measurement features for the collection of power consumption traces. We consider these two I/O-intensive workflows:

- *Epigenomics* [12]: A bioinformatics workflow that maps the epigenetic state of human cells on a genome-wide scale by processing multiple sets of genome sequences in parallel. We consider an Epigenomics instance with 577 tasks.
- *SoyKB* [11]: A bioinformatics workflow that re-sequences soybean germplasm lines selected for desirable traits such as oil, protein, soybean cyst nematode resistance, stress resistance, and root system architecture. We consider a SoyKB instance with 676 tasks.

We profiled these workflows when executed with Pegasus [5], a state-of-the-art workflow management system. Pegasus monitors and logs fine-grained profiling data such as I/O operations, runtime, memory usage, and CPU utilization [13].

The workflows were executed on the *taurus* cluster at the Grid'5000 Lyon site, which is instrumented at the node level with power meters. We used a single node to run the workflow tasks and collect power measurements (although not efficient, it allowed us to collect non-biased measurements). Each node is equipped with two 2.3 GHz hexacore Intel Xeon E5-2630 CPUs, 32GB of RAM, and standard magnetic hard drives. Power measurements are collected every second from power meters (with an accuracy of 0.125 Watts[1]) that are connected to a data collector via a serial link. We are interested in identifying relationships between power consumption, task duration, CPU utilization, and volume of I/O. Detailed execution profiles (but without power/energy data) and performance analysis for both workflows can be found in [25].

Table 1 shows the execution profiles of Epigenomics and SoyKB tasks, with one row per task type. Since most Epigenomics tasks require 1 CPU core, power measurements were collected from a resource where only a single core was enabled (i.e., only 1 CPU slot is advertised by the resource manager). Only the `pileup` task requires 2 cores, but there is only one such task in the workflow. For SoyKB, many tasks require 2 CPU cores. Therefore, we collected power measurements from a resource configured with two cores. The last two columns in Table 1 show the average power consumption per task and the energy consumption to compute all tasks of that type in sequence. As power measurement were collected every second, tasks with very short runtimes (e.g., `sol2sanger` in the Epigenomics workflow) may not allow accurate power measurements, and are not emphasized in our upcoming analyses.

3 Workflow Energy Consumption Analysis

Energy-aware workflow scheduling studies [8, 17, 20, 23, 30] typically assume that the power consumed by the execution of a task at time t, $P(t)$, is linearly related to the task's CPU utilization, $u(t)$, as:

$$P(t) = (P_{\max} - P_{\min}) \cdot u(t) \cdot \frac{1}{n}, \tag{1}$$

[1] Manufactured by OMEGAWATT: http://www.omegawatt.fr/gb/index.php.

Table 1. Execution and energy profiles of the Epigenomics (top) and SoyKB (bottom) workflow tasks. Energy measurements are for running all tasks of that type in sequence. Runtimes are shown in seconds, I/O operations in MB, and power in W. (μ is the mean, and σ the standard deviation.)

Task	Count	#cores	Runtime		CPU util.		I/O Read		I/O Write		Power		Energy
			μ	σ	μ	σ	μ	σ	μ	σ	μ	σ	(Wh)
fastqSplit	7	1	5.8	1.9	99.8%	0.0	508.1	173.2	254.1	86.6	126.9	5.48	1.4
filterContams	140	1	1.2	0.2	99.1%	0.0	25.4	3.7	12.7	1.8	100.9	5.6	4.6
sol2sanger	140	1	0.4	0.1	95.7%	0.2	66.9	9.8	29.0	4.3	98.5	3.8	1.4
fast2bfq	140	1	0.8	0.1	97.8%	0.1	35.5	5.2	6.4	0.9	98.3	3.6	2.9
map	140	1	57.9	5.0	99.9%	0.0	437.9	2.4	2.6	0.6	126.8	0.9	285.7
mapMerge	8	1	5.9	6.9	99.5%	0.0	171.2	205.6	84.0	103.4	113.5	7.7	1.5
maqIndex	1	1	33.5	–	99.9%	–	511.7	–	338.3	–	125.1	–	1.2
pileup	1	2	38.4	–	80.8%	–	559.3	–	264.1	–	135.5	–	1.4
Task	Count	#cores	Runtime		CPU util.		I/O Read		I/O Write		Power		Energy
			μ	σ	μ	σ	μ	σ	μ	σ	μ	σ	(Wh)
align_reference	25	2	1.8	0.0	53.9%	0.0	2609.7	0.0	186.6	0.01	134.8	4.7	1.6
sort_sam	25	2	1.3	0.1	61.9%	0.0	901.5	0.0	187.2	1.6	101.7	1.9	0.9
dedup	25	2	2.0	0.0	60.7%	0.0	901.9	0.0	186.9	0.2	106.2	4.3	1.5
add_replace	25	2	1.3	0.0	62.0%	0.0	901.5	0.0	186.9	0.0	102.6	1.7	0.9
realign_creator	25	2	133.1	2.6	75.9%	0.0	3230.8	8.7	189.6	2.8	135.3	0.3	125.1
indel_realign	25	1	34.3	0.0	18.9%	0.0	953.8	5.8	187.0	0.0	123.2	0.6	25.9
haplotype_caller	500	1	79.3	6.9	66.7%	0.0	1149.8	24.2	186.9	0.0	130.8	1.0	1329.5
genotype_gvcfs	20	1	263.8	29.6	95.9%	0.0	1058.0	16.2	187.6	0.1	126.6	0.3	185.5
comb_variants	1	1	35.5	–	26.5%	–	958.0	–	186.9	–	108.9	–	1.1
variants_indel	1	2	48.6	–	23.7%	–	1699.5	–	454.4	–	114.0	–	1.5
filtering_indel	1	1	34.7	–	20.3%	–	955.2	–	186.9	–	109.1	–	1.0
variants_snp	1	2	48.6	–	23.2%	–	1699.5	–	454.4	–	115.4	–	1.5
filtering_snp	1	2	34.7	–	10.2%	–	955.3	–	186.9	–	109.6	–	1.0
merge_gcvf	1	1	46804.5	–	99.9%	–	3061.2	–	238.8	–	128.9	–	1675.3

where P_{max} is the power consumption when the compute node is at its maximum utilization, P_{min} is the idle power consumption (i.e., when there is no or only background activity), and n is the number of cores on the compute node. Therefore, the energy consumption of the task, E, is defined as follows:

$$E = r \cdot P_{\min} + \int_0^r P(t)\mathrm{d}t, \tag{2}$$

where r denotes the task's runtime. To determine the idle power consumption P_{min}, we collected power measurements on one node of our cluster at every second whenever no activity was performed on that node over a 2-month period (for a total of 216,000 measurements). The average idle power consumption from these measurements is 98.08W (standard deviation 1.77W).

The power model in Eq. 1 does not consider the energy consumption of I/O operations, and hereafter we quantify the extent to which this omission makes the model inaccurate. Figure 1 shows scatter plots of the power consumption versus CPU utilization for all task types of both workflows. We observe very low Pearson's correlation coefficient values between power consumption and CPU utilization (0.38 for Epigenomics, −0.02 for SoyKB). This means that no linear increase is observed in the power consumption as CPU utilization increases. For example, the `align_reference` SoyKB task has an average CPU utilization at about 108% and consumes about 135W, while the `sort_sam` task from that same workflow has a CPU utilization at about 124% but consumes only 102W. This difference in power consumption is mostly explained by volumes of I/O (reads and writes). Figure 2 shows scatter plots of the power consumption versus I/O read volumes per task and computational resource. In contrast to the CPU utilization analysis, Pearson's correlation coefficient values are 0.86 for Epigenomics, and 0.64 for SoyKB.

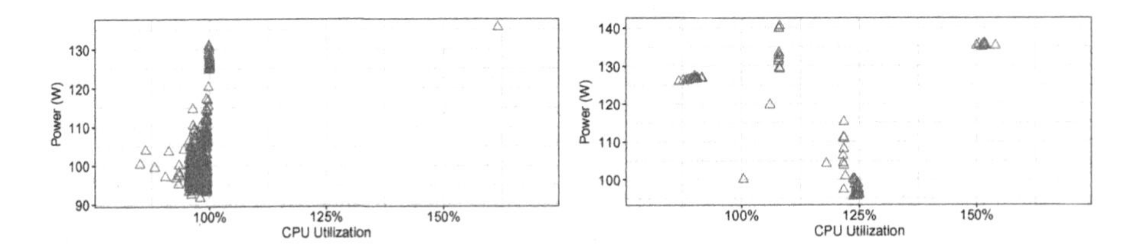

Fig. 1. Task power consumption vs. CPU utilization for the Epigenomics (left) and SoyKB (right) workflows.

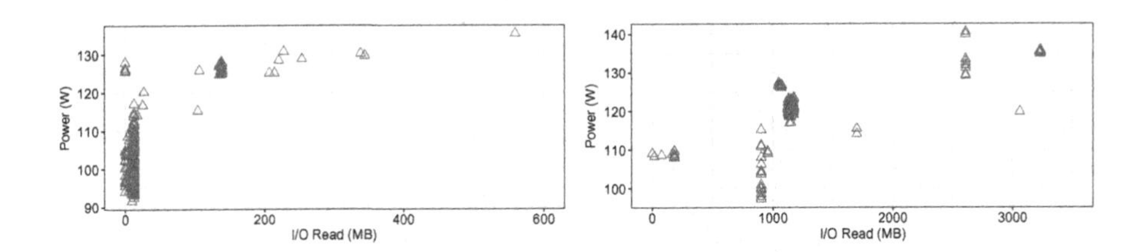

Fig. 2. Task power consumption vs. I/O read for the Epigenomics (left) and SoyKB (right) workflows.

These results show that power consumption is not strictly dependent, or even mainly influenced, by CPU utilization $u(t)$ (Eq. 1), but that it depends significantly on I/O volumes. Hence, we conduct a principal component analysis (PCA) to evaluate the variance of each parameter (CPU utilization, I/O reads, and I/O writes) and their impact on the power consumption. In this analysis, we aim to understand how CPU utilization and I/O operations are influencing (positively or negatively) power consumption, and consequently quantify the weight of each parameter. From the principal components, set of values of linearly uncorrelated variables, we obverse the *loadings* (the weight by which each standardized original variable should be multiplied to get the component score), which contain the data variance.

Table 2 shows the principal component (PC) loadings (rotations) for each parameter. For Epigenomics, the first two PCs explain most of the variability (85.3%). All parameters present similar variance for PC1, with the I/O read and I/O write parameters dominating, while CPU utilization has greater impact on PC2. Since PC1 explains most of the variance (64.3%), the power consumption of the Epigenomics workflow is also significantly impacted by the number of I/O operations (in particular I/O reads) as shown in Fig. 2-left. Similarly, the first two PCs for SoyKB explain most of the variability (85.4%). I/O read has greater impact on PC1, while PC2 is mostly impacted by CPU utilization and I/O write. Although I/O read has significant impact on PC1, this component only explains 49% of the variance, thus I/O read has less influence on the power consumption for SoyKB (Fig. 2-right). Note that the impact of I/O read on PC2 is minimal.

Overall, these results provide motivation and quantitative bases for developing a more accurate power model that captures the implications of I/O operations on power consumption in addition to that of CPU utilization.

Table 2. Principal component (PC) loadings (rotations) for the Epigenomics and SoyKB workflows.

Parameter	Epigenomics			SoyKB		
	PC1	PC2	PC3	PC1	PC2	PC3
CPU utilization	0.53	0.84	−0.03	−0.55	−0.62	0.56
I/O Read	0.59	−0.35	0.71	−0.73	0.04	−0.67
I/O Write	0.59	−0.40	−0.69	−0.39	0.78	0.47

4 Analysis of Power and Energy Consumption for Concurrent Task Execution

The power consumption model in Eq. 1 assumes that the consumed power is simply the CPU utilization divided by the number of cores. To evaluate the validity of this assumption we collected and analyzed power measurements for solitary and concurrent workflow task executions.

Since our cluster nodes are all equipped with dual, hexacore CPUs, we performed task executions with two schemes for core allocation (see Fig. 3): (1) *unpaired*—cores are enabled in sequence on a single socket until all cores on that socket are enabled, and then cores on the next socket are enabled in sequence; and (2) *pairwise*—cores are enabled in round-robin fashion across sockets (i.e., each core is enabled on a different socket than the previously enabled core). We report on results for only a subset of workflow tasks because (1) some tasks are unique; (2) some task runtimes are very short and overheads in Pegasus, such as releasing the next task, make the benefit of running these tasks in parallel negligible; or (3) energy measurements may not be accurate for tasks with very short runtimes due to the measurements interval of 1s. Finally, all our results report average runtime, power and energy measurements for concurrent executions of instances of the same task type.

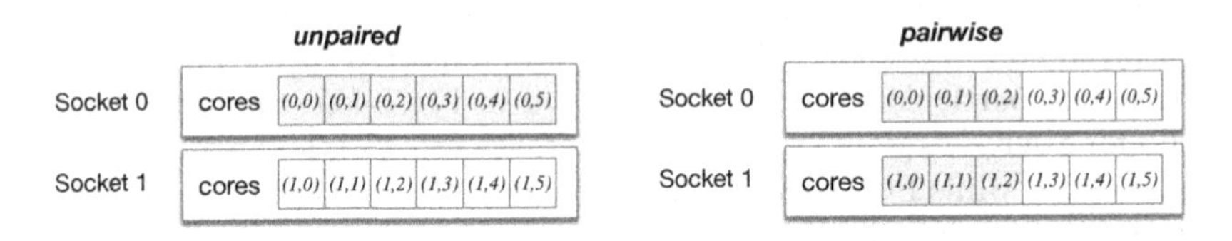

Fig. 3. Example of CPU core usage for the *unpaired* (left) and *parwise* (right) schemes when 6 cores are enabled.

Epigenomics: Figure 4 shows the average task runtime, average task power consumption, and total energy consumption (i.e., to run all 140 tasks) when running `map` tasks concurrently for different numbers of CPU cores. Task performance is significantly impacted when multiple cores are used within a single socket. For example, when 2 cores are enabled in different sockets (*pairwise*), no performance decrease is observed. However, a performance degradation of about 25% occurs when both cores are within a single socket (*unpaired*). The above is due to the fact that each socket has a single L3 cache shared between its cores.

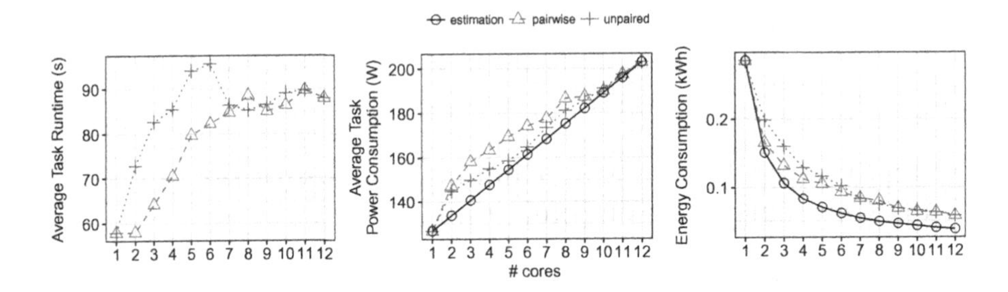

Fig. 4. Average task runtime (left), average task power consumption (center), and energy consumption to run all 140 `map` tasks from Epigenomics (right). Power and energy consumption computed using Eqs. 1 and 2 are shown as `estimation`.

While the use of multiple cores within a single socket limits performance, it consumes less power per unit of time: on the order of 10% (Fig. 4-center). According to Eq. 1, power consumption should grow linearly. Instead, we observe that power consumption is not equally divided among the number of cores per CPU. Equation 1 thus underestimates the energy usage per unit of time—root mean squared error (RMSE) for *pairwise* is 10.64, and 4.92 for *unpaired.*

The energy profile shown in Fig. 4-right accounts for the execution of all 140 `map` tasks. Although power consumption is lower when using a single socket, the total energy consumption is higher due to higher task runtimes. Workflow task executions may benefit from single socket CPU usage if task runtimes are very short. In this case, the performance loss is negligible and the difference of power consumption may save energy (e.g., the `filterContams` task in Epigenomics). The energy consumption for the set of `map` tasks presents a logarithmic decrease as a function of the number of cores. This logarithmic behavior is due to the increase in power consumption. The estimation errors propagated by Eq. 1 into Eq. 2 leads to energy consumption estimation errors up to 23% (RMSEs are 0.02 for *pairwise* and 0.03 for *unpaired*).

SoyKB: Figure 5 shows the average task runtime, the average task power consumption, and total energy consumption (i.e., to run all 500 tasks) when running `haplotype_caller` tasks concurrently using 2 up to 8 CPU cores. Due to disk space quota on Grid'5000, we were unable to run workflow instances that used more than 8 cores concurrently. We only report on results for more than 2 cores because the workflow cannot be executed on a single core. Task runtime differences between *unpaired* and *pairwise* is minimal regardless the number of cores used. A small degradation in runtime is observed when the number of cores increase from 2 to 4. However, there is a significant performance decrease when the number of cores exceeds 4. This is because `haplotype_caller` performs substantial I/O operations (it only has 67% of CPU utilization on average). The performance degradation is due to simultaneous I/O operations, which cause tasks to idle due to I/O resources being unavailable and/or saturated. This idle time (IOWait) is reported in the logs generated by Pegasus.

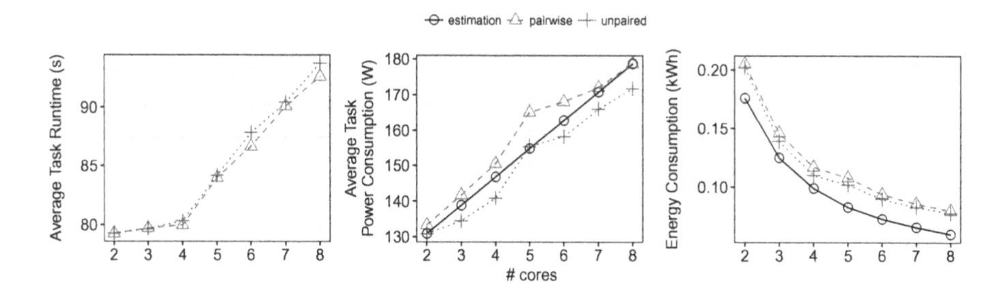

Fig. 5. Average task runtime (left), average power consumption (center), and energy consumption to run all 500 `haplotype_caller` tasks from SoyKB (right). Power and energy consumption computed using Eqs. 1 and 2 are shown as `estimation`.

Similar to Epigenomics, the *unpaired* scheme consumes slightly less power (about 5%, as see in Fig. 5-center. The power consumption estimated by Eq. 1 lies between the real-world consumption with the two schemes, with prediction errors up to 10% (RMSE up to 4.85 for *pairwise*). In Fig. 5-right, we see that the actual energy values are well above the estimated values (up to 22% higher). The main factor for this discrepancy is I/O operations, including the time spent waiting for I/O to complete (as indicated by IOWait values in the Pegasus logs).

5 Modeling and Simulating Energy Consumption of I/O-Intensive Workflows

In this section, we present an augmented model for power consumption that accounts for I/O in addition to CPU utilization. This model also accounts for the number of cores and the way in which they are activated (*unpaired* or *pairwise* schemes), as well as for the time spent waiting for I/O operations to complete (IOWait).

5.1 Model

We model $P(t)$, the power consumption of a compute node at time t, as:

$$P(t) = P_{\text{CPU}}(t) + P_{\text{I/O}}(t), \tag{3}$$

where $P_{\text{CPU}}(t)$, resp. $P_{\text{I/O}}(t)$, is the power consumption due to CPU utilization, resp. I/O operations. In what follows, we detail the model for both these terms.

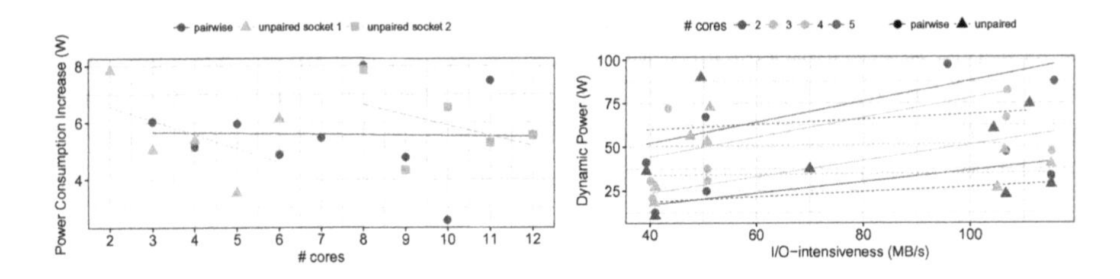

Fig. 6. Linear regression models. *Left:* power consumption increase in function of number of cores enabled for the Epigenomics `map` tasks. *Right:* dynamic power consumption vs. I/O-intensiveness for the SoyKB `realign_creator`, `indel_realign`, `haplotype_caller`, and `genotype_gvcfs` tasks.

CPU – Let s denote the number of sockets on the compute node, and n the number of cores per socket, so that the total number of cores on the compute node is $s \cdot n$. Let K the set of tasks that use at least one core on the compute node. We have:

$$P_{\text{CPU}}(t) = \textstyle\sum_{k,i,j} P_{\text{CPU}}(k,i,j,t), \tag{4}$$

where $P_{\text{CPU}}(k, i, j, t)$ is the power consumption of CPU utilization at time t due to the execution of task k ($k \in K$) on socket i ($0 \leq i < s$) at core j ($0 \leq j < n$).

In previous sections, we examined the impact of cores/socket allocation on power consumption in addition to CPU utilization. We have seen that the power consumption does not have constant increase as cores are enabled on sockets, and the behavior depends on the scheme used to enable further cores (*pairwise* or *unpaired*). Figure 6-*left* shows a scatter plot of power consumption increase for each additional enabled core for the `map` task of the Epigenomics workflow. The increase for the *unpaired* scheme can be approximated by linear regression with negative slope. For the *pairwise* scheme, an approximation by linear regression leads to nearly constant increase (noting that the RMSE is relatively high). Although this figure is for a particular task of the Epigenomics workflow, very similar results are obtained for all tasks for both production workflows considered in this work. Therefore, we derive a model that is only dependent on the task's CPU utilization and the hardware platform.

Based on the above, we now define our model for $P_{\mathrm{CPU}}(k, i, j, t)$ as:

$$P_{\mathrm{CPU}}(k,i,j,t) = \begin{cases} (P_{\max} - P_{\min}) \cdot \frac{u(t)}{s \cdot n} & \text{if } j = 0 \text{ (first core on a socket)} \\ 0.881 \cdot P_{\mathrm{CPU}}(k,i,j-1,t) & \text{if } j > 0 \text{ and } \textit{pairwise} \\ 0.900 \cdot P_{\mathrm{CPU}}(k,i,j-1,t) & \text{if } j > 0 \text{ and } \textit{unpaired} \end{cases} \quad (5)$$

where $u(t)$ is the task's CPU utilization at time t (which can be computed by benchmarking the task on a dedicated compute node). The model is written recursively as the power consumption due to enabling a core on a socket depends on the power consumption due to previously enabled cores on that socket. The 0.881 and 0.900 constants above are obtained from the aforementioned linear regressions. Finally, note that $P_{\mathrm{CPU}}(k, i, j, t)$ does not depend on i since only the rank (j) of a core in a socket matters.

I/O – Similarly to the above model for power consumption due to CPU utilization, we have:

$$P_{\mathrm{I/O}}(t) = \textstyle\sum_{k,i,j} P_{\mathrm{I/O}}(k,i,j,t), \quad (6)$$

where $P_{\mathrm{I/O}}(k, i, j, t)$ is the power consumption of I/O operations at time t due to the execution of task k ($k \in K$) on socket i ($0 \leq i < s$) at core j ($0 \leq j < n$) on the compute node.

Figure 6-*right* shows dynamic power consumption (i.e., power consumption beyond $P_{\min}$) vs. I/O-intensiveness for 4 tasks of the SoyKB workflow (**`realign_creator`**, **`indel_realign`**, **`haplotype_caller`**, and **`genotype_gvcfs`**). We define the I/O-intensiveness as the I/O volume (for reads and writes) in MB divided by the time the task spends performing solely computation (i.e., the runtime minus the time for performing and waiting for I/O operations). A higher value indicates a more I/O-intensive task, as it represents I/O overhead per second of CPU usage. We are able to compute the I/O-intensiveness of each task based on profiling data in Pegasus logs. The four task types in Fig. 6 exhibit a range of CPU utilizations, with relatively high volumes of data read/written. As for the results in Fig. 6, similar results are obtained for all tasks in the workflows we consider. We use a linear regression, shown in the figure, which has positive slope regardless of the core allocation scheme (with a steeper slope for the *pairwise* scheme). Based on these results, we model $P_{\mathrm{I/O}}(k, i, j, t)$ as follows:

$$P_{\mathrm{I/O}}(k,i,j,t) = \begin{cases} 0.486 \cdot (1 + 0.317 \cdot \omega(t)) \cdot P_{\mathrm{CPU}}(k,i,j,t) & \text{if } \textit{pairwise} \\ 0.213 \cdot (1 + 0.317 \cdot \omega(t)) \cdot P_{\mathrm{CPU}}(k,i,j,t) & \text{otherwise} \end{cases} \quad (7)$$

where the 0.486 and 0.213 values above come from the linear regressions, and $\omega(t)$ is 0 if I/O resources are not saturated at time t, or 1 if they are (i.e., idle time due to IOWait). $\omega(t)$ is equal to 1 whenever the volume of I/O requests placed by concurrently running tasks exceeds some platform-dependent maximum I/O throughput. When using this model, e.g., to drive simulations of workflow task executions so as to evaluate energy-efficient workflow scheduling algorithms, it is then necessary to keep track of simulated I/O requests so as to set the $\omega(t)$ value accordingly. It turns out that, in our result, the impact of IOWait does not show any strong correlation with the features of different task types. This is why $\omega(t)$ in Eq. 7 is weighted by a single factor (0.317). We computed this factor as the average of the most accurate such factor values we computed individually for each task type. Our evaluation of the model (see Sect. 5.2) shows that it achieves high accuracy across task types. It is thus tempting to claim that the impact of the IOWait effect on power consumption can be captured reasonably well using a single, application-independent value for the above factor. Providing a definitive answer as to whether this claim is general would require a comprehensive set of experiments with more workflow applications running under this condition.

5.2 Experimental Evaluation

To evaluate the accuracy of our model, we extended a simulator [29] of the state-of-the-art Pegasus [5] workflow management system (WMS), which is the WMS we used to perform the experiments described in Sect. 2. This simulator is built using the WRENCH simulator framework [28], which can be used to build simulators of WMSs that are accurate, can run scalably on a single computer, and can be implemented with minimal software development effort [4]. We extended the simulator by replacing its simulation model for power consumption (the traditional model in Eq. 1) by the model proposed in Sect. 5.1. We provide the simulator with a description of the hardware specifications of the *taurus* Grid'5000 cluster and with traces from individual Epigenomics and SoyKB workflow task executions. As a result, our simulator can simulate the exact procedure used for obtaining all real-world experimental results described in previous sections, making it possible to draw direct comparisons between real-world and simulated results. The simulator code, details on the simulation calibration procedure, and experimental scenarios used in the rest of this section are all publicly available online [29].

Figure 7 shows the simulated power and energy consumption measurements as well as with the traditional model based on Eqs. 1 and 2 (shown as `estimation`) and with our proposed model (shown as `wrench-*`). Due to space constraints, we only show results for the `map` Epigenomics task, and the `haplotype_caller` and `indel_realign` SoyKB tasks. For the `map` tasks, the RMSE for *pairwise* is 4.24, and 3.49 for *unpaired*, which improves over the traditional model by about two orders of magnitude for the former, and a half for the later. Similarly, RMSEs for the `haplotype_caller` tasks are 2.86 and 2.07 for *pairwise* and *unpaired* respectively, or improvements of about two orders of magnitude for both schemes. Last, RMSEs for the `indel_realign` tasks are 0.59

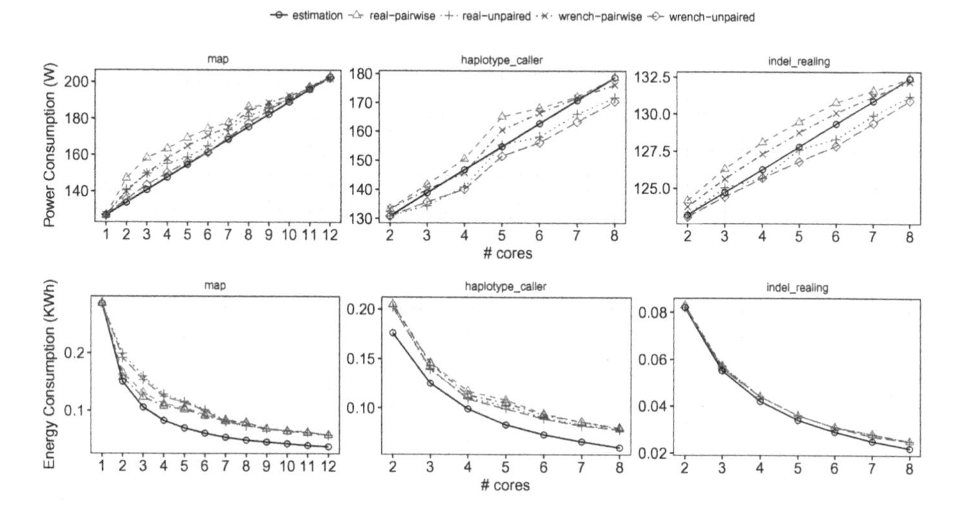

Fig. 7. Per-task power (top) and total energy (bottom) consumption measurements for the Epigenomics `map` task and the SoyKB `haplotype_caller` and `indel_realign`, as well as estimated with Eq. 1 and 2 (`estimation`) and our proposed model (`wrench-*`).

for *pairwise* and 0.47 for *unpaired*, or improvements by about an order of magnitude. Predicted energy consumption based on our proposed model nearly match the actual measurements for both schemes for all task types (RMSEs $\ll 0.01$).

6 Related Work

In the past few years, green computing has become a major topic of discussion in the scientific computing community. Many recent studies have addressed green solutions, in particular on distributed computing platforms. Research efforts in this field commonly include powering off or putting idle machines into low power states based on predictions of future workloads. On the application side, efforts are mainly focused on the optimization of resource provisioning and workload scheduling constrained by budgets and application deadlines.

A recent survey [19] of techniques for improving energy-efficiency describes methods to evaluate and model the energy consumed by resources on distributed systems. The survey presents taxonomies of compute node and network energy-aware techniques classified according to the technology employed. These techniques include adjustment of the processor's frequency and power consumption through DVFS [9], workload consolidation by running multiple tasks on the same physical machine in order to reduce the number of nodes that are powered on [16], energy-aware task scheduling [14,27], virtual machine migration [3,18], the coordination of network protocols [10], etc. These strategies often model energy consumption as a function of runtime, or do not consider the performance loss of running multiple tasks within a socket.

Several models have been developed to predict the power consumption of distributed system workloads. Most of them focus on measuring the resource utilization of distributed systems [6,15,22]. In [7], an integrated power consumption model, which incorporates previous approaches into a single model, describes a distributed system where several clients issue requests to a central storage server. Most of these models are limited to single-core and energy consumption is related to CPU usage. A few models consider data transfers, but as a separate operation (I/O operations during task execution are not considered).

In the context of scientific workflows, several works [8,17,20,23,30] have proposed energy-aware algorithms for task scheduling or resource provisioning. These algorithms are often designed to meet energy budget or deadline constraints. Their model assumes that the total energy usage is equal to the integral of the consumed power, which is linearly related to the resource utilization. In this work, we have shown that I/O operations also have significant impact on the power consumption, and thereby the energy. To the best of our knowledge, this is the first work that profiles and analyzes power and energy consumption of real scientific workflow applications at a fine-grained level, and proposes a model that also accounts for cores/sockets allocation and I/O usage.

7 Conclusion and Future Work

In this work, we have profiled and analyzed the power consumption of two production scientific workflow applications executed on a distributed platform. We have investigated the impact of resource utilization and I/O operations on the energy usage, as well as the impact of executing multiple tasks concurrently on multi-socket, multi-core compute nodes. In contrast to traditional power consumption model used in the energy-efficient workflow scheduling literature, we find that power consumption is impacted non-linearly by the way in which cores in sockets are allocated to workflow tasks. Furthermore, our experimental results show that I/O operations have significant impact on power consumption. Based on these results, we proposed a power model for I/O intensive workflows that accounts for the above behaviors. Experimental evaluation of this model shows that it accurately captures real-world behavior, with order of magnitude improvement over the traditional model.

In future work, we plan to instantiate and validate our proposed model for other workflows and platform configurations. In particular, we hope to use power-metered platforms in which compute nodes have SSDs instead of HDDs. With SSDs, the impact of I/O on power consumption may exhibit different behaviors that could mandate augmenting our model. The power consumption of I/O could also be smaller relative to that of computation, but note that platforms that target extreme-scale computing also often employ low-power compute nodes (i.e., equipped with ARM processors). Another future work goal is to extend the synthetic workflow generator in [24], which produces realistic synthetic workflow configurations based on profiles extracted from workflow execution traces. The objective is to extend the generated workflow descriptions to include data obtained from real-world power profiles that is sufficient to instantiate the power consumption model proposed in this work.

Acknowledgments. This work is funded by NSF contracts #1642369 and #1642335, "SI2-SSE: WRENCH: A Simulation Workbench for Scientific Workflow Users, Developers, and Researchers", and CNRS under grant #PICS07239; and partly funded by NSF contract #1664162, and DOE contract #DE-SC0012636.

References

1. Baliga, J., et al.: Green cloud computing: balancing energy in processing, storage, and transport. Proc. IEEE **99**(1), 149–167 (2011)
2. Balouek, D., et al.: Adding virtualization capabilities to the grid'5000 testbed. In: Ivanov, I.I., van Sinderen, M., Leymann, F., Shan, T. (eds.) CLOSER 2012. CCIS, pp. 3–20. Springer, Cham (2013). https://doi.org/10.1007/978-3-319-04519-1_1
3. Beloglazov, A., et al.: A taxonomy and survey of energy-efficient data centers and cloud computing systems. In: Advances in Computers, vol. 82 no. (2) (2011)
4. Casanova, H., et al.: WRENCH: a framework for simulating workflow management systems. In: 13th Workshop on Workflows in Support of Large-Scale Science (WORKS 2018) (2018)
5. Deelman, E., et al.: Pegasus, a workflow management system for science automation. Future Gener. Comput. Syst. **46**, 17–35 (2015)
6. Enokido, T., Takizawa, M.: An extended power consumption model for distributed applications. In: IEEE International Conference on Advanced Information Networking and Applications (AINA), pp. 912–919 (2012)
7. Enokido, T., Takizawa, M.: An integrated power consumption model for distributed systems. IEEE Trans. Ind. Electron. **60**(2), 824–836 (2013)
8. Ghose, M., et al.: Energy efficient scheduling of scientific workflows in cloud environment. In: IEEE International Conference on High Performance Computing and Communications (HPCC), pp. 170–177 (2017)
9. Guérout, T., et al.: Energy-aware simulation with DVFS. Simul. Model. Pract. Theory **39**, 76–91 (2013)
10. He, S., et al.: Energy-efficient capture of stochastic events under periodic network coverage and coordinated sleep. IEEE Trans. Parallel Distrib. Syst. **23**(6), 1090–1102 (2012)
11. Joshi, T., et al.: Next generation resequencing of soybean germplasm for trait discovery on XSEDE using pegasus workflows and iPlant infrastructure. In: XSEDE (2014). poster
12. Juve, G., et al.: Characterizing and profiling scientific workflows. Future Gener. Comput. Syst. **29**(3), 682–692 (2013)
13. Juve, G., et al.: Practical resource monitoring for robust high throughput computing. In: Workshop on Monitoring and Analysis for High Performance Computing Systems Plus Applications (HPCMASPA 2015), pp. 650–657 (2015)
14. Kliazovich, D., Bouvry, P., Khan, S.U.: Dens: data center energy-efficient network-aware scheduling. Cluster Comput. **75**(1), 65–65 (2013)
15. Lee, Y.C., Zomaya, A.Y.: Energy efficient utilization of resources in cloud computing systems. J. Supercomput. **60**(2), 268–280 (2012)
16. Lefevre, L., Orgerie, A.C.: Towards energy aware reservation infrastructure for large-scale experimental distributed systems. Parallel Process. Lett. **19**(03), 419–433 (2009)

17. Li, Z., et al.: Cost and energy aware scheduling algorithm for scientific workflows with deadline constraint in clouds. IEEE Trans. Serv. Comput. **11**(4), 713–726 (2018)
18. Liu, H., et al.: Performance and energy modeling for live migration of virtual machines. Cluster Comput. **16**(2), 171–182 (2013)
19. Orgerie, A.C., et al.: A survey on techniques for improving the energy efficiency of large-scale distributed systems. ACM Comput. Surv. (CSUR) **46**(4), 47:1–47:31 (2014)
20. Pietri, I., Sakellariou, R.: Energy-aware workflow scheduling using frequency scaling. In: International Conference on Parallel Processing Workshops (ICCPW) (2014)
21. Romanus, M., et al.: The anatomy of successful ECSS projects: lessons of supporting high-throughput high-performance ensembles on XSEDE. In: XSEDE, pp. 1–9 (2012)
22. Samak, T., et al.: Energy consumption models and predictions for large-scale systems. In: International Parallel and Distributed Processing Symposium Workshops (IPDPSW), pp. 899–906 (2013)
23. Shepherd, D., et al.: Workflow scheduling on power constrained VMS. In: IEEE/ACM 8th International Conference on Utility and Cloud Computing (2015)
24. Ferreira da Silva, R., et al.: Community resources for enabling and evaluating research on scientific workflows. In: IEEE International Conference on e-Science. eScience 2014 (2014)
25. Ferreira da Silva, R., et al.: Online task resource consumption prediction for scientific workflows. Parallel Process. Lett. **25**(3), 1541003 (2015)
26. Taylor, I., et al.: Workflows for e-Science. Springer, London (2007). https://doi.org/10.1007/978-1-84628-757-2
27. Wang, L., et al.: Energy-aware parallel task scheduling in a cluster. Future Gener. Comput. Syst. **29**(7), 1661–1670 (2013)
28. The WRENCH Project (2019). http://wrench-project.org/
29. WRENCH Pegasus Simulator (2019). https://github.com/wrench-project/pegasus
30. Wu, T., et al.: Soft error-aware energy-efficient task scheduling for workflow applications in DVFS-enabled cloud. J. Syst. Architect. **84**, 12–27 (2018)

Exploratory Visual Analysis of Anomalous Runtime Behavior in Streaming High Performance Computing Applications

Cong Xie[1], Wonyong Jeong[1], Gyorgy Matyasfalvi[2], Hubertus Van Dam[2], Klaus Mueller[1,2], Shinjae Yoo[2], and Wei Xu[2](✉)

[1] Stony Brook University, Stony Brook, NY 11790, USA
[2] Brookhaven National Laboratory, Upton, NY 11973, USA
xuw@bnl.gov

Abstract. Online analysis of runtime behavior is essential for performance tuning in streaming scientific workflows. Integration of anomaly detection and visualization is necessary to support human-centered analysis, such as verification of candidate anomalies utilizing domain knowledge. In this work, we propose an efficient and scalable visual analytics system for online performance analysis of scientific workflows toward the exascale scenario. Our approach uses a call stack tree representation to encode the structural and temporal information of the function executions. Based on the call stack tree features (e.g., execution time of the root function or vector representation of the tree structure), we employ online anomaly detection approaches to identify candidate anomalous function executions. We also present a set of visualization tools for verification and exploration in a level-of-detailed manner. General information, such as distribution of execution times, are provided in an overview visualization. The detailed structure (e.g., function invocation relations) and the temporal information (e.g., message communication) of the execution call stack of interest are also visualized. The usability and efficiency of our methods are verified in a real-world HPC application.

Keywords: Anomaly detection · High Performance Computing · Streaming analysis · Trace events · Visual analytics

1 Introduction

Performance analysis is a critical task for the diagnosis of parallel High Performance Computing (HPC) applications. In particular, domain scientists are typically interested in detecting abnormal runtime behaviors during the execution of HPC applications. Since the supercomputer resources in use are limited and costly, the timely identification of the causes of adverse performance events (e.g., abnormal communication latencies) is essential. We have been working

J. M. F. Rodrigues et al. (Eds.): ICCS 2019, LNCS 11536, pp. 153–167, 2019.
https://doi.org/10.1007/978-3-030-22734-0_12

with a group of chemists who use an HPC cluster to solve complex molecular equations. The development of the system presented in this paper was driven by their need to monitor and identify computation latencies at runtime.

Anomalous function executions are usually identified by examining the detailed traces collected in the HPC cluster. A trace is essentially a log of a sequence of specific events (determined by program instrumentation) that occur in a computing core during execution (e.g. function entry, function exit, or message passing). Figure 1(a) shows an example sequence of trace events inside one execution of the `main` function in an HPC core. It represents the call stack information (Fig. 1(b)) inside the execution of the root function. Domain scientists typically detect anomalous function executions based on the extracted temporal information [2,18] (e.g., execution time and exit timestamps) or structural information [22] (e.g., call relations from parent function to children) from the trace events. While existing detection approaches achieve good performance and scalability in offline analysis, detecting abnormal runtime behavior in online analysis remains challenging. One reason for this is that the complexities of most feature extraction and anomaly detection algorithms are too high to support online training. It would be prohibitively slow to use a standard algorithm like Local Outlier Factor (LOF) [3] to update the learning result every time a new trace event is generated. Conventional anomaly detection models do not support continuous updates, rather they need to be re-trained in each time window. Furthermore, in order for human experts to stay aware and gain insights into the performance data an interactive visual interface will be most appropriate. However, offline visual analysis tools are not designed to provide visualizations sufficiently responsive in a streaming data environment. For example, standard visualization algorithms such as Multidimensional Scaling (MDS) [12] or t-Distributed Stochastic Neighbor Embedding (t-SNE) [13] are not fast enough to plot tens of thousands of points at the rate of streaming data.

We recently devised a visual analysis approach for the detection of abnormal HPC runtime behavior [22]. This tool, however, was designed for offline analysis and cannot be used for online analysis of streaming data. In the current work

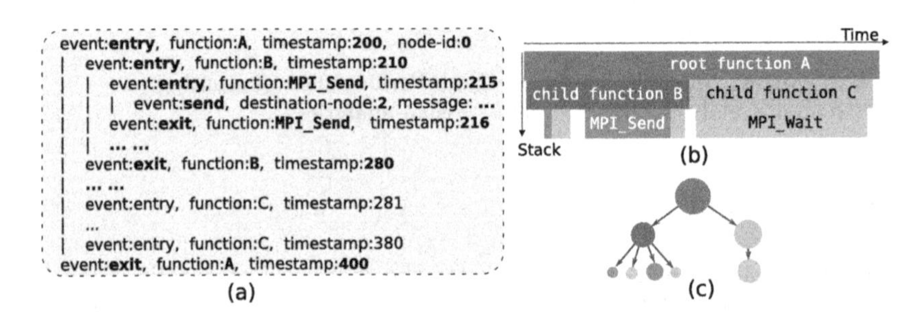

Fig. 1. (a) Example trace events during an execution of `main` in a compute core. (b) The call stack of the execution reconstructed from the trace events. (c) The call stack represented as a Call Stack Tree (CSTree), which is a directed tree with vertex weights.

we build on this approach to create a new method for streaming trace event data. Similar to our earlier work we make use of the call stack tree (CSTree) representation we devised there (Fig. 1(c)), but we now use it to encode the structural as well as the temporal information inside a function execution via a directed tree. Abnormal behavior of function execution can then be detected by identifying abnormal trees in a call stack forest. We also modify our original tree feature extraction process to now support online tree representation and online anomaly detection. We propose a set of new online visualizations to assist the scientist in the understanding and exploration of the function executions. In particular, we visualize the feature vectors and the learned anomaly labels of the function executions in an overview projection. The structural and temporal information of the CSTree are visualized in a structure and timeline visualization, respectively, capable of dealing with large amounts of data in real time.

The remainder of our paper is structured as follows. Section 2 reviews related work. Section 3 defines the problem and gives an overview of our approach. Section 4 introduces our online algorithms for vector representation and anomaly detection of CSTrees. Section 5 describes our visual analytics approach for anomalous CSTree exploration. Section 6 presents a case study to validate our approach. And Sect. 7 concludes the paper and discusses the future work.

2 Related Work

Domain scientists usually employ instrumentation and measurement tools [1,5,10,16] to generate performance data, such as trace events, and associated metrics. Many techniques have been proposed for the evaluation and diagnosis of these data [4]. In this paper, we focus on the most important two tasks in the diagnosis of the HPC application performance: detection of abnormal runtime behavior and visualization of performance.

2.1 Abnormal Runtime Behavior Detection

For anomalous runtime behavior detection, most existing approaches identify candidates based on temporal information [7]. For example, a disproportionately large function execution time has a very high probability of being abnormal since it has a large negative impact on the downgraded performance. However, diagnosis without the execution context makes it difficult to determine the source of this major latency. For example, the cause in the delay may be triggered by a child function or by another node in the HPC system (due to delayed communication). Furthermore, semantics of the executions are also necessary for the diagnosis. For example, the initialization function call may execute more computations and communicates more than other functions and as result takes more time to complete than other function calls. However, this phenomenon should not be identified as abnormal runtime behavior.

In contrast, the Call Stack Tree representation (CSTree) [22] can encode the temporal information as well as the context structure obtained from the call stacks to identify the potential anomalous executions. The embedding vectors are then generated from the CSTrees. Each vector encodes the structural information from the call stack of one function execution. A conventional anomaly detection algorithm such as the One-Class Support Vector Machine (OCSVM) [15] is then able to take the vectors and identify the anomalies.

However, both the tree representation and the anomaly detection are too complex for online analysis. To deal with that, in this paper, we modify the training strategy of these learning algorithms for the analysis of streaming data.

2.2 Performance Visualization

Performance visualization makes problem detection and diagnosis of HPC applications [7,21] more transparent to domain experts. Most visualization approaches support the comprehension of different levels and aspects of the performance data, including the trace timeline, call stack structures, and the messages.

For the trace events, a common practice is to visualize the events along a time axis, as is done in Vampir [9] and Jumpshot [23]. Most existing temporal visualizations provide level-of-detail explorations. Users can zoom into different time window granularities to see detailed events [21]. Other temporal visualizations are also capable of presenting the relationships between threads, such as SyncTrace [8].

Call path visualization (e.g., the call relationship between parent and children functions) is critical for understanding the behavior of the runtime execution. Existing approaches employ a directed tree or graph to present the structure in a call stack, such as Vampir [9]. These visualizations usually use the visual properties of the tree to encode detailed information from the call path. For example, CSTree [22] utilizes the color and size of a tree node to encode the type and the execution duration of a function.

Communication delay is usually the main reason for application latency. A straightforward approach to encode the message passing is to draw a directed line between the sending and receiving functions, which is adopted in the Jumpshot [23] implementation. Since the messages can also be regarded as directed edges, the communication between threads or processes can also be summarized in terms of an adjacency matrix [9].

One major issue with the existing visualizations is their limited capability for online performance evaluation. Some offline analysis tools focus on the complete event or message passing structure which is available only when the communication is finished. In addition, most visualization paradigms, such as MDS are too complex for real time streaming data. On the other hand, we also need to deal with online data reduction and sampling to prevent overdraw. In our approach, we adjust the common projection, timeline, and tree visualizations for streaming to facilitate incremental updates and data visualizations.

3 Problem Formulation and Approach Overview

We focus on the following problem: Given a set of functions of interest (FOIs) {A} and all of their invoked executions in an HPC cluster, determine which executions are associated with anomalous runtime behavior. An FOI A is usually a key function for computation or communication, which is specified by domain experts. An anomalous runtime behavior is then in most cases indicated by temporal or structural features. For example, a deadlock will cause large execution time and an infinite loop will generate unexpected call path structure.

3.1 Call Stack Tree Representation and Problem Formulation

As mentioned in Sect. 2, the Call Stack Tree (CSTree) (Fig. 1(c)) representation provides a comprehensive way to encode the execution of an FOI A since it takes advantage of the execution's context information. Each execution of A is converted to a CSTree T where a vertex in T is a function invoked in the call stack and a directed edge represents the call from the parent to the child function. All executions of A collectively give rise to a forest $\mathcal{T} = \{T_i\}$ where each tree T represents a single execution of A. The runtime behavior can be observed directly from the features of a CSTree. For example, if the volume of a vertex in the CSTree is proportional to the execution time of that function, then a very large vertex in the tree can represent a delayed function execution. Furthermore, a large set of child vertices of the same function type indicates that the parent function invokes the child function multiple times in a loop.

Given this representation the detection of anomalous behavior is then formulated as the problem of finding anomalous tree structures in the call stack forest. Our visual interface exposes the candidate anomalies, which are those CSTrees whose structures differ most.

3.2 Approach Overview and System Architecture

Based on the CSTree representation, our online visual analytics approach for detecting anomalous CSTrees uses the following four steps with the architecture[1] in Fig. 2:

Step 1 Data processing: First, the data analysis server (Fig. 2(b)) pairs and orders trace events (Fig. 2(a)) generated during the execution of an HPC application by their type (i.e., message or function event). Second, given a set of functions of interest {A}, the data analysis server can also generate the call stack trees rooted at A and insert them into the call stack forest $\mathcal{T} = \{T_i\}$.

Step 2 Tree feature extraction and anomaly detection: Each new CSTree in the forest is converted into a feature vector. For this paper the feature vector consists of the temporal features of the root functions (i.e. total duration of execution). However, other options for feature extraction, such as the graph

[1] Please visit https://github.com/CODARcode/ChimbukoVisualization to see the project page and the source code.

kernel, are also provided in the code. Anomaly detection is then performed based on these features, resulting in a set of candidate anomalous CSTrees.

Step 3 Overview visualization: An overview projection is calculated by the visualization server (Fig. 2(c)). The visualization platform is the web browser (Fig. 2(d)) showing the feature vector distribution and their learned labels.

Step 4 Detailed visual exploration: The detailed visualizations enable the user to investigate the execution's context and make a decision whether a candidate anomalous CSTree is truly anomalous. In specific, the user can select a feature vector from the overview to view the associated CSTree structure and communication patterns (both provided by the visualization server) to understand the execution's context.

Steps 3 and 4 can be performed repeatedly for better insights into the anomaly detection results.

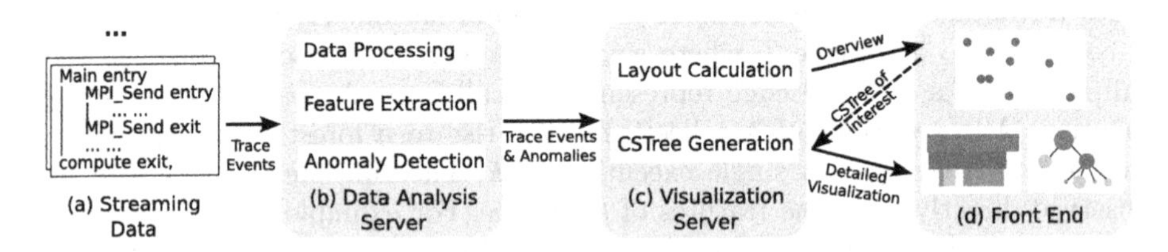

Fig. 2. Our system architecture for online anomalous CSTree detection. (a) The streaming trace events. (b) Feature vectors of the new CSTrees are generated in the data analysis server. Then the anomaly labels (normal or abnormal) of the CSTrees are learned online based on their feature vectors. (c) The visualization server calculates the layout of the overview and generates the CSTree of interest for detailed exploration. (d) On the browser end, the user is allowed to interact with the overview of the tree distribution in a 2D view. The candidates of interest in the overview can be further investigated via the detailed visualizations provided by the server.

4 Online Anomaly Detection of Call Stack Trees

Our approach begins with updating the call stack forest in the data server. For a FOI `A`, a new CSTree is generated and inserted into the call stack forest when an exit event of `A` is received in the trace stream. Since the user will focus on the recent data, the CSTrees older than a given threshold (e.g., 1 h) will be removed from the forest. Then the feature extraction and anomaly detection are performed for the updated call stack forest in the data analysis server.

4.1 Feature Vector Representation

Our approach provides different options for temporal and structural feature extraction from the CSTree for anomaly detection, including time-based and tree-based extraction methods.

Time-Based Representation. A straightforward way to represent runtime behavior is to utilize the information from the root function of the CSTrees. The domain expert is able to customize the specific set of features to be extracted, such as the execution duration, message frequency, and the exit time of the execution. With this pre-designed feature set, a vector of each CSTree is constructed and used as the input for anomaly detection.

Tree-Based Representation. Thus far there is no contextual information in the above temporal feature construction, however, structural representation is critical for the diagnosis of the runtime behavior. To provide an option for structural feature extraction, we follow the Graph Kernel [22] approach for the vector representation, which uses an analogy to document analysis. A tree is analogous to a document while the subtrees are analogous to the words in a document. The subtrees (Fig. 3(b)) are extracted using Weisfeiler-Lehman Graph Kernels [17]. In the initialization of the algorithm, each node is considered as a subtree of depth 0 (e.g., `A` - `E` in Fig. 3(b)). Then in the following iteration, each subtree is expanded towards the children to find sub-structures of larger depths, such as `F` - `J` in Fig. 3(b). At last, the CSTree (Fig. 3(a)) is represented as a bag-of-subtrees (Fig. 3(c)). The duration of each subtree in a CSTree is synonymous to "word frequency" in a document.

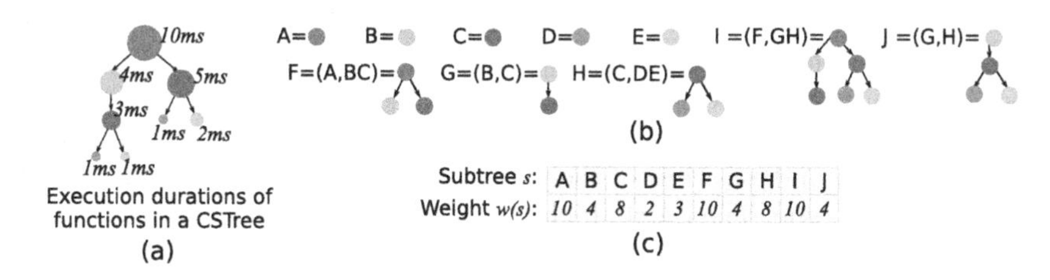

Fig. 3. (a) An example CSTree with the function execution time of the vertices. (b) An example of the result substructure generated from a CSTree using Weisfeiler-Lehman Graph Kernels. (c) The bag-of-subtree representation of the CSTree.

4.2 Online Anomaly Detection

The extracted feature vectors of the CSTrees are used as the input of the anomaly detection. Conventional algorithms such as Local Outlier Factor can be too complex for online training. For better online performance in the presence of a very large amount of streaming data we chose a fairly straightforward statistical approach. Using a Gaussian distribution to model the feature vector distribution of the CSTrees, we calculate the rolling mean (μ) and standard deviation (σ) of the feature vectors in the call stack forest using Welford's method [14]. We then label all CSTrees at 3σ (i.e., 0.003%) of the data distribution as anomaly candidates. In our own experiments we found this to be a good threshold, but the confidence level can be user-adjusted based on the estimated percentage of anomalies in the dataset.

5 Visual Exploration of Anomalous Call Stack Trees

It is necessary that human experts can verify the learning results to make sure the identified candidates are true anomalies. In addition, exploration of the temporal and structural patterns of the candidates also helps the user understand why they are potentially anomalous. In this section, we describe our level-of-detail visualization system designed for the exploration and verification of the CSTrees in different granularities.

5.1 Overview Visualization

The overview visualization is generated by the visualization server and displayed in a browser-based interface. The overview shows the general distribution of the feature vectors of CSTrees. As noted above, the most common approach available for this purpose is a low-dimensional embedding method such as MDS. However, after some experimentation we determined, as we also mentioned above, that these methods are not sufficiently fast and scalable for large streaming datasets.

To deal with this problem, we resorted to a standard bivariate projective scatterplot approach. In this visualization the x and y positions are calculated by the basic attributes of the CSTrees specified the user. For example, in Fig. 4(a), the CSTrees are visualized as points in the scatter plot. They are projected by the execution time and exit time of their root functions. The color of each points represent the FOI type. The points highlighted with thick borders are candidate anomalies which are detected in the previous step.

Display scalability can still be a problem when the data volume is massive. After discussing this with a domain expert we decided to adopt a negative down sampling strategy to reduce the data that are displayed. Since users are typically mainly interested in the candidate anomalies, they can set a down sampling rate (see the dashed box in Fig. 4(a)) to keep only a portion of the normal points in the projection. At the extreme, when the down sampling rate is set to 0, the projection will only visualize the candidate anomalies. In addition, users can also filter the displayed data by FOI type. This helps users to put more focus on a subset of more important FOIs.

In an online analysis, as opposed to a postmortem analysis, users will usually want to focus on the latest trace data. Hence, the visualization server will only maintain a subset of the data, namely those collected within a given time period (e.g., 1 h). The scatter plot will also be updated whenever new CSTrees are processed and generated from the server end. On the other hand, even in online analysis, keeping a historical perspective is desirable. We provide this view by visualizing the distribution of all history data in form of a heatmap (see Fig. 4(b)). In this map the user is free to specify the color encoding used for highlighting the number of anomalies or trace events.

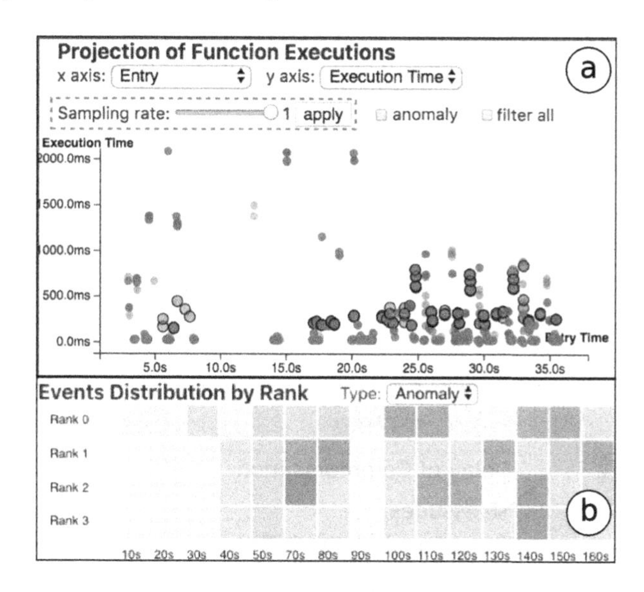

Fig. 4. (a) The CSTrees are projected by the execution durations and the entry timestamps of their root functions. The user can specify other axis-encoding schema for projection. Filtering and negative down sampling are supported to reduce visual clutter. (b) The heatmap visualizes the temporal distribution of all the history data.

Further, the user can specify the points of interest in the scatter plot, which will send requests to the visualization server for more detail on these selections. The structure and timeline visualizations will then provide the corresponding CSTree structures and events sequences for further exploration.

5.2 Structure Visualization

The details of the selected CSTree are visualized in the Structure Visualization, as shown in Fig. 5(a). The tree vertex size and color respectively represent the execution time and function name of the corresponding tree node. The directed edge shows the call relationship from a parent to a child function. Force-directed Layout [11] is employed to calculate the position of the tree vertices. For clearer representation, we set a maximum depth to only keep important parent functions which influence the tree the most. This reduces the visual clutter, which is helpful especially for CSTrees with a large number of descendant functions. On the other hand, in order to provide complete anomaly information, the abnormal substructures are preserved in the visualization even if they are beyond the maximum depth limit.

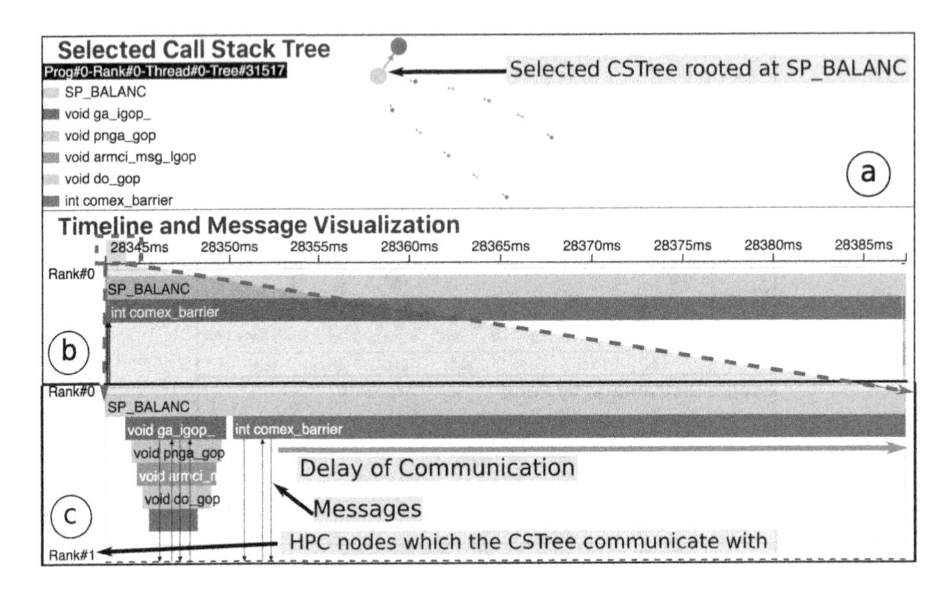

Fig. 5. (a) The call stack tree structure visualization shows the tree at a limited depth except for abnormal substructures. (b) The timeline visualize the event sequence and message communications. (c) The timeline can be zoomed in for detailed exploration.

5.3 Timeline Visualization

While the structure visualization is effective, the message communication between different HPC cores is not visualized. To deal with this problem, we propose the Timeline Visualization (Fig. 5(b)), which basically follows the visual design of Vampir [9] and Jumpshot [23]. Our visualization shows the event sequence as well as the message passing in a stack timeline. The x and y axes of the timeline encode the timestamp and the growing direction of the call stack. Since communication is one of the main reasons for HPC application delays, messages are also visualized, via lines between different cores, as indicated in Fig. 5(b). Users can zoom in and out of the x-axis (see red box in Fig. 5(b)) and so explore the detailed call stack in different time ranges, as shown in Fig. 5(c).

6 Case Study

We conducted a case study[2] with an NWChem [19] developer at Brookhaven National Laboratory (BNL). NWChem is a massively-parallel computational chemistry application deployed on BNL's HPC cluster. Our performance analysis study presented here focuses on analyzing NWChem's the molecular dynamics functions.

[2] Please see our video demo here and case study design details here.

6.1 Experiment Settings and Online Performance Evaluation

Our participant is a chemist and not an expert in visualization or machine learning and sought to use our system to find anomalous runtime behaviors in the function executions. In the onset, we had a number of thorough discussions with this domain scientist to learn about functions of interest and possible anomalous behavior patterns. We also held a short training session to introduce our visualization system.

We applied the temporal feature extraction described in Sect. 4.1 since it was faster for large scale datasets. We also employed the anomaly detection by standard deviation described in Sect. 4.2. After studying the dataset in our interface, the participant set the significance level defining a candidate anomaly to be roughly 3σ (i.e., 0.003%) to reduce false positives.

The online analysis will not affect the HPC application execution since it only takes the trace events output. The streaming trace events of NWChem were generated by SOSFlow [20] and ADIOS [6] in real time. We tested different NWChem application settings with different molecular system sizes (small and large in the first columns of Table 1) as well as the scalability in different HPC settings (the second columns of Table 1). The details for the six datasets are shown in Table 1. We found that our feature extraction and anomaly detection algorithms did not cause significant delays for the streaming analysis. The throughput of our system was acceptable according to our user.

Table 1. Summary of experiment datasets and the throughput.

NWChem Setting	Number of HPC cores	Number of trace events	Number of trace events per second	Number of anomalies	Throughput (MBps)
Small	2	393,542	58.5k	350	5.2
Small	4	1,201,533	128.1k	445	11.5
Small	8	4,024,651	224.2k	2235	20.5
Large	2	784,122	52.9k	818	4.7
Large	4	2,386,634	101.0k	1121	9.1
Large	8	7,972,872	172.9k	3683	15.8

6.2 Case 1: Delay of Communication

During the online analysis, our participant first examined the overview distribution of the CSTrees. He was interested in the `SP_BALANC` function, which was designed to redistribute the work over the processors to minimize the time spend waiting in communication functions. He suggested that it was a critical factor to the overall performance of the code. In the scatter plot, he noticed that most of the points in the projection were normal. To put more focus on the potential abnormal CSTrees, he reduced the negative sampling rate (Fig. 4(a)). He noticed an abnormal point, which was execution #31517 of the Thread #0 in Rank #0 of the program #0, as shown in the upper left of Fig. 5(a).

From the Structure Visualization, he found that there was a big green function of comex_barrier which spent the majority of the time in SP_BALANC. comex_barrier was a function responsible for the communication between different HPC cores. He made an assumption that the barrier function was the major reason for the latency of this execution.

To learn about the temporal pattern and message communication, our participant examined the Timeline Visualization. He found that barrier function invoked some communications, as shown in the message passing visualizations with Rank #1 in Fig. 4(c). After comparing with other regular executions of SP_BALANC, he concluded that this execution waited for a long time for the response of Rank #1. As a result, the communication delay to other computing cores made this execution a candidate anomaly. He concluded that our system helped him understand one of the reasons for the performance fluctuation of SP_BALANC, which provided insights of how to optimize the source code to improve the overall program performance.

6.3 Case 2: Delay of Computation

Our participant continued to explore other functions of interest. In the scatter plot, he only visualized CF_CENMAS function, which computed the center of mass coordinates of individual molecules in the NWChem simulation. He located another candidate anomaly and showed its detailed structure and timeline. From the structure visualization (Fig. 6(a)) he learned that it was abnormal since the root node in the CSTree was very huge. He zoomed into the timeline (Fig. 6(b)) and found that all of the child functions were executed as expected; however, they were invoked after waiting a long time for CF_CENMAS.

He expressed that this could happen when the computation in CF_CENMAS took a long time. The descendant functions (e.g., comex_barrier) had to wait for the computation to finish to communicate with other HPC cores. Our expert noted that the visual exploration our system provided was a critically important supplement to the learned labels (i.e., normal or abnormal) of the automatic algorithm. He stated that our system provided the much needed comprehensive analysis support for understanding and diagnosing the runtime behavior.

6.4 Feedback and Discussion

After the case study with the chemist, we invited additional 5 scientists from BNL to use our system. We conducted an interview session where we asked them to rate our system and give feedback. Their average ratings were 4.8 for usability (1 = not useful, 5 = very useful) and 4.7 for learning cost (1 = hard to learn, 5 = easy to learn). One participant mentioned that the visualizations were easy to understand since they are also commonly used by the existing performance analysis tools [4,7]. He also compared our system with the commonly available tools

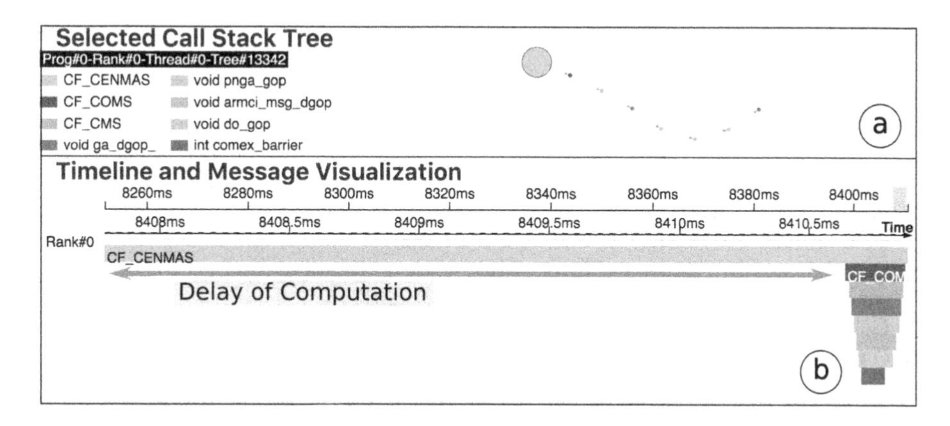

Fig. 6. (a) The CSTree structure of the `CF_CENMAS` function. (b) The user zoomed in the timeline to see the detailed temporal sequence at the beginning of the execution, as shown in the gray range in the time axis.

in his community. He indicated that with current tools such as Jumpshot [23] he would only be able to learn about the executions after a lengthy session with the system. He would first manually locate the time window of an anomaly candidate and then look at a detailed view on the respective call paths and messages. Conversely, he said that our interface was more intuitive, enabling him to quickly discover the anomaly and succinctly explain it by using the three linked views.

7 Conclusion and Future Work

We described a visual analytics approach for the online detection of anomalous function executions in HPC clusters and their visual exploration. Our approach is based on the CSTree representation. It provides effective anomaly detection and visualization tools that address challenges with streaming performance evaluation for parallel computation at scale. We demonstrated our approach with a real world NWChem application.

In the case study, we learned that our bag-of-subtree vector can be too sparse since the subtree corpus will be massive. To cope with this problem, a Stack2vec embedding [22] can be a viable option to generate the embedding vectors from the sparse bag-of-subtree vector. We also plan to integrate machine log analysis and source code examination into our system which will provide more insights into the execution scheduling and code design optimization.

Acknowledgments. This research was partially supported by NSF grant IIS 1527200, BNL LDRD grant 16-041 and 18-009, ECP CODAR project 17-SC-20-SC, and the MSIP (Ministry of Science, ICT and Future Planning), Korea, under "IT Consilience Creative Program (ITCCP)" supervised by NIPA.

References

1. Adhianto, L., et al.: HPCTOOLKIT: tools for performance analysis of optimized parallel programs. Concurr. Comput. Pract. Exp. **22**(6), 685–701 (2010)
2. Arnold, D.C., Ahn, D.H., De Supinski, B.R., Lee, G.L., Miller, B.P., Schulz, M.: Stack trace analysis for large scale debugging. In: IPDPS, pp. 1–10. IEEE (2007)
3. Breunig, M.M., Kriegel, H.P., Ng, R.T., Sander, J.: LOF: identifying density-based local outliers. ACM SIGMOD **29**, 93–104 (2000)
4. Ezzati-Jivan, N., Dagenais, M.R.: Multi-scale navigation of large trace data: a survey. Concurr. Comput. Pract. Exp. **29**(10) (2017)
5. Geimer, M., Wolf, F., et al.: The scalasca performance toolset architecture. Concurr. Comput. Pract. Exp. **22**(6), 702–719 (2010)
6. Gu, J., Klasky, S., Podhorszki, N., Qiang, J., Wu, K.: Querying large scientific data sets with adaptable IO system ADIOS. In: Yokota, R., Wu, W. (eds.) SCFA 2018. LNCS, vol. 10776, pp. 51–69. Springer, Cham (2018). https://doi.org/10.1007/978-3-319-69953-0_4
7. Isaacs, K.E., et al.: State of the art of performance visualization. In: EuroVis 2014 (2014)
8. Karran, B., Trumper, J., Dollner, J.: SYNCTRACE: visual thread-interplay analysis. In: VISSOFT 2013, vol. 00, pp. 1–10 (2013)
9. Knüpfer, A., et al.: The Vampir performance analysis tool-set. In: Resch, M., Keller, R., Himmler, V., Krammer, B., Schulz, A. (eds.) Tools for High Performance Computing, pp. 139–155. Springer, Heidelberg (2008). https://doi.org/10.1007/978-3-540-68564-7_9
10. Knüpfer, A., et al.: Score-P: a joint performance measurement run-time infrastructure for periscope, scalasca, TAU, and Vampir. In: Brunst, H., Müller, M., Nagel, W., Resch, M. (eds.) Tools for High Performance Computing 2011, pp. 79–91. Springer, Heidelberg (2011). https://doi.org/10.1007/978-3-642-31476-6_7
11. Kobourov, S.G.: Spring embedders and force directed graph drawing algorithms. arXiv preprint arXiv:1201.3011 (2012)
12. Kruskal, J.B.: Multidimensional scaling by optimizing goodness of fit to a nonmetric hypothesis. Psychometrika **29**(1), 1–27 (1964)
13. van der Maaten, L., Hinton, G.: Visualizing data using t-SNE. J. Mach. Learn. Res. **9**, 2579–2605 (2008)
14. Salonen, J.: (2013). http://jonisalonen.com/2013/deriving-welfords-method-for-computing-variance/
15. Schölkopf, B., Williamson, R., Smola, A., Shawe-Taylor, J., Platt, J.: Support vector method for novelty detection. In: NIPS, pp. 582–588 (2000)
16. Shende, S.S., Malony, A.D.: The TAU parallel performance system. Int. J. High Perform. Comput. Appl. **20**(2), 287–311 (2006). https://doi.org/10.1177/1094342006064482
17. Shervashidze, N., Schweitzer, P., van Leeuwen, E.J., Mehlhorn, K., Borgwardt, K.M.: Weisfeiler-lehman graph kernels. J. Mach. Learn. Res. 12, 2539–2561 (2011)
18. Sigovan, C., et al.: A visual network analysis method for large-scale parallel I/O systems. In: IEEE IPDPS, pp. 308–319 (2013)
19. Valiev, M., et al.: NWChem: a comprehensive and scalable open-source solution for large scale molecular simulations. Comput. Phys. Commun. **181**(9), 1477–1489 (2010)
20. Wood, C., et al.: A scalable observation system for introspection and in situ analytics. In: Proceedings of the 5th Workshop on Extreme-Scale Programming Tools, ESPT 2016, Piscataway, NJ, USA, pp. 42–49. IEEE Press (2016)

21. Xie, C., Xu, W., Ha, S., et al.: Performance visualization for TAU instrumented scientific workflows. In: VISIGRAPP (3: IVAPP), pp. 333–340. SciTePress (2018)
22. Xie, C., Xu, W., Mueller, K.: A visual analytics framework for the detection of anomalous call stack trees in high performance computing applications. IEEE Trans. Vis. Comput. Graph. (2018)
23. Zaki, O., Lusk, E., Gropp, W., Swider, D.: Toward scalable performance visualization with Jumpshot. Int. J. High Perform. Comput. Appl. **13**(3), 277–288 (1999). https://doi.org/10.1177/109434209901300310

Analysis of the Construction of Similarity Matrices on Multi-core and Many-Core Platforms Using Different Similarity Metrics

Uxía Casal, Jorge González-Domínguez(✉), and María J. Martín

CITIC, Computer Architecture Group, Universidade da Coruña,
Campus de Elviña, 15071 A Coruña, Spain
{uxia.casal.baldomir,jgonzalezd,mariam}@udc.es

Abstract. Similarity matrices are 2D representations of the degree of similarity between points of a given dataset which are employed in different fields such as data mining, genetics or machine learning. However, their calculation presents quadratic complexity and, thus, it is specially expensive for large datasets. MPICorMat is able to accelerate the construction of these matrices through the use of a hybrid parallelization strategy based on MPI and OpenMP. The previous version of this tool achieved high performance and scalability, but it only implemented one single similarity metric, the Pearson's correlation. Therefore, it was suitable only for those problems where data are normally distributed and there is a linear relationship between variables. In this work, we present an extension to MPICorMat that incorporates eight additional metrics for similarity so that the users can choose the one that best adapts to their problem. The performance and energy consumption of each metric is measured in two platforms: a multi-core platform with two Intel Xeon Sandy-Bridge processors and a many-core Intel Xeon Phi KNL. Results show that MPICorMat executes faster and consumes less energy on the many-core architecture. The new version of MPICorMat is publicly available to download from its website: https://sourceforge.net/projects/mpicormat/

Keywords: Similarity matrix · High Performance Computing · Intel Xeon Phi · Performance evaluation · Energy consumption

1 Introduction

The construction of similarity matrices is a fundamental step for many applications of different areas such as bioinformatics, data mining, text mining or machine learning. For instance, they are usually necessary when constructing gene co-expression networks, as they can represent the similarity between genes. However, the calculation of these 2D matrices is highly time-consuming due to

J. M. F. Rodrigues et al. (Eds.): ICCS 2019, LNCS 11536, pp. 168–181, 2019.
https://doi.org/10.1007/978-3-030-22734-0_13

its quadratic complexity. In the Big Data era the size of datasets is continuously increasing in many fields, and thus finding fast and scalable solutions is a highly important task.

We have recently developed the tool MPICorMat [3], a High Performance Computing (HPC) framework that accelerates the construction of similarity matrices with Message Passing Interface (MPI) and OpenMP routines. It presents high performance and scalability on multi-core clusters, but has the following limitations:

- It only includes Pearson's correlation as similarity metric, as it is suitable for data with linear relationship (very common in genetic scenarios, the focus of the previous work). However, there are in literature a number of metrics to build similarity matrices and the choice of the best one depends on the application field and the input data. For instance, if the linearity hypothesis cannot be assumed, Spearman's or Kendall's tau-b correlations may be more useful than Pearson's because they identify both linear and non-linear relationships [13].
- The calculation of the similarity metric for each pair of attributes relies on the GNU Scientific Library [4], which forces the users to install and tune that library in their systems.
- The experimental evaluation is limited to performance measures in traditional multi-core clusters, without taking into account energy consumption and/or the use of many-core platforms.

This work overcomes these limitations by presenting a new version of MPICorMat (v3) that includes eight additional metrics. The choice of the metric is made by the users through a command line parameter in order to adapt the execution to the characteristics of their data. The eight metrics were implemented using C++, MPI and OpenMP in order to avoid any dependency with external libraries and thus improve the portability of the tool. We also provide a highly detailed experimental evaluation that compares the performance of the different metrics on a multi-core platform and an Intel Xeon Phi Knights Landing (KNL) many-core system, not only in terms of performance but also of energy consumption.

The rest of the paper is organized as follows. Section 2 presents the related work. Section 3 describes the behavior of the extended application and the metrics included. Section 4 shows the results of the experimental evaluation in both the multi-core and many-core systems. Finally, conclusions are discussed in Sect. 5.

2 Related Work

There are in the literature a number of works than compare the behavior of different similarity measures for the reconstruction of gene co-expression networks [5,15,17]. These studies show that the choice of the most appropriate measure depends on the nature of the gene interactions to be analyzed.

On the other hand, similarity matrices also play a crucial role in other fields such as clustering, image retrieval, or recommending systems. For instances, in [11] authors compare seven popular similarity measures for the clustering of patients. They include the Euclidean distance, as well as the Pearson, Spearman and Kendall correlations, among others. Authors conclude that an absolute best similarity measure does not exist, but it strongly depends on data.

However, most of the available software tools for the calculation of similarity matrices on parallel architectures focus on only one similarity metric. For instance, TINGe [20] and CUDA-MPI [12] are parallel approaches based on Mutual Information (MI) for clusters (implemented with MPI), and GPUs (implemented with CUDA), respectively. In [8] TINGe is adapted for the first generation of the Intel Xeon Phi (KNC) architecture. The construction of Pearson's correlation-based similarity matrices was addressed for MPICorMat [3], LightPCC [7] and FastGCN [6] for multicore clusters (implemented with MPI and OpenMP), the Intel Xeon Phi KNC coprocessor and NVIDIA GPUs, respectively.

MPICorMat_v3, in contrast, allows the user to choose among several similarity metrics to better adapt to the characteristics of the problem in hand. Moreover, up to our knowledge, our performance evaluation is the first one focused on using the KNL generation of Intel Xeon Phi to accelerate the construction of similarity matrices.

3 MPICorMat Version 3

As previously explained, MPICorMat is a parallel tool to accelerate the construction of similarity matrices on HPC systems. It receives as input a file that contains a 2D matrix with dimensions $n \times m$, where n is the number of attributes and m the number of samples. It returns a file with an $n \times n$ similarity matrix with the similarity values for each pair of attributes.

The third version of this tool increments its usefulness by including eight additional similarity metrics (besides the Pearson's correlation already available in the previous versions of the tool). The users should indicate its desired metric using a command line parameter. Information about the input parameters, as well as installation and execution instructions, are available in the reference manual of the tool.

The implementations of the nine metrics have been integrated into the parallel approach already available in previous versions of MPICorMat, as it was proved efficient in our previous work [3]. MPICorMat follows a hybrid parallel approach with MPI and OpenMP that is able to exploit the computational capabilities of multi-core clusters, with hybrid distributed/shared memory architecture. As will be shown in Sect. 4, the focus of the experimental evaluation of this work consists in testing the suitability of this parallel implementation in the Intel Xeon Phi KNL many-core processor, compared to an Intel multi-core system. Only the OpenMP parallelization is necessary for both platforms, as they are shared-memory machines. OpenMP is a parallel programming interface based on compiler directives that follows a fork-join model, where a master or parent thread creates a number of slaves or children that are able to access to the same shared memory and perform different tasks to complete a work.

As the similarity metric must be calculated for all gene pairs, the MPICorMat workload can be seen as a 2D matrix, where each point represents one pair of attributes. Only half of the matrix (upper or lower triangular) must be calculated as all metrics are symmetric. Concretely $\frac{n \cdot (n-1)}{2}$ pairs. MPICorMat divides the workload (pairs) among the threads, which do not need any synchronization as computation is completely independent among pairs. Pairs are assigned by rows (the whole row to the same thread) in order to reuse data (one attribute is repeated in all the pairs of the row). However, due to the triangular nature of the problem, the most intuitive static block distribution, with the same number of rows per thread, would lead to unbalanced workload (some rows have more pairs than other, as can be seen in Fig. 1). Instead, MPICorMat uses a dynamic OpenMP distribution, where each (still not computed) row is assigned to a thread once it has finished all its previously assigned work. We refer to [3] for more information.

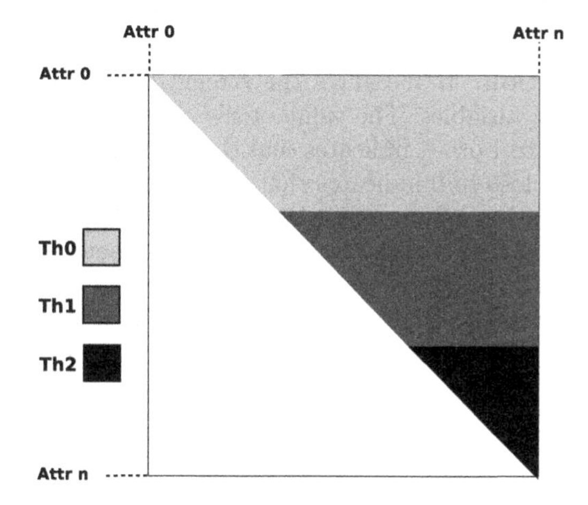

Fig. 1. Not efficient distribution of pairs among threads.

Algorithm 1 shows the pseudocode of the parallel OpenMP implementation in MPICorMat, the one tested in the experimental evaluation of Sect. 4. After reading the file with the input values for each attribute and sample (Line 1) and allocating memory for the output similarity matrix (Line 2), several OpenMP threads are launched to parallelize the loop that traverses the rows (Lines 3 and 4). As previously explained, each thread is in charge of a complete row. The thread starts calculating the position of the row in the output triangular matrix (Line 5). Then, it calculates the metric for all the pairs of the row (Lines 6 and 7). Next, the parallel region finishes and the output is written by the master thread (Line 8).

Algorithm 1. Pseudo-code of the OpenMP parallelization in MPICorMat.

```
1 Read input matrix M with the values of the attributes
2 Initialize matrix of scores S
3 #pragma omp parallel for schedule(dynamic)
4 for each row i from 0 to n do
5 |   rowPos = CalculatePos(i)
6 |   for each column j from i to n do
7 |   |   S[rowPos + j - i] := CalculateMetric(i, j)
  |   end
  end
8 Write S in the output file
```

3.1 Similarity Metrics Included in MPICorMat_v3

A different implementation of `CalculateMetric()` (Line 7 in Algorithm 1) is performed for each metric. The following nine similarity metrics are available for MPICorMat users since its third version.

Pearson's Correlation. It measures the strength of the linear relationship between two random variables. The value of the correlation is between -1 and 1. A correlation close to 1 or -1 indicates that the relationship is almost perfectly linear while a value close to 0 indicates that the two variables are uncorrelated. The Pearson's correlation assumes the data are normally distributed and there is a linear relationship between the two variables. It is sensitive to outliers and requires the data to be measured on interval or ratio scale. Assume that X and Y are two random variables with n observations (x_i, y_i with $i = 1, 2, \ldots, n$) and $\overline{x}$ and $\overline{y}$ are the means of X and Y, respectively. Then, Pearson's correlation is defined as:

$$\frac{\sum_{i=1}^{n}(x_i - \overline{x}) \cdot (y_i - \overline{y})}{\sqrt{\sum_{i=1}^{n}(x_i - \overline{x})^2 \cdot \sum_{i=1}^{n}(y_i - \overline{y})^2}} \tag{1}$$

Spearman's Correlation. It is equal to the Pearson's correlation between the rank values of the variables, being the rank value of an observation its relative position within all the values of the variable. While Pearson's correlation assesses linear relationships, Spearman's correlation assesses monotonic relationships (whether linear or not). It takes values between -1 and 1. A positive correlation implies that the ranks of both variables increase together, while a negative correlation implies that the rank of one variable increases as the rank of the other decreases. A definition of the Spearman's correlation able to deal with tied ranks (elements that have the same rank value) is:

$$\frac{\frac{1}{n} \cdot \sum_{i=1}^{n}(R(x_i) - \overline{R(x)}) \cdot (R(y_i) - \overline{R(y)})}{\sqrt{(\frac{1}{n} \cdot \sum_{i=1}^{n}(R(x_i) - \overline{R(x)})^2) \cdot (\frac{1}{n} \cdot \sum_{i=1}^{n}(R(y_i) - \overline{R(y)})^2)}} \tag{2}$$

where $R(x_i)$ and $R(y_i)$ are the ranks of the observation i in the variables X and Y, respectively, while $\overline{R(x)}$ and $\overline{R(y)}$ are the means of the ranks. The procedure to calculate the ranks usually consists in sorting the observations of the variable.

Euclidean Distance. This is probably the simplest metric, indicating the straight-line distance between two points in Euclidean space. The Euclidean distance between two attributes X and Y, with x_i and y_i denoting their value for sample i, is measured as:

$$\sqrt{\sum_{i=1}^{n}(x_i - y_i)^2} \tag{3}$$

Mutual Information (MI). It quantifies the amount of information that one random variable provides about another. MI can only take positive values. High MI indicates a large reduction in uncertainty, while low MI indicates a small reduction in uncertainty, and MI equal to 0 means that the variables are independent. MI is a metric that only works over discrete values. If the input data are real values (either in simple or double precision), a preliminary step that discretizes the values, grouping similar elements into the same bucket, is required. The number of buckets is indicated by the user as an argument of the application through the command line. The accuracy of the metric usually increases with the number of buckets, but also its complexity. Mi is defined as:

$$\sum_{i=1}^{n}\sum_{i=j}^{n} p(x_i, y_j)\log_2 \frac{p(x_i, y_j)}{p(x_i)\cdot p(y_j)} \tag{4}$$

where $p(x_i)$ and $p(y_j)$ are the probabilities of the buckets that contain the values x_i and y_i, and $p(x_i, y_j)$ is the joint probability of the buckets associated to x_i and y_j.

Kendall's Tau-b. It is a non-parametric metric of association based on the number of concordances and discordances in paired observations. It is an alternative method to Spearman's correlation, i.e., it also identifies monotonic relationships. Suppose two pairs (x_i, y_i) and (x_j, y_j), they are concordant if they are in the same order with respect to each variable. That is, if $x_i < x_j$ and $y_i < y_j$, or if $x_i > x_j$ and $y_i > y_j$. Otherwise, they are discordant. The value of this coefficient ranges from -1 (one ranking always reverses the other) to 1 (the ranks of the two attributes are the same). If the two variables are independent, the value is approximately equal to 0. Assume that P is the number of concordant pairs, Q is the number of discordant pairs, X_0 the number of tied pairs on X and Y_0 the number of tied pairs on Y. Then, Kendall's tau-be is defined as:

$$\frac{P - Q}{\sqrt{(P + Q + X_0)\cdot(P + Q + Y_0)}} \tag{5}$$

Goodman & Kruskal Gamma Coefficient (G&K). It is another widely-used rank-based coefficient that ranges between −1 and 1. As Kendall's tau-b, a value −1 indicates 100% perfect inversion, value 1 indicates 100% perfect agreement, and value 0 indicates the absence of association. It is defined as:

$$\frac{P - Q}{P + Q} \tag{6}$$

Maximal Information Correlation (MIC). It is based on the idea that if a relationship between two variables exists, then a grid that partitions the data to encapsulate that relationship can be drawn on the scatterplot of the two variables [10]. Its value ranges between 0 and 1 and it takes the value 0 if the variables are independent. The MIC for two attributes X and Y is defined as:

$$\frac{MI(X,Y)}{H(X)} \tag{7}$$

where $MI(X,Y)$ is the mutual information between the variables X and Y and it can be obtained from Eq. 4, and $H(X)$ is the entropy of the attribute X, which can be calculated as follows (being $p(x_i)$ the probability of the bucket that contains the variable x_i):

$$-\sum_{i=1}^{n} p(x_i) \cdot \log_2(p(x_i)) \tag{8}$$

Hoeffding D Test. This metric approximates a weighted sum over observations in order to test the independence of two datasets. In this work, each attribute is seen as a dataset. The statistic D is defined as:

$$30 \cdot \frac{(n-2) \cdot (n-3) \cdot D_1 + D_2 - 2 \cdot (n-2) \cdot D_3}{n \cdot (n-1) \cdot (n-2) \cdot (n-3) \cdot (n-4)} \tag{9}$$

where:

$$D_1 = \sum_{i=1}^{n} (Q_i - 1) \cdot (Q_i - 2) \tag{10}$$

$$D_2 = \sum_{i=1}^{n} (R(x_i) - 1) \cdot (R(x_i) - 2) \cdot (R(y_i) - 1) \cdot (R(y_i) - 2) \tag{11}$$

$$D_3 = \sum_{i=1}^{n} (R(x_i) - 2) \cdot (R(x_i) - 2) \cdot (Q_i - 1) \tag{12}$$

being $R(x_i)$ and $R(y_i)$ the ranks as in Spearman's correlation and Q_i the bivariate rank, which refers to the number of points j ($j = 1, 2, ..., n$) with both x_j and y_j values lower than the ith point. Hoeffding's D lies on the interval $[-0.5, 1]$, with larger values indicating a stronger relationship.

Table 1. Characteristics of the datasets used for evaluation.

Name	Number of attributes	Number of samples
GDS5037	41,000	108
GDS3795	61,170	160
GDS3244	54,675	200

Weighted Rank Correlation. It is a variation of the Spearman's rank correlation but giving weight to the distance between two ranks by using a linear function of those ranks (more weight to higher ranks than to lower ones). Assume that $R(x_i)$ and $R(y_i)$ are the ranks as in Spearman's correlation, then the weighted rank correlation metric can be calculated as:

$$1 - \frac{90}{g(n)} \cdot \sum_{i=1}^{n} (R(x_i) - R(y_i))^2 \cdot (2 \cdot (n+1) - (R(x_i) + R(y_i)))^2 \quad (13)$$

where:

$$g(n) = n \cdot (n-1) \cdot (n+1) \cdot (2 \cdot +1) \cdot (8 \cdot n + 11) \quad (14)$$

4 Experimental Evaluation

Three datasets, with a different number of attributes and samples, were used in the evaluation of the nine metrics included in the third version of MPICorMat. The datasets were downloaded from the Geo Expression Omnibus (GEO) Dataset Browser available at the National Center for Biotechnology Information (NCBI) website [9]. Table 1 shows their characteristics. As they contain genetic information, the attributes represent genes of a population.

Although MPICorMat includes support for MPI parallelization, all the experiments were carried out with only one MPI process and several OpenMP threads, as the two platforms are shared memory architectures. The scalability of the hybrid MPI/OpenMP parallel approach has not been tested again in a multicore cluster as it has not been modified since [3].

4.1 Performance Evaluation on an Intel Xeon Phi KNL

Knights Landing (KNL) is the code name for the second-generation Intel Xeon Phi product family [14]. It is a many-core processor that delivers massive thread and data parallelism with high memory bandwidth. Concretely, it provides features such as four threads per core, deeper out-of-order buffers, higher cache bandwidth, new instructions, better reliability, larger translation look-aside buffers (TLBs), and larger caches. Additionally, it introduces the new Advanced Vector Extensions instruction set, AVX-512 [19], in order to fully exploit its 512-bit vector registers, which can hold 16 single precision or 8 double precision floating-point numbers. In this project, we used the Intel Xeon Phi KNL

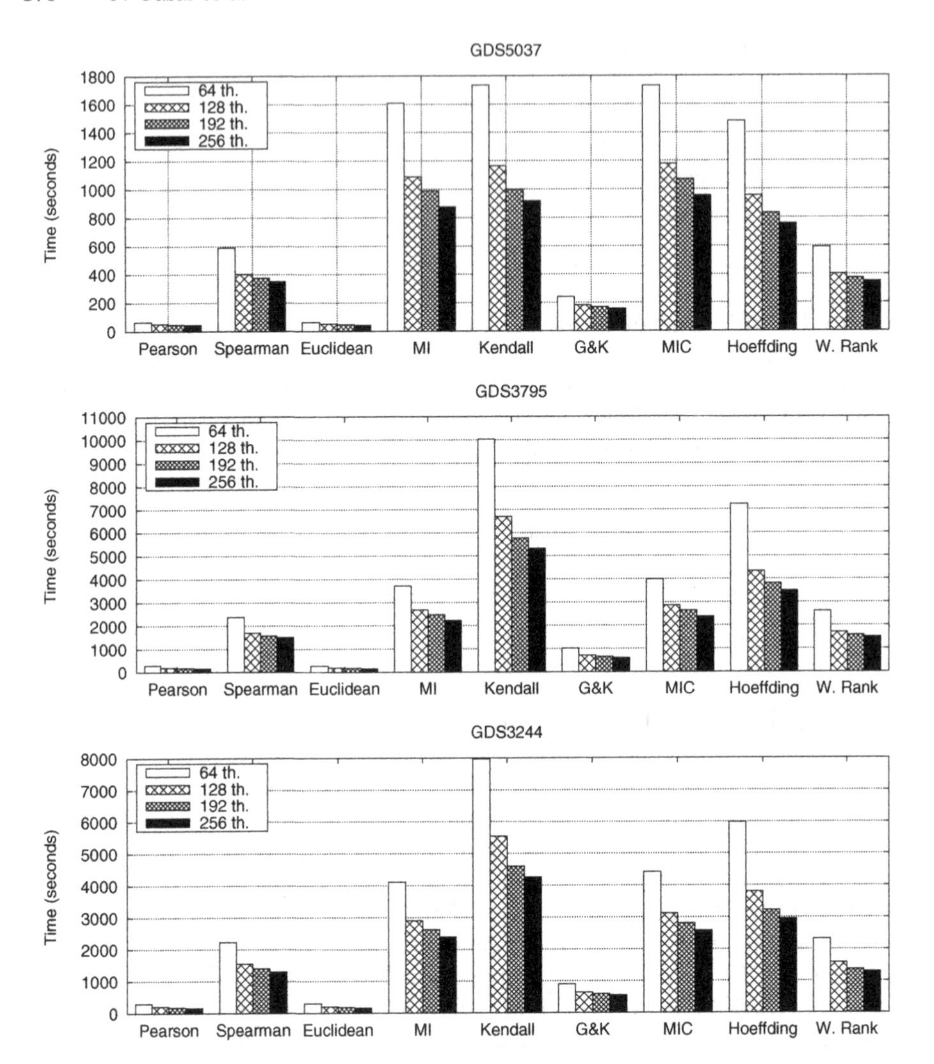

Fig. 2. Runtime of MPICorMat_v3 on the Intel Xeon Phi KNL for different metrics and number of threads. Automatic vectorization with Intel AVX-512 instructions has been used.

processor 7210. It has 64 active cores at 1.30 GHz, allows up to four threads per core (256 total threads) and it is configured in the quadrant clustering [16] and the flat memory modes [1]. MPICorMat has been compiled with the Intel ICPC compiler version 18.0.3 activating the automatic vectorization with Intel AVX-512 instructions (`-xMIC-AVX512` flag). Remark that all the runtimes shown in this section were obtained with the many-core system in exclusive mode, i.e., no other works were executed at the same time.

Figure 2 shows the runtime for the three datasets, the nine metrics and different number of threads (from 64 threads or one thread per core to 256 threads or four threads per core). 32 buckets are used for MI and MIC. The first conclusion that can be obtained is that the runtime heavily depends on the metric. Pearson's correlation and Euclidean distance are the simplest metrics, while Kendall's tau-b is the most complex one. The results shown in these graphs also indicate that hyperthreading is beneficial for MPICorMat. Using two threads per core reduces the runtime on average 1.41, 1.47 and 1.45 times compared to the single-thread execution with the GDS5037, GDS3795 and GDS3244 datasets, respectively. This average speedup increases to 1.71, 1.77, 1.80 if fully exploiting the hyperthreading, with four threads per core. 256 threads will be used from now on for all the experiments in the Intel Xeon Phi KNL, as this configuration obtains the best runtime for all scenarios (combination of dataset and metric).

As previously mentioned, the runtimes shown in the graphs of Fig. 2 were obtained by activating the automatic vectorization with the Intel AVX-512 instructions. Table 2 shows the speedups compared to an execution with 256 threads but without vectorization (`-no-vec` flag in the compiler). Its impact depends on the characteristics of the metrics, being especially beneficial for the ranking procedure necessary for Spearman's correlation, Hoeffding D test and weighted rank correlation (see Sect. 3.1).

Table 2. Speedup obtained thanks to the use of automatic vectorization with Intel AVX-512 instructions in the Xeon Phi KNL, compared to not vectorized versions of the metrics. All executions are carried out with 256 threads.

	GDS5037	GDS3795	GDS3244
Pearson	1.50	1.43	1.41
Spearman	10.59	14.65	13.63
Euclidean	1.36	1.31	1.30
MI	3.66	4.36	4.26
Kendall	1.46	1.49	1.49
G&K	1.56	1.90	1.81
MIC	3.55	4.27	4.15
Hoeffding	7.33	9.52	9.06
Weighted rank	10.59	14.71	13.66

4.2 Performance and Energy Consumption Comparison Between Intel Architectures

The experimental evaluation has also been performed in an Intel multi-core platform in order to compare its performance to the Intel Xeon Phi KNL many-core. Concretely, a machine with two eight-core Intel Xeon E5-2660 Sandy Bridge-EP

Table 3. Speedup of the execution of MPICorMat_v3 in the Xeon Phi KNL for each metric (with 256 threads and automatic vectorization using Intel AVX-512 instructions) compared to the execution on the Sandy Bridge-based multi-core platform (with 16 threads and automatic AVX 256-bit vectorization).

	GDS5037	GDS3795	GDS3244
Pearson	1.54	1.13	1.15
Spearman	3.92	4.94	4.74
Euclidean	1.73	1.07	1.34
MI	5.63	6.79	6.59
Kendall	2.97	2.91	2.82
G&K	1.56	1.50	1.47
MIC	5.41	6.58	6.36
Hoeffding	4.36	5.56	5.39
Weighted rank	3.93	4.96	4.75

processors (i.e., a total of 16 cores) and 64 GB of memory. The Intel ICPC compiler has also been used in this machine (in this case, version 18.0.1) activating the automatic vectorization with AVX instructions. Remark that the impact of vectorization should be lower than in the Intel Xeon Phi KNL as the length of the vector registers is 256 bits, instead of 512 bits as in the many-core.

The execution in the Intel Xeon Phi KNL is faster than in the Sandy Bridge-based multi-core system using 16 threads (one per core) for all combinations of dataset and metric. Table 3 shows the speedup for each scenario. The highest the speedup, the fastest the execution in the many-core compared to the multi-core systems (speedup equal to 1 would mean same execution time). The magnitude of the benefit thanks to running on the Intel Xeon Phi KNL depends again on the metric. Speedups are higher for those metrics that require a ranking of the data (such as Spearman's correlation, Hoeffding D test and weighted rank correlation, with average speedups of 4.53, 5.10 and 4.55, respectively), as well as for those based on probabilities (MI and MIC, with average speedups of 6.34 and 6.12, respectively).

Nowadays, reduction of energy consumption is key in order to develop and maintain large HPC infrastructures. In this sense, many-core systems are expected to accelerate the execution at the same time that save energy. The Performance API (PAPI) analysis library [2,18], together with the Running Average Power Limit (RAPL) of the Intel architectures, has been used to measure and report energy values on both platforms when calculating the similarity matrices with different metrics. Figure 3 shows the energy consumption (in Joules) for each metric and platform using the GDS5037 dataset. On average, the Intel Xeon Phi KNL consumes 4.46 times less energy than the multi-core platform, reaching factors of 5.31 and 5.85 for MI and MIC, respectively.

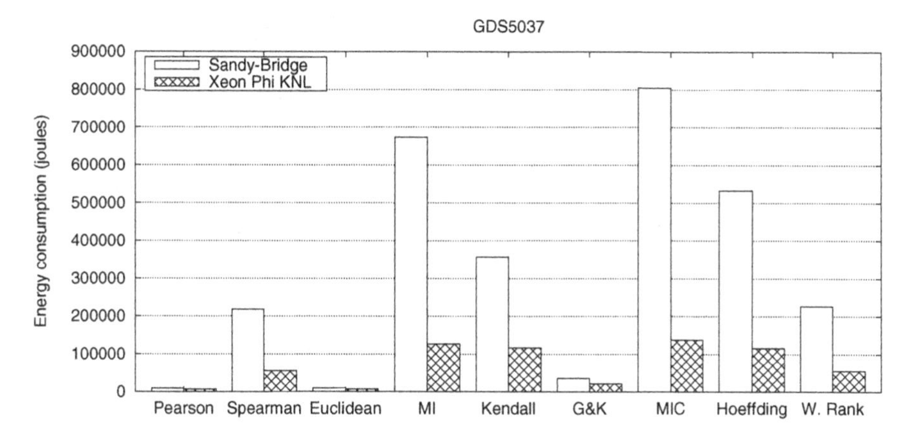

Fig. 3. Energy consumed by MPICorMat_v3 on the Intel Xeon Phi KNL (with 256 threads and automatic vectorization using Intel AVX-512 instructions) and the Sandy Bridge-based multi-core machine (with 16 threads and automatic AVX 256-bit vectorization) for different metrics.

5 Conclusions

The construction of similarity matrices is a bottleneck for many algorithms of different areas due to its quadratic complexity with the number of attributes. MPICorMat is a publicly available tool that helps to alleviate this problem by efficiently exploiting HPC resources. However, previous versions of this tool were only able to calculate similarity matrices based on Pearson's correlation, which limited its interest for many researchers. In this work, we have presented a new version of MPICorMat that includes a total of nine different similarity metrics so that the users can choose the one most suitable for their applications. The implementations of the new metrics were integrated into the framework of MPICorMat so all of them can benefit from the parallel implementation.

The experimental evaluation has focused on testing the adequacy of the metric implementations to the hardware characteristics of the Intel Xeon Phi KNL, as the scalability in multi-core clusters had been effectively tested in a previous work [3]. The use of this and other kind of many-core accelerators (such as GPUs) is gaining popularity in the last years as they provide high performance at low power consumption. Our experimental evaluation using three datasets from genetic scenarios with different characteristics has led to several conclusions:

- The best performance is obtained in all cases with four threads per core (256 threads per Intel Xeon Phi KNL). For instance, the runtime of applying Kendall's tau-b metric to the GDS3422 dataset is reduced from around 29 min with only one thread per core, to around 15 min when hyperthreading with four threads per core is used.

- Automatic vectorization with Intel AVX-512 instruction should be applied in order to improve performance. The magnitude of this performance improvement depends on the metric, varying from an overall speedup of 1.32 for Euclidean distance to 12.99 for weighted rank correlation.
- Execution times in the Intel Xeon Phi KNL are lower than in a multi-core platform with two octa-core Sandy Bridge processors for every combination of metric and dataset. The overall performance improvement is 3.74, being more significant for metrics based on probabilities such as MI and MIC, with an overall speedup of 6.34 and 6.12, respectively.
- The energy consumption is lower in the many-core architecture for all the experiments, needing on average 4.46 times less energy.

As future work, we plan to implement a GPU version of the code and compare the results with the ones obtained in the Intel Xeon Phi architecture.

Acknowledgments. This work was supported by the Ministry of Economy, Industry and Competitiveness of Spain and FEDER funds of the European Union [grant TIN2016-75845-P (AEI/FEDER/UE)], as well as by Xunta de Galicia (Centro Singular de Investigacion de Galicia accreditation 2016–2019, ref. EDG431G/01).

References

1. Asai, R.: MCDRAM as High-Bandwidth Memory. HBM) in Knights Landing Processors, Developers Guide. Colfax Research (2016)
2. Browne, S., Dongarra, J., Garner, N., Ho, G., Mucci, P.: A portable programming interface for performance evaluation on modern processors. Int. J. High Perform. Comput. Appl. **14**(3), 189–204 (2000)
3. González-Domínguez, J., Martín, M.J.: Fast parallel construction of correlation similarity matrices for gene co-expression networks on multicore clusters. In: 17th International Conference on Computer Science (ICCS 2017), Zurich, Switzerland, vol. 108, pp. 485–494 (2017)
4. Gough, B.: GNU Scientific Library Reference Manual, 3rd edn. Network Theory Ltd., Bristol (2009)
5. Kumari, S., et al.: Evaluation of gene association methods for coexpression network construction and biological knowledge discovery. PloS one **7**(11), e50411 (2012)
6. Liang, M., Zhang, F., Jin, G., Zhu, J.: FastGCN: a GPU accelerated tool for fast gene co-expression networks. PloS one **10**(1), e0116776 (2015)
7. Liu, Y., Pan, T., Aluru, S.: Parallel pairwise correlation computation on Intel Xeon Phi clusters. In: 28th International Symposium on Computer Architecture and High Performance Computing (SBAC-PAD 2016), Los Angeles, CA, USA, pp. 141–149 (2016)
8. Misra, S., Pamnany, K., Aluru, S.: Parallel mutual information based construction of genome-scale networks on the Intel Xeon Phi coprocessor. IEEE/ACM Trans. Comput. Biol. Bioinform. **12**(5), 1008–1020 (2015)
9. National Center for Biotechnology Information (NCBI): Geo Expression Omnibus (GEO) Dataset Browser. https://www.ncbi.nlm.nih.gov/sites/GDSbrowser. Accessed Dec 2018
10. Reshef, D.N., et al.: Detecting novel associations in large data sets. Science **334**(6062), 1518–1524 (2011)

11. Serra, A., Greco, D., Tagliaferri, R.: Impact of different metrics on multi-view clustering. In: 2015 IEEE International Joint Conference on Neural Networks (IJCNN), pp. 1–8 (2015)
12. Shi, H., Schmidt, B., Liu, W., Müller-Wittig, W.: Parallel mutual information estimation for inferring gene regulatory networks on GPUs. BMC Res. Notes **4**(1), 189 (2011)
13. de Siqueira Santos, S., Takahashi, D.Y., Nakata, A., Fujita, A.: A comparative study of statistical methods used to identify dependencies between gene expression signals. Brief. Bioinform. **15**(6), 906–918 (2013)
14. Sodani, A., et al.: Knights landing: second-generation Intel Xeon Phi product. IEEE Micro **36**(2), 34–46 (2016)
15. Song, L., Langfelder, P., Horvath, S.: Comparison of co-expression measures: mutual information, correlation, and model based indices. BMC Bioinform. **13**(1), 328 (2012)
16. Vladimirov, A., Asai, R.: Clustering Modes in Knights Landing Processors: Developers Guide. Colfax International (2016)
17. Wang, Y.R., Huang, H.: Review on statistical methods for gene network reconstruction using expression data. J. Theor. Biol. **362**, 53–61 (2014)
18. Weaver, V.M., et al.: Measuring energy and power with PAPI. In: 41st International Conference on Parallel Processing Workshops (ICPPW 2012), pp. 262–268 (2012)
19. Zhang, B.: Guide to Automatic Vectorization with Intel AVX-512 Instructions in Knights Landing Processors. Colfax International (2016)
20. Zola, J., Aluru, M., Sarje, A., Aluru, S.: Parallel information-theory-based construction of genome-wide gene regulatory networks. IEEE Trans. Parallel Distrib. Syst. **21**(12), 1721–1733 (2010)

High Performance Algorithms for Counting Collisions and Pairwise Interactions

Matheus Henrique Junqueira Saldanha(✉) and Paulo Sérgio Lopes de Souza

Institute of Mathematics and Computer Sciences, University of São Paulo, São Paulo, Brazil
mhjsaldanha@gmail.com, pssouza@icmc.usp.br

Abstract. The problem of counting collisions or interactions is common in areas as computer graphics and scientific simulations. Since it is a major bottleneck in applications of these areas, a lot of research has been carried out on such subject, mainly focused on techniques that allow calculations to be performed within pruned sets of objects. This paper focuses on how interaction calculation (such as collisions) within these sets can be done more efficiently than existing approaches. Two algorithms are proposed: a sequential algorithm that has linear complexity at the cost of high memory usage; and a parallel algorithm, mathematically proved to be correct, that manages to use GPU resources more efficiently than existing approaches. The proposed and existing algorithms were implemented, and experiments show a speedup of 21.7 for the sequential algorithm (on small problem size), and 1.12 for the parallel proposal (large problem size). By improving interaction calculation, this work contributes to research areas that promote interconnection in the modern world, such as computer graphics and robotics.

Keywords: Collision count · Pairwise interaction · GPU · High performance computing · Parallel computing · Algorithm

1 Introduction

As the performance growth of a single processor core decreases, attention has been shifting towards other means to decrease execution time of programs. Given the lower price of primary and secondary memory, compared to processors, exchanging memory usage for a reduction on algorithm complexity is an interesting option. Another commonly explored alternative is to exploit the increasing parallelism available in hardware, which can be accomplished by parallelizing existing sequential algorithms. This does not always work well, and a good parallel algorithm might only be brought forth by a complete re-analysis of the problem and design of new parallel algorithms. In this article, solutions that

J. M. F. Rodrigues et al. (Eds.): ICCS 2019, LNCS 11536, pp. 182–196, 2019.
https://doi.org/10.1007/978-3-030-22734-0_14

explore these two options are provided to accelerate the problem of calculating interactions (e.g. forces, contacts) in N-body environments.

The problem of interaction counting is widely present in areas that promote interconnection in the modern world: computer graphics and virtual reality, that connect people through games and link computer science to medicine, for example, where virtual environments can be used for practicing surgery; robotics, that interconnects objects and robots; and scientific simulations, that connect computing with areas that need to simulate complex natural phenomena such as protein folding and planet motion. Despite being largely used, interaction counting is a major bottleneck [12] in many applications. It may come in two flavors: (1) interactions might only matter within subsets of objects (e.g. collisions, where only neighbor objects are relevant), and it is common to use strategies such as spatial partitioning [6] or bounding volume hierarchies [18] to find these subsets before performing interaction counting; or (2) all interactions matter (e.g. gravitational forces), where it is common to use techniques that allow a set of objects to be regarded as a single object, such as the Fast Multipole Method [8]. Besides that, it is also often the case where the objects in question are in motion and collision detection must be done every small time steps, such as in a robot calculating collision-free paths. For such situations, besides the aforementioned strategies, time and space coherence is also used, that is, the fact that objects won't move too much in a short time span is exploited [9]. In any of these cases, however, there will still be a phase where smaller sets of objects undergo interaction calculation, and the usual way is to iterate over each object and compare it with the others, giving a $O(N^2)$ complexity.

Even though a lot of research has been carried out on alternative strategies to amortize the cost of this pairwise-comparisons $O(N^2)$ approach, not much has been done regarding this brute-force algorithm itself. This paper proposes a new parallel algorithm for calculating interactions that is designed to make better use of architectural resources on GPUs. The algorithm is mathematically proved to be correct, and results show a speedup of 1.12 over parallelization of the straightforward approach. We also propose a sequential algorithm for counting collisions among punctual objects, which has a $O(N)$ complexity at the cost of high virtual memory usage, and experiments show a speedup of 21.7 for limited size objects. By accelerating the pairwise-comparisons algorithm, we hope to facilitate smoother animations and virtual reality environments, simulations that take less time (or provide better results in the same amount of time), and robots that can better avoid collisions with its surroundings.

The document is organized as follows. An overview of research in collision counting is discussed in Sect. 2. In Sects. 3 and 4, two algorithms for collision counting are proposed. Experiments with the proposed algorithms are discussed in Sect. 5. Finally, Sect. 6 concludes the paper.

2 Related Work

Calculating interactions is a common problem that has been the subject of a vast amount of interesting research. Covering all of its facets is not our objective here, so in the following a short overview of the subject is provided.

I-COLLIDE [5] is a well-known suite of algorithms that perform fast collision detection among a large number of rigid or deforming objects, such as an environment with tens of thousands of triangles (that might compose an object). They focus on pruning potentially colliding sets (PCS) multiple times before performing the exact collision detection. CULLIDE [7] follows the same line and is parallelized for GPUs. Specializations of these algorithms for haptic systems, where collision detection must be performed thousands of times per second, were done in [9,10].

Spatial partitioning and bounding volumes hierarchies (BVHs) are broadly used for defining reduced sets of potentially colliding objects. A comparison among such techniques is found in [6], and k-d trees and octrees are notable in such problem. Reference [18] gives an efficient approach to construct BVH trees borrowing techniques from spatial partitioning, and [21] focuses on fast re-building of these trees on situations where there are moving objects.

Collision detection often involves calculating trajectories of objects, which is not always simple. In [20] the authors present fast and accurate methods for evaluating collision between triangulated models in such circumstances; their method involves a series of *coplanarity* and *inside* tests among elements (e.g. edges, vertices). In [17], where hair simulation is explored, there is a large number of hair-body collisions and hair-hair interactions (collisions and other forces such as friction and static attraction). In either case, the GPU approach we propose in this paper could be used to accelerate the calculation of interactions, such as the coplanarity tests. Similar studies are found in [3,4,15,16,19].

More related to engineering, in [22] is proposed a parallel algorithm for contact detection using spatial partitioning strategies, which is then experimented on simulation of concrete. Similarly, a framework that uses GPU for evaluating forces among particles of sand is given in [13]. Finally, in [14] the authors show an efficient parallelization, in the CUDA programming model, for the problem of evaluating gravitational forces among all bodies in an N-body system. They parallelized the straightforward sequential algorithm, where each body is compared with every other body. For summing these forces, in this paper is proposed a different sequential algorithm whose parallelization manages to make better use of GPU resources. To the best of our knowledge, this is the first work that brings an alternative to the pairwise-comparisons approach (for summing interactions) which provides benefits when parallelized.

3 A Sequential *O(N)* Approach

Consider the problem of counting the total number of collisions or contacts among beads (i.e. punctual objects) in a space $\mathcal{S}^3$ such that

$$\begin{cases} \mathcal{S}^3 = \mathcal{S} \times \mathcal{S} \times \mathcal{S} \\ \mathcal{S} = \{-a,\ -a+1,\ \ldots,\ a-1,\ a\} \subset \mathbb{Z} \end{cases}$$

that is, $\mathcal{S}^3$ is a three-dimensional, discrete and finite space whose axes are symmetric around the 0. Problems that do not suit such description are those that involve objects that have length, area or volume, possess non-discrete coordinates, or the range of possible coordinates is not bound by some known value a. Although this excludes a large number of problems, there still remains some that fit in the given conditions. For example, in [1] proteins are represented as a chain of beads in $\mathbb{Z}^3$, so each bead is located at a limited distance from the previous one, which limits the coordinates to some interval, therefore allowing us to define a space $\mathcal{S}^3$ for the problem.

If the problem at hand can be made to fit into the aforementioned conditions, then the brute-force counting algorithm can be replaced by one with lower time complexity. First, the space $\mathcal{S}^3$ must be computationally represented in a way that each coordinate in $\mathcal{S}^3$ maps uniquely to a number stored in memory. This can be accomplished with a three-dimensional array of some numeric data type, which can be indexed using the coordinates of $\mathcal{S}^3$ themselves; however, to avoid indexing with negative numbers the space $\mathcal{S}^3$ must be translated to have its origin in (a, a, a). For simplicity, in what follows we consider that three-dimensional arrays can be indexed with negative numbers.

Algorithm 1. Counting number of collisions among a vector of beads.

```
int countColls(point3D beads[],
               int space[][][]){
  int collisions = 0;
  for (b in beads){
    int beadCnt = space[b.x][b.y][b.z];
    collisions += beadCnt;
    space[b.x][b.y][b.z] += 1;
  }
  return collisions;
}
```

Algorithm 2. Counting number of contacts among a vector of beads.

```
int countContacts(point3D beads[],
                  int space[][][]){
  int contacts = 0;
  for (b in beads)
    space[b.x][b.y][b.z] += 1;
  for (b in beads){
    contacts += space[b.x+1][b.y][b.z];
    contacts += space[b.x-1][b.y][b.z];
    contacts += space[b.x][b.y+1][b.z];
    contacts += space[b.x][b.y-1][b.z];
    contacts += space[b.x][b.y][b.z+1];
    contacts += space[b.x][b.y][b.z-1];
  }
  return contacts / 2;
}
```

Having defined the array that represents $\mathcal{S}^3$ in memory, we may then perform the counting algorithm. Algorithm 1 shows the main procedure for counting, with linear time complexity, the number of collisions among a vector of beads, each of which has three integer coordinates. In this algorithm, the number associated with a point of $\mathcal{S}^3$ represents the number of beads in that spot so far (assume, for now, that it is initialized with zeros). We begin by initializing the number of collisions with zero, and then iterate over the vector of beads. For each bead, we access the $\mathcal{S}^3$ space array using the bead's coordinates as the index, retrieving from memory the number *beadCnt* of beads in that place. We are on the process of adding a bead to a place that already contains *beadCnt* beads, thus generating *beadCnt* extra collisions that are added into *collisions*. Finally, we increment the number associated with that place in the space, to effectively add one new bead there. For initializing the space array, we would also iterate over the vector of beads, initializing only elements at the coordinate of each bead.

With another similar algorithm (see Algorithm 2), we can also calculate the total number of contacts among beads, that is, count how many pairs of beads are neighbors. As in the previous algorithm, the *space* array holds how many beads are placed in each coordinate of $\mathcal{S}^3$ so far, and we begin by initializing a *contacts* variable with zero. The first loop "places" beads in the *space* array, such that its element (x, y, z) has the number of beads with coordinates (x, y, z). Afterwards, we iterate over the vector of beads again, and for each bead b we fetch from *space* the number of beads in each of the six spots that are neighbors of b, then add them into *contacts*. There is still a problem to deal with: if beads b_1 and b_2 are neighbors, their contact is counted twice (in iterations for b_1 and b_2); because of this, the function returns *contacts/2*.

Determining the initialization pattern for the problem of contact counting follows a reasoning similar to the one used with collisions. The elements of *space* that are accessed are all neighbors of each bead, including the bead's own position, so they must be initialized to zero.

The presented algorithms involve 2 steps: initializing *space* and counting either contacts or collisions. Both steps consist of N iterations, one for each bead, and in each iteration we perform a fixed number of $O(1)$ instructions: sum, subtraction and memory load/store. Hence, the algorithms have $O(N)$ time complexity, so they tend to perform better than the quadratic approach for large enough N. However, experiments show that these algorithms are faster even for small N (< 64), which are elaborated in Sect. 5.

On the other hand, the algorithms make use of the three dimensional array *space*, whose size depends on the cardinality of $\mathcal{S}^3$. As defined earlier, $\mathcal{S} = \{-a,\ -a+1,\ \ldots,\ a-1,\ a\}$, so each axis of $\mathcal{S}^3$ has $2a+1$ elements, resulting in a total of $(2a+1)^3$ elements and consequently a $O(a^3)$ complexity of memory consumption. The a might be known on compilation time, in problems where the beads are confined in a box of known edge length. Another possibility is that a is a function of N. For example, in the case of proteins modelled as a chain of beads that begins in $(0, 0, 0)$, a chain of size N would require each axis to span from $-N$ to N, meaning a would be a function $a(N) = N$, and the memory usage complexity can be rewritten as $O(N^3)$.

Although there are cases in which the memory complexity is $O(N^3)$, this concerns only virtual memory. When counting collisions, for example, virtual pages are mapped physically only if they contain beads. In the worst case, each bead will be in a different page and N pages will be mapped, making it $O(N)$ in physical memory usage. An important consequence of this is that swap memory tends to be depleted before physical memory, so by the time swap depletes the counting program still will not have begun to access disk for swapping, which would greatly impact performance.

4 An Efficient Parallel Algorithm

The $O(N)$ approach tends to incur high memory consumption, which poses an obstacle to handle large problem sizes, in which case it is reasonable to use parallel computing to distribute memory usage or computation among processing nodes. Applying this technique is possible for both the $O(N)$ and the $O(N^2)$ approaches, but due to the seemingly higher difficulty in using it for the $O(N)$ one, we analyze and propose an efficient parallelization of the $O(N^2)$ algorithm for counting any kind of symmetric pairwise interactions (SPI) among objects, not limited to collisions. The proposed algorithm is aimed mainly at GPUs, since it makes better use of its architectural characteristics, and experiments were performed using them. However, results are not limited to GPUs as there might be parallel architectures with similar characteristics and could be better exploited with the algorithm presented in the following.

In Algorithm 3, we present the standard sequential code for calculating SPI. For each object, we accumulate its interaction with all subsequent objects. Analyzing the nested loops in search for parallelization, we notice that they are not completely data-parallel due to the *interactions* variable, which is read and written in every iteration. Such variable is a reduction variable, which implies that parallelizing the iterations is still viable, provided that there is a reduction phase that agglomerates the intermediate results calculated by each parallel execution unit (denoted as *threads* from here on).

A second aspect of the standard sequential algorithm is that the outer loop is not balanced in terms of work executed per iteration. The first outer iteration executes $N - 1$ inner iterations, whereas the last outer iteration executes none. When we try to parallelize the problem, this could cause threads to be assigned different amounts of work, resulting in idle threads waiting for others to finish their larger burden. One way to promote balancing is to assign outer iterations to GPU threads in a round-robin fashion such that each thread performs at least two outer iterations. However, by agglomerating outer iterations this approach reduces the number of threads we can launch, and it also arguably reduces memory locality, which are undesirable properties for GPU algorithms.

Algorithm 3. Standard algorithm for calculating SPI.

```
for (i = 0 to N-1)
  for (j = i+1 to N-1)
    interactions += interact(obj[i],
                             obj[j]);
```

Algorithm 4. Proposed algorithm for calculating SPI.

```
for (i = 0 to N-1)
  for (j = 1 to (N-1)/2)
    interactions += interact(obj[i],
                             obj[(i+j)%N]);
```

For balancing the loop iterations in a way optimized for GPU, we offer an alternative solution whose main portion of code is shown in Algorithm 4. We now explain the idea behind it, prove that it works when N is odd and finally elaborate on how to make it work with even N too.

In the proposed algorithm, the outer loop can be seen as follows. Each bead i (we will use the term *bead* for simplicity, but it can be any kind of object) evaluates the interaction of itself with subsequent beads in a circular fashion. This circularity can be mathematically modelled by working in the universe of integers *modulo* N [11], which has interesting properties that we will use later:

$$a \equiv_N b \implies \forall k \in \mathbb{Z} \quad a + k \equiv_N b + k \tag{1}$$

$$a \equiv_N 0 \iff \exists c \in \mathbb{Z} \quad a = c.N \tag{2}$$

For each bead i we can now define two functions: $reach(s)$, that returns the index of the bead being evaluated in step s; and $reached(s)$ that returns the index of the bead evaluating bead i in step s. The algorithm begins in step $s = 1$ and goes forward until some stopping condition that we discuss now. For any bead i, we have:

$$\begin{array}{lll} s = 1 & reach(1) \equiv_N i + 1 & reached(1) \equiv_N i - 1 \\ s = 2 & reach(2) \equiv_N i + 2 & reached(2) \equiv_N i - 2 \\ & \ldots & \\ s & reach(s) \equiv_N i + s & reached(s) \equiv_N i - s \end{array}$$

where bead i evaluates beads j with j increasing, and bead i is evaluated by bead k with k decreasing as s advances (see Fig. 1).

An undesired situation here is that bead i evaluates bead j when bead j has already evaluated bead i, so the same interaction would be evaluated twice. This is illustrated in Fig. 1, where this situation happens in step 3, where bead 3 reaches bead 1, but bead 1 had already reached bead 3 in step 2. For an odd number N of beads, this hazard happens on the step where bead i reaches bead $j+1$ on the same step that bead j reaches bead i. Mathematically, the violation happens when $reach(s) - 1 \equiv_N reached(s)$, which for an arbitrary bead i gives

$$\begin{array}{lll} i + s - 1 \equiv_N i - s & & \\ 2s - 1 \equiv_N 0 & & \textbf{using (1)} \\ 2s - 1 = c.N & c \in \mathbb{Z} & \textbf{using (2)} \\ s = \dfrac{c.N + 1}{2} \; \therefore \; s = \dfrac{N + 1}{2} & & \end{array} \tag{3}$$

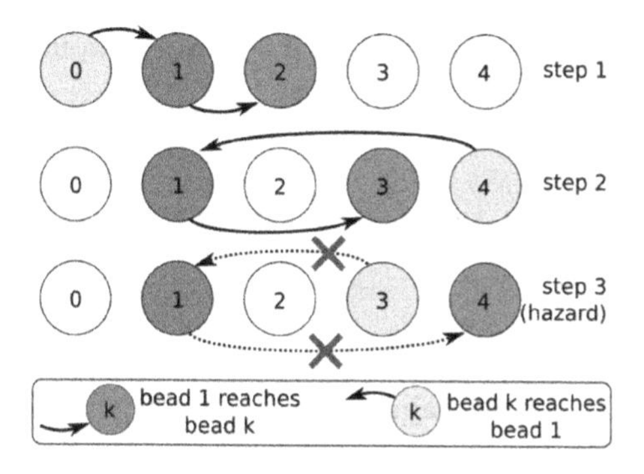

Fig. 1. Algorithm execution from the perspective of bead 1. Bead 1 reaches beads 2 and 3, and it is reached by beads 0 and 4 as s increases.

Note that modular arithmetic gives a set of answers instead of a unique one. It shows that the step s where the violation occurs is $(c.N+1)/2$ for some integer c. If c is 0, then $s = 1/2$ which is invalid because, as stated earlier, the first step is $s = 1$; any $c < 0$ gives $s < 0$ which also does not make sense; $c = 1$ gives $(N + 1)/2$, which is in fact the first step where the violation occurs (N is odd, so the division results in an integer). Taking $c = 2, 3$, and so on yields higher steps where the violation would occur, but they do not matter for us because we will stop the algorithm before the first violation. Finally, since the first violation occurs at $s = (N + 1)/2$, we need to stop at the previous iteration, that is, at $s = (N - 1)/2$.

We still need to prove the equivalence between the proposed and straightforward algorithms. First, the number of different interactions among beads amount to $N.(N - 1)/2$, which is precisely the amount of interactions evaluated by the straightforward approach. We now prove that the proposed algorithm performs this same amount of work, and that all interactions evaluated are mutually different, hence proving the equivalence of both approaches.

Proposition 1. *The proposed algorithm evaluates $N.(N - 1)/2$ interactions among beads, for odd N.*

Proof. As stated before, each of the N beads evaluates its interaction with $(N-1)/2$ subsequent beads, which amounts to a total of $N.(N - 1)/2$ evaluations. □

Proposition 2. *All $N.(N - 1)/2$ interactions evaluated by the proposed algorithm are different from each other, for odd N.*

Proof. Take two arbitrarily different beads i and j, with $i < j$. Each bead evaluates the interaction between itself and subsequent beads, so for beads i and j one side of the interactions they evaluate is inherently different since $i \neq j$. Consequently, both beads would only evaluate the same interaction if it was the interaction among beads i and j themselves. Let us see when this happens.

For bead i to evaluate the interaction of i with j, it would be necessary that

$$\begin{aligned} reach(s) &\equiv_N j \qquad \text{from bead i's perspective} \\ i + s &\equiv_N j \\ i + s - j &\equiv_N 0 \\ i + s - j &= c.N \qquad c \in \mathbb{Z} \\ s &= c.N + (j - i) \ \therefore \ s = j - i \end{aligned} \tag{4}$$

For $c < 0$, s would be a negative value because $j - i$ (the distance between beads i and j) cannot be higher than $N - 1$, and this is a contradiction because negative s never happens. If $c > 0$, then $s = c.N + (j - i) > N$, but s only gets values $1, ..., (N - 1)/2$. Finally, taking $c = 0$ makes $s = j - i$, which happens as long as $j - i$ also resides in interval $1, ..., (N - 1)/2$, that is, the distance between j and i is lower than or equal to $(N - 1)/2$. On the other side, bead j will evaluate its interaction with i when its $reach(s) \equiv_N i$, and developing this expression in a similar way as before we obtain $s = c.N - (j - i)$, but the only valid value for c is 1. This gives $s = N - (j - i)$, and $s \in \{1, ..., (N - 1)/2\}$ only if $j - i \geq (N + 1)/2$.

Therefore, let the distance between i and j be called $d = j - i$, then i will evaluate j only if $d \leq (N - 1)/2$, and j will evaluate i only if $d \geq (N + 1)/2$. This means that i and j do not mutually evaluate each other, so the interaction among beads i and j is evaluated only once. This proves that all interaction evaluations are mutually different, completing the proof of equivalence between this approach and the straightforward one, for odd N. □

When N is even, a slight modification is needed[1]. The stopping condition of the outer loop is derived in a similar way as done before, but in this case the problem is not that the $reach()$ and $reached()$ arrows cross themselves; instead, they reach the same value. That is, for a given bead i we have $reach(s) \equiv_N reached(s)$, which will result, following the same mathematical steps as before, in $s = N/2$. This is the step in which bead i is evaluating some bead j while bead j is also evaluating bead i. To prevent this situation, we allow execution of $N/2 - 1$ steps normally, and the first half of the beads are made to execute one more. This works because in step $N/2$ the beads being reciprocally evaluated are on different halves of the vector of beads, since they are within a distance of $N/2$ from each other. The consequence of this modification is that the algorithm is not completely balanced any longer; some outer iterations execute one extra inner iteration.

This concludes the formulation of the alternative, balanced algorithm. This proposed approach has some nice theoretical properties, based on the concepts of *depth* and *work* [2]. Depth is the largest amount of work done sequentially by a single thread, while work is the total amount of work done by all threads launched. In the straightforward algorithm, we have a work-complexity of $N.(N - 1)/2$ because that is the number of interactions that need to be evaluated, and a depth-complexity of $N - 1$ because the thread that performs the

[1] Full formulation is available in mjsaldanha.com/articles/1-hpc-sspi/.

first outer loop evaluates that many interactions. Note that we are ignoring the work and depth of the reduction phase because it is performed in the exact same way in both the straightforward and the proposed approaches. In our proposed algorithm, the work-complexity is maintained (seen in Proposition 1), while the depth-complexity becomes $(N-1)/2$ for odd N, and $N/2$ for even N. Therefore, we reduced the depth while preserving the amount of work, which indicates that if we had infinite physical processing units, the proposed approach could be faster.

In practice, both approaches can be parallelized by assigning each outer iteration to one thread, followed by a reduction phase where the threads cooperate to accumulate the intermediate results obtained. As was already mentioned, it is possible to parallelize the straightforward approach in a balanced way by distributing the outer loops over a smaller number of threads in a round-robin fashion. However, although this might perform well in distributed or multicore systems, it increases the algorithm's depth, reduces the number of threads that can be launched and degrades memory locality, which are not good properties for GPUs. Besides that, it also makes it considerably more difficult to manage usage of shared memory, which is a fast memory shared only by a block of threads. For these reasons, we have implemented in the CUDA programming model only the straightforward parallelization and the proposed one, and we show in Sect. 5 that the proposed approach is slightly better.

5 Experiments and Results

In order to evaluate the performance of the $O(N)$ sequential counting algorithm of Sect. 3, we implemented both approaches, linear and quadratic, for counting collisions[2]. We designed the implementations so that they had similar characteristics to the protein structure prediction program (from [1]) analyzed and implemented in the broader context of this research. Hence, in each execution we perform the counting procedure upon multiple bead vectors, the space array is allocated only once, and each bead vector is generated by placing the first bead at $(0, 0, 0)$ and positioning the next bead in the neighborhood of the previous bead (similar to a protein), choosing any of the 6 directions randomly.

Figure 2 (left side) shows the experimental results. Each program was executed with varying problem sizes and each execution comprised counting the number of collisions for 1000 different bead vectors. For each problem size, we collected the wall clock time for each of 100 executions and took their mean. The total vertical length of the black error bars equals four standard deviations of the samples. These experiments were run using an Intel i7-4790 3.6 GHz and 32 GB of primary memory. For a problem size of 1920, which was the largest problem size that could be run in the system, the speedup was 21.7 and the linear approach required 52.82 GB of virtual memory, as pointed in Fig. 2 (left).

[2] In mjsaldanha.com/articles/1-hpc-sspi/ the reader can find the source code for all experimented programs mentioned in this article.

With the results shown in Fig. 2 (left), the gain in execution time provided by the linear approach is clear. However, this comes at the cost of high consumption of virtual memory, which greatly limits the largest problem size that can be supported by one's system. In the case of the protein structure prediction (PSP) program we implemented[2] during the research project that revolves this paper, the number of beads among which collisions are calculated rarely exceeds 1000 (proteins rarely have that number of amino acids), and for this problem size the virtual memory usage is 7.5 GB, a feasible amount. Experiments with the PSP program showed speedups of 11.8 and 72.4 for proteins with 128 (a common size) and 768 amino acids, respectively, so by using the proposed algorithm the program was accelerated significantly. Possible reasons for the higher speedup are related to factors that are discussed below.

Some considerations must be made regarding the generation of beads in resemblance to the PSP algorithms. By allocating the space array only once in each execution, we reduce the cost of requesting memory from the operating system, which is present only in the linear approach; on the other hand,

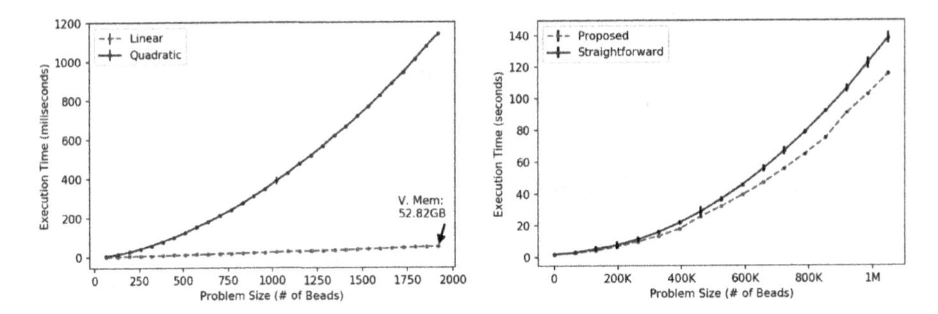

Fig. 2. To the left, execution time for linear and quadratic approaches for counting collisions. To the right, for counting collisions in GPU. All vertical black error bars have a length of four standard deviations of 100 samples taken.

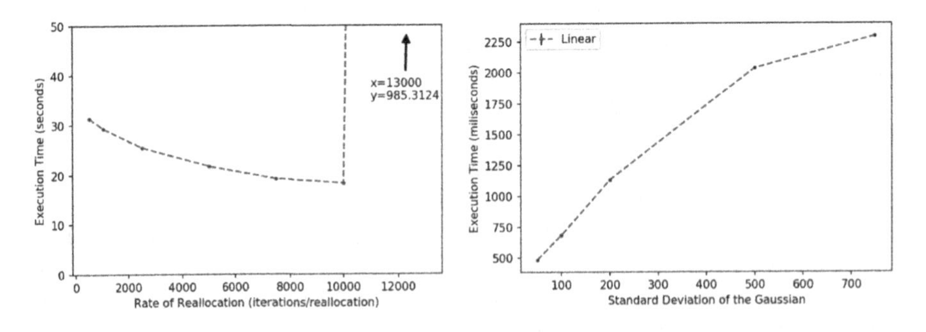

Fig. 3. Linear approach when: (to the left) varying the rate by which memory is reallocated, for a problem size of 13 000 vectors of beads; and (to the right) generating bead coordinates with a normal distribution with mean 0 and varying standard deviation, for a problem size of 1000 bead vectors.

this means we do not reclaim virtual memory after using it for a single vector of beads, causing physical pages that were mapped to remain mapped until the end of the program execution. For bigger problem sizes, memory should be reallocated every K iterations so as to free unused allocated virtual pages and prevent the program to swap memory. In Fig. 3 (left), we show what happens to the execution time when varying such number K of iterations. Notice that performance improves as we reallocate memory less often (due to lower memory management overhead), reaching an optimal point at a rate of one reallocation every 10 000 iterations. After that, the program begins to swap memory and performance degrades significantly.

A second consideration is that, by generating bead vectors as "random proteins", we are statistically confining beads to a smaller region around the origin of the space, hence improving cache usage and reducing the number of memory pages required. For better illustrating this, we show in Fig. 3 (right) what happens to execution time if we generate the beads' coordinates using a normal distribution with mean 0 and varying standard deviation. As expected, performance degrades at higher deviation due to a lower rate of cache hits.

It follows from these two considerations that the frequency of reallocation of the space array, as well as the regularity of bead positions, must be taken into consideration when using the linear approach. There is a lot of room for analyzing such factors and how they apply to real applications; as that is not the objective of this paper, it is left as future work.

The parallel algorithms for SPI presented in Sect. 4 were implemented[2] in the CUDA programming model and evaluated using NVIDIA GPUs. Experiments used an NVIDIA Tesla P100 with 16 GB of memory and 3584 CUDA cores spread over 56 multiprocessors. Both the straightforward and the proposed approaches were optimized to achieve high occupancy and memory bandwidth, and all optimizations were applied to both approaches alike. Then, we executed each program 100 times for each problem size and took the mean of their wall clock execution times, which are shown in Fig. 2 (right side); four standard deviations are represented by the black vertical error bars. Each program execution comprised calculating collisions for 100 randomly generated vectors of floating point spheres with a diameter of 1. Even though comparisons with sequential approaches could have been made, it is quite clear that parallel algorithms yield lower execution times and the objective here is mainly to show the benefits of the proposed parallelization compared with the straightforward one.

Figure 2 (right) shows that the proposed approach seems to perform well on GPU. In fact, for all problem sizes larger than 525 000, the proposed approach is more than 12% faster than the straightforward one ($p < 0.01$ using a t-test assuming unknown and different variances). The speedup is mainly due to two factors. First, in the straightforward approach each CUDA block is responsible for calculating interactions of a group of 1024 spheres with *all* subsequent spheres; because of this, some blocks perform more work than others, and in the last moments not all of the 56 multiprocessors are used, and the program is waiting for a few lengthy blocks to end their part of the work. Second, besides this block-level unbalance, threads within a warp (group of 32 threads that execute simultaneously) are also unbalanced, so when the warp is nearly concluding its work the 32nd thread ends its job earlier and becomes idle waiting for the other threads to finish. In our proposed balanced approach, threads or multiprocessors that would otherwise be idle are put to perform useful work and contribute to finishing the interaction counting more quickly. Finally, if the problem in hand allowed the CUDA kernels to be launched in parallel, the GPU could hide a lot of the block-level unbalance; however, the warp-level one would remain negatively impacting the program speed.

6 Conclusion

Interaction counting is a usual problem, and when it needs to be performed, the quadratic pairwise-comparisons approach immediately comes to mind. For a long time this has been a significant bottleneck [12] on important areas such as computer graphics and scientific simulations; in the former, collision counting must be performed enough times per second to allow image frames to be delivered in a visually fluid way, and more frames mean more fluidity, so every millisecond matters; in the latter, simulations of galaxies or proteins may need a large number of iterations if a high level of reality is desired (possibly taking weeks to execute), so if the interaction counting performed every iteration is accelerated, either less simulation time would be required or more iterations could be performed in order to achieve better results. In either case, performing interaction counting in less time yields great benefits, which is why a lot of research has been done on the subject. However, research walked toward algorithms that focus on reducing the number of objects among which interaction counting must be performed using the usual $O(N^2)$ approach, often in parallel.

In this paper two algorithms are proposed that aim at improving the pairwise-comparisons approach itself: a sequential approach with $O(N)$ complexity that works well for punctual objects in a limited discrete space, and a parallel approach that runs more efficiently on GPUs than the pairwise-comparisons algorithm's straightforward parallelization. These approaches can, of course, be used together with the algorithms that focus on pruning sets of interacting objects, in order to accelerate the phase where brute-force interaction counting must be performed. By using the $O(N)$ approach, interaction counting can be made significantly faster, at the cost of high memory consumption; our experience with accelerating a protein structure prediction algorithm, as already mentioned, shows a speedup of 72.4 using the same hardware. The proposed parallel algorithm may be used on large problems, with objects of any shape, for counting any kind of interactions, and experiments using GPU show that it can yield a 1.12 speedup.

A possible direction for future research is to evaluate possible benefits of parallelizing the proposed $O(N)$ algorithm in order to share the memory consumption among nodes, which would allow the algorithm to be used for bigger problem sizes. Also, the $O(N)$ algorithm is sensitive to cache effects and to how frequently memory is allocated and deallocated, so these factors should be further investigated, especially when applied to real problems. Besides that, since the proposed parallel algorithm has nice properties it could be analyzed on different parallel platforms so as to understand the limitations of the algorithm on each architecture; examples are Intel Xeon Phi processors and FPGAs.

Acknowledgement. We thank São Paulo Research Foundation (FAPESP) for funding this research project (grant 2017/25410-8, associated to 2013/07375-0), and the Center for Mathematical Sciences Applied to Industry (CeMEAI) for providing access to powerful computational resources.

References

1. Benítez, C.M.V., Lopes, H.S.: Parallel artificial bee colony algorithm approaches for protein structure prediction using the 3DHP-SC model. In: Essaaidi, M., Malgeri, M., Badica, C. (eds.) Intelligent Distributed Computing IV, vol. 315, pp. 255–264. Springer, Heidelberg (2010). https://doi.org/10.1007/978-3-642-15211-5_27
2. Blelloch, G.E.: Programming parallel algorithms. Commun. ACM **39**(3), 85–97 (1996)
3. Bridson, R., Fedkiw, R., Anderson, J.: Robust treatment of collisions, contact and friction for cloth animation. ACM Trans. Graph. (ToG) **21**(3), 594–603 (2002)
4. Brochu, T., Edwards, E., Bridson, R.: Efficient geometrically exact continuous collision detection. ACM Trans. Graph. (TOG) **31**(4), 96 (2012)
5. Cohen, J.D., Lin, M.C., Manocha, D., Ponamgi, M.: I-COLLIDE: an interactive and exact collision detection system for large-scale environments. In: Proceedings of the 1995 Symposium on Interactive 3D Graphics, pp. 189–ff. ACM (1995)
6. Elseberg, J., Magnenat, S., Siegwart, R., Nüchter, A.: Comparison of nearest-neighbor-search strategies and implementations for efficient shape registration. J. Softw. Eng. Robot. **3**(1), 2–12 (2012)
7. Govindaraju, N.K., Redon, S., Lin, M.C., Manocha, D.: CULLIDE: interactive collision detection between complex models in large environments using graphics hardware. In: Proceedings of the ACM SIGGRAPH/EUROGRAPHICS Conference on Graphics Hardware, pp. 25–32. Eurographics Association (2003)
8. Greengard, L., Rokhlin, V.: A fast algorithm for particle simulations. J. Comput. Phys. **73**(2), 325–348 (1987)
9. Gregory, A., Lin, M.C., Gottschalk, S., Taylor, R.: A framework for fast and accurate collision detection for haptic interaction. In: ACM SIGGRAPH 2005 Courses, p. 34. ACM (2005)
10. Gregory, A., Mascarenhas, A., Ehmann, S., Lin, M., Manocha, D.: Six degree-of-freedom haptic display of polygonal models. In: Proceedings of the Conference on Visualization 2000, pp. 139–146. IEEE Computer Society Press (2000)
11. Knuth, D.E., Graham, R.L., Patashnik, O.: Concrete Mathematics: A Foundation for Computer Science. Adison Wesley, Boston (1989)
12. Lin, M., Gottschalk, S.: Collision detection between geometric models: a survey. In: Proceedings of IMA Conference on Mathematics of Surfaces, vol. 1, pp. 602–608 (1998)
13. Longmore, J.P., Marais, P., Kuttel, M.M.: Towards realistic and interactive sand simulation: a GPU-based framework. Powder Technol. **235**, 983–1000 (2013)

14. Nguyen, H.: Fast n-body simulation with CUDA. In: GPU Gems 3, chap. 31. Addison-Wesley Professional, New Jersey (2007)
15. Provot, X.: Collision and self-collision handling in cloth model dedicated to design garments. In: Thalmann, D., van de Panne, M. (eds.) Computer Animation and Simulation 1997, pp. 177–189. Springer, Vienna (1997). https://doi.org/10.1007/978-3-7091-6874-5_13
16. Redon, S., Kheddar, A., Coquillart, S.: Fast continuous collision detection between rigid bodies. Comput. Graph. Forum **21**, 279–287 (2002)
17. Selle, A., Lentine, M., Fedkiw, R.: A mass spring model for hair simulation. ACM Trans. Graph. (TOG) **27**(3), 64 (2008)
18. Stich, M., Friedrich, H., Dietrich, A.: Spatial splits in bounding volume hierarchies. In: Proceedings of the Conference on High Performance Graphics 2009, pp. 7–13. ACM (2009)
19. Tang, M., Kim, Y.J., Manocha, D.: Efficient local planning using connection collision query. Ewha Womans University, Korea, Technical report (2010)
20. Tang, M., Tong, R., Wang, Z., Manocha, D.: Fast and exact continuous collision detection with Bernstein sign classification. ACM Trans. Graph. (TOG) **33**(6), 186 (2014)
21. Wald, I.: On fast construction of SAH-based bounding volume hierarchies. In: IEEE Symposium on Interactive Ray Tracing, RT 2007, pp. 33–40. IEEE (2007)
22. Zheng, J., An, X., Huang, M.: GPU-based parallel algorithm for particle contact detection and its application in self-compacting concrete flow simulations. Comput. Struct. **112**, 193–204 (2012)

Comparing Domain Decomposition Methods for the Parallelization of Distributed Land Surface Models

Alexander von Ramm[1(✉)], Jens Weismüller[1], Wolfgang Kurtz[1], and Tobias Neckel[2]

[1] Leibniz Supercomputing Centre (LRZ), Boltzmannstraße 1, 85748 Garching, Germany
alexander.vonramm@lrz.de
[2] Department of Informatics, Technical University of Munich, Boltzmannstraße 3, 85748 Garching, Germany
http://www.lrz.de

Abstract. Current research challenges in hydrology require high resolution models, which simulate the processes comprising the water-cycle on a global scale. These requirements stand in great contrast to the current capabilities of distributed land surface models. Hardly any literature noting efficient scalability past approximately 64 processors could be found. Porting these models to supercomputers is no simple task, because the greater part of the computational load stems from the evaluation of highly parametrized equations. Furthermore, the load is heterogeneous in both spatial and temporal dimension, and considerable load-imbalances occur triggered by input data. We investigate different domain decomposition methods for distributed land surface models and focus on their properties concerning load balancing and communication minimizing partitionings. Artificial strong scaling experiments from a single core to 8,192 cores show that graph-based methods can distribute the computational load of the application almost as efficiently as coordinate-based methods, while the partitionings found by the graph-based method significantly reduce communication overhead.

Keywords: Load-balancing · Graph-partitioning · Hydrology · High-perfomance computing

1 Introduction

Predicting hydrological phenomena is of great importance for various fields such as climate change impact studies and flood prediction. An outline of the principles and structure of physically-based hydrological models is given in [8]. Experimental hydrology often provides the scientific basis for hydrological catchment models. However, the great complexity, size and uniqueness of hydrological

J. M. F. Rodrigues et al. (Eds.): ICCS 2019, LNCS 11536, pp. 197–210, 2019.
https://doi.org/10.1007/978-3-030-22734-0_15

catchments[1] makes a methodology involving physical experiments unfeasible if not impossible for catchments exceeding a few hectars. Hydrologists remedy by performing in silico experiments, simulating the hydrological processes in the basins.

Different classes of hydrological models exist. In this study, we focus on distributed land-surface models (dLSMs). dLSMs focus on detailed modelling of vertical processes and the correct simulation of mass and energy balance at the land-surface. To this end, they solve the hydrological water balance equation globally for the entire domain and locally for the subdomains dictated by the discretization.

A key characteristic of dLSMs is their focus on vertical surface-processes, such as evapotranspiration, snow processes, infiltration and plant growth. Lateral processes such as discharge concentration and lateral subsurface flow are simulated in a simplified way, if they are simulated at all. For this investigation we will consider vertical and lateral processes in a de-coupled way. The reason for this is the fact that vertical processes can, in general, be computed concurrently for all points in the domain. Lateral processes, in contrast, require information from the surrounding area and can therefore not be computed concurrently. They consist mainly of simple fluid dynamics simulations, and are computationally less demanding than the vertical processes. However, the structure of communication is dictated by the lateral processes, so for a scalable, parallel dLSM, an integrated approach that considers both aspects is required.

Various computer codes exist that implement different aspects of the hydrological theory behind dLSMs. WaSiM [22] for example is a hydrological model that solves the one-dimensional Richards's equation to simulate vertical soil water movement. PROMET [16] features a detailed simulation of plant growth. The open-source code WRF-Hydro [9] was designed to link multi-scale process models of the atmosphere and terrestrial hydrology. These models are the basis for a number of different applications ranging from flood prediction [10], climate change impact studies [17] to land use scenario evaluation [19].

In contrast to dLSMs, which focus on surface processes, integrated hydrological models such as Parflow [18] or HydroGeoSphere [6] focus on the simulation of subsurface flow. They solve the three-dimensional Richards's equation fully coupled to the surface runoff. Parallel implementations of integrated hydrological models, scaling to supercomputer capabilities are already available [7]. However, for a number of reasons integrated hydrological models can not be used substitutively for dLSMs.

The hydrological modelling community using dLSMs is currently moving from desktop models to small-size distributed clusters. Currently, dLSMs do not scale well beyond 64 processors [24]. The authors of [4] list four examples related

[1] In [25], the hydrological catchment is defined as "the drainage area that contributes water to a particular point along a channel network (or a depression), based on its surface topography." As such it makes a suitable logical unit of study in hydrology, which can be modelled by physically-based models.

to global environmental change that require high-resolution models of terrestrial water on a global scale.

Advanced methods for Bayesian inference and uncertainty quantification such as e.g. [3] are often based on Markov-Chain Monte Carlo methods, which require a great amount of sequential model evaluations. In order to obtain results for dLSM-based hydrology analyses in a reasonable amount of time, the execution time of a single model evaluation needs to be reduced. Parallelization and deployment on supercomputers is one possibility to achieve this. In summary, better parallelization schemes and dLSMs scaling to the abilities of modern supercomputers would greatly advance the capabilities of hydrologists.

Parallelization efforts have been undertaken for a number of hydrological models. However, the employed strategies rarely aim for an ideal load balancing and minimized communication efforts. In [14], a hydrological model is parallelized under the assumption that downstream cells require information of upstream cells. The catchment is interpreted as a binary tree, which is partitioned and distributed among processors, using a master-slave approach. With this approach, the authors decrease the execution time of the serial algorithm by a factor of five on multiple processors. WRF-Hydro partitions the domain into rectangular blocks [9], yet we are not aware of any scaling experiments for this implementation. In WaSiM [22], the domain is partitioned by distributing different rows to different processors. The parallelization strategies of WRF-Hydro and WaSiM both do not take the communication structure into account that is necessary to compute the flow-equations in parallel. It is likely that this limits the scaling potential to a limited amount of processors. In [24], the TIN-based hydrological model tRIBS is parallelized. A graph-based domain decomposition method is employed to produce a communication efficient partitioning. Scaling experiments indicate scalability up to about 64 processors. No contribution except [24] attempts to minimize the required point-to-point communication. Therefore, the hydrological community still lacks a parallelized dLSM which efficiently scales to the capabilities of modern supercomputers.

In this paper, we re-visit some of the graph-based domain decomposition methods suggested by [24], adapt it for a regular grid and conduct a more thorough analysis of its communication patterns. We compare different coordinate-based domain decomposition methods with a graph-based method, and investigate the impact on the necessary point-to-point communication. Furthermore, we perform an artificial strong scaling experiment and compute theoretical peak values for parallel speed-up for an example application on up to 8,192 processors. As we want to conduct a methodological study of the suitability of different domain-decomposition methods for dLSMs we limit the evaluation to a simplified communication and work-balance model and do not measure performance of an actual application.

We will start by giving an overview over the functionality of dLSMs. Subsequently, we introduce the governing equations of the lateral processes and show how they dictate a specific communication structure. We then introduce the different domain decomposition methods. Finally, we present theoretical peak values of speed-up and efficiency and evaluate the potential for the minimization of point-to-point communication.

2 Theory

2.1 The Computational Domain of dLSMs

The focus of this study lies on the efficient decomposition of the computational domain of dLSMs. The computational domain commonly comprises one hydrological catchment as defined in the introduction, is denoted by $\Omega \in \mathbb{R}^2$ and fulfills the following assumptions:

1. Any point $x \in \Omega$ has an elevation h(x).
2. There is exactly one outlet O_Ω located on the boundary $\partial\Omega$ of the domain Ω, where the minimum h_0 of h is located.
3. Ω does not contain any sinks, i.e., there is a monotonously decreasing path from any point $x \in \Omega$ to the outlet O_Ω.

In practice, the domain is commonly derived from a digital elevation model of the basin and a given outlet O_Ω with a gauging station where discharge measurements are taken.

For Ω as well as for any subdomain ω of Ω, the hydrological water balance equation holds. Commonly the domain is discretized into a regular grid. However, other discretization methods also exist. The water budget equation is then solved for any cell of the grid individually. This step usually requires the greatest computational effort. The structure of point-to-point communication is dictated by the lateral processes. In dLSMs these processes are commonly simulated under the following assumptions:

1. The flow direction is dependent on the topography given by h.
2. Flow follows the steepest gradient of h on the domain Ω.
3. Flow is one-dimensional along the deterministic flow paths derived under the previous two assumptions.

The exact drainage network is commonly derived from the digital elevation model of the domain by the D8-Algorithm [20]. This algorithm assigns a flow direction to each cell under the consideration of the altitude $h(x)$ of all eight neighbouring cells. The flow is always directed towards the neighbouring cell with the smallest altitude. In Fig. 1, a simple example of a domain for a dLSM is given.

2.2 Governing Equations

In hydrological catchments, a number of lateral processes occur physically. These include, but are not limited to, channel flow, surface runoff and subsurface flow. In this paper, we focus on channel flow, but the method can be extended to include other lateral processes. Given the assumption of one-dimensional flow,

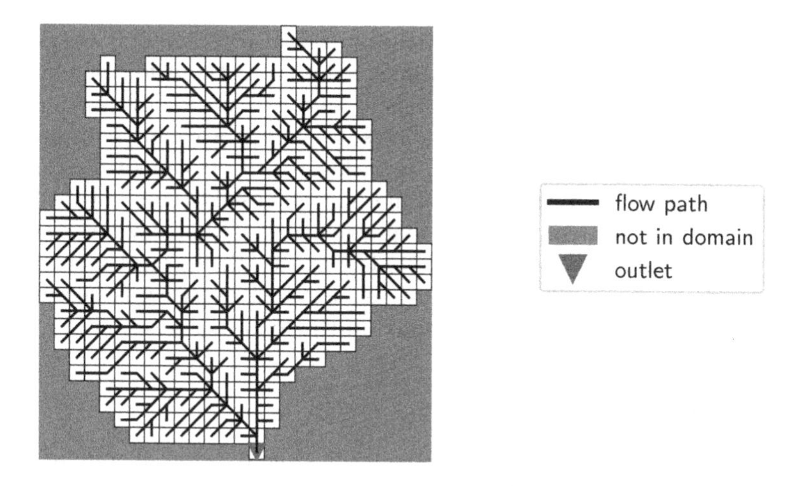

Fig. 1. Typical domain for the solution of the Saint-Venant Equations in hydrological land surface models. White cells show cells for which the hydrological water balance equation is solved. Black lines indicate the flow structure, derived by the D8-Algorithm, on which the Saint Venant equations are solved. Grey cells are not part of the catchment to be simulated and thus not part of the domain Ω.

the governing equations of the channel flow are the Saint-Venant equations, with spatial dimension x and time t:

$$\frac{\delta Q(x,t)}{\delta t} + \frac{\delta A(x,t)}{\delta t} = f(x,t) \tag{1}$$

$$\frac{1}{A(x,t)}\frac{\delta Q(x,t)}{\delta t} + \frac{1}{A(x,t)}\frac{\delta}{\delta x}\left(\frac{Q(x,t)^2}{A(x,t)}\right) + g\frac{\delta y}{\delta x} - g(S_0 - S_f) = 0 \tag{2}$$

Here, Q is the discharge measured in $[m^3/s]$, A represents the cross-sectional area given in $[m^2]$, g is gravitational acceleration and y is the water level in [m]. Finally, S_0 represents the slope of the channel, S_f is the friction slope and $f(x,t)$ is a source term describing the runoff generated on every point x for every time step t. The source term $f(x,t)$ represents the result of the hydrological water budget equation, and as such its computation comprises the simulation of all vertical processes. The one-dimensional Saint-Venant equations are solved on the flow structure derived by the D8-Algorithm. For the flow structure graph, the following holds:

1. All cells of the discretized domain Ω_h are nodes of the graph.
2. A cell can only be connected by an edge to its neighbouring cells.
3. The root of the flow structure is located at the outlet O_Ω.

Note that whenever the flow structure (i.e. the graph) is cut during domain decomposition, point-to-point communication is induced. Hence, for an efficient communication the graph should be cut as little as possible.

Solutions of the Saint-Venant equations in dLSMs are commonly obtained by kinematic or diffusive wave approximations. Solution methods for the kinematic wave usually involve looping over the cells in an upstream to downstream order to compute a discharge for each cell. For these kinematic wave methods, only the discharge of upstream cells is required, as backwater effects cannot be simulated with this method [23]. Diffusive wave methods can be used to simulate backwater effect. Thus, they also require additional information from downstream cells, which induces additional communication. The exact parallel implementation of these methods is not part of this study, but a rough overview is necessary in order to comprehend the method proposed in the following section.

2.3 Domain Decomposition

At the core of most parallelization strategies lies the division of computational work among computational resources. We investigate an approach which involves the decomposition of the domain Ω into subdomains ω to be computed on different processors. For the domain decomposition, we consider four different algorithms, three coordinate-based and one graph-based. For all algorithms we use the implementation provided by the Zoltan library [5].

The first algorithm considered is called "Block-Partitioning". The algorithm considers the unique ID of each cell and assigns each processor a block of ids. Therefore, this algorithm is heavily dependent on the cell-ID, which is application dependent. We include this algorithm in the analysis, as it is the most commonly used domain decomposition method included in current parallel implementations of dLSMs. WaSiM [22], for example, uses an altered version of this algorithm.

Secondly, we consider Recursive Coordinate Bisection [2]. This algorithm determines partitions by recursively dividing the domain along its longest dimension. This method reduces communication by minimizing the length of partition borders.

Additionally, a method employing the Hilbert curve was considered. Here, discrete iterates of the space-filling Hilbert curve[2] are constructed. Partitioning is done by cutting the preimage of the discrete iterates of the Hilbert curve into parts of equal size and assigning the resulting 2D subdomains to different processors [5].

Finally, we investigate the method also used by [24]. Here, the flow direction graph is considered in order to minimize the necessary communication. We employ graph-partitioning methods, which attempt to partition the given graph into sub-graphs of almost equal size, while minimizing the amount of edges cut. We use the parallel graph-partitioning algorithm described in [11,13] and parallelized in [12,21].

[2] Space-filling curves are surjective maps of the unit interval onto the 2D unit square (or a general rectangle) which provide decent surface-to-volume ratios of the resulting 2D subdomains when used for parallel partitioning; see [1] for details on and properties of space-filling curves.

3 Application Example

In order to avoid the overhead of the parallelization of an entire dLSM, we first measured the execution time per cell and time step of a PROMET model of the Upper-Danube catchment with 76,215 cells. These measurements were then used to perform an artificial strong scaling experiment from one processor to 8,192 processors. The outlet of this model is located at the gauging station in Passau, Germany. A more detailed description of the catchment, the model and its validation is given in [15]. The catchment was chosen for its heterogeneous domain, which includes cells in alpine regions as well as cells with agricultural use in the Alpine Foreland. These characteristics suggest a heterogenous load behaviour. Furthermore, it represents a typical model size of current dLSMs.

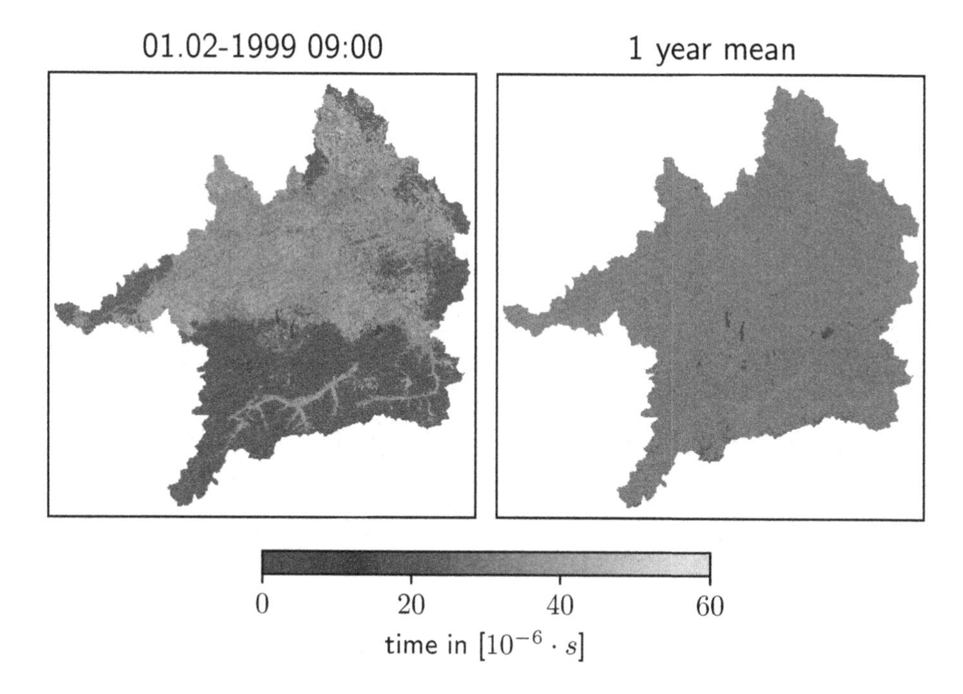

Fig. 2. Measured execution times for individual cells for one time step (left) and mean execution time over one year (right) of the hydrological Model PROMET [16]. The individual time step demonstrates the heterogeneity of computational load in one time step. The one-year mean shows the homogeneity of the total computational effort over the entire simulation period.

The measured cell executions times of the Upper-Danube model are displayed in Fig. 2. While the total computational effort is homogeneously distributed, the computational effort for individual time steps can be quite heterogeneous.

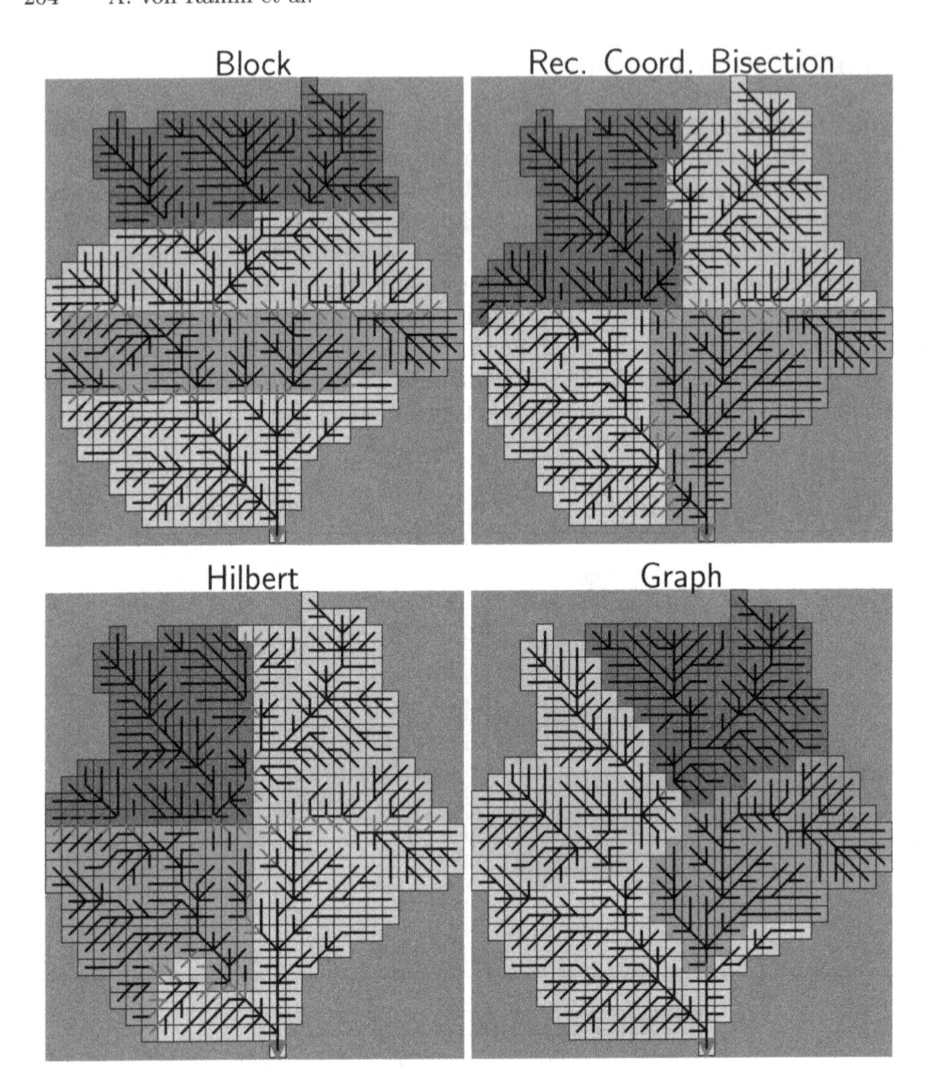

Fig. 3. Resulting partitioning of the different algorithms applied to a small head-catchment. Cells with identical colour are computed on the same processor. Flow paths in red correspond to cut edges in the graph.

The results of the domain decomposition algorithms introduced in the previous section for a small, overseeable head catchment are displayed in Fig. 3.

The illustration of the block-wise domain decomposition (top-left) shows PROMET's scheme to determine the cell-ID by sequential assignment from the top left to the bottom right corner of the domain.

The scaling experiments in Sect. 4 were performed using the measured cell execution times. Rather than performing dummy-calculations to simulate the load generated by the source-term $f(x, t)$ of Eq. (1), we decided to estimate the runtime by computing the sum of execution times on each processor. So with exception of the single cell measures, no real application code is executed. This allowed us to consider more methods without the overhead of implementing them into a dLSM and wasting CPU-hours on dummy calculations. Consequently, the reported metrics are theoretical peak values, which only take into account the load balance.

In order to evaluate the scaling experiments, we computed theoretical peak values of parallel speed-up S_p and the parallel efficiency E_p based on the estimated runtimes. These are:

$$S_p = \frac{T_1}{T_p} \tag{3}$$

$$E_p = \frac{S_p}{p}, \tag{4}$$

where T_1 is the total runtime on a single processor, and T_p is the runtime on p processors. Furthermore, the edge-cut count E_C was computed, which is the sum of all edges spanning over two processors. We use the edge-cut count as an estimate for the communication overhead introduced by a parallel implementation of the lateral processes.

We considered two synchronization scenarios, which should represent the lower and upper boundary in terms of synchronization requirements. The first scenario (unsynchronized) assumes no synchronization during the simulation and represents the lower boundary. The second scenario (synchronized) assumes that all cells need to be synchronized at the end of every time step and therefore represents the upper boundary, and would be required for a diffusive wave model for the lateral processes.

4 Results

The theoretical speed-up values computed during the strong scaling experiment are displayed in Fig. 4.

Figure 4 shows that, for synchronized simulations, all methods solve the load-balancing task equally well, and produce a partitioning which yields a theoretical speed up of 5,740. This corresponds to a parallel efficiency of 0.7. For unsynchronized simulations, the coordinate-based methods outperform the graph-based method by 6%, with theoretical speed up values of 7,200 and 6,722 respectively. This can be explained by the more constant partition size produced by the coordinate-based methods. Without a synchronization barrier at the end of each time step, the total computational load over the whole simulation time needs to be balanced, rather than the more heterogeneous load at each individual time step. The distribution of partition sizes is less relevant for the synchronized case, because load-imbalances are realized at the end of every time step without the possibility to be damped over the remaining simulation.

Much more severe differences between the methods can be found in the communication overhead potentially introduced by a parallel implementation

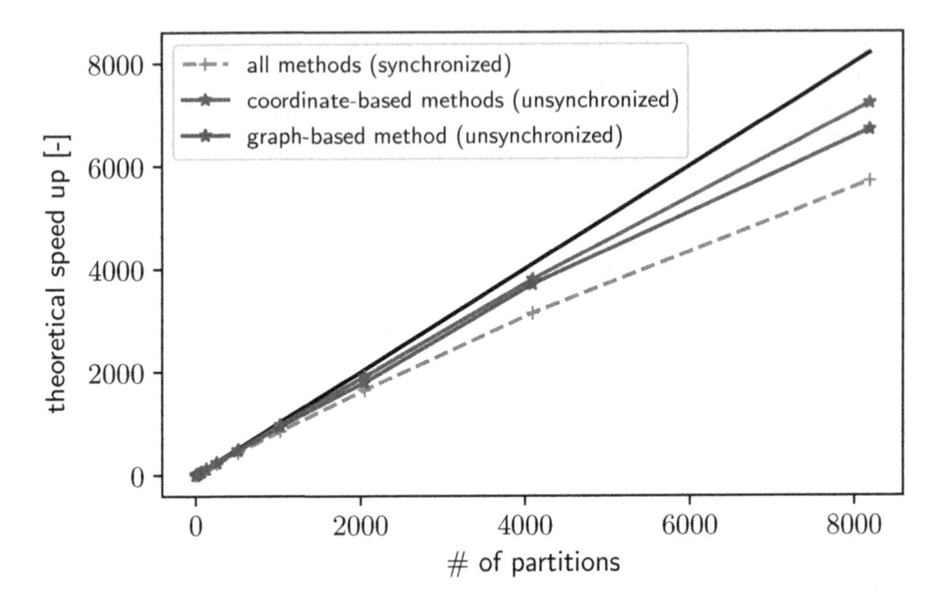

Fig. 4. Theoretical peak values of parallel speed-up computed for the four different domain decomposition algorithms for the synchronized and unsynchronized scenario. Runtime was estimated by summing the execution times of all cells on given processor. At 8,192 processors, each partition contains approximately nine cells.

of lateral processes. In Fig. 5, the total amount of flow paths cut is displayed. Here, the graph-based method shows significant advantages over the coordinate-based methods. The superiority of graph-partitioning becomes even more apparent if the flow paths cut per partition are considered. Here, the graph-based method is capable of producing partitionings with approximately 2.1 flow path cuts per partition, which mostly corresponds to one inlet and one outlet per partition. This holds true over the entire range of processors considered. For coordinate-based methods the number of flow paths cut is strongly dependant on the length of the partition border. Thus, for decreasing partition sizes the methods approach again. However, for future problems with significantly more than 76,000 cells and therefore bigger partitions the superiority of the graph-based method will be even more severe.

5 Discussion

For unsynchronized simulations, there is a strong connection between the number of cells computed on a processor and the workload of this processor. This is caused by the homogeneous distribution of the total computational effort of the simulation (see Fig. 2 (right)). For synchronized simulations, this connection is significantly weaker, because load-imbalances that are present in a given time step are realized immediately. In unsynchronized simulations, these imbalances can dissipate over the remaining simulation. Therefore, the disadvantage

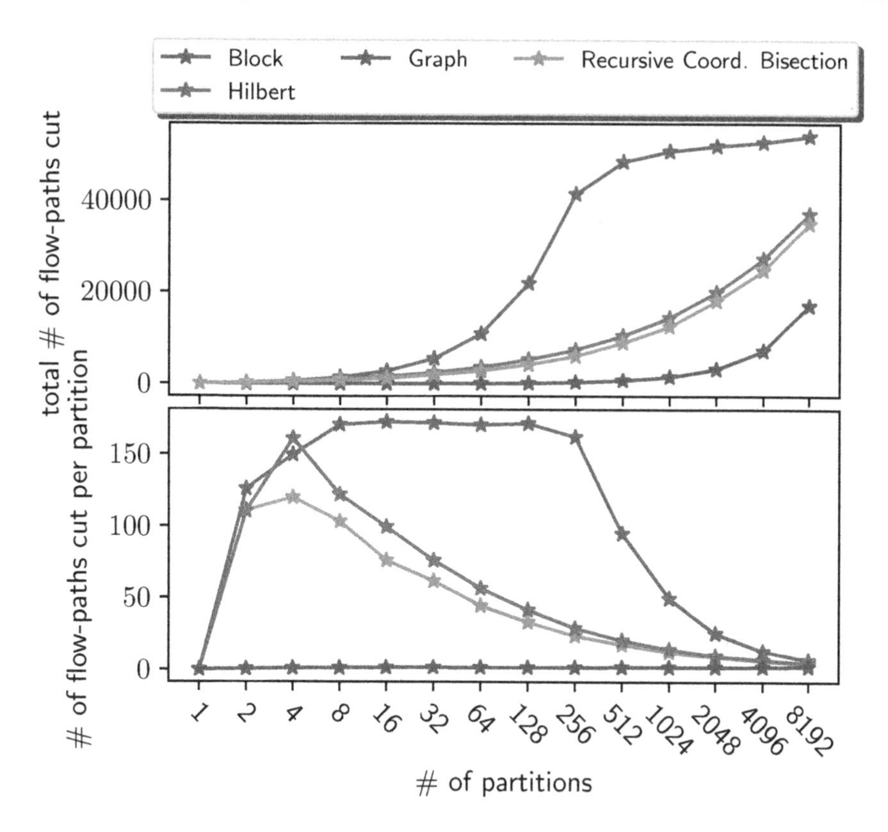

Fig. 5. Total number of flow paths cut and number of paths cut per partition for four different domain decomposition algorithms on different numbers of processors. The edge-cut count is determined by counting the edges that connect nodes on different partitions. It serves as a proxy for the necessary point-to-point communication. At 8, 192 processors, each subdomain contains approximately 9 cells.

of the graph-based method seen for unsynchronized simulations is not relevant for synchronized simulations.

Furthermore, the great drop between unsynchronized and synchronized simulations indicates that speed-up could be further improved by relaxing the synchronization requirements of the different algorithms for the simulation of the lateral processes. Of course, this has to be done under the consideration of the underlying physical processes.

Subsequently, the superiority of the of the graph-based domain decomposition in terms of the number of flow paths cut per partition over the entire range of processors needs to be noted. This result emphasizes the advantages of using graph-based domain decomposition for parallel implementations of the lateral processes in dLSMs most convincingly.

The artificial strong scaling experiment was conducted up to 8, 192 processors and yielded parallel efficiences between 0.70 and 0.88. These are, in comparison to integrated hydrological model such as Parflow, rather poor efficiencies on a small amount of processors. The catchment model of the Upper-Danube in the considered resolution simply is not big enough to be scaled to more processors. Models with more cells need to considered for scaling experiments with more processors. It is assumed that hydrologists will want to compute such models in the near future. Therefore, larger test scenarios with the possibility to be scaled up to the capabilities of current supercomputers will become available soon.

Furthermore, these results have to be seen in the context of current parallel dLSMs. The implementation of [24] reaches a parallel efficiency of approximately 0.3 on 64 processors.

Other codes scale even worse. For hyper-resolution global modelling as described in [4] this will not suffice. Also, for catchment studies at higher spatial resolutions, scalability beyond 64 processors is advantageous and would support scientific progress in hydrology.

6 Conclusion

We investigated several domain decomposition strategies for dLSMs. Strategies based purely on the coordinates of cells as well as a strategy acknowledging the special nature of the domain of the lateral processes were considered. We performed simplified scaling experiments to evaluate the suitability of these methods, and computed theoretical peak values for speed-up and parallel efficiency. For synchronized simulations, graph-based and coordinate-based domain decomposition yield similar speed-up values, which suggests that the advantageous communication structure of the graph-based methods would lead to a more scalable solution. For unsynchronized simulations, coordinate-based methods scale about 6% better. It remains to be demonstrated that the advantageous communication structure makes up for this gap.

One of the shortcomings of this study, the disregard of the induced communication effort, will be addressed by a parallel implementation of the methods used to solve the Saint-Venant equations in dLSMs. Furthermore, for dLSMs inducing greater and more heterogenous computational effort dynamic load-balancing strategies should be researched. Examples for such dLSMs include WaSiM, where the vertical soil water movement is simulated by the 1D-Richards's equation.

Acknowledgements. This work has in parts been funded by the German Federal Ministry of Education and Research under the grant reference 02WGR1423F ("Virtual Water Values" project), and by the Bavarian State Ministry of the Environment and Consumer Protection under the Hydro-BITS grant. The responsibility for the content of this publication lies with the authors.

References

1. Bader, M.: Space-Filling Curves. Springer, New York (2010)
2. Berger, M.J., Bokhari, S.H.: A partitioning strategy for nonuniform problems on multiprocessors. IEEE Trans. Comput. **C–36**(5), 570–580 (1987)

3. Betz, W., Papaioannou, I., Beck, J.L., Straub, D.: Bayesian inference with subset simulation: strategies and improvements. Comput. Methods Appl. Mech. Eng. **331**, 72–93 (2018). https://doi.org/10.1016/j.cma.2017.11.021
4. Bierkens, M.F.P., et al.: Hyper-resolution global hydrological modelling: what is next? Hydrol. Process. **29**(2), 310–320 (2014). https://doi.org/10.1002/hyp.10391
5. Boman, E., et al.: Zoltan 3.0: parallel partitioning, load-balancing, and data management services; user's guide. Sandia National Laboratories, Albuquerque, NM. Technical report SAND2007-4748W (2007). http://www.cs.sandia.gov/Zoltan/ug_html/ug.html
6. Brunner, P., Simmons, C.T.: HydroGeoSphere: a fully integrated, physically based hydrological model. Ground Water **50**(2), 170–176 (2011). https://doi.org/10.1111/j.1745-6584.2011.00882.x
7. Burstedde, C., Fonseca, J.A., Kollet, S.: Enhancing speed and scalability of the parflow simulation code. Comput. Geosci. **22**(1), 347–361 (2018). https://doi.org/10.1007/s10596-017-9696-2
8. Freeze, R., Harlan, R.: Blueprint for a physically-based, digitally-simulated hydrologic response model. J. Hydrol. **9**(3), 237–258 (1969). https://doi.org/10.1016/0022-1694(69)90020-1
9. Gochis, D.J., et al.: The WRF-Hydro modeling system technical description, (Version 5.0) (2018). https://ral.ucar.edu/sites/default/files/public/WRF-HydroV5TechnicalDescription.pdf
10. Karsten, J., Gurtz, J., Herbert, L.: Advanced flood forecasting in alpine watershed by coupling meteorological and forecasts with a distributed hydrological model. J. Hydrol. **269**, 40–52 (2002)
11. Karypis, G., Kumar, V.: Multilevel algorithms for multi-constraint graph partitioning. In: SC 1998: Proceedings of the 1998 ACM/IEEE Conference on Supercomputing, p. 28 (1998). https://doi.org/10.1109/SC.1998.10018
12. Karypis, G., Kumar, V.: A coarse-grain parallel formulation of multilevel k-way graph partitioning algorithm. In: Parallel Processing for Scientific Computing. 8th SIAM Conference on Parallel Processing for Scientific Computing (1997)
13. Karypis, G., Kumar, V.: Multilevelk-way partitioning scheme for irregular graphs. J. Parallel Distrib. Comput. **48**(1), 96–129 (1998). https://doi.org/10.1006/jpdc.1997.1404
14. Li, T., Wang, G., Chen, J., Wang, H.: Dynamic parallelization of hydrological model simulations. Environ. Model. Softw. **26**, 1736–1746 (2011). https://doi.org/10.1016/j.envsoft.2011.07.015
15. Mauser, W.: Global Change Atlas - Einzugsgebiet Obere Donau - online, chap. E4 - Validierung der hydrologischen Modellierung in DANUBIA. GLOWA-Danube-Projekt (2011)
16. Mauser, W., et al.: PROMET - Processes of Mass and Energy Transfer An Integrated Land Surface Processes and Human Impacts Simulator for the Quantitative Exploration of Human-Environment Relations Part 1: Algorithms Theoretical Baseline Document (2015). http://www.geographie.uni-muenchen.de/department/fiona/forschung/projekte/promet_handbook/index.html
17. Mauser, W., Bach, H.: Promet - large scale distributed hydrological modelling to study the impact of climate change on the water flows of mountain watersheds. J. Hydrol. **376**(3–4), 362–377 (2009)
18. Maxwell, R.M., et al.: PARFLOW User's Manual
19. Niehoff, D., Fritsch, U., Bronstert, A.: Land-use impacts on storm-runoff generation: scenarios of land-use change and simulation of hydrological response in a meso-scale catchment in sw-germany. J. Hydrol. **267**(1–2), 80–93 (2002)

20. O'Callaghan, J.F., Mark, D.M.: The extraction of drainage networks from digital elevation data. Comput. Vis. Graph. Image Process. **28**(3), 323–344 (1984). https://doi.org/10.1016/S0734-189X(84)80011-0
21. Schloegel, K., Karypis, G., Kumar, V.: Parallel multilevel algorithms for multi-constraint graph partitioning. In: Bode, A., Ludwig, T., Karl, W., Wismüller, R. (eds.) Euro-Par 2000. LNCS, vol. 1900, pp. 296–310. Springer, Heidelberg (2000). https://doi.org/10.1007/3-540-44520-X_39
22. Schulla, J.: Model Description WaSiM. Hydrology Software Consulting J. Schulla (2017). http://www.wasim.ch/de/products/wasim_description.htm
23. Todini, E.: A mass conservative and water storage consistent variable parameter Muskingum-Cunge approach. Hydrol. Earth Syst. Sci. **11**(5), 1645–1659 (2007). https://doi.org/10.5194/hess-11-1645-2007
24. Vivoni, E.R., et al.: Real-world hydrologic assessment of a fully-distributed hydrological model in a parallel computing environment. J. Hydrol. **409**(1–2), 483–496 (2011). https://doi.org/10.1016/j.jhydrol.2011.08.053
25. Wagener, T., Sivapalan, M., Troch, P., Woods, R.: Catchment classification and hydrologic similarity. Geogr. Compass **1**(4), 901–931 (2007). https://doi.org/10.1111/j.1749-8198.2007.00039.x

Analysis and Detection on Abused Wildcard Domain Names Based on DNS Logs

Guangxi Yu[1,2], Yan Zhang[1,2](✉), Huajun Cui[1], Xinghua Yang[1], Yang Li[1,2], and Huiran Yang[1]

[1] Institute of Information Engineering, Chinese Academy of Sciences, Beijing, China
zhangyan80@iie.ac.cn
[2] School of Cyber Security, University of Chinese Academy of Sciences, Beijing, China

Abstract. Wildcard record is a type of resource records (RRs) in DNS, which can allow any domain name in the same zone to map to a single record value. Former works have made use of DNS zone file data and domain name blacklists to understand the usage of wildcard domain names. In this paper, we analyze wildcard domain names in real network DNS logs, and present some novel findings. By analyzing web contents, we found that the proportion of domain names related to pornography and online gambling contents (referred as *abused* domain names in this work) in wildcard domain names is much higher than that in non-wildcard domain names. By analyzing behaviors of registration, resolution and maliciousness, we found that abused wildcard domain names have remarkably higher risks in security than normal wildcard domain names. Then, based on the analysis, we proposed GSCS algorithm to detect abused wildcard domain names. GSCS is based on a domain graph, which can give insights on the similarities of abused wildcard domain names' resolution behaviors. By applying spectral clustering algorithm and seed domains, GSCS can distinguish abused wildcard domain names from normal ones effectively. Experiments on real datasets indicate that GSCS can achieve about 86% detection rates with 5% seed domains, performing much better than BP algorithm.

Keywords: DNS · Abused wildcard domain name · Analysis · Detection

1 Introduction

The Domain Name System (DNS) is an important part of critical Internet infrastructure, which aims to translate domain names into IP addresses. In fact, the mappings are recorded in different record types, called resource records (RRs). One of these record types is *wildcard* record, which can allow any domain name in the same zone to map to a single record value (i.e. IP or domain name). Wildcard RRs are original used to forward mail to the same zone [1]. But today, with the development of Internet applications and services, wildcard RRs are used widely. Besides normal applications and services, some malicious attacks also take advantage of wildcard RRs.

J. M. F. Rodrigues et al. (Eds.): ICCS 2019, LNCS 11536, pp. 211–225, 2019.
https://doi.org/10.1007/978-3-030-22734-0_16

To understand the use of wildcard domain names, especially the malicious usage, some works have been presented. In paper [2], based on DNS zone files, researchers found that 17.8% domain names were wildcard domain names and 19.1% of them were involved in blackhat SEO. In paper [3], based on zone file data and domain name blacklists, researchers found that wildcards are popular among all types of Internet domains. And among malicious users, spammers use wildcards the most. All these works have made rich achievements, but so far, there is no comprehensive study to analyze wildcards usage based on user request data (e.g. passive DNS data or DNS logs), which can directly express real query behaviors of wildcard domains.

In this paper, we first perform an analysis on wildcard domain names from perspectives of normal domain names and abused domain names. In this study, we regard the domain names related to pornography and online gambling as ABUSED domain names, because these contents are illegal in China[1]. And we regard all wildcard domain names except for abused ones as normal wildcard domain names. Being different from prior studies constructing dataset from zone files or known malicious domain lists, we analyze wildcard domain names in real network DNS logs, which are collected from a large ISP network containing millions of hosts. In addition, we collect auxiliary data including WHOIS information and web content information to gain an insight from original DNS logs. On the whole, we analyzed 919,939 domain names. We found that 153,163 (17% of all) domain names are wildcard domain names. Then by analyzing the 66.4% wildcard domain names with web contents, we found that 22.5% of them are abused domain names (related to pornography and online gambling). What's more, by analyzing wildcard domain names' behaviors of registration, resolution and maliciousness, we also found that the abused wildcard domain names have remarkably higher risks in security than normal ones.

Then, based on the analysis of wildcard domain names, we propose a machine learning based algorithm named GSCS (Graph based Spectral Clustering with Seeds) to distinguish abused wildcard domain names from normal ones. Our GSCS algorithm includes the following steps. First, to discover the similarity of resolution behaviors, we build a bipartite DNS graph and its projection graph for abused domain names. Then, by applying simple and efficient spectral clustering algorithm on the similarity matrix of the projection graph, we can divide wildcard domain names into different clusters. Finally, based on seed domain names, we can discover inherent clustered groups of abused wildcard domain names. Our experiment results based on real datasets show that GSCS can detect abused wildcard domain names more effectively than BP (belief propagation) algorithm.

Our main contributions in this paper include:

- We found that the proportion of abused domain names (i.e., domain names related to pornography and online gambling contents) in wildcard domain names is much higher than that in non-wildcard domain names. Specifically, 22.5% versus 4.4%.

[1] For illegal contents, we should note that various countries hold different laws. For example, all pornographic contents in the Internet are illegal in China, but only the contents with child pornography are illegal in U.S.

- We found that, compared with normal wildcard domain names, abused wildcard domain names have remarkably higher risks in security, including a higher proportion of domains related to malicious activities (10% versus 2.7%), a lower proportion of domains deploying SSL (2.3% versus 14%), and being more likely to avoid regulation (be registered out of China, in bulk and in recent years).
- We propose an effective algorithm GSCS to detect abused wildcard domain names. Compared with the BP algorithm which can get only 72% detection rate, GSCS can improve the detection rate to 86%.

The rest of this paper is structured as follows. In Sect. 2, we provide background information on DNS and wildcard domain names. Section 3 describes the analysis of our dataset. In Sect. 4, we make a comprehensive analysis of wildcard domain names based on web content and WHOIS information. And we propose an abused wildcard domain names detection algorithm in Sect. 5. Section 6 summarizes the related work. Finally, Sect. 7 concludes the paper's work.

2 Background

Domain Name System. The domain name system is a hierarchical system, which contains local DNS servers, authority name servers and root servers. Correspondingly, a domain name is also a hierarchical string with each level related to a zone. In detail, a domain name d consists of a set of labels separated by dots; they are called top-level domain (TLD), second-level domain (2LD), third-level domain (3LD), etc., from right to left. TLDs are managed by registries such as CNNIC (for *cn*) and Versign (for *com* and *net*), and 2LDs are offered to public by registrars such as Alibaba and GoDaddy. Before using a domain name in the Internet, domain owner should get its 2LD from a registrar. Then, the WHOIS information of this domain name is updated to database. In general, for a domain name with a benign website, the meaning of domain name is related to content of website. However, malicious domains are usually not.

Wildcard Domain Names. Wildcard domain names are domain names starting with an asterisk label (*) to match non-existing subdomain names. Note that, names beginning with other labels are never wildcard domain names, and the asterisk at other places in the domain will also not work as a wildcard. As mentioned before, wildcard RRs can allow any domain name in the same zone to map to a single record value, and simple examples are shown in Table 1.

Table 1. Simple examples of wildcard RR.

`*.example1.com` 3600 IN MX 10 `a.example1.com`
`*.example2.com` 3600 IN A `1.2.3.4`

In the beginning, wildcard domain names are used to forward mails [1]. As shown in Table 1, the MX RR would cause any MX query for any domain name ending in *example1.com* to return an MX RR pointing at *a.example1.com*. In the following, because many DNS implementations diverge from the original definition of wildcards, some other record types are extended [4]. In addition, several domain name registrars have also deployed wildcard records for TLDs to provide a platform for advertising. Some of these TLDs are country code TLDs (ccTLDs) such as *.fm*, *.la*, and there are also some Internationalized TLDs, for example the wildcard domain name *.*中国* has been resolved to an IP address *218.241.116.40*, which belongs to CNNIC. Because wildcard TLD domain names are usually maintained by domain name registrars, in this paper, we ignore these cases and only consider the wildcard domain names with 2LDs.

3 Data Collection

Previous works about wildcard domain names analysis collected dataset from some zone files or some malicious domain lists. In this paper, we collect data from real network DNS logs, and analyze the usages of wildcard domain names comprehensively. Additionally, we utilize auxiliary data like WHOIS and web content, etc. Below we elaborate our data.

DNS Logs. We measure wildcard domain names by analyzing DNS real logs, which are generated by local DNS servers operated by a large ISP in China. These logs record the interactive information between local DNS servers and client hosts. As shown in Table 2, each record in the logs consists of five fields. For the log data size, take a middle level province as an example, it is over 1.9 TB per day. In this paper, we collected DNS logs over five days, from January 1 to January 5, 2018. Note that we only considered the normal queries with NOERROR response. Finally, we obtained 919,939 distinct domain names with different 2LD zones. Next, we make a comprehensive analysis based on these domain names.

Table 2. The form of a record in DNS logs

Source IP	Domain name	Query time	Destination IP	RCODE

Wildcard Domain Name. For each of 919,939 domain names, we queried its wildcard domain using the *dig* tool, and collected their responses together. For example, for *google.com*, we directly queried the wildcard domain name **.google.com*. In our study, we focus on A and CNAME records, because these two types of records are the main part of host queries. Finally, we collected 153,163 wildcard domain names, which accounts for about 17% of the total number of collected domain names. The result is similar to that obtained in paper [2].

WHOIS Information. To obtain the registration information of the collected wildcard domain names, we leverage the WHOIS records published by registrars. We used the

Ruby whois[2] tool to obtain WHOIS information of 135,785 (88.7% of all) wildcard domain names and used a python script to sparse them. For the missing of remaining domain names, the major reasons are request block and incomplete information provided by registrars.

Web Content. We implemented a crawler to visit the websites of the collected wildcard domain names. Meanwhile, we also recorded the HTMLs and URLs for further analysis. We finally extracted 101,763 (66.4% of all the wildcard domain names) HTMLs. The two major reasons for missing web contents of the remaining domain names are the request timed out and websites lacking (i.e., Websites not exist).

4 Analysis on Abused Wildcard Domain Names

Wildcard domain names offer DNS administrators the convenience of changing host names. However, problems do exist. In this section, we analyzed the usages of wildcard domain names through a series of automated and manual experiments, and then gave quantitative analysis based on these experiments. In detail, first, we grouped domain names into several categories according to text of HTMLs crawled in Sect. 3. Second, we analyzed the registration characteristics based on WHOIS data of the collected wildcard domain names. Next, we analyzed the resolution behaviors of these domain names. Finally, we checked the maliciousness and SSL deployment of the collected wildcard domain names.

4.1 Content Categories

Based on web content data from 101,763 (66.4% of all) wildcard domain names, we grouped these collected wildcard domain names into several categories using a semi-automatic method. We first manually looked into the title, page text of a few HTMLs and summarized seven main categories according to the key words of websites. For example, adult websites usually contain some descriptive words, such as porn, sex, gay, etc. Descriptions of seven main categories are as follows:

(1) Porn. We define domain names linked to adult content like pornographic pictures, videos and novels as porn domain names;
(2) Gambling. It refer to domains related to online gambling;
(3) Parking. It refer to domains linked to ads constructed by domain-parking agency, based on the words included in a domain name;
(4) Sale. Domain names sold over the Internet by domain agency are regarded as domains for sale;
(5) Business & Gov. Domain names serve as normal business and government;
(6) Entertainment. Domain names serve as entertainment content like games;
(7) Error. Web pages of domain show an error caused by web servers.

[2] https://whoisrb.org/.

Then, based on these key words belonging to different categories, we created generic content-signatures to automatically categorize the remaining pages into each category. Finally, we automatically classified all crawled webpages, and the results are shown in Table 3.

Table 3. Content categories

Categories	Number	Proportion	Description
Porn	19028	18.7%	Adult/Pornographic domain
Gambling	3888	3.8%	Online gambling domain
Parking	2032	2%	Parking Domain
Sale	5860	5.8%	Domain for sale
Business & Gov	28303	27.8%	Business/Government related domain
Entertainment	7206	7.1%	Entertainment/Game/Lottery, etc.
Error	9268	9.1%	Server error
Unclassified	26178	25.7%	Unclassified domain

Finding 1. Looking over the website categories, pornographic and online gambling websites take up a remarkable proportion of the wildcard domain names. Notably, about 22.5% crawled webpages contain adult and gambling contents, which are referred as ABUSED domain names. As a comparison, using the same method, we analyzed 100K non-wildcard domain names (i.e., domains without wildcard RRs). Finally we found that only about 4.4% crawled webpages contain pornographic or online gambling information. In China, pornographic and online gambling websites are banned by the Internet regulators. In other words, the relatively large proportion of websites of abused domain names suggests that wildcard domain names used to spread illegal information have not been regulated efficiently.

4.2 Registration Characteristics

As mentioned before, we obtained WHOIS information of 135,785 (88.7% of all) wildcard domain names. Based on the WHOIS data, in this subsection, we made a comprehensive analysis of wildcard domain name registration from perspectives of registrars, registrants and registration time windows. Specially, we studied registration behaviors by correlating domain names with their content categories. In the following, we summarize our findings.

Finding 2. Compared with normal wildcard domain names, abused wildcard domain names were much more likely to be registered out of China. We identified more than 2,100 registrars. For abused wildcard domain names and normal ones, the detailed distributions of the top 5 registrars are shown in Table 4. Especially, Godaddy and Alibaba are two dominant registrars in domain market, and Alibaba plays an important role in China domain market. From Table 4 we can see that, for abused wildcard domain names, only one registrar (Alibaba) in top 5 is from China, while for normal wildcard domain names, only one registrar (GoDaddy) in top 5 is not from China.

Table 4. Categories of registrars (Top 5)

Abused wildcard domain names	Ratio	Normal wildcard domain names	Ratio
GoDaddy.com, LLC	18.3%	Alibaba Cloud Computing Co., Ltd.	25.5%
NameCheap Inc.	9.6%	GoDaddy.com, LLC	10.1%
Alibaba Cloud Computing Co., Ltd.	9.3%	Xin Net Technology Corporation	6.1%
NameSilo, LLC	6.6%	Chengdu West Dimension Digital Technology Co., Ltd.	5.7%
Danesco Trading Ltd.	3.9%	eName Technology Co.,Ltd.	4.3%

This suggests that registrars out of China may hold loose conditions for registration, thus abused wildcard domain name owners like to register illegal domain names from them to avoid regulation.

Finding 3. Compared with normal wildcard domain names, abused wildcard domain names were registered more recently. About 40% abused wildcard domain names were registered in recent one year, and about 53% were registered in recent two years. However, registration dates of domain names in other categories were scattered. Totally, about 26% normal domain names were registered in recent one year, and 35% were registered in recent two years. The cause of this phenomenon may be that in recent years more and more people try to register abused domain names for great economic benefit, and they also try to avoid regulation by using a large number of new domain names.

Finding 4. Compared with normal wildcard domain names, abused wildcard domain names were more likely to be registered in bulk. As shown in Fig. 1, based on the data of created date, we compared the differences of registration characteristics between abused wildcard domain names and normal ones. Finally, by counting the days that have more than 20 registered domain names, we found that there are 8 days for normal wildcard domain names while 103 days for abused ones. Correspondingly, only 175 (0.2% of all) normal wildcard domain names were registered in these 8 days,

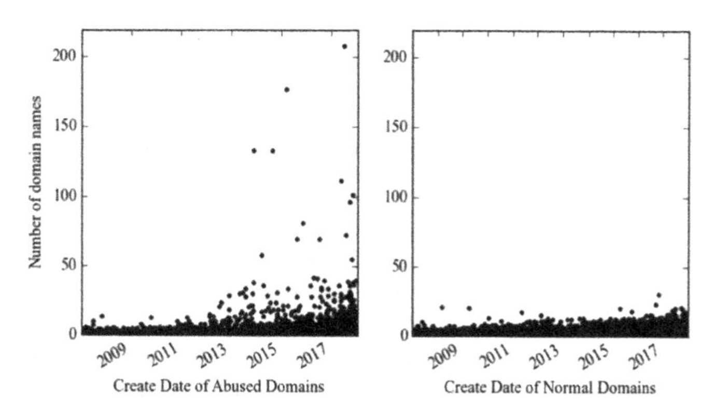

Fig. 1. Comparisons of abused and normal domains based on their created date data

and 4,239 (18.5% of all) abused wildcard domain names were registered in those 103 days. As a case, registrant *Li xiaoyu* registered 203 domain names in June 14, 2017, and resolved all of them to pornographic websites.

4.3 Resolution Behaviors

To understand the resolution behaviors of wildcard domain names, we analyzed wildcard records of destination IPs and name servers. In this paper, we obtained destination IP and name server for each wildcard domain name by using the *dig* tool. We summarize our findings as follows.

Finding 5. For abused wildcard domain names, their destination IP addresses are relatively concentrated than those of normal wildcard domain names. We collected 90,897 IP addresses used by wildcard domain names and analyzed the IP distributions of abused wildcard domain names and normal ones. By analyzing/24 IP addresses, we found that the IP addresses of abused wildcard domain names are relatively concentrated than the normal ones. As shown in Fig. 2(a), we could find that about 20% abused wildcard domain names were resolved to top 10 IP addresses, and top 100 IP addresses held about 50% abused wildcard domain names. We also collected 87,546 IP addresses used by non-wildcard domain names and compare their distributions with those of wildcard domain names. Results are shown in Fig. 2(b), which indicates that the IP addresses of wildcard domain names are relatively concentrated than those of non-wildcard ones.

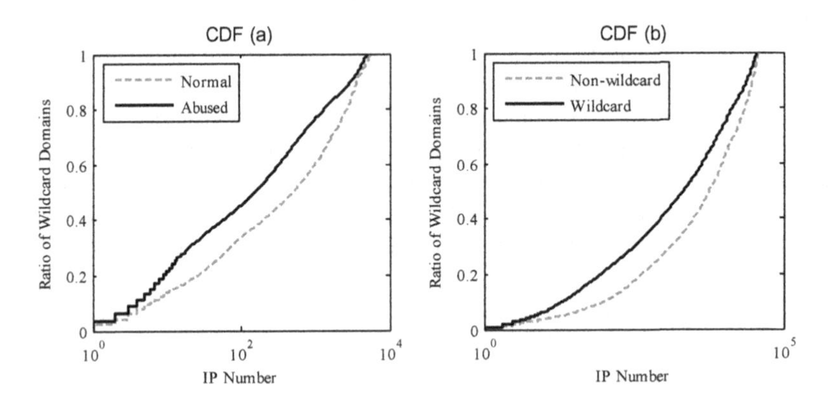

Fig. 2. CDF of IP addresses (IP numbers are shown in logarithmic coordinate)

Finding 6. For wildcard domain names resolved to the same IP, the name servers of abused ones are more concentrated than those of normal ones. To analyze the usage of name servers, we first grouped the wildcard domain names by each destination IP, and then we analyzed the number of distinct name servers used in each group.

As shown in Fig. 3, the X-axis is the ratio of the number of name servers over the number of wildcard domain names in each group. A lower value of this ratio means more wildcard domain names are resolved by the same name server. And the Y-axis is

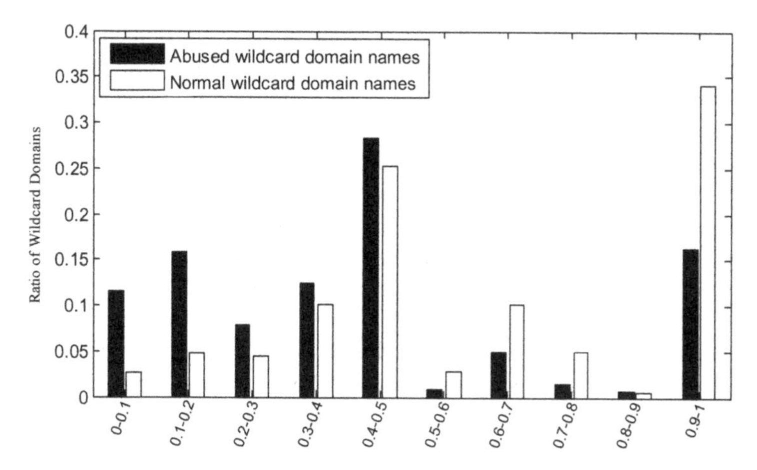

Fig. 3. Usage of name servers. The black bars describe the case of name servers used by abused domain names, and the white bars describe the case of name servers used by normal ones.

the relative number of wildcard domain names within different ratio ranges. The results show that the proportions of abused wildcard domain names are always higher than those of normal ones when the value of X-axis is lower than 0.5, suggesting the name servers of abused wildcard domain names are more concentrated than those of normal ones.

4.4 Maliciousness and Security of Domain Names

To analyze the malicious use of collected wildcard domain names, we checked these domain names with VirusTotal[3] and malicious domain lists, including DNS-BH[4], Malware Domain List[5], etc. Besides the malicious use, to analyze security of the collected wildcard domain names, we also checked SSL configuration of them.

Finding 7. The proportion of malicious domain names in abused wildcard domain list is apparently higher than that in normal wildcard domain list. Totally, we found 4,155 domain names were involved in malicious activities. When we looked into the categories of these malicious domain names, we found that about half of them were abused wildcard domain names. In other words, about 10% abused wildcard domain names were involved in malicious activities, however, only 2.7% normal wildcard domain names were related to malicious activities. This finding suggests a higher risk of being compromised for users when accessing websites with abused wildcard domain names. Obviously, pornography and online gambling contents provided by abused wildcard domain names are easy to allure victims.

[3] https://www.virustotal.com.

[4] http://www.malwaredomains.com.

[5] https://www.malwaredomainlist.com.

Here, we also made a blackhat SEO analysis of wildcard domains based on our DNS logs. For blackhat SEO testing, we use the method proposed in paper [2], which only considers the difference of hyperlinks in webpages between two visits for a domain name. We randomly selected 5K domain names and we finally found that 77 (1.5%) domain names are suspicious blackhat SEO domain names, which is much lower than 19.1% mentioned in paper [2]. In detail, 24.7% of these SEO domain names are abused domain names and 34.5% domain names are parking related domain names.

Finding 8. Only 2.3% abused wildcard domain names have adopted wildcard certificates to secure Internet traffic between users and web servers. As a contrast, about 14% normal wildcard domain names have adopted wildcard certificates. In today's Internet, HTTPS is a popular and effective information security protection method. Usually, web administrators adopt HTTPS only for several detailed domain names. For wildcard domain names, wildcard certificates can secure entire domains under the same zone with a single, flexible certificate. To analyze the SSL deployment of wildcard domain names, we extracted URLs of these domain names. Finally, discarding redirection, we found 11,306 URLs among all 101,763 domains with web contents adopted wildcard certifications. In detail, only 527 URLs belonged to abused domain names. To make a comparison, we also analyzed the SSL deployment of normal wildcard domain names. Finally, we found the application rate of SSL deployment in these domain names is higher than that in abused ones. This finding suggests that owners of abused wildcard domains rarely concern the security of transportation between their websites and users.

According to the above findings, we can see that abused wildcard domain names not only are related to illegal contents but also have higher risks in security than normal wildcard domain names. So it's necessary to distinguish wildcard domain names from normal ones. In the next section, we propose the GSCS algorithm to detect abused wildcard domain names.

5 Abused Domain Detection Based on DNS Graph

5.1 The GSCS Algorithm

In this section, to mine the relationships among abused wildcard domain names and detect them, we propose a graph-based method. In fact, graph-based method has already been used in malicious domain names detection [5–7]. Different from the former works, we exploit the inner relationships among abused wildcard domain names based on information of name servers and WHOIS. In addition, we avoid using traditional classification algorithms, which will be heavily influenced by unbalanced dataset of abused wildcard domain names.

We first describe the DNS graph model. Given a bipartite DNS graph $G = (D, I, E)$, the vertex set D and I consists of wildcard domain names and destination IPs, and the edge set E represents the connections between domain names and IPs. We then build a projection graph (named P) of bipartite G to extract hidden information between nodes in

```
Algorithm GSCS (Graph based Spectral Clustering with Seeds)
Input:   Wildcard domains
         IPs: destination IPs of wildcard domains
         Seeds: Abused domain name seeds
Output: Abused Clusters: Clusters of abused domain names
1. G = Bulid_graph(Wildcard domains, IPs)
2. P = Projection(G)
3. for each connected component of P do
4.       if ns_consistency_score < confidence_threshold then
5.            sub_component = Spectral_cluster(component)
6.            for each element of sub_component do
7.                 if seed in element then
8.                      Move element to Abused Clusters
9.                 end
10.           end
11.      else
12.           Move component to Abused Clusters
13.      end
14. end
15.return Abused Clusters
```

vertex set D. Here, we show the GSCS algorithm and introduce the detailed steps as follows:

- Using step 1 and 2, we transform records of wildcard domain names into a graph. In our analysis, we use/24 IPs to construct the bipartite graph, because IP addresses of collected wildcard domain names are relatively concentrated and/24 is a common block size of BGP routing prefixes [8].
- Through systematic analysis of the graph, we find that the projection graph consists of a large number of isolated components, so we analyze each of them separately (step 3–14).
- Based on *finding 6*, we use a consistency score to filter out components. The score is defined by

$$ns\_consistency\_score = \frac{max_{ns}}{num_{ns}}$$

  Where, max_{ns} refers to the number of name server, which is used by the largest number of domain names in one component, and num_{ns} is the total number of name servers in one component. For example, if all domain names in one component are resolved by the same name server, the score is equal to 1.
- For other components, we first apply a spectral clustering algorithm to decompose each of them into sub-components. Then, using seeds of abused domain names, we filter out all sub-components with seeded domain names.

Next, we simply describe the spectral clustering algorithm used in algorithm GSCS. The key step in the spectral clustering algorithm is computing similarity matrix. In this paper, we construct similarity matrix using weight of edge, and the weight is defined by

$$weight = \begin{cases} 1 - \frac{|D_1 - D_2|}{T} & |D_1 - D_2| < T \\ \frac{1}{|D_1 - D_2|} & |D_1 - D_2| \geq T \end{cases}$$

Where, $|D_1 - D_2|$ is the interval of registration date between every two domain names. In detail, we extract the information of registration data from WHOIS data. Additionally, we set $T = 30$ based on results of several experiments and use X-means to cluster nodes of domain names.

5.2 Evaluation

Based on the information of wildcard domain names, we first built a domain resolution graph and its corresponding projection graph. Totally, we found 29,492 connected components in the projection graph. Next, to evaluate the effectiveness of our GSCS algorithm, we varied confidence threshold of the consistency score and seed size. In experiments, we set the threshold to 0.6, 0.7 and 0.8 respectively. Under the condition of each value, we randomly selected seeds from the abused wildcard domain name list, and set the seed size range from 1% to 10% with a step length of 1%. Then we calculated the true positive rate (TPR) and the false positive rate (FPR) based on different groups of seeds that are arranged in order of size. Finally, we found that both TPR and FPR increase with the size of seeds, and we drew the ROC of GSCS with different thresholds in Fig. 4. We can see that the performance when threshold is set to 0.8 is better than the performance when threshold is set to the other two values. Especially, when setting threshold to 0.8 with 5% seed domain names, we can get 86% TPR with 3% FPR. We can also see that, when we set a small seed size, we get low TPR and FPR. Because the graph is composed of many isolated components, the smaller seed size we set, the more components are discarded. Conversely, the smaller threshold we set, the more components are considered, so FPR will go higher. However, when the seed size goes higher than 5%, TPR will nearly not increase while FPR will still get higher. So, threshold 0.8 and seed size 5% should be appropriate choices for our GSCS algorithm.

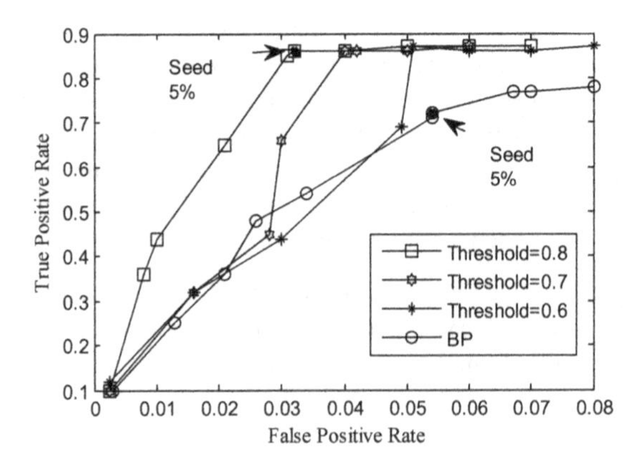

Fig. 4. ROC of various confidence thresholds

For false positives, we found that most of them belonged to *Error* or *Unclassified* categories mentioned in Subsect. 4.1. For example, when setting threshold to 0.8 with 5% seeds, there were 2,171 false positives. However, we found about 70% of them belonged to *Error* or *Unclassified* categories, which suggests that these domain names would have been used as abused domains before they were discarded.

To further analyze the effectiveness of our detection method, we make a comparison with BP algorithm, which is often used in the field of graph analysis and has also been used to detect malicious domain names [6, 9]. In this paper, we used the abused wildcard domain names and normal wildcard domain names collected from our data as ground truth, and we assigned priors to graph nodes and edge potential matrices according to [6], which are shown in Tables 5 and 6.

Table 5. Priors assigned to a domain node

Node	P (Abused)	P (Normal)
Abused	0.99	0.01
Normal	0.01	0.99
Unknown	0.5	0.5

Table 6. Edge potential matrices

	Abused	Normal
Abused	0.51	0.49
Normal	0.49	0.51

In our comparison experiments, we also set the seed size range from 1% to 10% with a step length of 1%, and we found both TPR and FPR increase with the size of seeds, as those in GSCS do. For the convenience of study, we show the results of BP in Fig. 4. We can see that our proposed method outperforms BP for the task of abused wildcard domain names detection. In detail, when using the same 5% seed domains used in GSCS, BP only obtained 72% TPR with 5.3% FPR. After analyzing, we found the factor of isolated components is a key reason leading to this inferior performance. Because we use a small number of seeds lying in several isolated components, other components without seeds cannot get information of propagation from these components with seeds.

From the above results and analysis, we can conclude that our GSCS algorithm is an effective solution to detect abused wildcard domain names.

6 Related Work

Wildcard Domain Names. Wildcard record is a type of RRs, which has been widely used in the Internet. Now, several studies have been proposed and focused on security implications of wildcard domain names. Du *et al.* [2] conducted the first comprehensive investigation on wildcard domain names used for blackhat SEO technique called "spider pool". Based on DNS zone files, their research shows 17.8% of all domain names are wildcard domains and 19.1% wildcard domain names are used for spider pool. Similarly, based on DNS zone files and known malicious domain lists, Kalafut *et al.* [3] studied the prevalence of wildcard DNS configuration and showed that it is broadly involved in malicious behaviors. In addition, several studies also briefly mentioned usage of wildcard domain names. Liu *et al.* [10] mentioned that wildcard records were also used to spawn shadowed domains, which are malicious subdomains under legitimate domains compromised by miscreants. Sharifnya *et al.* [11] found wildcard records used in botnet to resolve to a C&C server.

Although wildcard domain name has been widely used, now the study has still not paid much attention to it. And there is no comprehensive study to analyze abused wildcards usage based on user request data. So our study can be regarded as a complementary research on wildcard domain names.

Graph-Based Detection Method. As DNS data has the characteristics of graph, graph-based methods have been proposed to detect malicious or abused domain names. In general, graph-based methods can be divided into two categories, including hosts-domains graph and domain resolution graph. For hosts-domains graph, Lee *et al.* [5] proposed *GMAD*, a graph expressing DNS query sequences, to detect infected clients and malicious domain names. Using event logs collected by enterprises, Manadhata *et al.* [6] constructed a host-domain graph to detect malicious domain names. For domain resolution graph, Berger *et al.* [7] proposed a detection system called *DNSMap* to detect malicious website using dynamic FQDN-to-IP address mappings. In addition, to infer the maliciousness of unknown node in graph, some researchers chose to use BP algorithm [6, 9, 12]. To distinguish an abused wildcard use from a benign one, we proposed a detection method referencing graph idea.

7 Conclusion

In this paper, we first performed comprehensive analysis on wildcard domain names. Being different with former works that use DNS zone file data and domain name blacklists, our work is based on real DNS query logs and information of web content and WHOIS. We found that 153,163 (17%) domain names in our dataset were wildcard domain names. Our important findings from the analysis include: (1) the proportion of abused domain names (i.e., domain names related to pornography and online gambling contents) in wildcard domain names is much higher than that in non-wildcard domain names (22.5% versus 4.4%); (2) abused wildcard domain names have remarkably higher risks in security than normal wildcard domain names. Then, based on the analysis, we proposed an effective algorithm named GSCS to detect abused wildcard domain names. GSCS first uses a domain graph to study the similarities of abused wildcard domain names' resolution behaviors, and then applies spectral clustering algorithm and seed domains to detect abused wildcard domain names. Experiments on real datasets indicate that GSCS can get about 86% detection rates with 5% seed domains, performing much better than BP algorithm. Future work will focus on further improving the detection rate by applying more machine learning methods with several datasets and more entries.

Acknowledgments. The work was supported in part by Scientific Research Foundation of the Institute of Information Engineering, Chinese Academy of Sciences (Grant No. Y6Z0011105 and J810091105).

References

1. Mockapetris, P.V.: Domain names-concepts and facilities. RFC 1034 (1987)
2. Du, K., Yang, H., Li, Z., Duan, H.-X., Zhang, K.: The ever-changing labyrinth: a large-scale analysis of wildcard DNS powered blackhat SEO. In: USENIX Security Symposium, pp. 245–262 (2016)
3. Kalafut, A., Gupta, M., Rattadilok, P., Patel, P.: Surveying DNS wildcard usage among the good, the bad, and the ugly. In: Jajodia, S., Zhou, J. (eds.) SecureComm 2010. LNICST, vol. 50, pp. 448–465. Springer, Heidelberg (2010). https://doi.org/10.1007/978-3-642-16161-2_26
4. Lewis, E.: The role of wildcards in the domain name system. RFC 4592. (2006)
5. Lee, J., Lee, H.: GMAD: graph-based malware activity detection by DNS traffic analysis. Comput. Commun. **49**, 33–47 (2014)
6. Manadhata, P.K., Yadav, S., Rao, P., Horne, W.: Detecting malicious domains via graph inference. In: Kutyłowski, M., Vaidya, J. (eds.) ESORICS 2014. LNCS, vol. 8712, pp. 1–18. Springer, Cham (2014). https://doi.org/10.1007/978-3-319-11203-9_1
7. Berger, A., D'Alconzo, A., Gansterer, W.N., Pescapé, A.: Mining agile DNS traffic using graph analysis for cybercrime detection. Comput. Netw. **100**, 28–44 (2016)
8. Xu, K., Wang, F., Gu, L.: Behavior analysis of internet traffic via bipartite graphs and one-mode projections. IEEE/ACM Trans. Netw. (TON) **22**, 931–942 (2014)
9. Zou, F., Zhang, S., Rao, W., Yi, P.: Detecting malware based on DNS graph mining. Int. J. Distrib. Sens. Netw. **11**, 102687 (2015)
10. Liu, D., Li, Z., Du, K., Wang, H., Liu, B., Duan, H.: Don't let one rotten apple spoil the whole barrel: towards automated detection of shadowed domains. In: Proceedings of the 2017 ACM SIGSAC Conference on Computer and Communications Security, pp. 537–552. ACM (2017)
11. Sharifnya, R., Abadi, M.: DFBotKiller: domain-flux botnet detection based on the history of group activities and failures in DNS traffic. Digit. Investig. **12**, 15–26 (2015)
12. Oprea, A., Li, Z., Yen, T.-F., Chin, S.H., Alrwais, S.: Detection of early-stage enterprise infection by mining large-scale log data. In: 2015 45th Annual IEEE/IFIP International Conference on Dependable Systems and Networks (DSN), pp. 45–56. IEEE (2015)

XScan: An Integrated Tool for Understanding Open Source Community-Based Scientific Code

Weijian Zheng[1], Dali Wang[2](✉), and Fengguang Song[1]

[1] Indiana University-Purdue University, Indianapolis, IN 46202, USA
zheng273@purdue.edu, fgsong@iupui.edu
[2] Oak Ridge National Laboratory,
PO Box 2008, MS 6301, Oak Ridge, TN 37831, USA
wangd@ornl.gov

Abstract. Many scientific communities have adopted community-based models that integrate multiple components to simulate whole system dynamics. The community software projects' complexity, stems from the integration of multiple individual software components that were developed under different application requirements and various machine architectures, has become a challenge for effective software system understanding and continuous software development. The paper presents an integrated software toolkit called X-ray Software Scanner (in abbreviation, *XScan*) for a better understanding of large-scale community-based scientific codes. Our software tool provides support to quickly summarize the overall information of scientific codes, including the number of lines of code, programming languages, external library dependencies, as well as architecture-dependent parallel software features. The XScan toolkit also realizes a static software analysis component to collect detailed structural information and provides an interactive visualization and analysis of the functions. We use a large-scale community-based Earth System Model to demonstrate the workflow, functions and visualization of the toolkit. We also discuss the application of advanced graph analytics techniques to assist software modular design and component refactoring.

Keywords: Application software analysis · Community-based code · Code modulation · Code refactoring

1 Introduction

Many scientific communities have employed community-based models to simulate complex dynamics [8,22]. These community-based models usually adopted

This research was funded by the U.S. Department of Energy, Office of Science, Advanced Scientific Computing Research (Interoperable Design of Extreme-scale Application Software).

J. M. F. Rodrigues et al. (Eds.): ICCS 2019, LNCS 11536, pp. 226–237, 2019.
https://doi.org/10.1007/978-3-030-22734-0_17

open modeling and open coupling infrastructure, and integrated many individual components to address community-driven scientific questions. Since these models are developed by multidisciplinary communities over a variety of high-performance computational facilities [8], close collaborations and continuous communications among many domain science groups and computational science groups are required for improving model understanding and development. To this end, we present a software toolkit to facilitate the understanding of these complex software systems and layout some considerations on further code migration and refactorization. We design and develop a static software analysis tool, which is named as X-ray Software Scanner (i.e., *XScan*). The XScan toolkit can collect an array of software specific information related to overall source code meta data, third-party library requirements/dependencies, major HPC software features, and detailed function relationship. XScan consists of four components (see Fig. 1): (1) an integration with CLOC [3] to provide information about the used high level programming languages and number of lines of code in each programming language (Sect. 2.1.1); (2) a CMake [13] based analysis component to reveal what external third-party libraries are required by an open-source community project and their dependency relationship (Sect. 2.1.2); (3) an HPC-specific query language (Feature Query Language FQL) and component to search for HPC hardware and architecture features that are required by an open-source project (Sect. 2.1.3); and (4) a Doxygen [24] based data collector to collect function-relevant information and build graphs to facilitate big graphs and networking analysis targeting software engineering problems (Sect. 2.2).

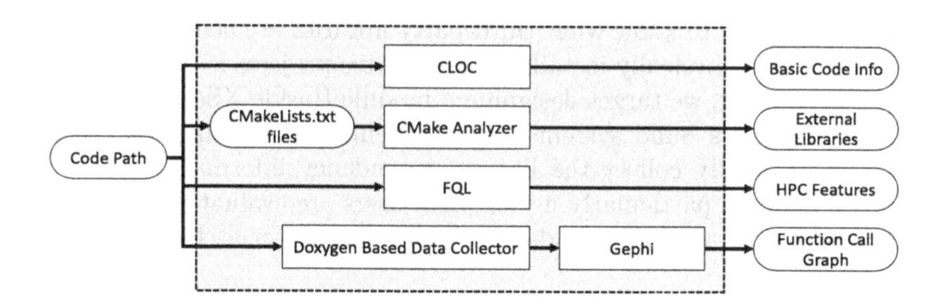

Fig. 1. Architecture of the XScan toolkit.

Finally, for the demonstration purpose, we apply the XScan toolkit to collect the overall information and structural relationship of an open-source community Earth System Modeling system, called Exascale Energy Earth System Model (E3SM). Mainly funded by US Department of Energy, E3SM is a computationally advanced coupled climate-energy model to investigate the challenges posed by the interactions of weather-climate scale variability with energy and related sectors.

2 Methodology

Figure 1 shows the architecture of our XScan toolkit, whose four components can be classified into two categories: (1) Components to show overall information of source code, and (2) Components to show function relationship information. The two categories will be described in the following two subsections, respectively.

2.1 Category 1: Overall Information of Source Code

2.1.1 Language and Size

The very first step of software system understanding is to present the information about a community project's size and programming languages. There are several existing tools available for code size evaluations, such as CLOC (Counting Lines Of Code) [3], Sonar [19] and SLOC [25]). We incorporates CLOC to XScan to provide the language and size information because CLOC consists of a single Perl file and can be executed on any machine. With CLOC, XScan can measure the lines of source code for a variety of languages, and can differentiate between the actual code, blank lines, code crossing multiple lines, and comment lines for each programming language. We consider it a convenient feature for XScan users.

2.1.2 External Library Dependency

In addition to the programming language and size information, it is important for community users to know what third-party libraries are needed by the open-source code. Since physically installing open-source projects is often challenging and time-consuming, we target designing a module (inside XScan) to parse and analyze the project's build systems (e.g., Makefile [20], CMake [13], Autoconf [12]) to automatically collect the library dependency information. This quick scan functionality is particularly useful when users are evaluating many choices of open source software and considering distinct computer architectures.

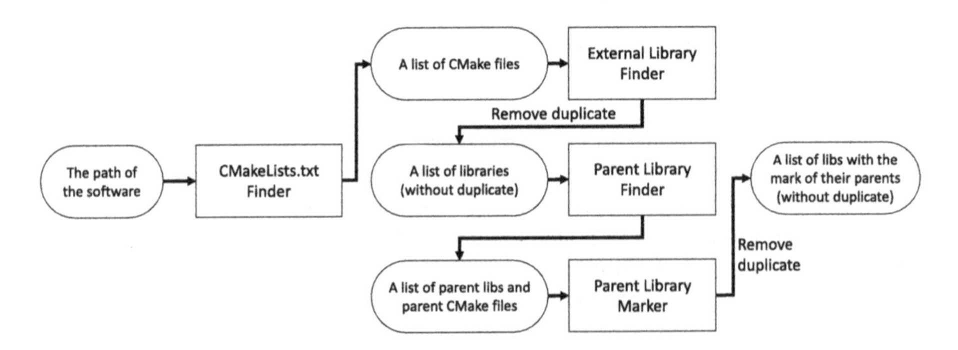

Fig. 2. Workflow of the CMake analyzer.

In this paper, we choose CMake as an example to achieve the objective without any physical software installation. CMake is an operating system independent tool for building and testing software. By modifying a configuration file (commonly named as *CMakeLists.txt*), developers are able to control the whole building process on most systems.

We design and develop a "CMake Analyzer" module in XScan, which scans all the *CMakeLists.txt* files in an open-source software project, and finds all the needed external libraries and identifies the dependency relationship among the detected libraries.

The design of CMake Analyzer is shown in Fig. 2, in which ellipses represent the data exchanged between software components meanwhile rectangles represent the software components. CMake Analyzer works in the following three steps: (1) Search for CMake files by using the *CMakeLists.txt Finder* component (see Fig. 2); (2) Scan the CMake files and identify the needed external libraries by the *External Library Finder* component; And (3) Use the *Parent Library Finder* and *Parent Library Marker* components to search and mark the parent of each needed library. The final output is a dependency graph for all the libraries.

2.1.3 Query Software Features

User community may want to understand special characteristics of the complex community projects. We defined these special characteristics as "features" of these community projects. For example, the user who are interested in high performance computing (HPC), may ask whether MPI or GPU is used by the code. To answer these questions, we created a new language, called *Feature Query Language* (FQL) [26] to describe the user's questions. It is an extensible and flexible language. Users can translate their questions to a query quickly when they know the keywords of a feature. Next, we design and implement a program parser to parse and execute the FQL queries. We also provide a number of predefined queries for commonly asked HPC related feature questions. By executing those queries, the FQL component can provide users with an overview of the HPC features in the community project.

2.2 Category 2: Function Relationship of the Source Code

It is important to gather function relationship information, from which users can gain insights into the internal structure of a community project.

XScan uses Doxygen [24] as a lower-level internal library to collect detailed information including code structure, function dependency, class inheritance relations as well as collaboration diagrams. Instead of using Doxygen as a GUI tool that generates documentation for projects, we call the Doxygen library and API directly to create an in-memory software-data repository, and perform in-situ graph partitioning, coloring, and data analysis.

Figure 3 shows an overview of the in-memory Doxygen-enhanced Analyzer, which consists of three major parts: (1) *Data Collection* (i.e., the large blue

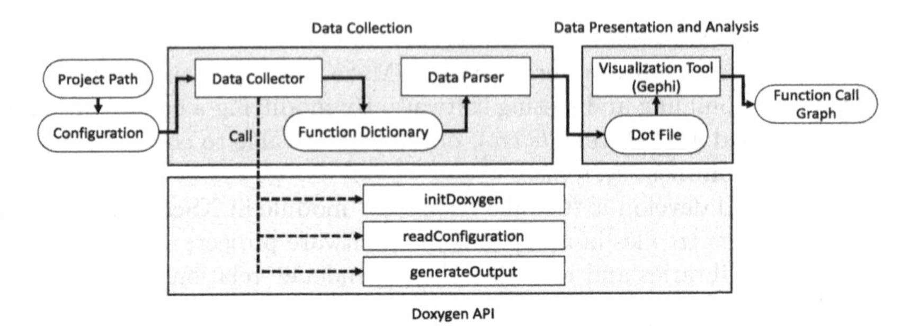

Fig. 3. Workflow of Doxygen-Based software analyzer (Color figure online)

rectangle on the upper left in Fig. 3), (2) *Doxygen API* (the large yellow rectangle at the bottom), and (3) *Data Presentation and Analysis* (the blue rectangle on the upper right). The following subsections will describe them in more details.

2.2.1 Data Collection and the Doxygen API

Data Collection: The input is a configuration file to configure Doxygen. Next, *Data Collector* calls the Doxygen APIs (shown in the yellow box) to obtain information of the code. In our implementation, *Data Collector* first calls *initDoxygen()*, then calls *readConfiguration* to read the configuration file, and finally calls *generateOutput* to generate the output.

Function Dictionary: The *generateOutput* function yields a *Function Dictionary*, which is stored in an in-memory data structure. In XScan, we need to know each function's function name, caller, module name and location. All of them are stored in this data structure. Because any information can be found by looking up the in-memory *Function Dictionary* data structure, XScan can perform fast graph and network data analytics efficiently in an in-situ manner.

Data Parser: Finally, XScan converts the code information stored in *Function Dictionary* to different graphs. The output graphs are stored in the Dot format [5]. Note that a Dot file can be read by many graph visualization tools.

2.2.2 Data Presentation and Analysis

This last part is responsible for reading the graphs generated from *Data Parser*, and using the open-source graph visualization tool Gephi [2] to visualize them with different layouts. These graphs will also be processed by using big graph/network analysis tools to facilitate software engineering code optimization, redesign and refactoring.

3 Case Study

In this section, we first introduce an Earth System Model. Then, we show the general code information of the model using XScan functions stated in Sect. 2.1.

After that, we present different function call graphs generated by the Doxygen-based Software Analyzer within XScan. Preliminary data analyses based on those graphs are also demonstrated at the end of the section.

3.1 The E3SM Application

The Energy Exascale Earth System Model (E3SM) is a computationally advanced coupled climate-energy model to investigate the challenges posed by the interactions of weather-climate scale variability with energy and related sectors [8]. E3SM is a 3D (plus time) computer-based general-circulation model that uses mathematical formulas to simulate the chemical and physical processes that drive Earth's climate. Being extraordinarily sophisticated, E3SM can be used to study phenomena ranging from the effect that an ocean surface temperature has on tropical cyclone patterns to the impact of land use on carbon dioxide concentration in the atmosphere.

Currently, E3SM contains several major community model components to simulate Earth systems: Atmosphere, Ocean, Land, Glacier, and Sea Ice. E3SM also provides a complicated system script tool, which allows science-oriented modelers to automatically reconfigure, compile, build, and submit jobs. It is an extremely valuable feature in helping E3SM users conduct computational experiments on various high-end computers.

3.2 Overall Source Code Information of E3SM

3.2.1 Language and Code Size

We use XScan/CLOC to check the overall information about E3SM's source code with respect to the source code size and programming languages. As listed in Table 1, the top ten programming languages used in E3SM (from the most to least) are FORTRAN 90, HTML, XML, C, Perl, Python, Tex, Fortran 77, Shell, and CMake. FORTRAN 90 is the most used programming language. There are 2184 FORTRAN files and nearly 1 million lines of FORTRAN code (excluding comments and empty lines), while HTML is mainly used for the documentation purpose.

3.2.2 External Library Dependencies

XScan is applied to scan the required third-party libraries for E3SM. Figure 4 shows the external libraries needed by E3SM. From the XScan-generated graph, we can see that E3SM needs more than ten external libraries, which include the data library NetCDF, message passing library MPI, numerical libraries of LAPACK, timing library GPTL, scientific application library Trillinos, scientific simulation library Zoltan, and program instrumentation library Extrae. We may also notice that MPI appears twice in the graph as it is used by both GPTL and the main program of E3SM.

Table 1. Top 10 programming languages inside the E3SM source code package

Detected language	Number of files	Number of lines of the code
Fortran 90	2,184	934,296
HTML	429	158,640
XML	307	85,785
C	124	46,245
Perl	156	37,860
Python	240	34,786
Tex	171	23,596
Fortran 77	70	21,118
Bourne Shell	236	20,598
CMake	213	6,125

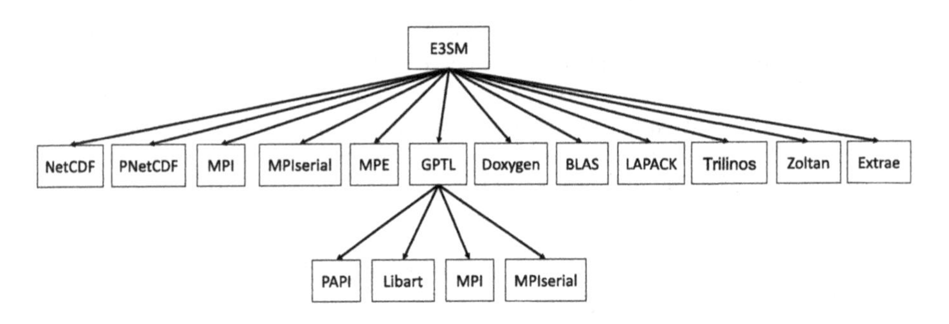

Fig. 4. CMake analyzer results of E3SM via XScan.

3.2.3 HPC Specific Features in E3SM

We apply XScan/FQL to extract HPC related features that are used and required by E3SM. The HPC results obtained by executing related FQL queries are listed in Table 2.

From the table, we find that MPI, OpenMP and OpenACC are all employed by the E3SM project. We can also find further information about how E3SM uses MPI, OpenMP, and OpenACC. For instance, E3SM uses the MPI Cartesian topology, one-sided communication and MPI I/O techniques. E3SM is written mainly in the FORTRAN language meanwhile using the C language for system level functions, such as parallel I/O and string handling. Also, please note that OpenMP is not listed in previous Fig. 4 since recent GCC compiler has the full support of OpenMP.

Table 2. FQL result of the E3SM.

MPI Yes	Min version required: 2.0	MPI one-sided communication: Yes
	MPI process topology: Cartesian	MPI I/O Yes
OpenMP Yes	Hybrid MPI/OpenMP: Yes	Task programming constructs: Yes
	OpenMP scheduling method: Dynamic	
CUDA No	Support multiple GPUs: –	Single/Double precision: –
OpenACC Yes	Atomic operation: No	Asynchronous operation: No
Language Support	Min required C compiler: C99	Fortran standard: Fortran 2003

3.3 Presentation and Exploration of Code Function Relationship in E3SM

In this section, we present two sets of graph results generated by XScan for the E3SM project: Partitioned function graphs computed by Doxygen-enhanced Software Analyzer, and Preliminary data analysis results on investigating software modularity.

3.3.1 Different Presentations of Source Code Function Graphs

Figure 5 presents the global function-call graph for the E3SM project. In this graph, different functions have different colors which are decided by their resident directories. As shown in the figure, the directory */components/cam* has the largest number of functions (i.e., the green nodes). Moreover, the functions located in directory */cime*—which are shown as the red nodes and working as the coupler to combine different models—have many connections with other colored nodes.

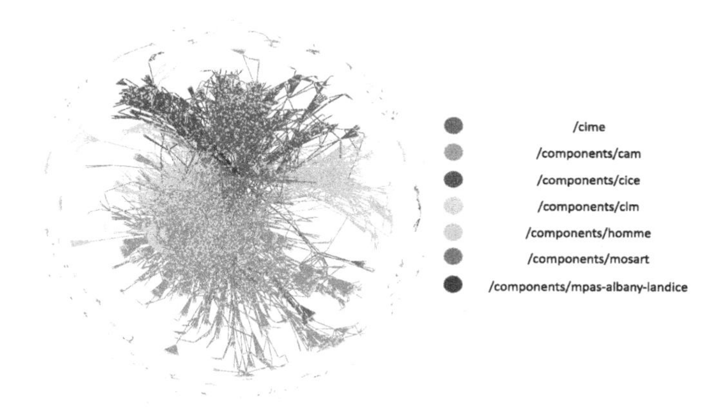

Fig. 5. Global function graph for the E3SM project. (Color figure online)

Next, we want to dive into the specific */components/cice* directory to investigate the function relationship in the CICE component. For example, Fig. 6a shows a colored function graph for the CICE component only, based on each function's subdirectory location. Since there are four sub-directories in CICE, the function graph has four colors.

Finally, we use XScan to study how a software module of interest can interact with other modules. By using the identical set of vertices and edges as Fig. 6a, Fig. 6b focuses on the module of *ice_forcing*, and reveals its interaction with other modules. In Fig. 6b, all the nodes that have no connection with the module *ice_forcing* are colored in the grey color. The rest of the nodes (i.e., with colors) are colored separately based on their corresponding module names. As shown in the figure, nine modules have interactions with the *ice_forcing* module.

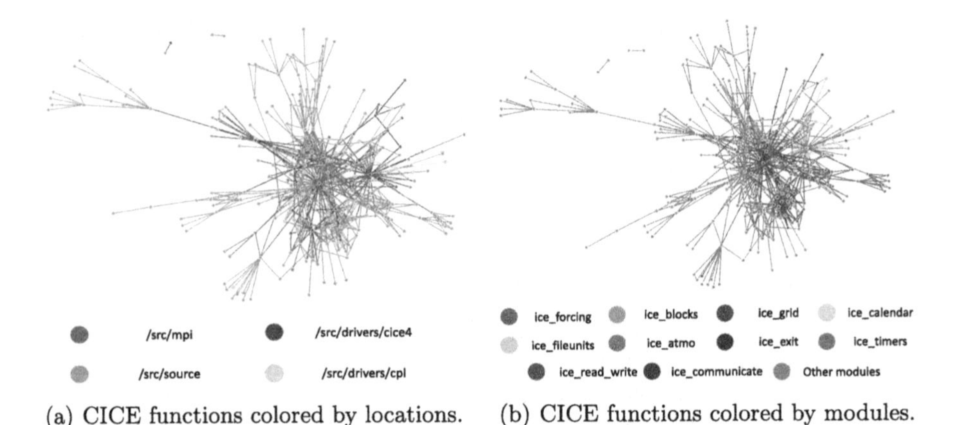

(a) CICE functions colored by locations. (b) CICE functions colored by modules.

Fig. 6. Different colored function graphs for the CICE component of E3SM. (Color figure online)

3.3.2 Preliminary Graph Analysis Results

Here, we present our preliminary big graph data analysis results based on the graph information collected by XScan. One purpose of the data analysis is to help users better understand the software modularity, and assist users to make good decision on software component refactoring. We use the CICE case again to illustrate our data analysis approach.

As shown in Fig. 6, the CICE component has more than 50 modules. For the purpose of demonstration, we abstract 13 modules as presented in Fig. 7. We propose and make a few modifications to those nodes, using common software engineering approaches, such as function encapsulation, merge, and function interface redesign. For all the modifications, we check how graph modularity score are changed. Graphs with higher modularity scores are more modular with denser connections inside the module and sparser connections between modules.

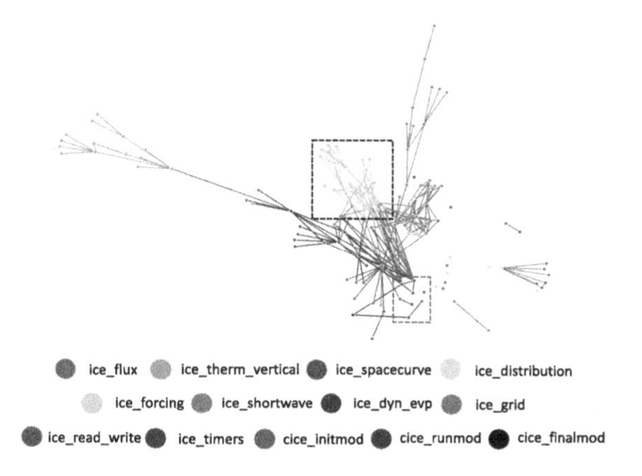

Fig. 7. Partial function graph for CICE before merging interface functions. (Color figure online)

In Fig. 7, there are a certain number of functions (presented as nodes in the graph) that have edges connected with other modules. We call them *interface functions* of a module. We can merge the interface functions of the module *ice_timers* as shown in the red rectangle box. It is an action to mimic the function encapsulation and interface redesign. As a result, the new graph is more modular with its modularity score improved from 0.649 to 0.66. On top of that, we further combine the interface functions of module *ice_forcing*, which is shown in the big rectangle box on the top of the figure. The modified new graph is even more modular with its modularity score rising to 0.71. Therefore, we can conclude that merging the interface functions of a module is able to make open source software more modular.

Currently, we only apply the modularity metric to evaluate the impact of source code modifications. Our next research plans to design new metrics and apply graphs analysis techniques to guide our further code developments, such as using k-component analysis [15] to estimate the difficulties of module refactoring.

4 Related Work

A number of tools have been developed to collect various kinds of overall code information. For example, a few tools can analyze a project's code size and languages by scanning the source code [3,19,25]. Certain tools like the ScanCode toolkit [18] and fossology [6] can provide the software license, copyright, dependency and other kinds of information of the code. By integrating other third party package managers (e.g., MAVEN, PIP, NPM) and code scanners (e.g., Licensee [11], ScanCode), OSS Review toolkit can tell user dependencies of different open-source libraries of the project [21].

Function call graph is commonly used to represent the calling relationship between different functions [17]. Many tools such as Doxygen [24], CFlow [16] and Egypt [4], are developed to statically extract the relationship. There are also many different ways to use the function-call graph. For example, some researchers analyze the software change impacts by checking the transitive closure of the function-call graph [1,9]. Function-call graph based reverse engineering work, such as the work of Mitchell et al. [14] and Vassilios et al. [23], can abstract the structure of software by applying clustering based machine learning algorithms to function call graphs.

5 Future Work

In the next phase, we will further enhance the in-memory software analyzer to collect more information (such as global variable, data context, data aliases) from static software analysis and to help users to evaluate the complexity of scientific software projects. We will apply graph algorithms and graph analysis tools (such as networkX [7] and SNAP [10]) to the software function graphs to estimate efforts to factorize a submodule as well as the workload to combine submodules from multiple different software to build an integrated software system. We will also look into the methods to estimate the effort of migrating software to a new platform. For instance, we can list modules, functions, and libraries (i.e., both internal and external) that need to be modified. Furthermore, with the vast collection of software information, we can transform the software understanding problem into software optimization problem. Then, we will apply machine learning approaches (such as reinforcement learning) to aid users in optimizing software structure and functional redesigns.

References

1. Arnold, R.S.: Software Change Impact Analysis. IEEE Computer Society Press, Los Alamitos (1996)
2. Bastian, M., Heymann, S., Jacomy, M., et al.: Gephi: an open source software for exploring and manipulating networks. ICWSM **8**(2009), 361–362 (2009)
3. CLOC (2018). https://github.com/AlDanial/cloc
4. Egypt (2018). http://www.gson.org/egypt
5. Gansner, E., Koutsofios, E., North, S.: Drawing graphs with dot (2006)
6. Gobeille, R.: The fossology project. In: Proceedings of the 2008 International Working Conference on Mining Software Repositories, pp. 47–50. ACM (2008)
7. Hagberg, A., Swart, P., Chult, D.S.: Exploring network structure, dynamics, and function using networkx. Technical report, Los Alamos National Lab. (LANL), Los Alamos, NM (United States) (2008)
8. Hurrell, J.W., et al.: The community earth system model: a framework for collaborative research. Bull. Am. Meteorol. Soc. **94**(9), 1339–1360 (2013)
9. Law, J., Rothermel, G.: Whole program path-based dynamic impact analysis. In: Proceedings of the 25th International Conference on Software Engineering, pp. 308–318. IEEE Computer Society (2003)
10. Leskovec, J., Sosič, R.: SNAP: a general-purpose network analysis and graph-mining library. ACM Trans. Intell. Syst. Technol. (TIST) **8**(1), 1 (2016)

11. Lisensee (2015). https://ben.balter.com/licensee
12. MacKenzie, D., Elliston, B., Demaille, A.: Autoconf: creating automatic configuration scripts. User Manual, Edition 2 (1996)
13. Martin, K., Hoffman, B.: Mastering CMake: A Cross-platform Build System. Kitware, Klifton-Park (2010)
14. Mitchell, B.S., Mancoridis, S.: On the automatic modularization of software systems using the bunch tool. IEEE Trans. Softw. Eng. **32**(3), 193–208 (2006)
15. Moody, J., White, D.R.: Structural cohesion and embeddedness: a hierarchical concept of social groups. Am. Soc. Rev. **68**, 103–127 (2003)
16. Poznyakoff, S.: GNU cflow (2005)
17. Ryder, B.G.: Constructing the call graph of a program. IEEE Trans. Softw. Eng. **3**, 216–226 (1979)
18. ScanCode (2016). https://github.com/nexB/scancode-toolkit
19. Sonar (2018). https://www.sonarqube.org/
20. Stallman, R.M., McGrath, R., Smith, P.: GNU Make: A Program for Directed Compilation. Free Software Foundation (2002)
21. OSS-Review Toolkit (2017). https://github.com/heremaps/oss-review-toolkit
22. Trolle, D., et al.: A community-based framework for aquatic ecosystem models. Hydrobiologia **683**(1), 25–34 (2012)
23. Tzerpos, V., Holt, R.C.: ACCD: an algorithm for comprehension-driven clustering. In: Proceedings of Seventh Working Conference on Reverse Engineering, pp. 258–267. IEEE (2000)
24. Van Heesch, D.: Doxygen: Source code documentation generator tool (2019). http://www.doxygen.org
25. Wheeler, D.A.: SLOC count user's guide (2004)
26. Zheng, W., Song, F., Wang, D.: FQL: an extensible feature query language and toolkit on searching software characteristics for HPC applications. arXiv preprint arXiv:1905.09364 (2019)

An On-Line Performance Introspection Framework for Task-Based Runtime Systems

Xavier Aguilar[1(✉)], Herbert Jordan[2], Thomas Heller[3], Alexander Hirsch[2], Thomas Fahringer[2], and Erwin Laure[1]

[1] Department of Computational Science and Technology and Swedish e-Science Research Center (SeRC), KTH Royal Institute of Technology, Lindstedvägen 5, 10044 Stockholm, Sweden
xaguilar@pdc.kth.se

[2] Department of Computer Science, University of Innsbruck, Technikerstrasse 21a, 6020 Innsbruck, Austria

[3] Department of Computer Science, Friedrich-Alexander-University Erlangen-Nuremberg, Martenstr. 3, 91058 Erlangen, Germany

Abstract. The expected high levels of parallelism together with the heterogeneity and complexity of new computing systems pose many challenges to current software. New programming approaches and runtime systems that can simplify the development of parallel applications are needed. Task-based runtime systems have emerged as a good solution to cope with high levels of parallelism, while providing software portability, and easing program development. However, these runtime systems require real-time information on the state of the system to properly orchestrate program execution and optimise resource utilisation. In this paper, we present a lightweight monitoring infrastructure developed within the AllScale Runtime System, a task-based runtime system for extreme scale. This monitoring component provides real-time introspection capabilities that help the runtime scheduler in its decision-making process and adaptation, while introducing minimum overhead. In addition, the monitoring component provides several post-mortem reports as well as real-time data visualisation that can be of great help in the task of performance debugging.

Keywords: Runtime system · Performance monitoring · HPX · Performance introspection · Real-time visualisation · AllScale

1 Introduction

The increase in size and complexity of upcoming High Performance Computing (HPC) systems pose many challenges to current software. Upcoming, and

Supported by the FETHPC AllScale project under Horizon 2020.

J. M. F. Rodrigues et al. (Eds.): ICCS 2019, LNCS 11536, pp. 238–252, 2019.
https://doi.org/10.1007/978-3-030-22734-0_18

already existing, HPC infrastructures contain heterogeneous hardware with multiple levels of parallelism, deep memory hierarchies, complex network topologies and power constrains that impose enormous programming and optimisation efforts. Thus, new high-productivity, scalable, portable, and resilient programming approaches are needed.

Task-based runtime systems have emerged as a good solution for achieving high parallelism, performance, improved programmability, and resilience. Proof of that is the amount of new task-based runtimes that have appeared in the HPC landscape in recent years, e.g. HPX [11], Legion [3], StarPU [2], or OmpSs [5] among others.

The AllScale project tries to settle in the HPC runtime landscape as a new solution that provides a unified programming interface for parallel environments. One key aspect differentiating AllScale against other existing runtimes is that it is heavily based on recursive parallelism to diminish the amount of global synchronisation present in classical parallel approaches. Reducing thereby one of the main factors that hinders scalability.

In this paper, we give an overview of the AllScale toolchain as well as an extensive characterisation of one of its key components, the AllScale Monitoring Component. This monitoring framework provides introspection capabilities such as real-time performance data visualisation and real-time performance feedback to the AllScale Runtime Scheduler. In addition, it can also provide several performance reports suited for post-mortem performance debugging. Furthermore, even though the monitoring component presented in this paper has been especially designed for the AllScale Runtime, it can easily be decoupled and adapted to any other task-based runtime system.

The paper is structured as follows: Sect. 2 provides an overview of the AllScale project; the AllScale Monitoring Component is described in depth in Sect. 3; Sect. 4 provides a detailed evaluation of the monitoring component; related work is described in Sect. 5, and finally, conclusions and future work are presented in Sect. 6.

2 AllScale Overview

2.1 Vision

The AllScale project pursues the development of an innovative programming model supported by a complete toolchain: an API, a compiler, and a runtime system. AllScale is strongly based on recursive programming in order to overcome inherent limitations that nested flat parallelism presents, for example, the use of global operations and synchronisation points that limit the scalability of parallel codes. Moreover, AllScale aims to provide a unified programming model to express parallelism at a high level of abstraction using C++ templates. In that way, problems imposed by the use of multiple programming models can be mitigated, for example, the need of expertise in complementary models such as MPI, OpenMP, or CUDA. Furthermore, by using a high level of abstraction to express parallism, problems such as the need to design the algorithm based on the underlying hardware can be mitigated too.

2.2 Architecture

The AllScale toolchain is divided into three major components: an API, a compiler, and a runtime system. The AllScale API, based on the prec operator [10], provides a set of C++ parallel operations such as stencils, reductions, and parallel loop operations, as well as a set of data structures (arrays, sets, maps, etc.) to express parallelism in an unified manner. AllScale applications can be compiled with standard C++ tools, however, in order to take advantage of all the benefits of the AllScale environment, the code has to be compiled with the AllScale Compiler. The compiler, based on the Insieme source-to-source compiler [8], interprets the API and generates the necessary code to execute the application in the most optimal manner by the AllScale Runtime System. Finally, the AllScale Runtime System [9] manages the execution of the application following customisable objectives that can be defined by the user. This multi-objective optimisation combines execution time, energy consumption, and resource utilisation, and thus, improves classical approaches where self-tuning during execution is mainly based in execution time. In addition, the AllScale Runtime provides transparent dynamic load balancing and data migration across nodes, and it is the basis for the resilience, scheduler and performance monitoring components. The AllScale Runtime System builds on top of the High Performance ParalleX (HPX) runtime [11].

3 AllScale Monitor Component

The AllScale environment includes a monitoring framework with real-time introspection capabilities. Its main purpose is to support the AllScale Scheduler in the management of resources and scheduling of tasks. To this end, the AllScale Monitor Component has been designed to introduce minimum overhead while being able to continuously monitor the system and provide in real time the performance information collected. In addition, the AllScale Monitoring Component can also generate several post-mortem reports to help developers in the process of performance debugging.

3.1 Performance Data

The AllScale Monitor Component collects several performance metrics on the execution of tasks (WorkItems [8,9] in AllScale jargon). These WorkItems, which can be hierarchically nested, are entities that describe the work that has to be performed by the runtime system. The AllScale Monitor Component takes control at the beginning and end of a WorkItem, and collects performance data such as execution time or WorkItem dependencies (what WorkItems have been spawned from within the current one). From these raw metrics, the monitoring component is also able to compute multiple derived metrics: execution time of a WorkItem and all its children; percentage of exclusive and inclusive time per WorkItem over total execution time; average execution time of all children of a WorkItem; or standard deviation of the execution time for all children of a WorkItem, among others. These derived metrics are computed on demand to minimise the amount of overhead introduced into the application when collecting the data. In other words, they are computed only if the scheduler requests them or if a performance report that includes them has to be generated.

The metrics described above are collected in an event action basis, that is, the collection is triggered by an event such as WorkItem start or stop for example. However, the AllScale Monitor Component also samples many other metrics on a per-node basis[1], for instance, WorkItem throughput, memory consumption, runtime idle rate, or CPU load, among others. The runtime idle rate is defined as the percentage of time spent by the runtime in scheduling actions against the time spent executing actual work. In addition, the AllScale Monitor Component is able to sample HPX counters (HPX internal metrics) and PAPI counters.

Power is becoming an important constraint in HPC, and therefore, the AllScale Monitoring Component can also provide power and energy consumed on x86 platforms, Power8, and Cray systems. If power and energy measurements are not available, the monitoring component is able to provide simple estimates.

3.2 Data Collection

The AllScale Monitor Component has been designed to introduce as minimum overhead as possible. When an application starts, the AllScale Runtime system creates a pool of worker threads that execute tasks as they are generated. These WorkItems, which are then monitored by the AllScale Monitor, are very small and thousands of them are executed in every run. Therefore, solutions where each thread collects and writes its data to shared data structures do not work due to heavy thread contention.

For this reason, the AllScale Monitor Component has been implemented following a multiple producer-consumer model with a double buffering scheme. All runtime worker threads measure and keep their raw performance data locally. Once their local buffers are full, they transfer the data to specialised monitoring threads that process and store such data into per-process global structures. The data have to be stored at a process level in order to facilitate the performance introspection later on. Only the specialised monitoring threads write into global data structures, and thereby, the contention is very low. Furthermore, producers and consumers utilise a double buffering scheme, that is, while producers create data in their buffers, consumers process data from an exclusive buffer that nobody else accesses. When a consumer finishes processing its data, it switches buffers with the next producer waiting to send data, and the process starts again. Even though our system has the possibility to have more than one

[1] Node defined as compute or cluster node.

consumer thread, we have seen in our experiments than one specialised thread per node is enough to process the amount of data generated.

The monitoring infrastructure also has another specialised thread responsible for collecting metrics sampled on a per-node basis. This thread awakes at fixed time intervals and collects performance metrics such as WorkItem throughput, node idle rate, memory consumption, or energy consumed, among others.

The AllScale Monitor Component is also able to select the type of tasks monitored, thereby, further reducing its overhead and memory footprint. In AllScale, there are two types of WorkItems derived from the recursive paradigm exploited in the runtime: splittable and processable. Splittable WorkItems are tasks that are split and generate more tasks, they correspond to the inductive case of the recursion. Processable WorkItems correspond to the base case of the recursion, and are WorkItems that are not split anymore. The user can configure the monitoring component to monitor only Splittable WorkItems, Processable WorkItems, or both. In addition, the AllScale Monitor can also monitor WorkItems to a certain level of recursion.

3.3 Execution Reports

The AllScale Monitor Component can generate text reports at the end of program execution. These reports include raw measurements per WorkItem such as exclusive time, i.e. execution time of the WorkItem, as well as several derived metrics such as inclusive time per WorkItem, or mean execution time for all children of a WorkItem.

The AllScale Monitoring Component can also generate task dependency graphs, in other words, graphs with all the WorkItems executed by the runtime and their dependencies. WorkItem dependencies are created when a WorkItem is spawned from within another WorkItem due to the recursive model used. In our task graphs, each node represents a WorkItem and contains WorkItem name, exclusive execution time, and inclusive execution time. WorkItem graphs are coloured by exclusive time from yellow to red, being yellow shorter execution time and red longer execution time.

In addition, the monitoring framework can also generate heat maps with processes in the Y-axis and samples in the X-axis. These plots allow the user to inspect the evolution of certain metrics across processes and time, for example, the node idle rate, the WorkItem throughput, or the power consumed.

3.4 Introspection and Real-Time Visualisation

The AllScale Monitor Component provides an API to access the collected performance data in real time, while the application runs. As previously explained, some performance measurements can be accessed directly whereas more complex metrics are calculated on-demand to reduce the overhead introduced into the application. Several of these on-demand metrics are computed recursively, taking advantage of the task dependency graph stored in memory, for example, the total time of a WorkItem and all its children, or the average execution time of the children for a certain WorkItem.

The AllScale Monitor Component provides a distributed global view of the performance, that is, there is an instance of the monitoring component per node, and each one of these instances can be directly asked for its performance data. The introspection data can also be pipelined in real time via TCP to a web-server based dashboard, the AllScale Dashboard. This dashboard can be accessed with any available web browser.

The AllScale Dashboard provides two views as shown in Fig. 1, a global view of the system, and a more detailed view per node. Figure 1a shows the overview page of the system performance. In the upper part of the picture, the dashboard presents the system wide total speed, efficiency, and power for all nodes. The speed is the ratio of time doing useful work (executing WorkItems) against total time at maximum processor frequency. Efficiency is the ratio of useful time against the time at the current CPU speed and number of nodes used. Finally, power is the ratio of current power consumed against total power at the highest CPU frequency. Under each metric there is a slider bar that allows the user to change the weight of each objective in the multi-objective optimiser, thereby, being able to change the behaviour of the runtime in real time. The global score in the right part of the frame indicates how well the three customisable objectives (time, resource, power) are fulfilled. Finally, the frame also contains information of the system task throughput.

The global view also contains plots that depict the speed, efficiency, and power per node across time. Remember that this dashboard presents live information so all these plots evolve in real time while the application runs. In the right lower quadrant of the dashboard, we have a view on how the work is distributed across the system. Its dropdown menu allows the user to change in real time the policy used by the runtime to distribute such work. Thereby, allowing the user to observe how the work (and data) gets redistributed by changing the scheduling policy. The distribution of data within the runtime is unfortunately beyond the scope of this paper. This topic is covered in detail in [9].

Figure 1b shows the detailed view where performance data is depicted per node. For every node we have the CPU load, the memory consumption, the amount of data transferred through the network (TCP and MPI), speed, efficiency and power, the task throughput, and the amount of data owned by the node.

The AllScale Dashboard is a very powerful and convenient tool that can speed up the development process, because it allows developers to see as soon as they run the code the effect of source code changes. Furthermore, it also allows developers to change in real time the behaviour of the runtime to explore how the system behaves when changing different parameters.

The current version of the AllScale Dashboard uses one server that collects information from all the nodes involved in the execution. We are currently working on more scalable solutions to be able to tackle high node counts, for example, collecting the data in a tree manner, or having several hierarchically distributed servers. We also want to explore how to improve the graphical interface for high

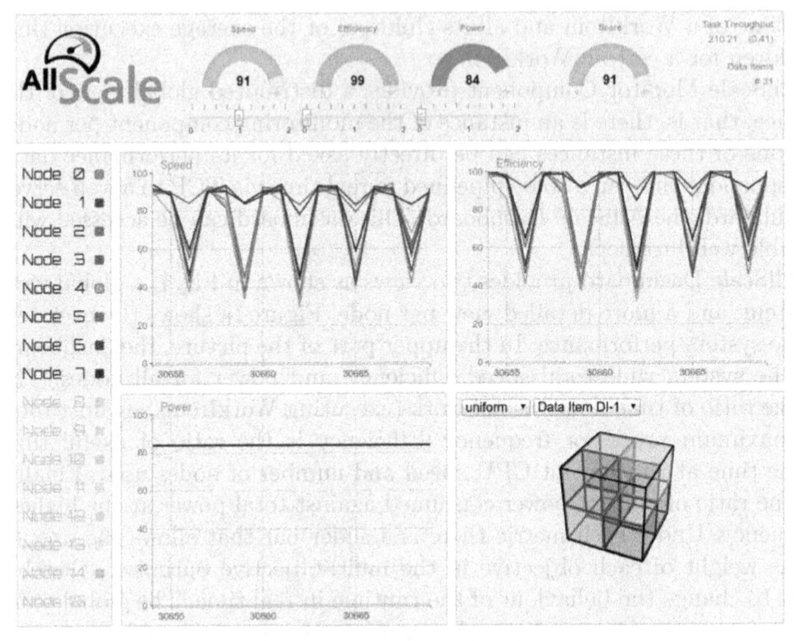

(a) Dashboard global system view.

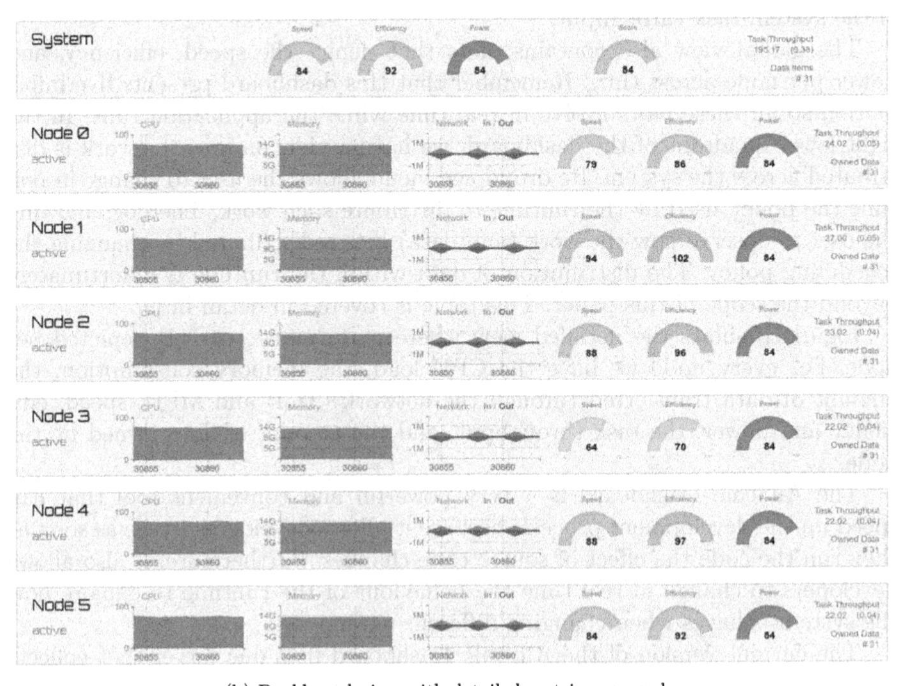

(b) Dashboard view with detailed metrics per node.

Fig. 1. The AllScale Dashboard showing performance data in real time while the application runs.

node counts, for instance, showing clusters of nodes instead of individual ones in the dashboard.

4 Evaluation

4.1 Overhead

The AllScale Monitoring Component has been designed with focus on having as minimum footprint in the system as possible to allow continuous performance introspection. We evaluated its overhead with several experiments. In the first set of experiments, we used a node with two 12-core E5-2690v3 Haswell and 512 GB of RAM memory, and run two benchmarks executing parallel loops that perform a simple stencil. The first benchmark (*chain* test) is a sequence of parallel loops where each loop waits for the completion of the previous one. The second benchmark (*fine_grain* test), on the other hand, computes loops in parallel by only waiting for the part of the previous loop it actually depends on. Both benchmarks were run with a grid of $128 \times 1024 \times 1024$ elements and for 100 time steps.

Figure 2a shows the overhead introduced by the monitoring framework into the stencil benchmarks running in 12, 24, and 48 logical cores (24 physical cores with hyper threading). Overhead is computed as the percentage difference in total execution time of the benchmark with and without the performance monitoring. In the experiments, the monitoring component collected execution time and dependencies for each WorkItem as well as node metrics every second. The picture depicts the overhead for two different versions of the monitoring component. First, an early naive implementation that we wrote where all threads process their own performance data and write it into per-process data structures. And second, our current producer-consumer implementation using a double buffering scheme.

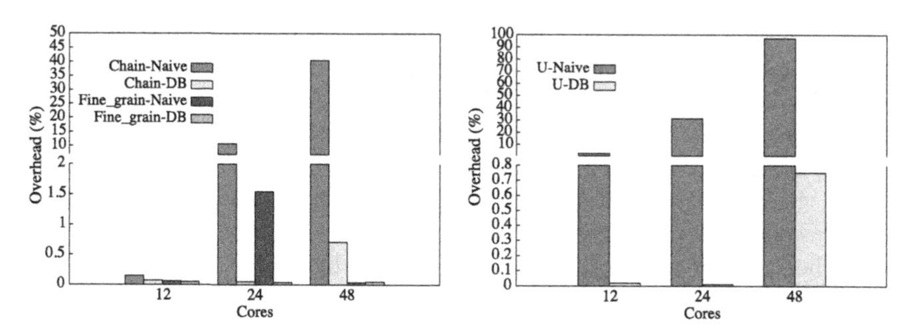

(a) Overhead with the stencil benchmarks

(b) Overhead with iPIC3D

Fig. 2. Overhead introduced by the AllScale Monitoring Component. Naive is an early implementation in which all threads process their own performance data. DB is the producer-consumer model with double buffering.

As can be seen in the picture, the overhead increases as we increase the number of cores, especially in the naive implementation due to its high thread contention. It is particularly bad for the chain benchmark, going from less than 0.5% to around 40%. In contrast, the overhead introduced when using our double buffering implementation is much lower, being always less than 1%. We can also see in the plot that there is a difference in the overhead between both benchmarks. This difference is caused by the nature of each benchmark. While the fine_grain benchmark executes all the iterations in parallel, the chain benchmark imposes threads to wait for the previous iteration to be finished. This synchronisation prevents the updates to the shared information to spread out, increasing the contention in global data structures.

We also evaluated the overhead of the monitoring component using a real-world application: iPIC3D [13], a Particle-in-Cell code for space plasma simulations that simulates the interaction of Solar Wind and Solar Storms with the Earth's Magnetosphere. In our evaluation we used sets of 8×10^6 particles uniformly distributed, and run 100 time steps of the simulation.

Figure 2b shows the overhead introduced by both implementations of the AllScale Monitoring Component. As can be seen in the picture, the naive implementation has a high overhead that increases rapidly with the number of threads used. This naive implementation introduces more than 90% of overhead with 48 logical cores. Thus, demonstrating that classical implementations where each thread processes its own data do not work in a scenario where lots of threads execute a huge amount of small tasks. For the double buffering implementation, the overhead is always smaller than 1%.

Figure 3 shows three different Paraver timelines of iPIC3D where we can see how the monitoring infrastructure interacts with the application. Figure 3a depicts two time steps of the application with the AllScale Monitoring turned off. We can see in Fig. 3b how our current AllScale Monitor Component implementation does not affect the application. In contrast, Fig. 3c serves to expose how the first naive approach we implemented affects the application execution a lot.

We also performed distributed experiments in a Cray XC40 equipped with two Intel Xeon E5-2630v4 Broadwell at 2.2 GHz and 64 GB of RAM per node. Nodes were connected with an Intel OmniPath interconnect. Figure 4 shows the running time of iPIC3D for different number of particles per node up to 64 nodes. The graph contains for each number of particles the simulation time with and without the monitoring infrastructure. The monitoring component used is the producer-consumer version with double buffering. As can be seen in the picture, the overhead of our monitoring component is negligible. It can also be observed that in some cases the execution time with the monitoring component activated is shorter than without monitoring. This can be explained by the fact that the overhead is so small that it gets absorbed by the natural execution noise and job execution time variability present in HPC systems.

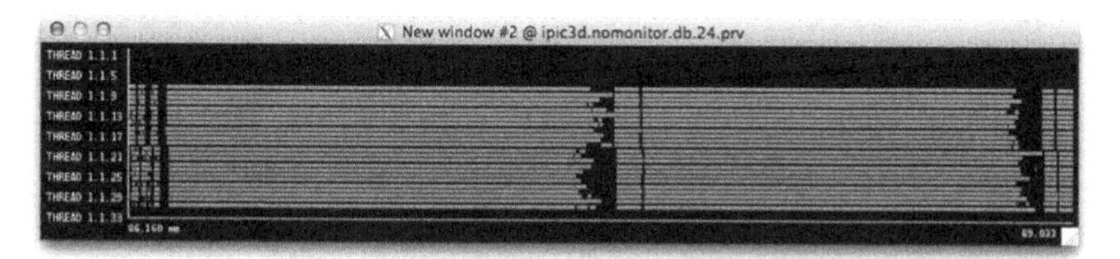

(a) WorkItem execution in iPIC3D without the AllScale Monitoring (24 cores)

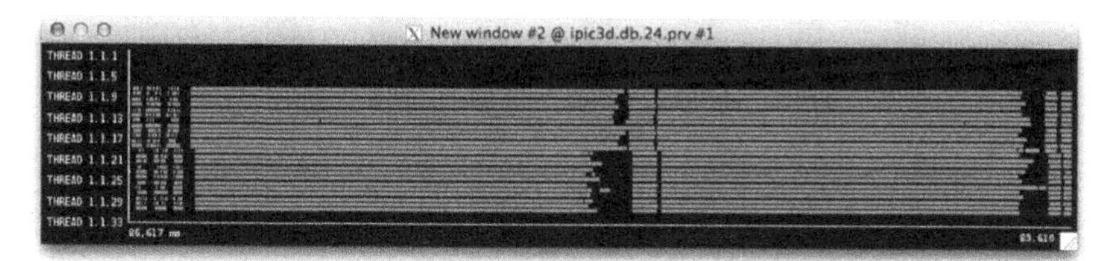

(b) WorkItem execution in iPIC3D with the AllScale Monitoring using double buffering (24 cores)

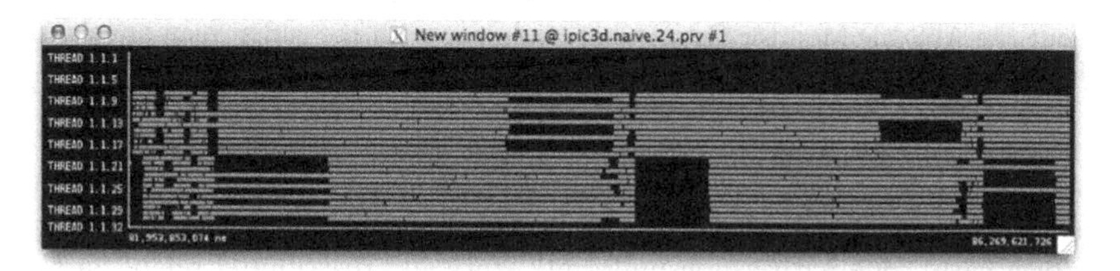

(c) WorkItem execution in iPIC3D with the naive implementation of the AllScale Monitoring (24 cores)

Fig. 3. Paraver timelines for two iterations of iPIC3D. Y-axis contains threads and X-axis time. Green is WorkItem execution. Black is outside WorkItem execution. (Color figure online)

4.2 Use Case

As previously explained, the AllScale Runtime can optimise the execution of applications regarding three different weighted objectives: time, resource, and energy. The runtime contains a prototype of an optimiser module that implements the Nelder-Mead method [14] to search for the best set of parameters that fulfil the objectives set. There are several parameters that the runtime can tune while trying to fulfil such objectives, for instance, number of threads and frequency of the processors used. The multi-objective optimiser uses the performance data provided by the AllScale Monitor Component to guide this parameter optimisation. It uses power consumed, system idle rate and task throughput among others.

In this section, we demonstrate how the AllScale Runtime uses in real time the introspection data collected by the AllScale Monitor Component to run an

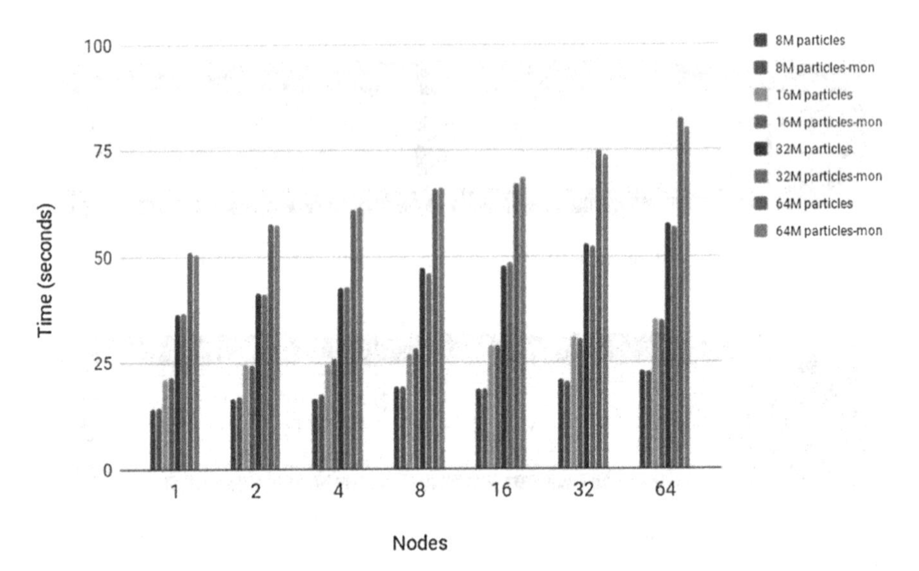

Fig. 4. Execution times for different configurations of particles per node with and without the AllScale Monitor Component. Columns with the suffix -mon are the runs with the monitoring infrastructure activated.

application as energy efficient as possible. To that aim, we run iPIC3D with three different tests cases: one with a uniformly distributed set of particles in a nebula (uniform case), one with a non-uniformly statically distributed set of particles (cluster case), and finally, one with a non-uniformly dynamically changing set of particles in an explosion (explosion case). We run the experiments in a node with 2 Intel Xeon Gold processors (36 cores in total), and 16×10^6 particles.

In the experiments we gave the maximum weight to the energy objective, that is, the runtime will try to optimise only by the energy consumed. To this end, the runtime will dynamically tune the number of threads and the frequency of the processors used. Figure 5 shows the power consumed for the three test cases while running with and without the optimiser. The green line (label *Power*) shows the power when running without any self-tuning using the *ondemand* governor for the CPU frequencies. With the *ondemand* governor the operating system manages the CPU frequencies. The purple line (label *Power-opt*) represents the power consumed when using the optimiser with the energy objective. The blue line shows the number of threads the runtime is using when the optimiser is tuning the run.

We can see several things in the figure. First, we can see how the optimiser starts always with the maximum number of threads and changes it until it finds an optimal value. Second, it is noticeable that even though the runs without optimisation (green line) consume much more power, they are much shorter. In terms of total energy consumed, in the explosion case, the version using the optimiser consumes a total of 4,066.97 J compared to 5,400.48 J without it. For the cluster case the numbers are 3,758.3 J with the optimiser against 3,696.75 J without it. Finally, for the uniform case, the run with the optimiser consumes 3,197.12 J against 2,064.93 J without it. As can be seen from the numbers, the current optimiser prototype does not find an optimal configuration for the cluster and uniform cases. Figure 5c for example, shows how the optimiser does not find an optimal thread number and keeps changing it during the whole execution. Thus, further investigation on why the current optimiser prototype does not find optimal solutions for these two test cases is needed. It is important to remark however that thanks to the AllScale Monitor Component we have been able to detect the bad behaviour of the optimiser module, thereby, being able to warn their developers and helping them to improve it.

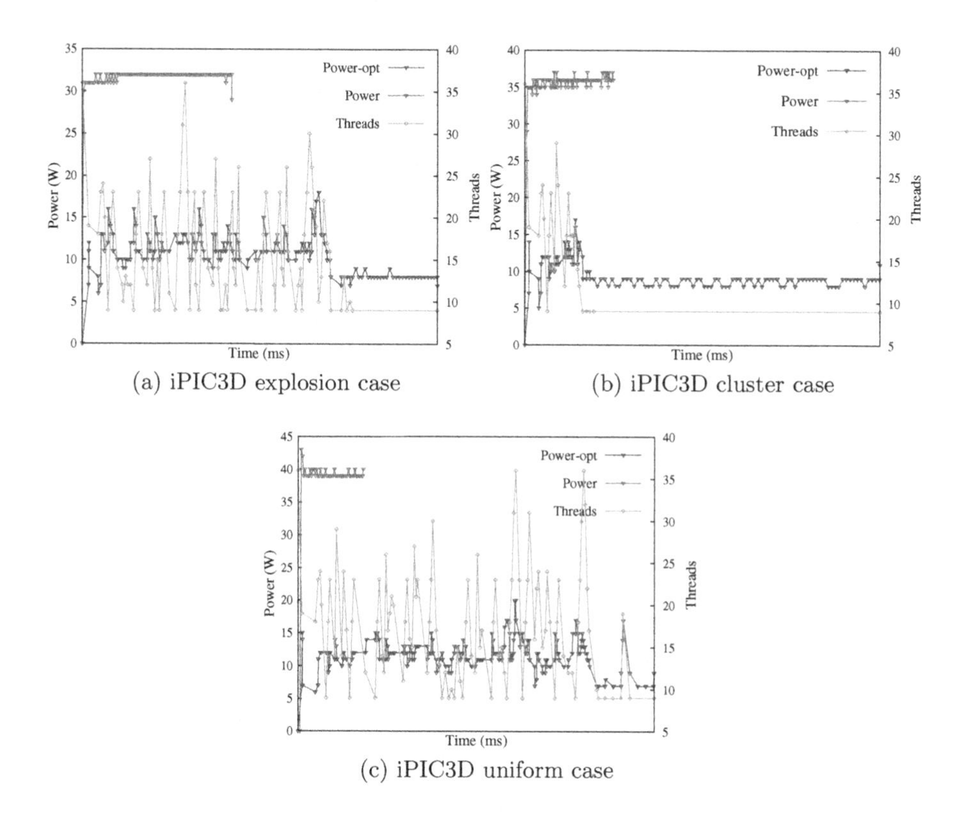

(a) iPIC3D explosion case

(b) iPIC3D cluster case

(c) iPIC3D uniform case

Fig. 5. Power consumed by iPIC3D when running with (power-opt line) and without (power line) the AllScale Multi-objective Optimiser.

5 Related Work

Performance observation and analysis is a topic extensively explored in HPC. There are many tools that can be used for post-mortem performance analysis of parallel applications, e.g. Paraver [15], Score-P [12], Scalasca [6], or TAU [17].

These tools are really useful to detect and optimise performance problems, however, they were designed for post-mortem analysis and are not sufficient for continuous monitoring introspection. In contrast, our monitoring component is not a stand-alone tool but a runtime system component, which cannot provide the same level of performance detail as the previous tools, but allows continuous monitoring introspection with very low overhead.

Most state-of-the-art runtime systems include means to generate performance reports for post-mortem analysis. StarPU [2] uses the FXT library [4] to generate Pajé traces and graphs of tasks. Legion [3] provides profiles and execution timelines with performance data of the tasks executed by the runtime. OmpsSs [5] generates Paraver [15] traces with Extrae. It also provides task dependency graphs in pdf format. APEX [7] generates performance profiles that can be visualised with TAU [17].

Nevertheless, classical post-mortem techniques will not suffice in the Exascale era. Real-time introspection capabilities are a requirement for performance tuning and adaptation. Thus, several performance tools and task-based runtimes implement introspection strategies. Tools such as the work of Aguilar et al. [1] or Score-P [12] include online introspection capabilities. However, these tools are mainly designed for MPI monitoring. The runtime Legion [3] contains task mappers that can request performance information to the runtime while the application runs. OmpSs has an extension to use introspection data to guide the process of task multi-versioning [16]. StarPU provides an API to access performance counters from the runtime.

6 Conclusions and Future Work

The path to Exascale brings new challenges to scalability, programmability, and performance of software. Future systems will have to include performance introspection to support adaptivity and efficient management of resources. In this paper we have presented the AllScale Monitor Component, the monitoring infrastructure included within the AllScale Runtime. This monitoring framework provides real-time introspection to the AllScale Scheduler with minimum overhead as we have demonstrated. We have also shown that real-time introspection capabilities are very useful to orchestrate application execution as well as to analyse the performance of the system in real time. Thus, speeding up the development process because performance deficiencies can be detected almost instantly, without the need to wait until the end of a long running job for example. In addition, the monitoring framework is self-contained, so it can be easily decoupled and used in other runtimes.

As future work, we want to further investigate the use of historical performance data together with real-time introspection data to help in the decision-making process of the scheduler. We are also working in improving the scalability of the dashboard server as well as the scalability of its graphical interface.

References

1. Aguilar, X., Laure, E., Furlinger, K.: Online performance introspection with IPM. In: 2013 IEEE 10th International Conference on High Performance Computing and Communications & 2013 IEEE International Conference on Embedded and Ubiquitous Computing (HPCC_EUC), pp. 728–734. IEEE (2013)
2. Augonnet, C., Thibault, S., Namyst, R., Wacrenier, P.A.: StarPU: a unified platform for task scheduling on heterogeneous multicore architectures. Concurr. Comput. Pract. Exp. **23**(2), 187–198 (2011)
3. Bauer, M., Treichler, S., Slaughter, E., Aiken, A.: Legion: expressing locality and independence with logical regions. In: Proceedings of the International Conference on High Performance Computing, Networking, Storage and Analysis, p. 66. IEEE Computer Society Press (2012)
4. Danjean, V., Namyst, R., Wacrenier, P.-A.: An efficient multi-level trace toolkit for multi-threaded applications. In: Cunha, J.C., Medeiros, P.D. (eds.) Euro-Par 2005. LNCS, vol. 3648, pp. 166–175. Springer, Heidelberg (2005). https://doi.org/10.1007/11549468_21
5. Duran, A., et al.: OmpSs: a proposal for programming heterogeneous multi-core architectures. Parallel Process. Lett. **21**(02), 173–193 (2011)
6. Geimer, M., Wolf, F., Wylie, B.J., Ábrahám, E., Becker, D., Mohr, B.: The scalasca performance toolset architecture. Concurr. Comput. Pract. Exp. **22**(6), 702–719 (2010)
7. Huck, K.A., et al.: An autonomic performance environment for exascale. Supercomput. Front. Innov. **2**(3), 49–66 (2015)
8. Jordan, H.: Insieme: a compiler infrastructure for parallel programs. Ph.D. thesis, Ph. D. dissertation, University of Innsbruck (2014)
9. Jordan, H., et al.: The Allscale runtime application model. In: 2018 IEEE International Conference on Cluster Computing (CLUSTER), pp. 445–455. IEEE (2018)
10. Jordan, H., Thoman, P., Zangerl, P., Heller, T., Fahringer, T.: A context-aware primitive for nested recursive parallelism. In: Desprez, F., et al. (eds.) Euro-Par 2016. LNCS, vol. 10104, pp. 149–161. Springer, Cham (2017). https://doi.org/10.1007/978-3-319-58943-5_12
11. Kaiser, H., Heller, T., Adelstein-Lelbach, B., Serio, A., Fey, D.: HPX: a task based programming model in a global address space. In: Proceedings of the 8th International Conference on Partitioned Global Address Space Programming Models, p. 6. ACM (2014)
12. Knüpfer, A., et al.: Score-P: a joint performance measurement run-time infrastructure for periscope, scalasca, TAU, and vampir. In: Brunst, H., Müller, M., Nagel, W., Resch, M. (eds.) Tools for High Performance Computing 2011, pp. 79–91. Springer, Heidelberg (2012). https://doi.org/10.1007/978-3-642-31476-6_7
13. Markidis, S., Lapenta, G., et al.: Multi-scale simulations of plasma with iPIC3D. Math. Comput. Simul. **80**(7), 1509–1519 (2010)
14. Nelder, J.A., Mead, R.: A simplex method for function minimization. Comput. J. **7**(4), 308–313 (1965)
15. Pillet, V., Labarta, J., Cortes, T., Girona, S.: PARAVER: a tool to visualize and analyze parallel code. In: Proceedings of WoTUG-18: Transputer and Occam Developments, vol. 44, pp. 17–31 (1995)

16. Planas, J., Badia, R.M., Ayguade, E., Labarta, J.: Self-adaptive OmpSs tasks in heterogeneous environments. In: 2013 IEEE 27th International Symposium on Parallel & Distributed Processing (IPDPS), pp. 138–149. IEEE (2013)
17. Shende, S.S., Malony, A.D.: The TAU parallel performance system. Int. J. High Perform. Comput. Appl. **20**(2), 287–311 (2006)

Productivity-Aware Design and Implementation of Distributed Tree-Based Search Algorithms

Tiago Carneiro[1](✉) and Nouredine Melab[1,2]

[1] INRIA Lille - Nord Europe, Lille, France
tiago.carneiro-pessoa@inria.fr
[2] Université de Lille, CNRS/CRIStAL, Lille, France
nouredine.melab@univ-lille.fr

Abstract. Parallel tree search algorithms offer viable solutions to problems in different areas, such as operations research, machine learning and artificial intelligence. This class of algorithms is highly compute-intensive, irregular and usually relies on context-specific data structures and hand-made code optimizations. Therefore, C and C++ are the languages often employed, due to their low-level features and performance. In this work, we investigate the use of Chapel high-productivity language for the design and implementation of distributed tree search algorithms for solving combinatorial problems. The experimental results show that Chapel is a suitable language for this purpose, both in terms of performance and productivity. Despite the use of high-level features, the distributed tree search in Chapel is on average 16% slower and reaches up to 85% of the scalability observed for its MPI+OpenMP counterpart.

Keywords: Tree search algorithms · High productivity · PGAS · Chapel · MPI+OpenMP

1 Introduction

Tree-based search algorithms are strategies that implicitly enumerate a solution space, dynamically building a tree. This class of algorithms is often used for the exact resolution of permutation combinatorial optimization problems (COP) and offers viable solutions to problems in different areas, such as operations research, artificial intelligence, bioinformatics and machine learning [15,19]. As the decision version of permutation COPs are usually NP-Complete, the size of problems that can be solved to optimality is limited, even if large-scale distributed computing is used [9,16]. In this sense, it is expected that exascale computers are willing to allow a significant decrease in the execution time required to solve COP instances to optimality. However, such large scale systems are going to be complex to program, and efforts towards programmability are crucial for better exploiting this future generation of computers [2,13].

J. M. F. Rodrigues et al. (Eds.): ICCS 2019, LNCS 11536, pp. 253–266, 2019.
https://doi.org/10.1007/978-3-030-22734-0_19

Tree-based search algorithms are compute-intensive and highly irregular, which demands hand-optimized data structures for efficient search and load balancing [6,14,17]. Thus, high-productivity languages are not often employed within the scope of tree search, as they historically suffer from severe performance penalties [11]. Instead, this kind of application is frequently written in either C or C++, due to low-level features present in both languages [5,9].

Chapel is a productivity-aware programming language for high-performance computing that is competitive to both C-OpenMP and C-MPI+OpenMP in terms of performance, considering different benchmarks [8]. The objective of the present research is to investigate Chapel's features to design and implement distributed tree search algorithms for solving permutation combinatorial problems. To the best of our knowledge, the present research is the first one that investigates the use of a high-productivity language for this purpose.

The experimental results show that Chapel is a suitable language for the design and implementation of distributed tree search algorithms, both in terms of performance and productivity. It is possible to conceive a distributed tree search algorithm starting from its multicore counterpart by adding few modifications. The distributed algorithm performs load balancing among different computer nodes and also uses all CPU cores that a computer node has. Despite the high level of its features, the distributed tree search in Chapel is on average 16% slower and reaches up to 85% of the scalability achieved by its C-MPI+OpenMP counterpart. Finally, the distributed load balancing strategies provided are effective: the dynamic load balancing version is up to 1.5× faster than its static counterpart.

The remainder of this paper is structured as follows. Section 2 brings background information and related works. The distributed tree-based search in Chapel is detailed in Sect. 3. Section 4 presents a performance evaluation. Then, Sect. 5 brings a discussion in terms of performance, programmability, and limitations of Chapel for programming distributed tree search algorithms. Finally, conclusions are outlined in Sect. 6.

2 Background and Related Works

2.1 The Chapel Programming Language

Chapel is an open-source parallel programming language designed to improve the programmability for high-performance computing. It incorporates features from compiled languages such as C, C++, and Fortran, as well as high-level elements related to Python and Matlab. The parallelism is expressed in terms of lightweight tasks, which can run on several locales or a single one. In this work, the term *locale* refers to a symmetric multiprocessing computer in a parallel system [10].

In Chapel, both *global view of control flow* and *global view of data structures* are present [8]. Concerning the first one, the program is started with a single task and parallelism is added through data or task parallel features. Moreover, a task can refer to any variable lexically visible, whether this variable is placed in the same locale on which task is running, or in the memory of another one. Concerning the second one, indexes of data structures are globally expressed, even in case the implementation of such data structures distributes them across several locales. Thus, Chapel is a language that realizes the Partitioned Global Address Space (PGAS) programming model [1].

Finally, indexes of data structures are mapped to different locales using *distributions*. Contrasting to other PGAS-based languages, such as UPC and Fortran, Chapel also supports user-defined distributions [7].

2.2 Tree-Based Search Algorithms

Tree-based search algorithms are strategies that implicitly enumerate a solution space, dynamically building a tree [15]. The internal nodes of the tree are incomplete solutions, whereas the leaves are solutions. Algorithms that belong to this class start with an initial node, which represents the root of the tree, i.e., the initial state of the problem to be solved. Nodes are branched during the search process, which generates children nodes more restricted than their parent node. Generated nodes are evaluated, and then, the valid and feasible ones are stored in a data structure called *Active Set*.

At each iteration, a node is removed from the active set according to the employed search strategy [19]. The search generates and evaluates nodes until the data structure is empty or another termination criterion is reached. If an undesirable state is reached, the algorithm discards this node and then chooses an unexplored (frontier) node in the active set. This action prunes some regions of the solution space, keeping the algorithm from unnecessary computation. The degree of parallelism of tree-based search algorithms is potentially very high, as the solution space can be partitioned into a large number of disjoint portions, which can be explored in parallel.

As these algorithms are compute-intensive, diverse strategies have been used for improving performance, such as instruction-level parallelism, architecture-specific code optimizations and problem-specific data structures [6,12,14,17]. Thus, parallel tree-based search algorithms are frequently written in C/C++, due to their low-level features and supported parallel computing libraries [5]. In the context of distributed algorithms, the same performance-aware strategies are combined with distributed programming libraries for implementing load balancing and explicit communication between processing nodes [9,16,18]. As a consequence, programming distributed tree search algorithms can be challenging and time-consuming.

3 Distributed Tree-Based Search Algorithms in Chapel

A major objective of Chapel concerning productivity is allowing distributed programming using concepts close to the ones of shared-memory programming [8]. In this section, a multicore single-locale tree-search algorithm is initially proposed. Then, it is extended using Chapel's productivity-aware features for distributed programming.

3.1 Algorithm Overview

This work focuses on permutation combinatorial problems, for which an N-sized permutation represents a valid and complete solution. Permutation combinatorial problems are used to model diverse real-world situations, and their decision versions are often NP-Complete [16,19].

This section presents two backtracking algorithms for enumerating *all* complete and feasible solutions of the N-Queens problem. Backtracking is a fundamental problem-solving paradigm that consists in dynamically enumerating a solution space in a depth-first fashion. Due to its low memory requirements and its ability to quickly find new solutions, depth-first search (DFS) is often preferred [19].

The N-Queens problem consists in placing N non-attacking queens on a $N \times N$ chessboard, and it is often used as a benchmark for novel tree-based search algorithms [3,12]. The N-Queens is easily modeled as a permutation problem: position r of a permutation of size N designates the column in which a queen is placed in row r. Furthermore, the concepts herein presented are similar to any permutation combinatorial problem and can be adapted for solving other problems of this class with straightforward modifications [6,14].

3.2 The Single-Locale Multicore Implementation

Algorithm 1 presents a pseudocode for the single-locale backtracking in Chapel. The algorithm starts receiving the problem to be solved (*line 1*) and the cutoff depth (*line 2*). Then, it is required to generate an initial load for the parallel search. For this purpose, *task* 0 performs backtracking from depth 1 (initial problem configuration) until the cutoff depth $cutoff$, storing all feasible, valid, and incomplete solutions at depth $cutoff$ in the active set A (*line 4*). After generating the initial load, the parallel search strategy begins through a `forall` statement (*line 5*).

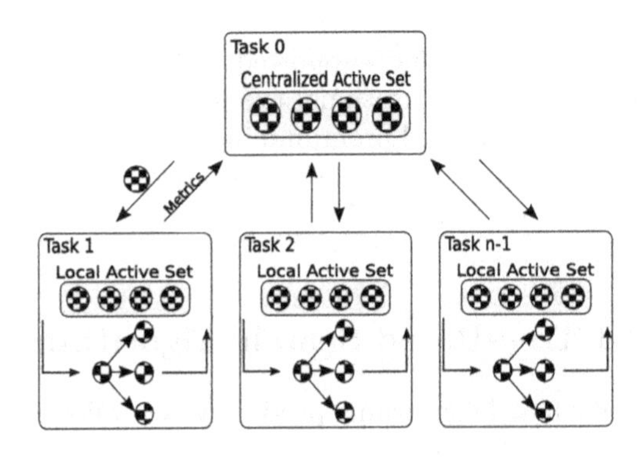

Fig. 1. Task 0 is responsible for managing the centralized active set A and performing load balancing. The searches are independent, and metrics are reduced using the *Reduce Intents* of Chapel (Own representation adapted from [9]).

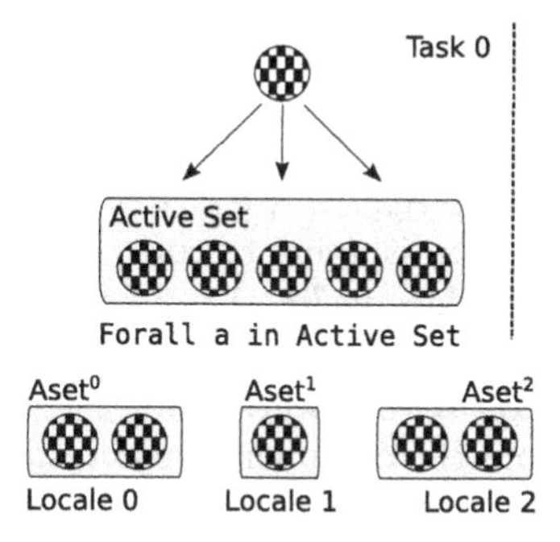

Fig. 2. Task 0 is responsible for distributing the active set across several locales. The distributed active set A_d consists of several sets $A_d^i, i \in \{0, ..., l-1\}$, where l is the number of locales on which the application is going to run.

As one can see in Fig. 1, nodes in the centralized active set A are assigned to tasks in chunks. Each task has its active set and executes a backtracking search strategy. In turn, nodes are used to initialize the backtracking, which enumerates the solution space rooted by a node. The load balancing is done through the iterator (`DynamicIters`) used to assign indexes of A to tasks, like in OpenMP. Metrics are reduced through *Reduce Intents*. In Chapel, it is possible to use the `Tuple` data type (equivalent to C-structs) and reduce all metrics at once (*line 6*). Differently from OpenMP, it is not required to define a tuple reduction. Finally, the parallel search finishes when the active set A is empty.

Algorithm 1. The multicore tree-based search algorithm.

```
1 I ← get_problem()
2 cutoff ← get_cutoff_depth()
3 A ← ∅
4 A ← generate_initial_active_set(cutoff, I)
5 forall node in A with(+ reduce metrics) do
6 |   metrics+ = tree_search(node, cutoff, I)
7 end
```

3.3 The Multi-locale Distributed Implementation

One can see in Algorithm 2 a pseudocode for the distributed tree-based search algorithm in Chapel. Thanks to Chapel's global view of control flow, the search also starts serially, with *task* 0 generating the initial load to populate the active set A (*line 4*). To make it possible to distribute A, it is required to define a domain (*line 5*) and define how this domain it is going to be distributed across different locales (*line 6*) [7]. In this work, only *standard distributions* are used[1]. Finally, the distributed active set A_d of type `Node` is defined over the domain D (*line 8*).

After the initial load generation, the nodes of A are distributed across several locales by using a parallel `forall` (*line 9*), which generates the distributed active

[1] https://chapel-lang.org/docs/modules/layoutdist.html.

set A_d. Thanks to Chapel's global view of A_d, the indexes of both active sets are directly accessed in *line 9*. The compiler is responsible for the communication code. Moreover, as shown in Fig. 2, A_d is an abstraction. The distributed active set A_d consists of several sets $A_d^i, i \in \{0, ..., l-1\}$, where l is the number of locales on which the application is going to run.

The parallel search takes place in *line* 12. As one can see in Algorithm 2, its `forall` is similar to the one of Algorithm 1. However, distributed iterators are used instead (`DistributedIters`). Additionally, the distributed search exploits two levels of parallelism, and the compiler is also responsible for generating the code that exploits all CPU cores a locale has. Finally, the metrics are reduced in the same way as in the single-locale algorithm.

Algorithm 2. The multilocale tree-based search algorithm.

```
1  I ← get_problem()
2  cutoff ← get_cpu_cutoff_depth()
3  A ← ∅
4  A ← generate_initial_active_set(cutoff, I)
5  Range ← 0..(|A| − 1)
6  D ← Range mapped according to a standard distribution
7  A_d ← ∅
8  A_d ← [D] : Node
9  forall s in Range do
10 |  A_d[s] ← A[s]
11 end
12 forall node in A_d following the iterator with(+ reduce metrics) do
13 |  metrics+ = tree_search(node, cutoff, I)
14 end
```

3.4 Search Procedure and Data Structures

The kernel of both parallel algorithms previously presented is based on a serial and hand optimized backtracking for solving permutation combinatorial problems, originally written in C [6]. The serial backtracking was then adapted to Chapel, obeying the handmade optimizations, instruction-level parallelism, data structures, and C-types. The data structure `Node` is similar to any permutation combinatorial problem. It contains an unsigned 8-bit integer vector of size $cutoff$, identified by *board*, and an unsigned integer variable. The vector *board* stores the feasible and valid incomplete solution. In turn, the integer variable, identified by *bitset*, keeps track of board lines by setting its bit n to 1 each time a queen is placed in the n-th line.

The search performed by the kernel is a non-recursive backtracking that does not use dynamic data structures, such as stacks. Initially *depth* receives the value of $cutoff$. Next, *board* and *bitset* are initialized with `Node[i].board` and `Node[i].bitset`, respectively. The semantics of a stack is obtained by trying to increment the value of the vector *board* at position *depth*. If this increment results in a feasible and valid incomplete solution, the *depth* variable is then incremented, and the search proceeds to the next depth. After trying all configurations for a given depth, the search backtracks to the previous one.

4 Performance Evaluation of a Multi-locale Backtracking

The objective of this section is to show that it is possible to use a high-productivity language for programming distributed tree-based search algorithms and achieve metrics similar to MPI+X.

4.1 Protocol

The following programs were conceived for enumerating all valid and complete configurations of the N-Queens problem.

- **Chapel:** implementation of the multi-locale backtracking search algorithm described in Algorithm 2, written in Chapel.
- **MPI+X:** single program - multiple data (SPMD) counterpart written in C of the program above introduced. This program uses MPI for communication and X means the use of OpenMP for exploiting all cores/threads a locale has.

Both implementations use the data structures and search procedure detailed in Sect. 3.4. In this section, it is investigated how the application scales according to the number of locales. Furthermore, the influence of the PGAS data structure distribution on the application execution time is also studied. Tree search algorithms for solving combinatorial problems are usually highly irregular applications. Therefore, the influence of the distributed load balancing strategies on the overall performance of the application is also investigated. Finally, all metrics collected for the implementation in Chapel are compared to the ones achieved by its MPI+X counterpart.

4.2 Parameters Settings

Problems of size (N) ranging from 15 to 20 are considered. The experiments take from few seconds to several hours of parallel processing. The number of locales ranges from 1 to 32, and the application is the same for either one or more than one computer node(s). The number of locales is passed to the application using Chapel's built-in command line parameter `-nl l` (`-np l` for MPI), where l is the number of locales on which the application is executed.

All computer nodes are symmetric and operate under Debian 4.9.130 − 2, 64 bits. They are equipped with *two* Intel Xeon X5670 @ 2.93 GHz (a total of 12 cores/24 threads), and 96 GB RAM. Thus, up to 384 cores/768 threads are used in the experiments. All locales are interconnected through an Infiniband network: Mellanox Technologies MT26428 (ConnectX VPI PCIe 2.0 5GT/s - IB QDR/10GigE).

Concerning the MPI+X implementation, OpenRTE 2.0.2 along with *gcc* 6.3.0 and OpenMP 4.5 were used for compilation and execution. The Chapel implementation was programmed in its current version (1.18), and the *default* task layer (qthreads) is the one employed. Chapel's multi-locale code runs on top of

Table 1. Summary of the environment configuration for multi-locale execution and compilation.

Variable	Value
`CHPL_RT_NUM_THREADS_PER_LOCALE`	24
`CHPL_TARGET_ARCH`	*native*
`CHPL_COMM`	*gasnet*
`CHPL_COMM_SUBSTRATE`	*ibv*
`GASNET_IBV_SPAWNER`	*mpi*

GASNet, and several environment variables should be set with the characteristics of the system the multi-locale code is supposed to run. One can see in Table 1 a summarization of the runtime configurations for multi-locale execution. The Infiniband GASNet implementation is the one used for communication (`CHPL_COMM_SUBSTRATE`) along with MPI, which is responsible for getting the executables running on different locales (`GASNET_IBV_SPAWNER`).

Chapel provides several standard distributions to map data structures onto locales. Different tests were also carried out to identify the best option in the context of this work. The one chosen was the one-dimension *BlockDist*, which horizontally maps elements across locales. For instance, in case $l = 3$ and $|A_d| = 8$, elements $0, ..., 2$ are on locale l_0, $3, .., 5$ on locale l_1, and $6, 7$ on locale l_2. In the scope of the present research, choosing a different standard distribution does not lead to performance improvements.

Experiments were carried out to choose a suitable cutoff depth. This parameter directly influences the size of A_d, and therefore the time spent in distributing the active set across locales. As observed in Fig. 3, the fastest data structure distribution is observed for $cutoff = 3$. However, such a cutoff value limits parallelism, resulting in a slow distributed search. In contrast, when the cutoff is set to 6, the distribution of A_d becomes $10\times$ slower than the search procedure itself. This behavior happens due to the combinatorial nature of N-Queens: a cutoff depth twice deeper results in an active set $725\times$ bigger. When choosing $cutoff = 5$, the search takes the same time as for $cutoff = 4$. Despite that, the distribution of A_d is on average $9\times$ slower for $cutoff = 5$. Thus, the cutoff depth chosen is 4. Preliminary experiments also show that $cutoff = 4$ is the best value for the MPI+X implementation.

Chapel also provides two different distributed load balancing iterators: *guided* and *dynamic*, which are also similar to OpenMP's schedules of the same name. Experiments were carried out to identify the best *chunk* for both guided and dynamic multi-locale load balancing strategies. Both strategies present the best performance using the *default* chunk size.

4.3 Results

First of all, concerning the distributed load balancing strategies provided by Chapel, using the *dynamic* iterator is from 1.17× to 1.51× times faster than using no load balancing (static version). Moreover, the *guided* iterator does not seem a suitable load balancing in the scope of this work: it shows benefits compared to the static version only for sizes ranging from 18 to 20. For these problem sizes, using the guided iterator makes the search up to 1.21× faster than its static counterpart. In turn, using the dynamic distributed iterator results in a search from 1.21× to 1.25× faster than using the guided one.

The benefits of using load balancing are not observed for the smallest solution space, i.e., for the problem of size $N = 15$. In such a situation, the static search performs better because there is no communication among locales during the search. As shown in Fig. 4, the overhead of data structure initialization

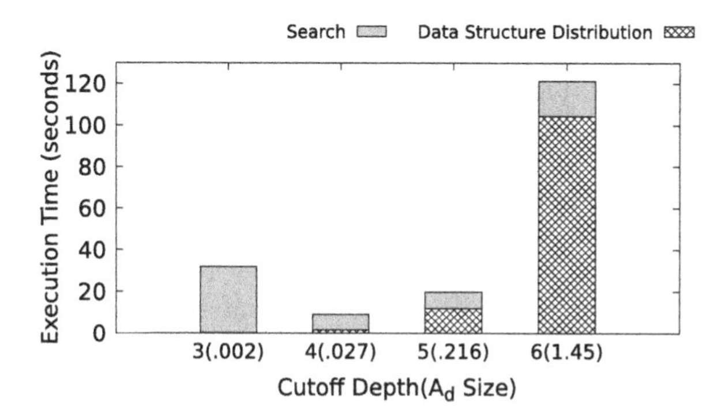

Fig. 3. Influence of the *cutoff* choice on the execution time. Values are for the N-Queens of size $N = 17$ and 32 locales (384 cores). On the X axis: the cutoff depth and the distributed active set size in parentheses (in 10^6 nodes).

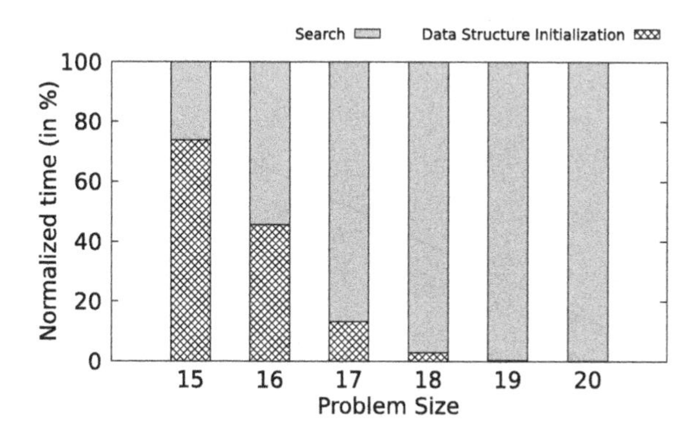

Fig. 4. Proportion of the initialization and distribution of A_d compared to the whole execution time. Results are for the N-Queens of sizes (N) ranging from 15 to 20 and executed on 32 locales (384 cores).

and distribution becomes less detrimental as the solution space grows, and the benefits of using distributed load balancing can be observed.

It is shown in Fig. 5 how the distributed search scales according to the number of locales. The worst scalability is observed for the smallest size ($N = 15$). In such a situation, the initialization and distribution of A_d amount for almost the whole execution time (see Fig. 4). For problem sizes ranging from 17 to 20, the dynamic version scales up to 20.5× ($N = 19$), whereas guided and static scale up to 16.8× and 16.9×, respectively (also for $N = 19$). The MPI+X version scales up to 25.4× ($N = 18$). Therefore, the distributed search in Chapel achieves up to 80% of the scalability observed for its MPI+X counterpart.

As shown in Fig. 4, running the Chapel program on multiple locales comes with the overhead of distributing A_d across several nodes. However, the scalability results previously discussed take into account the search running on 1 locale (`-nl 1`). In such a situation, the active set is not distributed across different nodes, and the search works similarly to a multicore and single locale one. One can see in Fig. 6 the best speedup reached for 4 to 32 locales compared to the search running on 2. Results closer to the linear speedup are observed: for 32 locales (16× more nodes) all three variations of the search written in Chapel and the MPI implementation reach a speedup of almost 13×, which corresponds to 81% of the linear speedup.

It is worth to mention that the time spent on distributing A_d does not grow linearly according to the number of locales. As one can see in Fig. 7, the time required to distribute A_d grows up to size $N = 16$, then it becomes almost constant. This behavior comes from the fact that the size of A_d is the same for one or more locale(s). Thus, as the number of locales grows, the number of messages sent grows as well, but their size decreases. Moreover, the A_d distribution is performed in parallel (Algorithm 2, *line 9*), and the Infiniband GASNet implementation supports one-sided communication.

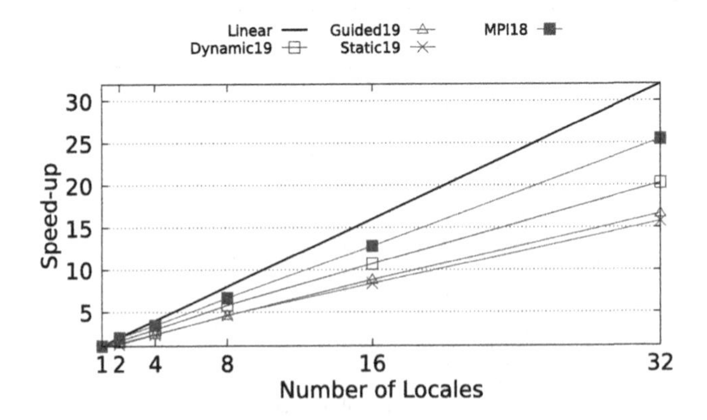

Fig. 5. The *highest* speedup achieved by Chapel and MPI+X implementations when executed on 2 (24) to 32 locales (384 cores).

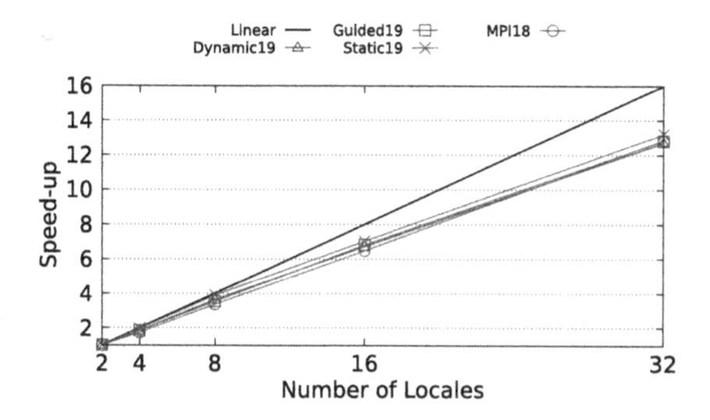

Fig. 6. The *highest* speedup achieved by Chapel and MPI+X implementations when executed on 4 (48) to 32 locales (384 cores) and compared to the execution on 2 (24).

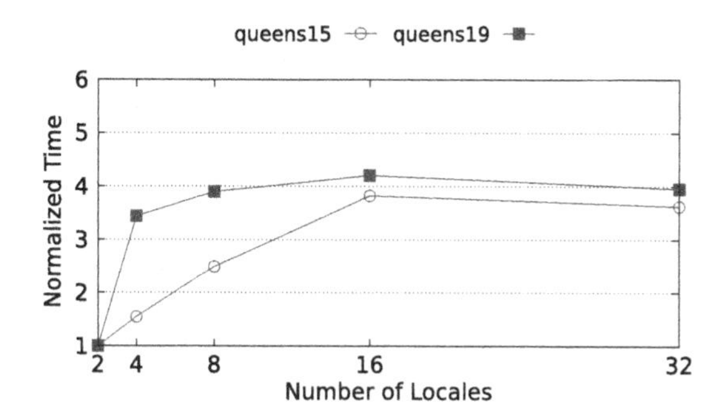

Fig. 7. Normalized time required to initialize and distribute A_d. Results are for 2 (24) to 32 locales (384 cores).

In terms of wall-clock time, Chapel is equivalent to MPI+OpenMP when running on one locale. For the smaller solution space (i.e., $N = 15$), Chapel stands out and it is up to 25% faster than MPI+X. In such a situation, A_d is not distributed, and the program behaves like a single-locale and multicore one. Moreover, MPI implements the SPMD programming model. This way, MPI is started, and its functions are called even for one locale. Additionally, it is worth to mention that Chapel is a compiled language and it is possible to program in Chapel both search strategy and data structures equivalent to the ones present in its counterpart written in C. In contrast, for multiple locales and bigger problem sizes, the Chapel distributed search is on average 16% slower than its MPI+X counterpart.

5 Discussion

Chapel's high-productivity features make it straightforward to design and implement a distributed backtracking based on a multicore and single-locale version. There is no need for dealing with communication, metrics reduction, or distributed load balancing. Moreover, differently from the classic MPI+X, there is no need for using a different parallel library for each level of parallelism.

Thanks to Chapel's global view of the control flow and data structures, the main difference between the multi- and single-locale versions lies mainly in the use of the PGAS data structures and distributed iterators for load balancing. As a consequence, the multi-locale version is only 8 *lines longer* than its single-locale counterpart, which results in a code 33% bigger. In contrast, it is required to add 24 lines to the backtracking written in C-OpenMP to use MPI, which almost doubles the program size.

Chapel also shows to be competitive to MPI+OpenMP regarding performance. It is possible to program the search and data structures in a way equivalent to C, resulting in efficient single-core use. Additionally, the compiler also generates code for exploiting all cores a locale has, and the load balancing provided by Chapel is effective. One would argue that it could be possible to program an MPI+X version faster than the one used, and the results of the last section could have been more advantageous for MPI+X. However, that is also the case for Chapel.

The multicore portion of the MPI+X implementation was programmed by hand, differently from its counterpart written in Chapel. Therefore, a first improvement in the Chapel implementation is programming the code for using all CPU cores a locale has. Additionally, It is worth to mention that Chapel also supports MPI and ZeroMQ libraries for programming inter-locale communication and the distributed load balancing. However, it does not seem necessary, as the use of high-productivity features resulted in performance competitive to MPI+OpenMP.

Concerning limitations, it took much more time to configure the GASNet library for running on a cluster than programming the multi-locale code itself. Moreover, Chapel's documentation is restricted to a few system configurations, e.g., Infiniband network with Slurm for job spawning. In our case, GASNet could not run on an MXM network with a non-default partition key, and a modification in the GASNet code was necessary. This problem would keep a not so enthusiastic user from Chapel. The bright side is that it was not a Chapel-only effort, as other PGAS libraries, such as UPC, Fortran, SHMEM use GASNet as communication layer.

Finally, graphics processing units are crucial for solving big and challenging combinatorial optimization problems [5,14]. The adoption of Chapel by the parallel optimization community, besides performance and productivity, also depends on the support of GPUs. According to Chapel's official documentation, the Xeon Phi accelerator is supported. However, there is no information concerning the support of GPUs.

5.1 Future Works

Permutation combinatorial optimization problems are commonly solved to optimality by using Branch-and-Bound search algorithms (B&B) [9]. Therefore, a first future research direction is to extend the proposed multi-locale backtracking into a distributed B&B. This way, it will be possible to solve challenging optimization problems, such as the Quadratic Assignment and the Flow-shop Scheduling Problem. Additionally, this future work will aim at larger scale clusters. Thus, it will be possible to investigate the limits of the productivity-aware features of Chapel concerning performance and scalability. A final future work is to compare Chapel to other PGAS-based libraries and high-productivity languages, such as SHMEM, UPC, and Julia.

6 Conclusion

This work investigated the use of Chapel high-productivity language for the design and implementation of distributed tree search algorithms. A distributed backtracking for enumerating all valid solutions of the N-Queens problem was conceived. The concepts herein presented can be adapted for solving other permutation combinatorial problems by performing straightforward modifications.

Programmers familiarized with OpenMP can easily conceive a distributed tree-based search in Chapel. Despite the high level of its features, the distributed search written in Chapel scales well and the distributed load balancing schemes are effective. Experimental results show that Chapel is competitive to C-MPI+OpenMP in terms of performance and scalability. The most significant drawbacks found do not concern performance nor scalability. Instead, they are related to the configuration of the communication layer, which can be more time consuming than programming the distributed application itself.

It is worth to point out that the parallel optimization community already possesses legacy code mainly written in C, C++. Therefore, programmers may be resistant to learn another language and translate programs to Chapel. The capacity of Chapel to include C code can be a partial solution for this situation. One can reuse C code for node bounding and search procedure, whereas Chapel distributed programming features are employed for load balancing and communication. Finally, the lack of support for accelerators may also limit the adoption of Chapel by the parallel optimization community.

Acknowledgments. The experiments presented in this paper were carried out on the Grid'5000 testbed [4], hosted by INRIA and including several other organizations (http://www.grid5000.fr). We thank Bradford Chamberlain, Elliot Ronaghan (from Cray inc.) and Paul Hargrove (Berkeley lab.) for helping us to run GASNet on GRID5000. Moreover, we also thank Paul Hargrove for the modifications in GASNet InfiniBand implementation necessary to run GASNet on GRID'5000 MXM InfiniBand networks.

References

1. Almasi, G.: PGAS (partitioned global address space) languages. In: Padua, D. (ed.) Encyclopedia of Parallel Computing, pp. 1539–1545. Springer, Boston (2011). https://doi.org/10.1007/978-0-387-09766-4_210

2. Asanovic, K., et al.: The landscape of parallel computing research: a view from Berkeley. Technical report, Technical Report UCB/EECS-2006-183, EECS Department, University of California (2006)
3. Bell, J., Stevens, B.: A survey of known results and research areas for n-queens. Discrete Math. **309**(1), 1–31 (2009)
4. Bolze, R., et al.: Grid'5000: a large scale and highly reconfigurable experimental grid testbed. Int. J. High Perform. Comput. Appl. **20**(4), 481–494 (2006)
5. Carneiro, T., de Carvalho Júnior, F.H., Arruda, N.G.P.B., Pinheiro, A.B.: Um levantamento na literatura sobre a resolução de problemas de otimização combinatória através do uso de aceleradores gráficos. In: Proceedings of the XXXV Ibero-Latin American Congress on Computational Methods in Engineering (CILAMCE), Fortaleza-CE, Brasil (2014)
6. Carneiro Pessoa, T., Gmys, J., de Carvalho Junior, F.H., Melab, N., Tuyttens, D.: GPU-accelerated backtracking using CUDA dynamic parallelism. Concurr. Comput. Pract. Exp. **30**, e4374-n/a (2017). https://doi.org/10.1002/cpe.4374
7. Chamberlain, B.L., Choi, S.E., Deitz, S.J., Navarro, A.: User-defined parallel zippered iterators in chapel. In: Proceedings of Fifth Conference on Partitioned Global Address Space Programming Models, pp. 1–11 (2011)
8. Chamberlain, B.L., et al.: Chapel comes of age: making scalable programming productive. Cray User Group (2018)
9. Crainic, T., Le Cun, B., Roucairol, C.: Parallel branch-and-bound algorithms. Parallel combinatorial optimization, pp. 1–28 (2006)
10. Cray Inc.: Chapel language specification, vol. 986. Cray Inc. (2018)
11. Da Costa, G., et al.: Exascale machines require new programming paradigms and runtimes. Supercomput. Front. Innov. **2**(2), 6–27 (2015)
12. Feinbube, F., Rabe, B., von Löwis, M., Polze, A.: NQueens on CUDA: optimization issues. In: 2010 Ninth International Symposium on Parallel and Distributed Computing (ISPDC), pp. 63–70. IEEE (2010)
13. Fiore, S., Bakhouya, M., Smari, W.W.: On the road to exascale: advances in high performance computing and simulations–an overview and editorial. Future Gener. Comput. Syst. **82**, 450–458 (2018)
14. Gmys, J., Mezmaz, M., Melab, N., Tuyttens, D.: IVM-based parallel branch-and-bound using hierarchical work stealing on multi-GPU systems. Concurr. Comput. Pract. Exp. **29**(9), e4019 (2017)
15. Grama, A.Y., Kumar, V.: A survey of parallel search algorithms for discrete optimization problems. ORSA J. Comput. **7** (1993). https://doi.org/10.1.1.45.9937
16. Mezmaz, M., Melab, N., Talbi, E.G.: A grid-enabled branch and bound algorithm for solving challenging combinatorial optimization problems. In: IEEE International Parallel and Distributed Processing Symposium, IPDPS 2007, pp. 1–9. IEEE (2007)
17. San Segundo, P., Rossi, C., Rodriguez-Losada, D.: Recent developments in bit-parallel algorithms. INTECH Open Access Publisher (2008)
18. Tschoke, S., Lubling, R., Monien, B.: Solving the traveling salesman problem with a distributed branch-and-bound algorithm on a 1024 processor network. In: 9th International Parallel Processing Symposium. Proceedings, pp. 182–189. IEEE (1995)
19. Zhang, W.: Branch-and-bound search algorithms and their computational complexity. Technical report, DTIC Document (1996)

Development of Element-by-Element Kernel Algorithms in Unstructured Implicit Low-Order Finite-Element Earthquake Simulation for Many-Core Wide-SIMD CPUs

Kohei Fujita[1,2](✉), Masashi Horikoshi[3], Tsuyoshi Ichimura[1,2,4], Larry Meadows[5], Kengo Nakajima[2,6], Muneo Hori[7], and Lalith Maddegedara[1,2]

[1] Earthquake Research Institute and Department of Civil Engineering, The University of Tokyo, Tokyo, Japan
{fujita,ichimura,lalith}@eri.u-tokyo.ac.jp
[2] Center for Computational Science, RIKEN, Kobe, Japan
[3] Core and Visual Computing Group, Intel K.K, Tokyo, Japan
masashi.horikoshi@intel.com
[4] Center for Advanced Intelligence Project, RIKEN, Tokyo, Japan
[5] Data Center Group, Intel Corporation, Hillsboro, USA
lawrence.f.meadows@intel.com
[6] Information Technology Center, The University of Tokyo, Tokyo, Japan
nakajima@cc.u-tokyo.ac.jp
[7] Japan Agency for Marine-Earth Science and Technology, Yokosuka, Japan
horimune@jamstec.go.jp

Abstract. Acceleration of the Element-by-Element (EBE) kernel in matrix-vector products is essential for high-performance in unstructured implicit finite-element applications. However, the EBE kernel is not straightforward to attain high performance due to random data access with data recurrence. In this paper, we develop methods to circumvent these data races for high performance on many-core CPU architectures with wide SIMD units. The developed EBE kernel attains 16.3% and 20.9% of FP32 peak on Intel Xeon Phi Knights Landing based Oakforest-PACS and Intel Skylake Xeon Gold processor based system, respectively. This leads to 2.88-fold speedup over the baseline kernel and 2.03-fold speedup of the whole finite-element application on Oakforest-PACS. An example of urban earthquake simulation using the developed finite-element application is shown.

Keywords: Finite-element method · Random data access · Many-core · SIMD

1 Introduction

Energy efficiency is one of the requirements for exascale computing, and many-core CPUs with wide SIMD units are considered to be one of the architectures

J. M. F. Rodrigues et al. (Eds.): ICCS 2019, LNCS 11536, pp. 267–280, 2019.
https://doi.org/10.1007/978-3-030-22734-0_20

towards this goal. For example, the Post-K supercomputer that is scheduled for operation around 2021 is announced to equip many-core CPUs with 512 bit wide SIMD cores [1]. However, some applications are difficult to benefit from wide SIMD and many-core architectures due to random data access and data recurrence, and thus algorithms enabling efficient use of these architectures are important towards exascale computing. The unstructured low-order implicit finite-element method is one of the applications that are not straightforward for efficient use of many-core wide SIMD CPUs. As the unstructured low-order implicit finite-element method is the de facto standard in structural simulations used in the manufacturing industry, the acceleration of this method is expected to lead to large ripple effects. In this paper, we focus on the unstructured low-order implicit finite-element method as a target for scalable algorithm development for many-core wide SIMD CPUs.

Many studies target implicit solvers on large scale systems. For example, the SC15 Gordon Bell Prize Paper [2] uses multi-grid solvers for simulation of extremely large mantle convection problems, and the SC16 Gordon Bell Prize Paper uses the Sunway Taihu Light Supercomputer to solve climate problems [3]. Also, an SC15 Gordon Bell Prize Finalist [4] solves an urban earthquake problem on the K computer. However, all of these simulations use structured or semi-structured mesh, which are both SIMD and multi-core friendly, for efficient computation of matrix-vector products for the implicit solvers. However, for solving general manufacturing or geoscience problems, use of pure unstructured mesh is required. The state-of-the-art of low-order unstructured finite-element method on CPU based systems is SC14 Gordon Bell Prize Finalist *GAMERA* [5]. Targeting architectures with high arithmetic capabilities and relatively low memory bandwidth, *GAMERA* uses a matrix-free kernel for computation of matrix vector products (i.e., the Element-by-Element kernel, EBE [6]). Here, instead of storing and reading the global matrix from memory, element matrices are generated and multiplied with the right hand side vector every time a matrix-vector product is called. This enables fast computation of matrix-vector products with good load balance on massively parallel systems with relatively low memory bandwidth, which lead to high application performance and high scalability. However, as *GAMERA* is targeted for the K computer which comprises 8 core CPUs with 2 wide SIMD cores [7], the computation algorithm used for the EBE kernel does not consider the wider SIMD and larger core counts of current and near future many-core CPUs. Thus, a new time-parallel computation algorithm *GHYDRA* was developed in a 2018 IPDPS paper [8], that enables use of uniformity of mesh in time domain to reduce random data access. This algorithm is suitable for recent wide SIMD CPUs; however, EBE kernel algorithms suitable for this type of architecture is required to fully exploit its performance. In this paper, we develop EBE kernel algorithms for efficient computation of the *GHYDRA* finite-element solver algorithm on many-core CPU systems, and compare with conventional EBE algorithms on the K computer, Intel Xeon Phi Knights Landing based Oakforest-PACS, and an Intel Skylake Xeon Gold CPU based system. We also show an urban earthquake simulation example using the unstructured finite-element solver accelerated by the developed EBE kernel algorithms.

2 Finite-Element Solver Algorithm

2.1 Target Problem

Earthquake simulations involve large domain nonlinear time-evolution problems with locally complex structures. For example, soft soil near surface have complex geometry and softens during strong ground motion under large earthquakes. Thus, we solve the target dynamic nonlinear continuum mechanics problem using a nonlinear dynamic 3D finite-element method with low-order solid elements because this method is suitable for modeling complex geometry and analytically satisfies the traction-free boundary condition at the surface. The target equation using Newmark-β method ($\beta = 1/4$, $\delta = 1/2$) for time integration is:

$$\mathbf{A}^n \, \delta\mathbf{u}^n = \mathbf{b}^n, \tag{1}$$

where

$$\begin{cases} \mathbf{A}^n = \frac{4}{dt^2}\mathbf{M} + \frac{2}{dt}\mathbf{C}^n + \mathbf{K}^n, \\ \mathbf{b}^n = \mathbf{f}^n - \mathbf{q}^{n-1} + \mathbf{C}^n\mathbf{v}^{n-1} + \mathbf{M}\left(\mathbf{a}^{n-1} + \frac{4}{dt}\mathbf{v}^{n-1}\right), \end{cases}$$

and

$$\begin{cases} \mathbf{q}^n = \mathbf{q}^{n-1} + \mathbf{K}^n\delta\mathbf{u}^n, \\ \mathbf{u}^n = \mathbf{u}^{n-1} + \delta\mathbf{u}^n, \\ \mathbf{v}^n = -\mathbf{v}^{n-1} + \frac{2}{dt}\delta\mathbf{u}^n, \\ \mathbf{a}^n = -\mathbf{a}^{n-1} - \frac{4}{dt}\mathbf{v}^{n-1} + \frac{4}{dt^2}\delta\mathbf{u}^n. \end{cases} \tag{2}$$

Here, $\delta\mathbf{u}$, $\mathbf{u}$, $\mathbf{v}$, $\mathbf{a}$, and $\mathbf{f}$ are incremental displacement, displacement, velocity, acceleration, and external force vectors, respectively. $\mathbf{M}$, $\mathbf{C}$, and $\mathbf{K}$ are the consistent mass, damping, and stiffness matrices, respectively. dt is the time increment and n is the time step. We use Rayleigh damping for $\mathbf{C}$, where the element damping matrix $\mathbf{C}_e^n$ is calculated using the element consistent mass matrix $\mathbf{M}_e$ and the element stiffness matrix $\mathbf{K}_e^n$ as $\mathbf{C}_e^n = \alpha\mathbf{M}_e + \beta\mathbf{K}_e^n$. Here, α and β are obtained by solving the following least squares equation:

$$\text{minimize}\left[\int_{f_{\min}}^{f_{\max}} \left(h - \frac{1}{2}\left(\frac{\alpha}{2\pi f} + 2\pi f\beta\right)\right)^2 \mathrm{d}f\right],$$

where, $f_{\max}$, $f_{\min}$, and h are the maximum and minimum target frequency and the damping ratio, respectively. Although arbitrary constitutive models can be used, we use the Ramberg Osgood model [9] and Masing rule [10] for the nonlinear constitutive modeling in urban earthquake simulations. In summary, time history response $\mathbf{u}^n$ is computed by repeating the following steps.

1. Read boundary conditions.
2. Evaluate $\mathbf{C}^n$ and $\mathbf{K}^n$ based on the constitutive relations and strain at time-step $n-1$.
3. Obtain $\delta\mathbf{u}^n$ by solving Eq. 1.
4. Update Eq. 2 using $\delta\mathbf{u}^n$.

Since most of the computation cost is incurred in solving Eq. 1, we explain the details of the solver in the next subsection.

Standard solver algorithm

1: set $x_{-1} \leftarrow 0$
2: for($i = 0; i < n; i = i + 1$){
3: guess $\bar{x}_i$ using standard predictor
4: set b_i using x_{i-1}
5: solve $x_i \leftarrow A^{-1}b_i$ with error tolerance $\frac{|Ax_i - b_i|}{|b_i|} < \epsilon$ using initial solution $\bar{x}_i$ **(Computed using iterative solver with EBE kernel (1 vector))**
6: }

Time-parallel solver algorithm

1: set $x_{-1} \leftarrow 0$ and $\bar{x}_i \leftarrow 0$ ($i = 0, \ldots, m-2$)
2: for($i = 0; i < n; i = i + 1$){
3: set b_i using x_{i-1}
4: guess $\bar{x}_{i+m-1}$ using standard predictor
5: $b_j = \bar{b}_j$
6: while ($\frac{|A\bar{x}_i - b_i|}{|b_i|} > \epsilon$) do {
7: guess $\bar{b}_j$ using $\bar{x}_{j-1}$ ($j = i+1, \ldots, i+m-1$)
8: refine solution $\{\bar{x}_j \leftarrow A^{-1}\bar{b}_j\}$ with initial solution $\bar{x}_j$ ($j = i, \ldots, i+m-1$) **(Computed using iterative solver with EBE kernel (m vectors))**
10: }
11: $x_i = \bar{x}_i$
11: }

Fig. 1. Standard and time-parallel solver algorithm. The same solution is obtained within error tolerance ϵ using both of the solver algorithms.

2.2 Overview of Time-Parallel Finite-Element Solver Algorithm

Solving Eq. 1 is a common target problem arising from low-ordered implicit finite-element methods used in solid continuum mechanics problems in engineering and science fields. In these low-ordered (e.g., second-ordered) implicit simulations, iterative solvers such as Conjugate Gradient (CG) methods are generally used. Since these solvers involve much random access and intensive memory access, it is difficult to attain performance on current systems.

The time-parallel algorithm in *GHYDRA* aims to accelerate this solver by reducing random access by using the uniformity of mesh in the time domain [8]. As shown in Fig. 1, instead of solving one time step at a time, several time steps are solved together in parallel. When solving m time steps together, the arithmetic count per iteration of the iterative solver becomes m times larger. However, as the solutions for future time steps can be used as accurate initial solutions, the total number of iterations becomes approximately $1/m$ (see [8] for detailed comparison of convergence of this method). Thus, the total arithmetic count does not change by using this algorithm. However, since the mesh connectivity does not change in time, the random access involved in the kernel used to compute matrix-vector products can be reduced, which leads to shorter time-to-solution of the application with the same solution within the solver error threshold ϵ. This time-parallel algorithm is different from parallel-in-time integration methods intended to improve the time-to-solution at the strong scaling limits, as the *GHYDRA* algorithm uses the uniformity of mesh in the time domain to reduce random data access and thus enables speedup of the application even within the strong scaling limits.

In *GHYDRA*, the time-parallel algorithm is combined with inexact preconditioned conjugate gradient method, multi-grid method, and mixed precision computation. Below we briefly summarize these components:

- Inexact preconditioned conjugate gradient method [11]: An inexact preconditioned conjugate gradient method is a preconditioned conjugate gradient method which uses another solver for solving the preconditioning matrix equation instead of multiplying a fixed preconditioning matrix. As it is common to use another iterative solver for solving the preconditioning matrix equation, we call the original CG loop as the "outer loop" and the preconditioning solver as the "inner loop". As the inner loop does not need to be solved exactly, this makes room for making improvements such as multi-grid and mixed precision arithmetic.
- Multi-grid method: We reduce the computation cost and communication size of the inexact preconditioned conjugate gradient method by using a multi-grid solver for the inner loop. Here we use a two step geometric multi-grid; we use the targeted second-ordered tetrahedral mesh FEmodel for the inner fine loop, and use the same mesh without intermediate nodes of each element for the inner coarse loop (i.e., a first-ordered tetrahedral mesh FEmodel_c). We use a 3×3 block Jacobi preconditioned conjugate gradient solver for solving the inner fine and inner coarse loops.
- Mixed precision arithmetic: Even for a target problem in FP64, we need only to solve the preconditioning matrix equation roughly. Thus, we compute the whole multi-grid preconditioner in FP32. This leads to halving memory access and communication size, and enable usage of high-performance FP32 arithmetic units equipped in recent CPUs.

When using the algorithm above, we can move most of the compute cost of the outer loop to the inner loops by using appropriate threshold values in the inner CG solvers. Thus, even though it may seem a complicated algorithm, the performance of the whole solver is dependent on kernels such as matrix-vector product kernels, inner product kernels, or saxpy kernels which are in common with standard CG solvers. In the following, we explain the most time consuming matrix-vector product kernel used in the inner fine loop, which is not straightforward to attain performance on many-core wide SIMD CPUs.

2.3 Baseline Element-by-Element Kernel Algorithm

As the relative memory transfer capability to floating point computation capability is becoming lower, using an algorithm that can reduce memory access is becoming important for fast time-to-solution. Thus, the Element-by-Element (EBE) method, which is a matrix-free matrix-vector multiplication method, is used for computation of $\mathbf{f} = \mathbf{A}\mathbf{u}$. Here, $\mathbf{u}, \mathbf{f}$ are displacement and nodal force vectors, and $\mathbf{A}$ is the global matrix that is generated by superimposing element matrices $\mathbf{A}_i^e$. $\mathbf{A}_i^e$ can be computed by using coordinates of nodes ($\mathbf{x}$) and element material properties. In standard matrix-vector product methods, $\mathbf{f} = \mathbf{A}\mathbf{u}$ is computed by reading the global matrix $\mathbf{A}$ from memory and multiplying it with $\mathbf{u}$. In the contrary, in the EBE method, matrix-vector products are computed by computing local matrix-vector products

$$\mathbf{f}_i^e = \mathbf{A}_i^e \mathbf{u}_i^e = \mathbf{A}_i^e \mathbf{Q}_i^{e\mathrm{T}} \mathbf{u}, \tag{3}$$

and adding them up as

$$\mathbf{f} = \sum_i \mathbf{Q}_i^e \mathbf{f}_i^e. \tag{4}$$

Here, $\mathbf{Q}_i^e$ are matrices for mapping element-wise nodal values to global nodal values. Since the coordinates, displacement and force vectors $\mathbf{x}, \mathbf{u}, \mathbf{f}$ can be kept on cache by reordering of nodes and elements, we can drastically reduce memory access when compared with methods reading the global matrix from memory. On the other hand, the EBE method involves more computation and consequently transfers memory access cost to computation cost.

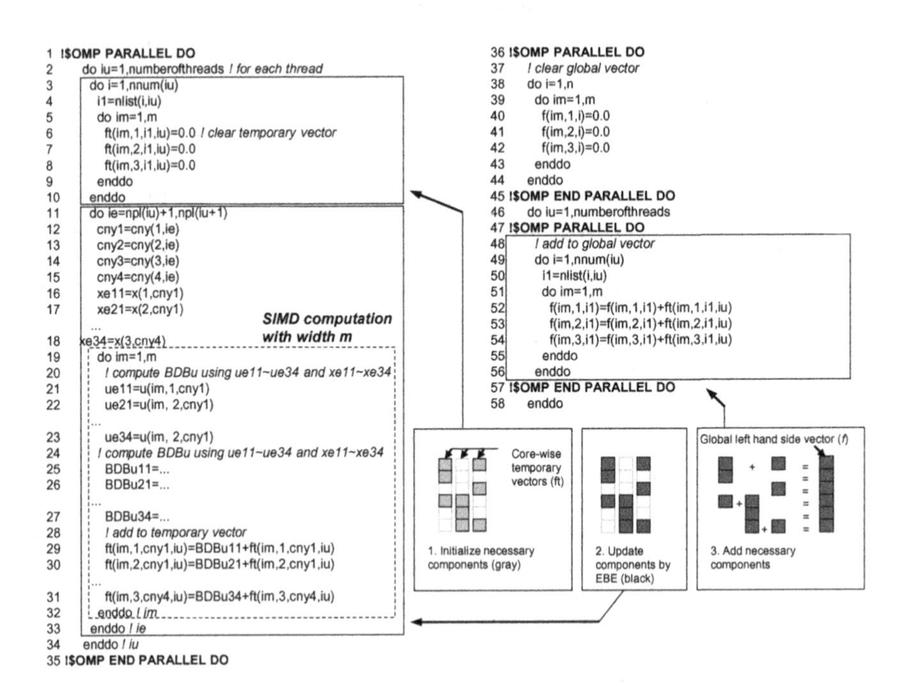

Fig. 2. Baseline EBE kernel with m vectors

Although the EBE method has low algorithmic Byte/FLOP and is potentially suitable for current computers, it is not straightforward for parallel computation due to the data recurrence for adding local force vectors of elements with shared nodes in Eq. 4. Figure 2 shows the multi-core EBE computation algorithm for *GHYDRA*. Here, temporary vectors `ft` are allocated for each core, initialized on lines 3–10, and the core wise results are added to `ft` in lines 11–33. Finally, the core-wise results are added to the global vector `f` in lines 46–58. The innermost loop for time-parallelism (m) can be computed using SIMD units. However, in practice, we use $m = 4$ in earthquake problems as increasing m leads to deterioration of the accuracy of the predicted solutions and thus increase in the total number of arithmetic counts. Thus, only 1/4 of the SIMD lanes of 512 bit SIMD registers (i.e., 4 out of the 16 FP32 lanes) can be used. Although this is better than the non-time-parallel algorithm in which SIMD cannot be used completely due to the data recurrence involved in lines 29–31, this EBE implementation is expected to lead to poor performance on recent many-core CPUs with wider SIMD units. Development of EBE kernel algorithms that uses SIMD in full width is required for improving performance.

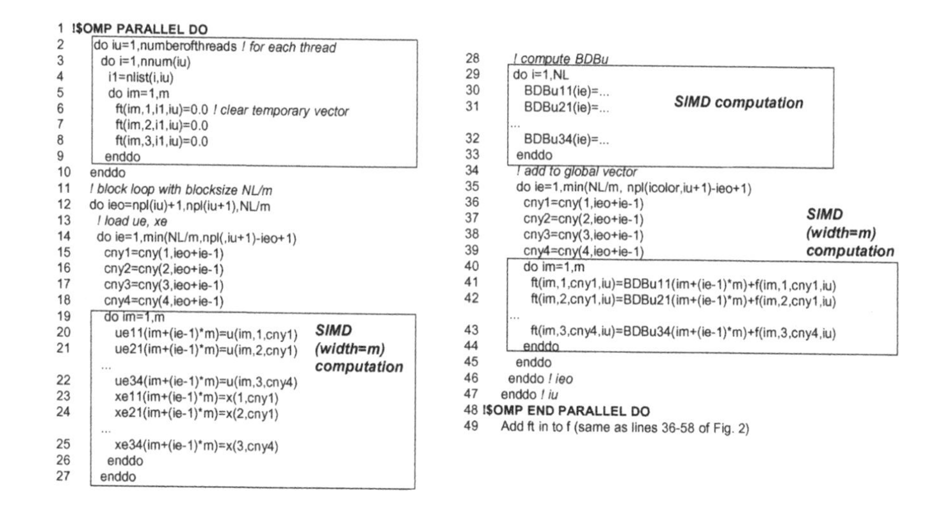
```
1  !$OMP PARALLEL DO
2    do iu=1,numberofthreads ! for each thread
3     do i=1,nnum(iu)
4      i1=nlist(i,iu)
5      do im=1,m
6       ft(im,1,i1,iu)=0.0 ! clear temporary vector
7       ft(im,2,i1,iu)=0.0
8       ft(im,3,i1,iu)=0.0
9      enddo
10   enddo
11   ! block loop with blocksize NL/m
12   do ieo=npl(iu)+1,npl(iu+1),NL/m
13    ! load ue, xe
14    do ie=1,min(NL/m,npl(,iu+1)-ieo+1)
15     cny1=cny(1,ieo+ie-1)
16     cny2=cny(2,ieo+ie-1)
17     cny3=cny(3,ieo+ie-1)
18     cny4=cny(4,ieo+ie-1)
19     do im=1,m
20      ue11(im+(ie-1)*m)=u(im,1,cny1)      SIMD
21      ue21(im+(ie-1)*m)=u(im,2,cny1)      (width=m)
    ...                                     computation
22      ue34(im+(ie-1)*m)=u(im,3,cny4)
23      xe11(im+(ie-1)*m)=x(1,cny1)
24      xe21(im+(ie-1)*m)=x(2,cny1)
    ...
25      xe34(im+(ie-1)*m)=x(3,cny4)
26     enddo
27    enddo
28    ! compute BDBu
29    do i=1,NL
30      BDBu11(ie)=...
31      BDBu21(ie)=...                      SIMD computation
    ...
32      BDBu34(ie)=...
33    enddo
34    ! add to global vector
35    do ie=1,min(NL/m, npl(icolor,iu+1)-ieo+1)
36     cny1=cny(1,ieo+ie-1)
37     cny2=cny(2,ieo+ie-1)                 SIMD
38     cny3=cny(3,ieo+ie-1)                 (width=m)
39     cny4=cny(4,ieo+ie-1)                 computation
40     do im=1,m
41      ft(im,1,cny1,iu)=BDBu11(im+(ie-1)*m)+f(im,1,cny1,iu)
42      ft(im,2,cny1,iu)=BDBu21(im+(ie-1)*m)+f(im,2,cny1,iu)
    ...
43      ft(im,3,cny4,iu)=BDBu34(im+(ie-1)*m)+f(im,3,cny4,iu)
44     enddo
45    enddo
46   enddo ! ieo
47  enddo ! iu
48 !$OMP END PARALLEL DO
49  Add ft in to f (same as lines 36-58 of Fig. 2)
```

Fig. 3. EBE kernel with m vectors for wide-SIMD CPUs

3 Developed EBE Kernel Algorithms for Many-Core Wide SIMD CPUs

In this section, we show the developed algorithms for faster EBE computation on many-core wide SIMD CPUs.

We first develop an algorithm to utilize SIMD units in the main computation loop (Fig. 3). As the data recurrence in addition of results into the `ft` vector was blocking the use of SIMD units, we split the innermost loop into two loops; i.e., the computation part of $\mathbf{f}_i^e = \mathbf{A}_i^e \mathbf{u}_i^e$ (Fig. 3 lines 14–33) and the summation part of $\mathbf{f}_i^e$ to $\mathbf{f}$ (lines 35–45). This leads to use of SIMD operations to the computationally rich first loop. Here, we reduce the overhead of loop splitting by

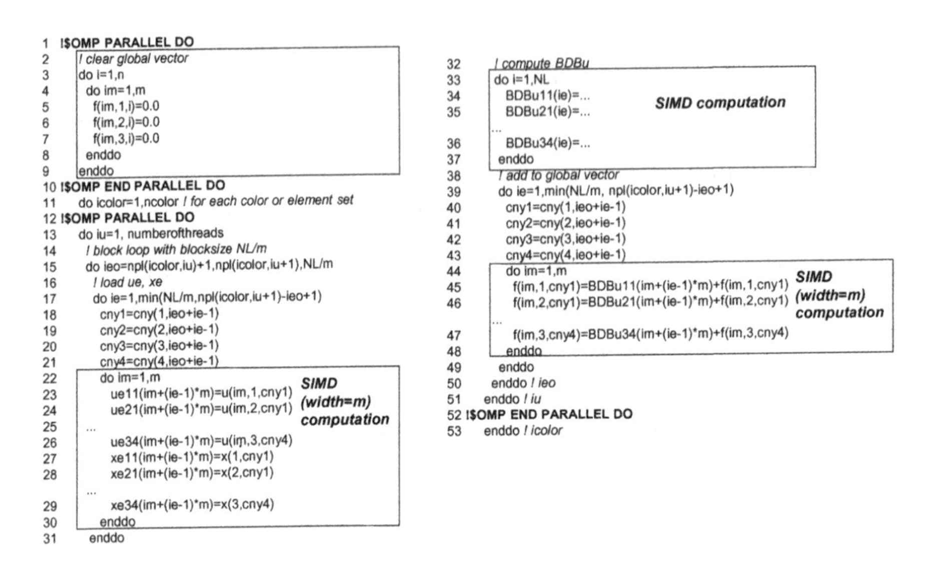

Fig. 4. Coloring/thread partitioning of EBE kernel with m vectors for wide-SIMD CPUs. The same code can be used for both coloring/thread partitioning methods shown in Fig. 5.

blocking the loop with small block size (`NL`, which is typically set to the SIMD vector length). This keeps the temporary buffers `BDBu11-34` on cache.

When using many cores, the size of the thread-wise temporary buffer `ft` becomes large. In addition, the overhead of initializing `ft` and adding the thread-wise results to the global vector `f` is not negligible in the total kernel cost. Thus, we developed a thread partitioning method to eliminate the use of thread-wise temporary buffers (Fig. 4). The procedure in standard coloring methods are to color the whole mesh such that elements in each color does not have shared nodes (Fig. 5a). Instead, we partition the domain with prescribed thread numbers (Fig. 5b). Here, we partition the overall mesh into the number of threads using a graph partitioning method (METIS [12]). Then, we remove elements that share nodes between the thread partitions. The remaining elements becomes the first set of elements to be computed (Set #1 of Fig. 5b). The removed elements are partitioned again to recursively decompose the mesh into sets (Set #2 and #3 of Fig. 5b). As will be shown in Sect. 4, the proposed thread partitioning method enables better cache reuse of nodal values `u`, `x` and `f` when compared with the case of standard coloring methods.

In practice, we use $m = 4$ in earthquake problems as increasing m leads to deterioration of the accuracy of the predicted solutions and thus increase in the total number of arithmetic counts. On the other hand, 16 FP32 floats can be packed in a single 512 bit SIMD register; thus, we can only use the first 1/4 of the SIMD capability in the vector load/store computation. In order to accelerate this part, we used 16 wide SIMD for loading from `u` while using 4 wide SIMD for storing to local vectors `ueXX` (Fig. 6a). Similarly, we used 4 wide SIMD for loading local vectors `BDBuXX` while using 16 wide SIMD for loading and storing `f` (Fig. 6b). This enables reduction of inefficient 4 wide SIMD accesses and thus improvement in overall kernel performance.

4 Performance Measurements

We measure performance of the developed kernels and the accelerated finite-element application on the K computer, Intel Xeon Phi Knights Landing based Oakforest-PACS, and an Intel Skylake Xeon Gold CPU based system. The latter two systems are examples of many-core systems with 512-bit SIMD. Table 1 summarizes the configuration of the systems. K computer (K) [7] consists of 82,944 compute nodes, each with a single eight-core SPARC64 VIIIfx CPU. Each core has two sets of SIMD FMA arithmetic units of width 2. Oakforest-PACS (OFP) [13] is a supercomputer system introduced by the Joint Center for Advanced HPC, which was established by the University of Tokyo and the University of Tsukuba. The system comprises 8,208 nodes with a 68-core Intel Xeon Phi 7250 (Knights Landing) CPU [14], 96 GB of DDR4 RAM, and 16 GB of stacked 3D MCDRAM. Intel Skylake Xeon Gold CPU based system comprises two 20-core Intel Skylake Xeon Gold 6148 CPU [15] and 192 GB of DDR4 RAM.

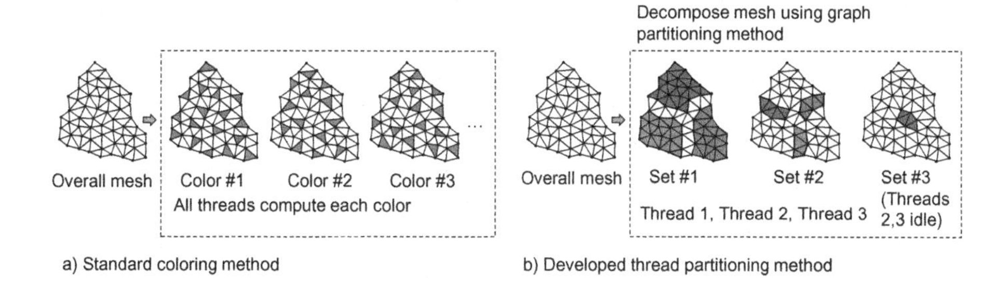

Fig. 5. Coloring/thread partitioning methods for the EBE kernel

Table 1. Performance measurement environment. Memory bandwidth on OFP is measured values of STREAM benchmark, while others are hardware peak values.

	K computer	Oakforest-PACS	Intel Skylake Xeon Gold based server
Nodes	8	1	1
Sockets/node	1	1	2
Cores/socket	8	68	20
FP32 SIMD width	2	16	16
Clock frequency	2.0 GHz	1.4 GHz	2.4 GHz
Total peak FP32 FLOPS	1024 GFLOPS	6092 GFLOPS	6144 GFLOPS
Total DDR bandwidth	512 GB/s	80.1 GB/s	255.9 GB/s
Total MCDRAM bandwidth	-	490 GB/s	-

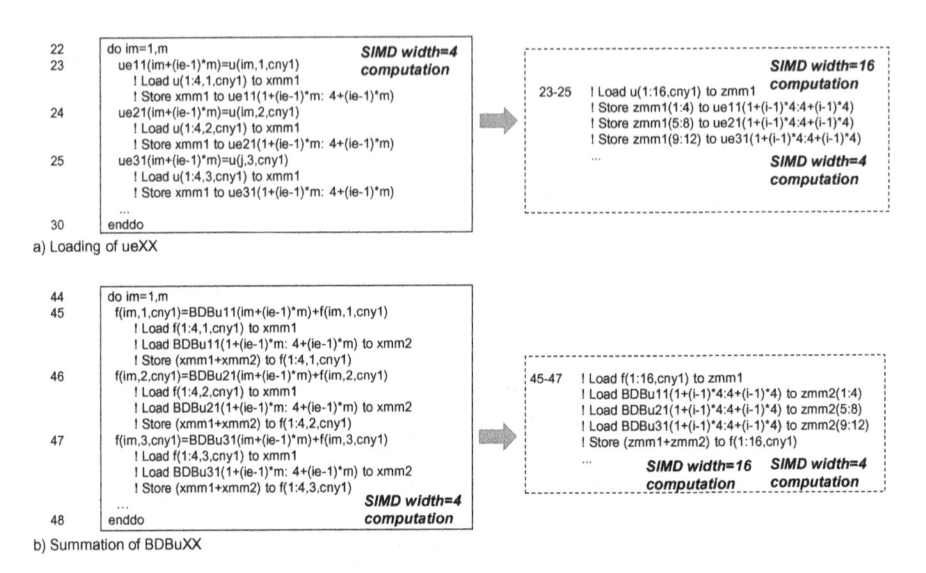

Fig. 6. Tuning of m vector EBE kernel for 512 bit SIMD architecture. Here, `xmm` indicate 128 bit FP32 registers and `zmm` indicate 512 bit FP32 registers.

Both the Xeon Phi 7250 CPU and Skylake Xeon Gold 6148 CPU support AVX-512 SIMD instructions.

We first measure performance of the EBE kernel. For the K computer, we use 8 nodes each with 1 MPI process with 8 OpenMP threads per process. For OFP, we use quadrant/flat mode, and run 8 MPI processes with 8 OpenMP threads per process on a single node. Each thread is bound to one CPU core (hyperthreading is not used), and we use `numactl --preferred=1` option such that memory is preferentially allocated to MCDRAM. For the Skylake system, we use 8 MPI processes with 5 OpenMP threads per process on a single node. Each MPI process on K/OFP/Skylake runs on a mesh with 6,534,144 second-order tetrahedral elements and 9,044,560 nodes and elapsed time for computing 529 times of matrix-vector products is measured. Figure 7 shows the EBE kernel performance. Comparing the K computer and OFP for the baseline kernel with one vector, we can see that high performance of 26.1% FP32 peak efficiency is attained for the K computer; however, only 1.98% is attained on OFP. The kernel performance is improved by using the time-parallel algorithm with $m = 4$; a $9.45\,\text{s}/7.52\,\text{s} = 1.25$-fold speedup was obtained on K computer and $20.1\,\text{s}/7.50\,\text{s} = 2.68$-fold speedup was obtained on OFP, respectively, for elapsed time per vector. This performance is expected to be further improved by using the developed EBE algorithms. First we see the effectiveness of the kernel algorithm enabling use of SIMD for the main computation part. By using SIMD by loop splitting and blocking for the main computation part of the EBE kernel, we can shorten elapsed time from 7.50 s to 3.75 s on OFP. Note that loop splitting and blocking is disabled for the K computer as the 2 wide FP32 SIMD of

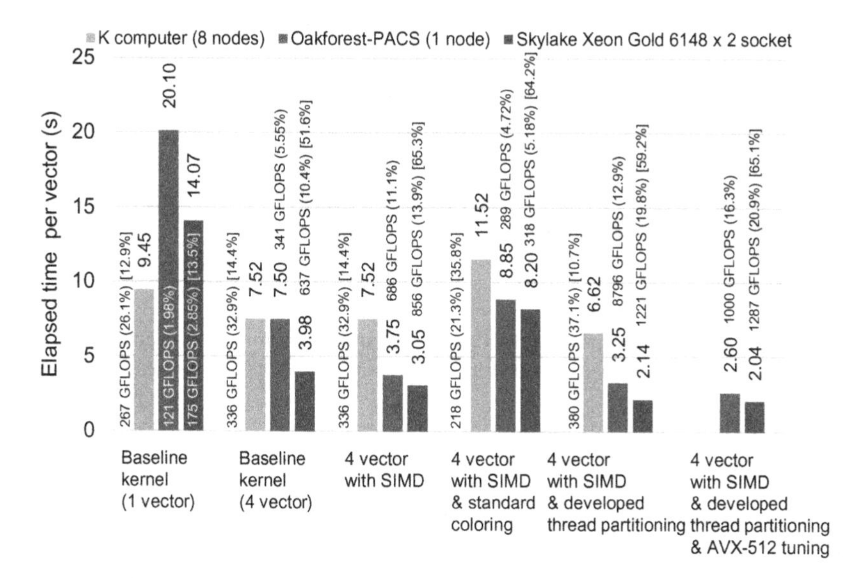

Fig. 7. EBE kernel performance. Elapsed time per vector is shown. Numbers in brackets indicate efficiency to FP32 peak, and numbers in square brackets indicate memory bandwidth efficiency to hardware peak of K computer and Skylake system. 8 nodes of the K computer (1 MPI process × 8 OpenMP threads per node), 1 node of OFP (8 MPI processes × 8 OpenMP threads), and a Skylake system (8 MPI processes × 5 OpenMP threads) is used for computation.

the K computer can be directly applied to the innermost time-parallel m loop. Next we see the effectiveness of using coloring and thread aware partitioning. When using standard coloring generated by serial greedy algorithm, the mesh was decomposed into 41 colors, resulting in total of 263.9 GB memory transfer per node incurred for computation of the EBE kernel on the K computer. This is more memory transfer than the baseline algorithm with 69.3 GB memory transfer, as the cache reuse for `u`, `x` and `f/ft` is disabled; which lead to longer elapsed time. When using the developed thread partitioning method, the mesh was decomposed into 5 colors. With more cache reuse, the total memory transfer was reduced to 45.3 GB, which resulted to 7.52 s/6.62 s = 1.13-fold speedup on the K computer and 3.75 s/3.25 s = 1.15-fold speedup on OFP. Furthermore, by using the AVX-512 tuning for reducing random data access, the elapsed time was decreased to 2.60 s on OFP. This lead to high performance of 16.3% of peak FP32 performance on 64 cores of OFP's Intel Xeon Phi Knights Landing CPU. We can see that the effective use of SIMD and circumventing random access to temporary vectors lead to significant speedup of 7.50 s/2.60 s = 2.88-fold from the baseline algorithm with $m = 4$ vectors on OFP. Compared with the baseline algorithm without time-parallelism ($m = 1$), this is 20.1 s/2.60 s = 7.73-fold speedup on OFP. The developed EBE algorithm is also effective for the Skylake system, and lead to 3.98 s/2.04 s = 1.95-fold from the baseline algorithm with

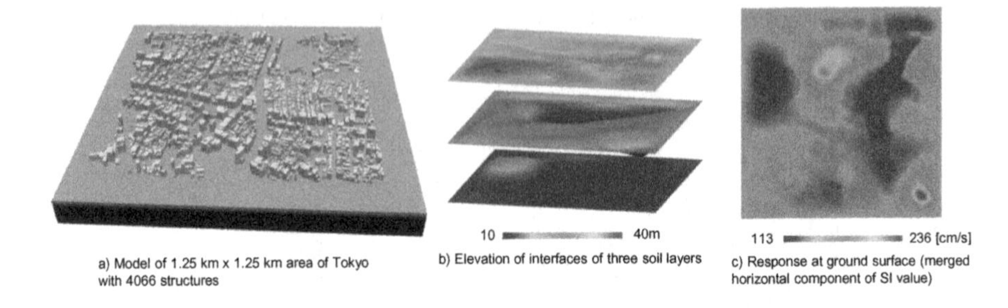

Fig. 8. Urban earthquake problem settings and computation results. Earthquake wave propagation in the three-layered soil structure is computed in this application example.

$m = 4$ vectors. Compared with the baseline algorithm without time-parallelism ($m = 1$), this is 14.1 s/2.04 s = 6.89-fold speedup on the Skylake system.

Next we see the performance of developed application on an urban earthquake problem targeting a 1.25 km × 1.25 km area of Tokyo. Using digital elevation map of [16] and soil information of [17], we constructed a soil model consisting of three soil layers. We discretized this problem with minimum element size of 1 m, which lead to a problem size of 1,022,620,536 degrees-of-freedom, 252,738,195 second-order tetrahedral elements and 340,873,512 nodes (Fig. 8a, b). We input the wave observed during 1995 Kobe Earthquake [18] with time stepping $dt =$ 0.01 s. The mesh is partitioned with METIS into 1,152 partitions and computed using 144 nodes of OFP (8 MPI processes per node). Figure 9 shows the elapsed time for solving the first 25 time steps using *GHYDRA* with the baseline EBE kernel and *GHYDRA* with the developed EBE kernel (both with $m = 4$). We also show performance of *GAMERA*, which can be considered as the non-time parallel version of *GHYDRA* with the baseline EBE kernel ($m = 1$). We can see that the application was accelerated by 125.6 s/61.9 s = 2.03-fold by using the developed EBE kernel algorithms leading to high peak performance of 11.6% of FP64 peak. Together with the 247.2 s/125.6 s = 1.97-fold speedup using time-parallelism, the application was accelerated by total of 3.99-fold when compared

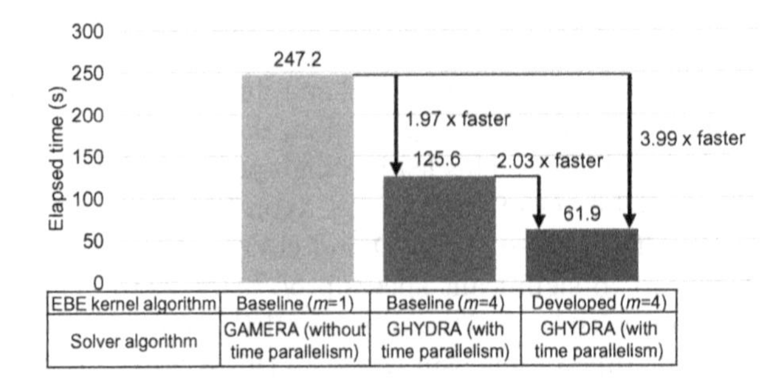

Fig. 9. Elapsed time for solving the first 25 time steps of application problem. Measured on 144 nodes of OFP.

to the SC14 Gordon Bell Prize Finalist solver *GAMERA*. We can see that the acceleration of the EBE kernel with algorithms suitable for wide SIMD and multi-cores are effective for attaining high performance and shorter time-to-solution of the overall application. With less energy-to-solution, we can afford to conduct more detailed simulations of larger areas of interest, which is expected to contribute to earthquake disaster mitigation.

5 Closing Remarks

In this paper we developed algorithms to accelerate the Element-by-Element kernel in unstructured low-order implicit finite-element methods on systems with many-core wide SIMD CPUs. By using the developed algorithm on the Intel Xeon Phi Knights Landing based Oakforest-PACS system, the elapsed time of the EBE kernel and total unstructured finite-element application was accelerated by 2.88-fold and 2.03-fold, respectively. The developed EBE kernel algorithms were also effective for the K computer and an Intel Skylake Xeon Gold CPU based system. These insights are expected to enable high performance on other large-scale many-core wide CPU based supercomputer systems, enabling energy efficient finite-element computation towards exascale computing.

Acknowledgments. Our results were obtained using the Oakforest-PACS at the Joint Center for Advanced HPC and the K computer at the Center for Computational Science, RIKEN (hp170249, hp180217). We acknowledge support from the Japan Society for the Promotion of Science (17K14719, 26249066, 25220908, 18H05239).

References

1. Outline of the Development of the Post-K computer. https://www.r-ccs.riken.jp/en/postk/project/outline
2. Rudi, J., et al.: An extreme-scale implicit solver for complex PDEs: highly heterogeneous flow in earth's mantle. In: Proceedings of the International Conference on High Performance Computing, Networking, Storage and Analysis (SC 2015), p. 5. ACM (2015)
3. Yang, C., et al.: 10M-core scalable fully-implicit solver for nonhydrostatic atmospheric dynamics. In: Proceedings of the International Conference on High Performance Computing, Networking, Storage and Analysis (SC 2016), p. 6. IEEE Press (2016)
4. Ichimura, T., et al.: Implicit nonlinear wave simulation with 1.08T DOF and 0.270T unstructured finite elements to enhance comprehensive earthquake simulation. In: Proceedings of the International Conference on High Performance Computing, Networking, Storage and Analysis (SC 2015), p. 4. ACM (2015)
5. Ichimura, T., et al.: Physics-based urban earthquake simulation enhanced by 10.7 BlnDOF x 30 K time-step unstructured FE non-linear seismic wave simulation. In: Proceedings of the International Conference on High Performance Computing, Networking, Storage and Analysis (SC 2014), pp. 15–26. IEEE Press (2014)

6. Winget, J.M., Hughes, T.J.: Solution algorithms for nonlinear transient heat conduction analysis employing element-by-element iterative strategies. Comput. Methods Appl. Mech. Eng. **52**(1–3), 711–815 (1985)
7. Miyazaki, H., Kusano, Y., Shinjou, N., Shoji, F., Yokokawa, M., Watanabe, T.: Overview of the K computer system. FUJITSU Sci. Tech. J. **48**(3), 302–309 (2012)
8. Ichimura, T., et al.: A fast scalable implicit solver with concentrated computation for nonlinear time-evolution problems on low-order unstructured finite elements. In: IEEE International Parallel and Distributed Processing Symposium (IPDPS), Vancouver, BC 2018, pp. 620–629 (2018). https://doi.org/10.1109/IPDPS.2018.00071
9. Idriss, I.M., Dobry, R., Sing, R.D.: Nonlinear behavior of soft clays during cyclic loading. J. Geotech. Eng. Div. **104**(ASCE 14265) (1978)
10. Masing, G.: Eigenspannungen und Verfestigung beim Messing. In: Proceedings of the 2nd International Congress of Applied Mechanics, pp. 332–335 (1926)
11. Golub, G.H., Ye, Q.: Inexact conjugate gradient method with inner-outer iteration. SIAM J. Sci. Comput. **21**(4), 1305–1320 (1997)
12. Karypis, G., Kumar, V.: A fast and high quality multilevel scheme for partitioning irregular graphs. SIAM J. Sci. Comput. **20**(1), 359–392 (1998)
13. Oakforest-PACS. http://www.cc.u-tokyo.ac.jp/system/ofp/index-e.html
14. Sodani, A.: Knights landing (KNL): 2nd Generation IntelR Xeon Phi processor. In: IEEE Hot Chips 27 Symposium (HCS), Cupertino, CA, pp. 1–24 (2015). https://doi.org/10.1109/HOTCHIPS.2015.7477467
15. Kumar, A.: The New Intel Xeon Processor Scalable Family (Formerly Skylake-SP). In: 2017 IEEE Hot Chips 29 Symposium (HCS), Cupertino, CA (2017)
16. 5m mesh digital elevation map, Tokyo ward area, Geospatial Information Authority of Japan. https://www.gsi.go.jp/MAP/CD-ROM/dem5m/index.htm
17. National Digital Soil Map, The Japanese Geotechnical Society. https://www.jiban.or.jp/?page_id=432
18. Strong ground motion of The Southern Hyogo prefecture earthquake in 1995 observed at Kobe JMA observatory, Japan Meteorological Agency. https://www.data.jma.go.jp/svd/eqev/data/kyoshin/jishin/hyogo_nanbu/dat/H1171931.csv

A High-Productivity Framework for Adaptive Mesh Refinement on Multiple GPUs

Takashi Shimokawabe[1(✉)] and Naoyuki Onodera[2]

[1] Information Technology Center, The University of Tokyo, Tokyo, Japan
shimokawabe@cc.u-tokyo.ac.jp

[2] Center for Computational Science and e-Systems, Japan Atomic Energy Agency, Chiba, Japan

Abstract. Recently grid-based physical simulations with multiple GPUs require effective methods to adapt grid resolution to certain sensitive regions of simulations. In the GPU computation, an adaptive mesh refinement (AMR) method is one of the effective methods to compute certain local regions that demand higher accuracy with higher resolution. However, the AMR methods using multiple GPUs demand complicated implementation and require various optimizations suitable for GPU computation in order to obtain high performance. Our AMR framework provides a high-productive programming environment of a block-based AMR for grid-based applications. Programmers just write the stencil functions that update a grid point on Cartesian grid, which are executed over a tree-based AMR data structure effectively by the framework. It also provides the efficient GPU-suitable methods for halo exchange and mesh refinement with a dynamic load balance technique. The framework-based application for compressible flow has achieved to reduce the computational time to less than 15% with 10% of memory footprint in the best case compared to the equivalent computation running on the fine uniform grid. It also has demonstrated good weak scalability with 84% of the parallel efficiency on the TSUBAME3.0 supercomputer.

Keywords: Adaptive mesh refinement · GPU · Stencil computation

1 Introduction

The stencil-based applications are important applications running on the GPU supercomputers. Thanks to the wide bandwidth and high computational power of GPU, various stencil applications have successfully achieved high performance [7,8,11]. Recently grid-based physical simulations with multiple GPUs require effective methods to adapt grid resolution to certain sensitive regions of simulations. An adaptive mesh refinement (AMR) method is one of the key technique to compute certain local regions that demand higher accuracy with higher resolution [1,3,6]. While GPU computation has the potential to achieve

J. M. F. Rodrigues et al. (Eds.): ICCS 2019, LNCS 11536, pp. 281–294, 2019.
https://doi.org/10.1007/978-3-030-22734-0_21

high performance, it forces the programmer to learn multiple distinctive programming models such as CUDA or OpenACC and introduce various complicated optimizations. For this reason, most of existing AMR libraries supporting GPU provide only several numerical schemes that optimized for GPU, or the programmer has to provide GPU optimized kernels written in CUDA [6].

In order to improve productivity and achieve high performance, various types of high-level programming models for GPU were proposed [2,4,5,9,10,13]. However, since these programming models focus on stencil computations on uniform grids, it is difficult to apply them to the AMR applications where additional data structures such as tree structures are essential. Although Daino was proposed as a directives-based programming framework for AMR on GPUs, it needed to use its own directives [14]. To enhance the portability and transparency of frameworks themselves and the user codes using them, the framework should be written in standard languages without language extension.

In this paper, we propose a high-productivity framework for a block-based AMR for grid-based applications running on multiple GPUs. In previous research, we proposed a high-productivity GPU programming environment for stencil computations on uniform grids [9,10]. By extending this framework and adding the AMR data structure with halo exchange functions and mesh refinement mechanisms, we construct this AMR framework. The framework is implemented in the C++ language with CUDA and can be used in the user code is written just in C++, which improves portability of both framework and user code and facilitates cooperation with the existing codes. The framework provides data structure suitable for AMR method and class which can easily express stencil calculation on grid with various resolutions.

2 Overview of AMR Framework

The proposed block-based AMR framework is designed to provide highly-productive programming environment for stencil applications with explicit time integration and adapting grid resolution to certain sensitive regions of simulations. The framework is intended to execute the user program on NVIDIA's GPU. The programmer can develop user programs just in the C++ language. The programmer simply describes a C++11 lambda expression that updates a grid point, which is applied to the entire grids with various resolution over a tree-based AMR data structure effectively.

The framework can locally change the resolution of the grids for arbitrary regions in the time integration loop of applications. An entire computational domain is divided into a large number of the small uniform grid blocks with the same size recursively. The computation for all grid blocks can be solved with a single execution of a conventional stencil calculation for Cartesian grid regardless of their resolutions. This strategy may be effective for performance improvement because GPU can often derive high performance when accessing contiguous memory. The framework also provides some functions and C++ classes to realize other processes required for the AMR computations, such as mesh refinement, exchanging data in halo regions between grid blocks with different resolutions, and data migration to maintain load balancing.

3 Implementation and Programming Model of AMR Framework

This section describes the implementation and programming model of this proposed framework.

3.1 Data Structure for AMR Framework

In order to realize AMR computations, this framework recursively divides a computational domain into a large number of uniform grid blocks and represent their spatial distributions by tree structures. Each leaf node of the tree structures has a uniform grid block per each physical variable. Each block contains the same number of cells regardless of the resolution to be expressed. A grid block, for example, contains 16^3 cells in 3D with halo regions, which size depends on the numerical schemes adopted by the application. Figure 1 shows a schematic diagram of the physical spatial distributions of grid blocks with trees and the memory space layout that holds the actual data. A quadtree or an octree tree is used as the tree structure in 2D or 3D, respectively. Since the GPU often achieves high performance when accessing consecutive memory areas, the grid blocks are allocated in one large contiguous memory area for each physical variable.

Each leaf node does not directly hold a grid block itself but holds an ID that specifies an assigned grid block. From these IDs, the position of the assigned grid block in the contiguous memory area can be determined. By based on ID mapping, a single tree structure can be associated with an arbitrary number of physical variables, which is important in developing a framework. Changing the positions and the number of grid blocks with time integration can be made only by changing the tree structure with varying the values of the IDs on the leaf nodes. It is unnecessary to allocate and deallocate the memory for grid blocks that may cause performance degradation especially on GPU. In order to express arbitrary shapes of the computational domains flexibly, the framework arranges multiple tree structures in an entire computational domain as shown in Fig. 2.

In order to represent the AMR data structure by the multiple tree structures, this framework provides `Field` class, which is used as follows.

```
Field field(3, {4, 4, 8}); // dimension and size of trees
field.grow(2);             // grow all trees by 2 levels
```

In the multi-GPU computation, the entire program is parallelized by MPI and each process handles a single GPU. Each process independently holds the same tree structures as an object of the `Field` class at all times. The change of the tree structures, which means the change of the spatial resolution of the computational domain, is determined by (1) instructions to change mesh resolution and (2) instructions to migrate grid blocks between GPUs. Only the instructions in (1) and (2) are synchronized with MPI without explicit synchronization of the tree structures themselves. By sharing all the instructions in all processes, every process can change the own tree structures in the same way, which allows us

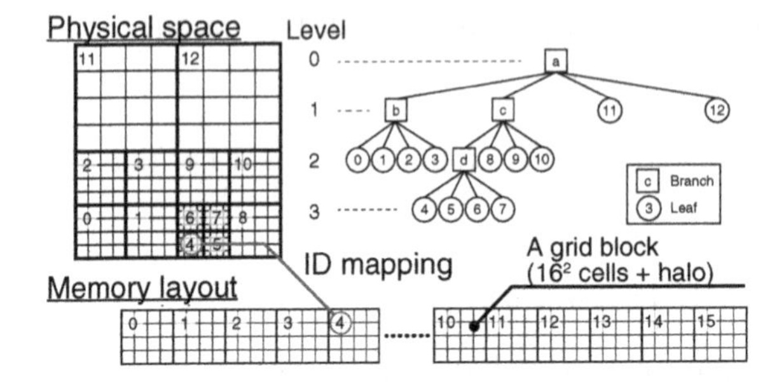

Fig. 1. AMR data structure.

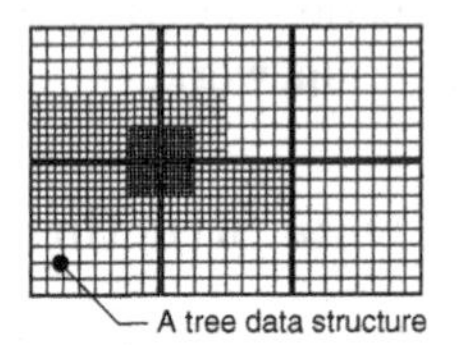

Fig. 2. Multiple trees to represent arbitrary shapes of the computational domains.

to always keep the same tree structures among all processes. In addition to the ID indicating the grid block, each leaf node holds a rank number of MPI that handles a GPU in which the grid block data are actually stored.

3.2 Array Structure for Multiple Grid Blocks

In order to represent the entire computational domain by a large number of grid blocks, the framework provides an unique data type `MArray` with `Range` type, which represents a 1D/2D/3D rectangular range. An object of `MArray` holds a single large array data, which is virtually divided into and used as multiple grid blocks as shown in Fig. 1, with the number of grid blocks and one object of `Range`. By using these types, the multiple grid blocks are allocated as follows:

```
unsigned int length[] = {16+2*mgn, 16+2*mgn, 16+2*mgn};
int begin [] = {-mgn, -mgn, -mgn};
int narrays = 4096;            // number of grid blocks
Range3D whole(length, begin); // size of each grid block
MArray<float, Range3D> f(whole, narrays, MemoryType::DEVICE);
```

`MArray` is initialized with parameters that specify a `Range` that represents the range of a grid block, the number of grid blocks the `MArray` contains, and a location of memory to allocate. This `Range` object is used to determine the halo regions of each grid block. These grid blocks are also exploited as temporary areas for storing data used for mesh refinement and halo exchange with MPI.

3.3 Writing and Executing Stencil Functions

In this framework, a stencil calculation must be defined as a C++11 lambda expression called a *stencil function* with `MArrayIndex` provided by the framework. The stencil function for 3D diffusion equation is defined as follows:

```
auto diffusion3d = [] __device__ (const MArrayIndex &idx,
  int level, float ce, float cw, float cn, float cs, float ct,
  float cb, float cc, const float *f, float *fn) {
  fn[idx.ix()] = cc*f[idx.ix()]
    + ce*f[idx.ix(1,0,0)] + cw*f[idx.ix(-1,0,0)]
    + cn*f[idx.ix(0,1,0)] + cs*f[idx.ix(0,-1,0)]
    + ct*f[idx.ix(0,0,1)] + cb*f[idx.ix(0,0,-1)]; }};
```

`MArrayIndex` holds the size of given grid block n^3 and represents a certain grid point (i, j, k), which is the coordinate of the point where this function is applied. It provides a function for accessing to the (i, j, k) point and its neighboring points for the stencil access; `idx.ix(-1, -2, 0)`, for example, returns the index representing $(i-1, j-2, k)$ point. Stencil functions can be defined as device (i.e., GPU) functions by using the qualifier `__device__` provided by CUDA.

To update `MArray` by the user-written stencil functions, the framework provides the `Engine` class, which is used to invoke the diffusion equation on the three-dimensional grid as follows:

```
Range3D inside; // where stencil functions are applied.
Engine_t engine;
engine.run(amrcontroller, inside, LevelGreaterEqual(1),
  diffusion3d, idx(f.range()), level(), ce,cw,cn,cs,ct,cb,cc,
  ptr(f), ptr(fn));
```

The parameters of `Engine::run` must begin with an object of `AMRController` that holds `Field` and another data structures required for AMR. The fourth parameter is a stencil function defined as a lambda expression, followed by any number of different types of additional parameters that are provided to this function. `f` and `fn` are `MArray` data. `Engine::run` applies a given stencil function to the grid blocks of the given `MArray fn` in the region represented by the second parameter `inside` and satisfying the condition for the AMR level given as the third parameter. Typically, `inside` specifies an inside region that is a region excluding the halo region from the computational domain as shown in Fig. 3. By specifying `LevelGreaterEqual(1)`, this stencil function is applied to the grid blocks on level 1 or higher. The `ptr` function provides the pointer pointing to (i, j, k) of the given grid block in the `MArray` to the user-defined stencil function. Similarly, `level` is used to obtain the AMR level of the applied given block inside the stencil function, which allows us to perform level-dependent computation. Since the grid blocks are allocated in the contiguous memory area as described above, the framework can apply a single stencil function to all grid blocks at various levels that are contained inside a single `MArray` simultaneously.

3.4 Data Transfer of Halo Regions

In this framework, each grid block on a leaf node has halo regions for stencil calculations. To advance the time step, it is necessary to exchange data in the halo regions between adjacent grid blocks with the same and different resolutions.

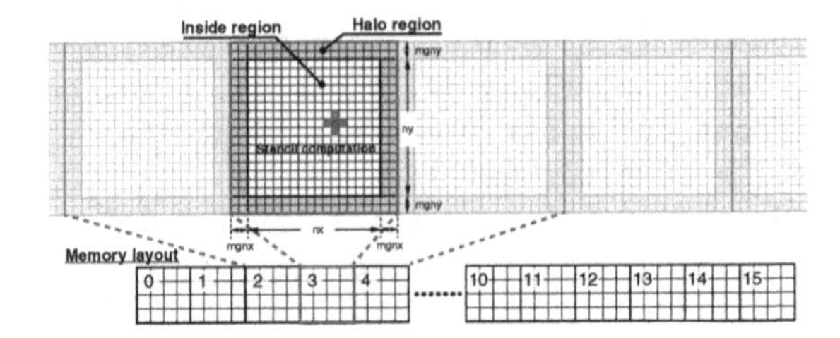

Fig. 3. Executing a stencil function with multiple grid blocks allocated in a large array.

Data exchange of the halo regions inside a GPU is performed in the following order. First, data exchange of the halo regions is performed between adjacent grid blocks with the same resolution (i.e., the same level), which do not need the interpolation of values. Next, the data of the halo regions are transferred from the high-resolution grid blocks to the low-resolution grid blocks. Finally, the data of the halo regions are transferred from the low-resolution grid blocks to the high-resolution grid blocks. The framework can handle values defined at cell center and node center points. It can copy values at the same physical location between high- and low-resolution with interpolation functions. Currently, the interpolation values are calculated by a linear function.

Figure 4 shows exchanging data in the halo regions between the grid blocks allocated in the different GPUs. First, the framework designates several pieces of current unused grid blocks from the continuous memory area as temporary regions in each process. They are placed in the surround area of the subdomain of each process. These temporary grid blocks are called the ghost blocks in our framework. Next, the data in the grid blocks that are necessary for the stencil computation are actually transferred from the adjacent GPU using the CUDA APIs with MPI and stored in the ghost blocks. Referring to these ghost blocks, the stencil functions are executed at each process independently.

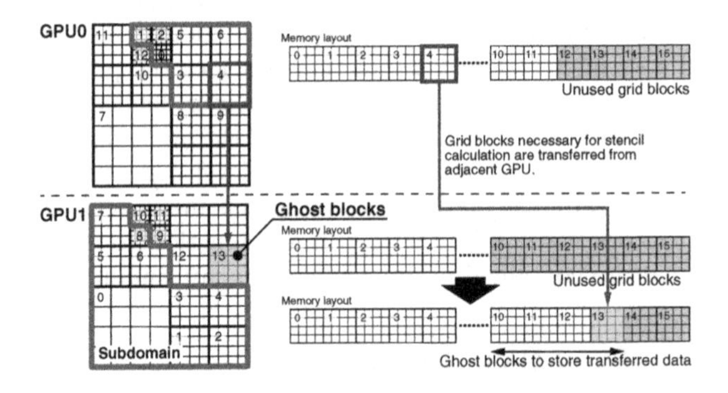

Fig. 4. Halo exchange between grid blocks allocated in the different GPUs.

To execute the stencil computations, only the halo regions of the ghost blocks are required. However, in this framework, the whole regions of the ghost blocks are transferred between neighboring GPUs instead of the halo regions of them. In order to make full use of transferred data of grid blocks, we exploit a temporal blocking method that is a well-known technique for locality improvement in stencil computations [12,15]. By using this method with several decomposed subdomains, several time steps can be advanced in each subdomain independently of the others. This also contributes to reducing the number of communications.

The framework exploits the temporal blocking based on the countdown proposed in our previous research [12]. Figure 5 shows the scheme of halo exchange using the temporal blocking with multiple GPUs. The number of executions of the function of halo exchange is counted. Based on this count, when it is expected that there will be no more effective data for performing the stencil calculation, actual communication will be carried out. Otherwise, the function of halo exchange does not perform any communication. As a result, the programmers can use the temporal blocking method without modifying their user codes.

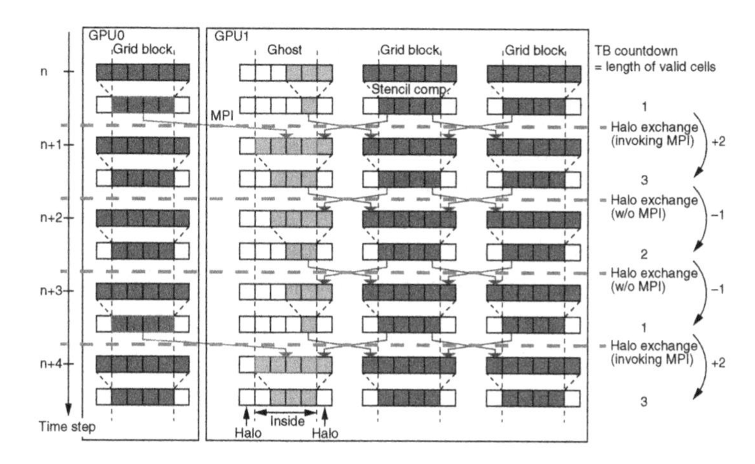

Fig. 5. Scheme of halo exchange using the temporal blocking with multiple GPUs. For the sake of simplicity, this figure is supposed to exchange halo regions at the same level.

In order to perform the halo exchange inside a GPU and between different GPUs with the temporal blocking method, this framework provides the `AMRController::exchange_halo`. This function is typically used as follows:

```
amrcontroller.exchange_halo(f, u, v, w);
```

`f`, `u`, `v` and `w` are `MArray` data. By using the C++11 variadic templates, this function can apply halo exchange between grid blocks to any number of different types of `MArray` data simultaneously. In this function, first inter-GPU communication with MPI is performed, which updates values on ghost blocks allocated each GPU. After that, inside each GPU, the halo exchanges between grid blocks including the ghost blocks are performed.

3.5 Mesh Refinement

Modifying the resolution of the grid blocks on the leaf nodes is not done automatically on the framework side because it is necessary to take care of the change of arbitrary physical quantities and variables in the user codes. To change the resolution of the grid blocks, the programmers explicitly specify the leaf nodes that having these grid blocks in the user code and issue the instructions of changing their resolutions by using the functions provided by the framework. These instructions issued to some leaf nodes in each process are shared by all processes before mesh refinement is actually executed.

After all instructions are shared by all processes, each process changes its own tree structure as follows. When a leaf node is specified to be fine resolution by an instruction for refining mesh, the framework forcibly raises its level by 1. To maintain a 2:1 balance of the resolution, the levels of its adjacent leaf nodes are also increased by 1 if necessary. When a leaf node is specified to be coarse resolution by an instruction, the framework decreases its level if it is able to continue to meet a 2:1 balance with its surrounding leaf nodes.

The resolution of the grid blocks is actually changed, after the new levels of the all leaf nodes after mesh refinement are determined on the tree structures. First, some of the unused grid blocks pooled in the continuous memory area are assigned to the grid blocks that store fine or coarse values after the mesh refinement. The framework assigns the grid blocks for this purpose in order from the smallest numeral to prevent fragmentation of memory. After that, the framework actually copies the values between grid blocks for the mesh refinement with interpolation in parallel. The unnecessary grid blocks that hold original values are returned to a group of the unused grid blocks for future use.

When several grid blocks with a high resolution that are not allocated on the same single GPU are changed to a single grid block with a low resolution, data migration is executed before mesh refinement in order to collect those original data on the same GPU.

3.6 Data Migration Between GPUs and Load Balancing

In the AMR method, the sizes and the physical positions of local regions with high resolution change in the time integration loop of applications. The load balancing among GPUs using data migration is necessary to make efficient use of computational resources and improve performance.

This framework provides a function to issue an instruction to migrate grid blocks from a GPU to another GPU. When migrating a grid block, the programmer first issues this instruction with specifying a new process for the leaf node handling this grid block. All instructions issued in each process are shared by all processes with MPI Allreduce. After that, the framework actually performs the migration of the grid blocks using MPI according to these instructions. The some of the unused grid blocks are assigned to store the migrated grid blocks.

By using this migration mechanism, dynamic load balancing is realized as follows. In our framework, a computational domain is represented by multiple trees (Fig. 2). While traversing trees in turn, the leaf nodes are assigned to each process in a depth-first search on each tree. The leaf nodes assigned to a certain process are typically owned by a few adjacent trees. By using this strategy, our applications can achieve localizing the distribution of the leaf nodes handled by each process and load balancing of them. Localizing their distribution contributes to making inter-process communication more effective. In our application, when the number of leaf nodes assigned to a certain process increases by 10% compared to the average number of leaf nodes assigned to each process, the redistribution of all leaf nodes based on the migration described above is carried out.

4 Performance Analysis and Discussion

This section presents the performance of compressible flow simulation based on the proposed framework on a NVIDIA Tesla P100 GPU and its weak scaling results obtained on TSUBAME3.0. TSUBAME3.0 is equipped with 2,160 P100 GPUs. The peak performance of each GPU in double precision is 5.3 TFlops. Each node of it has four P100 attached to the PCI Express bus 3.0 ×16 (15.8 GB/s), four Intel Omni-Path Architecture HFI (12.5 GB/s) and two sockets of the Intel CPU Xeon E5-2680 V4 2.4 GHz 14-core.

4.1 Application: 3D Compressible Flow

We perform 3D compressible flow computation written by this framework and show computational results of the Rayleigh-Taylor instability. To simulate this, we solve 3D Euler equations described as follows:

$$\frac{\partial U}{\partial t}+\frac{\partial E}{\partial x}+\frac{\partial F}{\partial y}+\frac{\partial G}{\partial z}=S,\quad U=\begin{bmatrix}\rho\\ \rho u\\ \rho v\\ \rho w\\ \rho e\end{bmatrix},\ E=\begin{bmatrix}\rho u\\ \rho uu+p\\ \rho vu\\ \rho wu\\ (\rho e+p)u\end{bmatrix},$$

$$F=\begin{bmatrix}\rho v\\ \rho uv\\ \rho vv+p\\ \rho wv\\ (\rho e+p)v\end{bmatrix},\ G=\begin{bmatrix}\rho w\\ \rho uw\\ \rho vw\\ \rho ww+p\\ (\rho e+p)w\end{bmatrix},\ S=\begin{bmatrix}0\\ 0\\ 0\\ \rho g\\ \rho wg\end{bmatrix},\tag{1}$$

where ρ is density, (u, v, w) are velocity, p is pressure, and e is energy. Here, g is gravitational acceleration. An advection term is solved using three-order upwind scheme with three-order TVD Runge-Kutta method. Time integration of five variables ρ, ρu, ρv, ρw, and ρe is solved, which requires 13 neighbor elements of each variable are used to update them on a center point of the grid.

4.2 Performance Evaluation on Single GPU

We show the performance results of the application on a single GPU by varying the size of the grid blocks assigned to leaf nodes. We change the number of cells that each grid block contains to 8^3, 12^3 or 16^3, and evaluate performance using 5 levels of resolution of AMR. In each grid block, halo regions having a width of 2 are added around those cells in this simulation due to the adopted numerical schemes. The maximum width of a grid block is 16 times the minimum one in physical space on the 5-level AMR. When the number of cells each grid block contains is increased, the volume occupied by one grid block becomes large and it is difficult to finely adjust the resolution locally. Then, we set the maximum value of the length of one side of a grid block to 16 in this measurement.

Table 1 shows the total performance of computational kernels themselves and the overall performance of an entire time step at a certain time step when the number of cells in each grid block is changed. The 15 different computational kernels are executed at each time step. The entire time step includes computational times for exchange of the halo regions and control of the AMR structure as well as the above 15 kernels. These results are evaluated by using NVIDIA GPU profiler nvprof. This table also shows the number of leaves from the coarsest level 1 to the finest level 5. Note that the length of the whole computational domain needs to be a constant multiple of the length of a grid block in the current implementation. Then, when the number of cells each grid block contains is 12^3, the size of whole computational domain is different from others.

As shown in Table 1, comparing the total performance of the 15 kernels, the performance is higher when the length of one side of a grid block is longer. This is because when the volume of the halo regions with respect to that of the inside region decreases in each grid block, the memory access needed for updating values in the inside region is reduced, resulting in performance improvement. When the length of one side of a grid block is 16, the total performance of the kernels is 914 GFlops, which is 65% of the performance obtained by the same computation on the normal structure grid with the size of 128^3 (i.e., 1.41 TFlops). Considering that the ratio of the inside region to the entire computational region including the halo regions is 51% in each grid block and the cache can be used as part of the memory access, the ratio of 65% is considered appropriate.

In this framework, each grid block has halo regions so that the stencil functions for the structure grid can be used without any modification. However, the cost of exchanging these halo regions is relatively high. Due to this overhead, the observed overall performance decreases to 246 GFlops in the above case.

Table 1. Performance on a single NVIDIA Tesla P100 GPU.

# of cells	Equivalent domain size	# of leaves (level 1/2/3/4/5)	Kernels' performance (GFlops)	Overall performance (GFlops)
8^3	$512 \times 512 \times 2048$	180/449/705/2385/17208	765.2	102.4
12^3	$576 \times 576 \times 2304$	20/176/210/786/4848	811.9	178.1
16^3	$512 \times 512 \times 2048$	16/80/220/718/4752	913.9	245.9

4.3 Time to Solutions

We evaluate the computation time of two versions of simulation codes. The first version uses the temporal blocking method to reduce the number of communications and the second one does not exploit it. The latter version is used as references for this performance evaluation. Both versions may migrate data on grid blocks every 200 steps to improve load balancing if necessary. We perform simulations on a physical volume equivalent to the finest uniform grid with the size of $2,048 \times 1,024 \times 4,096$ using 5 levels of AMR by using 4 GPUs on each node and total 32 GPUs on TSUBAME 3.0. We use grid blocks having 16^3 cells with 2-width halo region from the results of the previous section.

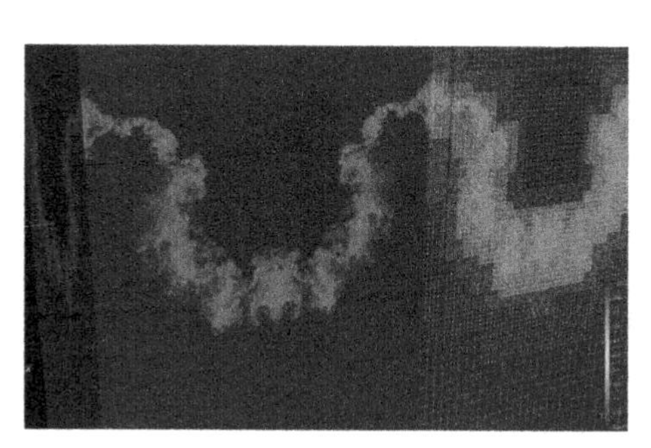

Fig. 6. A snapshot of density distribution results obtained by the simulation of 3D compressible flow. The boundary lines of the grid blocks are also shown in part.

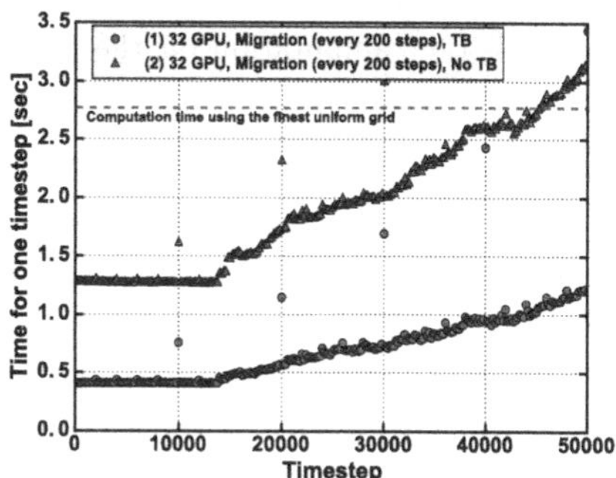

Fig. 7. Computational times for each time step using 32 GPUs. (Color figure online)

Figure 6 shows a snapshot of computational results of the Rayleigh-Taylor instability obtained by 3D compressible flow computation written by this AMR framework. By applying the AMR method to fluid simulation, we have succeeded in simulating with a fine structure around the interface of two fluids.

Figure 7 shows the computation time taken for the calculation of each time step in the above two versions. At the 10,000th step, the first version takes 0.41 s for the computation on this time step, while the second version takes 1.28 s for the same computation. With the benefits of the framework, programmers can easily introduce the temporal blocking to this application and achieve approximately 3.1 times speedup without any additional development cost. When the finest uniform grid is used over entire computational domain instead of AMR, the computational time of 2.7 s per each time step is required, which is depicted as a blue dashed line in Fig. 7. It indicates that the first version is 6.7 times faster than the same computation on the finest uniform grid. Since the restart files are output every 10,000 steps, the computational times for every 10,000 steps is longer than the those required for other time steps.

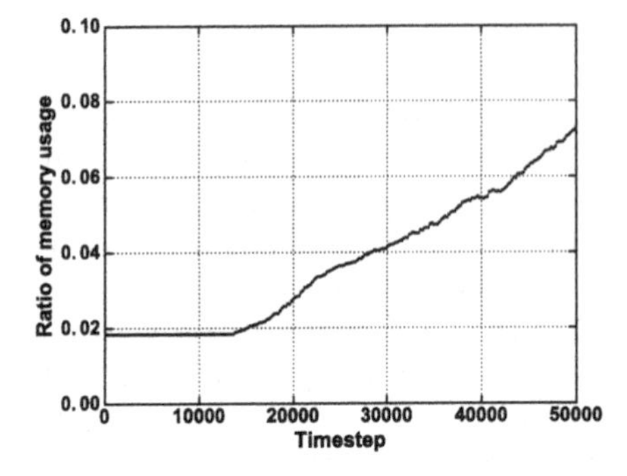

Fig. 8. Ratio of memory usage of AMR simulation for each time step in comparison with the computation on the finest grid.

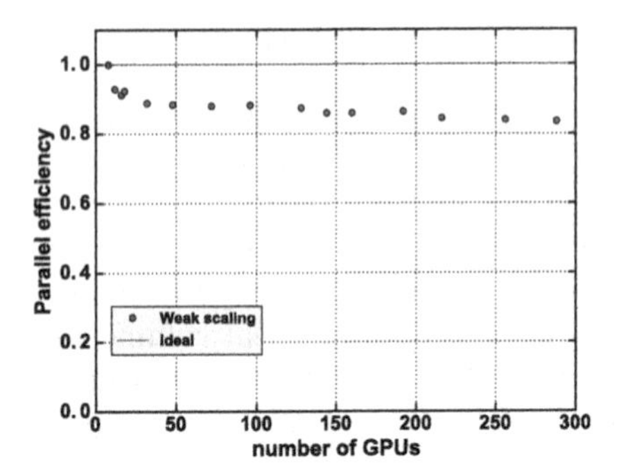

Fig. 9. Weak scaling on TSUBAME 3.0.

Figure 8 shows the memory consumption ratio of this AMR simulation at each time step, compared to the simulation performed using the finest uniform grid over the physical volume having the same size. By using 5 levels of AMR, this memory consumption rate is kept to be less than 10% in overall runtime.

4.4 Weak Scaling Results

We show the weak scaling results of AMR applied simulation for the Rayleigh-Taylor instability using multiple GPUs on TSUBAME3.0. Figure 9 shows the performance results of the simulation using 5-level AMR with the temporal blocking method and the data migration to improve load balancing. We use 4 GPUs per each node for this simulation. We assign a physical volume equivalent to the finest uniform structure grid with the size of 1024^3 to each GPU. As shown in this figure, the weak scaling efficiency is above 84% for a physical volume equivalent to the finest uniform grid with the size of $6144 \times 6144 \times 8192$ on 288 GPUs with respect to the 8-GPU performance.

In order to further analyze the weak scaling results, Fig. 10 shows the breakdown results of the computation time using 8 and 288 GPUs at the 1000th step. The computation time obtained by the stencil functions and the time taken by the halo exchange inside each GPU are almost the same in both figures. On the other hand, the communication time among GPUs with MPI is greatly affected by the number of GPUs to be used. Because of the complex geometry of the subdomains, each GPU needs to communicate with more number of GPUs in AMR than in the case of computation using a structure grid with multiple GPUs. When the number of GPUs used increases, the number of GPUs each GPU communicates increases, resulting that the communication time takes longer. In the refinement and data migration, MPI Allreduce is used to share the instructions among all processes to update the tree structures held by each process. As the number of GPUs increases, the communication between all processes increases, resulting in increasing the total computation time in one time step.

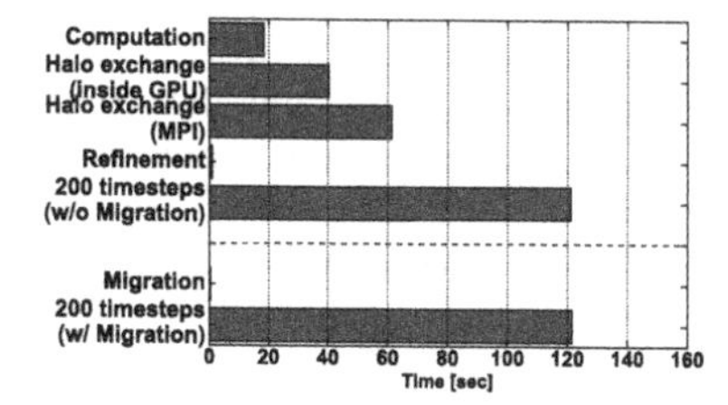

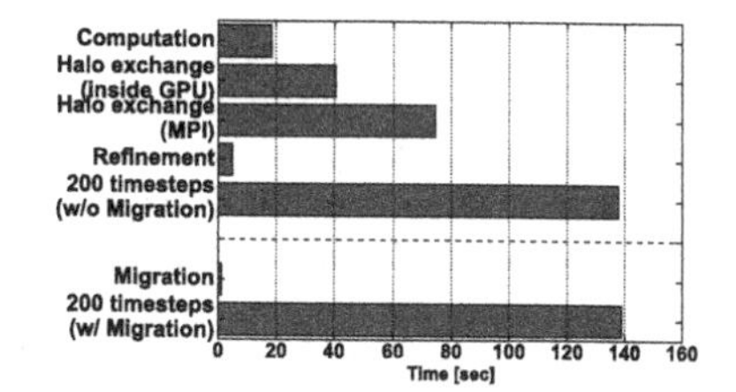

Fig. 10. Breakdown of the computation time at one time step using 8 GPUs (*left figure*) and 288 GPUs (*right figure*).

5 Conclusion

This paper has presented the programming model and implementation of the high-productivity framework for a block-based AMR for stencil applications, and evaluation of 3D compressible flow based on the proposed framework performed on a supercomputer equipped with multiple GPUs. The framework can execute the user-written stencil functions that update a grid point on Cartesian grid over a tree-based AMR data structure effectively. This framework also provides mesh refinement mechanism and data migration that are required for AMR applications. The countdown based temporal blocking method, which is applied to the user codes without any modification, are contributes to reducing the number of communications and making full use of transferred data. With our proposed framework, we have conducted performance studies of the framework-based compressible flow simulation on a single GPU and using multiple GPUs on TSUBAME 3.0. The framework-based compressible flow simulation has achieved to reduce the computational time to less than 15% with 10% of memory footprint compared to the equivalent computation running on the fine uniform grid. The good weak scaling is obtained using 288 GPUs of TSUBAME 3.0 with the efficiency reaching 84%.

Acknowledgments. This research was supported in part by JSPS KAKENHI Grant Number JP17K00165, JP26220002 and in part by "Joint Usage/Research Center for Interdisciplinary Large-scale Information Infrastructures" and "High Performance Computing Infrastructure" in Japan (Project ID: jh180061-NAH, jh180041-NAH).

References

1. Berger, M.J., Oliger, J.: Adaptive mesh refinement for hyperbolic partial differential equations. J. Comput. Phys. **53**, 484 (1984)
2. Christen, M., Schenk, O., Burkhart, H.: PATUS: a code generation and autotuning framework for parallel iterative stencil computations on modern microarchitectures. In: 2011 IEEE International Parallel Distributed Processing Symposium (IPDPS), pp. 676–687 (2011)
3. Fryxell, B., et al.: FLASH: an adaptive mesh hydrodynamics code for modeling astrophysical thermonuclear flashes. Astrophys. J. Suppl. Ser. **131**(1), 273 (2000)

4. Gysi, T., Osuna, C., Fuhrer, O., Bianco, M., Schulthess, T.C.: STELLA: a domain-specific tool for structured grid methods in weather and climate models. In: Proceedings of the International Conference for High Performance Computing, Networking, Storage and Analysis, SC 2015, pp. 41:1–41:12. ACM, New York (2015)
5. Maruyama, N., Nomura, T., Sato, K., Matsuoka, S.: Physis: an implicitly parallel programming model for stencil computations on large-scale GPU-accelerated supercomputers. In: Proceedings of 2011 International Conference for High Performance Computing, Networking, Storage and Analysis, SC 2011, pp. 11:1–11:12. ACM, New York (2011)
6. Schive, H.Y., Tsai, Y.C., Chiueh, T.: GAMER: a graphic processing unit accelerated adaptive-mesh-refinement code for astrophysics. Astrophys. J. Suppl. Ser. **186**(2), 457 (2010)
7. Shimokawabe, T., Aoki, T., Ishida, J., Kawano, K., Muroi, C.: 145 TFlops performance on 3990 GPUs of TSUBAME 2.0 supercomputer for an operational weather prediction. Proc. Comput. Sci. **4**, 1535–1544 (2011). Proceedings of the International Conference on Computational Science, ICCS 2011
8. Shimokawabe, T., et al.: An 80-fold speedup, 15.0 TFlops full GPU acceleration of non-hydrostatic weather model ASUCA production code. In: Proceedings of the 2010 ACM/IEEE International Conference for High Performance Computing, Networking, Storage and Analysis, SC 2010, pp. 1–11. IEEE Computer Society, New Orleans (2010)
9. Shimokawabe, T., Aoki, T., Onodera, N.: High-productivity framework on GPU-rich supercomputers for operational weather prediction code ASUCA. In: Proceedings of the 2014 ACM/IEEE International Conference for High Performance Computing, Networking, Storage and Analysis, SC 2014, pp. 1–11. IEEE Computer Society, New Orleans (2014)
10. Shimokawabe, T., Aoki, T., Onodera, N.: High-productivity framework for large-scale GPU/CPU stencil applications. Proc. Comput. Sci. **80**, 1646–1657 (2016)
11. Shimokawabe, T., et al.: Peta-scale phase-field simulation for dendritic solidification on the TSUBAME 2.0 supercomputer. In: Proceedings of the 2011 ACM/IEEE International Conference for High Performance Computing, Networking, Storage and Analysis, SC 2011, pp. 1–11. ACM, Seattle (2011)
12. Shimokawabe, T., Endo, T., Onodera, N., Aoki, T.: A stencil framework to realize large-scale computations beyond device memory capacity on GPU supercomputers. In: 2017 IEEE International Conference on Cluster Computing (CLUSTER), pp. 525–529, September 2017
13. Unat, D., Cai, X., Baden, S.B.: Mint: realizing CUDA performance in 3D stencil methods with annotated C. In: Proceedings of the International Conference on Supercomputing, ICS 2011, pp. 214–224. ACM, New York (2011)
14. Wahib, M., Maruyama, N., Aoki, T.: Daino: a high-level framework for parallel and efficient AMR on GPUs. In: Proceedings of the International Conference for High Performance Computing, Networking, Storage and Analysis, SC 2016, pp. 53:1–53:12. IEEE Press, Piscataway (2016)
15. Wolf, M.E., Lam, M.S.: A data locality optimizing algorithm. In: Proceedings of the ACM SIGPLAN 1991 Conference on Programming Language Design and Implementation, PLDI 1991, pp. 30–44. ACM, New York (1991)

Harmonizing Sequential and Random Access to Datasets in Organizationally Distributed Environments

Michał Wrzeszcz[1,2], Łukasz Opioła[1,2(✉)], Bartosz Kryza[2], Łukasz Dutka[2], Renata G. Słota[1], and Jacek Kitowski[1,2]

[1] AGH University of Science and Technology, Faculty of Computer Science, Electronics and Telecommunications, Department of Computer Science, Kraków, Poland
{wrzeszcz,lukasz.opiola,renata.slota,jacek.kitowski}@agh.edu.pl
[2] Academic Computer Centre CYFRONET AGH, Kraków, Poland
bkryza@agh.edu.pl, lukasz.dutka@cyfronet.pl

Abstract. Computational science is rapidly developing, which pushes the boundaries in data management concerning the size and structure of datasets, data processing patterns, geographical distribution of data and performance expectations. In this paper we present a solution for harmonizing data access performance, i.e. finding a compromise between local and remote read/write efficiency that would fit those evolving requirements. It is based on variable-size logical data-chunks (in contrast to fixed-size blocks), direct storage access and several mechanisms improving remote data access performance. The solution is implemented in the Onedata system and suited to its multi-layer architecture, supporting organizationally distributed environments – with limited trust between data providers. The solution is benchmarked and compared to XRootD + XCache, which offers similar functionalities. The results show that the performance of both systems is comparable, although overheads in local data access are visibly lower in Onedata.

Keywords: Random access · Variable-size block · Distributed file system · Organizationally distributed environment

1 Introduction

Recent years have brought significant advances in computational science and rapid development of data centers, which keep growing and employing modern, distributed storage technologies. Big institutions are getting dedicated network links and the throughput of network infrastructures is increasing. This technological progress aligns well with the trends in computational science that push towards globalization and distributed computing. The idea of e-Science [11] allows scientists from different fields and organizations to cooperate without

J. M. F. Rodrigues et al. (Eds.): ICCS 2019, LNCS 11536, pp. 295–308, 2019.
https://doi.org/10.1007/978-3-030-22734-0_22

borders, performing parallel, distributed analysis on large, shared datasets. However, current data access and sharing solutions can only partly fulfill this vision. The reason behind the lack of suitable solutions is the challenging nature of data access globalization. Some relevant issues are: autonomy of data providers, geographical distribution of vast datasets, complicated maintenance of network-based communication, data security and privacy or decentralized authorization. Among them there is efficient and cost-effective access and processing of distributed datasets in such decentralized environments.

Analysis of different use-cases show that scientists use very diverse methods to process their datasets. Quite often the data is stored in giant (sometimes sparse) files, which are read or written by variable-size chunks in a seemingly arbitrary order – consider for example the popular HDF5 [9] format that can hold multi-dimensional data. While sequential access is usually well handled, the case of random read & write is a pitfall for most of network-based file systems. When data is located on remote storage systems or distributed, these operations trigger transfers of whole files or large blocks between storage clusters that can generate unnecessary costs. Transfer management and optimization can be very challenging, especially when files are accessed in a random manner.

In this paper we present our solution for harmonizing performance of sequential and random access to local and remote datasets in organizationally distributed environments. The solution was implemented as a part of data access system called Onedata [16,17], evaluated and compared to commonly used XRootD virtual filesystem.

2 Related Work

Below is a summary of existing solutions related to efficient access to large, distributed datasets: distributed data access systems, solutions optimizing random access performance of network-based storage and tools for large data transfers.

The need of unified data access is apparent as more and more initiatives [22,25] and products appear, trying to fulfill those requirements. For example, IBM offers Active File Management (AFM) [12] as an additional layer over their GPFS storage to achieve caching of data originating from remote sites (home-and-cache model) with support for data modifications. By creating associations between data clusters, one can implement a single namespace view across sites around the world, though this requires full trust between them. XRootD [3] is a commonly used, open-source alternative to GPFS + AFM, which embraces a very similar model when coupled with XCache [7]. XRootD can be used to unify access to different storage systems into a single virtual endpoint, accessible from anywhere. XCache is essentially an XRootD service employing a caching plugin, which manages a local cache of data read from remote sites for faster consecutive reading. Like in IBM's solutions, XRootD/XCache requires all the sites to be federated. However, in contrast to AFM, XCache does not support remote write operations. DataNet Federation Consortium (DFC) [5] aims to implement a national data management infrastructure to streamline scientific development. The prototype supports three types of federation mechanisms: (1) tightly coupled federations, realized by federating iRODS data grids, (2) loosely coupled federations, i.e. external services offering certain datasets for retrieval and querying and (3) asynchronous federations where queries to external services are processed in an asynchronous (message queue based) manner. This approach promotes (read-only) integration with open data services, rather than unifying data access to distributed data providers. DFC employs iRODS [18] to create a single federation, but it can be also used to achieve cross-federation data access. However, it does not implement location transparency of the stored data. The files must be manually moved/copied between iRODS Zones. It requires certain administrative effort to set up cross-federation data access – user accounts pointing to their home Zone must be created in remote Zones.

As mentioned before, random access performance is a weakness of most file systems, and especially problematic in network-based storage systems. There have been attempts to overcome these limitations or introduce optimization mechanisms. For example, in [29] the authors propose three methods to optimize random queries on HDFS [20] and guarantee the performance of sequential access. All the methods are based on network-level optimizations and yield satisfactory results. Another attempt to adjust HDFS to random access profile is presented in [15], where the authors introduce some low-level modifications to make the filesystem better suited for computations in the field of High Energy Physics (HEP). The next example is VarFS [10] – a filesystem build on Ceph [24] especially for the purpose of random write operations. The general idea is that instead of using objects or blocks of fixed-size as most file systems do, this layer uses variable-size objects while remaining POSIX compatible. This way, random write performance can be greatly increased and the overheads of sequential write operations are acceptable. The conclusion is that it is possible to achieve reasonable random access performance in a commonly used distributed data access systems such as HDFS or Ceph, however these solutions are designed for federated environments.

Data transfers are an inherent aspect of distributed data access systems. Whenever a file is accessed remotely, it must be pushed through the network between data sites. In the great majority of data access systems, fixed-size block is used as the basic unit of data and only the required blocks are transferred. The stability and latency of the network link have a significant impact on efficiency – hence various optimization techniques are employed. They include prefetching of blocks, caching the data locally, tuning the network etc. Pre-staging used to be a reasonable choice in some scenarios, but with the growth of data volumes, replication of whole datasets is becoming unviable. Still, there are many cases where large files are moved between data clusters as part of the scientific process. The choice of tools for managing data transfers is wide, see products from Signiant [21], Axway [2], IBM [13] or Serv-u [19] as commercial examples. Non-commercial solutions include mdtmFTP [28], FDT [8] or GridFTP [1], which is extensively used in scientific communities. A common approach is to embrace parallel network links between clusters to speed up file moving – this idea was formalized in the Parallel FTP protocol [4].

The choice of tools for data access and transfers is wide, but there is a lack of integrated solutions for efficient data access in organizationally distributed environments. Such solution should hide away the complexity of manual data management between autonomous sites and offer a unified, transparent view on all user's datasets, at the same time ensuring the performance sufficient for scientific computing.

3 Data Access in Organizationally Distributed Environments

Our solution for harmonizing performance of sequential and random access to local and remote datasets is a part of Onedata – an eventually consistent distributed data access system. Onedata aims to provide access to distributed data under a single namespace [27]. Its main goal is to achieve truly transparent, efficient, scalable and cost-effective data access to autonomous data providers, despite the inherent lack of trust between them [17].

The Onedata system is based on a multi-layer architecture (see Fig. 1). Onezones provide an Authentication and Authorization Infrastructure, and mediate in cooperation of Oneproviders, which realize access to datasets stored in different organizations. Oneclients, subject to Oneproviders, employ FUSE (filesystem in userspace [23]) to implement POSIX data access and seamless integration with filesystems. While Onezones are key to overcome the lack of trust in organizationally distributed environments, Oneproviders and Oneclients are responsible for handling data access.

Users access their data through the Oneclient software using logical paths pointing at logical files. To provide efficient data access without significant overheads, Oneclient accesses the data directly on the storage system whenever possible. Otherwise, proxy mode is used – the data is read/written through a network connection to Oneprovider. Prior to direct data access, Oneclient gains knowledge about logical files from metadata managed by Oneprovider. The metadata includes such information as logical filenames, permissions, access types and a registry of data-chunks – as decribed below.

3.1 Data-Chunks

Onedata introduces data-chunk as the basic unit of data. The content of each logical file consists of one or more data-chunks, each representing a range of bytes. Data-chunks have similar role as blocks in a standard filesystem, although they can correspond to a series of blocks or other entities (e.g., objects) on the underlying storage system. Thus, data-chunk handling is a vital factor in data access scalability, performance and cost-effectiveness. Without appropriate models for metadata consistency and synchronization [26], file metadata that includes the registry of data-chunks can become a bottleneck of the whole data access system (e.g. [6,14,24]).

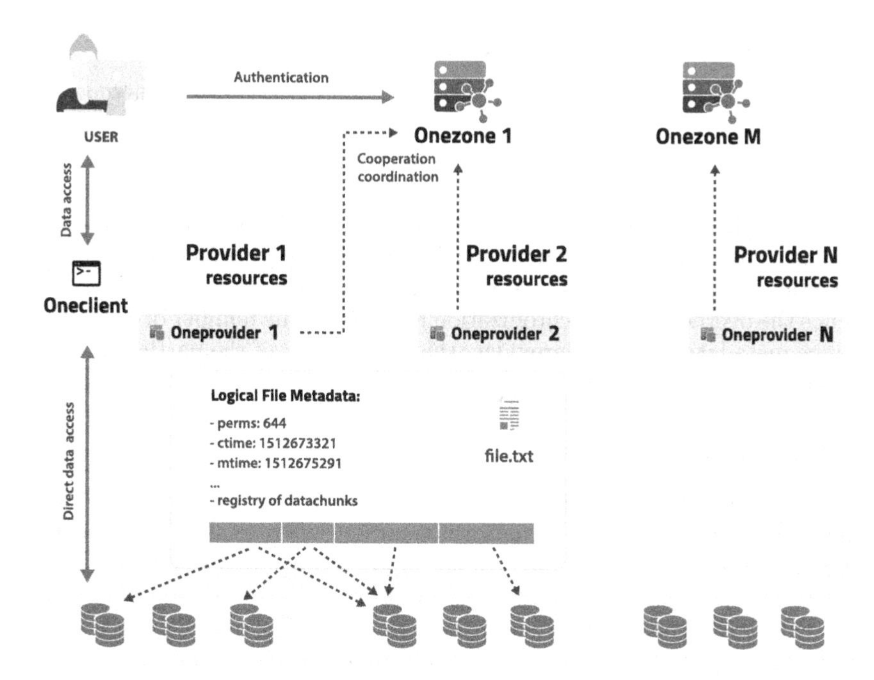

Fig. 1. Multi-layer Onedata architecture and logical file matadata with data-chunks.

Datasets differ in characteristics – in extreme cases the files may be small and numerous or large and sparse. While small data-chunk size would result in creation of numerous data-chunks for big files, large data-chunk size would cause synchronization of large data pieces even when a single byte is read. Thus, Onedata uses variable-size data-chunks – in specific cases, a data-chunk can represent a single byte or the whole file.

Oneprovider services synchronize file metadata including the registry of data-chunks (see Fig. 1). When a data-chunk is overwritten by a Oneprovider, the remaining Oneproviders mark the modified data-chunk as invalid. As the actual file content is not exchanged, this is a lightweight mechanism even for large files. Data transfers are performed only when a data-chunk being read is absent from the Oneprovider that handles the reading. As needed, data-chunks are split on the fly to limit the transfer size to the missing data range only. Therefore, the variable-size data-chunks minimize the cost of remote data access.

All things considered, there are several advantages of variable-size block management in the context of highly distributed systems, which align well with our concept of data-chunks:

- universal fit for small and large files,
- limiting the network and storage cost to possible minimum,
- flexibility and ability to dynamically adapt to circumstances,
- seamless integration with different underlying storage systems, no matter their blocksize or type (file/object-based etc.).

3.2 Models for Metadata Consistency and Synchronization

Metadata access overheads can be related to round trip times, which are often impossible to reduce. For this reason, the metadata is replicated between Oneproviders and cached by Oneclients. Whenever possible, it is processed locally and/or asynchronously. As a result, most of the metadata (including the registry of data-chunks) is eventually consistent and adopts last-write-wins conflict resolution. The causal consistency model is applied only to metadata managed by Onezones that is crucial for cooperation and security.

Since the overheads of synchronization grows with the number of entities that exchange metadata, only Oneproviders that store parts of the particular dataset are involved in processing and replication of the corresponding metadata.

3.3 Data Access Performance

One of the basic assumptions for Onedata is direct access to storage systems whenever possible to retain the performance they offer. Scalability is achieved by handling multiple underlying storage systems in parallel and limiting the metadata processing overheads as much as possible, by using appropriate consistency and synchronization models (see Sect. 3.2).

Data access performance can be further improved by employing specialized mechanisms that support particular data access patterns when data is not accessible directly. However, the sequential and random data access patterns require different optimization strategies to limit the negative impact of the network and data access latency.

4 Harmonizing Random and Sequential Access

There are three main factors when considering efficient data access:

- operation: read/write,
- data location: local/remote,
- access pattern: sequential/random.

The write operation on the side of Oneclient works in the same way, regardless if local or remote. The data is written directly to the local storage system and events are produced that update the data-chunk registry asynchronously. This process is depicted in Fig. 2 – initially, the file is stored only in the second Oneprovider. Oneclient overwrites a part of the file content within the first Oneprovider, a registry update is broadcasted and the data-chunk is invalidated in the second Oneprovider.

Similarly to the write operation, local data read is performed directly on the storage system. In both cases, the only overheads are caused by fetching the file metadata, which stays cached for faster consecutive operations. Essentially, the efficiency of write and local read operations depends roughly on the local storage system performance. Thus, to harmonize the performance of data access, we focused mostly on mechanisms that support sequential or random data reading from remote sites, discussing local operations for reference only.

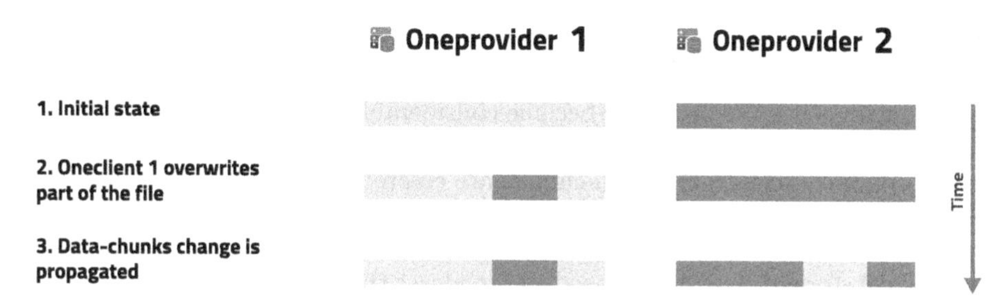

Fig. 2. Remote write causing a data-chunk invalidation.

4.1 Sequential Remote Read

During sequential remote reading, Oneclient requests its Oneprovider to transfer the missing data-chunks to the local storage as needed and reads the file block by block. While certain data-chunks of the file may be distributed between many remote Oneproviders, the key factor of sequential remote read performance is reduction of the delay in access to non-local data-chunks. It is achieved using a prefetching mechanism. For each opened file handle, Oneclient continuously detects the access pattern (sequential vs. random), based on comparing the read offsets on consecutive read operations. When a file is detected to be accessed sequentially, Oneclient requests transfer from remote Oneprovider of more data in advance to be immediately accessible as reading proceeds. Data transfers are prioritized so that the prefetch requests do not hinder the transfers of data needed instantly. Moreover, all Oneclients operating on a particular logical file are asynchronously notified of any prefetched data-chunks in order to minimize the number of transfer requests. In summary, Onedata employs three mechanisms that support sequential remote read: prefetching, prioritization of transfer requests and broadcasting of information about synchronized data blocks.

4.2 Random Remote Read

Random read performance is cumbersome to optimize in any file system, especially when data is stored in a remote location. The introduction of data-chunks greatly limits the network traffic caused by transfers, but reading a file remotely (especially by small blocks) results in creation of numerous small data-chunks. This causes the data-chunks registry to grow, increasing the costs of processing and synchronization between Oneproviders. Moreover, the prefetching mechanism in Oneclient is undesirable during random read, as it hampers the performance by transferring unneeded bytes. For these reasons, we introduced several optimizations to data-chunk management and Oneclient behaviour to suite them to remote data access.

The data-chunks registry includes information about data-chunks stored in local and remote Oneproviders. The registry is updated whenever one of the following takes place: an event is received from Oneclient reporting data modification, an update of certain data-chunks appears from another Oneprovider or a transfer request is completed. As random read can result in thousands of small transfer requests per second, the overheads of synchronizing the data-chunks registry with other Oneproviders become considerable. For this reason, we introduced the distinction of public (instantly advertised) and private (stored only locally) data-chunks. Public data-chunks are created as a result of data modifications, so that other Oneproviders are quickly informed about any changes in the file content. Private data-chunks are a result of replicating fragments of data to the local storage. There is no need to broadcast them quickly – it is done only after they are merged to a larger data-chunk and made public.

To minimize the costs of processing and synchronizing the data-chunks registry, it is based on a tree structure with fast offset-based access. Consequently, the registry processing time grows logarithmically with size, and upon any modification, only the changed parts of the tree are broadcasted to other Oneproviders.

In case Oneclient determines that the file is not read in sequential access pattern, it assumes that the file is accessed randomly and the prefetching algorithm works differently. Rather than requesting consecutive parts of data, it discovers which fragments (if any) of the logical file are accessed frequently and prefetches them (see Fig. 3). Such behavior is beneficial for two reasons. Firstly, it is probable that further read operations will appear within such fragment and will be handled much faster. Secondly, such aggregation merges several data-chunks into a larger public one, decreasing the overall data-chunks number and triggering a broadcast of the aggregated data-chunk.

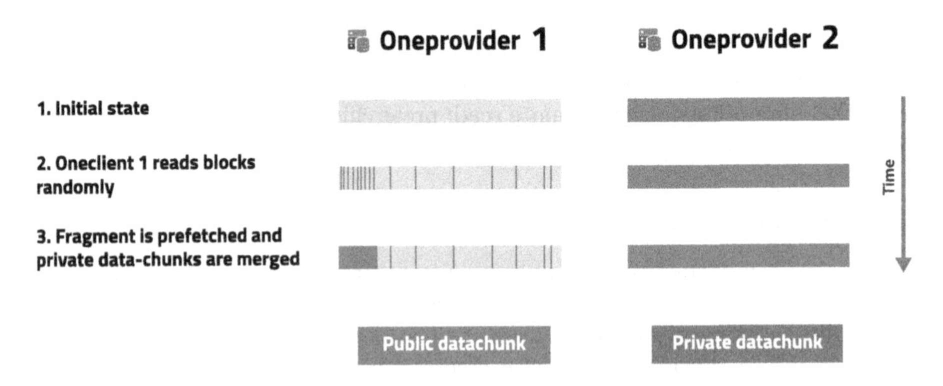

Fig. 3. During random read, private data-chunks are merged into a bigger one, prefetched and published.

To summarize, the following mechanisms support random remote read in Onedata: private data-chunks, data-chunk merging, selective tree-based processing and broadcasting of the data-chunk registry, automatic discovery of random access pattern and frequently accessed fragments of logical files that trigger prefetching of the whole fragment.

4.3 Influence of Random and Sequential Read on One Another

Due to their nature, sequential and random read require different optimizations. Some of the mechanisms dedicated for one read type have an opposite effect on the other. For this reason, Oneclient performs continuous detection of data access pattern and adjusts its behaviour accordingly.

Besides access pattern recognition, Onedata harmonizes random and sequential access when many Oneclients operate on logical files sequentially and randomly in parallel. In such case, randomly reading Oneclients trigger merging of smaller data-chunks into larger ones, which makes certain fragments of files better suited to sequential read. On the other hand, sequentially reading Oneclients trigger prefetching, which does not block random read (due to lower priority) but increases the chances of hitting already prefetched data during further reading.

5 Evaluation

We have performed benchmarks to verify the read and write performance and estimate overheads introduced by the Onedata software. For reference, an installation of XRootD and XCache with standard settings has been tested using the same underlying storage and benchmarks. The environment consisted of two identical virtual machines: 12 CPU × 2 GHz and 40 GB RAM. All test cases were based on a 64 KB block and the test file size was 200 GB. The network link between the machines yielded about 5.2 Gb/s. The presented results have been obtained from several runs with repeatable measurements.

5.1 Local Data Access

The purpose of testing local data access was mainly to estimate the overheads of virtualization. Data was read or written to a test file by one process on one host in three cases:

- directly on the storage system,
- via Oneclient connected to the Oneprovider on the host and with direct access to the storage,
- via XRootDFS (FUSE-based client for XRootD) connected to the XRootD server on the host.

Results of the tests are presented in Figs. 4 and 5. Thanks to the fact that Oneclient operates directly on the storage system and communicates with Oneprovider only to fetch the required metadata, the measurements are close to the underlying file system performance. XRootDFS uses a network link to the XRootD server to read/write file data and yields lower results, despite the fact that both client and server were located on the same machine. As a consequence, when scalability is concerned, XRootDFS depends on the network capacity, and Oneclient depends on the underlying storage scalability.

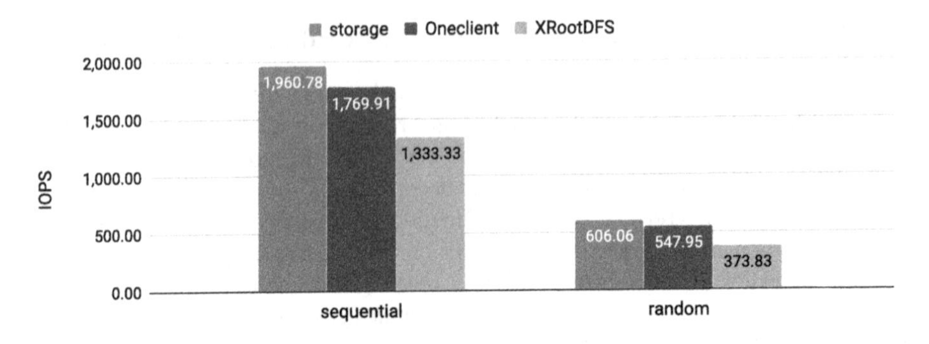

Fig. 4. Local read performance.

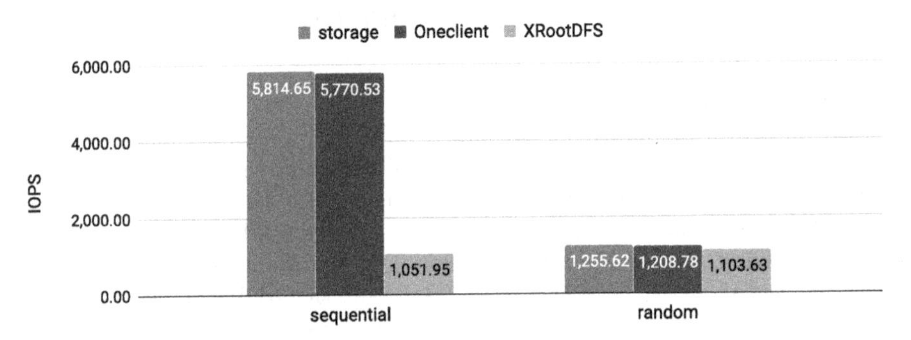

Fig. 5. Local write performance.

5.2 Remote Data Access

The environment for remote data access tests consisted of two machines, hosting one of the following setups:

- Oneprovider on the first host, Oneprovider + Oneclient (with direct storage access) on the second host,
- XRootD server on the first host, XCache + XRootDFS on the second host.

The tests included only remote read benchmarking. Remote write was not tested, because it is not supported by XCache, and in Oneclient it works the same way as local write and yields the same performance – the data is written locally and overwritten data-chunks are invalidated in remote providers.

The test file was placed on the first host, and read by the client software on the second host, via the caching layer (second Oneprovider/XCache). Effectively, reading the file caused data trasfers between the Oneproviders or XRootD and XCache. The file was read following three different patterns: sequential, random and hybrid (starting from a random offset every time, a 20 MB fragment was read). The results are shown in Fig. 6.

Figure 6a shows that the prefetching mechanism in Oneclient works effectively. The chart adopts a stairstep-like shape, depicting where the prefetching

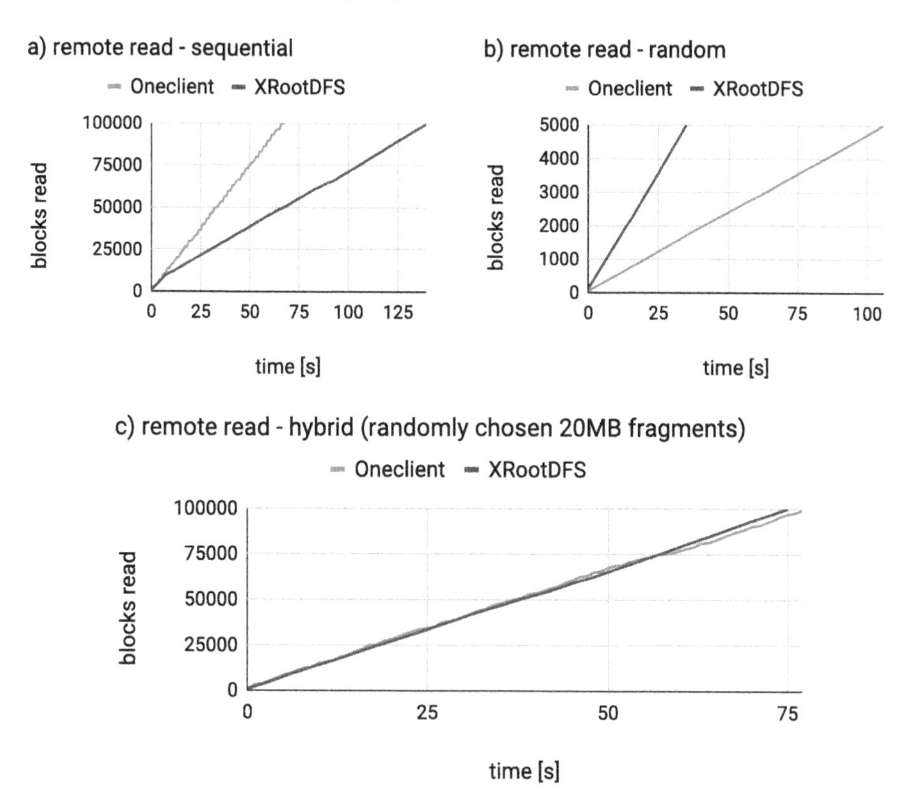

Fig. 6. Remote read performance – (a) sequential, (b) random, (c) hybrid.

or reading proceeded ahead of each other – the data is transferred by Oneprovider to the local storage and then read directly by Oneclient. XCache was configured with prefetching enabled and served the file in a slower, but stable manner.

Random remote read (Fig. 6b) is the most pessimistic case for any filesystem, as no prediction-based optimizations can be done when the access is completely random. This is where XCache is faster, thanks to its simpler architecture. The data is fetched from remote XRootD, served to the client and cached locally at the same time. On the other hand, Oneprovider transfers the data and writes it to the storage before informing the client that it is ready to be read. Nevertheless, consecutive reading of the same blocks (if required) would be faster via Oneclient (as shown in local read tests), and the total transfer size is about 15 times less for Oneclient thanks to variable-size data-chunks (64 KB vs. 1 MB default block size in XRootD). In this case, prefetching in XCache was disabled – it is worth mentioning that while Oneclient detects the reading pattern automatically, XCache needs to be restarted when prefetching settings change, making it less universally applicable. Moreover, the case of complete random read is quite rare, usually there is a pattern that causes some file fragments to be read more frequently than others.

Considering the above-mentioned, we have tested the two solutions in a more likely scenario when the file is read by bigger fragments (20 MB), but randomly chosen every time (Fig. 6c). Here, Oneclient and XRootDFS yielded similar measurements – a result of a compromise between prefetching and random read performance.

5.3 Discussion

The presented tests were performed on an elementary environment, where both systems were running on default settings and without any tuning. The purpose was to assess the impact of using variable-size data-chunks in comparison to a fixed-size block approach in XRootD, as well as the mechanisms introduced in Onedata in order to harmonize sequential and random data access. That said, these benchmarks should not be perceived as an absolute comparison of Onedata and XRootD + XCache performance. The results show that our data-chunk based solution achieves performance comparable to a state-of-the-art virtual filesystem based on a fixed block size, while offering additional features such as remote write and ensuring lower overheads in local data access. These tests should be treated as a proof-of-concept and their satisfying results are our incentive to further refine and optimize the proposal. We plan to perform tests in larger scale, on more complex environments, taking into account various parameters and other state-of-the-art filesystems for reference.

6 Conclusions and Future Work

In this paper we present our approach for harmonized data access in organizationally distributed environments. The solution has been implemented in a data access system called Onedata. The purpose is to provide a universal solution for data access that offers a satisfactory compromise between sequential and random read/write performance. It is achieved by combining a multi-layer architecture with Onezones with a novel approach to logical mapping of distributed file content – variable-size data-chunks, a layer over physical file blocks, objects etc. The system achieves good performance owing to support for direct storage access, which minimizes overheads during local data read and write, and a series of optimizations and mechanisms for efficient data-chunk management that boost remote data access: prefetching, prioritization of data transfers, event-based notifications of data-chunk registry changes, reading pattern recognition, detection of popular file fragments and data-chunks merging.

The universal character of Onedata means that it will not perform better than solutions dedicated for certain use-cases. Nevertheless, scientific computing is constantly developing, along with the diversity of datasets and data access patterns, which promotes integrated solutions that can cover various use-cases with sufficient performance. Furthermore, the sizes of datasets are ever-growing, which gradually makes pre-staging and full data replication obsolete, and encourages data access based on smaller file fragments. The proposed data-chunks are well suited to those needs, and have the great advantage of minimizing network and storage costs. The performance is comparable to the state-of-the-art XRootD virtual filesystem, coupled with XCache, at the same time the Onedata system offers more features, such as support for organizationally distributed environments and remote write, which are desired in scientific collaboration.

The future work includes further tests in larger scale and comparing more filesystems, following further optimizations: lowering the overheads of data transfer management, better access pattern recognition and increasing adaptability of the prefetching mechanisms to minimize the idle time during sequential reading.

Acknowledgments. This work has been partially supported by the funds of Polish Ministry of Science and Higher Education assigned to AGH University of Science and Technology and 2018–2020's research funds in the scope of the co-financed international projects framework (projects no. 3958/H2020/2018/2 and no. 3905/H2020/2018/2).

References

1. Allcock, W., Bester, J., Bresnahan, J., et al.: GridFTP: protocol extensions to FTP for the Grid. Glob. Grid Forum GFD-RP **20**, 1–21 (2003)
2. Axway AMPLIFY (2019). https://www.axway.com/en/products/amplify
3. Bauerdick, L., et al.: Using XRootD to federate regional storage. In: Journal of Physics: Conference Series, vol. 396, no. 4, p. 042009. IOP Publishing (2012)
4. Bhardwaj, D., Kumar, R.: A parallel file transfer protocol for clusters and grid systems. In: First International Conference on e-Science and Grid Computing, 2005. pp. 7–254. IEEE (2005)
5. DataNet Federation Consortium. http://datafed.org/
6. Dong, D., Herbert, J.: FSaaS: file system as a service. In: IEEE 38th Annual Computer Software and Applications Conference (2009)
7. Fajardo, E., Tadel, A., Tadel, M., Steer, B., Martin, T., Würthwein, F.: A federated XRootD cache. In: Journal of Physics: Conference Series, vol. 1085, p. 032025. IOP Publishing, September 2018
8. Fast Data Transfer. http://monalisa.cern.ch/FDT/
9. Folk, M., Heber, G., Koziol, Q., Pourmal, E., Robinson, D.: An overview of the HDF5 technology suite and its applications. In: Proceedings of the EDBT/ICDT 2011 Workshop on Array Databases, AD 2011, pp. 36–47. ACM, New York (2011). https://doi.org/10.1145/1966895.1966900, http://doi.acm.org/10.1145/1966895.1966900
10. Gong, Y., Hu, C., Xu, Y., Wang, W.: A distributed file system with variable sized objects for enhanced random writes. Comput. J. **59**(10), 1536–1550 (2016)
11. Hinrich, P., Grosso, P., Monga, I.: Collaborative research using eScience infrastructure and high speed networks. Futur. Gener. Comput. Syst. **45**(C), 161 (2015)
12. IBM Active File Management. https://www.ibm.com/support/knowledgecenter/en/STXKQY_4.1.1/com.ibm.spectrum.scale.v4r11.adv.doc/bl1adv_afm.htm
13. IBM MFT. https://www.ibm.com/customer-engagement/supply-chain/managed-file-transfer
14. Leong, D.: A new revolution in enterprise storage architecture. IEEE Potentials **28**(6), 32–33 (2009)

15. Li, Q., Sun, Z., Wei, Z., Sun, G.: A new data access mechanism for HDFS. In: Journal of Physics: Conference Series, vol. 898, p. 062018. IOP Publishing, October 2017
16. Onedata. https://onedata.org
17. Opioła, Ł., Dutka, Ł., Słota, R.G., Kitowski, J.: Trust-driven, decentralized data access control for open network of autonomous data providers. In: 16th Annual Conference on Privacy, Security and Trust (PST), pp. 1–10. IEEE, August 2018. https://doi.org/10.1109/PST.2018.8514209
18. Röblitz, T.: Towards implementing virtual data infrastructures - a case study with iRODS. Comput. Sci. (AGH) **13**(4), 21–34 (2012). http://dblp.uni-trier.de/db/journals/aghcs/aghcs13.html#Roblitz12
19. Serv-u. https://www.serv-u.com
20. Shvachko, K., Kuang, H., Radia, S., Chansler, R.: The hadoop distributed file system. In: 2010 IEEE 26th Symposium on Mass Storage Systems and Technologies (MSST), pp. 1–10. IEEE (2010)
21. Signiant. https://www.signiant.com
22. Słota, R., et al.: Storage management systems for organizationally distributed environments PLGrid PLUS case study. In: Wyrzykowski, R., Dongarra, J., Karczewski, K., Waśniewski, J. (eds.) PPAM 2013. LNCS, vol. 8384, pp. 724–733. Springer, Heidelberg (2014). https://doi.org/10.1007/978-3-642-55224-3_68
23. Szeredi, M.: Fuse: Filesystem in Userspace (2010). http://fuse.sourceforge.net
24. Weil, S.A., Brandt, S.A., Miller, E.L., Long, D.D., Maltzahn, C.: Ceph: a scalable, high-performance distributed file system. In: Proceedings of the 7th Symposium on Operating Systems Design and Implementation, pp. 307–320. USENIX Association (2006)
25. Wrzeszcz, M., Kitowski, J., Słota, R.: Towards trasparent data access with context awareness. Comput. Sci. **19**(2), 201–221 (2018)
26. Wrzeszcz, M., et al.: Consistency models for global scalable data access services. In: Wyrzykowski, R., Dongarra, J., Deelman, E., Karczewski, K. (eds.) PPAM 2017. LNCS, vol. 10777, pp. 471–480. Springer, Cham (2018). https://doi.org/10.1007/978-3-319-78024-5_41
27. Wrzeszcz, M., et al.: Metadata organization and management for globalization of data access with `Onedata`. In: Wyrzykowski, R., Deelman, E., Dongarra, J., Karczewski, K., Kitowski, J., Wiatr, K. (eds.) PPAM 2015. LNCS, vol. 9573, pp. 312–321. Springer, Cham (2016). https://doi.org/10.1007/978-3-319-32149-3_30
28. Zhang, L., Wu, W., DeMar, P., Pouyoul, E.: mdtmFTP and its evaluation on ESNET SDN testbed. Futur. Gener. Comput. Syst. **79**, 199–204 (2018)
29. Zhou, W., Han, J., Zhang, Z., Dai, J.: Dynamic random access for hadoop distributed file system. In: 32nd International Conference on Distributed Computing Systems Workshops (ICDCSW), pp. 17–22. IEEE (2012)

Towards Unknown Traffic Identification Using Deep Auto-Encoder and Constrained Clustering

Yongzheng Zhang[1,2], Shuyuan Zhao[1,2], and Yafei Sang[1,2](✉)

[1] Institute of Information Engineering, Chinese Academy of Sciences, Beijing, China
{zhangyongzheng,zhaoshuyuan,sangyafei}@iie.ac.cn
[2] School of Cyber Security, University of Chinese Academy of Sciences, Beijing, China

Abstract. Nowadays, network traffic identification, as the fundamental technique in the field of cybersecurity, suffers from a critical problem, namely "unknown traffic". The unknown traffic refers to network traffic generated by previously unknown applications (*i.e.*, zero-day applications) in a pre-constructed traffic classification system. The ability to divide the mixed unknown traffic into multiple clusters, each of which contains only one application traffic as far as possible, is the key to solve this problem. In this paper, we propose the *DePCK* to improve the clustering purity. There are two main innovations in our framework: *(i)* It learns to extract bottleneck features via deep auto-encoder from traffic statistical characteristics; *(ii)* It uses the flow correlation to guide the process of pairwise constrained k-means. To verify the effectiveness of our framework, we make contrast experiments on two real-world datasets. The experimental results show that the clustering purity rate of DePCK can exceed 94.81% on the ISP-data and 91.48% on the WIDE-data [1], which outperform the state-of-the-art methods: RTC [20], and k-means with log data [15].

Keywords: Unknown traffic · Deep auto-encoder · Bottleneck features · Pairwise constrained k-means

1 Introduction

The performance of network traffic identification directly affects network security and controllability, because it is a basic tool for network management tasks such as network monitoring, quality of service, traffic priority [20]. With the explosion of network applications, network traffic identification suffers from a critical problem, namely "unknown traffic". The unknown traffic is defined as network traffic generated by previously unknown applications (*i.e.*, zero-day applications) in a traffic classification system. The network traffic statistics of the Internet2 organization to the North American backbone network shows that nearly 50% of the traffic belongs to unknown traffic [16].

J. M. F. Rodrigues et al. (Eds.): ICCS 2019, LNCS 11536, pp. 309–322, 2019.
https://doi.org/10.1007/978-3-030-22734-0_23

The methods of fine-grained unknown traffic identification can be generally divided into three stages. First, extracting mixed unknown traffic from raw network traffic (including known traffic, and unknown traffic) [10,20,21]. Then, dividing the mixed unknown traffic into multiple clusters, each of which contains only one application traffic as far as possible [8,15,20]. Finally, identifying clusters through manually labeling [20] or association information (*e.g.*, DNS). To solve this problem, machine learning methods based on typical flow-statistical-features (*e.g.*, packet size, packet-interval) have been widely applied in unknown traffic identification, but most of them are aimed at solving the key problems of the first stage [7,9,10,14,21]. Previous research of the second stage has the following shortcomings: (*i*) Previous studies cannot perform beneficial feature selection just using unlabeled dataset [8,15,20]. (*ii*) Flow correlation is not entirely utilized to guide the clustering method [8,15,20]. All these issues will reduce the accuracy of clustering and affect the efficiency of unknown traffic identification.

In this paper, an unsupervised framework, which we call *DePCK*, is proposed to improve the dividing power of mixed unknown traffic (focusing on the second stage). To achieve traffic information embeddings without labels, it uses deep auto-encoder to build a self-supervised feature extraction model. To improve clustering performance, it fully uses flow correlation to guide the process of pairwise constrained clustering.

The major contributions can be summarized as follows:

- We propose the *DePCK*, an unsupervised framework for unknown traffic identification problem.
- We first use the bottleneck features (by mean of deep auto-encoder) to model unknown traffic.
- We use flow correlation (*i.e.*, 3-tuple of flow) to guide pairwise constrained clustering.
- The experiments of *DePCK* on two real-world datasets: ISP, and WIDE [1], show that the clustering purity rate of DePCK can exceed 94.81% on the ISP-data and 91.48% on the WIDE-data, which outperform the state-of-the-art methods: RTC [20], and k-means with log data [15].

The rest of this paper is structured as follows. A novel framework for network unknown traffic identification is proposed in Sects. 2, 3 and 4. Section 5 describes the datasets and evaluation metrics. Section 6 reports a large number of experiments and experimental results. Section 7 discusses related work in unknown traffic identification. Finally, Sect. 8 concludes the paper.

2 The *DePCK* Framework

Figure 1 provides the details of the *DePCK*. This framework includes two main modules: features extraction module and clustering module. In the features extraction module, according to the demonstrated capabilities [6] of feature learning and the theoretical function approximation properties [11] of deep neural networks (*DNNs*), we use a deep auto-encoder to train a self-supervised deep neural network and learn bottleneck features from unlabeled samples. This part is described in detail in Sect. 3. In the clustering module, we first extract the constrained relation between traffic flow and then use the *MPCKMeans* algorithm to match the unknown traffic identification scenario. The training data of this module is the bottleneck features of the features extraction module. This part is described in detail in Sect. 4.

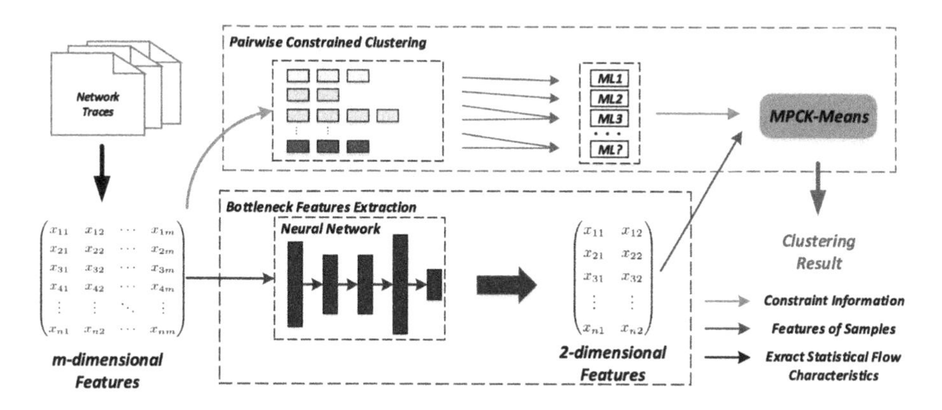

Fig. 1. The *DePCK* framework

3 Bottleneck Features Extraction

In this section, we describe the feature extraction module of *DePCK* based on deep auto-encoder, which can automatically train an unsupervised deep neural network and obtain the bottleneck features of the samples.

3.1 Deep Embedding

Network traffic classification schemes based on flow statistics generally train a classification model from a set of labeled data, which is composed of multiple statistical characteristics (*e.g.*, packet size, packet-interval) and class labels. Based on labels, most schemes usually first use supervised feature selection methods (*e.g.*, correlation coefficient, and covariance) to remove redundant and unrelated features.

Since unknown traffic has no labels, supervised feature selection methods are invalid to solve the unknown traffic identification problem. To tackle this problem, we use a neural network, which has demonstrated feature learning capabilities [18], to transform the feature with non-linear mapping. The non-linear transformation of features is to map the feature *space* from X to Z:

$$f_\theta : X \rightarrow Z \tag{1}$$

where Z is the latent *feature space*, and its dimensionality is smaller than the *space* X, and θ are parameters that can be automatically learned based on a deep neural network.

3.2 Training a Bottleneck Network

Deep neural networks have multiple hidden layers, which can train the input data through non-linear mapping and obtain hidden feature sets of samples. In our scene, without labeled data, we use an unsupervised deep auto-encoder to train the deep neural networks. Deep auto-encoder is an unsupervised neural network, which composed of multilayer auto-encoders.

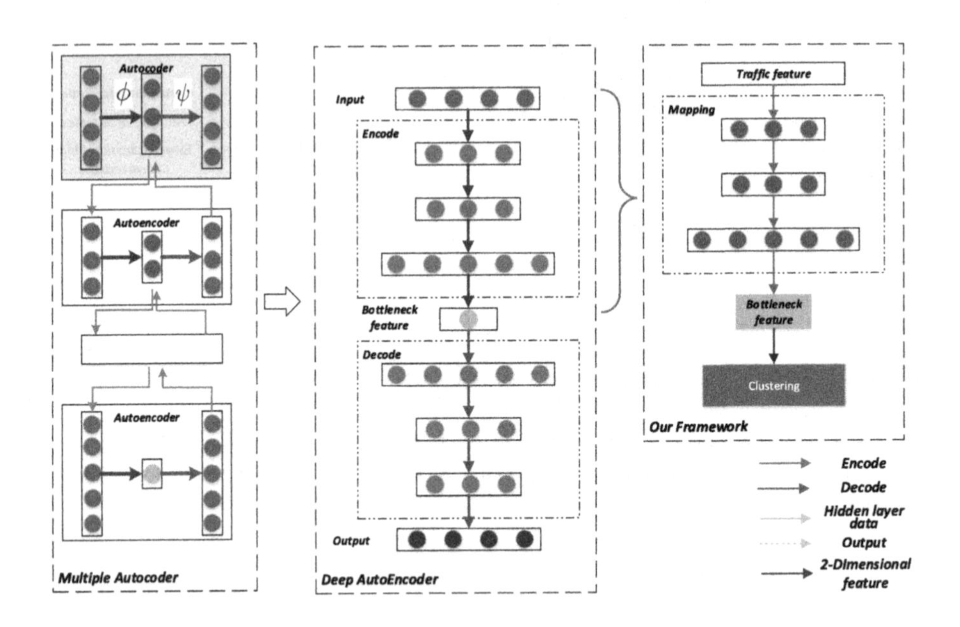

Fig. 2. Deep auto-encoder structure

An auto-encoder is a type of artificial neural network used to learn efficient data codings in an unsupervised manner [12]. The aim of an auto-encoder is to learn a representation (encoding) for a set of data. Architecturally, the form of an auto-encoder is a feedforward, non-recurrent and two-layer neural network. As shown in the green part of Fig. 2, an auto-encoder always consists of two parts, the encoder and the decoder, which can be defined as:

$$\psi : \alpha \rightarrow \beta \tag{2}$$

$$\phi : \beta \rightarrow \alpha \tag{3}$$

$$\psi, \phi = argmin_{\psi,\phi} ||\alpha - (\psi \circ \phi)\alpha||^2 \tag{4}$$

where the encoder stage of an auto-encoder takes the input $x \in R^d = \alpha$ and maps it to $z \in R^P = \beta$, and the decoder stage of the auto-encoder maps z to the reconstruction x' of the same shape as x:

$$z = r_1\{W_1 x + b^1\} \tag{5}$$

$$x' = r_2\{W_2 z + b^2\} \tag{6}$$

where z is latent representation, r_1 and r_2 are element-wise activation function such as a sigmoid function or a rectified linear unit. W_1 and W_2 are weight matrix, b_1 and b_2 are bias vector. In the model, all activation functions are rectified linear units (ReLUs).

The training process of the auto-encoders is to minimize the loss function J. The loss function is defined as:

$$J(x, x') = \sum_{x \in D} L^p(x, x') \tag{7}$$

where L^p is reconstruction errors, here we use the square of Euclidean norm: $||x - y||^2$. D is the dataset.

As shown in the middle part of Fig. 2, the deep auto-encoder is a deep neural network with multiple layers. After training auto-encoders by greedy layer-wise training, we connect all the encoders in series and then combine all the decoders in the opposite direction to form a deep auto-encoder. When designing the deep network structure, we set the middle layer with the minimum dimension in all layers to build the bottleneck features, because the bottleneck features have recently found success in a variety of speech recognition tasks [18]. Then, we get the final model by discarding and use the model as the initial mapping between the *raw feature space* and the *bottleneck feature space*.

4 Pairwise Constrained Clustering

In this section, we first describe the constrained relation between network traffic and then introduce the modified pairwise constrained clustering algorithm based on flow correlation.

4.1 Correlations Between Network Traffic

In Transmission Control Protocol/Internet Protocol (TCP/IP) model, IP flow, a series of data packets transferred between two programs, is the basic unit for end-to-end data transfer. In the program, the system determines an IP flow through the IPv4 five-tuple. A five-tuple refers to a set of five different values that uniquely identifies a UDP/TCP session. It includes a source IP address/port number, destination IP address/port number and transport protocol. An Internet Protocol address (IP address) is a numerical label assigned to each device connected to a computer network that uses the Internet Protocol for communication. A port is an endpoint of communication in an operating system, which identifies a specific process or a type of network service running on that system. Hence, a port can be used with an IP address of a host and the transport layer protocol for communication. For example, to transfer a file to a remote computer, one could specify the machine itself by IP address, use TCP for transport, and the FTP file server service on that computer on port 20.

We assume that the service provided by a particular port lasts for a certain period. In this example, the flows that have the same three-tuple (service IP address, service port number, and transport protocol) can be considered to belong to the same protocol. This assumption is typically valid for the Internet because of the port-reuse restriction rule enforced by operating systems, in which a local port number will become unavailable for some time after closing unless a particular program is bound to it [17]. Therefore, we can use the 3-tuple of flows to obtain constrained dataset.

The flows' constrained relation can be used to guide the clustering process when the unknown traffic is identified based on the clustering algorithm. In the clustering process, if there is a large number of associated flows between two independent clusters, the clustering algorithm can determine that the correlation between the two clusters is strong. Based on this idea, we propose an unknown traffic identification method based on pairwise constrained clustering algorithm.

4.2 Modified PCKMeans

In previous studies, unknown traffic clustering methods based on statistical characteristics cannot make good use of flow correlation [8,15,20]. To make full use of the flows' constrained relation, we propose the modified pairwise constraint clustering algorithm based on PCKMeans [5].

Pairwise Constraint Conditions. PCKMeans, an improved k-means algorithm, uses the prior knowledge of the data to guide the clustering process and gets better clustering results. Consequently, in addition to the distance between samples in the data, this algorithm uses pairwise *must-link* (ML) and *cannot-link* (CL) constraints to guide clustering. ML is a set of must-link pairs and CL is a set of cannot-link pairs. if $(x_i, x_j) \in ML, x_i$ and x_j should be assigned to the same cluster. Conversely, if $(x_i, x_j) \in CL, x_i$ and x_j should be assigned to the different cluster.

In our *DePCK*, we can use the 3-tuple of network flows to construct the ML set but can not build the CL set.

Modified PCKMeans. The PCKMeans algorithm implements the use of ML set and CL set by adding a constraint violation penalty term to the objective function of the k-means algorithm. In the case of a given dataset D, a set of must-link constraints ML, a set of cannot-link constraints CL, the PCKMeans algorithm can minimize objective function by giving the weights corresponding to the ML and CL respectively. The objective function J of PCKMeans can be computed as:

$$J_{ML} = \sum_{(x_i,x_j)\in ML} w_{ij} I(C_i \neq C_j) \tag{8}$$

$$J_{CL} = \sum_{(x_i,x_j)\in CL} \overline{w}_{ij} I(C_i = C_j) \tag{9}$$

$$J = \frac{1}{2}\sum_{x_i\in D} ||x_i - c_i||^2 + J_{ML} + J_{CL} \tag{10}$$

where x_i and x_j are the single sample of the dataset; C_i and C_j are the assigned cluster of x_i and x_j respectively, w_{ij} and $\overline{w}_{ij}$ are two sets that give weights

---

Algorithm 1. Modified PCKMeans

Input: $D = \{x_i\}_{i=1}^{n}$:set of samples; $ML = \{(x_i, x_j)\}$: set of must link samples; k: number of clusters; w: weight of constraints;
Output: J_{min}:divide the dataset into k clusters and have the smallest J value; $C = C_1, C_2, ..., C_n$:the set of clusters.
1: initialize the Centroids $\{c_i\}_{i=1}^{k}$ of k clusters at random
2: **repeat**
3: **for** $x_i \in D$ **do**
4: **for** $C_j \in C$ **do**
5: Calculate the objective function: $J^j = \sum_{x_i\in D} ||x_i - c_j||^2$
6: **end for**
7: assign sample x_i to the cluster j where $J^j{=}argmin(J^*)$
8: **end for**
9: **for** $c_i \in \{c_i\}_{i=1}^{k}$ **do**
10: recalculate the Centroids $\{c_i\}_{i=1}^{k}$ of k clusters: $c_i = \frac{\sum_{x_i\in D} x_i}{|C_i|}$
11: **end for**
12: **until** none of the Centroids $\{c_i\}_{i=1}^{k}$ of k clusters changes
13: **repeat**
14: **for** $x_i \in D$ **do**
15: **for** $C_j \in C$ **do**
16: Calculate the objective function: $J_{ML} = \sum_{(x_i,x_j)\in ML} w_{ij} I(C_i \neq C_j)$
17: Calculate the objective function: $J^j = \sum_{x_i\in D} ||x_i - c_i||^2 + J_{ML}$
18: **end for**
19: assign sample x_i to the cluster j where $J^j{=}argmin(J^*)$
20: **end for**
21: **for** $c_i \in \{c_i\}_{i=1}^{k}$ **do**
22: recalculate the Centroids $\{c_i\}_{i=1}^{k}$ of k clusters: $c_i = \frac{\sum_{x_i\in D} x_i}{|C_i|}$
23: **end for**
24: **until** none of the Centroids $\{c_i\}_{i=1}^{k}$ of k clusters changes

---

corresponding to the ML and CL respectively, $I(\cdot)$ is the indicator function, with $I(true) = 1$ and $I(false) = 0$. In Eq. (10), c_i is centroid of C_i.

Because there is no CL set of network traffic, we do not need to add J_{CL} to the objective function J. The objective function J of our model is defined as:

$$J = \frac{1}{2} \sum_{x_i \in D} ||x_i - c_i||^2 + J_{ML} \tag{11}$$

In the PCKMeans algorithm proposed by Basu *et al.* [5], the initialization phase strategy is designed as follows: Firstly, using the *must-link* set to construct λ neighborhood sets $\{N_p\}_{p=1}^{\lambda}$, then using the information of neighborhood sets to initialize the center centroids of k clusters as much as possible. The most significant advantage of this strategy is that the *must-link* set directly determines the distribution of the clusters. However, in the unknown traffic identification scenario, this advantage will have the opposite effect because it prevents the algorithm from discovering new traffic protocol or application from mixed traffic.

In order to solve the above problem, we propose an improved pairwise constraint clustering algorithm (MPCKMeans) Algorithm 1.

In this algorithm, we first use the k-means algorithm to complete the clustering of data D and obtain the centroid of the clustering clusters as the initial centroid of the next stage. Then, based on the result of k-means, we use the *must-link* constraints to guide the clustering process. The improved algorithm not only makes full use of the pairwise constraints, but also has excellent ability to discover unknown protocols.

5 Preliminaries

In this section, we first introduce how we build the ground truth dataset using traffic traces. Then, we show the assessment criteria.

5.1 Dataset

In this paper, two Internet traffic traces, WIDE [1] and ISP, are used for our experimental study. Table 1 shows the main detail of traffic traces.

Table 1. Traffic traces

Trace	Data time	Duration	Type	Volume
ISP-data	2015-08-17	1 day	Edge	130.7 GB
WIDE-data	2012-03-30	5 hours	Backbone	482.8 GB

The ISP trace was collected from our routers in the edge of a campus network on August 17, 2015 from 1 am to 12 pm. This trace consists of 3 million flows with full packet payload. The WIDE trace was captured by MAWI Working Group in March 2012 that was during 5 h. In this trace, all the IP addresses are anonymized and each packet just includes forty bytes of application layer payload.

We used two steps to obtain the ground truth dataset. Firstly, we used an open source tool nDPI [2] to label the ISP trace. Besides we used the port-based method to enhance the reliability of the dataset. Because the WIDE trace does not include the full payload, we directly use the port-based approach to label this dataset. Then we used tool Netmate to extract statistical flow characteristics. This tool's job is to classify packets into flows and to calculate the statistics of flows.. When building the experimental dataset, we will calculate as many features as possible. Finally, we select 28 flow features, which are described in Table 2.

From the ISP trace, eight protocols, BT, DNS, HTTP, IMAP, NTP, SSDP, SSL, LLMNR, were extracted and constituted the ISP-data. Our sampling rules

Table 2. Network flow statistical features

Category of features	Description of feature	No. of feature
Packets	Number of packets transferred in unidirection	2
Bytes	Volume of bytes transferred in unidirection	2
Packets size	Min, Max, Mean and Standard deviation of packets size in unidirection	8
Inter packet time	Min, Max, Mean and Standard deviation of inter packet time in unidirection	8
Connection duration	Min, Max, Mean and Standard deviation of subflow activity time	4
Idle time	Min, Max, Mean and Standard deviation of subflow idle time	4
Total		**28**

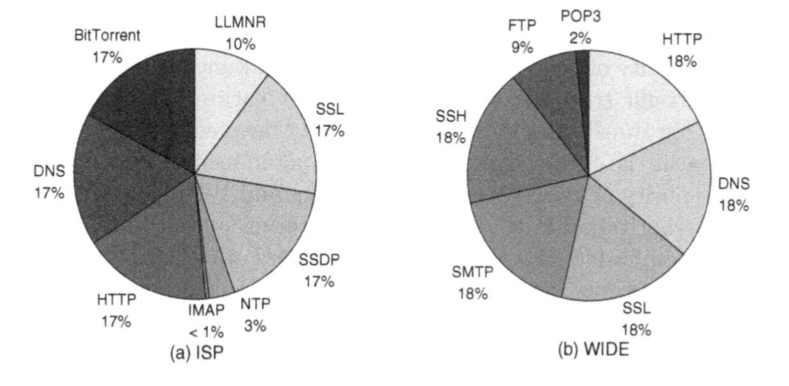

Fig. 3. Class distribution of the dataset

are that: randomly sampling 40K flows from each protocol if it contains more than 40k flows, otherwise sampling all the flows. The experimental dataset consists of 233k flow randomly sampled from the initial traffic dataset, which is described in Fig. 3(a). From the WIDE trace, seven protocols, POP3, FTP, SSH, SMTP, SSL, DNS, HTTP, were extracted and constituted the WIDE-data. This experimental dataset consists of 56k flows with the rule of randomly sampling up to 10K flows from each protocol, which are described in Fig. 3(b). During experiments, we simulated the problem of unknown applications. Both ISP-data and WIDE-data represent mixed unknown traffic datasets.

5.2 Assessment Criteria

To evaluate the effectiveness of our method, we focus on clustering purity. The clustering purity is defined as the average percentage of the dominant class label in each cluster [3]. To calculate the purity, each cluster is assigned to the category which is most frequent in the cluster. The definition of clustering purity is shown in Eq. (12).

$$P(C,S) = \frac{1}{|D|}\sum_{i=1}^{k}\max_{j}|c_i \cap s_j| \tag{12}$$

where k is the number of clusters, $C = \{c_1, c_2, ..., c_i\}$ is the set of clusters and $S = \{s_1, s_2, ..., s_j\}$ is the set of classes.

6 Performance Results

In this section, we first compare our feature extraction method with traditional approaches to explain why we use deep embedding. Secondly, we prove the effectiveness of the modified PCKMeans. Finally, we present and discuss the comprehensive experiments. To evaluate the effectiveness of the method, we use labeled data to simulate unknown traffic. In the experiment, the number of clusters is the only input parameter, which ranges from 10 to 100. Every experiment is repeated 100 times to ensure the reliability of the results.

6.1 Why We Use Deep Embedding

To show the validity of deep embedding, we used the k-means algorithm to identify unknown traffic traces based on three kinds of feature sets: initial statistical features [20], log transformation features [15], and deep embedding features. For each k, we repeat the clustering with different random seeds. Figures 4(a) and (b) illustrate the purity of clustering on the ISP-data and WIDE-data, respectively.

Our clustering target is to obtain high clustering purity with small cluster number. The results indicate that when the number of clusters is 10, that deep embedding outperforms initial features and log transformation features clustering purity on two datasets. This satisfies the original intention of our algorithm design. Besides, with the increase of cluster number, the clustering purity of deep embedding is almost always higher than that of the other two methods on two datasets.

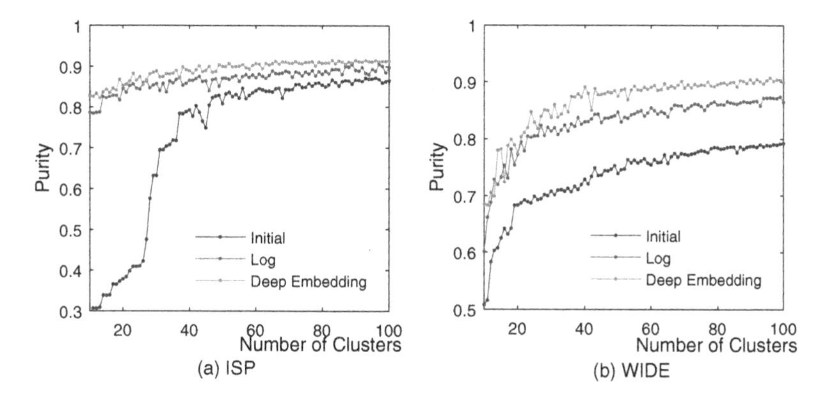

Fig. 4. Clustering purity comparison of different data preprocessing methods

6.2 Benefits of Modified PCKMeans

To show the ability of MPCKMeans, we perform the following experiments. In the case bottleneck features as the original input, we compare the effectiveness of the MPCKMeans algorithm and the k-means algorithm for unknown traffic identification.

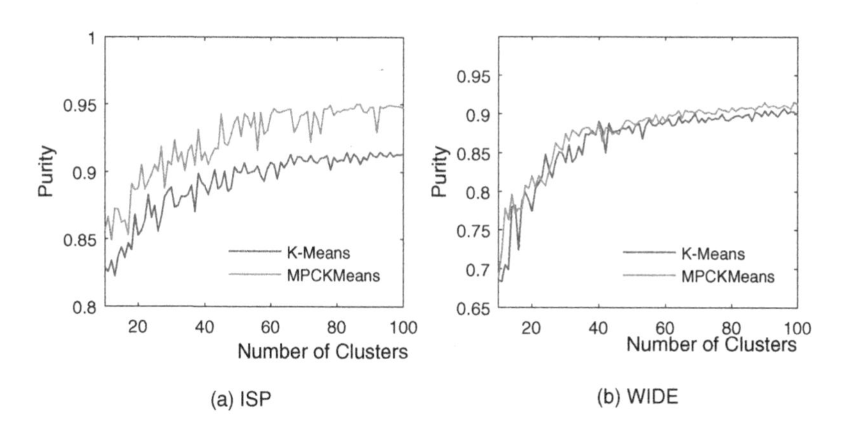

Fig. 5. Clustering purity comparison of different algorithms

The experimental results are shown in the Fig. 5. Viewing the trend as a whole, the experimental results of *DePCK* are almost always better than the results of k-means. Individually, Fig. 5(a) shows the results on the ISP-data. No matter how the parameter changes, the clustering purity of *DePCK* is always higher than that of k-means by about 5%. In Fig. 5(b), although the advantage of *DePCK* is not as obvious as in Fig. 5(a), the overall result is that our method is superior to the k-means. The results show that the MPCKMeans algorithm can get more pure clusters and is stable in different datasets.

6.3 Performance of *DePCK*

We perform a series of experiments on real-world traces to evaluate the proposed framework and the state-of-the-art methods (k-means [20]and log k-means [15]). All methods use the same datasets and parameters. The parameter ranges from ten to one hundred: $k = 10, 20, ..., 50, 60, ..., 100$. Table 3 shows the experimental results. When the parameters are the same, our method *DePCK* can always get the best experimental results on both datasets. The *DePCK* achieves over 94.81% clustering purity on the ISP-data and 91.48% clustering purity on the WIDE-data, which is obviously superior to other methods.

Table 3. Experiment results (the best results are in bold)

K	ISP-data			WIDE-data		
	k-means	log k-means	*DePCK*	k-means	log k-means	*DePCK*
10	29.50%	78.60%	**85.64%**	50.82%	63.12%	**82.02%**
20	37.80%	84.63%	**88.74%**	68.35%	80.26%	**82.02%**
30	63.19%	85.75%	**90.49%**	70.79%	84.36%	**87.49%**
40	79.21%	86.59%	**91.50%**	72.89%	85.98%	**88.74%**
50	80.79%	87.09%	**94.09%**	73.91%	87.68%	**88.83%**
60	83.97%	87.23%	**94.18%**	75.42%	88.43%	**89.09%**
70	84.18%	88.59%	**94.27%**	77.29%	87.94%	**90.59%**
80	85.40%	88.12%	**94.52%**	78.43%	88.96%	**90.88%**
90	86.66%	89.09%	**94.81%**	78.69%	89.39%	**91.22%**
100	86.44%	89.73%	**94.73%**	79.22%	89.44%	**91.48%**

7 Related Work

How to classify network traffic into known protocols and applications was extensively focused in past studies, but few discussed the identification of unknown traffic. We briefly review the work related to the unknown traffic identification.

In previous studies, (N+1)-Class traffic classification model was proposed to solve the problem of unknown traffic. N represents the number of known classes, and one represents all unknown classes. Erman *et al.* [9] proposed a semi-supervised classification method that can accommodate both known and unknown applications. Then Casas *et al.* [7] use the ensemble clustering technique to improve the semi-supervised method. Later, Erman's work is introduced to classify encrypted traffic by using the composite feature set and combining the first 40-B payload with statistical features of the flow level [14]. Besides, Fu *et al.* construct multiple one-class classifiers to divide traffic into N classes and an unknown class [10]. Although the way that all unknown traffic is identified as one class helps to increase the accuracy of a traffic classification system, the system's ability to efficiently achieve fine-grained identification of unknown traffic is feeble.

In theory, unsupervised methods can solve network traffic identification problem. Unsupervised methods have been widely applied in the network traffic classification, which deserves our reference. In [19], Zander *et al.* presented a classification method based on AutoClass program that uses the Expectation-Maximization (EM) algorithm and mixture models. Then Erman *et al.* [8] applied two unsupervised clustering algorithms, namely k-means and DBSCAN, to classify network traffic and compare them to the previously used AutoClass algorithm. The experimental results showed that both k-means and DBSCAN work very well and much more quickly than AutoClass. Similarly, in Liu *et al.* [15], the author adopted feature selection to find an optimal feature set and log transformation to improve the k-means accuracy. The report of this method showed that overall accuracy was up to 80%, and, after a log transformation, the accuracy was improved to 90% or more. This approach is superior to the previous methods. Zhang's work [20] is based on Erman's semi-supervised classification method. For fine-grained unknown traffic, an update module, based on k-means algorithm, is proposed to finely classify unknown traffic in the system. Besides, some other well-known unsupervised algorithms, such as Fuzzy C-means [13] and hierarchical clustering [4], were also used for traffic classification.

8 Conclusion and Future Work

In this paper, we proposed a robust scheme, the *DePCK*, for unknown traffic identification based on deep auto-encoder and modified pairwise constrained k-means. To the best of our knowledge, this is the first application of bottleneck features and pairwise constrained k-means in this area. Extensive experimental results reveal that *DePCK* achieves better performance compared to the state-of-the-art methods on real-world datasets. In addition, the *DePCK* can deal with textual protocols and binary protocols. To further solve the unknown traffic identification problem, our future research will focus on how to automatically determine the number of traffic clusters.

Acknowledgment. This work was supported by the National Natural Science Foundation of china (No. 61572496 and No. U1736218).

References

1. http://mawi.wide.ad.jp/mawi/
2. https://www.ntop.org/products/deep-packet-inspection/ndpi/
3. Aggarwal, C.C.: A human-computer interactive method for projected clustering. IEEE Trans. Knowl. Data Eng. **16**(4), 448–460 (2004)
4. Bacquet, C., Zincir-Heywood, A.N., Heywood, M.I.: Genetic optimization and hierarchical clustering applied to encrypted traffic identification. In: 2011 IEEE Symposium on Computational Intelligence in Cyber Security (CICS), pp. 194–201. IEEE (2011)

5. Basu, S., Banerjee, A., Mooney, R.J.: Active semi-supervision for pairwise constrained clustering. In: Proceedings of the 2004 SIAM International Conference on Data Mining, pp. 333–344. SIAM (2004)
6. Bengio, Y., Courville, A., Vincent, P.: Representation learning: a review and new perspectives. IEEE Trans. Pattern Anal. Mach. Intell. **35**(8), 1798–1828 (2013)
7. Casas, P., Mazel, J., Owezarski, P.: Minetrac: Mining flows for unsupervised analysis & semi-supervised classification. In: Proceedings of the 23rd International Teletraffic Congress, pp. 87–94. International Teletraffic Congress (2011)
8. Erman, J., Arlitt, M., Mahanti, A.: Traffic classification using clustering algorithms. In: Proceedings of the 2006 SIGCOMM Workshop on Mining Network Data, pp. 281–286. ACM (2006)
9. Erman, J., Mahanti, A., Arlitt, M., Cohen, I., Williamson, C.: Offline/realtime traffic classification using semi-supervised learning. Perform. Eval. **64**(9–12), 1194–1213 (2007)
10. Fu, N., Xu, Y., Zhang, J., Wang, R., Xu, J.: FlowCop: detecting "stranger" in network traffic classification. In: 2018 27th International Conference on Computer Communication and Networks (ICCCN), pp. 1–9. IEEE (2018)
11. Hornik, K.: Approximation capabilities of multilayer feedforward networks. Neural Netw. **4**(2), 251–257 (1991)
12. Liou, C.Y., Cheng, W.C., Liou, J.W., Liou, D.R.: Autoencoder for words. Neurocomputing **139**, 84–96 (2014)
13. Liu, D., Lung, C.H.: P2P traffic identification and optimization using fuzzy c-means clustering. In: 2011 IEEE International Conference on Fuzzy Systems (FUZZ), pp. 2245–2252. IEEE (2011)
14. Liu, H., Wang, Z., Wang, Y.: Semi-supervised encrypted traffic classification using composite features set. J. Netw. **7**(8), 1195 (2012)
15. Liu, Y., Li, W., Li, Y.C.: Network traffic classification using k-means clustering. In: Second International Multi-Symposiums on Computer and Computational Sciences, IMSCCS 2007, pp. 360–365. IEEE (2007)
16. Statistics, I.N.: Internet2 netflow statistics (2011)
17. Wang, Y., Xiang, Y., Zhang, J., Zhou, W., Wei, G., Yang, L.T.: Internet traffic classification using constrained clustering. IEEE Trans. Parallel Distrib. Syst. **25**(11), 2932–2943 (2014)
18. Yaman, S., Pelecanos, J., Sarikaya, R.: Bottleneck features for speaker recognition. In: Odyssey 2012-The Speaker and Language Recognition Workshop (2012)
19. Zander, S., Nguyen, T., Armitage, G.: Automated traffic classification and application identification using machine learning. In: The IEEE Conference on Local Computer Networks, 30th Anniversary, pp. 250–257. IEEE (2005)
20. Zhang, J., Chen, X., Xiang, Y., Zhou, W., Wu, J.: Robust network traffic classification. IEEE/ACM Trans. Netw. (TON) **23**(4), 1257–1270 (2015)
21. Zhao, S., Zhang, Y., Chang, P.: Network traffic classification using tri-training based on statistical flow characteristics. In: 2017 IEEE Trustcom/BigDataSE/ICESS, pp. 323–330. IEEE (2017)

How to Compose Product Pages to Enhance the New Users' Interest in the Item Catalog?

Nicollas Silva[1(✉)], Diego Carvalho[2(✉)], Adriano C. M. Pereira[1(✉)], Fernando Mourão[3(✉)], and Leonardo Rocha[2(✉)]

[1] Universidade Federal de Minas Gerais, Belo Horizonte, Brazil
{ncsilvaa,adrianoc}@dcc.ufmg.br
[2] Universidade Federal de São João del Rei, São João del Rei, Brazil
{dcarvalho,lcrocha}@ufsj.edu.br
[3] Seek AI Labs, Belo Horizonte, Brazil
fernando.mourao@catho.com

Abstract. Converting first-time users into recurring ones is key to the success of Web-based applications. This problem is known as Pure Cold-Start and it refers to the capability of Recommender Systems (RSs) to provide useful recommendations to users without historical data. Traditionally, RSs assume that non-personalized recommendation can mitigate this problem. However, several users are not interested in consuming just biased-items, such as popular or best-rated items. Then, we introduce two new approaches inspired by user coverage maximization to deal with this problem. These coverage-based RSs reached a high number of distinct first-time users. Thus, we proposed to compose the product's page by mixing complementary non-personalized RSs. An online study, conducted with 204 real users confirmed that we should diversify the RSs used to conquer first-time users.

Keywords: Non-personalized RS · Pure Cold-Start problem · Users coverage

1 Introduction

Recommender Systems (RSs) have assumed a prominent role in Web-based applications, affecting decisively distinct business phases, such as the acquisition and retention of users. In the retention phase, the performance of current prediction models is extremely satisfactory [2]. A recent study highlighted RSs as the main responsible for 35% of sales on Amazon, 2/3 of the movies watched on Netflix and 38% more click-through on Google News [9]. However, the user acquisition phase has not received much attention in recent years. In this phase, RSs help to consolidate the users' first impression about the item catalog, which may influence the conversion rate of first-time users into clients [11].

J. M. F. Rodrigues et al. (Eds.): ICCS 2019, LNCS 11536, pp. 323–338, 2019.
https://doi.org/10.1007/978-3-030-22734-0_24

In the literature, this problem is called Pure Cold-Start and it remains poorly exploited by researchers who just consider the Cold-Start problem [16]. Despite this, the Pure Cold-Start problem has grown in real domains since several users became to reach systems through incognito navigation or with social networks disable due to privacy issues [20]. In this context, it is not easy to capture personal information from cookies, social networks or browsing history. For this reason, the users are always unknown, and the system always faces the challenge of recommending useful items for them who do not have any information [6].

In this work, we identify an opportunity for improvements on state-of-the-art non-personalized RSs that address the Pure Cold-Start. The literature assumes that items biased by popularity, recency or positive ratings are enough to attract first-time users. We show that a non-negligible portion of these users is not interested in consuming such items in some domains. Hence, exploiting biased-items RSs to compose product pages is not the best method to conquer distinct first-time users. This work aims to answer a promising research question: *How to compose product pages to attract the maximum number of first-time users?*

We hypothesize that to satisfy distinct first-time users, RSs should balance recommendations that suit distinct user profiles. Aiming to validate this hypothesis, we evaluated three state-of-the-art RSs and two novel strategies, proposed by this work. Traditional RSs are inspired by the utility of biased-items [14] - (1) *Most Popular*; (2) *Best-Rated*; and (3) *Recent Items*. We propose two novel non-personalized RSs inspired by user coverage maximization, already exploited to address other RSs related problems: (1) *Max-Coverage*: selects items that cover a large number of distinct users, such as addressed in [15]; and (2) *Niche-Coverage*: selects items that cover distinct user profiles [13]. Complementary of our last work [20], we propose an extension of the Niche-Coverage method and deeper analyzes than previous ones to consolidate their practical application.

Offline assessments on four popular datasets from e-commerce and entertainment domains evinced that the methods are complementary. While traditional RSs retrieved potentially relevant items, obtaining high utility, the new RSs enhanced diversity. Further, the new RSs reached a higher number of distinct first-time users. Therefore, mixing these complementary RSs to compose product pages is a promising answer for our research question in real scenarios. To confirm this assumption, we conducted an online study with 204 real users. We build an A/B test comparing traditional RSs (scenario A) against complementary RSs (scenario B). For each scenario, we asked the users to select movies of their interest and answer questions about the list of items. The results highlighted as main contribution a clear message: we should combine complementary non-personalized RSs in product pages.

2 Related Work

In the literature, the term Pure Cold-Start refers to a subtask of the Cold-Start problem [10]. Despite being closely related, both problems should be addressed differently. Whereas in the Cold-Start problem exists a lot of strategies to deal with small consumption history of users, in the Pure Cold-Start there are few strategies to handle first-time users [1]. We identified three main categories of RSs designed to deal with the Pure Cold-Start problem: (1) Knowledge RSs; (2) Social Filtering RSs; and (3) Non-Personalized RSs.

Knowledge RSs try to acquire user information using small questionnaires in user-web interaction. So, several studies have been proposed to improve the classical RSs with this information [5,21]. However, *He et al.* [5] argue that the quality of recommendations depends on information provided by users, who may not be able to define clearly their preferences. In turn, Social Filtering RSs exploit 'external' information about users, such as social or demographic data. In general, these RSs use hybrid methods to mitigate the Cold-Start problem [16,18]. Despite the advantages obtained, these approaches are not commonly used in e-commerce scenarios, because many users are not willing to provide demographic information before buying products.

Non-Personalized RSs are the predominant solution in real-world scenarios due to simplicity, domain independence, and efficiency. These RSs derive global information about items and users [2], exploiting key features related to consumption, such as popularity, ratings, and release/consumption recency [14]. However, these strategies are targeted to specific profiles, biasing users interested in items that satisfy a large portion of a population. To balance the recommendations for all users, the concept of result diversification has been introduced from the field of IR [23]. In general, the items recommended are re-ordered on the basis of a given diversification objective [22]. In this work, to attract more first-time users, we propose to diversify the items with user-coverage.

3 Handling First-Time Users

The Pure Cold-Start problem occurs when the system does not have any information about users. For this reason, first, we simulate these scenarios and, next, discuss the main approaches that address this problem.

3.1 First-Time Users Definition

First, we select the MovieLens 1M and 10M, and the CiaoDVD and Amazon datasets, described in Table 1, to simulate entertainment and e-commerce scenarios. Next, we simulate the first-time users in our datasets as follows. We sort the users considering the timestamp from the first item consumed in their historical data. Then, we selected the last 20% of users as the first-time ones, since they present the most recent actions in each collection. So, we used all historical data of the selected users to compose test sets and removed them from the training sets used as inputs by the evaluated RSs. The number of users selected from each dataset is available on the last column of Table 1.

Table 1. Datasets - general information

Datasets	Users	Items	Sparsity	Genres	First-time
ML-1M	6,040	3,952	95.82%	18	1,277
ML-10M	69,878	10,283	98.60%	20	10,633
CiaoDVD	17,615	16,621	99.97%	17	3,523
Amazon	8,057	26,729	99.92%	471	1,612

3.2 Biased-Item Models

In Pure Cold-start problem, the state-of-the-art RSs are based on biased-items recommendations. These models assume that items biased by popularity, recency or positive ratings are useful to first-time users. For this reason, we implement and evaluate these non-personalized RSs, popularly used in real domains:

- **Popularity (Pop):** selects the k most popular items in the domain. The popularity is estimated by the number of distinct users who consumed an item i.
- **Best-Rated (BestR):** recommends the k best evaluated items in the domain. Basically, we sum the items' ratings and divide its by the number of users.
- **Recent Items (RecItems):** recommends the k last items consumed by users, calculated based on timestamp.

Generally, items recommended by these RSs are concentrated in the *head* of popularity distribution. However, several studies have discussed the *long tail* phenomenon in real scenarios such as Amazon and Netflix [7]. In these scenarios, tail products generate a significant fraction of the total revenue in aggregate and can boost head sales by offering consumers both their mainstream and specific tastes. For this reason, we suppose that there are many users interested in other items beyond the recommended by these state-of-the-art RSs. Hence, for every dataset, we select the top-100 items from Popularity, Best-Rated, and Recent Items, and count the number of biased-items in each user's consumption history. The values of each RS are normalized by the history size of each user and plotted in Fig. 1. Values close to 100% indicate that user consumption is strongly biased by the items recommended and values close to 0% show that user consumption is formed by other items.

In each ranking, we observe three user behaviors: (1) users who prefer biased-items (bias more than 70% - head of distribution); (2) users who prefer other items (bias less than 30% - tail of distribution); and (3) users who mix biased-items and others (bias around 30% and 70% - middle). These results show a non-negligible portion of users with (2) and (3) behaviors, i.e., interested in other items beyond the selected by these RSs. Specifically, in the e-commerce domains, around 40% to 60% of the users do not have any biased-item in your consumption history. Therefore, these results point out an opportunity for improvements on state-of-the-art non-personalized RSs that address the Pure Cold-Start problem.

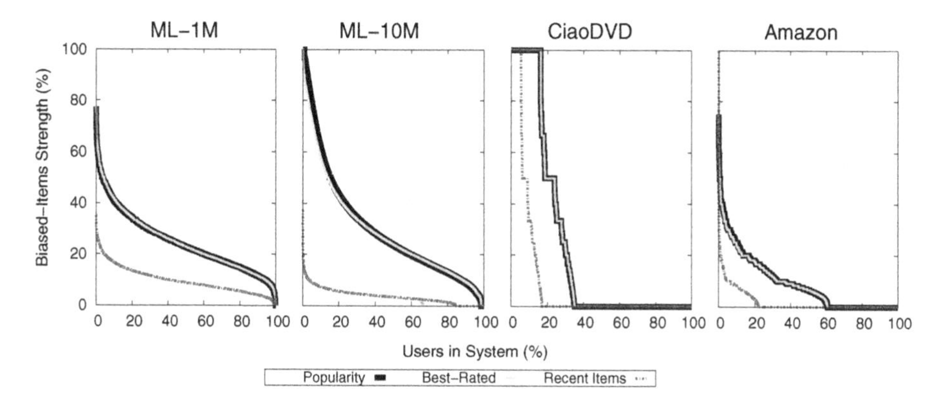

Fig. 1. Percentage of popular items consumed by all users

3.3 Coverage-Based Models

Exploring the improvements opportunity, we propose two non-personalized RSs based on user coverage maximization. Max-Coverage is inspired by a NP-hard problem (Maximum k-Coverage), already exploited to address other RSs related problems [15]. In turn, we propose a new method, called Niche-Coverage, which aims to apply Max-Coverage in a distinct niche of users found by any clustering approach. Both methods consider that maximizing user coverage is a relevant approach to handle the Pure Cold-Start problem.

Max-Coverage: This strategy models the recommendation domains in sets of items and users, and applies the Maximum k-Coverage problem to find the items for first-time users. Formally, considering a universe of elements $U = \{u_1, ..., u_m\}$, a family of sets $F = \{S_1, ..., S_n\}$, where each set S_i is a subset of U and an integer k, the *Maximum k-Coverage* consists to find a subfamily $F^* \subseteq F$ such that $|F^*| \leq k$ and the number of covered elements $|\bigcup_{S \in F^*} S|$ is maximized, i.e. using up to k sets, cover as many elements as possible.

In a recommendations domain, we model the domain based on the users-items interaction, creating sets of users and items. Then, let $U = \{u_1, ..., u_m\}$ as the users that previously have consumed items, we create the set $S = \{S_1, .., S_n\}$, where each element S_i is a subset of users who consumed the item i. Therefore, the objective is to find the subset $S^* \subseteq S$, such that $|S^*| \leq k$ and the number of distinct covered users $|\bigcup S_i|$ is maximized. In another viewpoint, Max-Coverage is modeled as a bipartite user-item graph, where the nodes are the users and items, and the edges represent the interactions of a given user to an item. Then, MaxCov aims to select k items that reach the maximum number of distinct users, as proposed in other RSs related problems [15].

The Maximum k-Coverage is a NP-hard problem and there is no optimal solution in polynomial time. Our RS is a greedy algorithm to select the item that maximizes the number of users covered at each iteration. k iterations are executed to evaluate every set S_i was not selected (i.e., $S \in F \setminus F^*$). In each iteration, the algorithm looks for the item that maximizes the intersection of users not covered yet ($|S \cap R|$). A superficial analysis of this strategy can conclude that the selected items are the most popular ones, considering that the goal is to find items related to many users. However, at each iteration, the set R (resting users) is constantly updated to exclude users covered by the selected S-set ($R \leftarrow R \setminus S$). For this reason, this strategy recovers increasingly less popular items. The algorithm ends when k items are selected or when there are no more users to be covered. The complexity of this algorithm is $O(kmn)$, where k is the number of items to be recommended, m is the number of users and n is the number of items.

Niche-Coverage: This model is inspired by users' behavior studies [13]. Since the first surveys in RSs, the main approaches are often implemented using collaborative filtering (CF) algorithm [14,19]. CF algorithms produce recommendations based on the assumption that similar users have similar tastes. Then, people who share common ratings are a good source of recommendations. However, these algorithms are not able to Pure Cold-Start problem, because it is impossible to find similar users to first-time users. Nevertheless, the assumption used still true for our problem and it is the premise used by Niche-Coverage. In this case, our approach intends to divide users into niches of common interests and identify items that cover the most users for each niche. We suppose that recommending items from distinct niches of users, the system can reach all distinct preference of first-time users because we present the things that appealed to all types of users. First, we find the k items used to cover all users from a specific niche through the Max-Coverage algorithm. Next, we merge the items selected based on the size of each niche to maximize the number of users covered. In a recommendation list R of size k, the biggest niche compose the most of items in R.

The definition of users niches is based on clustering methods. In recommendations domains, the most famous clustering methods are the traditional *k-means* and *Bisecting k-means* [4]. These methods use the ratings assigned by users-items interactions to group users in sets with common interests (i.e., niche of users). In this work, we compare both clustering methods looking for the most suitable and the number of clusters to be used. Then, we should find the number c of clusters with Maximum Rate (CMR), oppositely to [13]. For this, we look for the number of clusters that maximizes the mean Hit Rate, a traditional metric of business performance often associated with sales [2]. Specifically, we are interested in the niche that maximizes the hit rate metric because is crucial that system shows at least one relevant item for users in this first interaction. This process is shown by the Equation $c = arg\ max \left[\sum_{n=1}^{N} \left(\frac{\sum_{u=1}^{U_{test}} |R_{list}(u) \cap I_{test}(u)|}{U_{test}} \right)\right]$, where N is the number of users niche, U_{test} the set of first-time users, $I_{test}(u)$ the items in test set consumed by u and $R_{list}(u)$ the recommendations generated by Niche-Coverage for the user u.

Therefore, the goal is to select the *representative items* from each niche of users, which are the items with the highest chances of matching the preference of any user from the niche. Initially, we classify the set of users U in c niches. Next, at each iteration, we analyze each niche of users. First, the set R is updated to contain only users from the niche evaluated. So, we select the subset S that maximizes the number of users covered. In this case, each element S_i in set $F = \{S_1, .., S_n\}$ is a subset of users from the cluster who consumed the item i. Next, we apply the Max-Coverage approach to find k items from each niche of users. Again, the set R (resting users) is constantly updated to exclude users covered by the selected S-set. A set of *Items* saves the k items selected for each niche. Then, finally, we execute a *merge* function to generate the final recommendation list. This function select items from each niche according to the Max-Coverage order. The complexity of this algorithm is divided into two steps, clusters computation, and Max-Coverage recommendation. The clustering complexity depends on the implementation. In general, the complexity is $O(ncdi)$ where n is the number of d-dimensional vectors, c the number of clusters and i the number of iterations needed until convergence. However, in order to mitigate the Pure Cold-Start problem, we need to compute the clustering algorithm just one time, before the recommendation process. Hence, Niche-Coverage complexity is related to the recommendation step. Basically, it consists in to compute the Max-Coverage for c times (one for each niche of users). So, this complexity-time is $O(ckmn)$.

4 Empirical Assessments

This analysis aims to compare biased-items and coverage-based RSs for addressing the Pure Cold-Start problem. We used all historical data of the selected users in Sect. 3.1 to compose test sets and removed all data information about them. The other users compose the training sets and are used as inputs by the evaluated RSs. So, first, we analyze the best parameters to the Niche-Coverage algorithm, comparing k-means and Bisecting k-means. Next, we evaluated the recommendation lists issued by each RS, considering the most famous quality requirements. We also analyze the users reached by the items recommended, in order to consolidate the complementarity of our approaches. To attract first-time users with different preferences, it is not enough to assume that strategies focus only on the usefulness of items to users [7]. Aspects such as diversity, coverage, and surprise are important to compose an interface that presents the best of items catalog available to first-time users. The usefulness of each advisor is evaluated by *Hit Rate*, *Precision* and *Recall* [2]. The diversity of the recommended items is evaluated by the metrics of *ILD* and *Genre Coverage* [15].

4.1 Niche-Coverage Definitions

To define the best Niche-Coverage performance, we analyze two clustering methods and look for the number of clusters that maximizes the rate (CMR).

Then, we compute 2^c clusters with the k-means and Bisecting k-means algorithms, where c range is $c = \{1, 2, 3, ..., 8\}$. Considering each number of clusters, we run the Niche-Coverage algorithm to recommend 10 items for first-time users and evaluate the Hit Rate metric. In the entertainment scenario, we find the best hit rate using k-means with 10 and 4 clusters, respectively in ML-1M and ML-10M. In the e-commerce domain, we find the best hit rate using k-means with 2 niches to CiaoDVD and Bisecting k-means with 93 niches in Amazon. Moreover, the results found are better than just using one cluster (i.e., the Max-Coverage approach). So, we confirm our premise that dividing users in niches and run the Max-Coverage locally is better than only run Max-Coverage with all users.

4.2 Quality of Recommendations

First, we simulate a real web-scenario, where users handle with 5, 10 and 20 items, and measure the RS's effectiveness. Then, we show the Hit Rate and F-measure metrics in Fig. 2(a) and (b), respectively. In the entertainment scenario (ML-1M and ML-10M datasets), the users usually watch famous movies, that attracted the attention of many domain users. The Popularity and Best-Rated approaches have satisfactory performance in these scenarios. However, Max-Coverage and Niche-Coverage also have a high effectiveness rate. For example, in the ML-10M dataset, with 10 items recommended, the Niche-Coverage have the highest hit rate. In the e-commerce scenario, the users are interested in buy specific products, frequently related to their personal preference. For this reason, approaches based on biased-items are not the best option to satisfy first-time users. In this case, our Max-Coverage and Niche-Coverage approaches have the best performance, as shown in the second row of Fig. 2. Specifically, in CiaoDVD dataset with just 10 items recommended, the Niche-Coverage has double the performance of state-of-the-art RSs. Statistically, we consolidated the results by Wilcoxon test for non-parametric distributions. In the entertainment scenario, the RSs' performance is not statistically different. In turn, in the

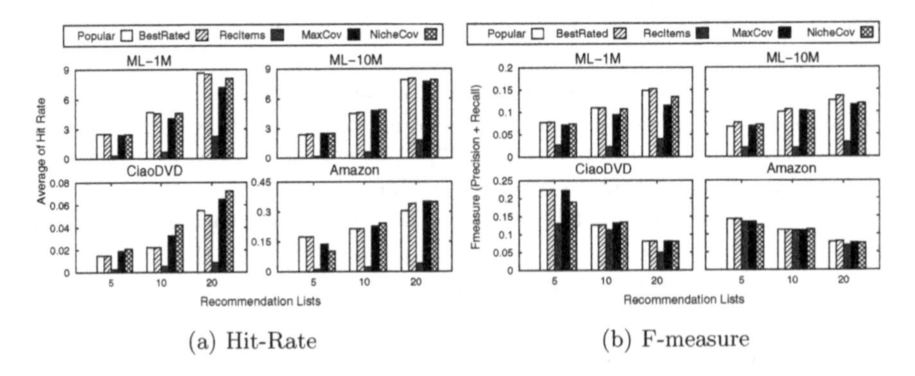

(a) Hit-Rate (b) F-measure

Fig. 2. Results of utility metrics on all domains.

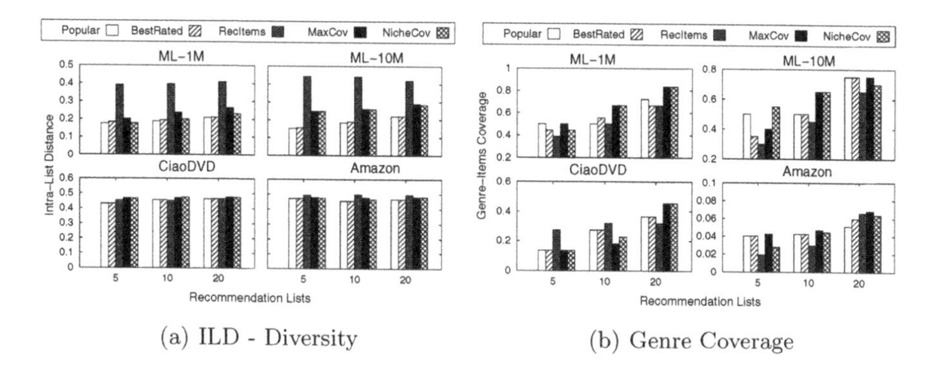

(a) ILD - Diversity

(b) Genre Coverage

Fig. 3. Results of diversity metrics on all domains.

e-commerce domains, the Niche-Coverage performance presents a statistical gain with 99% of confidence interval and *p-value* = 0.01.

Furthermore, these gains obtained by our approaches are related to distinct items. Basically, due to the assumption of maximizing the coverage of users on the domain, Max-Coverage and Niche-Coverage recover items related to most of the users profiles. For this reason, these RSs are also high values of diversity and item-genre coverage, as shown in Fig. 3(a) and (b). The Recent Items RS has the best value of diversity because it recommends just the last items consumed. However, these results are not efficient due to the low accuracy obtained (Fig. 2). On the other hand, the diversity presented by Max-Coverage and Niche-Coverage is achieved through potentially relevant distinct items. Specifically, in ML-10M dataset, our approaches have almost 50% of more diversity than traditional RSs with ILD metric. The same occurs in Genre Coverage metric, which Max-Coverage covers a greater number of distinct genres. We applied the Wilcoxon test, confirming the superiority of our approaches with a *p-value* = 0.001. These results show that Max-Coverage and Niche-Coverage are effective RSs, due to the high values of accuracy even gaining in terms of diversity.

4.3 Analysis of Complementarity

The last analysis point out to a complementary behavior between our approaches and the traditional ones. Our supposition is that Max-Coverage and Niche-Coverage do not recommend items biased by its rating or high influence. A straightforward analysis of the popularity of the first 10 items recommended by each strategy confirm this assumption. Our analysis demonstrates that all methods, except Best-Rated, recommend less popular items. The Max-Coverage recommends items that are less popular than the previous item. Niche-Coverage also diversifies the items from traditional RSs with less popular items. Hence, these analyses highlight that: (1) traditional approaches are much similar because they recommended biased-items; and (2) the new approaches are complementary to traditional methods because they recommended items based on the coverage.

Table 2. This table shows the number of users conquered and covered by each RS, denoted by the set <*conquered/covered*>. The cells marked with a color ■ mean a higher number of users conquered by RS. The symbol ▲ denotes significant positive gains, • non significant gains and ▼ significant negative losses. These gains are obtained concerning the best state-of-the-art RS, located in the first three rows, and applying a Chi-square test with 95% of a confidence interval.

Entertainment Scenario

	ML-1M			ML-10M		
RecLists	*top-5*	*top-10*	*top-20*	*top-5*	*top-10*	*top-20*
Popular	61.1% / **87.4%**	79.4% / **93.6%**	90.6% / **98.2%**	60.5% / **77.2%**	72.3% / **86.6%**	85.1% / **92.6%**
BestRated	64.1% / **88.1%**	78.7% / **93.8%**	90.4% / **98.1%**	62.0% / **82.1%**	73.1% / **87.0%**	84.7% / **92.7%**
RecItems	10.7% / **19.7%**	36.3% / **52.7%**	61.3% / **81.3%**	12.5% / **22.8%**	25.9% / **41.9%**	28.9% / **45.4%**
Max-Cov	69.6% / **89.3%** ▲	82.4% / **95.9%** ▲	91.9% / **99.3%** ▲	61.0% / **84.6%** •	76.1% / **91.6%** ▲	86.6% / **96.3%** ▲
Niche-Cov	66.1% / **87.7%** •	80.1% / **94.6%** ▲	91.3% / **98.9%** •	62.9% / **84.6%** •	76.6% / **91.5%** ▲	85.3% /**95.8%** •

E-commerce Scenario

	CiaoDVD			Amazon		
RecLists	*top-5*	*top-10*	*top-20*	*top-5*	*top-10*	*top-20*
Popular	5.29% / **7.01%**	8.03% / **11.0%**	12.6% / **16.5%**	10.9% / **14.7%**	15.6% / **21.1%**	23.5% / **31.7%**
BestRated	5.29% / **7.01%**	8.03% / **11.0%**	12.6% / **16.5%**	10.9% / **14.7%**	15.6% / **21.1%**	24.5% / **31.9%**
RecItems	0.05% / **0.09%**	0.15% / **0.25%**	0.21% / **0.33%**	0.54% / **0.91%**	1.16% / **1.86%**	1.88% / **2.87%**
Max-Cov	5.56% / **7.43%** •	8.92% / **11.7%** •	13.1% / **17.3%** •	10.9% / **14.7%** •	16.3% / **23.0%** •	25.6% / **34.4%** ▲
Niche-Cov	4.83% / **6.65%** ▼	8.50% / **11.2%** •	12.9% / **17.1%** •	8.12% / **12.0%** ▼	15.1% / **20.9%** •	24.1% / **32.6%** •

However, in real web-scenarios, the system owners are interested in the user's satisfaction. If the users watch/buy their products, their profit will be higher. For this reason, we develop a metric to evaluate the number of users conquered by each RS, based on user satisfaction in real scenarios. We consider that a user is conquered by the system if s/he consumed and liked at least one item. In this work, we define that users like an item when they provide a rating greater than their personal average. We analyze each recommendation list, counting the number of users conquered by the items. Note that, this method is more complex than simple coverage. We measure RS's coverage just considering if the user watched or bought the item recommended. Here, the ability to conquer is related to the rating assigned by users that watched or bought the item. In Table 2, we color the cases that the number of users conquered is higher than baselines and mark it with symbols of statistical significance. The Max-Coverage approach covers more users than other RSs due to its greedy algorithm. Moreover, we observe that in 9/12 cases, the Max-Coverage approach also conquers more users than baselines. In the other 3 cases, Niche-Coverage conquers more users.

Table 3. Number of users conquered exclusively by one RS.

RecList	Exclusive users of each RS			
	ML-1M	ML-10M	CiaoDVD	Amazon
Popularity	1.19%	0.41%	0.00%	0.00%
BestRated	1.92%	0.84%	0.00%	0.00%
RecItems	2.20%	1.17%	0.14%	0.71%
Max-Cov	**3.84%**	**9.80%**	1.57%	2.91%
Niche-Cov	0.56%	8.74%	**1.72%**	**3.68%**

We also count the number of users conquered exclusively by one RS. We observe in Table 3 that: (1) Popularity and Best-Rated do not aggregate users than those already conquered by other approaches; (2) Recent-Items conquers some different users, but it does not present items potentially relevant to first-time users; and (3) Max-Coverage and Niche-Coverage conquers more first-time users, which are distinct from others. These results point out a room for improvements, which are explored in the next section.

5 Construction of Product Pages

Mixing different recommendation lists on a product page is a common practice of real systems. However, we argue that these lists usually reach a similar subset of first-time users, since all of them are based on biased-items. To verify this behavior, we evaluate product pages composed by mixing three top-10 recommendation lists issued by distinct combinations of RSs. We restrict each page to have only three lists for working in smartphone scenarios, characterized by small screens. We evaluate five different combinations (Table 4). For each combination, we evaluate the number of users who rated positively at least one item from the three RSs, obtaining the percentage of *Users Conquered* for each combination. The results show that by mixing <BestR, MaxCov, NiCov> we can reach a high number of first-time users. In other words, in the evaluated scenarios, product pages should be composed by Best-Rated, Max-Coverage, and Niche-Coverage. This work suggests that systems incorporate our strategies to be used side by side, changing from the traditional approach for our suggestion.

Table 4. Percentage of users conquered mixing three RS

Approaches	Users conquered			
	ML-1M	ML-10M	CiaoDVD	Amazon
Pop, BestR, RecItems	87.35%	76.27%	8.10%	16.34%
Pop, BestR, MaxCov	88.99%	84.35%	9.60%	18.54%
Pop, BestR, NiCov	84.88%	83.75%	9.76%	19.33%
Pop, MaxCov, NiCov	87.25%	83.83%	9.76%	20.10%
BestR, MaxCov, NiCov	**89.15%**	**84.89%**	**9.76%**	**20.10%**

5.1 Online User-Centered Study

In order to evaluate our new approach for mixing RSs, we perform an experiment with volunteer users of different ages and preferences to evaluate the recommendations. Once the focus of this work is the first-time users, for who we do not have any information, a Web interface that presents the recommended items is able to simulate real scenarios. We follow the main guidelines of online evaluations presented in the literature [12]. We chose the movie scenario of *ML-Latest*, updated in August 2017. This dataset has 26M ratings assigned by 270K users to 45K movies on a scale from 1 to 5. The user-centered study was released during 8 days (from 09/07 to 09/14/2018), reaching 204 users that interacted with an online system. The users selected are 71% men and 29% women, from 11 to 63 years old. Moreover, 85% of users are frequent users of movies streaming systems. Initially, the participants are instructed to fill in a consent form. In the next three steps, users answer questions, selecting or ordering their favorite movies. In the end, the users answer questions about personal information. We are concerned in the three middle steps:

1. **A/B Test:** users have to choose one movie to watch or the option "None of the Movies". In this case, some users interact with a side A (traditional approach) and others with the side B (our approach).
2. **User Satisfaction:** users answer three questions about all movies presented in the first step. Basically, these questions are related to classical concepts, such as *unexpectedness*, *novelty* and *utility*.
3. **Ideal Ranking:** users have to build their ideal ranking between all movies presented in the first step. In this case, users can choose how many movies s/he wants. We suggest that users choose at least 5 movies.

The first step aims to compare side A (traditional approach) against the side B (our approach). Specifically, we present 10 movies of each RS, similarly to the current Web-scenarios. In this step, 102 participants interact only with the side A and the other half with the side B. Then, we ask for each user to select a movie to watch or the "None of the Movies" option. We are simulating real scenarios, where users have to make a decision: watch any movie or ignore the options. In this case, the labels have not a biasing effect because this step aims to highlight the most promising scenario instead of comparing the lists. Moreover, the users who interact with side A do not know about side B.

Table 5. User choices in the A/B test interface

Recommendation list	Percentage of users	
	Side A	Side B
Page top	39.21%	42.15%
Page middle	25.49%	17.64%
Page bottom	30.39%	33.33%
None	4.9%	6.8%

Table 5 shows the percentage of users that selected a movie from the list on top, middle or bottom of the page. In both sides, most users select movies from the list on the page top, related to Popularity (side A) and Best-Rated (side B). Despite the effectiveness of these RSs, this result may be related to the list position on the page. However, there is a high percentage of users who roll down the page to select movies from the bottom lists, related to Recent Items and Niche-Coverage methods. Probably, this behavior is related to the complementarity of these RSs in the movies domain. This result reinforces the assumption that complementary RSs should be used to compose web pages.

The second step assesses the quality perceived by real users. In this step, we follow the questionnaire used in [17]. We present each movie with a short synopsis and ask users: (1) *Did you already know about this movie before this recommendation? (Yes, No, Dont know)*; (2) *Have you already watched this movie? (Yes, No)*; and *Would you like to watch this movie (for the first time or once again)? (Yes, No, Dont know)*. We create a ranking with users answers. In the first question, we count the number of users who said "no" to simulate a feeling of surprise by something *unexpected*. In the second, we count users who said "no" to measure the RS *novelty*. For the third question, we count the users who said "yes" to discover the RS *utility*. Table 6 summarize these rankings with the area under the curve (AUC) normalized by its highest value.

Table 6. Quality perception of real users

Recommender	Real user satisfaction		
	Unexpected	Novelty	Utility
Popular	0.1929	0.4162	0.7196
BestRated	0.2570	0.5043	0.7688
RecItems	0.4798	0.6269	0.7177
Max-Cov	0.3394	0.4776	0.6647
Niche-Cov	0.3394	0.4912	0.7049

Indeed, all five RS are useful for real users in web-scenarios. Conversely the offline results, Recent Items is also useful for real users because it recommends distinct items from other RSs. Moreover, Max-Coverage and Niche-Coverage also recommended unexpected items, increasing novelty for users. These results are reinforced in the next step. Specifically, the third step aims to compare each recommendation list with the ideal ranking built by users. We use a traditional pooling strategy from the IR field to create a ground-truth about users. Basically, we select top-10 results from each RS, removing the items duplicates and present these movies to participants in a random order. The users have to order the movies according to their preference. Then, we can measure three ranking metrics to compare the recommendations and the feedback provided by the user: *Jaccard Similarity*; *Mean Reciprocal Rank (MRR)* [3]; and *Normalized Discounted Cumulative Gain (nDCG)* [8]. These metrics measure, respectively: (1) the similarity of each recommendation list with the ideal ranking; (2) the position of the first relevant item recommended; and (3) the effectiveness of each RS.

Table 7 shows the metric's average for 118 participants that built their ground-truth. We consider only the five-first movies to create a fair analysis.

Table 7. Quality of RSs based on users' ranking

Recommender	Ranking metrics		
	Jaccard	MRR	nDCG
Popular	**0.35**	0.32	**0.49**
BestRated	**0.35**	0.27	0.46
RecItems	0.02	0.04	0.03
Max-Cov	0.33	0.32	0.47
Niche-Cov	**0.35**	**0.33**	**0.49**

The first column shows the Jaccard's similarity and does not highlight any difference between the rankings. The reason for this result may be there are a few distinct items to be selected in this step. In turn, the MRR shows a higher value to Max-Coverage and Niche-Coverage than baselines approaches. In other words, these RSs are more likely to present relevant items in the first positions than the other RSs. In addition, *nDCG* metric confirms the effectiveness of our RSs, comparing with the baselines approaches. Thus, this user-centered study points to improvement possibilities for owners of web-applications.

6 Conclusions

Web applications assume that items biased by popularity, recency or positive ratings are enough to suit most of the first-time user's profiles. However, this work shows that it is not always true because there are many users not interested only in biased-items. Conversely, we introduce two new methods inspired in user coverage maximization: Max-Coverage and Niche-Coverage. While traditional RSs retrieved potentially relevant items, obtaining just high accuracy, the new RSs keep accuracy and enhance diversity. Our experiments show a statistical gain in both concepts. Further, the new RSs match the interest of a higher number of distinct first-time users. Thus, these results highlight complementary behaviors between our RSs and traditional approaches and show an opportunity for improvements to compose product pages. We assume that to enhance the interest of first-time users on the item catalog, the web-applications should mix these complementary RSs. An online user-centered study with 204 participants reinforces this assumption with metrics related to user satisfaction.

Acknowledgments. This work was partially funded by the INWeb (no. 573871/2008-6), MASWeb (FAPEMIG/PRONEX APQ-01400-14), CAPES, CNPq, Finep, and Fapemig.

References

1. Barjasteh, I., Forsati, R., Masrour, F., Esfahanian, A.-H., Radha, H.: Cold-start item and user recommendation with decoupled completion and transduction. In: Proceedings of the 9th ACM RecSys, pp. 91–98 (2015)

2. Bobadilla, J., Ortega, F., Hernando, A., Gutiérrez, A.: Recommender systems survey. Knowl.-Based Syst. **46**, 109–132 (2013)
3. Chapelle, O., Metlzer, D., Zhang, Y., Grinspan, P.: Expected reciprocal rank for graded relevance. In: Proceedings of the 18th ACM CIKM, pp. 621–630 (2009)
4. Ghazanfar, M.A., Prügel-Bennett, A.: Leveraging clustering approaches to solve the gray-sheep users problem in recommender systems. Expert. Syst. Appl. **41**(7), 3261–3275 (2014)
5. He, C., Parra, D., Verbert, K.: Interactive recommender systems: a survey of the state of the art and future research challenges and opportunities. Expert. Syst. Appl. **56**, 9–27 (2016)
6. Hernando, A., Bobadilla, J., Ortega, F., Gutiérrez, A.: A probabilistic model for recommending to new cold-start non-registered users. Inf. Sci. **376**, 216–232 (2017)
7. Ho, Y.-C., Chiang, Y.-T., Hsu, J.Y.-J.: Who likes it more?: mining worth-recommending items from long tails by modeling relative preference. In: Proceedings of the 7th ACM WSDM, pp. 253–262 (2014)
8. Järvelin, K., Kekäläinen, J.: Cumulated gain-based evaluation of IR techniques. ACM Trans. Inf. Syst. (TOIS) **20**(4), 422–446 (2002)
9. Lee, D., Hosanagar, K.: Impact of recommender systems on sales volume and diversity (2014)
10. Lika, B., Kolomvatsos, K., Hadjiefthymiades, S.: Facing the cold start problem in recommender systems. Expert. Syst. Appl. **41**(4), 2065–2073 (2014)
11. Majumdar, A., Jain, A.: Cold-start, warm-start and everything in between: an autoencoder based approach to recommendation. In: IEEE IJCNN, pp. 3656–3663 (2017)
12. O'Donovan, J., Tintarev, N., Felfernig, A., Brusilovsky, P., Semeraro, G., Lops, P.: Joint workshop on interfaces and human decision making for recommender systems. In: Proceedings of 9th ACM RecSys, pp. 347–348 (2015)
13. Pereira, A.L.V., Hruschka, E.R.: Simultaneous co-clustering and learning to address the cold start problem in recommender systems. Knowl.-Based Syst. **82**, 11–19 (2015)
14. Poriya, A., Bhagat, T., Patel, N., Sharma, R.: Non-personalized recommender systems and user-based collaborative recommender systems. Int. J. Appl. Inf. Syst **6**(9), 22–27 (2014)
15. Puthiya Parambath, S.A., Usunier, N., Grandvalet, Y.: A coverage-based approach to recommendation diversity on similarity graph. In: Proceedings of the 10th ACM Conference on Recommender Systems, pp. 15–22. ACM (2016)
16. Rosli, A.N., You, T., Ha, I., Chung, K.-Y., Jo, G.-S.: Alleviating the cold-start problem by incorporating movies facebook pages. Clust. Comput. **18**(1), 187–197 (2015)
17. Rossetti, M., Stella, F., Zanker, M.: Contrasting offline and online results when evaluating recommendation algorithms. In: Proceedings of the 10th ACM Conference on Recommender Systems, pp. 31–34. ACM (2016)
18. Safoury, L., Salah, A.: Exploiting user demographic attributes for solving cold-start problem in recommender system. Lect. Notes Softw. Eng. **1**(3), 303 (2013)
19. Sedhain, S., Sanner, S., Braziunas, D., Xie, L., Christensen, J.: Social collaborative filtering for cold-start recommendations. In: Proceedings of the 8th ACM Conference on Recommender systems, pp. 345–348. ACM (2014)
20. Silva, N., Carvalho, D., Pereira, A.C., Mourão, F., Rocha, L.: The pure cold-start problem: a deep study about how to conquer first-time users in recommendations domains. Inf. Syst. **80**, 1–12 (2019)

21. Steck, H., van Zwol, R., Johnson, C.: Interactive recommender systems: tutorial. In: Proceedings of the 9th ACM RecSys, pp. 359–360 (2015)
22. Vargas, S., Baltrunas, L., Karatzoglou, A., Castells, P.: Coverage, redundancy and size-awareness in genre diversity for recommender systems. In: Proceedings of the 8th ACM Conference on Recommender Systems, pp. 209–216. ACM (2014)
23. Vargas, S., Castells, P., Vallet, D.: Intent-oriented diversity in recommender systems. In: Proceedings of the 34th International ACM SIGIR Conference on Research and Development in Information Retrieval, pp. 1211–1212. ACM (2011)

Rumor Detection on Social Media: A Multi-view Model Using Self-attention Mechanism

Yue Geng[1,2], Zheng Lin[1(✉)], Peng Fu[1], and Weiping Wang[1]

[1] Institute of Information Engineering, Chinese Academy of Sciences, Beijing, China
{gengyue,linzheng,fupeng,wangweiping}@iie.ac.cn

[2] School of Cyber Security, University of Chinese Academy of Sciences, Beijing, China

Abstract. With the unprecedented prevalence of social media, rumor detection has become increasingly important since it can prevent misinformation from spreading in public. Traditional approaches extract features from the source tweet, the replies, the user profiles as well as the propagation path of a rumor event. However, these approaches do not take the sentiment view of the users into account. The conflicting affirmative or denial stances of users can provide crucial clues for rumor detection. Besides, the existing work attaches the same importance to all the words in the source tweet, but actually, these words are not equally informative. To address these problems, we propose a simple but effective multi-view deep learning model that is supposed to excavate stances of users and assign weights for different words. Experimental results on a social-media based dataset reveal that the multi-view model we proposed is useful, and achieves the state-of-the-art performance measuring the accuracy of automatic rumor detection. Our three-view model achieves 95.6% accuracy and our four-view model using BERT as a view also reaches an improvement of detection accuracy.

Keywords: Rumor detection · Multi-view model · Self-attention · Deep learning

1 Introduction

Nowadays, social media enable not only journalists but also ordinary individuals to post ongoing events. As social media provide citizens with an ideal platform to stay abreast of momentous events, it is also eligible for broadcasting rumors. Rumors that are ultimately proven false often have damaging consequences in view of the fact that they negatively impact citizens' life and sometimes even trigger public panic. For instance, a rumor claiming that iodized salt could prevent radiation was posted in 2011 and a great number of citizens stripped supermarkets of salt in the belief that it could ward off radiation poisoning [6]. This piece of fake news caused panic in population as well as a huge market disorder.

J. M. F. Rodrigues et al. (Eds.): ICCS 2019, LNCS 11536, pp. 339–352, 2019.
https://doi.org/10.1007/978-3-030-22734-0_25

Therefore, it is crucial to identify rumors in social media where large amounts of information are easily spread. This emphasizes the need for studies that can assist in analyzing the veracity of news. Rumors can give rise to shock, suspicion or protest in public since misinformation affects individuals' perception of events and causes harmful consequences which tend to be more severe over time.

To alleviate these problems, studies have been conducted from different perspectives ranging from psycholinguistic analysis to deep learning techniques. Early studies extracted groups of related features (i.e., message, topic) and built a machine learning classifier to evaluate the credibility of social media posts [16]. One of the drawbacks of this kind of method is that hand-crafted features hardly explore the inner relationship among replies. Recently, Ma et al. [12] exploited recurrent neural networks (RNN) to represent the content of the source tweet and its replies/retweets. RNNs automatically learn both temporal and textual features and thus yield outstanding performance. This work detected rumor events mainly based on contents whereas the stances of users were not concentrated on. Since those users who read the rumors may have common sense, and they may share opinions or raise questions on suspicious posts, we introduce a new sentiment view for rumor detection. Specifically, since source tweet and replies are proven useful in previous work, a supplementary sentiment view is adopted, and then the multi-view model is constructed. Besides, the previous work tokenizes the posts and equally treat each word while we train a self-attention layer which pays more attention to significant words in the source tweet. We train and test our model based on a Weibo dataset[1]. This dataset incorporates 4664 events and the posts in the event are sorted by time. In each event, a source tweet is associated with a number of replies, retweets, and user profiles. We utilize the dataset to understand how the lexical content and the users' reactions are related to its veracity. Our work indicates that the content of all the posts and the users' sentiment are capable of better exploiting representations of rumors. Our research also reveals that GRU with self-attention mechanism [19] can provide strong assistance for social-media-based rumor detection. The source code is available at GitHub[2].

The main contributions of our research include:

- We develop a multi-view network which analyzes features related to a specific event in social media. This model enables deep neural networks to learn representations containing adequate information from three different perspectives, including the source tweets, replies/retweets, and the latent sentiment semantics. All of the views we proposed are useful for rumor detection based on experimental results.
- We also apply the Gated Recurrent Unit (GRU) with self-attention mechanism to better capture the features of the content as well as the propagation path of a certain event and automatically assign a weight for each reply/retweet corresponding to its significance.

---

[1] http://alt.qcri.org/~wgao/data/rumdect.zip.
[2] https://github.com/crystalyue/multi-view-rumor-detection.

- Our model demonstrates the state-of-the-art performance of detecting rumors in social media on Weibo dataset. Our three-view model outperforms baseline PPC_RNN+CNN [10] by 3.5%. We also test the performance of the Bidirectional Encoder Representations from Transformers (BERT) model [5] as the fourth view, and the combination of our proposed model with BERT model can achieve an even better result.

The rest of this paper is organized as follows. We begin with an overview of related work in Sect. 2. Section 3 presents the rumor detection task and a detailed description of our multi-view system including the preprocessing methods and the feature sets. Section 4 introduces the datasets used in this paper and provides the experimental settings. We also analyze the detection performance of our model. The purpose of the evaluation experiments is to compare the predictive capacity with that of the prevailing methods. Besides, we conduct experiments on assessing the contributions of different parts of our model to the overall performance. Finally, we conclude and present directions for future work in Sect. 5.

2 Related Work

Related work on rumor detection can be roughly classified into four categories, content-based, knowledge-based, propagation-path-based, and hybrid methods.

Qian et al. [17] introduced a Two-Level Convolutional Neural Network with User Response Generator (TCNN-URG) where Two-Level Convolutional Neural Network (TCNN) captures underlying semantic information at both word and sentence levels. User Response Generator (URG) is based on Conditional Variational Autoencoder (CVAE) which generates user responses to new articles with the assistance of historical user responses. Sarkar et al. [4] proposed a different idea which is to build a hierarchical neural network architecture. First, they took a sequence of weighted average word embeddings as inputs to generate a sentence embedding. Second, they created a document embedding taking the sentence embeddings as inputs. The resulting document embeddings contain semantic information at both sentence and document levels.

In addition, there are other groups of approaches. Knowledge-based rumor detection methods mainly focus on information retrieval or knowledge graph. By extracting basic elements of the document and searching them from websites, Wu et al. measure the quality of the query results. The results are accumulated to obtain a final score for a given document [21]. The other possible means is to build a Wikipedia Knowledge Graph and evaluate the veracity of news by calculating the truth value of the shortest path between entities in the knowledge graph. A subject-predicate-object statement's veracity amounts to both the path length between the two target entities and the generality of the entities [3].

Studies also found that temporal features could strengthen the predictive power of models. Kwon et al. [9] proposed a time-series-fitting model representing a rumor's spreading pattern. Ma et al. extracted more time-sensitive features and explored how they vary in time [13]. Subsequently, Jin et al. discovered conflicting viewpoints in tweets by constructing a credibility propagation network of tweets. Based on a topic model method, the credibility propagation finally generates a credit score for each piece of news [7]. Kernel-based methods are also capable of automatically modeling the propagation path of an event. Propagation trees giving clues for how tweets are transmitted in their life span. Ma et al. then identified rumors by comparing the similarities between different tweets' propagation tree structures [14].

In the meanwhile, researchers introduced hybrid deep learning models which drastically improve the accuracy of rumor detection. Apart from linguistic features, user profiles including party affiliation, speaker title, location and credit history can also be used as additional information. Long et al. [11] included user profile information in attention layers. Volkova et al. showed that a joint learning neural network model based on social network interactions and news contents advanced lexical models since syntax and grammar features did not make any contribution to evaluate the veracity of rumors [20]. Liu et al. modeled the propagation path using ensemble learning and encoded eight kinds of social-media-based user characteristics. They then built a classifier that incorporates both RNN and CNN to utilize textual contents as well as user characteristics along the propagation path [10]. Moreover, Ma et al. proposed a bottom-up and a top-down tree-structured neural network both of which naturally model the propagation paths of a rumor [15].

However, existing network-based methods did not focus on the crucial opinions in the replies as well as the different extent of information that different words in the source post can provide. In this work, we use a multi-view representation model to exploit contents, replies and the supporting and opposing sentiment to improve the performance of rumor detection.

3 Method

In this section, we present the details of our proposed model for classifying rumor events. First, we describe the overall structure of our model and introduce a method that assigns different weights for each word in the source tweet. Then, we describe each part of our model and explain how the sentiment view is constructed.

3.1 Problem Statement

Let $E = \{e_1, e_2, ..., e_n\}$ be an event where e_1 is the source tweet and $\{e_2, ..., e_n\}$ are the posts related to the source tweet. Each event is associated with a label L indicating whether the source tweet in this event is a rumor or not. Note that $L = 0$ denotes the event is a non-rumor while $L = 1$ denotes the event is a rumor. Our rumor detection task can be defined as automatically distinguishing the veracity of an event given its corresponding label L.

3.2 The Proposed Model

Overview. We propose a multi-view neural network model to classify Weibo posts into two categories—rumor and non-rumor. The architecture was presented in Fig. 1. Three views are incorporated in our model which are named as content view, reply view and sentiment view respectively. All the source tweets and the reply/retweet sentences are tokenized using *Jieba* tools[3]. We initialize our embedding layer with pre-trained 200-dimensional word embeddings for Chinese words and phrases following the setup as in [18]. We implement and train the proposed model using *PyTorch*[4].

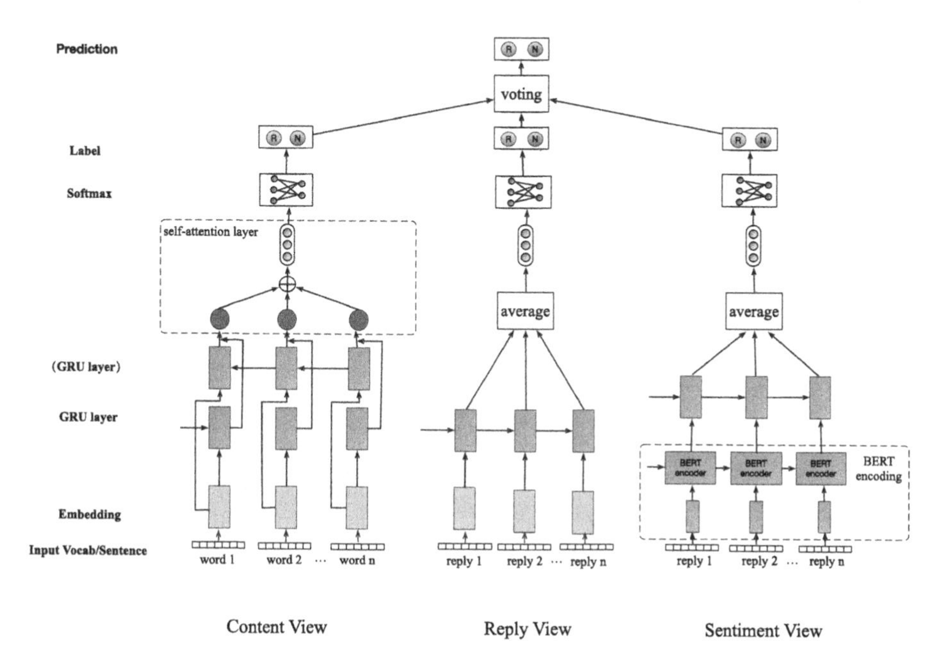

Fig. 1. The framework of the proposed multi-view model. The content view combines bidirectional GRU and a self-attention layer to evaluate the veracity of the source tweet in an event. The reply view generates representations for each reply/retweet in the event through the GRU layer. The sentiment view extracts sentiment embeddings via fine-tuned BERT encoders. The vote layer integrates the results of three views and finally predicts a label by majority voting.

The content sub-network consists of an embedding layer and a GRU network with a self-attention layer. For the content view, GRU takes each word's embedding in the source post as input. In the embedding layer, the source tweet S_n with a length l is represented as a vector $[S_1, S_2, ..., S_l]$ where Si is the word

[3] https://pypi.org/project/jieba/.
[4] https://pytorch.org/.

embedding of the l-th word. The network is followed by a self-attention layer for the reason that not all the words in a given post have the same significance. Thus, we calculate the similarities among the words in the post and calculate the weight for each word. Since it is unlikely to accurately distinguish rumors from true news only depending on its source tweet, we proposed other views providing more information to assist in promoting the performance of our model. For the reply view, we average the word embeddings to produce a post embedding which enables GRU to receive a reply/retweet at a time. GRU is used for excavating the lexical features and temporal features of events. In addition to content and reply views, we also generate sentiment embedding for each post in the event with the assistance of a fine-tuned BERT model and then send the embeddings to GRU. Each view is followed by a softmax layer which predicts a result, i.e., 1 for rumor and 0 for non-rumor. Take the three-view model as an example, the results of the three views are combined as a majority vote classifier. If there are two or more views generate the same result, then that result will be regarded as the final prediction of our model. The voting procedure can be described using the following equation:

$$Predict_{output}(event) = u(\sum_{v \in V} Predict_v(event) - \frac{|V|}{2}), \quad V \subseteq \{content, reply, sentiment\}, V \neq \varnothing \tag{1}$$

$$u(x) = \begin{cases} 1 & x \geq 0, \\ 0 & x < 0. \end{cases} \tag{2}$$

where $Predict_{output}(event)$ indicates the prediction label of the current processing event based on each view's output. V is a non-empty subset of {content, reply, sentiment}. The prediction of each view can be 0 or 1 and if the sum of all the predictions is greater than the threshold $\frac{|V|}{2}$, then we consider the current event as a rumor event.

Propagation Path Modelling via GRU. In terms of framework, we employ the gated recurrent units in order to better capture lexical and temporal information since the standard recurrent neural network is a biased model, where the earlier inputs are more likely to be abandoned during training. GRU receives each word embedding in content view and receives document embedding in reply view, and post-based sentiment embedding is used in sentiment view. GRU [2] is designed to dynamically remember and forget the information flow. There are two types of gates controlling how information is updated, i.e., reset gate and update gate. A single layer GRU accepts input vectors $<x_1, x_2, ..., x_N>$, computes the corresponding hidden states $<h_1, h_2, ..., h_N>$. Specifically, let $\odot$ denote the element-wise product of two vectors, the single layer GRU computes the hidden state h at time t and the corresponding output as:

$$h_t = (1 - z_t) \odot h_{t-1} + z_t \odot \widetilde{h_t} \tag{3}$$

$$\widetilde{h_t} = tanh(W_h x_t + U_h(r_t \odot h_{t-1}) + b_h) \tag{4}$$

$$r_t = \sigma(W_r x_t + U_r h_{t-1} + b_r) \tag{5}$$

$$z_t = \sigma(W_z x_t + U_z h_{t-1} + b_z) \tag{6}$$

where r_t and z_t are reset and update gates respectively. h_t is the hidden state of GRU and $\widetilde{h_t}$ is the candidate output. $\sigma(\cdot)$ are element-wise sigmoid functions. Mean square error (MSE) is used as the loss function for model training. We choose Adam algorithm [8] for updating network parameters.

Self-Attention-Based Content Representation. Self-attention [19] is a special kind of attention mechanism where the query Q is replaced by an embedding x_j from the source input itself. d_k is the dimension of Q, K, V. Specifically, the query Q, key K and value V are the same in this scenario. Self-attention computes the attention between elements at different positions in the sequence. For each embedding pair x_i and x_j, we calculate the Scaled Dot Product as the attention weight. Then we adopt a softmax function to normalize these weights. Finally, we weighted sum the weights and the corresponding value to obtain the final attention scores.

$$Attention(Q, K, V) = softmax(\frac{QK^T}{\sqrt{d_k}})V \tag{7}$$

The adoption of the self-attention mechanism is based on the assumption that not all the posts make equal sense. In the self-attention module, query Q, key K and value V represent the concatenation of each output of the hidden units in GRU. Hence, the resulting probabilities can be regarded as the weight of each post. In this way, the significance of each post can be automatically reflected by its weight and the weighted sum of all the posts encodes a better representation for the sentiment view. Moreover, it requires a small number of parameters and has a very fast computation speed. The Scaled Dot-Product Attention we adopt is presented in Fig. 2.

Sentiment View. One of the main challenges of constructing sentiment view is to obtain sentiment embeddings of each post. The usual practice is to supervised train a sentiment classifier with annotated labels. However, in this task, there are no sentiment labels indicating the sentiment polarity of the posts. To overcome this obstacle, it is reasonable to fine-tune a pre-trained sentiment classifier and take the output of its hidden layers as sentiment embeddings. Specifically, we employ a pre-trained Bidirectional Encoder Representations from Transformers (BERT) model [5] which advances the state-of-the-art model in eleven Natural Language Processing (NLP) tasks. In the original paper which proposes BERT, the authors report the experimental results on two model sizes: $\text{BERT}_{\text{BASE}}$ with 12 transformer blocks and 768 hidden size, and $\text{BERT}_{\text{LARGE}}$ with 24 transformer blocks and 1024 hidden size. In our paper, we adopt the first model since $\text{BERT}_{\text{BASE}}$ can already achieve an outstanding result and it is costly to

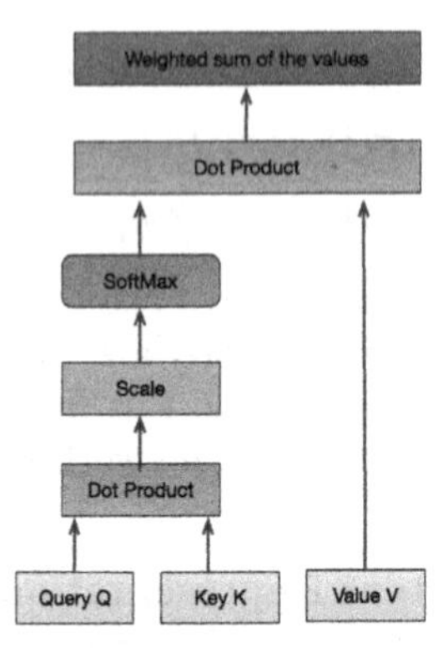

Fig. 2. Scaled dot-product attention adopted in our model.

train $\mathrm{BERT_{LARGE}}$. The architecture of the $\mathrm{BERT_{BASE}}$ model in our paper is the same as that in the original paper. The $\mathrm{BERT_{BASE}}$ model is fine-tuned by Weibo sentiment corpus[5] for better excavating the sentiment features of Weibo contents. All the sentences in the posts are fed into the $\mathrm{BERT_{BASE}}$ model and we take the outputs of the last hidden layer as our sentiment embeddings. The 768-dimensional vectors are then used as the inputs of GRU sentiment view. Figure 3 shows an example of $\mathrm{BERT_{BASE}}$ model. Since our reply view mainly focuses on contents, it is unlikely for reply view to capture much sentiment information. As a result, we adopt the third view which specifically extract sentiment embeddings. In the sentiment view, we adopt the fine-tuned $\mathrm{BERT_{BASE}}$ as a sentiment extractor which specifically captures sentiment features. Since $\mathrm{BERT_{BASE}}$ is a strong state-of-the-art model which outperforms other models in so many NLP tasks, we would like to see if our proposed model can improve its performance. We first use $\mathrm{BERT_{BASE}}$ without fine-tuning as a single view to test its performance and then combine our 3-view model with the $\mathrm{BERT_{BASE}}$ model to test the 4-view models performance. Results indicate that this view is not only able to help detect rumor by itself but also enables our 3-view and 4-view model to improve detection accuracy.

4 Experiment

4.1 Experimental Settings

In the experiment, we empirically set the size of hidden units as 300 and the maximum training epoch as 200. The training process finishes when the number of training epoch meets the restriction or the validation loss converges. The input dimension of content/reply view is 200. For Sentiment view, the output dimension of Bert is 768 as the configuration of the pre-trained model and the GRU input embedding dimension is also 768. The output vector size of all the

[5] https://github.com/baidu/Senta.

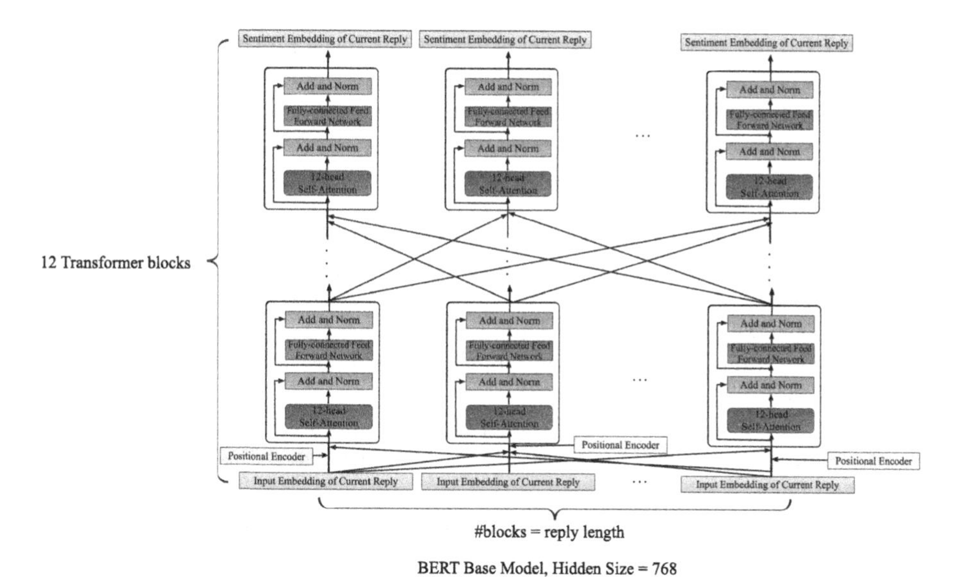

Fig. 3. A sample of BERT model with 12 transformer blocks.

views is set to 300. Batch size is set to 1 and the learning rate is 0.001. We restrict the number of replies in an event to 4096.

4.2 Dataset

We conduct experiments on an available public dataset: Weibo [12]. Posts including a source tweet and its relevant replies/retweets form an event. The dataset is comprised of 4664 events with 2,313 rumors and 2,351 non-rumors. In the same event, posts are sorted by published time and hence a propagation path is naturally constructed. Propagation path represents the extent to which each event is retweeted. Our dataset contains binary labels, i.e., rumor and non-rumor. Table 1 summarizes the statistics of the dataset. To get an overall comparison, we divide our dataset into three subsets by strictly following the same partition configuration as the previous papers. The validation set incorporates 10% of the total events. The remaining rumor events are split in a ratio of 3:1 and are used for model training and testing respectively.

4.3 Baseline Models

We carefully select a series of previous work on rumor classification as baselines, some of which are classical and others are state-of-the-art:

- DTC [1] A decision-tree-based classifier that utilizes a series of hand-crafted features.

- SVM-RBF [22] An SVM classifier with Radial Basis Function (RBF) kernel that also utilizes hand-crafted features.
- RFC [9] A random forest classifier that fit the utilizes user, linguistic and structure characteristics.
- SVM-TS [13] A linear SVM model that utilizes time-series to model how each kind of features vary in time.
- DT-Rank [23] A ranking method based on the decision tree. Searching for enquiry phrases and ranking the clustered results enable this method to detect rumors.
- GRU-RNN [12] An RNN-based model that learns long-distance dependencies among different time steps, which utilizes more information from user comments.
- PPC_RNN [10] A time series classifier that incorporates recurrent neural networks which combine tweet texts and the user characteristics along the propagation path to detect rumors.
- PPC_CNN [10] A time series classifier that incorporates convolutional neural networks which combine tweet texts and the user characteristics along the propagation path to detect rumors.
- PPC_RNN+CNN [10] A classifier that utilizes RNN and CNN to respectively represent the propagation path, and integrate two paths to detect rumors at the early stage of propagation.

Table 1. Statistics of the Weibo dataset

Statistic	Weibo
# Events	4664
# Rumors	2313
# Non-rumors	2351
# Users	2,746,818
# Posts	3,805,656
Avg. #of posts/event	816

4.4 Results and Discussion

This section presents the experimental results that demonstrate the state-of-the-art performance of our rumor detecting model. Table 2 shows the experimental results of our proposed model and that of baseline models. The three-view model achieves 95.6% accuracy on Weibo dataset. The baseline models listed in the table are carefully selected. So they are representative and classical methods for classification tasks. There are also three recently proposed state-of-the-art models that already achieve a great result, while our proposed model outperforms these baseline models. The reason why our model outperforms PPC_RNN+CNN is that our model introduces sentiment information. Besides, PPC_RNN+CNN adopts ensemble learning with CNNs and RNNs learning features, so that it's time-consuming. It is clear that our model achieves state-of-the-art performance based on all the evaluation indicators, including the overall accuracy, and the precision, recall, F1 score for rumor and non-rumor classes.

Table 2. Fake news detection results on Weibo dataset

Method	Class	Acc.	Prec.	Recall	F1
DTC	R	0.831	0.847	0.815	0.831
	N		0.815	0.847	0.830
SVM-RBF	R	0.818	0.822	0.812	0.817
	N		0.815	0.824	0.819
RFC	R	0.849	0.786	0.959	0.864
	N		0.947	0.739	0.830
SVM-TS	R	0.857	0.839	0.885	0.861
	N		0.878	0.830	0.857
DT-Rank	R	0.732	0.738	0.715	0.726
	N		0.726	0.749	0.737
GRU-RNN	R	0.910	0.876	0.956	0.914
	N		0.952	0.864	0.906
PPC_RNN	R	0.912	0.878	0.958	0.916
	N		0.944	0.866	0.908
PPC_CNN	R	0.919	0.899	0.958	0.922
	N		0.946	0.880	0.916
PPC_RNN+CNN	R	0.921	0.896	0.962	0.923
	N		0.949	0.889	0.918
Content+Reply+Sentiment	R	**0.956**	0.944	**0.966**	0.955
	N		**0.968**	0.947	**0.957**

According to the results in Table 3, we can find that models with two views yield better accuracy than the model with one view. For the models with a different number of views, the superiority of the 3-view model with a voting mechanism is explicit. This result reveals that the selected views make considerable sense and they can capture more useful information from different perspectives. Besides, other kinds of views can also be used in our model. As it was mentioned before, we adopted the pre-trained BERT model to construct a sentiment view. We are also interested in how accurate BERT can achieve in this task. Thus, we combine all the posts in the event as input and train $\text{BERT}_{\text{BASE}}$ model using veracity labels. Obviously, BERT reaches a very great result. We combine four views together, i.e., content, reply, sentiment, Bert using vote classification and assign weights for each view, and we are happy to find that the combined model achieves an even better result on this dataset. This implies that our proposed views do make contributions in the task and can improve BERT's prediction performance.

Table 3. Control experiment on different views of our model

View	Class	Acc.	Prec.	Recall	F1
Content	R	0.894	0.890	0.888	0.889
	N		0.898	0.899	0.899
Reply	R	0.929	0.908	0.948	0.928
	N		0.950	0.912	0.931
Sentiment	R	0.931	0.935	0.920	0.928
	N		0.928	0.941	0.935
Content+Reply	R	0.953	0.956	0.946	0.951
	N		0.951	0.960	0.955
Content+Sentiment	R	0.938	0.916	0.958	0.937
	N		0.960	0.920	0.939
Content+Reply+Sentiment	R	0.956	0.944	0.966	0.955
	N		0.968	0.947	0.957
Bert	R	0.961	0.941	0.980	0.960
	N		0.981	0.943	0.962
Content+Reply+Sentiment+Bert	R	**0.965**	0.946	**0.982**	0.964
	N		**0.983**	0.949	**0.966**

Figure 4 plots the relevance between hyperparameters and detection accuracy. It is clear that when the dimension of the hidden layer is 300, the model performs better than that using an other hidden size. Besides, the model's detection

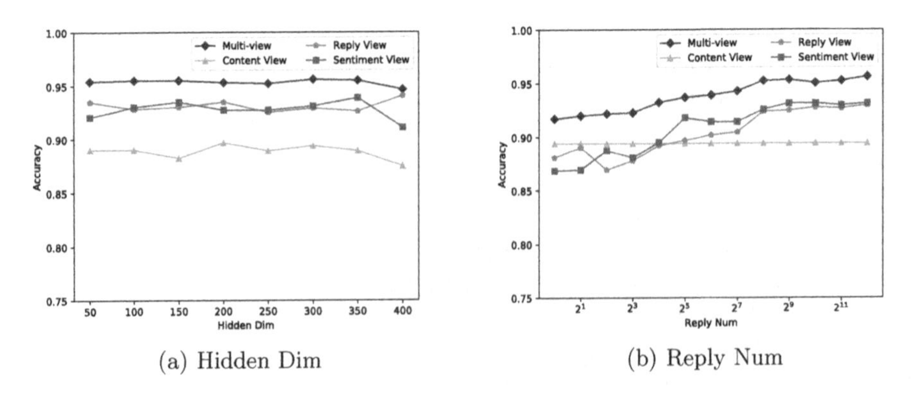

(a) Hidden Dim (b) Reply Num

Fig. 4. Rumor detection accuracy when different hyperparameter values are taken.

accuracy steadily ascends when the number of replies is increasing. This implies a larger number of replies provides more adequate information thus yield better performance.

The overall experiments demonstrate that each view can learn event representations from different perspectives and the fine-tuned pre-trained $\text{BERT}_{\text{BASE}}$ model is able to capture sentiment information expressed in replies and retweets, which can be utilized to guide the prediction for rumor events.

5 Conclusion and Future Work

In this paper, we provide insights into detecting real-world rumors. We created deep neural networks for automatically predicting the veracity of a rumor and using sentiment embeddings to help better distinguish rumors from true news. Our model achieves higher accuracy than existing baseline models in the task of rumor detection on Weibo dataset. The multi-view model we proposed comprehensively consider source tweet, reply and sentiment information. Despite that our model has a simple structure, it can be a hard-to-beat baseline since the model has already achieved 96.5% accuracy in the defined task. Since our model is generalizable and robust, other insightful views may also be added into it. In addition, we introduced the BERT model into our view, which assists in improving the final detection performance. In the future, we would like to exploit other views and build a more efficient model which has a faster speed but still demonstrates promising results.

Acknowledgments. This work was supported by the National Key Research and Development Program of China under Grant No. 2016YFB1000604.

References

1. Castillo, C., Mendoza, M., Poblete, B.: Information credibility on twitter. In: Proceedings of the 20th International Conference on World Wide Web, pp. 675–684. ACM (2011)
2. Chung, J., Gulcehre, C., Cho, K., Bengio, Y.: Empirical evaluation of gated recurrent neural networks on sequence modeling. arXiv preprint arXiv:1412.3555 (2014)
3. Ciampaglia, G.L., Shiralkar, P., Rocha, L.M., Bollen, J., Menczer, F., Flammini, A.: Computational fact checking from knowledge networks. PloS ONE **10**(6), e0128193 (2015)
4. De Sarkar, S., Yang, F., Mukherjee, A.: Attending sentences to detect satirical fake news. In: Proceedings of the 27th International Conference on Computational Linguistics, pp. 3371–3380 (2018)
5. Devlin, J., Chang, M.W., Lee, K., Toutanova, K.: Bert: pre-training of deep bidirectional transformers for language understanding. arXiv preprint arXiv:1810.04805 (2018)
6. The Guardian: Chinese panic-buy salt over Japan nuclear threat (2011). https://www.theguardian.com/world/2011/mar/17/chinese-panic-buy-salt-japan
7. Jin, Z., Cao, J., Zhang, Y., Luo, J.: News verification by exploiting conflicting social viewpoints in microblogs. In: AAAI, pp. 2972–2978 (2016)

8. Kingma, D.P., Ba, J.: Adam: a method for stochastic optimization. arXiv preprint arXiv:1412.6980 (2014)
9. Kwon, S., Cha, M., Jung, K., Chen, W., et al.: Prominent features of rumor propagation in online social media. In: International Conference on Data Mining. IEEE (2013)
10. Liu, Y., Wu, Y.F.B.: Early detection of fake news on social media through propagation path classification with recurrent and convolutional networks. In: AAAI (2018)
11. Long, Y., Lu, Q., Xiang, R., Li, M., Huang, C.R.: Fake news detection through multi-perspective speaker profiles. In: Proceedings of the Eighth International Joint Conference on Natural Language Processing (Volume 2: Short Papers), vol. 2, pp. 252–256 (2017)
12. Ma, J., et al.: Detecting rumors from microblogs with recurrent neural networks. In: IJCAI, pp. 3818–3824 (2016)
13. Ma, J., Gao, W., Wei, Z., Lu, Y., Wong, K.F.: Detect rumors using time series of social context information on microblogging websites. In: Proceedings of the 24th ACM International on Conference on Information and Knowledge Management, pp. 1751–1754. ACM (2015)
14. Ma, J., Gao, W., Wong, K.F.: Detect rumors in microblog posts using propagation structure via kernel learning. In: Proceedings of the 55th Annual Meeting of the Association for Computational Linguistics (Volume 1: Long Papers), vol. 1, pp. 708–717 (2017)
15. Ma, J., Gao, W., Wong, K.F.: Rumor detection on twitter with tree-structured recursive neural networks. In: Proceedings of the 56th Annual Meeting of the Association for Computational Linguistics (Volume 1: Long Papers), vol. 1, pp. 1980–1989 (2018)
16. Mendoza, M., Poblete, B., Castillo, C.: Twitter under crisis: can we trust what we RT? In: Proceedings of the First Workshop on Social Media Analytics, pp. 71–79. ACM (2010)
17. Qian, F., Gong, C., Sharma, K., Liu, Y.: Neural user response generator: fake news detection with collective user intelligence. In: IJCAI, vol. 3834, p. 3840 (2018)
18. Song, Y., Shi, S., Li, J., Zhang, H.: Directional skip-gram: explicitly distinguishing left and right context for word embeddings. In: Proceedings of the 2018 Conference of the North American Chapter of the Association for Computational Linguistics: Human Language Technologies, Volume 2 (Short Papers), vol. 2, pp. 175–180 (2018)
19. Vaswani, A., et al.: Attention is all you need. In: Advances in Neural Information Processing Systems, pp. 5998–6008 (2017)
20. Volkova, S., Shaffer, K., Jang, J.Y., Hodas, N.: Separating facts from fiction: linguistic models to classify suspicious and trusted news posts on twitter. In: Proceedings of the 55th Annual Meeting of the Association for Computational Linguistics (Volume 2: Short Papers), vol. 2, pp. 647–653 (2017)
21. Wu, Y., Agarwal, P.K., Li, C., Yang, J., Yu, C.: Toward computational fact-checking. Proc. VLDB Endow. **7**(7), 589–600 (2014)
22. Yang, F., Liu, Y., Yu, X., Yang, M.: Automatic detection of rumor on Sina Weibo. In: Proceedings of the ACM SIGKDD Workshop on Mining Data Semantics, p. 13. ACM (2012)
23. Zhao, Z., Resnick, P., Mei, Q.: Enquiring minds: Early detection of rumors in social media from enquiry posts. In: Proceedings of the 24th International Conference on World Wide Web, pp. 1395–1405. International World Wide Web Conferences Steering Committee (2015)

EmoMix: Building an Emotion Lexicon for Compound Emotion Analysis

Ran Li[1,2], Zheng Lin[1], Peng Fu[1(✉)], Weiping Wang[1], and Gang Shi[1,2]

[1] Institute of Information Engineering, Chinese Academy of Sciences, Beijing, China
{liran,linzheng,fupeng,wangweiping,shigang}@iie.ac.cn
[2] School of Cyber Security, University of Chinese Academy of Sciences, Beijing, China

Abstract. Building a high-quality emotion lexicon is regarded as the foundation of research on emotion analysis. Existing methods have focused on the study of primary categories (i.e., anger, disgust, fear, happiness, sadness, and surprise). However, there are many emotions expressed in texts that are difficult to be mapped to primary emotions, which poses a great challenge in emotion annotation for big data analysis. For instance, "despair" is a combination of "fear" and "sadness," and thus it is difficult to divide into each of them. To address this problem, we propose an automatic building method of emotion lexicon based on the psychological theory of compound emotion. This method could map emotional words into an emotion space, and annotate different emotion classes through a cascade clustering algorithm. Our experimental results show that our method outperforms the state-of-the-art methods in both word and sentence-level primary classification performance, and also offer us some insights into compound emotion analysis.

Keywords: Emotion lexicon · Compound emotion · Text emotion analysis · Natural language processing

1 Introduction

With the rise of opinion-rich content in social media, such as Twitter, Sina Microblog, Facebook and WeChat, there is an increasing desire for individuals to express their feelings, mood, and emotions rather than just browsing and receiving information. These emotions, expressed in sentences, contain useful information for understanding public opinion, opinion mining, business decisions, mood management, and information forecasting [5]. Different from mature sentiment lexica labeled with sentiment polarity, emotion lexicon should contain multiple emotions with different intensity [21].

Analyzing and summarizing the emotions in texts has become an area of focus for many studies. An emotion lexicon, which reflects the unstructured characteristics of texts, plays an important role in emotion analysis and provides the advantage of high speed and easy understanding. In addition, it is also a fundamental task for affective modification and mood management, which are essential

J. M. F. Rodrigues et al. (Eds.): ICCS 2019, LNCS 11536, pp. 353–368, 2019.
https://doi.org/10.1007/978-3-030-22734-0_26

for research institutions, information consulting organizations, and government decision-making departments.

Much work has been done on emotion lexicon construction [15–17,20,21,23]. Xu *et al.* [21] manually labeled a large Chinese lexicon with part-of-speech(POS), emotion, and intensity. In addition, Staiano *et al.* [17] and Song *et al.* [16] used a crowd-based method for emotion lexicon construction. Song *et al.* [15] synchronously used seed words and emoticons to build an emotion lexicon. However, their lexica all focus on primary emotions(i.e. happiness, sadness, like, anger, disgust, fear, and surprise), which are not enough for representing our daily lives.

The emotions of humans are complex, and many emotions are derived from two or more primary emotions, which are called compound emotions in the field of psychology. Compound emotions express the complex relationship between people and objective things that are widely used in our daily lives. A Chinese Microblog regarding the 2014 Malaysia airlines event can be considered as an example, "你患马航焦虑症 了吗？患了，时刻关注马航事件，迫切想知道结果。" (which translates to: "Do you suffer from Malaysia airlines anxiety disorders? Yes, I'm keeping track of Malaysia airlines and eager to know about the result"). Both fear and anticipation are expressed in this microblog, and thus anxiety, a compound emotion, can be directly used to make a detailed analysis. Learning about compound emotions can allow us to accurately understand the tendency of hot social issues. In another context, compound emotions can also useful in a test for depression. There are many different types of depression, and each type of depression has its own emotional characteristics. Anxiety is a major symptom of menopausal depression [14], other kinds of depression, such as postpartum depression (guilt and inferiority) [2] and major depressive disorder (sad, anxious, guilty, sentimentality, pessimism, and in the worst case, insomnia or even try to suicide) all have their own characteristics of different compound emotions. Effectively identifying the compound emotions can provide a foundation for detecting and predicting depression. Therefore, it is necessary to build a compound emotion lexicon, which can be used as a fundamental tool for sophisticated applications of emotion analysis.

In this paper, we construct a novel compound emotion lexicon called EmoMix, for complex emotion analyzing. Specifically, the candidate words are mapped into an emotion space that is built based on Plutchik's emotion wheel [11]. Then a cascade clustering algorithm is used to tag the candidate words with compound emotion labels. The major contributions of this work are as follows:

- We propose a novel building method of emotion lexicon to meet the demands of compound emotion analysis. To the best of our knowledge, our EmoMix is the first method for creating a lexicon with the psychological theory of compound emotion.
- We propose a unified emotion labeling scheme based on a cascade clustering algorithm to address the requirements of compound emotions. Our method uses an emotion space to help find the most relevant emotion words. Although we built a Chinese lexicon as an example, the building method is language independent.

- We conduct experiments for lexicon quality, state-of-the-art emotion lexica are used for evaluation, and show that EmoMix is competent in the emotion classification task. Additionally, the results of case studies demonstrate that EmoMix outperforms the conventional emotion lexica in compound emotion analysis.

The rest of this paper is organized as follows: In Sect. 2, we review the related work. Section 3 presents some preliminaries. Section 4 describes the details of our proposed EmoMix lexicon. Section 5 presents performance evaluation results. Finally, Sect. 6 concludes the paper.

2 Related Work

This paper involves two research focuses: emotion models and emotion lexicon construction.

Emotion Models. In general, psychologists classify emotion models into two categories, discrete [4,18] or dimensional models [11,13]. Discrete models introduce the basic cognition of human emotions from psychologists. In previous work, Tomkins [18] classified human emotions into eight primary emotions: surprise, interest, joy, rage, fear, disgust, shame, and anguish. Subsequently, Ekman [4] put forward another primary emotion model with six emotions, namely, anger, disgust, fear, happiness, sadness, and surprise. However, it is difficult for these models to describe complex emotions within the constraints of the number of emotions and their independence characteristic. Human emotions are considered to lie in two or three dimensions in the dimensional models. Plutchik offered a 3-dimensional model, called the wheel of emotions [11], which contains primary emotions, intensity dimensions, and compound emotions. Among them, compound emotions are acquired by adding two primary emotions together.

It is widely accepted that human expressions cannot be attributed to only primary emotions. We built a compound emotion lexicon based on Plutchik's wheel of emotions, which is more suitable for the variety of human expressions.

Emotion Lexicon Construction. Emotion lexicon plays an important role in opinion mining and emotion analysis. Unlike a sentiment lexicon, an emotion lexicon has more granular classifications. Existing emotion lexicon construction methods can be roughly classified into manual and automated labeling. Xu *et al.* [21] constructed an affective lexicon ontology (ALO) in Chinese, which leveraged sentiment lexica and wordnet to obtain candidate words. They manually labeled the candidate words with emotional categories and intensity, and divided them into seven primary categories, and 21 subclasses. Then, Staiano *et al.* [17] presented a crowd-sourced approach DepecheMood (DPM) for emotion lexicon generation by combining the document-frequency distributions of words and the emotion distributions over documents. This method used the score in each emotion dimension to represent the strength of the emotion. As an extension to their work, Song *et al.* [16] analyzed the emotions of Rappler news articles at the topic level. By developing a non-negative matrix factorization model, a fine-grained emotion lexicon was built that associated words with emotions based on the hidden topics obtained from the factorization process. An approach based on manual labeling is labor-intensive, slow, and costly; therefore, researchers have focused more attention on automatic lexicon construction.

Xu *et al.* [20] proposed an automatic method for building a Chinese emotion lexicon by using WordNet-Affect. Their method translated WordNet-Affect into Chinese and filtered all non-emotion words. Then, synonym words were expanded to obtain the Chinese emotion lexicon. To address domain dependence, Yang *et al.* [23] proposed an Emotion-aware LDA (EaLDA) model to automatically build a domain-specific lexicon with six primary emotions. To solve the data scarcity problem, Song *et al.* [15] adopted a multi-label random walk algorithm based on a three-layer heterogeneous graph to build the emotion lexicon. They combined the effects of seed words and emoticons co-occurring with candidate words to capture the fine-grained emotion of candidate words. However, these lexica ignore the compound emotion at the word level, which results in a loss of valuable information for emotion analysis in complex situations.

In this paper, we introduce a cascade clustering algorithm into lexicon construction, and generate a compound lexicon with advantages as shown in Table 1.

Table 1. A comparison of emotion lexicons

Lexicon	Language	Psychology emotion model				
		Usage	Model	Primary	Secondary	Compound
Lex_xu [20]	Chinese	√	Ekman	6	×	×
EaLDA [23]	English	√	Ekman	6	×	×
Lex_song [15]	Chinese	×	-	7	×	×
DPM [17]	English	×	-	8	×	×
Lex_song [16]	English	×	-	8	×	×
ALO [21]	Chinese	×	-	7	√(21)	×
EmoMix (ours)	**Chinese**	√	**Plutchik**	**8**	**√(24)**	√

3 Preliminaries

3.1 Plutchik's Wheel of Emotions and Compound Emotion

The psychologist Robert Plutchik created the wheel of emotions [11], which is widely accepted for illustrating the various relationships among the emotions. As shown in Fig. 1, there are three important concepts to his theory: (1) **Primary Emotions.** There are eight primary emotions that are indecomposable (joy, trust, fear, surprise, sadness, anticipation, anger, and disgust). (2) **Emotion Intensity.** Each primary emotion has different degrees of intensity, which range from very light to very intense. (3) **Compound Emotions.** All emotions other than primary emotions occur as a result of a combination of the primary emotions. Recently, some work in natural language processing (NLP) has started to pay attention to this theory; however, most of the focus is on primary emotions [3] and emotional intensity [1], which do not fully express the complex feelings of users. If primary emotion lexica were used to analyze the 2014 Malaysia airlines event in Sect. 1, one might only obtain the anticipation of a compound emotion "anxiety" and miss the component of fear.

Fig. 1. Plutchik's emotion wheel

Table 2. The extension of compound emotions

Human feelings	Emotions	Opposite feelings	Emotions
Optimism	Anticipation + Joy	Disapproval	Surprise + Sadness
Hope	Anticipation + Trust	Unbelief	Surprise + Disgust
Anxiety	Anticipation + Fear	Outrage	Surprise + Anger
Love	Joy + Trust	Remorse	Sadness + Disgust
Guilt	Joy + Fear	Envy	Sadness + Anger
Delight	Joy + Surprise	Pessimism	Sadness + Anticipation
Submission	Trust + Fear	Contempt	Disgust + Anger
Curiosity	Trust + Surprise	Cynicism	Disgust + Anticipation
Sentimentality	Trust + Sadness	Morbidness	Disgust + Joy
Awe	Fear + Surprise	Aggressiveness	Anger + Anticipation
Despair	Fear + Sadness	Pride	Anger + Joy
Shame	Fear + Disgust	Dominance	Anger + Trust

Different from existing work, to meet the demand of complex emotion analysis, this study focuses on the third concept, "Compound Emotions." To build such a compound emotion lexicon, the combinations of two adjacent primary emotions are not enough. Following the definition of "primary dyad" in Plutchik's theory (the combinations of two adjacent primary emotions), Turner, another psychologist, extended the rest of the compound emotions [19]. He defined "secondary dyad" (mixed emotions that are one step apart on the wheel) and "tertiary dyad" (emotions that are two steps apart on the wheel) emotions.

Table 3. Chinese microblog examples

Weibo_id	Date	Weibo	Emotion
走饭	2012-2-28	不管谁对我说重话我都无所谓不过十分钟后我会躲到厕所里哭。 (which translates to: "I don't care who says something bad to me, but I'll be crying in the bathroom in ten minutes.")	sentimentality (sadness+trust)
	2012-3-18	我有抑郁症，所以就去死一死，没什么重要的原因，大家不必在意我的离开。拜拜啦。 (which translates to: "I'm suffering from depression, thus I try to die. There is no important reason. Don't care about my departure. Bye.")	despaired (sadness+fear)
你不管哇	2016-4-20	人的一切痛苦，本质上都是对自己的无能的愤怒。(which translates to: "All the pain of a human is essentially anger at one's own incompetence.")	contempt (anger+disgust)
	2018-1-08	锁上我的记忆，锁上我的忧伤，不再想你，怎么可能再想你，快乐是禁地，生死之后，找不到进去的钥匙。 (which translates to: "Lock my memory, lock my sadness, I'll no longer think of you. How can I think of you? Happiness is forbidden, after life and death, I cannot find the key to go in.")	despaired (sadness +fear)
末离_moli	2017-1-17	每一个到来的人，我都知道他们终要走，这次我要走了，没有送别的人。 (which translates to: "Everyone who comes, I know they are going to leave, this time I want to go, no one to see off.")	despaired (sadness+fear)
	2017-1-19	这次大概是真的要结束了。(which translates to: "This time I'm really coming to the end.")	despaired (sadness+fear)

Additionally, he pointed out that the emotions on the opposite side of the wheel are in conflict, and they can't generate compound emotions. As shown in Table 2, we can discover that the extended compound emotions also appear as pairs of opposites.

The above-mentioned research has provided the theoretical foundation for building a compound emotion lexicon. In this work, the rules of compound emotions and opposite emotions function as theoretical guides for emotion lexicon construction.

3.2 Motivation and Basic Idea

This work is supported by using a set of observations from users' microblogs who suffer from depression. We collected over 1,600 microblogs from 57 users who committed suicide due to depression, as shown in Table 3. For example, consider the following Chinese microblog, "我有抑郁症，所以就去死一死，没什么重要的原因，大家不必在意我的离开。拜拜啦。" (which translates to: "I am suffering from depression, thus I try to die. There is no important reason. Don't care about my departure. Bye."), which expresses pessimism and despair. It is very difficult to express these feelings by any single primary emotion. As bloggers' delicate emotions move far beyond primary classes and are much more complicated, a fundamental tool for compound emotion analysis is urgently required.

It is a common phenomenon in our daily lives that our expressions are often complex and overlapping. Even more than one primary emotion can be expressed in a single word. After statistical analysis of these microblogs, we find that there are over 1,800 compound emotions, accounting for 63.7% of the total emotional words. This implies that a primary classification only gives limited information on the actual intentions of users' expressions. However, most existing Chinese lexica classify emotional words into a single class from six to eight primary emotion categories. A few works [21] can be found where emotional words are labeled with multiple primary emotions; however, they are only a tiny proportion (14.3%) of the total number of words in the lexicon. Based on these observations from microblogs, the idea behind our EmoMix is to explore new methods for building a compound lexicon that can provide a foundational resource in emotion applications. Two primary emotions can evolve into a new emotion, and the new emotion category is no longer a part of primary categories. Each emotion, including primary and compound emotions, could combine with negation words to convert it to the opposite emotion. With such a compound emotion lexicon, we can provide many personalized applications could be provided that are difficult to accomplish using only a primary emotion lexicon. In addition to emotion management and depression detection, a compound emotion lexicon could be applied to recommend products, track hotpots, and forecast trends.

Above all, our goal is to build a compound emotion lexicon with the following characteristics: (1) The lexicon should be in accordance with modern psychological theory. (2) The lexicon should have a computational model of emotions. (3) The lexicon should directly reflect the compound emotion category of each word.

To meet the design principles above, we follow the steps described below to build the lexicon. First, we select the domain-independent and widely-applicable corpus to train word vectors. Word embeddings trained from different corpus are compared and we choose the best fit for our work. Second, we construct an emotion space accompanied by Plutchik's wheel of emotions and map the emotional words into the emotion space. The emotion space is built to refine the candidate words so that they could be closer to both semantically and emotionally similar words and further away from emotionally dissimilar words. Third, similar words are classified through cluster analysis. According to the position of word embedded in the emotion vector space, the emotional words could be divided into different compound emotion categories. In this way, a lexicon is obtained that has compound emotions in the word level.

4 Emotion Lexicon Construction

4.1 Dataset

To build a compound emotion lexicon, we gather different kinds of resources, such as dictionaries (Chinese synonym dictionary[1], Contemporary Chinese Dictionary, and NTUSD[2]) and semantic networks (HowNet and WordNet). From these resources, candidate words are selected that are related to emotion, such as psychology, feeling, affection, character, and attitude. The unlabeled corpora of different domains (Microblogs, News, Literature, Encyclopedias, etc.) are compared to train the word embeddings, and Baidubaike (an online Chinese encyclopedia)[3] is chosen for our work due to its wide coverage and accordance with the rules of NLP. We use SGNS (skip-gram model with negative sampling) [9] to train word embeddings on Baidubaike [8], which includes 0.745G word tokens. We train 300-dimensional word vectors with n-gram features from the corpus.

4.2 The Emotion Space

Generally, existing word embeddings can capture semantic and syntactic information from an unlabeled corpus, but they cannot acquire sufficient emotion information, which makes it difficult to apply the word embedding directly to emotion analysis. Therefore, we propose an Emotion Space to refine the word embedding. The emotion space can transfer semantic similarity into the emotion similarity. As discussed in Sect. 3.1, one compound word can be composed of multiple primary emotions. Each of these primary emotions and its opposites on Plutchik's wheel form an emotion pair. These emotion pairs can be regarded as axes to build a 4-*dimensional* emotion space. To better represent the compound emotion character of words, the cosine similarity between candidate words and

[1] http://ir.hit.edu.cn/.
[2] http://nlg.csie.ntu.edu.tw/download.php.
[3] https://baike.baidu.com/.

primary emotions is computed, and then these words are mapped into the emotion space.

Formally, the set of candidate words is denoted as C, and the seed words of the primary emotions that are used to predict the emotion similarity are denoted as S, respectively. The i-th primary emotion pair can be presented as $\langle s_i^+, s_i^- \rangle \in S$, where $i \in \{1, 2, 3, 4\}$ represents the index of the primary emotion pair. For example, we use ⟨高兴(*Joy*),悲伤(*Sadness*)⟩ as $\langle s_1^+, s_1^- \rangle$, which is presented as the positive and negative of one axis of the emotion space. For each candidate word $c^* \in C$, 3COSMUL [7] is exploited to calculate the similarity between the word embedding of the candidate words and the seed words of every four emotion pairs. The cosine similarity measurement can be defined as $\cos(u, v) = \frac{u \cdot v}{\|u\| \|v\|}$. Thus, we can obtain the emotion similarity e_i that represents how much closer the word is to a primary emotion and further away from its opposite.

$$e_i = \arg\max_{c^* \in C} \frac{\prod_{j=1}^{m} \cos(c^*, s_{i,j}^+)}{\prod_{k=1}^{n} \cos(c^*, s_{i,k}^-) + \varepsilon}, \tag{1}$$

where $\varepsilon = 0.001$ is used to prevent division by zero, m is the number of seed words in s_1^+, and n is the number of seed words in s_1^-.

After calculating the four emotion similarities with each emotion pair, we can express the emotion vector of the candidate word $c^* \in C$ as:

$$E_{c^*} = (e_1, e_2, e_3, e_4), \tag{2}$$

When the candidate word is more similar to the opposite emotion, then the value of the emotion similarity e_i will below the zero. If each opposite emotion pair is regarded as the two extremes of a single axis, then all candidate words can be mapped into the emotion space by using E_{c^*} as the coordinate values. As shown in Fig. 2, there are four quadrants representing different compound emotions in the hyperplane composed of every two emotion axes.

4.3 Clustering Algorithm for Building the Lexicon

In this section, a cascade clustering algorithm is used to build the compound emotion lexicon. Since the emotion similarities are calculated by different seed words, the values on different axes are not in the same ratio. We cannot simply rank the four values of the axes and select the top two of them to mix into a compound emotion. To obtain more accurate results, a density-based clustering algorithm is leveraged to group candidate words into subclasses naturally. Then, all subclasses are divided into primary or compound emotions with the help of a modified k-means method, as shown in Fig. 3.

Specifically, let $C = \{E_{c_1}, E_{c_2}, \cdots, E_{c_n}\}$ be the set of pre-processed emotion vectors corresponding to the candidate words in emotion space. The measure

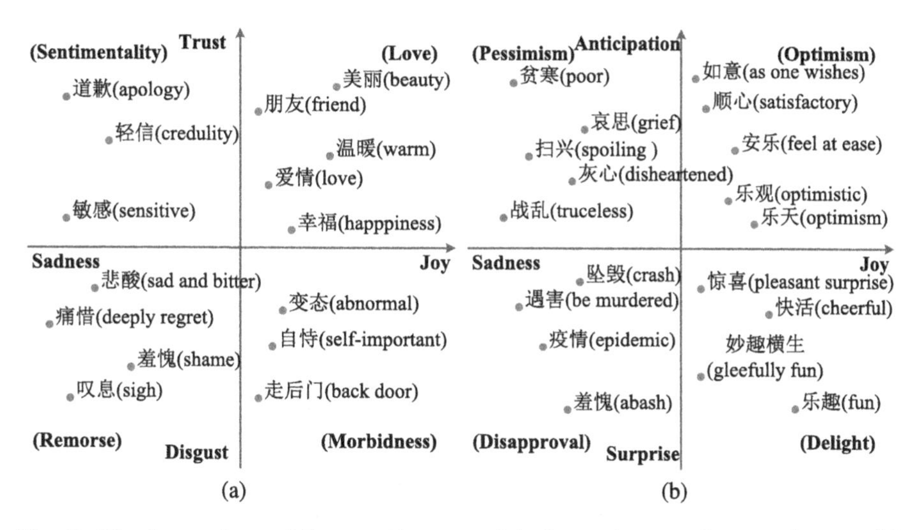

Fig. 2. The hyperplane of the emotion axes (a): Joy-sadness and trust-disgust; (b): Joy-sadness and anticipation-surprise

used in literature [12] is used to find out the density peaks of candidate words. The local density ρ_{c_i} of the candidate word c_i is defined as

$$\rho_{c_i} = \sum_{j} \chi(d_{c_i,c_j} - d_{cut-off}), \tag{3}$$

where $\chi(x) = 1$ if $x < 0$; otherwise, $\chi(x) = 0$; and $d_{cut-off}$ is a cutoff distance. A parameter t is used to adjust the size of $d_{cut-off}$. The minimum distance between the candidate word c_i and any other word with a higher density can be denoted as follows:

$$\delta_{c_i} = \min_{j:\rho_{c_i} > \rho_{c_j}} (d_{c_i,c_j}), \tag{4}$$

where $\delta_{c_i} = \max_j(d_{c_i,c_j})$ for the highest density word. After comparing the quantity $\gamma_{c_i} = \rho_{c_i}\delta_{c_i}$, the highest density word is obtained, which is the center of the emotion subclass $Sub = \{sub_1, sub_2, \cdots, sub_n\}$, where n is the number of subclasses.

The idea of k-means is used in the next step. Eight typical subclasses ($k = 8$) were selected as the initial primary emotion classes, which are denoted as $Pri = \{pri_1, pri_2, \cdots, pri_8\}$, and then the similarity between the rest of the subclasses sub_j is calculated, where $j \in \{1, 2, \cdots, n-8\}$ and each primary emotion class pri_i, where $i \in \{1, 2, \cdots, 8\}$. The similarity function can be presented as follows:

$$similarity^{(i)} := \arg\min_{j} \left\| x^{(i)} - \mu_j \right\|^2. \tag{5}$$

Specially, another constraint is added to limit the k-means method to consolidated emotion subclasses. The distribution of eight similarities is normalized into a range of [0, 1], and then they are ranked and the ratio of the top two emotions is calculated. Thus, the constraint of the similarity ratio can be defined as:

$$Constraint = \begin{cases} Consolidated & \text{if } ratio_{x,y} \geq \lambda, \\ Not\ Consolidated & \text{otherwise.} \end{cases} \quad (6)$$

If the similarity ratio is greater than a threshold λ, then the subclass is classified into the nearest primary emotion; otherwise, it retains the two primary emotions labels, which are regraded as compound emotions. Then, the center of a new primary emotion is relocated, and all similarities are recalculated. The consolidated process is repeated until a global stable state is achieved and all the primary and compound emotions remain unchanged.

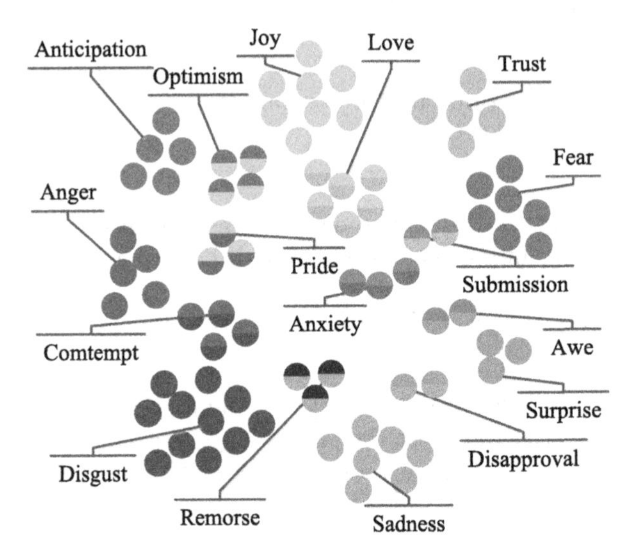

Fig. 3. Emotion classify based on clustering

5 Experiments

In this section, we describe the empirical evaluations of the proposed EmoMix. Since there is no existing method to metric the quality of compound lexica, its performance is verified in following ways: (1) we use a standard task of primary emotion analysis to test EmoMix with traditional lexica on the word and sentence levels; (2) we conduct several case studies of compound emotion analysis to show the potential of EmoMix in different applications.

5.1 Quality of Lexicon on the Word Level

Baseline: To the best of our knowledge, this method is the first method for creating compound emotion lexica. Thus, we can only compare Precision, Recall, and F-measure of the lexica generated through state-of-the-art methods with the lexicon ALO [21]. We choose ALO as a baseline because it is a large scale manually crafted lexicon and it is widely regarded as a standard in Chinese emotion analysis. The Precision P and Recall R are defined as follows:

$$P = \frac{\sum_{e \in E} |W_{ALO}(e) \cap W_{TestLex}(e)|}{\sum_{e \in E} |W_{TestLex}(e)|}, \tag{7}$$

$$R = \frac{\sum_{e \in E} |W_{ALO}(e) \cap W_{TestLex}(e)|}{\sum_{e \in E} |W_{ALO}(e)|}, \tag{8}$$

where E is the set of seven emotions, $W_{ALO}(e)$ is the word set in emotion e of ALO, $W_{TestLex}(e)$ is the word set with emotion label e in the testing lexicon, and the F-measure F is defined as $F = \frac{2 \cdot P \cdot R}{P+R}$.

As the emotion categories present in ALO do not exactly match with the ones provided by Plutchik's wheel (there are seven main classes in ALO while there are eight in Plutchik's). To meet the baseline ALO, EmoMix is adjusted by merging "trust" and "anticipation" into the emotion "like," and mapping "joy" to "happiness," as shown in Table 4. For a better comparison, we also use the same seed words in different lexicon construction methods.

Tuning the Parameter: There are two adjustable parameters in our method, including the size of emotion subclasses parameter t and the threshold of the top-2 primary emotions λ. We investigate how the parameter t and threshold λ affect the final result of clustering. Generally, to obtain enough numbers of the emotion subclasses, a smaller parameter t is better; however, if we regard each word as a subclass under extreme conditions, then the natural distribution of emotion classes is lost. Finally, we set parameter $t = 0.02$. λ is also an important parameter for determining whether it is a primary or compound emotion, and we chose a threshold $\lambda = 7$. Therefore, the words in primary emotion classes would have a domination of emotion distribution.

Lexicon Evaluation: Experiments are conducted on the following emotion lexicon generation methods for comparison: (1) The DPM lexicon [17] is built by document-by-emotion matrix M_{DE} and word-by-emotion matrix M_{WE}. M_{DE} is built from crowd-sourced emotions data and M_{WE} is obtained by using compositional semantic method over M_{DE}. (2) The PMI lexicon [10] uses pointwise mutual information (PMI)-based scores and starts from a small set of seed words for emotion lexicon construction. (3) The Lex_yang lexicon [22] is built from a variation of PMI to calculate the collocation strength between candidate word and the emoticon. Every word entry of the lexicon contains several emotion senses ordered by the collocation strength. (4) The Lex_song lexicon [15] is obtained from a multi-label random walk algorithm combining the seed words and emoticons to construct the emotion lexicon. (5) The EmoMix_s lexicon is generated as a special case of our method that uses only emotion space to label words. (6) EmoMix lexicon is built using the full configuration of our method, which combines the emotion space and the cascade clustering algorithm.

The results are shown in Table 5, and we can easily find that EmoMix has obvious advantages over DPM and Lex_yang. Although our Precision is only a

Table 4. Mapping of EmoMix on ALO and Lex_song

ALO [21]	Lex_song [15]	EmoMix	ALO	Lex_song	EmoMix
Disgust	Disgust	Disgust	Like	Like	**Trust**
Sadness	Sadness	Sadness	Like	Like	**Anticipation**
Anger	Anger	Anger	Fear	Fear	Fear
Happiness	Happiness	**Joy**	Surprise	Surprise	Surprise

Table 5. The quality of lexica on emotions

Methods	DPM [17]	PMI [10]	Lex_yang [22]	Lex_song [15]	EmoMix_s	**EmoMix**
Precision	0.202	0.282	0.361	0.541	0.520	**0.545**
Recall	0.060	0.083	0.106	0.159	0.236	**0.352**
F-measure	0.092	0.128	0.164	0.246	0.308	**0.428**

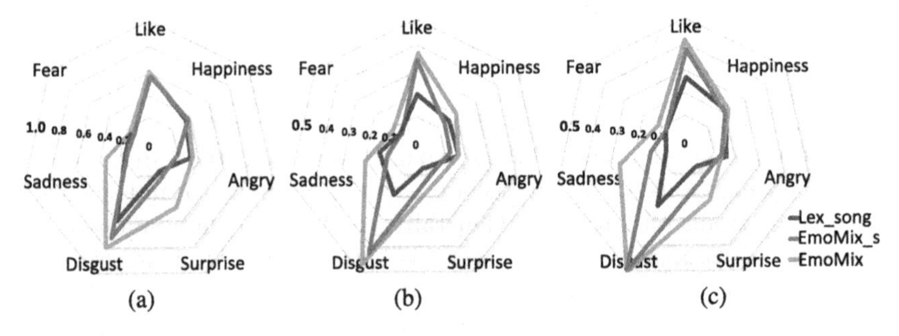

Fig. 4. Emotion performance of lexica (a): Precision; (b): Recall; (c): F-measure

bit more than Lex_song, our Recall and F-measure were higher, which means it could detect and identify more emotion words. The advantage of EmoMix comes from its emotion expression and the natural distribution of emotion classes.

In addition, we further study the performance of each class in the three best lexica, as shown in Fig. 4. The "like" and "disgust" emotions are higher than other emotions, and "fear" is much lower. In fact, the taxonomy of ALO is not very appropriate for emotion analysis. The reason for this is that their classes of "like" and "disgust" are more similar to "positive" and "negative" in sentiment lexica. For example, the subclasses "trust" and "wish" in "like" are not as fine-grained as "like," and "suspicion" in "disgust" is the same. We expect that EmoMix can offer more elaborate categories in both emotion classes and subclasses for emotion analysis applications.

5.2 Emotion Classification on the Sentence Level

Sentence level experiments are conducted based on the public dataset, **Emotion Analysis in Chinese Weibo Texts (EACWT)**. This dataset has been widely used to verify primary emotion classification, which is provided for the emotion analysis shared task in NLP&CC2013. The emotional sentences in EACWT are annotated with emotion labels from anger, disgust, fear, happiness, like, sadness, or surprise (the same as in ALO). In this experiment, we compare the performance between the best performance on the sentence level and EmoMix. Furthermore, our results are compared with those obtained by a supervised learning TextCNN [6]. For the unbalanced distribution of the emotion classes, class-by-class classification results are provided as well as the overall performance of experiments.

A straight-forward voting-based algorithm is used for each lexicon to assign emotion labels for sentences, in the same manner as in [15]. Note that our main idea is to compare the quality of lexicon building methods rather than improve the performance of emotion classification on the sentence level; therefore, we only consider the negation and degree words in this task. We also use a basic CNN algorithm with raw training data for the same goal. The results for single-label classification are summarized in Table 6. As shown, EmoMix not only outperforms the state-of-the-art lexicon, Lex_song, but it also provides a competitive performance with the supervised learning method, TextCNN. It performs poorly when the data is lacking or imbalanced because the results of supervised leaning methods are closely impacted by the training dataset. It is worth stressing that Lex_song is a domain-specific lexicon of Weibo, while EmoMix is not; therefore, we expect EmoMix has general applicability in other domains.

Table 6. Sentence level emotion classification

Methods	TextCNN [6]			Lex_song [15]			EmoMix		
Indicators	Precision	Recall	F-measure	Precision	Recall	F-measure	Precision	Recall	F-measure
Anger	0.216	0.423	0.286	0.258	0.024	0.043	0.286	0.026	0.047
Disgust	0.243	0.197	0.218	0.504	0.113	0.184	0.502	0.127	0.202
Fear	0.017	0.011	0.013	0.043	0.005	0.010	0.125	0.008	0.016
Happiness	0.263	0.243	0.252	0.501	0.086	0.147	0.579	0.095	0.163
Like	0.419	0.436	0.427	0.601	0.320	0.417	0.619	0.345	0.443
Sadness	0.347	0.371	0.359	0.526	0.009	0.017	0.636	0.009	0.018
Surprise	0.009	0.014	0.011	0.400	0.015	0.029	0.444	0.017	0.033
Total	0.227	**0.227**	0.227	0.552	0.139	0.222	**0.581**	0.148	**0.236**

5.3 Case Studies on the Paragraph Level

In this paper, the research question is to raised as to whether primary emotion classes can be used to reliably evaluate complex emotions in multiple kinds of applications. The advisable emotion categories can help reduce error classification, improve the quality of analysis, and increase the consistency of results. Because of that, we analyze four different kinds of media as a case study, including headline (news), microblog, blogs, and fairy tales (literature). As shown in Table 7, the different emotions expressed in several types of texts are analyzed.

Table 7. Case studies

ID	Text	Type	Media	ALO	EmoMix
1	有他们的地方就有平安！(which translates to: "Where they are, there is peace!")	Sentence	Headline	Like	Admiration
2	深圳虐童事件背后：社会该如何预防父母虐童？(which translates to: "Behind the event that parents abused their daughter in Shenzhen: how to prevent child abuse?")	Sentence	Microblog	Anger	Anxiety
3	我只能把这份不能爱的爱，深藏心底，我只能这样静静的去品其中的酸甜苦辣。......每一次的忆起，都让我身心震撼，每一次的忆起，都叫我无法去控制自己那颗跳动的心！(which translates to: "I can only put this love which is not love, deep in my heart, I can only quietly taste the bittersweet....... Every time I think of it, it will make me shocked and I cannot control my beating heart!")	Paragraph	Blogs	Surprise	Sentime-ntality
4	《丑小鸭》节选:他拍拍翅膀，发现翅膀比以前有力得多，他试了两下，竟然可以飞起来了!他飞出沼泽，飞过森林，到了一个美丽的湖边，这时，他又看见了那群美丽的天鹅，正在水面上游来游去。......"杀死我吧，哦，杀死我吧！" 他把头低低地垂到水上，只等待着死。(which translates to: "《The Ugly Duckling》 Then, all at once, the duckling raised his wings and they flapped with much greater strength than before. Before he knew it, he found himself in a large garden where the apple trees were in full blossom and long branches of lilacs overhung the shores of the lake...."Kill me! Oh, kill me!" said the poor duckling, and he waited for his death bowing his head towards the water.")	Chapter	Fairy Tales	Sadness	Despair

Headlines. Headlines are an important text that are used to attract readers, and they directly affect the subsequence output of news media. Moreover, headlines are not just about the topics, but they are also about the emotion expressed regarding the topic. Readers often determine whether to read on according to the emotional orientation of news headline, and hence, fine-grained emotion analysis cannot be ignored. As an example, the emotion "admiration" has a higher distinction over "like," and applying this study to headlines would help build more personalized recommendation services.

Microblogs. A traditional microblog is limited to 140 words, which is called a short-text. Users often express their feelings with one or a few sentences. However, they might not realize that the emotions in microblogs could affect their relationship with friends. Mood management, a burgeoning psychological assistant application, could help users improve the control ability of emotions. In particular, some special psychological assistant application, like the detection of autism and depression, could effectively help users and family monitor their moods and avoid tragedies. During this process, fine-grained emotions can assist doctors to find out about different psychological characteristics and provide more options for the treatment or prevention.

Blogs. A blog is another type of writing for self-expression that can be a paragraph or an article. Bloggers have enough space to express their feelings, and thus the emotions in blogs are usually exquisite, complex, and multiple. A blogger may have buried an inexpressible feeling or sentimentality, at the bottom of her heart; traditional primary emotions are limited by the number of categories, and they would be insufficient at expressing such complex feelings.

Fairy Tales. Literature is a much larger category of text that can be as long as a chapter level or a document level. Authors like to write many foreshadowing plots to highlight protagonists' emotions. For example, in The Ugly Duckling, after so much suffering, the ducking's feeling of despair is very difficult to express with only a primary emotion. It is worth mentioning that the emotional analysis of literature has a business value in Text-to-Speech(TTS), which could make the voice of machine more emotional.

Our findings indicate that the value of compound emotions in question is indeed competitive. These results pave the way to the practical analysis emotions in real life with simple, and comprehensible analysis benchmarks.

6 Conclusion and Future Work

In this paper, we presented EmoMix, a compound emotion lexicon, which was built using a novel construction method. Based on Plutchik's wheel of emotions, we first constructed the emotion space by using four pairs of primary emotions and mapped the word embeddings into it. Then, to group the words naturally, a cascade clustering algorithm was applied, and the candidate words were classified into primary and compound emotion classes. Experimental results showed that EmoMix was competitive to the current state-of-the-art lexica on primary emotion classification. Additionally, it had a promising performance on compound emotion analysis over other lexica.

In future work, we would like to add a combination for the intensity and extend the method to the other languages. In addition, we will further enhance the quality and capacity of our compound emotion lexicon, EmoMix.

Acknowledgement. This work is supported by the National Key R&D Program of China (No. 2016YFB1000 604), and the National Natural Science Foundation of China (No. 61602467).

References

1. Abdul-Mageed, M., Ungar, L.: EmoNet: fine-grained emotion detection with gated recurrent neural networks. In: ACL, vol. 1, pp. 718–728 (2017)
2. Brockington, I.F.: Postpartum depression: causes and consequences. Am. J. Psychiatry **155**(2), 303–304 (1998)
3. Deyu, Z., Zhang, X., Zhou, Y., Zhao, Q., Geng, X.: Emotion distribution learning from texts. In: EMNLP, pp. 638–647 (2016)
4. Ekman, P.: An argument for basic emotions. Cogn. Emot. **6**(3–4), 169–200 (1992)
5. Kannangara, S.: Mining twitter for fine-grained political opinion polarity classification, ideology detection and sarcasm detection. In: WSDM, pp. 751–752 (2018)
6. Kim, Y.: Convolutional neural networks for sentence classification. In: EMNLP, pp. 1746–1751 (2014)
7. Levy, O., Goldberg, Y.: Linguistic regularities in sparse and explicit word representations. In: CoNLL, pp. 171–180 (2014)

8. Li, S., Zhao, Z., Hu, R., Li, W., Liu, T., Du, X.: Analogical reasoning on Chinese morphological and semantic relations. In: ACL, vol. 2, pp. 138–143 (2018)
9. Mikolov, T., Yih, W.t., Zweig, G.: Linguistic regularities in continuous space word representations. In: NAACL, pp. 746–751 (2013)
10. Perrie, J., Islam, A., Milios, E., Keselj, V.: Using Google n-grams to expand word-emotion association lexicon. In: Gelbukh, A. (ed.) CICLing 2013. LNCS, vol. 7817, pp. 137–148. Springer, Heidelberg (2013). https://doi.org/10.1007/978-3-642-37256-8_12
11. Plutchik, R.: The nature of emotions: human emotions have deep evolutionary roots, a fact that may explain their complexity and provide tools for clinical practice. Am. Sci. **89**(4), 344–350 (2001)
12. Rodriguez, A., Laio, A.: Clustering by fast search and find of density peaks. Science **344**(6191), 1492–1496 (2014)
13. Russell, J.A., Barrett, L.F.: Core affect, prototypical emotional episodes, and other things called emotion: dissecting the elephant. J. Pers. Soc. Psychol. **76**(5), 805 (1999)
14. Schmidt, P.J.: Mood, depression, and reproductive hormones in the menopausal transition. Am. J. Med. **118**(12), 54–58 (2005)
15. Song, K., Feng, S., Gao, W., Wang, D., Chen, L., Zhang, C.: Build emotion lexicon from microblogs by combining effects of seed words and emoticons in a heterogeneous graph. In: HT, pp. 283–292 (2015)
16. Song, K., Gao, W., Chen, L., Feng, S., Wang, D., Zhang, C.: Build emotion lexicon from the mood of crowd via topic-assisted joint non-negative matrix factorization. In: SIGIR, pp. 773–776 (2016)
17. Staiano, J., Guerini, M.: Depeche mood: a lexicon for emotion analysis from crowd annotated news. In: ACL, vol. 2, pp. 427–433 (2014)
18. Tomkins, S.: Affect Imagery Consciousness: Volume I: The Positive Affects. Springer, New York (1962)
19. Turner, J.H.: On the origins of human emotions. Q. Rev. Biol. **30**(4), 483–484 (2000)
20. Xu, J., Xu, R., Zheng, Y., Lu, Q., Wong, K.-F., Wang, X.: Chinese emotion lexicon developing via multi-lingual lexical resources integration. In: Gelbukh, A. (ed.) CICLing 2013. LNCS, vol. 7817, pp. 174–182. Springer, Heidelberg (2013). https://doi.org/10.1007/978-3-642-37256-8_15
21. Xu, L., Lin, H., Pan, Y., Ren, H., Chen, J.: Constructing the affective lexicon ontology. J. China Soc. Sci. Tech. Inf. **27**(2), 180–185 (2008)
22. Yang, C., Lin, K.H.Y., Chen, H.H.: Building emotion lexicon from weblog corpora. In: ACL, pp. 133–136 (2007)
23. Yang, M., Peng, B., Chen, Z., Zhu, D., Chow, K.P.: A topic model for building fine-grained domain-specific emotion lexicon. In: ACL, pp. 421–426 (2014)

Long Term Implications of Climate Change on Crop Planning

Andrew Lewis[1(✉)], Marcus Randall[1,2], Sean Elliott[2], and James Montgomery[3]

[1] School of Information and Communication Technology, Institute for Integrated and Intelligent Systems, Griffith University, Nathan, QLD, Australia
{a.lewis,m.randall}@griffith.edu.au

[2] Bond Business School, Bond University, Robina, QLD, Australia
{mrandall,selliott}@bond.edu.au

[3] School of Technology, Environments and Design, University of Tasmania, Hobart, TAS, Australia
james.montgomery@utas.edu.au

Abstract. The effects of climate change have been much speculated on in the past few years. Consequently, there has been intense interest in one of its key issues of food security into the future. This is particularly so given population increase, urban encroachment on arable land, and the degradation of the land itself. Recently, work has been done on predicting precipitation and temperature for the next few decades as well as developing optimisation models for crop planning. Combining these together, this paper examines the effects of climate change on a large food producing region in Australia, the Murrumbidgee Irrigation Area. For time periods between 1991 and 2071 for dry, average and wet years, an analysis is made about the way that crop mixes will need to change to adapt for the effects of climate change. It is found that sustainable crop choices will change into the future, and that large-scale irrigated agriculture may become unviable in the region in all but the wettest years.

Keywords: Crop planning · Climate change · Water management · Optimisation · Differential Evolution

1 Introduction

Climate change is having a large impact on all aspects on animal and plant life across the planet [11,14,16]. Emissions and other anthropomorphic effects have seen large changes in political positions and industrial practices [6]. One such industry that is more susceptible to the effects of climate charge, and has a significant impact on the human condition, is agriculture [17]. Given an amount of arable land and a set of climate conditions, the question becomes what may sensibly be planted and how will this change over time? To address this, one

J. M. F. Rodrigues et al. (Eds.): ICCS 2019, LNCS 11536, pp. 369–382, 2019.
https://doi.org/10.1007/978-3-030-22734-0_27

particular food growing region in Australia, the Murrumbidgee Irrigation Area (MIA), is examined with a set of climate models, and an optimisation approach tailored towards crop planning.

The tool used to derive which crop should be planted, and the areas in which to do this, is Differential Evolution (DE) [12,13]. This has been found to be an effective tool for this problem by Montgomery, Fitzgerald, Randall and Lewis [9]. DE uses recombination and mutation to transform vectors so that they improve over time. This is done by gradually reducing the scale of changes made by the algorithm. At the beginning of the search, where the vectors are usually highly distributed, there is scope for poor solutions to make significant improvements in quality as they move towards their better counterparts. As time passes, and vectors become closer and improve. The self-scaling will mean smaller, finer-grained improvements are made. This behaviour helps the population to converge on a good region of state space. In its multi-objective form, DE's standard solution comparison and replacement mechanism (in which a new solution is compared against a single 'parent') is typically replaced by non-dominated sorting over the union of the current population and current iterations' candidate solutions produced using DE's mutation mechanism. While this works against overall population convergence, multi-objective DEs have been effective (see, e.g., Montgomery, Randall and Lewis [10]).

The remainder of this paper is organised as follows. Section 2 describes the use of optimisation tools in crop planning, particularly the recent modelling work that has been done for the MIA. Section 3 explains the problem model and describes the DE implementation. Section 4 examines climate change models, especially NARCLiM that develops predictive models for New South Wales (NSW) in Australia (thus incorporating the area of interest). This section also discusses the processes necessary to handle this data and convert it to a form that is usable by the optimisation model. Section 5 puts all of this together and applies the required computing to determine workable crop mixes in the decades to come. Comment is made on the changes observed. Finally, Sect. 6 concludes and describes other projects that are now possible because of this work.

2 The Optimisation of Crop Planning

The research area of crop planning has seen the application of various optimisation techniques with the aim of generating optimal crop mixes under different climate-related scenarios that maximise revenue while minimising water use. There is a very wide range of research material covering optimisation techniques applied to water management. A good summary of these may be found in Singh [15]. This paper focuses on the water management issues faced by farmers and regional planners in the Murrumbidgee Irrigation Area in south east Australia. The papers that have focused on the MIA have applied a range of computational techniques to generate optimal solutions to the multi-objective problem. The most recent research [7–9] has applied the optimisation techniques Non-dominated Sorting Genetic Algorithm (NSGA-II) and DE to generate good attainment surfaces.

The multi-criteria problem forming the underlying basis of this paper was given initially by Xevi and Khan [18]. Lewis and Randall [7] redefined this as a multi-objective problem in their work and applied the NSGA-II optimisation technique to generate the Pareto-efficient solutions to the problem. This work optimised the results of two key output variables. These were to maximise the net revenue generated by the chosen combination of crops and to minimise the corresponding environmental flow deficit. This work also added cotton to the range of crops considered when solving the multi-objective problem. This work was further extended by Montgomery et al. [9]. They considered the application of DE as the optimisation technique. One of their findings was that the attainment surfaces generated by DE tended to improve on those by the NSGA-II approach.

The previous work considered only current climate conditions. Effectively, the water requirements of each crop have been based on static underlying assumptions relating to temperature and rainfall, where only potential seasonal variation has been incorporated. The purpose of this paper is to project forward in time, on the basis of potential changes to future temperatures and rainfall. This introduces the application of climate change models to show potential future changes to temperatures and rainfall in the MIA. This will make allowance for potential future changes in the water requirement for each crop, and the aim is to generate robust optimal solutions that can withstand potential changes in climate conditions.

3 Problem Definition and Optimisation

In the following subsections the problem model and the DE solver used to optimise it are described.

3.1 Problem Model

As previously stated, the problem seeks to determine appropriate areas of land for a selection of crops such that a net revenue function is maximised, but an environmental water deficit is minimised. This model is for one year. In the instances used here, the crops that can be selected from are: rice, wheat, barley, maize, canola, oats, soybean, Winter pasture, Summer pasture, lucerne, vines, Winter vegetable, Summer vegetables, citrus, stone fruit and cotton. The multi-objective model used in this paper, reproduced from Lewis and Randall [7] (pp. 181–182) is given in Eqs. 1–7.

$$\begin{aligned} \text{Maximise } NR = & \textstyle\sum_{c=1}^{C} TCI(c) \times X(c) \\ & - C_w \times \textstyle\sum_{m=1}^{M} \left(\left(\sum_{c=1}^{C} WREQ(c,m) \times X(c) \right) - P(m) \right) \\ & - C_p \times \textstyle\sum_{m=1}^{M} P(m) \\ & - \textstyle\sum_{c=1}^{C} Vcost(c) \times X(c) \end{aligned} \quad (1)$$

$$\text{Minimise } EFD = \sum_{m=1}^{M} (Tenv\_f(m) - Env\_f(m)) \cdot [Env\_f(m) < Tenv\_f(m)] \quad (2)$$

s.t.

$$\sum_{m=1}^{M} P(m) \leq 50\,\text{GL} \quad (3)$$

$$\sum_{c=1}^{C} X(c) \leq T_{Area} \quad (4)$$

$$X(c) \leq Y(c) \qquad 1 \leq c \leq C \quad (5)$$

$$Allocation(m) = Inflow(m) - Env\_f(m) \quad (6)$$

$$P(m) = \left(\sum_{c=1}^{C} WREQ(c,m) \times X(c) \right) - Allocation(m) \quad (7)$$

Where:

NR is the net revenue,
C is the number of crops,
$TCI(c)$ is the total crop income for crop c,
$X(c)$ is a decision variable which is the area of crop c (in hectares),
C_w is total cost of water per unit volume \$/ML,
M is the number of months, i.e., 12,
$WREQ(c, m)$ is the water required for crop c in month m (in ML),
$P(m)$ is the groundwater pumped in month m,
C_p is the cost of groundwater pumping and delivery (in \$/ML),
$Vcost(c)$ are all other variable costs associated with crop c,
EFD is the deficit in environmental flow,[1]
$Tenv\_f(m)$ is the target environmental flow for month m,
$Env\_f(m)$ is a decision variable that is the environmental flow for month m,
T_{Area} is the total cropping area available,
$Y(c)$ is a the maximum allowable area for crop c,
$Allocation(m)$ is the amount of surface water available for irrigation of crops in month m and
$Inflow(m)$ is the total surface (river) water available in month m.

3.2 The DE Solver

As solutions to this problem are pseudo-continuous it may be solved using continuous solvers operating on integer-valued solution vectors. Differential Evolution (DE) [13], an exemplar continuous solver, has previously been shown to be effective [9] so is used again here. It is adapted to the multiobjective setting using a similar approach to Montgomery, Randall and Lewis in their DE for RFID

[1] Equation 2 is expressed in Iverson bracket notation.

antenna design [10]: a DE/rand/1/bin algorithm is used to generate new solutions from the current population, after which the next generation is selected by applying the non-dominated sorting algorithm from NSGA-II [3] to the union of these solution sets. Feasible solutions are compared using standard Pareto-dominance rules, while a feasible solution dominates any infeasible solution, and infeasible solutions are compared based on the amount they violate the pumped water constraint (to provide some selection pressure toward feasible space).

The population (hence, archive) size is set to 100 members, and the algorithm is executed for 10,000 iterations (one million function evaluations). Appropriate values of difference vector scale F and crossover probability Cr are 0.8 and 0.5, respectively.

Solutions are represented as vectors of $C + 12$ integer values (i.e., 28 in the current problem as there are 16 crops) corresponding to the areas allocated to the C crops and environmental flows for each month of the year. Certain highly lucrative crops (vines, summer and winter vegetables, citrus, and stone fruit) have restrictive upper bounds on their areas, set at 10% of Australian national production. The ranges of the remaining 'unbounded' crops are 0 to the size of the farming land (120,000 ha), while the bounded crops are restricted to 0 to their nominated maximum. Further details of these values may be found in Lewis and Randall [7]. Environmental flow variables range between 0 and the total inflow for their respective months (which will often exceed the target environmental flow, although in some months they are less than the target).

Initial solutions for the DE are created by the following steps:

1. Generate C uniform random values r_c in $[0, 1]$.
2. Allocate each bounded crop c with $r_c \geq 0.5$ its maximum area.
3. Normalise the r_c values for all remaining crops with $r_c \geq 0.5$, then allocate those crops space from the remaining area in proportion to their normalised r_c values.
4. Generate 12 randomised integer values in the range $[0, Inflow(m)]$ for each month m to set the solution's environmental flows.

4 Using Climate Predictions to Inform Crop Planning Models

As may be noted, in order to model climate impact on optimal crop selections, some knowledge of crop water requirements and available water sources are needed. Our previous work made use of published data (rainfall, reference evapotranspiration, and surface water supplies – streamflows in rivers used to supply water for irrigation) for typical years (wet, dry and average) for the MIA. This data had been accumulated from meteorological observations over a number of years, but the specific sources were not supplied. In any case, data for future years cannot be sourced from historical observations.

In order to compile the required data, recourse was made to the NSW and ACT Regional Climate Modelling (NARCliM) Project [4,5]. This is a research partnership between the NSW and ACT state governments in Australia and the Climate Change Research Centre at the University of NSW. It has generated projections from four global climate models dynamically downscaled by three regional climate models, for three time periods: 1990 to 2009 (base), 2020 to 2039 (near future), and 2060 to 2079 (far future). Meteorological data are available on a 10 km grid across the NARCliM domain, which covers most of S.E. Australia.

For the purposes of the work described in this paper, data were extracted from the R3 physical configuration of the Weather Research and Forecasting model downscaling of the CCCMA 3.1 global climate model. This combination was chosen for this preliminary investigation because its results lie closest to the mean of the 12 model ensemble generated by the NARCliM project. Use of data from other models is planned for future work investigating the extremes of climate predictions, and uncertainties of future climate outcomes.

Data were extracted from the CCCMA3.1-R3 output data sets covering a 120,000 ha region north and west of the approximate location of the Berembed Weir on the Murrumbidgee River. This approximates the area of the MIA modelled in previous work. The data used were monthly averages of mean, maximum and minimum daily temperatures, and precipitation. The data were aggregated to give monthly mean values over the entire region.

In addition, data were aggregated over an 800,000 ha area covering the Snowy Mountains region. Precipitation data were averaged over this region to give an approximation to the available water sources in the catchment for the Murrumbidgee River.

From inspection of the annual aggregate precipitation data for the MIA, three years were selected from each modelled time period to represent wet, dry and average years. "Wet" years and "dry" years were those with the greatest and least aggregate annual rainfall in the relevant period, respectively. Apart from these two, obvious extremes, the "average" years were chosen as those with an aggregate annual rainfall closest to the 20 year mean annual rainfall. The chosen years are shown in Table 1.

Table 1. Representative years selected for modelling

Wet	Average	Dry
1991	2002	2003
2038	2037	2033
2063	2071	2064

From each of the selected years, average minimum and maximum daily temperatures were used to calculate reference evapotranspiration data by way of a temperature-based application of the FAO Penman-Monteith equation [1]. The calculated monthly ET_0 are shown in Table 2. As previously noted, monthly average rainfall was aggregated over the area to be modelled. These data are shown in Table 3.

Table 2. Reference Evapotranspiration (ET_0)

		Jan	Feb	Mar	Apr	May	Jun	Jul	Aug	Sep	Oct	Nov	Dec
Wet	1991	1.96	1.53	1.27	0.81	0.55	0.38	0.43	0.54	0.80	1.23	1.53	1.93
	2038	1.92	1.48	1.37	0.89	0.57	0.39	0.42	0.56	0.78	1.31	1.73	1.93
	2063	2.17	1.46	1.39	0.88	0.61	0.40	0.41	0.56	0.77	1.30	1.83	2.05
Average	2002	1.93	1.47	1.45	0.85	0.57	0.43	0.44	0.51	0.86	1.43	1.61	1.99
	2037	2.05	1.39	1.35	0.95	0.59	0.41	0.42	0.59	0.87	1.39	1.84	2.00
	2071	2.21	1.80	1.55	0.94	0.60	0.42	0.45	0.63	0.84	1.45	1.78	1.99
Dry	2003	2.03	1.83	1.47	1.05	0.78	0.43	0.43	0.63	0.94	1.65	2.03	2.05
	2033	1.90	1.54	1.48	0.95	0.65	0.48	0.44	0.66	1.03	1.42	1.90	1.97
	2064	2.03	1.74	1.45	1.03	0.75	0.40	0.46	0.69	0.89	1.53	1.97	2.03

Table 3. Average monthly precipitation for MIA (mm)

		Jan	Feb	Mar	Apr	May	Jun	Jul	Aug	Sep	Oct	Nov	Dec
Wet	1991	146.8	9.8	27.1	18.8	101.9	37.6	6.8	94.7	51.5	95.7	43.0	27.7
	2038	95.9	94.4	20.5	56.7	40.8	61.1	46.4	84.6	111.1	34.3	37.3	74.9
	2063	44.7	185.5	8.9	64.1	69.7	53.4	68.9	54.8	64.4	68.2	7.0	13.2
Average	2002	11.6	37.5	2.44	61.6	28.8	26.6	72.3	76.7	62.1	11.1	47.8	27.0
	2037	2.4	140.1	0.7	16.4	43.8	46.8	38.4	46.2	77.8	40.3	2.0	18.5
	2071	6.1	23.5	12.5	41.0	55.8	23.9	23.7	55.7	113.3	47.6	15.6	38.3
Dry	2003	18.3	1.0	20.8	41.9	27.0	6.5	9.8	28.6	35.7	5.8	38.3	22.9
	2033	32.3	1.4	2.2	0.0	38.1	8.6	33.6	28.3	16.4	20.8	36.3	24.8
	2064	0.7	36.8	18.0	1.9	5.3	102.8	28.3	37.6	54.4	9.1	1.5	7.3

Finally, it was necessary to calculate some approximation to the streamflow in the Murrumbidgee River, to provide estimates of surface water available for irrigation. Limited data were available for Berembed Weir directly, for the baseline years (1990 to 2009), but a more comprehensive series of data was available for the Murrumbidgee River at Wagga Wagga (Station 410001, Lat: –35.10, Long: 147.37, NSW Department of Industry – Lands and Water [2]) As there is no major intervention in streamflow between the Wagga Wagga station and the main canal off-take at Berembed Weir, streamflow data at Wagga Wagga was taken as a reasonable proxy for water available for irrigation at the weir. This was adequate for the years selected in the baseline, for which there were data available in the historical record, but more problematic for future years, in the timeframes of the climate predicted data. A median ratio was determined between the aggregate monthly precipitation in the Snowy Mountains catchment area, and the measured streamflow at Wagga Wagga, for the 20 years of the baseline. This can only be a gross approximation to actual streamflow, which is often determined by human intervention (water releases from Burrinjuck Dam on the Murrumbidgee River and Blowering Dam on the Tumut River). The median ratio was then applied to the aggregate monthly precipitation extracted from the climate model outputs in the Snowy Mountains catchment for the years selected in future time periods. These data are shown in Table 4.

With this accumulated data, it is possible to calculate the water required, per hectare, for cultivation of a crop, c, in a particular month, m, from Eq. 8:

$$WREQ(c, m) = k_c(c, m) \times ET_0(m) - Rain(m) \quad (8)$$

where $k_c(c, m)$ is the crop coefficient for that particular point in the crop's growing season, and $ET_0(m)$ and $Rain(m)$ are determined from Tables 2 and 3, respectively. Taken together with the streamflow (surface water available for irrigation) from Table 4, modelling of optimal crop selections can proceed.

Table 4. Monthly streamflow - Murrumbidgee River at Wagga Wagga (GL)

		Jan	Feb	Mar	Apr	May	Jun	Jul	Aug	Sep	Oct	Nov	Dec
Wet	1991	513	333	416	202	138	97	1011	546	869	447	333	351
	2038	522	701	286	179	357	571	481	529	969	365	312	419
	2063	237	1021	53	321	642	355	443	506	608	469	82	145
Average	2002	459	269	318	148	108	78	115	279	199	360	256	255
	2037	80	805	78	163	237	339	352	394	508	341	86	150
	2071	33	232	103	316	317	355	376	506	662	356	230	495
Dry	2003	331	212	127	91	53	40	62	173	141	276	258	289
	2033	251	97	45	23	261	146	628	532	324	195	237	133
	2064	50	206	145	193	110	591	193	567	342	233	145	247

5 Computational Experiments and Results

The DE solver described above was applied to determine optimal crop selection and environmental flow data, using the model previously described to evaluate the twin objectives of maximum net revenue and minimum "environmental flow deficit", the difference between the monthly environmental flow set as one of a trial solution's parameters and a "target", monthly, environmental flow to be released downstream. For the purposes of this preliminary study, an arbitrary monthly target of 100 GL was set for all months. Ten optimisation runs were performed using random seeds for each of the selected years in the three time periods, and median data inspected. The solution sets produced for each instance were highly consistent across trials (yet highly distinct between instances), suggesting that the solver was able to approximate the true Pareto fronts. (For reasons of simplicity and consistency, results reported in this paper are for single runs that closely approach median values.) The results are presented as objective attainment surfaces in following subsections.

Preliminary runs indicated some problems with environmental flow releases in future scenarios. In search of revenue, many trial solutions were generated that reduced environmental flow to zero, i.e., the river downstream was entirely dry. This was obviously considered unreasonable, so an arbitrary constraint was applied to maintain a minimum, monthly environmental flow of 30 GL.

In order to achieve the minimum environmental flows, it was necessary to make alterations to some projected inflows, as in some cases these were below the 30 GL limit. A decision was made to "top up" the monthly inflow to 50 GL whenever it fell below the limit, subtracting a corresponding amount from the following month. This was intended to model dam releases, and the 50 GL figure chosen to supply at least some water for irrigation before the minimum was reached.

As all nine scenarios include months where the inflow is less than the 100 GL target, the minimum achievable environmental deficit is always greater than zero. Across all scenarios and trials the optimisation algorithm was able to find solutions to meet these minima. Therefore, the lower extent of the attainment surfaces is a realistic indication of the extent to which the flow targets can be achieved. Ideally, these targets should be based on environmental cease-flow thresholds for downstream ecosystems, instead of arbitrarily imposed limits, and future work is to be directed towards this.

5.1 Wet Years

From Fig. 1 it may be seen that, in the baseline period, the optimisation algorithm has converged to a single solution. This represents 100% use of available land, with maximum allowed allocation to vegetable crops and citrus, and the remainder planted with cotton. There was sufficient water to achieve minimal environmental flow deficit and the highest net revenue achieved.

In the near-future scenario, it was no longer possible to maximise net revenue without incurring deficits in environmental flows, though they are relatively small in scale. It was possible to minimise these deficits by reducing cotton cultivation by 30% and replacing it with canola.

In the far future, net revenue achievable was near halved, compared to baseline, and flow deficits increased. No solutions were capable of using all available land, and to reach minimum flow deficits less than half the land was cultivated. In all solutions it proved infeasible to grow cotton, its place being taken by canola, with maximal crops of vegetables and citrus.

5.2 Average Years

From Fig. 2 it may be seen that there are several similar trends in results for average years, compared to wet years. The baseline year did not have enough water to converge to a single solution, instead spanning from almost as much net revenue but with 100 GL flow deficit, down to minimal flow deficit with 9% less net revenue. In the near future, net revenues were again nearly halved, and flow deficits increased. To achieve minimal flow deficits, cultivated land had to be

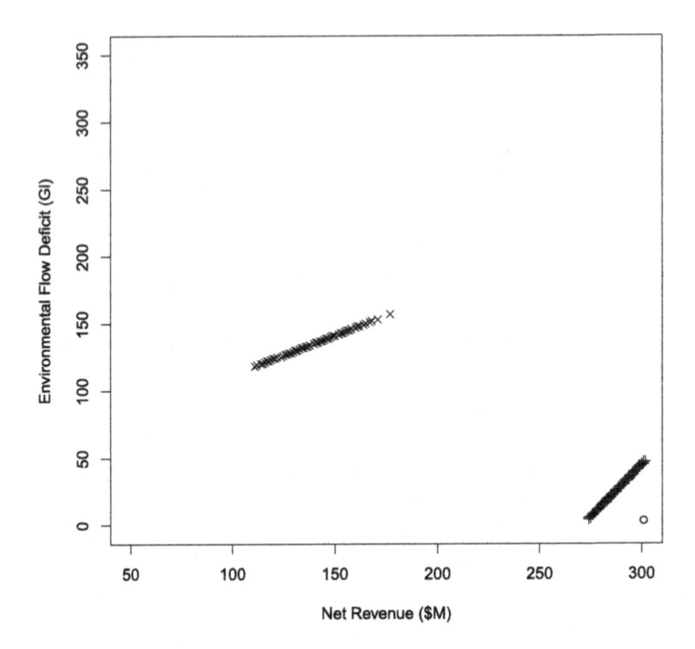

Fig. 1. Objective attainment surfaces for "wet" years, 1991 (o), 2038 (+) and 2063 (×).

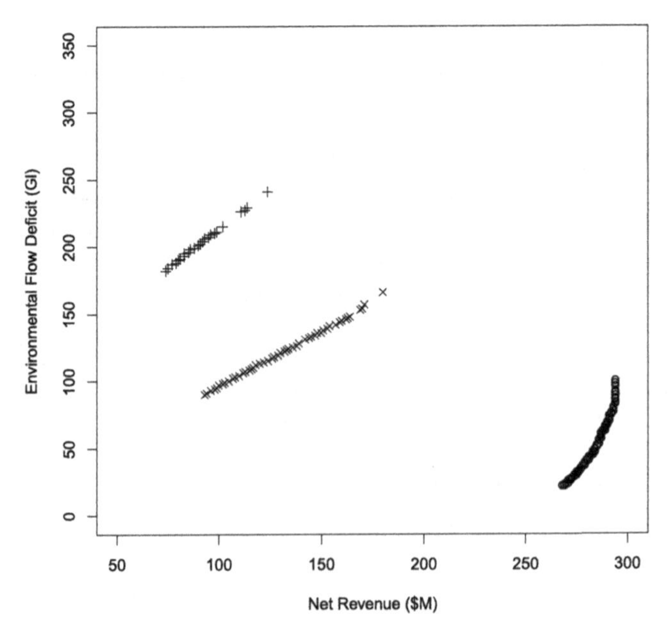

Fig. 2. Objective attainment surfaces for "average" years, 2002 (o), 2037 (+), and 2071 (×).

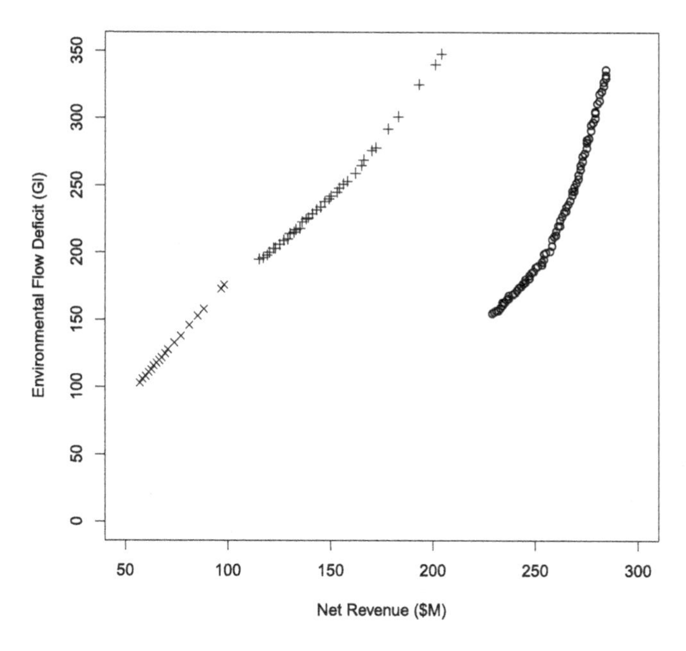

Fig. 3. Objective attainment surfaces for "dry" years, 2003 (o), 2033 (+), and 2064 (×).

severely curtailed. Crop selection still always included maximum vegetables and citrus, with a mix of cotton and canola filling the remainder. In the far future, it was once more infeasible to grow cotton, and again it was replaced by canola. To achieve minimal flow deficits, less than 40% of available land was cultivated.

5.3 Dry Years

Results for "dry" years are shown in Fig. 3. All years, from each of the periods, show significant environmental flow deficits – there is simply not enough water to achieve the (arbitrary) 100 GL targets each month in a dry year. In the baseline year, to achieve the maximum net revenue required an aggregate flow deficit of over 300 GL – targets were not reached for two thirds of the year. Flow deficits could be halved, with a corresponding 19% reduction in net revenue (and land cultivated). Maximum vegetable and citrus crops were included with a varying mix of cotton and canola.

For the near future, similar flow deficits were experienced, but with significantly reduced net revenue – at best 71% compared to the baseline, reducing to 40% if flow deficits are minimised. All solutions were unable to use all available land, ranging from a maximum of 85% down to 40%. Vegetables and citrus remained close to maximums across solutions. The cotton crop remained fairly stable at about 30,000 ha, with canola ranging from a minimum of 10,000 ha to a maximum of 45,000 ha, making an increasing contribution to an increasing net revenue.

In the far future, no cotton crop was possible. Maximum vegetable and citrus crops underpinned net revenue, with an increasing area of canola across solutions as net revenue increased. Net revenue only reached a maximum of one third of that achieved in the baseline years. Less than half the land available was cultivated, reducing to only 17% if environmental flows were to be maintained.

To summarise, net revenues can be seen generally to decline to half, in the near future, or a third or less in the far future. Crops forming the basis for revenues in the baseline, e.g., cotton, become increasingly non-viable as climate changes progress over the near and far future. Inspection of solution details showed that by the far future period, in a wet year the area cultivable decreased to, at best, 90% of the total if environmental flow requirements were ignored and 50% if they were taken fully into account. In a dry year, less than 50% of the total area was cultivable, or less than 20% if environmental flows were to be maintained.

All solutions included crop allocations that approach the (arbitrary) maximum limits placed on perishable commodities – it is possible that closer consideration of these limits, and the potential influence of population and market demand, might significantly change optimal choices. For this reason the results reported here should not be considered as recommending particular courses of action.

From these preliminary results, it appears that if sufficient water is released for aggregate downstream needs – environmental, agricultural and to support human populations – it may be that large-scale, irrigated agriculture may become unsustainable or uneconomic in the region in all but the wettest years.

It should be noted that this study only considers the impact of projected water availability on irrigated agriculture. No attempt has yet been made to incorporate changing crop yields as ambient temperatures and humidity change, or other foreseeable impacts on agro-economic and environmental systems.

Furthermore, the discussion of results is centred around characteristic "wet", "average" and "dry" years – there also has been no attempt yet to determine the (possibly changing) proportions of years in different time periods that can be characterised by these different labels. A brief inspection of the predicted climate data on which the modelling is based shows that by the far future period, 60% of years have less aggregate annual rainfall that the baseline average for the MIA, and 70% have less than baseline average precipitation in the Snowy Mountains catchment. Of perhaps greater concern is that the baseline period, 1990–2009, brackets the "Millenium drought" in Australia, recognised as the worst drought on record. For available water for agriculture in the MIA to be predicted to be predominantly *less* than available during this period is gravely concerning, and deserves further investigation.

6 Conclusions

Changes to the world climate will have profound implications for human centred activities, particularly the growing of crops to ensure food security. Using recognised climate models, the research reported in this paper has sought to shed light on some of the potential impacts of, in particular, future water availability on large-scale, irrigated agriculture, using the Murrumbidgee Irrigation Area (MIA) as a case study. It is important to note that the current and future, predicted conditions are specific to the MIA, and any conclusions and observations are solely applicable to this region.

The results obtained are preliminary, with many factors as yet unaccounted for. However, the framework and modelling approach described provide a foundation for future extension and development – it is easy to see how further detail of climate-related impacts, agronomic considerations of soil pathogen control, and commonly employed variations in sowing, crop species and irrigation strategies could be incorporated, to ensure modelling is more realistic and relevant to real-world practice. Collaborative research ties are being pursued to achieve this. The results reported are sufficiently concerning in their implications – that in the future large-scale, irrigated agriculture may become unsustainable in the region – that extensive future work is planned and underway.

The use of time in this paper is consistent with a snapshot approach (i.e., examining different time periods in isolation, and assuming complete knowledge of available water at the start of the season). An ongoing practical consideration for farmers and regional planners, though, is what crops should be grown over a certain timeframe (such as a decade) and how the crop mix should change over that time. To that end, a temporal model is currently being developed which attempts to maximise overall revenue while minimising cumulative water usage over extended timeframes.

References

1. Allen, R.G., Pereira, L.S., Raes, D., Smith, M., et al.: Crop evapotranspiration-guidelines for computing crop water requirements-FAO irrigation and drainage paper 56. Food Agric. Organ. Rome **300**(9), D05109 (1998)
2. Bureau of Meteorology (Australia): Data owner: NSW Department of Industry - Lands and Water. http://www.bom.gov.au/waterdata/
3. Deb, K., Pratap, A., Agarwal, S., Meyarivan, T.: A fast and elitist multi objective genetic algorithm: NSGA-II. IEEE Trans. Evol. Comput. **6**, 182–197 (2002)
4. Evans, J., Ji, F., Lee, C., Smith, P., Argüeso, D., Fita, L.: Design of a regional climate modelling projection ensemble experiment - NARCliM. Geosci. Model Dev. **7**(2), 621–629 (2014)
5. Evans, J., McCabe, M.: Regional climate simulation over Australia's Murray-Darling Basin: a multi-temporal assessment. J. Geophys. Res. Atmos. **115** (2010). https://doi.org/10.1029/2010JD013816
6. Kolk, A., Pinkse, J.: Multinationals' political activities on climate change. Bus. Soc. **46**(2), 201–228 (2007)
7. Lewis, A., Randall, M.: Solving multi-objective water management problems using evolutionary computation. J. Environ. Manag. **204**, 179–188 (2017)
8. Lewis, A., Randall, M., Capon, S., Jackwitz, E.: Constrained optimisation of agricultural water management with parameter-sensitive objectives. In: Proceedings of the 15th International Conference on Computer Applications, pp. 79–85 (2017)

9. Montgomery, J., Fitzgerald, A., Randall, M., Lewis, A.: A computational comparison of evolutionary algorithms for water resource planning for agricultural and environmental purposes. In: Proceedings of the 2018 IEEE Congress on Evolutionary Computation, pp. 1–8 (2018)
10. Montgomery, J., Randall, M., Lewis, A.: Differential evolution for RFID antenna design: a comparison with ant colony optimisation. In: Genetic and Evolutionary Computation Conference, Dublin, Ireland, pp. 673–680 (2011)
11. Parmesan, C., Yohe, G.: A globally coherent fingerprint of climate change impacts across natural systems. Nature **421**(6918), 37 (2003)
12. Price, K.: An introduction to differential evolution. In: Corne, D., Dorigo, M., Glover, F. (eds.) New Ideas in Optimization, pp. 79–108. McGraw Hill (1999)
13. Price, K., Storn, R., Lampinen, J.: Differential Evolution: A Practical Approach to Global Optimization. Springer, Heidelberg (2005). https://doi.org/10.1007/3-540-31306-0
14. Rosenzweig, C., Parry, M., et al.: Potential impact of climate change on world food supply. Nature **367**(6459), 133–138 (1994)
15. Singh, A.: Review: computer-based models for managing the water-resource problems of irrigated agriculture. Hydrogeol. J. **23**, 1–11 (2015)
16. Stern, N., et al.: Stern Review: The Economics of Climate Change, vol. 30. HM Treasury, London (2006)
17. Turral, H., Burke, J., Faurès, J.: Climate change, water and food security. Food and Agriculture Organization of the United Nations Rome, Italy (2011)
18. Xevi, E., Khan, S.: A multi-objective optimisation approach to water management. J. Environ. Manag. **77**(4), 269–277 (2005)

Representation Learning of Taxonomies for Taxonomy Matching

Hailun Lin[1(✉)], Yong Liu[1], Peng Zhang[1], and Jianwu Wang[2]

[1] Institute of Information Engineering, Chinese Academy of Science, Beijing, China
linhailun@iie.ac.cn
[2] Department of Information Systems, University of Maryland, Baltimore County, Baltimore, USA

Abstract. Taxonomy matching aims to discover categories alignments between two taxonomies, which is an important operation of knowledge sharing task to benefit many applications. The existing methods for taxonomy matching mostly depend on string lexical features and domain-specific information. In this paper, we consider the method of representation learning of taxonomies, which projects categories and relationships into low-dimensional vector spaces. We propose a method to takes advantages of category hierarchies and siblings, which exploits a low-dimensional semantic space to modeling categories relations by translating operations in the semantic space. We take advantage of maximum weight matching problem on bipartite graphs to model taxonomy matching problem, which runs in polynomial time to generate optimal categories alignments for two taxonomies in a global manner. Experimental results on OAEI benchmark datasets show that our method significantly outperforms the baseline methods in taxonomy matching.

Keywords: Taxonomy matching · Representation learning · Category embedding · Relation embedding · Maximum weight matching

1 Introduction

Taxonomy is used to annotate entity semantic information in knowledge base, which contains a hierarchy of categories. Categories in a taxonomy can be described as multi-relational data and represented as triples (h_c, r, t_c), where h_c denotes the head category, h_t denotes the tail category, r expresses the direct relationship between h_c and t_c. r has three kinds of value: *subclass*, *superclass* and *sibling*. Specifically, if the value of r is *subclass*, it indicates that h_t is the parent of h_c; if the value of r is *superclass*, it indicates that h_t is the child of h_c; if the value of r is *sibling*, it indicates that h_t would be the sibling of h_c. As is known to all, different taxonomies sometimes contain both overlapping and complementing data. In order to implement knowledge sharing from two

J. M. F. Rodrigues et al. (Eds.): ICCS 2019, LNCS 11536, pp. 383–397, 2019.
https://doi.org/10.1007/978-3-030-22734-0_28

taxonomies, we study the problem of taxonomy matching to discover categories alignments between them.

For taxonomy matching, the key step is to calculate the category pair relevance between two taxonomies. After all the category pairs relevant scores have been calculated, we can obtain the most relevant category pairs between two taxonomies. In recent years, taxonomy matching has received a lot of research interests, and many approaches have been proposed (e.g., [4,9,14–16]). These works mostly depend on string lexical features or domain-specific information to predict the relevance score between categories. Although there are many studies on taxonomy matching, most of those approaches have been demonstrated to achieve good performance only on fairly domain-specific taxonomies [1,13].

Therefore, we present a representation learning based taxonomy matching approach, which exploits a unified semantic model where we can learn to place categories, supercategories, and siblings as points in a hypothetical common semantic space, i.e., a continuous low-dimensional vector space. The category representation vector can significantly promote taxonomy matching. This method follows the assumption in TransE [3] (designed for learning knowledge graph representations), modeling categories relationships by translating operations between two categories in the semantic space.

Our methods works in three stages: it firstly embeds taxonomies including both categories and relations into a continuous low-dimensional vector space. Secondly, it creates a weighted bipartite graph to model the candidate relevant category pairs between two taxonomies. Thirdly, it performs a maximum weight matching algorithm to generate an optimal matching for two taxonomies in a global manner. Key aspects of our method are: (1) it automatically learns category and relation feature representations in semantic space to calculate the relevance between categories, without external data resources except taxonomies themselves; (2) it proposes category matching in a global manner, by finding matching with maximum weight in a bipartite graph. In general, the main contribution of this paper is three-fold:

- We show a multi-relational data modeling formulation for taxonomy matching that learns a unified semantic space for categories, supercategories and siblings, while drawing relations between them.
- We present a maximum weight matching algorithm for matching taxonomies, which can obtain a global optimal matchings between two taxonomies.
- We show from the experiments that the multi-relational data modeling with the maximum weight matching algorithm helps improve taxonomy matching accuracy.

The rest of this paper is organized as follows. Section 2 discusses the related work. Section 3 formulates the problem of taxonomy matching and proposes TransC framework. Section 4 introduces our model for taxonomy matching, and discusses its implementation. Section 5 introduces the experimental results. Finally, the paper is concluded in Sect. 6.

2 Related Work

The problem of taxonomy matching has its roots in the problems of identifying duplicate entities, which is also known as record linkage, duplicate detection, or coreference resolution. There are a lot of research work has been proposed for taxonomy matching (e.g., [2,5,7,9,12,14,16]). In this paper, we try to analyze the studies on this problem from the perspective of their measures for calculating category relevance [5].

Specifically, we simply classify the studies into the following six categories. (1) Lexicon-based measure: They perform taxonomy matching task according to the mention forms (i.e., words representation) of the categories in taxonomies. (2) Semantic-based measure: They adopt semantic dictionaries to complete taxonomy matching according to the meaning of the words. (3) Structure-based measure: They adopt category hierarchical information, including supercategories and subcategories. (4) Instance-based measures: They use the instances of categories. (5) Context-based measure: The adopt the descriptive text of categories. (6) Hybrid-based measure: They adopt various combination of different types of information, i.e., lexicon, semantic, structure, instance and context.

PARIS [16] adopted instance based measure for taxonomy matching. It considered that the category structure in one taxonomy may be more fined-grained than the category in the other taxonomy, so it aimed to find subclass matching relationships between two taxonomies. RiMOM [12] exploited a dynamic multistrategy for finding categories alignments, which automatically combined the measures based on two estimated factors, i.e., the lexicon similarity factor and the structure similarity factor. RiMOM adopted similarity flooding technique on a relationship graph between two taxonomies to enhance the structural information contributing to taxonomy matching. ServOMap [14], designed for biomedical ontologies, took advantage of lexicon and context based measure to calculate the relevance scores of categories. It exploited an inverted index used in information retrieval to reduce the number of candidate categories to consider. LogMap [7], also designed for biomedical ontologies, combined lexicon and semantic as well as structure measures to aligning categories between taxonomies.

In addition, Chen et al. [5] proposed FFCA technique, a combination of the fuzzy theory and formal concept analysis (FCA), to match taxonomies with the same domain. FFCA enriched each category from the source taxonomy by information obtained form WordNet. Demidova et al. presented a Markov Logic Network (MLN) based semi-supervised method for matching task [6]. They mentioned serval heuristic rules based on first-order logic to capture the similar semantic elements. Lee et al. [11] presented a Monte Carlo algorithm for finding greedy cuts to entity resolution problem. They adopted a combination of properties and instances based measures. SiGMa [9] adopted a simple greedy matching algorithm with a combination of lexicon and structure measures, which finds aligned categories in a greedy local search manner. The greedy based method can be seen as an efficient method to the task of large-scale taxonomy matching. However, due to its greedy nature, it can not correct previous mistakes in making decisions. Therefore, the greedy based method could not guarantee obtaining a global optimal matching for two taxonomies.

Based on the above analysis, the studies on taxonomy matching either focus on specific domains, or aim at providing a general way across various domains. Furthermore, we can see that most of existing studies employ the combinational strategies. Extensive experiments also show that the combination method outperforms the single strategy based method [9,12]. However, they are mostly capturing the linguistic features and structural features to predict the relevance score between categories, there is no single dominant taxonomy matcher that performs the best, regardless of its application domain [1].

3 Taxonomy Matching

In this section, we will study the problem of automatically taxonomy matching. For this purpose, we will firstly give some notations and formulate the problem of taxonomy matching in Sect. 3.1. Subsequently, the overall framework of TransC will be introduced in Sect. 3.2.

3.1 Notations and Problem Formulation

Suchanel et al. defined a taxonomy as a set of a formal collection of knowledge, including categories, relations, and the instances with their assertions [16]. In this paper, we describe a taxonomy as multi-relational data with numerous triple facts $T = \{(h_c, r, t_c)\}$. Given a triple $(h_c, r, t_c) \in T$, where $h_c, t_c \in C$ denote categories and $r \in R$ denotes the relationship between h_c, t_c. C is categories set and R is relation set, where $R = \{subclass, superclass, sibling\}$. Each category $c \in C$ contains an attributes set A_c. Each category and relation embedding in the hypothetical common semantic space takes value in $\mathbb{R}^k$.

Structure-Based Embeddings: $\mathbf{h_c^s}$ and $\mathbf{t_c^s}$ are the embeddings of category h_c, t_c, which are learned from the hierarchical structure of taxonomies. These embeddings are learned from translation-based method TransE [3].

Attribute-Based Embeddings: $\mathbf{h_c^a}$ and $\mathbf{t_c^a}$ are the attribute-based embeddings of category h_c, t_c, which are learned based on category attributes. In the following, we will elaborate on an encoder to learn attribute-based embeddings for taxonomy categories.

We note that, in a taxonomy, the categories set C is global, which means that some categories maybe identical across different taxonomies. Moreover, in addition to the equivalent ($\equiv$) relationship between two categories c and $c^{'}$, the relationship between c and $c^{'}$ could be subcategory relationship $c \subseteq c^{'}$ or supercategory relationship $c \supseteq c^{'}$. As the subcategory relationship or supercategory relationship can be inferred by equivalent relationship between categories, we aim to find out whether one category c of one taxonomy is equivalent to another category $c^{'}$ of another taxonomy. Since the set C is global in a taxonomy, we consider the one-to-one matching of categories between two taxonomies.

Definition 1. Given two taxonomies $T = \{(h_c, r, t_c)\}$ and $T^{'} = \{(h_c^{'}, r^{'}, t_c^{'})\}$, the goal of the taxonomy matching is to obtain a one-to-one (1-1) equivalent matching M from the categories set C of T to the categories set $C^{'}$ of $T^{'}$, which contains all semantically equivalent categories between two taxonomies.

3.2 The TransC Framework

Based on the problem definition, we propose a method called TransC, to address the task of taxonomy matching using two modules as follows:

Taxonomy Representation Learning. To supplement lexical representation in measuring category relevance scores, this module exploits a unified semantic model where we can learn to place categories, supercategories, and siblings as points (or vectors) in a hypothetical common semantic space.

Taxonomy Matching Generation. Based on taxonomy representations, this module exploits a weighted bipartite graph to model the candidate relevant category pairs between two taxonomies, and performs a maximum weight matching algorithm to generate an optimal matching for two taxonomies.

In the following sections, we will introduce those modules in details.

4 Methodology

In this section, we introduce our method that obtains the categories alignments for taxonomies. In what follows, we first introduce how to learn the representations of taxonomies, and then elaborate on the process for finding an optimal matching for two taxonomies based on the representations.

4.1 Taxonomy Representation Learning

To exploit both triple facts $(h_c, r, t_c) \in T$ and category attributes, we follow a representation learning method DKRL for knowledge graphs [17], and propose structure-based category embeddings and attribute-based category embeddings. These embeddings adopt energy-based model, which learns category representation vectors in low-dimensional vector space. The structure-based category embeddings are used to capture information in triple facts of taxonomies, and the attribute-based category embeddings are used to capture meta information in category attributes. We use the same embedding vector space to learn the two types of category representations. The energy function of our method is then defined as follows:

$$F(h_c, r, t_c) = F_S(h_c, r, t_c) + F_A(h_c, r, t_c), \tag{1}$$

where $F_S(h_c, r, t_c) = \|\mathbf{h_c^s} + \mathbf{r} - \mathbf{t_c^s}\|$ is the part of structure-based category embedding energy function, $F_A(h_c, r, t_c)$ is the part of attribute-based category embedding energy function. In this paper, we define $F_A(h_c, r, t_c)$ as follows:

$$F_A(h_c, r, t_c) = F_{AA}(h_c, r, t_c) + F_{AS}(h_c, r, t_c) + F_{SA}(h_c, r, t_c) \tag{2}$$

where $F_{AA}(h_c, r, t_c) = \|\mathbf{h_c^a} + \mathbf{r} - \mathbf{t_c^a}\|$ in which $\mathbf{h_c^a}$ and $\mathbf{t_c^a}$ are the attribute-based embeddings of category h_c, t_c. In this paper, we also define $F_{AS}(h_c, r, t_c) = \|\mathbf{h_c^a} + \mathbf{r} - \mathbf{t_c^s}\|$ and $F_{SA}(h_c, r, t_c) = \|\mathbf{h_c^s} + \mathbf{r} - \mathbf{t_c^a}\|$.

In the following subsection, we present a continuous bag-of-words encoder to build attribute-based category representation.

Continuous Bag-of-Words Encoder. For each category, a set of attributes are used to denote the meta information of the category. We assume that if categories are similar, their attributes should be similar. In the encoder, we take the words in the attributes for each category as input. Firstly, we sum up the words representation vectors to obtain the attribute representation vector. Secondly, we sum up the representation vectors of attributes to obtain the category representation vector:

$$\mathbf{c^a} = \mathbf{a}_1 + \cdots + \mathbf{a}_k, \tag{3}$$

where $\mathbf{a}_i$ is the i-th attribute representation vector belonging to the attributes set A_c of category c; $\mathbf{a}_i = \mathbf{x}_1 + \cdots + \mathbf{x}_m$, where $\mathbf{x}_j$ is the j-th word representation vector belonging to the words set of attribute $a \in A_c$. In this paper, $\mathbf{x}_j$ can be obtained by Word2Vec. $\mathbf{c^a}$ will be used to minimize F_A. The encoder framework is illustrated in Fig. 1.

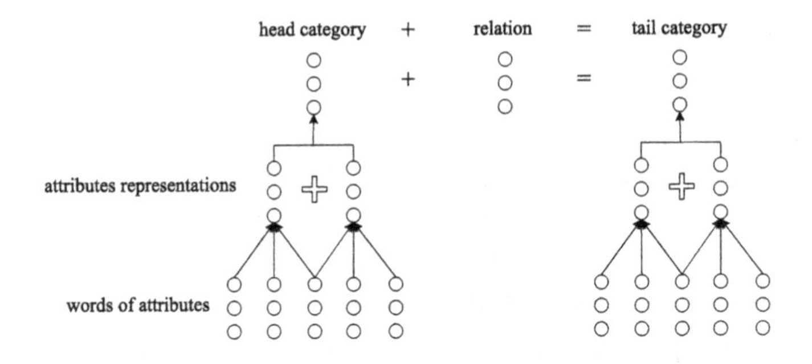

Fig. 1. The CBOW encoder

Training. Now we introduce how to learning category embeddings with our proposed models. Given a training set $S \subseteq T$ of triple facts (h_c, r, t_c), we adopt a margin-based ranking criterion as objective:

$$\mathcal{L} = \sum_{(h_c,r,t_c)\in S} \sum_{(h_c^{'},r^{'},t_c^{'})\in T'} \max(0, \gamma + F(h_c, r, t_c) - F(h_c^{'}, r^{'}, t_c^{'})) \tag{4}$$

where γ denotes a margin hyperparameter, $\gamma > 0$; $F(h_c, r, t_c)$ denote the dissimilarity score function between $\mathbf{h_c} + \mathbf{r}$ and $\mathbf{t_c}$, which we take to be either the L_1-norm or the L_2-norm; $S^{'}$ is the negative sampling set, generated according to Eq. 5. As we define two representation types for categories, h_c and t_c in the Eq. 4 could be either of these two types representations.

$$S^{'} = \{(h_c^{'}, r, t_c) | h_c^{'} \in C\} \cup \{(h_c, r, t_c^{'}) | t_c^{'} \in C\} \tag{5}$$

More specifically, during constructing corrupted triples, we follow the method described in [17], which sets different probabilities for replacing the head or tail entity for corrupting the golden triple. The loss function (Eq. 4) favors lower values for similar triples than for dissimilar triples. The input of the CBOW Encoders is the category attributes words, and its output is category representation vectors. The categories and relations are initialized by the random procedure proposed in [3]. We take stochastic gradient descent in minibatch mode to optimize the objective function.

4.2 Taxonomy Matching Generation

In this section, we will describe the process of our method to obtain a most suitable one-to-one equivalent categories matchings with highest confidence for a pair of taxonomies based on the categories embeddings.

Bipartite Graph Creation. In order to efficiently encode the complicated relationships between the categories C of T and the categories $C^{'}$ of $T^{'}$, we choose the bipartite graph model as our representation model, because that the bipartite graph can encode categories from taxonomies as the vertices, and encode the candidate matching relationships between these vertices explicitly.

Before we begin to construct a bipartite graph to model the candidate matching relationships of categories between two taxonomies, we firstly introduce the score function which measure the suitability of a matching between categories. Given a pair of categories $c, c^{'}$, their corresponding embeddings are $\mathbf{c}, \mathbf{c}^{'}$, the relevance score between $c, c^{'}$ is defined as:

$$w(c, c^{'}) = \cos(\mathbf{c}, \mathbf{c}^{'}) = \frac{\mathbf{c} \cdot \mathbf{c}^{'}}{\|\mathbf{c}\| \|\mathbf{c}^{'}\|} \tag{6}$$

We build a weighted bipartite graph $G = (V, E, W)$ exploiting the score function, where V denotes the vertices set, consisting of $|C|$ left vertices and $|C^{'}|$ right vertices; E denotes the edges set, including all the candidate links between categories from C and $C^{'}$; W: $E \rightarrow \mathbf{R}$ is the weight function. The graph G is defined as an undirected graph and its generation algorithm can be described in Algorithm 1, where V, E, W is initialized as a zero set, respectively.

Specifically, Algorithm 1 works in two steps as follows:

- Candidate matching categories selection. In this step, for each category c from taxonomy T, we pair it with each category $c^{'}$ contained in $T^{'}$. All categories $c^{'} \in C^{'}$ likely matching to the category c are selected (see Lines 2–6).
- Vertex connection. In this step, we assign the matching edge to the bipartite graph. For each category vertex c from taxonomy T in the graph, we add an edge between it and each of its candidate matching category $c^{'}$ from taxonomy $T^{'}$; the weight w of the edge $(c, c^{'})$ is set according to Eq. (6) (see Lines 7–12).

After those two steps, a weighted bipartite graph has been generated (see Line 13).

Algorithm 1. The algorithm for bipartite graph creation

Input: $T = \{(h_c, r, t_c)\}, T' = \{(h_c', r', t_c')\}, G = (V, E, W)$ and the embeddings of all the categories of T and T'.
Output: $G = (V, E, W)$.
1: Initialize graph $G = (V, E, W)$: $V = \varnothing, E = \varnothing, W = \varnothing$.
2: **for all** $c \in C$ in T **do**
3:   **for all** $c' \in C'$ in T' **do**
4:     Compute the likelihood w that c is equivalent matching to c' based on the embeddings of those two categories via Equation (6).
5:     **if** $w > 0$ **then**
6:       Add class c and c' to V.
7:       Add weight function $W(c, c') = w$ to W.
8:       Add edge (c, c') to E.
9:       Assign weight to edge (c, c') with w.
10:     **end if**
11:   **end for**
12: **end for**
13: **return** $G = (V, E, W)$.

Maximum Weight Matching in Bipartite Graph. In this section, we will introduce how to find an optimal one-to-one equivalent categories matching M for a pair of taxonomies. The goal of taxonomy matching is to find a most suitable one-to-one equivalent categories matching M for a pair of taxonomies with highest confidence, and the goal of maximum weight matching is to find a set of vertex-disjoint edges with maximum weight. Therefore, the taxonomy matching problem can be converted to a maximum weight matching problem.

Specifically, given a weighted bipartite graph $G = (V, E, W)$, we can use integer linear program (ILP) to model the matching problem:

$$\begin{aligned} &\max \sum_{e \in E} w(e)x(e) \\ &\text{s.t.} \\ &\qquad \sum_{e=(v,v')} x(e) \le 1 \qquad \forall v \in V \\ &\qquad 0 \le x(e) \le 1 \qquad \forall e \in E \\ &\qquad x(e) \text{ is an integer} \quad \forall e \in E, \end{aligned} \tag{7}$$

where x is the matching's incidence vector; w represents the likelihood of the categories matching as specified in Eq. 6. The dual of the Problem (7) is vertex cover problem. The dual problem is defined as:

$$\begin{aligned}
&\min \sum_{v \in V} y(v) \\
&\text{s.t.} \\
&\qquad y(e) \geq w(e) \quad \forall e \in E \\
&\qquad y(v) \geq 0 \qquad \forall v \in V,
\end{aligned} \tag{8}$$

where we define $y(v, v^{'}) \stackrel{def}{=} y(v)+y(v^{'})$. According to complementary relaxation condition, the matching M and y are optimal iff $\forall e \in M, y(e) = w(e)$ and for all free vertices $v, y(v) = 0$.

The essential idea of acquiring maximum weight matching is to repeatedly find an augmenting path in the bipartite graph and augment over it, until there are no augmenting paths left. The maximum weight matching problem has been extensively studied (e.g., [8,10]). The most efficient general algorithm for this problem is Hungarian algorithm [8]. Since the weights in the bipartite graph we constructed are real numbers, so we adjust the Hungarian algorithm improved by Lawler [10], to our taxonomy matching problem. In what follows, we describe our algorithm in details.

Maximum Weight Matching Algorithm. Given a weighted bipartite graph $G = (V, E, W)$ constructed from taxonomies T and $T^{'}$, we repeatedly find augmenting paths and augment over it, to find a set of vertex-disjoint edges with maximum weight. Our algorithm consists of four steps. It firstly initializes the dual variables in Problem (8). Secondly, it finds an augmenting path and augments over it. Thirdly, it computes the dual variable augmentation value and updates the dual variables in the fourth step. Repeating steps 2–4, we can finally obtain a matching with maximum weight according to the bipartite graph. In this paper, we model taxonomy matching problem as maximum weight matching problem. Therefore, we can guarantee to generate an optimal matching with highest confidence for two taxonomies in a global manner.

Here, we use $V_L = C$ and $V_R = C^{'}$ to represent the left vertices set and right vertices set in G, respectively. Let LF and RF denote the left and right free vertices (not matched vertex) in G, respectively, which is initialized as $LF = V_L$ and $RF = V_R$. Let $y(v)$ denote the dual variable value for each vertex $v \in V$, δ denote the dual variable augmentation value, whose initial value is $\delta_0 = \max\{W(e)|e \in E\}$. Let τ denote the current iteration times. Let M denote the set of vertex-disjoint edges with maximum weight, which is initialized as a zero set, i.e., $M = \varnothing$. The maximum weight matching algorithm can thus be described in Algorithm 2.

Algorithm 3 is to find an augmenting path by performing a bread-first-search (BFS) on a modified graph. Specifically, the algorithm firstly judges whether it has an augmenting path or not, using the left and right free vertices sets in G (see Lines 1–2). Secondly, the algorithm directs all edges in G (see Lines 4) and performs BFS algorithm to find an augmenting path (see Lines 5–6). Thirdly, augments the free vertices sets (see Lines 8). Finally, we acquire an augmenting path (see Lines 9–10).

Algorithm 2. The algorithm for maximum weight matching

Input: $G =< V, E, W >, V_L, V_R, LF, RF, \delta_0, M$.
Output: $M = \{(v, v^{'})|v \in V_L, v^{'} \in V_R\}$.
1: **for all** $v \in V$ **do**
2: **if** v is a left vertex in G **then**
3: $y(v) = \delta_0$.
4: **else**
5: $y(v) = 0$.
6: **end if**
7: **end for**
8: Set $\tau = 0$.
9: **repeat**
10: Find an augmenting path AP, using the method described in Algorithm 3.
11: Augment the matchings M: $M = M \oplus AP$.
12: $\tau = \tau + 1$.
13: For all edge $(v, v^{'}) \in E$, get minimum left vertex dual variable value $l\_y(v) = \min\{y(v)\}$ and minimum right vertex dual variable minus value $r\_y(v^{'}) = \min\{y(v) + y(v^{'}) - W(v, v^{'})\}$.
14: **if** $l\_y(v) < r\_y(v^{'})$ **then**
15: halt.
16: **else**
17: Set dual variable augmentation value $\delta_\tau = l\_y(v)$.
18: **end if**
19: **for all** $v \in V$ **do**
20: **if** v is a left vertex in G **then**
21: $y(v) = y(v) - \delta_\tau$.
22: **else**
23: $y(v) = y(v) + \delta_\tau$.
24: **end if**
25: **end for**
26: **until** $(AP = \varnothing)$
27: **return** M.

5 Experiments

In this section, we test the performance of our method TransC on two benchmark datasets. We will compare the accuracy of our method with the baseline methods.

5.1 Experimental Settings

In this paper, we employ LogMap, AML, YAM-BIO, XMap and SiGMa as the baseline methods. We compared the accuracy of the final category matchings in terms of precision, recall and F1-measure on the number of categories correctly matched. We used two datasets from OAEI 2017[1].

[1] http://oaei.ontologymatching.org/2017/.

Algorithm 3. The augmenting path searching algorithm

Input: $G=<V,E,W>, M, \{y(v)|v \in V\}, V_L, V_R, LF, RF$.
Output: AP.
1: **if** $LF = \varnothing$ or $RF = \varnothing$ **then**
2: $AP = \varnothing$.
3: **else**
4: Direct unmatched edges from $V_L \rightarrow V_R$, matched edges $V_R \rightarrow V_L$.
5: Add vertices s, t and connect them to free vertices in LF and RF, respectively.
6: Run BFS algorithm on G to find an augmenting path AP containing only edges $\{(v, v^{'})|v \in LF, v^{'} \in RF\}$ for which $y(v) + y(v^{'}) = W(v, v^{'})$.
7: **end if**
8: Augment the free vertices set LF and RF based on the augmenting path AP, respectively.
9: $AP = AP \setminus \{s, t\}$.
10: **return** AP.

The first dataset was derived from the large BioMed track (denoted as DS_{bio}). The dataset contains three biomedical ontologies: FMA, SNOMED-CT and NCI. These ontologies are semantically rich and contain tens of thousands of categories. Large BioMed track contains three matching problem: FMA-NCI, FMA-SNOMED and NCI-SNOMED, and each matching problem contains two tasks involving "small" largebio dataset (denoted as DS_{bio_s}) and "whole" large-bio dataset (denoted as DS_{bio_w}). The dataset DS_{bio} is used to find matchings between large ontologies with rich semantics.

The second dataset was derived from the conference track (denoted as DS_{conf}), The DS_{conf} dataset contains 16 different ontologies, which aims at finding all category matchings in ontology set describing the domain of organising conferences [13]. The DS_{conf} dataset contains 867 categories and 534 attributes in total.

For experiments of TransC, the parameters we used to measure the matching likelihood between two categories is experimentally set to $\lambda = 0.01$, $\gamma = 2$, $k = 50$, $d = L_1$, which yields the best accuracy, where λ is the learning rate for stochastic gradient descent (SGD), γ is the margin, k is the dimensions of category and relation embedding, d is the dissimilarity measure. We found reasonable values for the parameters by exploring its accuracy on the DS_{conf} dataset alignments, and then kept them fixed for all the experimental comparisons over the DS_{bio} and DS_{conf} datasets.

5.2 Experimental Analysis

In the following, we will test the performance of TransC and the baseline methods on taxonomies with different size and domains. Firstly, we tested the accuracy of all the methods on the DS_{bio} datasets, then tested the accuracy of all the methods on the DS_{conf} dataset.

In order to see how the accuracies of all the methods change with the increase of the size of ontologies, we test all the methods on the DS_{bio_s} dataset and

DS_{bio_w} dataset. Tables 1 and 2 show the results for the DS_{bio_s} dataset and DS_{bio_w} dataset, respectively. From the results, it can be seen that our method TransC achieves the best precision, recall and F1 measure on the datasets. In addition, we averaged the precision, recall and F1 measure of each method over the DS_{bio_s} and DS_{bio_w} datasets in Tables 1 and 2, respectively. The average results for these methods are shown in Table 3. From Table 3, we can notice that TransC achieves the best accuracy among all the methods. In summary, the experimental results show that TransC can obtain better accuracy over the baseline methods on the DS_{bio} dataset.

In the following, we test how the accuracies of all the methods change across ontologies from different domains. Firstly, we conducted experiments on the DS_{bio} dataset. Secondly, we conducted experiments on the DS_{conf} dataset. In order to test the performance of our method and the baseline methods on the DS_{bio} dataset, we averaged the accuracy on the DS_{bio_s} and DS_{bio_w} datasets (Table 3). The results is shown in Table 4. From Table 4, we can see that TransC achieves the best performance on the DS_{bio} dataset.

In the following, we conducted experiments on the DS_{conf} dataset. The results on the DS_{conf} dataset is presented in Table 5. From Table 5, it can be seen that TransC performs better than any of the baseline methods.

Table 1. Comparison over the DS_{bio_s} dataset

Method	Task								
	FMA-NCI			FMA-SNOMED			NCI-SNOMED		
	Precision	Recall	F1	Precision	Recall	F1	Precision	Recall	F1
SiGMa	0.841	0.654	0.735	0.959	0.692	0.804	0.896	0.647	0.751
XMap	0.977	0.901	0.937	0.974	0.847	0.906	0.894	0.566	0.693
YAM-BIO	0.969	0.896	0.931	0.966	0.733	0.834	0.899	0.677	0.772
AML	0.958	0.910	0.930	0.923	0.762	0.835	0.871	0.746	0.804
LogMap	0.944	0.897	0.920	0.947	0.690	0.798	0.947	0.690	0.798
TransC	**0.981**	**0.928**	**0.954**	**0.979**	**0.854**	**0.912**	**0.961**	**0.749**	**0.842**

Table 2. Comparison over the DS_{bio_w} dataset

Method	Task								
	FMA-NCI			FMA-SNOMED			NCI-SNOMED		
	Precision	Recall	F1	Precision	Recall	F1	Precision	Recall	F1
SiGMa	0.826	0.628	0.714	0.945	0.625	0.752	0.873	0.466	0.608
XMap	0.884	0.847	0.865	0.774	0.843	0.807	0.819	0.553	0.66
YAM-BIO	0.818	0.888	0.852	0.887	0.728	0.800	0.827	0.698	0.757
AML	0.838	0.872	0.855	0.882	0.687	0.772	0.904	0.668	0.768
LogMap	0.856	0.808	0.831	0.840	0.645	0.730	0.868	0.597	0.707
TransC	**0.896**	**0.892**	**0.894**	**0.948**	**0.849**	**0.896**	**0.911**	**0.708**	**0.797**

Table 3. Average comparison over the DS_{bio} dataset

Tasks	Approaches	Precision	Recall	F1
DS_{bio_s}	SiGMa	0.899	0.664	0.763
	XMap	0.948	0.770	0.851
	YAM-BIO	0.945	0.770	0.848
	AML	0.917	0.810	0.858
	LogMap	0.946	0.760	0.842
	TransC	**0.974**	**0.844**	**0.904**
DS_{bio_w}	SiGMa	0.881	0.573	0.691
	XMap	0.826	0.750	0.785
	YAM-BIO	0.844	0.770	0.806
	AML	0.875	0.740	0.803
	LogMap	0.855	0.680	0.759
	TransC	**0.918**	**0.816**	**0.864**

Table 4. Comparison over the DS_{bio} datasets

Approaches	Precision	Recall	F1
SiGMa	0.890	0.619	0.727
XMap	0.887	0.760	0.818
YAM-BIO	0.894	0.770	0.828
AML	0.896	0.770	0.831
LogMap	0.900	0.720	0.801
TransC	**0.946**	**0.830**	**0.884**

Overall, from Tables 4 and 5, we can see that TransC can obtain the best performance both on the DS_{bio} dataset and the DS_{conf} dataset. The results show that our method TransC can performs well on taxonomies from different domains. Based on the experimental results and analysis, we can see that TransC can performs well on taxonomies regardless of their scales and domains. This shows that TransC has good adaptability.

Table 5. Comparison over the DS_{conf} dataset

Approaches	Precision	Recall	F1
SiGMa	0.512	0.334	0.404
XMap	0.840	0.570	0.680
AML	0.840	0.660	0.740
LogMap	0.820	0.590	0.690
TransC	**0.862**	**0.690**	**0.766**

6 Conclusion

This paper presents a representation learning method for taxonomy matching. As our method models the taxonomy matching problem as an optimization problem on bipartite graphs, with the global nature of maximum weight matching, our method can obtain a global optimal matching for two taxonomies. Currently, our method mainly focuses on one-to-one equivalent category matchings between two taxonomies and runs in polynomial time. For future work, we plan to address the subclass and superclass matchings at the same time.

References

1. Achichi, M., Cheatham, M., Dragisic, Z., et al.: Results for the ontology alignment evaluation initiative 2017. In: Proceedings of the 12th International Workshop on Ontology Matching (2017)
2. Asprino, L., Presutti, V., Gangemi, A., Ciancarini, P.: Frame-based ontology alignment. In: Proceedings of 31st AAAI Conference on Artificial Intelligence, pp. 4905–4906 (2017)
3. Bordes, A., Usunier, N., García-Durán, A., Weston, J., Yakhnenko, O.: Translating embeddings for modeling multi-relational data. In: Proceedings of 27th Annual Conference on Neural Information Processing Systems, pp. 2787–2795 (2013)
4. Bouraoui, Z., Schockaert, S.: Learning conceptual space representations of interrelated concepts. In: Proceedings of 27th International Joint Conference on Artificial Intelligence, pp. 1760–1766 (2018)
5. Chen, R.C., Bau, C.T., Yeh, C.J.: Merging domain ontologies based on the wordnet system and fuzzy formal concept analysis techniques. Appl. Soft Comput. **11**(2), 1908–1923 (2011)
6. Demidova, E., Oelze, I., Nejdl, W.: Aligning freebase with the YAGO ontology. In: Proceedings of 22nd ACM International Conference on Conference on Information and Knowledge Management, pp. 579–588 (2013)
7. Jiménez-Ruiz, E., Grau, B.C.: LogMap: logic-based and scalable ontology matching. In: Proceedings of 10th International Semantic Web Conference, pp. 273–288 (2011)
8. Kuhn, H.W.: The hungarian method for the assignment problem. Naval Res. Logistics Q. **2**(1–2), 83–97 (1955)
9. Lacoste-Julien, S., Palla, K., Davies, A., Kasneci, G., Graepel, T., Ghahramani, Z.: SiGMa: simple greedy matching for aligning large knowledge bases. In: Proceedings of 19th SIGKDD, pp. 572–580 (2013)
10. Lawler, E.L.: Combinatorial Optimization: Networks and Matroids. Courier Dover Publications (1976)
11. Lee, T., Wang, Z., Wang, H., won Hwang, S.: Web scale taxonomy cleansing. Proc. VLDB Endowment **4**(12) (2011)
12. Li, J., Tang, J., Li, Y., Luo, Q.: RiMOM: a dynamic multistrategy ontology alignment framework. TKDE **21**(8), 1218–1232 (2009)
13. Lin, H., Wang, Y., Jia, Y., Xiong, J., Zhang, P., Cheng, X.: An ensemble matchers based rank aggregation method for taxonomy matching. In: Cheng, R., Cui, B., Zhang, Z., Cai, R., Xu, J. (eds.) APWeb 2015. LNCS, vol. 9313, pp. 190–202. Springer, Cham (2015). https://doi.org/10.1007/978-3-319-25255-1_16

14. Ba, M., Diallo, G.: Large-scale biomedical ontology matching with ServOMap. IRBM **34**(1), 56–59 (2013)
15. Ochieng, P., Kyanda, S.: A statistically-based ontology matching tool. Distrib. Parallel Databases **36**(1), 195–217 (2018)
16. Suchanek, F.M., Abiteboul, S., Senellart, P.: PARIS: probabilistic alignment of relations, instances, and schema. Proc. VLDB Endowment **5**(3), 157–168 (2011)
17. Xie, R., Liu, Z., Jia, J., Luan, H., Sun, M.: Representation learning of knowledge graphs with entity descriptions. In: Proceedings of 30th AAAI Conference on Artificial Intelligence, pp. 2659–2665 (2016)

Creating Training Data for Scientific Named Entity Recognition with Minimal Human Effort

Roselyne B. Tchoua[1(✉)], Aswathy Ajith[1], Zhi Hong[1], Logan T. Ward[2], Kyle Chard[2,3], Alexander Belikov[6], Debra J. Audus[4], Shrayesh Patel[5], Juan J. de Pablo[5], and Ian T. Foster[1,2,3]

[1] Department of Computer Science, University of Chicago, Chicago, IL, USA
roselyne@uchicago.edu
[2] Globus, University of Chicago, Chicago, IL, USA
[3] Data Science and Learning Division, Argonne National Lab, Lemont, IL, USA
[4] Materials Science and Engineering Division, National Institute of Standards and Technology, Gaithersburg, MD, USA
[5] Institute for Molecular Engineering, University of Chicago, Chicago, IL, USA
[6] Knowledge Lab, University of Chicago, Chicago, IL, USA

Abstract. Scientific Named Entity Referent Extraction is often more complicated than traditional Named Entity Recognition (NER). For example, in polymer science, chemical structure may be encoded in a variety of nonstandard naming conventions, and authors may refer to polymers with conventional names, commonly used names, labels (in lieu of longer names), synonyms, and acronyms. As a result, accurate scientific NER methods are often based on task-specific rules, which are difficult to develop and maintain, and are not easily generalized to other tasks and fields. Machine learning models require substantial expert-annotated data for training. Here we propose polyNER: a semi-automated system for efficient identification of scientific entities in text. PolyNER applies word embedding models to generate entity-rich corpora for productive expert labeling, and then uses the resulting labeled data to bootstrap a context-based word vector classifier. Evaluation on materials science publications shows that the polyNER approach enables improved precision or recall relative to a state-of-the-art chemical entity extraction system at a dramatically lower cost: it required just two hours of expert time, rather than extensive and expensive rule engineering, to achieve that result. This result highlights the potential for human-computer partnership for constructing domain-specific scientific NER systems.

Keywords: Scientific named entities · Word embedding · Natural Language Processing · Crowdsourcing · Polymers

J. M. F. Rodrigues et al. (Eds.): ICCS 2019, LNCS 11536, pp. 398–411, 2019.
https://doi.org/10.1007/978-3-030-22734-0_29

1 Introduction

There is a pressing need for automated information extraction and machine learning (ML) tools to extract knowledge from the scientific literature. One task that such tools must perform is the identification of entities within text. Many rule-based, ML, and hybrid named entity recognition (NER) approaches have been developed for particular entity types (e.g., people and places) [20,23].

Scientific NER remains challenging due to non-standard encoding and the use of multiple entity *referents* (terms used to refer to an entity). For example, in materials science, polymers are encoded in text using various representations (conventional or commonly-used), acronyms, synonyms, and historical terms. Further challenges arise when trying to distinguish between general and specific references to members of polymer families, or recognizing references to blends of two polymers, etc. Such challenges are not unique to polymer science. However, while NER is a well-studied topic in medicine and biology [5,16], it has only recently become a focus in materials science [10,17,29,36]. Many approaches applied in other domains rely on large, carefully annotated corpora of training data, a luxury not yet available in domains like polymer science. Here we introduce polyNER, a hybrid computer-human system for semi-automatically identifying scientific entity referents in text. PolyNER operates in three phases, first applying a fully automated analysis to produce an entity-rich set of candidates for labeling; then engaging experts to approve or reject a modest number of proposed candidates; and finally using the resulting labeled candidates to train a classifier. In both the first and third phases, it uses word embedding models to capture shared contexts in which referents occur. PolyNER thus seeks to substitute the labor-intensive processes of either assembling a large manually labeled corpus or defining complex domain-specific rules with a mix of sophisticated automated analysis and focused expert input.

We evaluate polyNER performance on polymer science publications. We compare the output of its first candidate enrichment phase against expert-labeled data, and find that it retrieves 61.2% of the polymers extracted by experts with a precision (26.0%) far higher than the ratio of target entities vs. non-entities in scientific publications (less than 2% in our experience). We evaluate the performance of polyNER overall by comparing its output polymer referents against both expert-labeled data and a state-of-the art rule-based chemical entity extraction system, ChemDataExtractor (CDE) [36], which we have previously enhanced with dictionary- and rule-based methods for identifying polymers [39]. We find that PolyNER can achieve either 52.7% precision or 90.7% recall, depending on user preference, a 10.5% improvement in precision or 22.4% improvement in recall over the enhanced CDE. These results highlight the potential for creating domain-specific scientific NER systems by combining sophisticated automated analysis with focused expert input.

The rest of this paper is as follows. We review scientific NER systems in Sect. 2. In Sect. 3, we motivate the need for identifying polymer names in text. We describe design and implementation in Sect. 4 and evaluate polyNER in Sect. 5. We summarize and discuss future work in Sect. 6.

2 Related Work

Natural Language Processing (NLP) is a way for computers to "read" human language. NLP tasks include automatic summarization, topic modeling, translation, named entity recognition (NER), and relationship extraction. NER systems, which aim to identify and categorize named entities (e.g., a person, organization, location), have been developed using both linguistic grammar-based techniques and ML models. While many ML models have been developed, their performance depends critically on the quantity and quality of training data.

Scientific domains such as molecular biology and medical NLP have long used NER for extracting symptoms, diagnoses, medications, etc. from text [5,16]. More recently, there has also been much interest in chemical entity and drug recognition [10,17,29,36]. However, even state-of-the-art NER systems do not typically perform well when applied to different domains [13]. Considerable effort is involved in selecting and (often manually) generating quality data for trainable statistical NER systems [14].

Various approaches have been proposed to address the lack of training data for NER and other information extraction tasks. *Distant supervision* maps known entities and relations from a structured knowledge base onto unstructured text [26,43]. However, many fields, including polymer science, lack such knowledge bases.

Data programming uses *labeling functions* (user-defined programs that provide labels for subsets of data) [27]. Errors due to differences in accuracy and conflicts between labeling functions are addressed by learning and modeling the accuracies of the labeling functions sets. Under certain conditions, data programming achieves results on par with those of supervised learning methods. But while writing concise scripts to define rules may seem to be a more reasonable task for annotators than exhaustively annotating text, it still requires expert guidance. Moreover, labeling functions typically rely on state-of-the-art entity taggers, such as CoreNLP [19], which recognizes persons, locations, organizations and more, and which itself has been trained using various corpora, including the Conference on Computational Natural Language Learning (CoNLL) dataset [32]. A user-defined function may be defined, for example, as: *if the word "married" appears between two PERSONs (as identified by a state-of-the-art named entity tagger), then extract the pair as potential spouses.* Eventually, we will explore using polyNER and data programming to extract polymer properties. For instance: *if a sentence contains a polymer name and the words "glass transition", then extract number(s) in the sentence as potential glass transition temperature(s) for that polymer.*

Other approaches use unskilled or semi-expert users to crowdsource the labeling task [7,15,38]. Nonetheless, domain expertise is often crucial for identifying and extracting complex scientific entities. Hence, we ask: how can we quickly generate annotated data for scientific named entity recognition?

3 Motivation

The complexity of scientific NER is primarily due to the fact that entities, for example biological [12] and chemical [14], can be described in different ways, with vocabularies often specialized to small communities. Such issues arise in the polymer science applications that we focus on here. In principle, International Union of Pure and Applied Chemistry (IUPAC) guidelines define polymer naming conventions [9]. However, such guidelines are not always followed in practice [37]. Polymer names may be reported as source-based names (based on the monomer name), structure-based names (based on the repeat unit), common names (requiring domain-specific knowledge), trade names (based on the manufacturer), and names based on chemical groups within the polymer (requiring context to fully specify the chemistry). Oftentimes, polymers are encoded using acronyms.

These different naming conventions arise in part because a desire for clarity in communications is at odds with the often complicated monomeric structures found in many polymers [1]. For example, sequence-defined polymers, where multiple monomers are chemically bound in a well-defined sequence as in proteins, often defy normal naming practices, as it is not possible to list concisely every monomer and their respective positions [18]. Another class of polymers that often suffer from complicated names are conjugated polymers, which exhibit useful optical and electrical properties. Conjugated polymers are complex due to the co-polymerization of multiple monomers (donor/acceptor units), the type and position of side chains along the polymer backbone, and the coupling between monomer units to control regioregularity [8].

Other challenges arise from the use of labels, structure referents (e.g., "micelles," "nanostructures"), and unusual author-coined acronyms. For example, one author defined the acronym DBGA for N,N-dibenzylglycidylamine and then used the string poly(DBGA) to represent poly(N,N-dibenzylglycidylamine). More naming variations result from typographical variants (e.g., alternative uses of hyphens, brackets, spacing) and alternative component orders.

These issues, which make identifying polymeric names a non-trivial exercise not only for computers but also for experts, arise in many fields with specialized vocabularies. Our long-term goal is to build a hybrid human-computer system in which we leverage both human and machine capabilities for the efficient extraction from text of properties associated with specialized vocabularies. In this work, we focus on the task of identifying polymer names.

4 Design and Implementation

As noted in the introduction, previous approaches to scientific NER have relied on large expert-labeled corpora to train NER tools. Our goal in polyNER is to slash the cost of NER training for new domains by using bootstrap methods to optimize the effectiveness and impact of minimal expert labeling. As shown in Fig. 1, rather than having experts review entire papers to identify entity referents, we use NLP tools to identify a set of promising candidate entity referents

(Candidate Generation), then in an Expert Labeling step employ experts to accept or reject those candidates, and finally in a Classifier Training step use the accepted candidates to train an entity classifier.

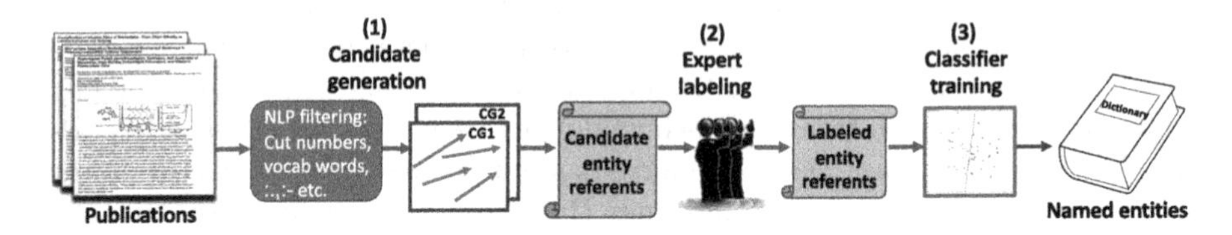

Fig. 1. PolyNER architecture: showing (1) Candidate Generation, which produces candidate named entities from word vectors, (2) Expert Labeling, and (3) Classifier Training, which uses labeled candidates to train supervised ML models for identifying referents.

Before turning to the details of the polyNER implementation, we define an NLP filtering process that is used in various places in polyNER to filter out words that are unlikely to be polymer referents. (1) We remove numbers. (2) Hypothesizing that names of scientific entities will not, in general, be English vocabulary words, we remove words found in the SpaCy and NLTK dictionaries of commonly used English words [2,4]. (We manually remove common polymer names, such as polystyrene and polyethylene, from the dictionaries.) (3) We use SpaCy's part-of-speech tagging functionality to remove non-nouns. (4) We remove unwanted characters (e.g. ':', '.', ',', ';', '-') from the beginning and the end of each candidate, allowing us to recognize, for example, *polyethylene;* (which fails the exact string comparison test against "polyethylene"). (5) We remove plurals (e.g., polyamides, polynorbornenes), as they can represent polymer family names.

4.1 Candidate Generation

This first phase uses word vector representations, context vector similarity measures, and minimal domain knowledge to identify a set of high-likelihood ("candidate") entity referents (names, acronyms, synonyms, etc.) in a supplied corpus of full-text documents (in the work presented here, scientific publications).

We first apply the NLP filtering process to reduce false positives. We next face the problem of determining whether a particular string is a polymer referent. String matching only gets us so far: for example, "polyethylene" names a polymer, but "polydispersity" does not. We need also to consider the context in which the string occurs. For example, the polymer name "polystyrene" in a sentence "The melting point of polystyrene is ..." suggests that X may also be a polymer in the sentence "The melting point of X is ...".

NLP researchers have developed a variety of *word embedding* methods for capturing this notion of context. A word embedding method maps each word in a sentence or document to a vector in an n-dimensional real vector space based on the linguistic context in which the word appears. (This mapping may be based, for example, on co-occurrence frequencies of words.) We can then determine the context-similarity between two words by computing the distance between their corresponding vectors in the feature space. Such vector representations can be created in many different ways [30,31]. Recently, the efficient neural network-based Word2Vec has become popular [21,22].

We consider two measures of context-similarity between word vector representations in this step. CG1 uses the Gensim implementation of the Word2Vec algorithm [28] to generate 100-dimension vectors. CG2 employs an alternative FastText word embedding method that considers sub-word information as well as context [3,11], allowing it to consider word morphology differences, such as prefixes and suffixes. Sub-word information is especially useful for words for which context information is lacking, as words can still be compared to morphologically-similar existing words. We set the length of the sub-word used for comparison—FastText's *n_gram* parameter—to five characters, based on our intuition that many polymers begin with the prefixes "*poly*" or "*poly(*". FastText produces 120-dimension vectors by default. We keep this slightly increased dimension (120 vs 100) as the embedding captures character information in addition to context information. Both CG1 and CG2 employ the continuous bag-of-words (CBOW) word embedding, in which a vector representation is generated for each word from an adjustable window of surrounding context words, in any order.

We compute a CG1 (Gensim) vector and a CG2 (FastText) vector for each NLP-filtered word in the input corpus, and also for a small set of representative polymer referents. Here we use polystyrene and its common acronym, PS, based on the assumption that polystyrene, as the most commonly mentioned polymer, provides a large number of example sentences in which polymers are mentioned. We can then determine, for each NLP-filtered word, the extent to which it occurs in a similar context to the representative polymers, by computing the similarities between the word's CG1 and CG2 vectors and those for polystyrene and PS. We discard the lower score for each of CG1 and CG2 to obtain two scores per word.

Having thus obtained scores, we then select as candidates, for each of CG1 and CG2, the N highest-scored words ($2N$ candidate total), with N selected based on the time available for experts. We also use a rule-based synonym finder to identify synonyms of generated polymer candidates [33]. For example, if polypropylene has been identified as a candidate, then the expression "*polypropylene (PP)*" leads to PP being added to the candidate list.

4.2 Expert Labeling

The previous step produces a set of candidate polymer referents: NLP-filtered strings that have been determined to occur in similar contexts to our representatives. We next employ an expert polymer scientist to indicate, for each such candidate, whether or not it is in fact a polymer referent. The expert simply approves or rejects each candidate via a simple web interface: a task that is more efficient than reading and annotating words in text.

The interface (see Fig. 2) provides the expert with example sentences as context for ambiguous candidates, and allows the expert to access the publication(s) in which a particular candidate appears when desired.

Name	Is polymer?	Notes	Submit notes	Bookmark	Example sentence	More Examples?
P(CL-co-PDSC)	✔	None	Add note		The resulting P(CL-co-PDSC) copolymer was isolated by precipitation in cold diethyl ether and dried in vacuo at room temperature.	?
TCLP	✔	None	Add note		Then DCLP was hydrolyzed to form a triple-chain ladder superstructure (TCLS), which was further converted into the target TCLP via subsequent in situ dehydration condensation.	?
ϕselfPS		None	Add note		Our study reveals that perturbations to PS Tg, which may be quantified by ϕselfPS calculations, correlate with partner fragility rather than partner Tg, with higher fragility partners resulting in higher ϕselfPS values.	?

Fig. 2. PolyNER web interface showing annotated candidates. Clicking on "?" delivers up to 25 more example sentences.

4.3 Candidate Discrimination

We next use the expert-labeled data to create a binary entity (polymer/not polymer) classifier. Many classification methods could be applied; we consider three in the work reported here: K Nearest Neighbor (KNN), Support Vector Classifier (SVC), and Random Forest (RF). Previous work has shown that KNNs perform reasonably well in text classification tasks [42]. SVC, an implementation of Support Vector Machines (SVMs), maps data into a feature space in which it can separate the data into two or more sets. RF groups decision trees ("weak learners") to form a strong learner; it produces models that are inspectable, and includes a picture of the most important features. In each case, we use the 100 (Gensim) + 120 (FastText) = 220 dimensions of the two word vectors as input features. Given limited training and testing data, we evaluate all three classifiers, as implemented within scikit-learn [24], in Sect. 5.3. We envision that with more annotated data, we will be able to use neural-network-based classifiers.

5 Evaluation

We report on studies in which we evaluate the performance of both the unsupervised Candidate Generation step and various classifiers trained on the labeled data that results from the Candidate Generation and Expert Labeling steps.

5.1 Dataset

We work with two disjoint sets of full-text publications in HTML format from the journal *Macromolecules*: P100 comprising 100 documents with 22 664 sentences and 508 391 (36 293 unique) words or "tokens," and P50 comprising 50 documents with 12 148 sentences and 270 514 (22 571 unique) tokens. For later use in evaluation, we engaged six experts to identify one-word polymer names in P100. They find 467 unique one-word polymer names.

5.2 Evaluation of Candidate Generation Methods

Recall that the polyNER candidate generation module employs two candidate generation methods, CG1 and CG2, plus a rule-based synonym finder. We evaluate the performance of both the complete polyNER candidate generation module and its CG1 and CG2 submodules by comparing the sets of candidates that they each generate from P100, both against the 467 one-word polymer names identified by experts in P100, and against two other polymer name extraction methods, CDE and CDE+, plus a sixth compound method formed by combining polyNER with CDE+. We use exact string-matching between candidates and expert-identified names (all lower cased). Results are in Table 1. For each, we evaluate extraction accuracy in terms of precision, recall, and F_1 score. Recall is the fraction of actual positives that are labeled correctly and precision the fraction of predicted positives that are labeled correctly; F_1, the harmonic average of precision and recall, reaches its best value at 1 and worst at 0.

The first two methods considered, CDE and CDE+, serve as baselines. CDE is a state-of-the-art Python package that extracts chemical named entities and associated properties and relationships from text [36]. As CDE aims to extract all chemical compounds, not just polymers, it serves only as a demonstration of an alternative approach in the absence of a polymer NER system. Its recall is high at 74.5% but its precision is, as expected, low at 8.7%. CDE+ extends CDE with manually defined polymer identification rules [39] to achieve a higher precision of 42.2% but a slightly decreased recall of 68.3%. These results emphasize the difficulty of automatically recognizing complex entities such as polymers.

Rows 3 and 4 show performance for CG1 and CG2 when employed independently. Recall that polyNER performs NLP filtering before applying CG1 and CG2. The filtering step eliminate all but 6878 of the 36 293 unique tokens in P100. Recall also that CG1 and CG2 each assign a score to each of the 6878 remaining words based on their context-based vector similarities to polystyrene and PS, and select the N highest scoring. In this evaluation, we set N=500. CG2, which takes word morphology into account, achieves higher precision and recall than does CG1 (41.8% vs. 15.6% precision and 44.8% recall vs. 16.7% recall for CG2 and CG1, respectively). CG2 retrieves more words starting with "poly" (67% of the 500 candidates vs. only 4% for CG1) while CG1 retrieves more acronyms (38% of the 500 candidates contained more upper than lower case letters, vs. 23% for CG2). CG1 returns more false positives. While character level information is useful for unseen words, or in this case for words lacking context information, we cannot dismiss the use of CG1. Authors often introduce polymer names and subsequently use acronyms more heavily, especially for long names. The facts that CG1 returns more acronyms and that there is likely more context information about acronyms, suggests that the performance of CG1, albeit lower, is solely based on context information.

Row 5 shows results for the complete polyNER candidate generator: that is, the combined CG1 and CG2 candidates plus their rule-based extracted synonyms. This method achieves 61.2% recall and 26.0% precision, producing an entity-rich set of candidates without any domain-specific rules and without any (tedious, time-consuming, and costly) expert-annotated corpus of polymer names. We are encouraged to observe that polyNER retrieves polymers not extracted by CDE: the combined recall for PolyNER $\cup$ CDE+ (row 6 in the table) is 81.6%—higher than CDE itself. This result suggests that polyNER's candidate generation module can be used not only to annotate automatically a diverse set of polymers based on context, but also to improve on the results of more sophisticated hybrid rule- and ML-based NER tools.

Figure 3, which shows every word in P100 in FastText vector space, illustrates the challenges and opportunities inherent in differentiating between polymer and non polymer word vectors. The polymer names (in red and green) form two rather diffuse clusters that overlap considerably with non polymers (in blue).

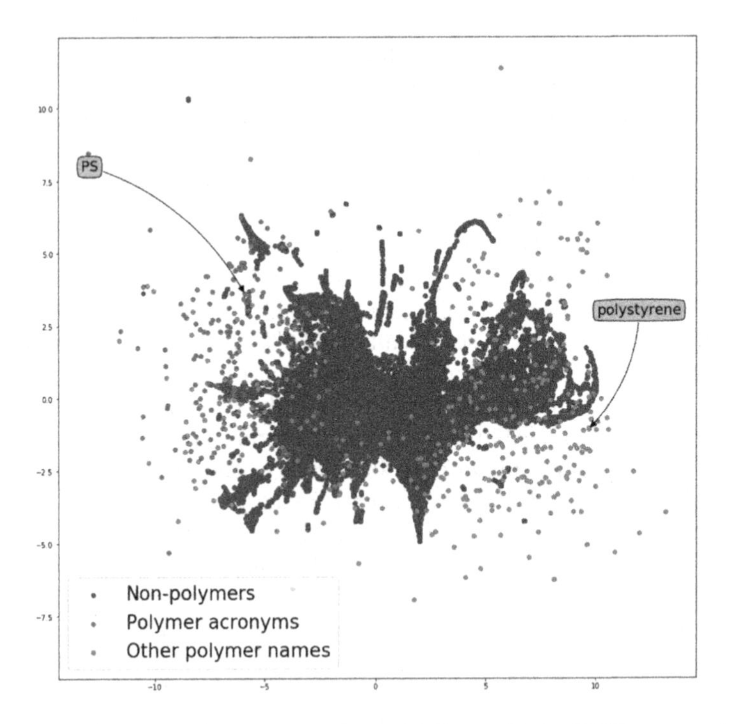

Fig. 3. A two-dimensional representation of all words in P100, generated with the scikit-learn implementation of t-distributed Stochastic Neighbor Embedding (t-SNE) [41]. Of the words identified by experts as polymers, we show acronyms in red and non-acronyms in green; all other words are blue. We label our two representative words. (The t-SNE plot, a dimensionality reduction technique used to graphically simplify large datasets, reduces the 120-dimensional vectors to two-dimensional data points. The axes have no "global" meaning.) (Color figure online)

Table 1. Results when polymer candidates are extracted from our test corpus, P100, via different methods. For each, we show true positives, false positives, false negatives, precision, recall, and F-score.

#	Method	Total	TP	FP	FN	Precision	Recall	F_1
1	CDE	3994	348	3646	119	8.7%	74.5%	15.6%
2	CDE+	755	319	436	148	42.2%	68.3%	52.2%
3	CG1	500	78	422	389	15.6%	16.7%	16.1%
4	CG2	500	209	291	258	41.8%	44.7%	43.2%
5	PolyNER	1099	286	813	181	26.0%	61.2%	36.5%
6	PolyNER ∪ CDE+	1495	381	1114	86	25.4%	81.6%	38.8%

Interestingly, the subset of polymer names that are acronyms (the red points) are clearly clustered.

5.3 Evaluation of Classifier Training Methods

The evaluation of Candidate Generation phase mainly illustrates the entity-richness of the candidate pool generated using the context-similarity criteria. The ultimate goal, however, is to train a classifier of context-aware vectors able to differentiate between polymer and not-polymer names in a set of test documents. Hence, we next evaluate how well classifiers trained on expert-labeled output from the Candidate Generation phase perform when applied to directly to a set of full-text documents. Here, we make use of our second dataset, P50, to train and test our classifiers, and P100 to validate the trained classifiers.

Before training our classifiers, we need a set of expert-labeled candidates. Thus we first apply the CG1 and CG2 methods of Sect. 4.1 to P50, generating a total of 897 unique candidates: 500 for CG2 and 466 from CG1, of which 69 overlapped. (We do not apply the rule-based synonym finder here.) Then we employ an expert to label as polymer or non-polymer each of those 897 candidates, producing a new dataset that we refer to as P50-labeled. Note that this task is quick work for the expert, as only 897 words need to be evaluated: the total time required was two hours. This expert review identifies 260 (29.0%) of the 897 as polymers.

Training and Validating the Classifiers: We next use the 897 expert-labeled words to train our three classifiers. We use 90% (807) for training and hold out 10% (90) for validation.

The left-hand side of Table 2 shows the performance of the different trained classifiers when applied to the P50 hold-out words. Note that performance here is defined with respect to how well the classifier does at predicting the expert labels assigned to the polyNER-generated candidates—not how well the classifier identifies *all* polymer referents in P50, as we do not have the latter information.

All three classifiers obtain between 66.7% and 100.0% precision and between 28.6% and 57.1% recall. SVC achieves the highest recall and F_1 score. The lower part of the table ("combined classifiers") shows that combined classifiers can improve performance. The 3-of-3 method achieves the highest precision (100.0%) but lowest recall, as one might expect. The ≥2 method also achieves 100.0% precision but with a higher recall (42.3% vs 28.6%). The ≥1 method has the lowest precision but the highest recall at 57.1%.

Table 2. Results when various classifiers (trained on expert-labeled P50 candidates) are applied to P50 holdouts (left) and P100 (right). The results in the bottom two rows are copied from Table 1 for ease of comparison.

Classifier	Validation on P50 holdouts			Testing on P100		
	Precision	Recall	F_1	Precision	Recall	F_1
KNN	75.0%	42.8%	54.5%	9.5%	77.8%	16.9%
SVC	66.7%	57.1%	61.5%	16.8%	76.5%	27.5%
RF	100.0%	28.6%	44.4%	51.0%	42.0%	46.1%
Combined						
3-of-3	100.0%	28.6%	44.4%	52.7%	39.1%	44.9%
≥2	100.0%	42.3%	60.0%	22.8%	66.6%	34.0%
≥1	57.1%	57.1%	57.1%	9.1%	90.7%	16.5%
CDE				8.7%	74.5%	15.6%
CDE+				42.2%	68.3%	52.2%

Testing the Trained Classifiers: We test the trained classifiers by applying each to all 6878 NLP-filtered nouns extracted from P100 and comparing the resulting polymer/non-polymer labels against our ground truth of polymer names extracted from P100 by experts. Results are in the right-hand side of Table 2. RF achieves the highest F_1 score (46.1%) with 51.0% precision and 42.1% recall. While recall is relatively low (fewer entities retrieved), precision is significantly better than that achieved by CDE+. We observe also that combined classifiers can improve precision (52.7%) at the expense of recall, or significantly increase recall (90.1%) at the expense of precision. Users can thus trade off precision and recall, in each case exceeding those achieved by the rule-based CDE+ system.

5.4 Discussion

These results are based on only limited training data: just 897 labeled words, of which 260 are polymers. We view the effectiveness of the classifiers trained with these limited data as demonstrating the feasibility of using small amounts of expert-labeled data to bootstrap context-aware word-vector classifiers. Importantly, this whole process was both inexpensive and generalizable to other domains. Candidate generation was fully automated and involved no domain knowledge besides the two representative words, polystyrene and PS. Labeling required just two hours of an expert's time. Classifier training was again automated and involved no domain knowledge.

6 Conclusion

Despite much progress in NLP, scientific named entity recognition (NER) remains a research challenge. A lack of labeled training data in fields such as polymer science limits the use of machine learning models for this task. PolyNER is a generalizable system that can efficiently retrieve and classify scientific named entities. It uses word representations and minimal domain knowledge (a few representative entities) to produce a small set of candidates for expert labeling; labeled candidates are then used to train named entity classifiers.

PolyNER can achieve either 52.7% precision or 90.7% recall when combining classifiers: a 10.5% improvement in precision or 22.4% in recall over a well-performing hybrid NER model (CDE+) that combines a dictionary, expert created rules, and machine learning algorithms. PolyNER's architecture allows users to tradeoff precision and recall by selecting which classifiers are used for discrimination. One out of every four candidates identified by our current polyNER prototype is in fact a polymer. This enrichment relative to the relative paucity of polymers in publications significantly reduces the effort required by experts. Considering that polyNER relies on simple distance from a known polymer(s), and default word embedding parameters, this result is encouraging.

An important issue to explore in future work is whether classifier performance can be improved by providing additional expert-labeled words. We plan to apply active learning [34] to select good candidates. As we generate more expert-labeled candidates, we will explore the use of neural network word vector classifiers to improve accuracy, and the use of polyNER-labeled data to annotate text for other NER approaches, such as bidirectional long short-term memory models. With a view to exploring generalizability, we are also working to apply polyNER to quite different problems, such as extracting dataset names from social science literature. We may explore more recent word representation models, which are pre-trained on large corpora [6,25].

Acknowledgments. We thank Mark DiTusa, Tengzhou Ma, and Garrett Grocke for contributing manually extracted polymer names. This work was supported in part by NIST contract 60NANB15D077, the Center for Hierarchical Materials Design, and DOE contract DE-AC02-06CH11357, and by computer resources provided by Jetstream [35,40]. Official contribution of the National Institute of Standards and Technology; not subject to copyright in the US.

References

1. Audus, D.J., de Pablo, J.J.: Polymer informatics: opportunities and challenges. ACS Macro Lett. **6**(10), 1078–1082 (2017)

2. Bird, S., Loper, E.: NLTK: the natural language toolkit. In: 42nd Annual Meeting of the Association for Computational Linguistics, p. 31 (2004)
3. Bojanowski, P., et al.: Enriching word vectors with subword information. arXiv:1607.04606 (2016)
4. Choi, J.D., et al.: It depends: dependency parser comparison using a web-based evaluation tool. In: 53rd Annual Meeting of the ACL, vol. 1, pp. 387–396 (2015)
5. Cohen, A.M., Hersh, W.R.: A survey of current work in biomedical text mining. Brief. Bioinform. **6**(1), 57–71 (2005)
6. Devlin, J., et al.: BERT: pre-training of deep bidirectional transformers for language understanding. arXiv:1810.04805 (2018)
7. Gao, H., et al.: Harnessing the crowdsourcing power of social media for disaster relief. IEEE Intell. Syst. **26**(3), 10–14 (2011)
8. Himmelberger, S., Salleo, A.: Engineering semiconducting polymers for efficient charge transport. MRS Commun. **5**(3), 383–395 (2015)
9. Hiorns, R.C., et al.: A brief guide to polymer nomenclature. Polymer **54**(1), 3–4 (2013)
10. Jessop, D.M., et al.: OSCAR4: a flexible architecture for chemical text-mining. J. Cheminform. **3**(1), 41 (2011)
11. Joulin, A., et al.: Bag of tricks for efficient text classification. arXiv:1607.01759 (2016)
12. Kim, J.-D., et al.: Introduction to the bio-entity recognition task at JNLPBA. In: International Joint Workshop on NLP in Biomedicine and its Applications, pp. 70–75 (2004)
13. Krallinger, M., et al.: Overview of the chemical compound and drug name recognition (CHEMDNER) task. In: BioCreative Challenge Evaluation Workshop, vol. 2, p. 2 (2013)
14. Krallinger, M., et al.: CHEMDNER: the drugs and chemical names extraction challenge. J. Cheminform. **7**(1), S1 (2015)
15. Krishna, R., et al.: Visual genome: connecting language and vision using crowdsourced dense image annotations. Intl. J. Comput. Vis. **123**(1), 32–73 (2017)
16. Leaman, R., Gonzalez, G.: BANNER: an executable survey of advances in biomedical named entity recognition. In: Pacific Symposium on Biocomputing, pp. 652–663 (2008)
17. Leaman, R., et al.: tmChem: a high performance approach for chemical named entity recognition and normalization. J. Cheminform. **7**(1), S3 (2015)
18. Lutz, J.-F.: Aperiodic copolymers. ACS Macro Lett. **3**(10), 1020–1023 (2014)
19. Manning, C.D., et al.: The Stanford CoreNLP natural language processing toolkit. In: ACL (System Demonstrations), pp. 55–60 (2014)
20. Marrero, M., et al.: Named entity recognition: fallacies, challenges and opportunities. Comput. Stand. Interfaces **35**(5), 482–489 (2013)
21. Mikolov, T., et al.: Efficient estimation of word representations in vector space. arXiv:1301.3781 (2013)
22. Mikolov, T., et al.: Distributed representations of words and phrases and their compositionality. In: Advances in Neural Information Processing System, pp. 3111–3119 (2013)
23. Nadeau, D., Sekine, S.: A survey of named entity recognition and classification. Lingvisticae Investig. **30**(1), 3–26 (2007)
24. Pedregosa, F., et al.: Scikit-learn: machine learning in Python. J. Mach. Learn. Res. **12**, 2825–2830 (2011)

25. Peters, M.E., et al.: Deep contextualized word representations. In: Conference of the North American Chapter of the Association for Computational Linguistics (2018)
26. Peters, S.E., et al.: A machine reading system for assembling synthetic paleontological databases. PLoS One **9**(12), e113523 (2014)
27. Ratner, A.J., et al.: Data programming: creating large training sets, quickly. In: Advances in Neural Information Processing Systems, pp. 3567–3575 (2016)
28. Rehurek, R., Sojka, P.: Software framework for topic modelling with large corpora. In: Workshop on New Challenges for NLP Frameworks (2010)
29. Rocktäschel, T., et al.: ChemSpot: a hybrid system for chemical named entity recognition. Bioinformatics **28**(12), 1633–1640 (2012)
30. Rumelhart, D.E.: Learning internal representations by back-propagating errors. Parallel Distrib. Process. **1**, 318–362 (1986)
31. Sabes, P.N., Jordan, M.I.: Reinforcement learning by probability matching. In: Advances in Neural Information Processing Systems, pp. 1080–1086 (1995)
32. Sang, E.F.T.K., De Meulder, F.: Introduction to the CoNLL-2003 shared task: language-independent named entity recognition. In: 7th Conference on Natural Language Learning, pp. 142–147 (2003)
33. Schwartz, A.S., Hearst, M.A.: A simple algorithm for identifying abbreviation definitions in biomedical text. In: Pacific Symposium on Biocomputing, pp. 451–462 (2002)
34. Settles, B.: Active learning. Synth. Lect. Artif. Intell. Mach. Learn. **6**(1), 1–114 (2012)
35. Stewart, C.A., et al.: Jetstream: a self-provisoned, scalable science and engineering cloud environment (2015). https://doi.org/10.1145/2792745.2792774
36. Swain, M.C., Cole, J.M.: ChemDataExtractor: a toolkit for automated extraction of chemical information from the scientific literature. J. Chem. Inf. Model. **56**(10), 1894–1904 (2016)
37. Tamames, J., Valencia, A.: The success (or not) of HUGO nomenclature. Genome Biol. **7**(5), 402 (2006)
38. Tchoua, R.B., et al.: A hybrid human-computer approach to the extraction of scientific facts from the literature. Proc. Comput. Sci. **80**, 386–397 (2016)
39. Tchoua, R.B., et al.: Towards a hybrid human-computer scientific information extraction pipeline. In: 13th International Conference on e-Science, pp. 109–118 (2017)
40. Towns, J., et al.: XSEDE: accelerating scientific discovery. Comput. Sci. Eng. **16**(5), 62–74 (2014)
41. van der Maaten, L.J.P., Hinton, G.: Visualizing data using t-SNE. J. Mach. Learn. Res. **9**, 2579–2605 (2008)
42. Yang, Y., Liu, X.: A re-examination of text categorization methods. In: 22nd Annual International ACM SIGIR Conference, pp. 42–49. ACM (1999)
43. Zhang, C., et al.: GeoDeepDive: statistical inference using familiar data-processing languages. In: ACM SIGMOD Conference, pp. 993–996 (2013)

Evaluating the Benefits of Key-Value Databases for Scientific Applications

Pol Santamaria[1(✉)], Lena Oden[2], Eloy Gil[1], Yolanda Becerra[1,3], Raül Sirvent[1], Philipp Glock[2], and Jordi Torres[1,3]

[1] Barcelona Supercomputing Center, 08034 Barcelona, Spain
{pol.santamaria,eloy.gil,yolanda.becerra, raul.sirvent,jordi.torres}@bsc.es

[2] Forschungszentrum Jülich, 52428 Jülich, Germany
{l.oden,p.glock}@fz-juelich.de

[3] Polytechnic University of Catalonia, 08034 Barcelona, Spain

Abstract. The convergence of Big Data applications with High - Performance Computing requires new methodologies to store, manage and process large amounts of information. Traditional storage solutions are unable to scale and that results in complex coding strategies. For example, the brain atlas of the Human Brain Project has the challenge to process large amounts of high-resolution brain images. Given the computing needs, we study the effects of replacing a traditional storage system with a distributed Key-Value database on a cell segmentation application. The original code uses HDF5 files on GPFS through an intricate interface, imposing synchronizations. On the other hand, by using Apache Cassandra or ScyllaDB through Hecuba, the application code is greatly simplified. Thanks to the Key-Value data model, the number of synchronizations is reduced and the time dedicated to I/O scales when increasing the number of nodes.

Keywords: Key-Value distributed databases · HPC · Big Data · NoSQL

1 Introduction

In recent years, the data produced in scientific workflows increased massively. One of the objectives of the Human Brain Project is the creation of a digital *brain atlas* identifying the regions of the human brain. The project support getting a better understanding of the microscopic structure of the human brain. For this purpose, Jülich researchers slice the brain into small tissues and obtain high-resolution images through 3-D Polarized Light Imaging. Since one scanner produces 1 TB of data per day, a whole brain imaged at 1 μm resolution generates one Petabyte worth of data.

The scans are analyzed with image processing techniques to identify cells, regions and, neuron densities. Only HPC resources are capable of processing

J. M. F. Rodrigues et al. (Eds.): ICCS 2019, LNCS 11536, pp. 412–426, 2019.
https://doi.org/10.1007/978-3-030-22734-0_30

the sheer amount of data. However, parallel file systems are the principal cause of performance and scalability issues. Besides, working with files enforces strict application workflows with synchronizations and complex code, hard to adapt under new requirements or hardware changes.

Key-Value *(KV)* databases are widely used in data analytics and represent a realistic alternative. They are well-suited for scientific applications such as time-series or spatial data. Furthermore, they allow analyzing partial results and react, for instance, by discarding a chemical simulation as soon as a certain event occurs.

In this work, we analyze the introduction of KV distributed databases in a high-performance data-analytic from the Human Brain Project. The use case originated in the *brain atlas* after facing a real problem driven by I/O, cells detection in high-resolution brain images. The Cell Segmentation Application *CSA* analyzes TIFF images through MPI-IO [3] and saves the results on large HDF5 files [8]. Post-execution queries will later refine the results.

The *brain atlas* is built on pipelines including sequential post-processing caused by the use of files. In this scenario, KV databases would allow processing the partial results in parallel. With this document, our goal is to discuss not only the performance but also the usability, advantages, and disadvantages of KV databases in HPC workflows.

The rest of the paper is organized as follows. In Sect. 2 we introduce the necessary knowledge to understand this work. Afterward, in Sect. 3 we discuss similar approaches and research works. Section 4 describes the problem we are trying to solve, the current procedure, and our proposed solution. We evaluate and discuss both approaches in Sect. 5. Finally, in Sect. 6 we present our conclusions and describe future lines of research.

2 Background

To perform parallel writes and to solve synchronization issues, HDF5 relies on buffering and aggregation of multiple POSIX operations. Its concurrency is based on the Single Writer Multiple Readers *(SWMR)* model which allows concurrent reads but enforces writes serialization. MPI-IO allows parallel non-contiguous writes in HDF5 but requires a POSIX-compliant file system.

Parallel File Systems *(PFS)* have performance issues on Big Data workloads even if they are HPC-oriented, such as GPFS. PFSs are typically deployed on a dedicated cluster with a set of raid disks to maximize the throughput. Often, these filesystems are located in a separate cluster, causing high-latencies and bottlenecks, and they rely on master - many slaves architectures that do not scale. Moreover, in these situations splitting a dataset for concurrent access or managing dynamic allocations is complicated. As a result, the I/O parallelism is low, and synchronizations are frequent. Besides, keeping POSIX metadata coherent is expensive.

To reduce the complexities of working with files and the associated performance, coherency, and availability issues, Key-Value *(KV)* distributed databases were introduced. They handle data identified by a key and an associated value comprised of multiple elements. For many years they have been present in business analytic deployments but seldom in HPC. These systems delay and optimize the indexation of keys for a better throughput.

Frequently, data analytics rely on Apache Cassandra [11], a distributed and highly scalable KV database with homogeneous architecture. An alternative is ScyllaDB [1], which targets applications with real-time requirements. They reduce the operations latency by increasing the concurrency degree through a lock-less design. The client-server communication protocol and configuration are inherited from Cassandra, which makes the transition instantaneous. Moreover, they designed the core architecture to overcome the CPU bottleneck of Cassandra with a thread-per-core and a shared-nothing architecture. Both solutions were developed to offer tuneable consistency and storage management for data analytics with commodity hardware.

To simplify the interaction with the distributed storage, we introduced Hecuba. It brings a set of tools and interfaces to simplify the interaction with distributed storage and improved performance. So far, support for Cassandra and Scylla has been built-in, enabling manipulation of distributed datasets transparently. Through Hecuba, Python applications access persistent data like regular objects stored in memory. To develop an application, the user describes the data model by extending a Hecuba class and instantiate as many objects as needed. The user can persist or retrieve the in-memory objects by giving an identifier to the constructor or calling the method *make_persistent*.

3 Related Work

There is an increasing requirement to save and analyze large amounts of data in real-time. With this aim, researchers have evaluated and proposed storage technologies that take advantage of hardware advancements. Based on the level of abstraction and data granularity, research has followed different directions.

On one side, there have been discussions and advancements in parallel file systems such as Lustre and GPFS. They have a considerable degree of complexity, translated into difficulties to configure, maintain and trace issues. Consequently, researchers often write and share knowledge about good practices as Cornell [17] did, or performance tuning as Latham et al. [12]. New HPC-oriented filesystems have been devised to overcome the limitations on data-intensive workloads. BeeGFS is a relevant technology that delivers great performance as in the work of Eekhoff et al. [5]. Another major topic is the convergence of data analytics with HPC-oriented file systems. For instance, Tantisiriroj et al. [16] adapted PVFS to match HDFS performance for Hadoop workloads.

Given the complexity of managing parallel access to bytes, Key-Value *(KV)* and Object stores were designed and implemented after studying common data access patterns. They abstract the data access with high-level APIs managing a set of values or complex objects. Ceph is an object storage which also provides a FileSystem API. It is widely used in the Cloud and has been deployed in supercomputing centers, but, HPC users often report integration issues. Alternatively, Intel's DAOS and Segates' MIRO were developed for HPC data analytic frameworks. Liu et al. [13] demonstrated that DAOS scales up to 32 nodes and outperforms Lustre and Ceph in different types of workloads.

On the other hand, research advanced in different ways for KV data stores. In 2015, Islam et al. [10] started by placing Memcached, an in-memory KV data store, between Lustre and HDFS to enable low latency and high throughput on reads. Their work continued with Shankar et al. [14,15] who replaced internal Memcached operations to improve the performance. This approach differs from our proposal because their storage is not durable data might be lost on failures.

Likewise, Wu et al. [18] proposed Anna, an in-memory KV data store. They claimed that Anna outperformed Redis thanks to a thread-per-core architecture that scales vertically. Recently, Wu et al. [19] introduced an adjustable replication factor based on access frequency with significant benefits. In this work, they also evaluated an integration with enterprise storage obtaining promising results. In our future work, we will consider Anna an alternative.

In 2015, Greenberg et al. [7] developed an HPC-oriented KV database to take advantage of specialized hardware which outperformed Apache Cassandra. However, further experimentation would be needed to test the behavior in a more realistic scenario since only a uniform distribution was used.

Our research on KV databases for scientific applications started with Hernandez et al. [9] where we evaluated scientific queries on Cassandra. We analyzed a molecular dynamics simulation in real-time with a significant speedup compared to traditional files. Then, Artigues et al. [2] proposed Qbeast based on our previous work on D8trees [4], a distributed multi-dimensional index generated at runtime which can be paired with Cassandra. These works were data analytics oriented while this paper focuses on data generation and usability.

Related to the cell segmentation on images in HPC, Gabriel et al. [6] tested the PVFS and NFS file systems to provide the images and preserve the results. They discovered that persistent storage limits the scalability while being one of the most expensive tasks. On the other hand, Zhang et al. [20] proposed a framework to reduce the time needed to process and analyze the cell segmentation results. They stated the need to move to short-cyclic analytical workflows with queries analyzing partial results.

4 Cell Segmentation on Large-Scale Brain Images

The segmentation and morphological characterization of neuronal cell bodies enable the study of the cells distribution, density, and other features. Retrieving and classifying the cell's information represents a large data analytical challenge. Even a small brain region covering a few millimeters contains several hundred thousands of neurons, while the number of neurons in a complete human brain is almost 90 billion. Each human brain is sliced in up to 2,500 sections, which must be individually scanned and processed. A single image has a size of around $80,000 \times 100,000$ pixels, is stored with one channel which results in 20–30 GB.

In this section, we give a short overview of the original cell segmentation application. Note that we use an already optimized version with a reduced number of synchronization and MPI-IO to allow a fair comparison. Finally, we describe the changes introduced to adapt the application to the Key-Value databases.

4.1 Original Application Using HDF5 Files

The brain-images are provided in the BigTIFF format, a TIFF extension that enables files larger than 4 GB. It has an internal multi-page hierarchy where each page has a copy of the image with different resolution. The page organizes the data as a bitmap divided into tiles or stripes. In this case, 2D tiles of 256×256 pixels with elements encoded as 16 bits are used, resulting in tiles of 128 KB. Big tiles (>4 KB) improve the throughput of sequential operations at the expenses of storage requirements because the edging tiles are filled with padding bytes.

The Cell Segmentation Application *(CSA)* is written in *Python* and parallelized with `mpi4py`, an interface to MPI. Figure 1a illustrates the original workflow based on HDF5. Steps *(1–5)* are inherent to the application and independent of the storage solution. Steps *(6–9)* indicate specific operations required by HDF5, and vary depending on the storage system as reviewed in Sect. 4.2.

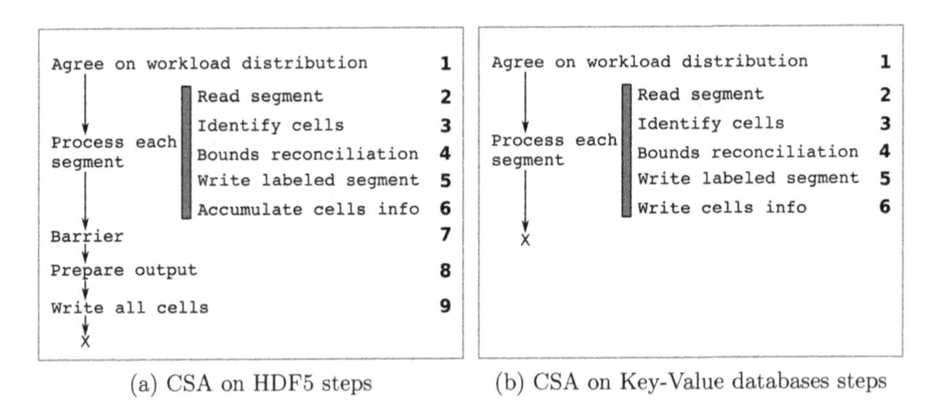

(a) CSA on HDF5 steps (b) CSA on Key-Value databases steps

Fig. 1. Cell Segmentation Application workflow

The application starts by creating an empty, uncompressed, HDF5 file through `h5py`. The *labeled image* array is the first dataset to be created. Each position of the *labeled image* array will have the identifier of the cell identified in the corresponding image, and zero otherwise. At the end of the execution, the *cells table* dataset is generated, which includes the identifiers, bounding boxes, and centroids of the cells. A resizable dataset for the *cells table* would allow partial writes. However, they become slower, and memory consumption is barely affected. Besides, it requires a chunk tunning process based on the underlying file system and future data access patterns. On the contrary, we know that the *labeled image* dataset will have the shape as the original image. Therefore, we preallocate the dataset for optimized write performance.

As a parallelization strategy, the application splits the image into multiple fix-sized segments *(Line 1 in Fig. 1a)* with a small overlap. The segment size represents a trade-off between memory consumption, performance and workload imbalance. We determined that memory consumption follows a quadratic increase with the size of segments. These segments are distributed evenly in a randomized manner to reduce the imbalance. After agreeing on the work distribution, each process gets a fixed range of cells IDs, which is wide enough to accommodate more cells than potentially found.

Next, every process works on its segments by fetching the data through the *Pytiff* module *(2)*. The cells are identified along with their bounding box and the rest of the characteristics *(3)* using *Sklearn*, `Scipy` and *OpenCV*. Then, the bounds reconciliation *(4)* discards the cells whose centroid does not belong to the segment. This step is necessary to remove the cells identified twice in the overlapping area by different processes. Then, the index of the current segment is written *(5)* into the *labeled image* through MPI-IO. Because the size of the *cells table* dataset is unknown in advance and it represents a small fraction of the output, the data is kept in memory *(6)* temporarily.

After completing the analysis, all processes wait on a barrier *(7)*. They need to reach an agreement *(8)* on the *cells table* layout to avoid overlapping data during the collective MPI-IO write *(9)*. Half the steps are only necessary to work with files, which requires complex APIs to synchronize or preallocate storage. We analyze these calls and their code in Sect. 4.3.

4.2 Implementation Using Key-Value Storage

The introduction of a Key-Value *(KV)* storage required minimal modifications to the original HDF5 application described in Sect. 4.1. The original workflow in Fig. 1a has been simplified to Fig. 1b.

In this approach, we save both datasets to the distributed KV database managed by Hecuba. We write the cells' characteristics *(6)* as soon as they are identified *(3)* and filtered by the bounds reconciliation phase *(4)*. Consequently, the *cells table* and *labeled image* are written one after the other and they can be accessed by other applications at this point. The operations *(6–9)* in Fig. 1a that introduced synchronizations are replaced with a simple step identified as *(6)* in Fig. 1b, greatly simplifying the code.

With regards to the data model, we considered saving the cell identifier for each pixel using the x-y coordinates as keys. Early experiments demonstrated that the number of small IOPS was huge and resulted in a performance impact, and we opted to store the labeled image as an array.

4.3 Interfacing with Hecuba Versus HDF5

We introduced some minor changes to replace the storage interface from HDF5 to Hecuba. First, the initialization of the HDF5 files shown in Listing 2 and the description of the data model is replaced with the code in Listing 1, resulting in cleaner code. The data model definitions can be defined on separate files and imported to reuse the data model in subsequent analytics.

```
from hecuba import StorageDict

class CellData(StorageDict):
  '''
  @TypeSpec <<rows:int,cols:int>,
            data:numpy.ndarray>
  '''
class Labels(StorageDict):
  '''
  @TypeSpec <<rows:int,cols:int>,
            data:numpy.ndarray>
  '''
# Create Hecuba persistent dictionaries
cell_table = CellData(ksp+'.cell_data')
data_table = Labels(ksp+'.labeled_data')
```

Listing 1: Hecuba objects in CSA

```
import mpi4py as MPI
import h5py

# Create HDF5 file with the MPI communicator
out_file = h5py.File(out, "w", driver="mpio",
                          comm=MPI.COMM_WORLD)
# Create and preallocate the indexing dataset
out_file.create_dataset("pyramid/00",
                          shape=img.shape,
                          dtype=np.int64)
# Create dataset after processing the image
out_file.create_dataset("cells",
                shape=(n_cells, n_features),
                            dtype=np.float32)
```

Listing 2: HDF5 objects in CSA

With Hecuba, we save each dataset as distributed Numpy NDArrays. Note that it is not necessary to explicit the internal type nor the shape since Hecuba extracts them at run-time. Describing the data model requires learning a simple and intuitive syntax. On the contrary, HDF5 requires more information to pre-allocate and optimize the file and has many optional arguments. Besides, MPI has to be initialized before the creation of HDF5 files.

```
# Persistent objects already instantiated
for (row, col, indexes) in local_slices:
  # ...
  # Write labeled data (numpy.ndarray)
  data_table[rows[0],cols[0]] = labels
  # Write cells information (numpy.ndarray)
  cell_table[rows[0], cols[0]] = table
```

Listing 3: Hecuba CSA main code

```
# Persistent objects already instantiated
total_table = None
for (row, col, indexes) in local_slices:
  # ...
  total_table = np.concatenate((total_table,
                          table), axis=0)
  # Write labeled data
  out_labeled[row[0]:row[1],
            col[0]:col[1]] = labels
table_length = np.array(total_table.shape[0],
                          np.uint64)
# Synchronize to unify the cells information
comm.Barrier()
len_cells = np.zeros(comm.size,
                      dtype=np.uint64)
comm.Allgather(table_length, len_cells)
offsets = [0] + np.cumsum(len_cells).tolist()
n_cells = len_cells.sum()
n_features = total_table.shape[1]
# Create and preallocate the cells information
out_file.create_dataset("cells",
          shape=(n_cells, n_features),
              dtype=np.float32)
# Write cells information
out_file["cells"][offsets[comm.rank]:
          offsets[comm.rank + 1]] = total_table
```

Listing 4: HDF5 CSA main code

The line of code used to pass the data to either Hecuba or HDF5 is almost identical. However, to do so, Hecuba does not require synchronizations or the collective initializations displayed in Listing 3. At the end of the execution, the HDF5 code introduces a global MPI barrier to unify the cells characteristics metadata as in Listing 4. It is necessary to allocate the dataset and to perform a collective write to non-overlapping areas.

5 Evaluation

In this section, we study the performance of the Cell Segmentation Application *(CSA)* supported by HDF5 files and Key-Value *(KV)* distributed databases. The objective is to identify the key aspects that limit the CSA from matching the data processing requirements of the Human Brain Project. Also, we demonstrate that KV distributed databases satisfy the HPC needs while offering a pleasant API for scientists.

First, we present the hardware setup and the experiments characteristics. Next, we describe how the experimentation evolved based on the results obtained. With this idea, we present a brief description and motivation of each experiment alongside the results obtained and an analysis.

5.1 Experimental Setup

The Juron Pre-Commercial Procurement (PCP) is one of the pilot systems granted to the Human Brain Project. The cluster features fast local NVMe disks facilitating the deployment of a wide variety of distributed storage technologies. For this reason, we selected the Juron PCP to run our experiments.

With regards to the storage technologies, we evaluate Apache Cassandra and ScyllaDB as Key-Value *(KV)* databases. Their results will be confronted with storing HDF5 files on a GPFS node pool. GPFS features a 2 GB pagepool on each node acting as a cache. On the contrary, Hecuba implements a client-side write queue for Cassandra and Scylla of 4 MB per process, resulting in 76 MB per node. However, both KV databases acknowledge writes only when they are persisted in disk.

The cluster features 19 high-density memory nodes with 2xPower8 processors clocked at 3.5 Ghz. Each processor has ten cores with a Simultaneous Multithreading (SMT) of 8, potentially executing 160 threads per node concurrently. Every node has a 1.6 TB Ultrastar SN100 NVMe local disk available to users. Additionally, 256 GB DDR4 and 4 NVIDIA Tesla P100 interconnected with NVLink are available on each node.

Regarding the inter-node communication, a 100 Gbps CoonectX-4 EDR Infiniband interconnect is used. Besides, the PCP system is connected to the Jülich GPFS system (JUST) to offer centralized storage to users. Likewise, multiple clusters rely on the storage provided by the JUST system, built on top of 22 Lenovo Distributed Storage Solution and three older GPFS Storage Servers, achieving a throughput of 220 GB/s. The connection between the Juron PCP and JUST is made through two Ethernet adapters, each with two 10 Gbps Ethernet ports, delivering a speed of 40 Gbps. The number of concurrent users is highly variable and the service does not offer on-demand scalability.

Early experiments performed best by running a process on each physical core and taking advantage of the SMT with underlying threads. For this reason, we run the original HDF5 application by launching 20 processes on each node with MPI. On the contrary, we left a physical core free when deploying Apache Cassandra or ScyllaDB resulting in 19 application processes on each node. The OS schedules the processes on cores, neither the application nor the database performs CPU pinning. This decision has been made to compare all solutions using the same amount of physical resources. The local NVMe disks are used to deploy the KV databases.

To evaluate the workflow, we selected a representative image from the dataset with a size of 24 GB and producing an output file of 87 GB in the HDF5 format containing the information of 16.6 Million cells individuated.

5.2 Performance Results

The first set of experiments illustrate the performance of the original Cell Segmentation Application *(CSA)* and the consequences of introducing Key-Value distributed databases. Figure 2a reports the CSA execution times when writing the output into HDF5 on GPFS, Scylla and Cassandra. As already mentioned, the input images are stored on the JUST GPFS, which also supports the HDF5 files.

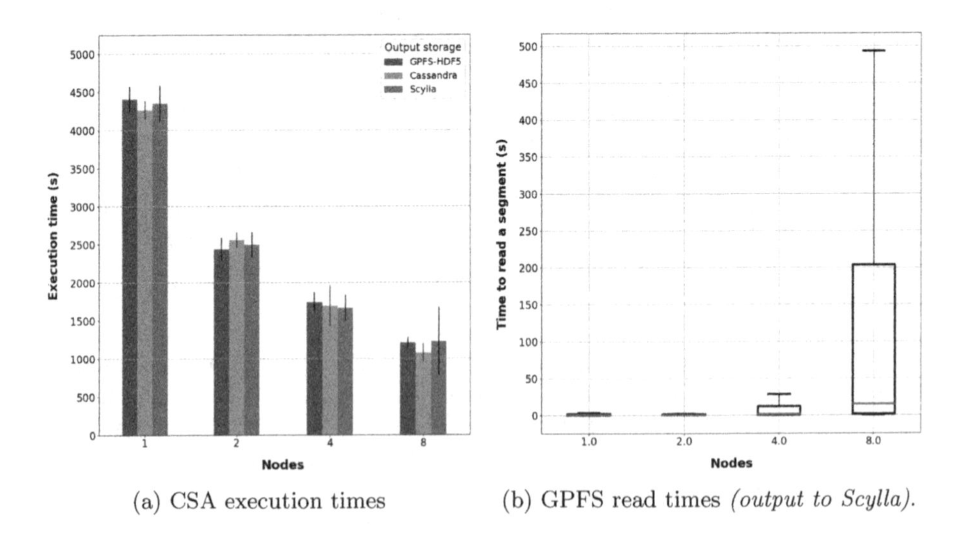

(a) CSA execution times (b) GPFS read times *(output to Scylla).*

Fig. 2. CSA: Input from GPFS and output to different storage solutions.

We evaluated the scalability from 1 to 8 nodes, running the HDF5 setup with 20 to 160 processes, and the KV counterpart with 19 to 152 application processes. We observed that the CSA does not scale, and it is subject to high variance without any visible performance pattern.

To understand these results, we measured the time dedicated to I/O. In every configuration, we found out that reading image segments from GPFS was the slowest operation and the main contribution to variability. For instance, the configuration whose performance varies the most in Fig. 2a is Scylla with 8 nodes. In Fig. 2b, we can appreciate the correlation between the variability and the GPFS read latencies of said execution. The measured read latencies grow exponentially when increasing the number of nodes. After a brief investigation, we found out that the GPFS itself caused the variance, and not the network, which is sufficiently overprovisioned.

We can compare the IOPS performed with the performance metrics obtained with the FIO benchmark. The CSA usually retrieves segments of 5200×5200 pixels made of tiles of 256×256 pixels. Therefore, each read operation accesses 441 tiles of 128 KB, and thus, fetches a total of ~55 MB. On the other hand, when the application starts, every process asks for a segment and produces more than 50 thousand read operations which the GPFS is unable to serve in a reasonable time. The read latencies observed for 8 nodes in Fig. 2b indicate that the lowest throughput obtained on GPFS is 106 KB/s and the average is 527 KB/s. The results are consistent with the FIO benchmark which reported 1337 IOPS of 4 KB or 1058 IOPS of 64 KB on GPFS.

Also, confronting Fig. 2b with 2a, we can see that the reading time would represent only 25.6% of the execution time if the workload were perfectly balanced. Since 92.5% of the processes analyze three slices and the rest only two, we can say that the average processing time of a segment, including reading and writing, is 409 s. However, we find read operations taking ~500 s, meaning that sometimes fetching a single segment from GPFS took longer than processing a complete segment on another process.

The GPFS performance issues are also related to the tiling size, which is too small for the GPFS to take full advantage of parallel reads. Synthetic tests demonstrated that tilling the image at 5200×5200 pixels, and thus, performing one read request, improved the read performance but not enough to compensate the task of reformatting the image and the extra space needed. Besides, since the input image segments are read while writing the *labeled image* segments, the GPFS pagepool is ineffective and evictions are continuously triggered. As a result, writing the *labeled image* dataset can result in slow operations which pollute the GPFS distributed pagepool and interfere with reads.

All in all, we can say that the GPFS introduces large amounts of variability and is unable to serve the input images. The causes are an inappropriate tilling of the input image, the large number of small IOPS and its inability to scale on-demand. On the other hand, the CSA itself introduces variability by randomly distributing the segments to processes.

After the initial tests with GPFS, we decided to obtain a baseline performance without the storage technologies, and we disabled the output in the CSA. Additionally, we distributed a copy of the image to all local NVMe disks.

We considered moving the image to a ramdisk, but it might have interfered with the memory usage. Besides, the NVMe already provided sub-microsecond read operations. Figure 3a reports the results illustrating a lower bound of the achievable performance of the provided CSA. The variability has been reduced in all scenarios but when storing HDF5 files on GPFS. Reading the images from the NVMe disks compared to GPFS show a speedup of ~2400 of the read latencies.

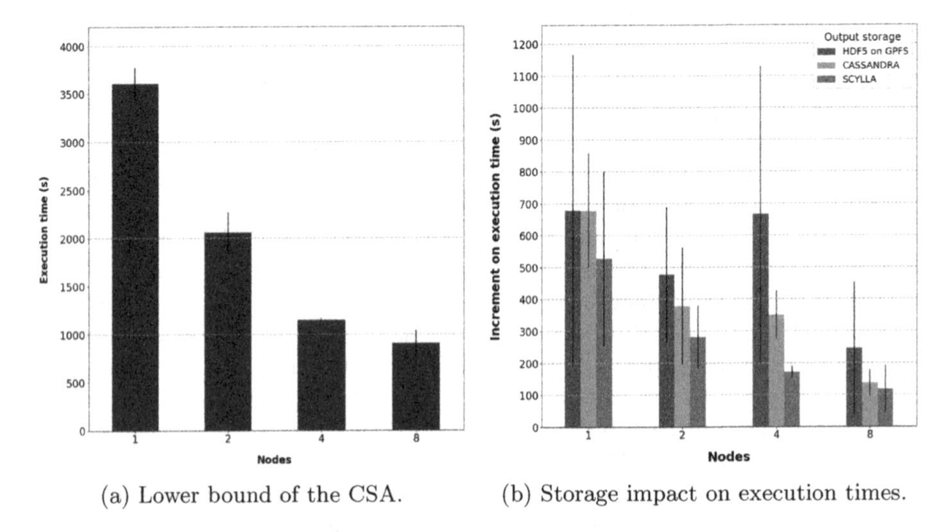

(a) Lower bound of the CSA. (b) Storage impact on execution times.

Fig. 3. Fetching images from local NVMe disks.

In Fig. 3a we can appreciate that the CSA does not scale even under ideal I/O conditions. The main cause is the trade-off between data granularity and cell reconciliation time. By having segments of 5200 × 5200 pixels, the data can be easily imbalanced even with a random distribution. However, if we reduce the size of the segment, the cells boundaries reconciliation time grows exponentially. As we add nodes, the imbalance has a greater impact on the execution times.

We repeated the cell segmentation tests by placing the input image on the local NVMe disks and storing the output to HDF5 on GPFS, Scylla and Cassandra. We took this decision after observing the reduction in the variability and the improved read latencies. As a result, we reduced the execution time and variability. To understand the influence of the I/O, Fig. 3b illustrates the difference between the cell segmentation when reading from NVMe and the lower bound obtained previously. We opted for this analysis because by only measuring the time dedicated to I/O we would not have considered the side effects of having threads doing background operations.

We can observe that the impact of I/O decreases with the number of nodes. Also, the variance of the results obtained with HDF5 on GPFS is significantly higher than the rest of the configurations. Indeed, the I/O time did not decrease with GPFS when running on four nodes; it took almost the same as with a single node. Cassandra behaves better than GPFS, but Scylla obtains the best results. We should keep in mind that reading the image now interferes with the KV databases since both rely on the NVMe disks. To analyze the performance differences, we decided to take a look at the average time dedicated to write.

The CSA performs 468 write operations to store the labeled segments with 32-bit elements. Around 400 operations write arrays of 5000 × 5000 elements resulting in ~100 MB. The remaining write operations write less information. In Fig. 4 we report the time dedicated by the user code to write the labeled segment at each iteration. For simplification, we only included the GPFS and Cassandra setups, since Scylla performed similarly to Cassandra but with lower average latencies.

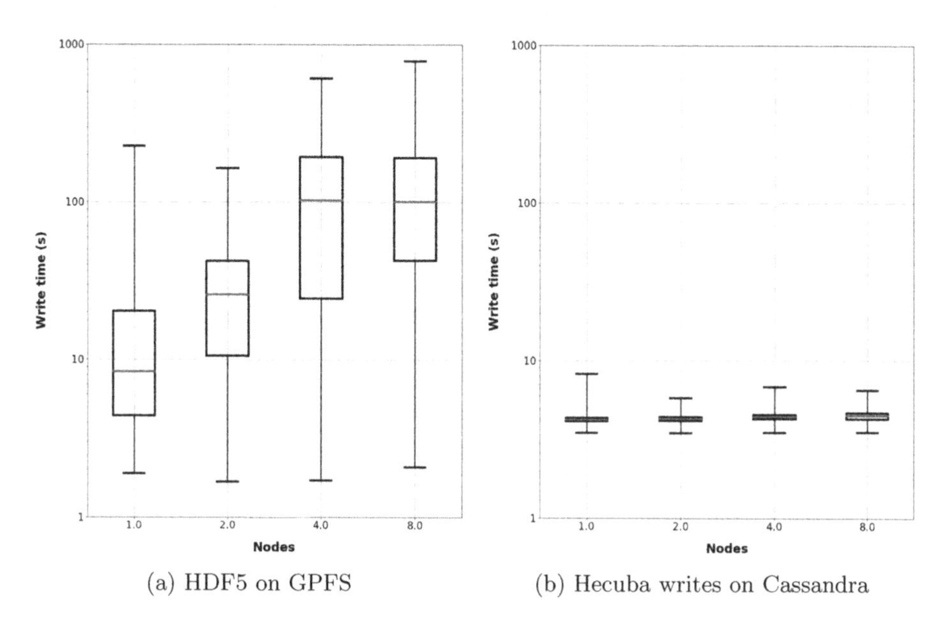

(a) HDF5 on GPFS

(b) Hecuba writes on Cassandra

Fig. 4. Labeled segment writes distribution *(Log10 scale)*.

In Fig. 4 we see that the mean write time on GPFS and its variance increase with the number of nodes. On the contrary, Cassandra achieves constant write times when scaling horizontally with little dispersion. The slowest write operations were observed on GPFS, reaching 600 s. Likewise, the fastest write operations were on GPFS, reaching subsecond write execution times. Cassandra reports write operation times in the order of 4 s. Since the interaction with Cassandra is done through the Hecuba prototype, we expect that further code optimizations should reduce the latencies. We also conclude that GPFS is unable to support the load with the addition of nodes.

The higher number of IOPS that Scylla achieves, thanks to its Thread Per Core architecture, explains the performance differences between KV databases in Fig. 3b. Consequently, client communication and database's processes interfere less with the application and execution times are shorter. However, HDF5 on GPFS can be faster sometimes for a variety of reasons. First, KV databases are optimized for queries rather than raw data ingestion. Moreover, they were not designed for HPC and, for instance, communication is done through TCP/IP over Infiniband. Finally, we use 4 KB working units with Hecuba, while the GPFS counterpart relies on 256 KB.

On the contrary, HDF5 uses MPI I/O to share data among the processes and write to storage efficiently through RDMA. In its turn, GPFS performs caching of 4 GB in the node's memory and writes asynchronously, unless instructed with the direct I/O flag. Hecuba was configured with a cache of 20 blocks per-process. Despite these differences, KV datastores achieve reasonable performance and notably better than the current workflow with HDF5 on GPFS. Besides, the latter is still subject to unpredictable performance despite only performing writes.

6 Conclusion and Future Work

The brain atlas use case provided by the Human Brain Project proved that deploying distributed Key-Value *(KV)* databases on HPC simplifies the programmers' task and achieves better performance than working with files. KV databases such as ScyllaDB and Apache Cassandra allowed horizontal scalability of the I/O with stable performance. Besides, the interface provided by Hecuba reduced the programming efforts and removed synchronizations.

We observed that the I/O does not scale with GPFS, which introduces high variability. For instance, running the application on eight nodes, the read operations on GPFS took ~61 s on average with a standard deviation of ~101 s. These latencies grow exponentially with the number of nodes. Besides, the GPFS is affected by other factors such as the coincidental load generated by separate applications. However, we proved that the I/O scales horizontally with distributed KV databases. We also identified the algorithmic factors that prevented the application to linearly scale, especially when running in more than four nodes. The work distribution policy, the large segments, and the cost of the bounds reconciliation strategy are the major impediments.

As a future work, we plan to enhance the integration of Cassandra and Scylla with HPC systems. In particular, replacing the TCP/IP communication with faster HPC protocols should provide significant benefits. In its turn, research on Hecuba should deliver performance improvements.

With regards to the Cell Segmentation Application, further research could be beneficial. First, by evaluating the effects of placing the input dataset to distributed storage and widen the set of technologies tested. Secondly, by designing and evaluating different mechanisms for the cell boundaries reconciliation would permit shrinking the image segments without an increase in execution time. As a positive collateral effect, smaller segments would reduce the imbalance.

To demonstrate the full potential of distributed KV databases queries should be implemented and evaluated. For instance, by performing analytics with data locality on real-time. Another field of interest consists of the benefits of persisting and recovering data in scientific workloads upon failures. Both are exciting research lines to explore through a data-driven approach. Finally, in the event of dynamic scheduling, the time needed to complete the workflow could be reduced by better load-balancing the application at runtime.

Acknowledgments. This project/research has received funding from the European Union's Horizon 2020 Framework Programme for Research and Innovation under the Specific Grant Agreement No. 720270 (Human Brain Project SGA1) and the Specific Grant Agreement No. 785907 (Human Brain Project SGA2). This work has also been supported by the Spanish Government (SEV2015-0493), by the Spanish Ministry of Science and Innovation (contract TIN2015-65316-P), and by Generalitat de Catalunya (contract 2017-SGR-1414).

References

1. Scylla Web Page. www.scylladb.com
2. Artigues, A., et al.: Paraview + Alya + D8tree: integrating high performance computing and high performance data analytics. Proc. Comput. Sci. **108**, 465–474 (2017)
3. Corbett, P., et al.: MPI-IO: a parallel file I/O interface for MPI version 0.3. Technical report, January 1995
4. Cugnasco, C., Becerra, Y., Torres, J., Ayguadé, E.: D8-tree: a de-normalized approach for multidimensional data analysis on key-value databases. In: Proceedings of the 17th ICDCN (2016)
5. Eekhoff, A., Tweddell, B., Pleiter, D.: BeeGFS benchmarks on JURON: streaming and metadata performance on OpenPOWER with NVMe. Technical report, ThinkParQ and Jülich Forschungszentrum, October 2017
6. Gabriel, E., Venkatesan, V., Shah, S.: Towards high performance cell segmentation in multispectral fine needle aspiration cytology of thyroid lesions. Comput. Methods Prog. Biomed. **98**, 231–240 (2010)
7. Greenberg, H.N., Bent, J., Grider, G.: MDHIM: a parallel key/value framework for HPC. In: 7th USENIX HotStorage (2015)
8. Group, T.H.: Hierarchical data format, version 5 (2014). www.hdfgroup.org/HDF5/
9. Hernandez, R., Cugnasco, C., Becerra, Y., Torres, J., Ayguadé, E.: Experiences of using cassandra for molecular dynamics simulations. In: 23rd Euromicro International Conference on PDP, March 2015
10. Islam, N.S., Shankar, D., Lu, X., Wasi-Ur-Rahman, M., Panda, D.K.: Accelerating I/O performance of big data analytics on HPC clusters through RDMA-based key-value store. In: 44th ICPP, September 2015
11. Lakshman, A., Malik, P.: Cassandra: a decentralized structured storage system. SIGOPS Oper. Syst. Rev. **44**, 35–40 (2010)
12. Latham, R., Ross, R., Thakur, R.: The impact of file systems on MPI-IO scalability. In: Kranzlmüller, D., Kacsuk, P., Dongarra, J. (eds.) EuroPVM/MPI 2004. LNCS, vol. 3241, pp. 87–96. Springer, Heidelberg (2004). https://doi.org/10.1007/978-3-540-30218-6_18
13. Liu, J., et al.: Evaluation of HPC application I/O on object storage systems. In: PDSW (2018)
14. Shankar, D., Lu, X., Islam, N., Wasi-Ur-Rahman, M., Panda, D.K.: High-performance hybrid key-value store on modern clusters with RDMA interconnects and SSDs: non-blocking extensions, designs, and benefits. In: IPDPS, May 2016

15. Shankar, D., Lu, X., Panda, D.K.: High-performance and resilient key-value store with online erasure coding for big data workloads. In: 37th ICDCS, June 2017
16. Tantisiriroj, W., Son, S.W., Patil, S., Lang, S.J., Gibson, G., Ross, R.B.: On the duality of data-intensive file system design: reconciling HDFS and PVFS. In: SC (2011)
17. Cornell, V.: How to ruin a perfectly good GPFS file system (2015). http://files.gpfsug.org/presentations/2015/CIUK-DDN.pdf
18. Wu, C., Faleiro, J., Lin, Y., Hellerstein, J.: Anna: A KVS for any scale, April 2018
19. Wu, C., Sreekanti, V., Hellerstein, J.: Eliminating Boundaries in Cloud Storage with Anna, August 2018
20. Zhang, X., Su, H., Yang, L., Zhang, S.: Fine-grained histopathological image analysis via robust segmentation and large-scale retrieval. In: CVPR, June 2015

Scaling the Training of Recurrent Neural Networks on Sunway TaihuLight Supercomputer

Ouyi Li[1,3], Wenlai Zhao[2,3](✉), Xuancheng Huang[2], Yushu Chen[3], Lin Gan[2,3], Hongkun Yu[2,3], Jiacheng Zhang[2], Yang Liu[2], Haohuan Fu[1,3], and Guangwen Yang[2,3]

[1] Ministry of Education Key Laboratory for Earth System Modeling, Department of Earth System Science, Tsinghua University, Beijing, China
loy16@mails.tsinghua.edu.cn
[2] Department of Computer Science and Technology, Tsinghua University, Beijing, China
zhaowenlai@tsinghua.edu.cn
[3] National Supercomputing Center in Wuxi, Wuxi, Jiangsu, China

Abstract. The recurrent neural network (RNN) models require longer training time with larger datasets and bigger number of parameters. Distributed training with large mini-batch size is a potential solution to accelerate the whole training process. This paper proposes a framework for large-scale training RNN/LSTM on the Sunway TaihuLight (SW) supercomputer. We take series of architecture-oriented optimizations for the memory-intensive kernels in RNN models to improve the computing performance. The lazy communication scheme with improved communication implementation and the distributed training and testing scheme are proposed to achieve high scalability for distributed training. Furthermore, we explore the training algorithm with large mini-batch size, in order to improve convergence speed without losing accuracy. The framework supports training RNN models with large size of parameters with at most 800 training nodes. The evaluation results show that, compared to training with single computing node, training based on proposed framework can achieve a 100-fold convergence rate with 8,000 mini-batch size.

Keywords: Neural machine translation · Recurrent neural networks · Large-scale training · Many-core architecture · Sunway TaihuLight supercomputer

1 Introduction

Deep learning has already proven its efficiency in many different tasks. Recurrent neural network (RNN) [13] plays an important role in deep learning. Neural machine translation (NMT) is one of the most successful application examples of RNN. Training RNN models in NMT task takes a lot of time.

J. M. F. Rodrigues et al. (Eds.): ICCS 2019, LNCS 11536, pp. 427–440, 2019.
https://doi.org/10.1007/978-3-030-22734-0_31

It is proved that larger datasets and bigger models lead to improvements in accuracy for many deep learning tasks. In NMT, as the number of parameters increases, the memory space required becomes larger. Taking a specific parameter setting in [2] as an example, attention based long short-term memory (LSTM) [7] with 1000 hidden units, 620 dimensional word embeddings and 75 words in a sentence with mini-batch size 40 needs 30 GB memory.

With the rapid development of deep learning, the size of dataset in related tasks gradually becomes larger. It takes more time to traverse the whole training dataset as one epoch during training models. For example, the size of NMT dataset grows from 1,200,000 pairs of sentences in source language (English) and target language (Chinese) in dataset NIST to 5,800,000 pairs of sentences in dataset WMT. In order to cope with the increasing size of the dataset, we need to ensure that the overall training time is controllable by reducing the time of traversing the dataset during training process.

Scaling the mini-batch size by synchronizing data parallel is a good solution. In some previous works, a large-scale training on distributed deep learning frameworks have been explored by [5,15]. They mainly focus on training CNN models on ImageNet [11]. Correspondingly, we explore the solution of large-scale training RNN/LSTM in NMT in this paper.

SW provides large number of computing nodes, we can explore training RNN/LSTM with large size of parameters with large mini-batch size by synchronizing data parallel. If we train attention based LSTM model with 2000 hidden units, 1440 dimensional word embeddings on one computing node, the allowed mini-batch size is 10. To explore training large models with large mini-batch size, 800 training workers are needed to reach the mini-batch size 8,000.

We propose architecture-oriented optimizations and communication optimizations customized for RNN/LSTM. We propose a distributed RNN/LSTM training framework based on SunwayCaffe [8,16]. We explore convergence acceleration ratio of training process and generalization of models under large-scale training scenarios. Our specific contributions include:

- We propose architecture-oriented optimizations for the memory-intensive kernels in RNN/LSTM, fully exploring the parallelism in SW26010 many-core architecture and improving the usage of memory bandwidth.
- We propose improved MPI_Allreduce implementation and lazy communication scheme customized for RNN/LSTM, so as to achieve high communication efficiency and high scalability for P2P-based distributed training.
- We propose a customized distributed framework design for NMT, which assigns the computing nodes with different tasks (training, testing and evaluating). The proposed framework provides environment for frequently validating models with little test interval. Training in the framework can overlap all testing time.
- We provide discussions and analysis of the convergence acceleration ratio of large-scale training RNN/LSTM and the generalization of models. Furthermore, we show an empirical training strategy, as well as the evaluation results of RNN/LSTM on a large dataset with up to 800 training nodes.

The rest of the paper is organized as follow. In Sect. 2, we describe the background of this paper. In Sect. 3.1, we propose our architecture-oriented optimizations for the memory-intensive kernels in RNN/LSTM on SW26010 processor. In Sect. 3.2, we propose our communication optimizations including the improved MPI_Allreduce implementation and lazy communication scheme. In Sect. 3.3, we describe our distributed framework designed for NMT. In Sect. 4, we explore large-scale training RNN/LSTM and we provide discussions and analysis of the convergence acceleration ratio of large-scale training RNN/LSTM and the generalization of the models. In Sect. 5, we show evaluation of the optimizations in our framework, and the evaluation results of large RNN/LSTM. In Sect. 6, we conclude the report and discuss about the future work.

2 Background

2.1 RNN Model and NMT Task

RNN is a class of artificial neural network and add sequential information to the artificial neural network model. RNN models can use their hidden units to process sequences of inputs. A finite RNN is a directed acyclic graph that can be unrolled. When it is unrolled, each layer deals with one element in the sequence. LSTM cell is a kind of RNN cell, and it adds gates to basic RNN cell to regulate the flow of information into and out of the cell.

The past several years have witnessed the rapid development of NMT, which aims to model the translation process using neural networks in an end-to-end manner. Most of the proposed NMT models belong to a family of encoder-decoder. The encoder-decoder system consists of an encoder and a decoder for a language pairs. They are jointly trained to maximize the probability of a correct translation given a source sentence. All neurons in encoders share their weights and all neurons in decoders share their weights.

There are three encoder-decoder models mentioned in this paper, the first model is encoder-decoder (RNNencdec) proposed in [14], the second one is attention-based encoder-decoder (RNNsearch) proposed in [2], the third one is attention-based encoder-decoder with twice number of hidden units and dimensional word embeddings of RNNsearch (RNNsearch-H).

We evaluate models on English-Chinese translation. We have two training datasets, one small dataset consisting of 1.25M pairs of sentences and one large dataset consisting of 5.8M pairs of sentences. We use the NIST 2002 dataset as validation dataset for the first training dataset and WMT dataset as validation dataset for the second one.

Bilingual evaluation understudy (BLEU) is an algorithm for evaluating the quality of text translated by machine from one natural language to another. BLEU considers the quality to be the correspondence between a machine's output and that of a human. It is now the normal standard for evaluating machine translation results. It is involved in our framework.

2.2 System Setup

One SW26010 many-core processor is composed of four core-groups (CGs), each CG consists of 65 cores: one management processor element (MPE) and 64 computing processor elements (CPEs). 64 CPEs are organized as a CPE cluster. Within a cluster, CPEs are connected in an 8 by 8 mesh. The MPE and CPEs are all based on 64-bit RISC architecture, but have different duties. The MPE supports the complete interrupt functions, memory management, superscalar and out-of-order issue/execution. It is good at handling management, task schedule, and data communications. CPE is designed to maximize the aggregated computing and minimize the complexity of the micro-architecture. Each of CGs owns 8 GB of DDR3 memory, shared by MPE and CPE cluster through the Memory Controller (MC). So one node has 32 GB memory. The on-chip network (NoC) connect the MPE/CPE chip with System Interface (SI).

2.3 Current Situation and Related Work

As for now, most of state-of-art RNN/LSTM in different tasks are trained with small mini-batch size performed in single-worker multi-GPU mode. In [14], they train RNNencdec model with mini-batch size 128. The bi-LSTM model has 4 layers with 1000 hidden units, 620 dimensional word embeddings. In [2], they proposed model RNNsearch, the encoder and decoder of RNNsearch both have 1000 hidden units, with mini-batch size 80, the length of the training sentences is 50.

To explore training RNN/LSTM with larger number of parameters requiring more memory space, we choose to train models on SW. Architecture-oriented optimizations for the memory-intensive kernels in RNN/LSTM are needed to improve the usage of memory bandwidth and to improve the overall computing performance.

As the size of dataset increases, the time consumption of training RNN/LSTM in NMT increases. There are two training datasets mentioned in this paper. The time consumption of training on WMT dataset for one epoch is five times of that on NIST dataset. Multi-server distributed training is a solution. SW provides a large number of computing nodes, we can scale the mini-batch size in data parallelism method to meet our need.

The support for the existing distributed frameworks of parameter solver (PS) architecture [12] on SW is not satisfactory. In contrast, open-sourced Caffe [8] deep learning framework on a single computing node has been well supported on SW. Our previous work consists of optimized math library SWDNN and SunwayCaffe [4,16]. SWDNN mainly optimized the convolutional layer and pooling layer according to the architecture of SW. SunwayCaffe provides a basic distributed framework for training CNN models on SW. This paper proposes a framework based on SunwayCaffe. Because of recurrent structure of RNN/LSTM, the communication mechanism in training process are optimized to achieve high communication efficiency and scalability.

In terms of large-scale training neural networks, many researchers have explored ways to reduce the training time under the premise of ensuring the accuracy. [5] proposed a scheme of large-scale training CNN models. They scale mini-batch size to 8k in AlexNet and ResNet with decline in testing accuracy. [15] proposed a layer-wise adaptive optimization algorithm LARS. They scale the mini-batch size of training ResNet to 32k with no decline in accuracy. As aspect of RNN/LSTM in NMT task, few people explore large-scale training.

Therefore, in this paper, for large-scale training RNN/LSTM according to current situation, we propose architecture-oriented optimizations for the memory-intensive kernels in RNN/LSTM for improving the usage of memory bandwidth and improving the computing performance. We propose an efficient implementation of MPI_Allreduce and a lazy communication scheme for high communication efficiency and high scalability. We propose a customized distributed framework designed for overlapping the testing time. We explore large-scale training RNN/LSTM and provide discussions and analysis of convergence acceleration ratio and generalization of models. Finally we show the results of large-scale training RNNsearch-H with an empirical training strategy.

3 Optimizations

3.1 Architecture-Oriented Optimization

In training process of RNN/LSTM, GEMM (General Matrix Multiply) is the most computation intensive operation. GEMM performs in the computing of gates in LSTM neurons, logit layers and attention module. Optimization of GEMM on SW26010 many-core processor has been discussed in [9,16]. We apply the implementation in our framework directly.

Besides the computation intensive kernels, the optimizations of the memory intensive kernels are also very important from the perspective of the overall performance. Gradient-based optimization algorithms (e.g. square root operation) adopt element-wise vector operations, for which performance is limited by memory bound. These layers can be efficiently implemented by DMA for large continuous data blocks and perform computation in CPE cluster.

Particularly, two kernels are major considerations in this paper: the exponential layer and the softmax layer.

Exponential Layer Optimization. In neural networks, activation layers are used to perform a nonlinear transformation of data. In RNN/LSTM, activation layers are mainly sigmoid layers and tanH layers, which both utilize nonlinear property of exponential function. In addition to activation layers, exponential function is also in the implementation of softmax layer.

On SW, the underlying implementation of the exponential function is included in SW basic math library. the specific implementation of the exponential function is method of look-up table with interpolation. The look-up table of exponential function is stored in memory of MPE, both computing in MPE and CPEs need to access data from main memory when exponential function is called. Data access from main memory takes more than 100 CPU cycles. Instead of SW basic math library, we implement the exponential function by Taylor Expansion using a small amount of LDM in CPEs. Our implementation avoids discrete data access from main memory. Efficiency of the operation is greatly improved while ensuring the precision.

Softmax Layer Optimization. In the decoding phase of NMT, models pass data through softmax layer at each time step, and the number of neurons in softmax layer is equal to the size of vocabulary, usually 50,000 or more. In ImageNet, the number of neurons in softmax layer is equal to the number of categories 1000. The categories of NMT is more than 50 times that of ImageNet, and the length of sentences in NMT is always 50 to 100. Therefore, in an training iteration, the amount of computation of softmax in NMT is 2500 to 5000 times that in ImageNet. The large number of neurons in softmax layer and the high occurrences make softmax layer a bottleneck in NMT when training on SW.

The softmax function is described as Eq. 1. K is the number of neurons in softmax layer. It maps the output value of j-th neuron to a new value in interval $(0, 1)$, which can be thought as the probability of the category represented by the output.

$$\alpha(x)_j = \frac{\exp(x_j)}{\sum_{k=1}^{K} \exp(x_k)} \qquad for\ j = 1\ to\ K \tag{1}$$

As the output of exponential function increases fast as x grows, the input of exponential function of each neuron in softmax layer must keep small while the

---

Algorithm 1. Implementation of softmax

Input
1: Vector<Data*> A; Batch size: B; Data count in one batch: N; Core group id: cg_id; CPE id: cpe_id;
output
2: Probability of each data of all batches: Vector<Data*> S
3: **function** PARALLEL_SOFTMAX(Vector<Data*> A, INT B, INT N)
4: $Sync\_4cg()$;
5: $start\_index = B/4 * cg\_id + B/(4 * 64) * cpe\_id$;
6: $local\_count = B/(4 * 64)$;
7: dma($local\_A, start\_index, local\_count$) $from$ A;
8: **for** each d in $local\_A_{cg\_id,cpe\_id}$ **do**
9: $M = max(d)$;
10: d = exp(d-M);
11: $SUM = \sum d$;
12: $local\_S = d \div SUM$;
13: **end for**
14: **return** ;
15: **end function**

---

output of softmax layer unchanged. Max value of all neurons is subtracted from value of each neuron first. This trick can be applied to make sure that none of the exponentials overflows [8].

According to Algorithm 1, the implementation of softmax layer on SW26010 mainly contains four parts of computations. Two parts in Line 10 and Line 12 are both element-wise operation, and their implementations are mentioned previously. Other two parts of computations are MAX operation in Line 9 and SUM operation in Line 11. We do data parallelism in batch level, each CPE computes max (summation) of vector data in different mini-batches. For the same data, two operations in Line 10 and Line 11 can be completed after one DMA operation according to the index of CPE in CGs. Through the above implementations on SW, we can accelerate the training process on SW26010 processor.

3.2 Communication Optimization

Communication Architecture. Parameter Server (PS) and P2P communication are two mainstream communication architecture for distributed deep learning [12]. PS follows a client-server scheme and can be easily scaled up on a distributed cluster, tolerating the imbalanced performance, unstable network bandwidth and unexpectable faults of the workers. However, for a supercomputer system, the sustaining performance and the stability of the computing nodes can be guaranteed, as well as the network condition. Therefore, a P2P communication architecture is more suitable.

We adopt two optimization methods to further improve the communication efficiency, including an improved MPI_Allreduce design and a lazy communication strategy for distributed RNN/LSTM training.

Improved Allreduce. The network topology of SW is a two-level fat tree, which consists of an intra super-node level and an inter super-node level. At the intro super-node level, 256 computing nodes are fully connected via a customized super-node network switch. At the inter super-node level, a central switching network is designed for the communication between different super-nodes. Generally, the intro super-node communication has a higher bandwidth and a lower latency than the inter super-nodes communication.

To improve the overall communication efficiency, we implement a kind of hierarchical Allreduce. [6] An improved Allreduce design is proposed with four stages, which include: (1) an inter super-node reduce stage; (2) an intra super-node reduce stage; (3) an intra super-node broadcast stage; (4) and an inter super-node broadcast stage. Compared with the original MPI_Allreduce operation, the improved Allreduce contains as less inter super-node communication requests as possible.

Besides, in the improved Allreduce operation, the computation operations (usually SUM operation is used to aggregate gradients) is accelerated using the CPEs, while it is handled only by MPE in the standard MPI_Allreduce implementation. With the optimizations on both computation and communication, the improved Allreduce operation can achieve about 20 times higher efficiency on average than the standard MPI_Allreduce operation.

Lazy Communication. In a training iteration, each layer invokes an Allreduce communication for the gradients, so that the number of communication requests is usually large for deep neural networks. In modern RNN/LSTM, the number of parameters in each RNN/LSTM layer is related to the length of the word vector. Usually if the size of hidden states is 1000, which is relatively large, there are about 1 million (1000×1000) parameters in a layer, and then the data size of the gradients involved in one Allreduce operation is about 4 MB. Hence we can see that, there are numerous communication requests with small data size, which is the main feature of the communication pattern in a distributed NMT training framework, and is not efficient in large-scale training tasks.

To address the above issue, we propose a lazy communication scheme in our framework. The basic design idea is that instead of executing the communication requests immediately, we *remember* them temporarily until the unsent data size is greater than a given $MAXSIZE$, which is set to 100 MB empirically in our framework.

The lazy communication design can merge multiple small-data-size communication into a large-data-size communication, which can improve the overall efficiency by lowering the launch cost and increasing the utilization rate of network bandwidth.

3.3 Framework Optimization

Considering the load balance and the overall training efficiency, we propose a new distributed framework design for large-scale training RNN/LSTM.

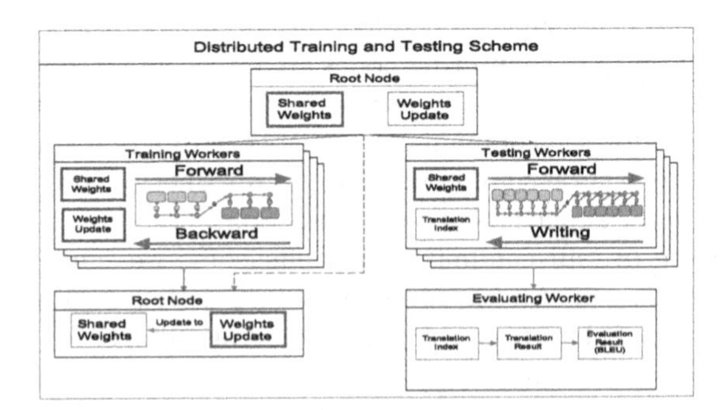

Fig. 1. The overview of distributed training and testing scheme

The training process in NMT contains three modules: training module, testing module and evaluating module. A training model and a testing model are involved in the training process. The model in training module and the model in testing module share all parameters, but not absolutely the same. There is clear relationship between three modules. Evaluating module runs after testing module, training module and testing module run in parallel.

In fact, except sharing parameters at beginning of training iteration and testing iteration, training module and testing module don't rely on each other. We propose a distributed training and testing scheme as shown in Fig. 1. A testing interval is set that testing process is performed after every testing interval of training iterations.

In NMT, even after a few training iterations, BLEU value changes quite a lot. So frequent validation help find the highest BLEU value. The model with the highest BLEU value on validation dataset is always the model with the best generalization in theory. Distributed training and testing scheme leads to complete overlap of testing process and evaluating process, we can save extra testing time. The scheme greatly improves efficiency of entire training process and also provides an opportunity to find the model with the best generalization.

4 Convergence and Generalization

4.1 Model Convergence Optimization Algorithm

The stochastic gradient based optimization algorithm applied in our training experiments is Adaptive Moment Estimation (Adam). Adam applies momentum on a per-parameter basis and automatically adapts step size subject to a user-specified maximum learning rate [10]. Adam's convergence speed and generalization made it a popular choice for NMT [1,3].

Learning Rate. Under large-scale training scenarios, when RNNsearch model is trained with Adam with same mini-batch size (100 nodes, 8000 mini-batch size and 50 nodes, 4000 mini-batch size) on NIST dataset, the convergence speed at different learning rates are shown in Fig. 2(a) and (b). Under large-scale training scenarios, although the learning rate can be adaptively adjusted with the first moment and the second moment in Adam, setting a higher or lower learning rate will result in a slower convergence speed. When learning rate is set too high (0.003 or more), the model does not converge.

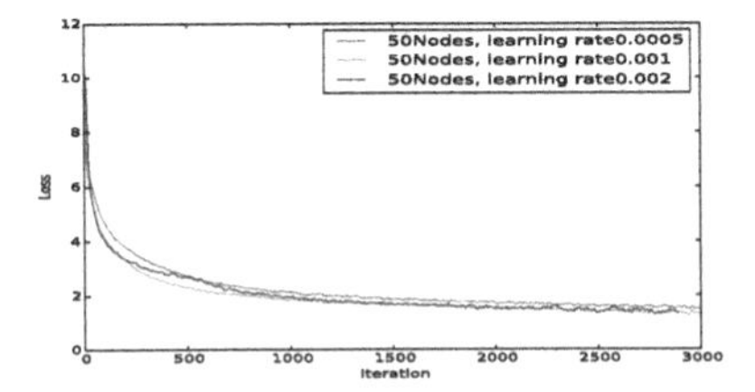

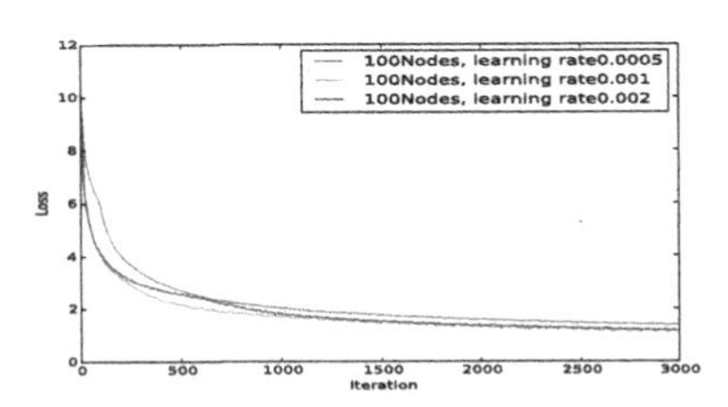

Fig. 2. Loss vs. learning rate

Parameter Setting. Under large-scale training scenarios, Adam is applied with momentum 0.9 and momentum2 0.999. The learning rate is 0.001. The length of sentences is set to 50 on dataset NIST and 75 on dataset WMT. When training models on single node, mini-batch size for RNNencdec model training on dataset NIST is 80, mini-batch size for RNNsearch model is 80 on NIST and 40 on WMT, mini-batch size for RNNsearch-H model training on dataset WMT is 10.

Exploring Large-Scale Training. The evaluation of models contains two parts. The first one is loss function, used to measure the degree of fit on training dataset. The second one is BLEU value of the models on the testing dataset in the training process, which represents the generalization of the trained models. We explore tradeoff between mini-batch size, accuracy and training convergence time.

We evaluate RNNsearch on dataset NIST. The convergence speed on training dataset is shown in Fig. 3(a) and (b). In Fig. 3(a), models trained with different mini-batch size fit the dataset very close after enough epochs. The increase of mini-batch size would not lead to a no-converge situation. As shown in Fig. 3(b), assessing on iterations, the model converge faster with the increase of mini-batch size. Large-scale training has a great advantage on the convergence speed for our framework provides a good scalability.

As shown in Fig. 3(c), training under large-scale training scenarios on WMT dataset, the trend of convergence speed is consistent with that on NIST dataset. Figure 3(d) shows that as mini-batch size increases, BLEU value increases faster. Under large-scale scenarios, the generalization of models increases faster when fitting faster on training dataset.

In terms of generalization of models trained with different mini-batch size, BLEU of RNNsearch on NIST assessing on epochs is shown in Fig. 3(e), with increase of mini-batch size, BLEU value increases slower assessing on epochs. From the perspective of training time, the convergence speed assessing on iterations is important. As shown in Fig. 3(f), assessing on iterations, when mini-batch size is above 32k (400 training nodes), the convergence speed of BLEU value does not increases. The convergence speed even decreases when mini-batch size reaches 64k (800 training nodes).

The highest BLEU results in Fig. 3(e) are shown in Table 1. Scaling mini-batch size to 4k leads no generalization gap. Scaling mini-batch size to 16k, leads 0.77 generalization gap, which is about 2.5% lose of BLEU value. As mini-batch size reaching 64k, we get 2.1 generalization gap, about 10% lose of BLEU value.

Training RNNsearch with 100 training nodes has an acceptable 0.3% lose of BLEU compared with the baseline of training with 10 training nodes. Training with 100 training nodes has a near convergence speed to the baseline assessing on epochs and converge quite faster than baseline assessing on iterations. So we select the mini-batch size of 8k as the scheme for training RNN/LSTM with large mini-batch size. **Training with mini-batch size 8k can bring a 100-fold convergence rate of training with a single node.**

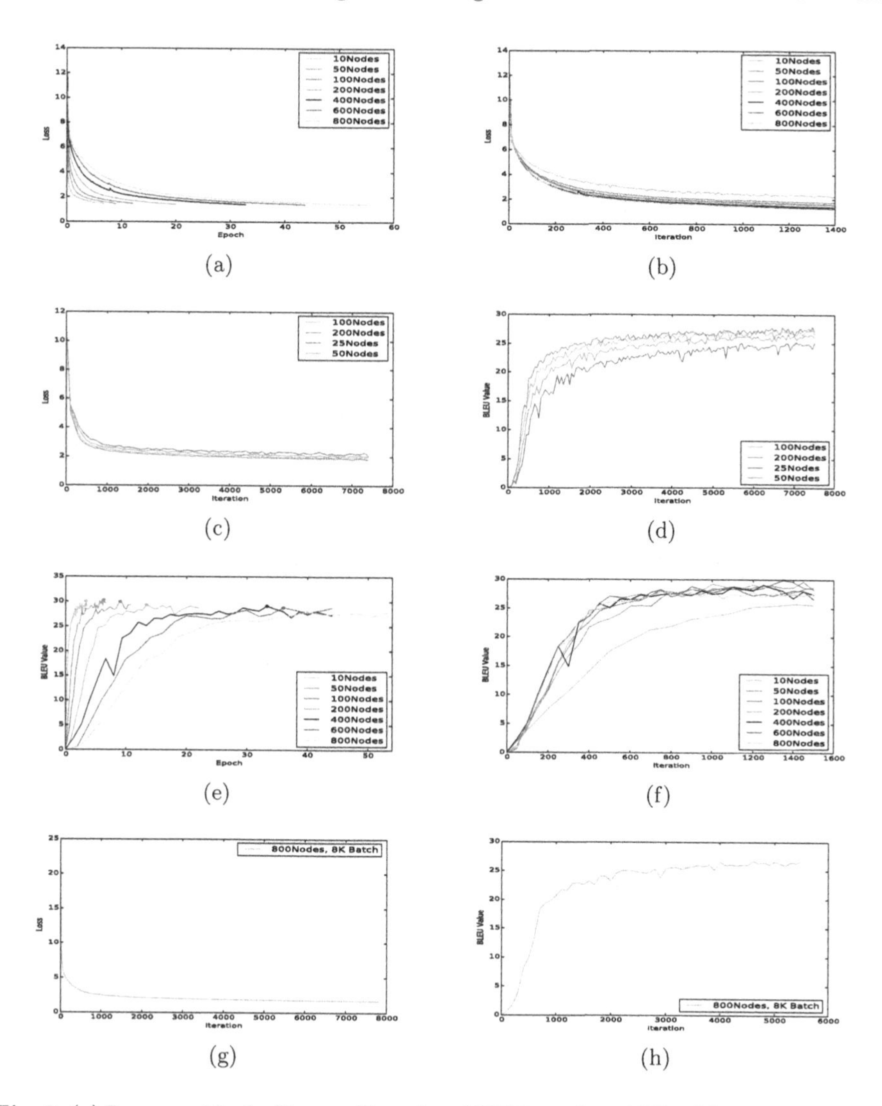

Fig. 3. (a) **Loss vs. Node Size** on **Epochs** of RNNsearch on NIST (b) **Loss vs. Node Size** on **Iterations** of RNNsearch on NIST (c) **Loss vs. Node Size** on **Iterations** of RNNsearch on WMT (d) **BLEU vs. Node Size** on **Iterations** of RNNsearch on WMT (e) **BLEU vs. Node Size** on **Epochs** of RNNsearch on NIST (f) **BLEU vs. Node Size** on **Iterations** of RNNsearch on NIST (g) Loss of RNNsearch-H with 800 Nodes (h) BLEU of RNNsearch-H with 800 Nodes

Table 1. Results of top BLEU of RNNsearch on NIST

Node size	10	50	100	200	400	600	800
Batch size	800	4000	8000	16000	32000	48000	64000
BLEU	30.04	30.31	29.93	29.27	29.12	28.81	27.95

5 Evaluation

In this section, we evaluate the performance of our framework in two ways firstly. One is to evaluate the computing performance of training on single worker and to compare the computing performance versus GPU. The other one is to evaluate the scalability of the framework. Secondly, we show the loss function and BLEU value both assessing on iterations of training model RNNsearch-H.

5.1 Performance

SW26010 vs. GPU. In terms of computing performance of single-worker training, we compare the performance of SW26010 with the performance of NVIDIA TITAN X. The single-precision computing capability of NVIDIA TITAN X is 11 TFlops, and the memory bandwidth of NVIDIA TITAN X is 505 GB/s. The double-precision computing capability of SW26010 is 3 TFlops. Compared with the double-precision computing capability, single-precision computing capability can reach 60% of double-precision computing capability [16], the memory bandwidth of SW26010 is 128 GB/s. As a result of the gap between NVIDIA TITAN X and SW26010, for one iteration training of RNNencdec, the average computing time on NVIDIA TITAN X is 0.71 s, the average computing time on SW26010 is 8.32 s. Actually, SW has a computing power of only 1/6 of NVIDIA TITAN X and 1/4 memory bandwidth of NVIDIA TITAN X, we can achieve 1/12 computing performance of NVIDIA TITAN X.

Table 2. Module time before and after architecture-oriented optimizations

Module	Time before	Percentage before	Time now	Percentage now
Total	66.765 s	100.0%	8.326 s	100.0%
Softmax	26.317 s	39.4%	0.583 s	7.00%
Activation function	34.363 s	51.4%	1.658 s	19.9%

The performance of training on SW is shown in Table 2. Data parallel optimization are operated on exponential layer. Optimizations described in Sect. 3.1 is operated on exponential layer and softmax layer. Training time in one iteration has decreases from 66.765 s to 8.326 s in RNNencdec after architecture-Oriented optimizations.

Scalability. After communication optimizations mentioned in Sect. 3.2, our framework provides high communication efficiency. When training RNNencdec under large-scale scenarios, scaling the size of training workers to 800 achieves 580x speedup, and parallel efficiency is 72.5%. When training RNNsearch under large-scale scenarios, scaling the size of training workers to 800 achieves 692x speedup, with parallel efficiency of 86.5%.

5.2 Experimental Results

As mentioned in Sect. 2, RNNsearch-H is with twice the size of hidden states and dimensional word embeddings of RNNsearch, the size of parameters is four times of RNNsearch, which needs about 30 GB memory space with mini-batch size 10. As mentioned in Sect. 4, we need 800 training nodes to reach the suitable mini-batch size 8k on SW. Training RNNsearch-H model under large-scale training scenarios on GPUs is unbearable for the memory bound and the limitation of the size of server cluster.

The loss function and BLEU value are shown in Figs. 3(g) and (h). As shown in Fig. 3(g), RNNsearch-H can converge to about the same level as model RNNsearch. RNNsearch-H can fit well on the training dataset. As shown in Fig. 3(g), BLEU value of RNNsearch-H can get to the same level as RNNsearch, after training enough time, the model can have a good generalization.

6 Conclusion

In this paper, we propose architecture-oriented optimizations for memory-intensive kernels in RNN/LSTM, exploring the parallelism in SW26010 many-core architecture. We propose a lazy communication scheme with improved MPI_Allreduce to achieve high communication efficiency and high scalability. We provide a distributed framework for large-scale NMT training to overlap all of testing time for frequently validation. At last, we provide discussions and analysis on convergence speed and generalization quality under different training mini-batch size and get a 100-fold convergence rate with 100 training nodes, 8k mini-batch size. We show an empirically training strategy, as well as the convergence and evaluation results, of training RNNsearch-H on a large dataset with 800 training nodes, 8k mini-batch size.

Acknowledgement. This work is supported in part by the National Key R&D Program of China (Grant No. 2017YFB0202204, 2017YFA0604500, 2016YFA0602200), by National Natural Science Foundation of China (Grant No. 91530323, 5171101179), and by the China Postdoctoral Science Foundation (Grant No. 2018M641359).

References

1. Arthur, P., Neubig, G., Nakamura, S.: Incorporating discrete translation lexicons into neural machine translation. arXiv preprint arXiv:1606.02006 (2016)
2. Bahdanau, D., Cho, K., Bengio, Y.: Neural machine translation by jointly learning to align and translate. arXiv preprint arXiv:1409.0473 (2014)
3. Britz, D., Goldie, A., Luong, T., Le, Q.: Massive exploration of neural machine translation architectures. arXiv preprint arXiv:1703.03906 (2017)
4. Fang, J., Fu, H., Zhao, W., Chen, B., Zheng, W., Yang, G.: swDNN: a library for accelerating deep learning applications on Sunway TaihuLight. In: 2017 IEEE International Parallel and Distributed Processing Symposium (IPDPS), pp. 615–624. IEEE (2017)

5. Goyal, P., et al.: Accurate, large minibatch SGD: training imageNet in 1 hour. arXiv preprint arXiv:1706.02677 (2017)
6. Hasanov, K., Lastovetsky, A.: Hierarchical optimization of MPI reduce algorithms. In: Malyshkin, V. (ed.) PaCT 2015. LNCS, vol. 9251, pp. 21–34. Springer, Cham (2015). https://doi.org/10.1007/978-3-319-21909-7_3
7. Hochreiter, S., Schmidhuber, J.: Long short-term memory. Neural Comput. **9**(8), 1735–1780 (1997)
8. Jia, Y., et al.: Caffe: convolutional architecture for fast feature embedding. In: Proceedings of the 22nd ACM International Conference on Multimedia, pp. 675–678. ACM (2014)
9. Jiang, L., et al.: Towards highly efficient DGEMM on the emerging SW26010 many-core processor. In: 2017 46th International Conference on Parallel Processing (ICPP), pp. 422–431. IEEE (2017)
10. Kingma, D.P., Ba, J.: Adam: a method for stochastic optimization. arXiv preprint arXiv:1412.6980 (2014)
11. Krizhevsky, A., Sutskever, I., Hinton, G.E.: ImageNet classification with deep convolutional neural networks. In: International Conference on Neural Information Processing Systems, pp. 1097–1105 (2012)
12. Li, M., et al.: Scaling distributed machine learning with the parameter server. In: OSDI, vol. 14, pp. 583–598 (2014)
13. Rumelhart, D.E., Hinton, G.E., Williams, R.J.: Learning representations by back-propagating errors. Nature **323**(6088), 533 (1986)
14. Sutskever, I., Vinyals, O., Le, Q.V.: Sequence to sequence learning with neural networks. In: Advances in Neural Information Processing Systems, pp. 3104–3112 (2014)
15. You, Y., Gitman, I., Ginsburg, B.: Scaling SGD batch size to 32K for imagenet training. arXiv preprint arXiv:1708.03888 (2017)
16. Zhao, W., Fu, H., Fang, J., Zheng, W., Gan, L., Yang, G.: Optimizing convolutional neural networks on the sunway taihulight supercomputer. ACM Trans. Arch. Code Optim. (TACO) **15**(1), 13 (2018)

Immersed Boundary Method Halo Exchange in a Hemodynamics Application

John Gounley[1,2(✉)], Erik W. Draeger[3], and Amanda Randles[1]

[1] Department of Biomedical Engineering, Duke University, Durham, NC, USA
amanda.randles@duke.edu
[2] Computational Science and Engineering Division,
Oak Ridge National Laboratory, Oak Ridge, TN, USA
gounleyjp@ornl.gov
[3] Center for Applied Scientific Computing,
Lawrence Livermore National Laboratory, Livermore, CA, USA
draeger1@llnl.gov

Abstract. In recent years, highly parallelized simulations of blood flow resolving individual blood cells have been demonstrated. Simulating such dense suspensions of deformable particles in flow often involves a partitioned fluid-structure interaction (FSI) algorithm, with separate solvers for Eulerian fluid and Lagrangian cell grids, plus a solver - e.g., immersed boundary method - for their interaction. Managing data motion in parallel FSI implementations is increasingly important, particularly for inhomogeneous systems like vascular geometries. In this study, we evaluate the influence of Eulerian and Lagrangian halo exchanges on efficiency and scalability of a partitioned FSI algorithm for blood flow. We describe an MPI+OpenMP implementation of the immersed boundary method coupled with lattice Boltzmann and finite element methods. We consider how communication and recomputation costs influence the optimization of halo exchanges with respect to three factors: immersed boundary interaction distance, cell suspension density, and relative fluid/cell solver costs.

Keywords: Red blood cell · Immersed boundary method · Parallel computing

This manuscript has been authored by UT-Battelle, LLC under Contract No. DE-AC05-00OR22725 with the U.S. Department of Energy. The United States Government retains and the publisher, by accepting the article for publication, acknowledges that the United States Government retains a non-exclusive, paid-up, irrevocable, world-wide license to publish or reproduce the published form of this manuscript, or allow others to do so, for United States Government purposes. The Department of Energy will provide public access to these results of federally sponsored research in accordance with the DOE Public Access Plan (http://energy.gov/downloads/doe-public-access-plan).

J. M. F. Rodrigues et al. (Eds.): ICCS 2019, LNCS 11536, pp. 441–455, 2019.
https://doi.org/10.1007/978-3-030-22734-0_32

1 Introduction

High-resolution computational simulations of blood flow have been employed to study biomedical problems such as malaria [7], thrombosis [28], and sickle-cell anemia [16]. However, as simulations are scaled from microvasculature to mesovasculature, the problem size demands efficient and scalable parallel fluid-structure interaction algorithms. As reviewed by [12], one of the most popular fluid-structure interaction algorithms in this space is the immersed boundary (IB) method. The IB method is often implemented as a partitioned fluid-structure interaction scheme, with separate solvers for the fluid and cells. Characterized by a time-invariant Eulerian fluid lattice and body-fitted Lagrangian meshes for the cells, the IB method transfers data between the fluid and cell grids using smoothed discrete delta functions [17,20]. While maintaining separate Eulerian and Lagrangian grids provides distinct advatanges (e.g., avoiding remeshing), it also complicates parallelization in a distributed-memory environment. In this study, we introduce a scalable IB framework for a hemodynamics application and explore how model parameters influence the cost of halo exchange and recomputation in the IB method.

Parallelization of the IB method for blood flow has several components. Depending on the method, the fluid solver requires at least a halo exchange. Likewise, the movement of blood cells across MPI domains must also be accounted for. Additionally, due to the diffusivity of the IB interface, the IB method interaction of the cell and fluid grids must also be parallelized. This halo exchange for the IB method is particularly interesting: because the IB method can transfer data between the fluid and cell grids, these Lagrangian and Eulerian data are effectively equivalent. Consequently, in principle, either could be communicated on the halo. For notational simplicity, we will denote as Lagrangian and Eulerian communication the transfer of the eponymous types of data.

Implementations of the IB method with distributed-memory parallelism originate with the work of [25] and [8]. While differences necessarily exist between continuous and direct forcing immersed boundary methods, the general challenges related to Lagrangian and Eulerian grids remain similar. In these and subsequent frameworks, the domain decomposition and requisite communication of IB-related data take various forms. To reduce or eliminate the movement of IB structures between tasks, [8] and [26] use separate domain decompositions for Eulerian and Lagrangian data and perform Eulerian communication of IB-related data. In contrast, the majority of implementations have used coincident domain decompositions for the fluid and structure. These schemes typically employ Lagrangian communication on a halo region (e.g., [18,24,25,27,29]). Eulerian communication over a halo region was judged prohibitively expensive for coincident domain decompositions [6]. More recently, a hybrid parallelization approach has improved load balance of the IB-related workload [19].

Intuitively, the optimal communication arrangement is expected to depend on particular details of the physical system being modeled. For instance, in the implementations discussed above, the IB structures being considered range from a suspension of point particles to a set of small cells to a single large membrane. Algorithmic choices would also seem to play a role: dynamic Lagrangian communication is inherently more complex than static Eulerian communication, but this could be offset by smaller message sizes. Further, other aspects of the simulations may already demand at least a basic level of Lagrangian or Eulerian communication. Moreover, choices about communicating Eulerian or Lagrangian data have implications for which aspects of the algorithm are fully parallelized versus involving some re-computation on overlap regions.

In this study, we investigate the relative parallel efficiency and scaling of Eulerian and Lagrangian communication frameworks applied to blood flow with coincident fluid and structural domain decompositions. Simulations are conducted with HARVEY, a parallel hemodynamics solver for flows in complex vascular geometries [22]. We describe an MPI+OpenMP implementation of the lattice Boltzmann and immersed boundary methods, coupled with finite element models for the cell mechanics. We explore the relative costs of Eulerian and Lagrangian communication for the force which is generated by the cell and spread onto the surrounding fluid. We investigate the dependence of the communication and recomputation costs on three factors: the support of the immersed boundary delta function, the density of the cell suspension, and the relative cost of the finite element method.

2 Methods

HARVEY performs the fluid-structure interaction with the immersed boundary method, coupling the lattice Boltzmann method for the fluid flow with a finite element method representing blood cells. An early version of this framework was presented in [9]. The present section extends that work by generalizing the IB method implementation and by discussing the parallelization schemes in depth. In the subsequent equations, we employ the convention of using lower- and upper-case letters for Eulerian and Lagrangian quantities, respectively.

2.1 Lattice Boltzmann Method for Fluid Flow

The Navier-Stokes equations governing continuum-level fluid flow are solved with the lattice Boltzmann method (LBM), which represents the fluid with a distribution function f of fictitious particles moving about a fixed Cartesian lattice [4]. The quantity f_i represents the component of the distribution with discrete velocity $\mathbf{c}_i$. For the D3Q19 velocity discretization used here, 18 of the 19 velocity vectors $\mathbf{c}_i$ point to nearest-neighbor lattice positions and remaining stationary velocity points to the same lattice position. The lattice Boltzmann equation for a fluid subject to an external force takes the form

$$f_i(\mathbf{x}+\mathbf{c}_i,t+1) = (1-\frac{1}{\tau})f_i(\mathbf{x},t) + \frac{1}{\tau}f_i^{eq}(\mathbf{x},t) + h_i(\mathbf{x},t) \tag{1}$$

for lattice position $\mathbf{x}$, timestep t, external force distribution h_i, equilibrium distribution f_i^{eq}, and relaxation time τ. Without loss of generality, we assume the LBM spatial (dx) and temporal (dt) steps equal to unity.

The external force field $\mathbf{g}(\mathbf{x}, t)$ is incorporated into the collision kernel – the right-hand side of Eq. 1 – in two steps [10]. First, the moments of the distribution function, density ρ and momentum $\rho\mathbf{v}$, are computed by the sums:

$$\rho = \sum_{i=1}^{19} f_i \qquad \rho\mathbf{v} = \sum_{i=1}^{19} \mathbf{c}_i f_i + \frac{1}{2}\mathbf{g}. \tag{2}$$

From these moments, the equilibrium Maxwell-Boltzmann distribution is approximated as

$$f_i^{eq}(\mathbf{x}, t) = \omega_i \rho \Big(1 + \frac{\mathbf{c}_i \cdot \mathbf{v}}{c_s^2} + \frac{\mathbf{v}\mathbf{v} : (\mathbf{c}_i \mathbf{c}_i - c_s^2 \mathbf{I})}{2c_s^4}\Big) \tag{3}$$

for the standard D3Q19 lattice weights ω_i and lattice speed of sound $c_s^2 = \frac{1}{3}$. Second, the external force $\mathbf{g}$ is converted into the force distribution h_i,

$$h_i = \Big(1 - \frac{1}{2\tau}\Big)\omega_i \Big[\frac{\mathbf{c}_i - \mathbf{v}}{c_s^2} + \frac{\mathbf{c}_i \cdot \mathbf{v}}{c_s^4}\mathbf{c}_i\Big] \cdot \mathbf{g}. \tag{4}$$

The lattice Boltzmann implementation in HARVEY is targeted at performing highly parallel simulations in sparse vascular geometries. To deal efficiently with this sparsity, the fluid points are indirectly addressed and an adjacency list for the LBM streaming operation is computed during setup. While the reference implementation of LBM stores two copies of the distribution function, we implement the AA scheme in HARVEY, which stores a single copy of the distribution function [1]. Other aspects of the lattice Boltzmann implementation, including grid generation and boundary conditions, may be found in previous work [9,21].

2.2 Finite Element Methods for Deformable Cells

Each cell is described by a fluid-filled triangulated mesh, derived from successive refinements of an icosahedron. Red blood cell membrane models include physical properties such as elasticity, bending stiffness, and surface viscosity [11]. For the sake of simplicity in this study, we model the cell surface as a hyperelastic membrane using a Skalak constitutive law. The elastic energy W is computed as

$$W = \frac{G}{4}\Big(I_1^2 + 2I_1 - 2I_2\Big) + \frac{C}{4}I_2^2 \tag{5}$$

for strain invariants I_1, I_2, and for shear and dilational elastic modul G and C, respectively [14]. We consider two common continuum-level finite element methods for the structural mechanics of deformable cells in blood flow. First, by assuming the displacement gradient tensor is constant over a given triangular element, the forces arising from the deformation of the triangular element can be computed using only the three vertices of the triangle [23]. This method is simple, efficient, and widely implemented but may be limited with respect to stability and extensibility. Second, subdivision elements have been used to develop more stable and extensible models, but require using a 'one-ring' of 12 vertices to compute the strain on a triangular element [3,5,15]. Compared with the simple model, the subdivision model is much more computationally expensive.

2.3 Immersed Boundary Method for Fluid-Structure Interaction

The Eulerian fluid lattice is coupled with the Lagrangian cell meshes by the immersed boundary method (IB) using a standard continuous forcing approach. Developed to model blood flow in the heart, the IB method uses discrete delta functions δ to transfer simulation data between the two grids [20]. Three computational kernels are involved in each timestep of the IB method: interpolation, updating, and spreading. At a given timestep t, the velocity $\mathbf{V}$ of the cell vertex located at $\mathbf{X}$ is interpolated from the surrounding fluid lattice positions $\mathbf{x}$:

$$\mathbf{V}(\mathbf{X},t) = \sum_{\mathbf{x}} \mathbf{v}(\mathbf{x},t)\,\delta(\mathbf{x}-\mathbf{X}(t)). \tag{6}$$

The position $\mathbf{X}$ of the cell vertices is updated using a forward Euler method

$$\mathbf{X}(t+1) = \mathbf{X}(t) + \mathbf{V}(t), \tag{7}$$

by the no-slip condition. With the cell having been translated and deformed by the fluid, the elastic response to cell deformation is computed according to either method discussed in the previous section. The Lagrangian force $\mathbf{G}$ is 'spread' from cell vertices onto the surrounding fluid lattice positions,

$$\mathbf{g}(\mathbf{x},t) = \sum_{\mathbf{X}} \mathbf{G}(\mathbf{X},t)\,\delta(\mathbf{x}-\mathbf{X}(t)) \tag{8}$$

which defines the external force $\mathbf{g}(\mathbf{x},t)$ acting on the fluid.

The support of the delta function, which we denoted by the symbol ϕ, determines the interaction distance between the Eulerian and Lagrangian grids. Support is measured by the number of Eulerian grid points in a given physical dimension which may influence or be influenced by a given IB point. For a given vertex, the delta function is computed for each dimension at each fluid point within the finite support, using the single-dimension distance $r \geq 0$ from the fluid point to the IB vertex. This corresponds to 8, 27, and 64 fluid points per IB vertex for delta functions with 2, 3, and 4 point support, respectively. The support of the delta function influences on simulation stability and accuracy, with certain supports being favorable for particular applications [14,20]. We consider three delta functions, where the index i indicates whether the distance $r \geq 0$ is taken in the x, y, or z direction.

Delta function support $\phi = 2$:

$$\delta_i(r) = \begin{cases} 1-r & \text{if } r \leq 1 \\ 0 & \text{if } r > 1 \end{cases} \tag{9}$$

Delta function support $\phi = 3$:

$$\delta_i(r) = \begin{cases} \frac{1}{3}\left(1+\sqrt{1-3r^2}\right) & \text{if } r \leq \frac{1}{2} \\ \frac{1}{6}\left(5-3r-\sqrt{-2+6r-3r^2}\right) & \text{if } \frac{1}{2} < r \leq \frac{3}{2} \\ 0 & \text{if } r > \frac{3}{2} \end{cases} \tag{10}$$

Delta function support $\phi = 4$:

$$\delta_i(r) = \begin{cases} \frac{1}{4}\left(1 + \cos(\frac{\pi}{2}r)\right) & \text{if } r \leq 2 \\ 0 & \text{if } r > 2 \end{cases} \quad (11)$$

We note that computational expense of the interpolation and spreading operations varies with the number of vertices, the complexity of the delta function and the number of fluid point within the support. The latter factor is exacerbated when the fluid points are not directly addressed, such as the indirect addressing in this study. Unlike the static adjacency list for LBM streaming, the dynamic set of fluid points falling within the support of a given IB vertex varies in time. Consequently, a lookup operation must be performed for each fluid point in the support to identify the memory location for the Eulerian velocity or force data with which it is associated. As the indirect addressing scheme is not random but has limited local patterns, it can be advantageous for larger ϕ to guess-and-check a subset of lookups and, if successful, interpolate between them.

2.4 General Parallelization Framework

The simulation domain is spatially decomposed among tasks into rectangular cuboid bounding boxes. Forming a partition of the vascular geometry, the bounding boxes for the Eulerian fluid domain and Lagrangian cell domain are coincident. The boundary between two bounding boxes is located exactly halfway between the last fluid point belonging to each bounding box. Based on [18], communication between tasks is governed by a hierarchy of overlapping halos on which Lagrangian or Eulerian communication is performed. When the halo of a task overlaps with the bounding box of another task, the latter task is considered a 'neighbor' of the former task with respect to this halo and vice versa. For linguistic convenience, fluid points and IB vertices which are and are not located within the task bounding box will be denoted as 'owned' and 'shared', respectively, from the task's perspective. Analogously, a cell is considered to be owned or shared based on the position of the unweighted average of its vertices. An example of the bounding box decomposition in shown in Fig. 1.

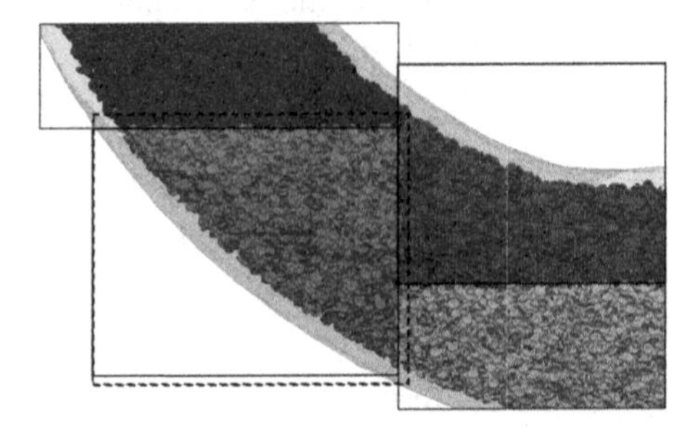

Fig. 1. Example of domain decomposition, with cells coloured by the bounding box to which they below. Dotted lines indicate halo for the green task.

Fluid Halo: A halo of fluid points is placed around the task bounding box, with a two-fold purpose. First, a single point-wide halo may be used to communicate LBM distribution components which will stream into the bounding box in the subsequent timestep. Second, by setting the halo width to $\lfloor \frac{\phi}{2} \rceil$, the IB interpolation operation may be computed locally for all owned vertices. As $\delta_i(\frac{3}{2}) = 0$ in Eq. 10, we have a single point-wide fluid halo for $\phi = 2$ and 3, but a two point-wide fluid halo for $\phi = 4$.

Cell Halo: A halo for cells is placed about the task bounding box in order to facilitate IB-related computation. In contrast to [18], a shared cell in a halo is a complete and fully updated copy of the cell. The width of this halo is set to $\lfloor \frac{\phi}{2} \rceil + r$, in which r is the largest cell radius expected in the simulation. This width ensures that all vertices which may spread a force onto a fluid point owned by the task are shared with the task. That is, if forces were known on cells within the halo, spreading may be computed locally for all owned fluid points.

2.5 Lagrangian and Eulerian Communication for IB Spreading

Algorithm 1 shows the basic coupling of fluid solver and finite element solver with the immersed boundary method for a serial code. To explore the options of communicating Eulerian or Lagrangian data, we focus on the parallelization of the last two steps: the finite element method (FEM) to compute forces at vertices of the cells and the IB spreading operation, in which forces defined at cell vertices are spread onto the fluid lattice. Two general approaches are possible for handling the communication at task boundaries.

Algorithm 1. FSI workflow

1 LBM: Collision and streaming
2 IB: Interpolate velocity of cell vertices
3 IB: Update position of cell vertices
4 FEM: Compute forces on cell vertices
5 IB: Spread forces onto fluid domain

First, Lagrangian data – the forces defined at cell vertices – can be communicated, as depicted in Fig. 2. This allows for the finite element method to be computed in a conservative manner. In our implementation, tasks run the finite element method over cells which they own. Forces defined at vertices within another task's halo are then communicated, which allows each task to perform the spreading operation locally. However, recomputation occurs when multiple tasks perform the spreading operation for vertices located near task boundaries.

Second, Eulerian data – the forces defined on the fluid grid – can be communicated instead, as depicted in Fig. 3. We compute the forces at all owned vertices, which leads to recomputation for finite elements which include vertices owned by two tasks. The forces of a task's owned vertices are spread onto owned

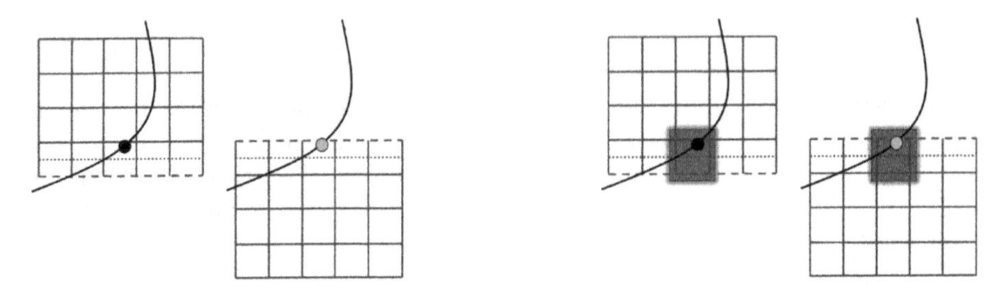

Fig. 2. Lagrangian communication for the IB spreading operation for $\phi = 2$. First (left), the Lagrangian force is computed on the owned IB vertex (black circle) by the upper task and communicated to the same vertex (yellow circle) on the lower task. Second (right), the IB spreading operation (red box) is performed for this vertex by both tasks. Solid blue lines indicate fluid grid points owned by the task, dash blue line denotes fluid points on the halo, and the dotted line represents the boundary between tasks. (Color figure online)

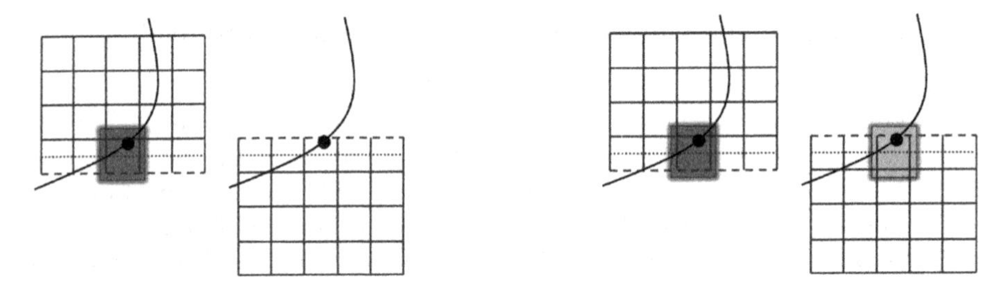

Fig. 3. Eulerian communication for the IB spreading operation for $\phi = 2$. First (left), the upper task computes the Lagrangian force on the owned vertex (black circle) and perform the IB spreading operation (red box). Second (right), the upper task communicates the Eulerian forces to the same fluid grid points (yellow box) on the lower task. Solid blue lines indicate fluid grid points owned by the task, dash blue line denotes fluid points on the halo, and the dotted line represents the boundary between tasks. (Color figure online)

and halo fluid points, which is a conservative operation. Finally, a halo exchange is performed for forces on fluid points adjacent to and located on the halo.

3 Results

3.1 Simulation Setup

The fluid domain is assumed to be cylindrical, representing an idealized blood vessel. A variety of approaches exist for generating dense suspensions of red blood cells or other suspended bodies. In the context of blood flow, the density of the suspension – the volume percentage of red blood cells in blood – is referred to as the hematocrit (Hct) level. Iterative schemes for packing rigid [30] and deformable [13] red blood cells have been demonstrated to achieve physiological hematocrit levels. To avoid the additional startup and parallelization cost of such a scheme, we perform dense packing of minimally enclosing ellipsoids in a cubic geometry using an external library [2]. The cubic arrangement is used to periodically 'tile' the vascular geometry during preprocessing and red blood cells meshes are initialized within ellipsoids. A warmup period is necessarily required before a well-developed flow is realized but other schemes incur similar costs [30]. In subsequent simulations, we completely tile the geometry with a dense red blood cell suspension and, if necessary, randomly remove cells until the desired hematocrit is achieved. A small example of a dense cell suspension in a vascular geometry is shown in Fig. 4.

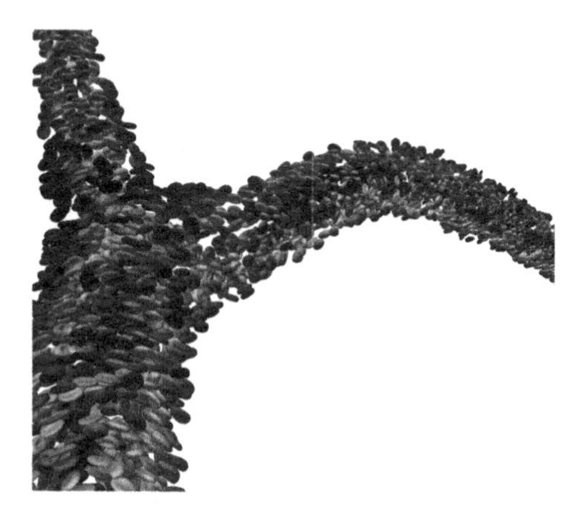

Fig. 4. Example image of red blood cells at a bifurcation in a vascular geometry. Cells are colored by vertex velocity. (Color figure online)

Runs are conducted on two different architectures, Intel Broadwell and IBM Blue Gene/Q. Part of the Duke Computer Cluster (DCC), the Broadwell system is a cluster with two Intel Xeon E5-2699 v4 processors per node and 56Gb/s Mellanox Infiniband interconnect, using 32 MPI ranks per node and 2 OpenMP threads per rank. The LLNL Blue Gene/Q system Vulcan has a Power BQC 16C on each node and custom interconnect, and is used with 16 MPI ranks and 4 OpenMP threads per rank. In the subsequent sections, we investigate single node performance and scaling across multiple nodes. For single node performance on Intel, we study a cylindrical geometry with a radius of 197 μm, which includes approximately 900,000 red blood cells when packed at 43% hematocrit. Due to the limited memory available on a Blue Gene/Q node, we use a scaled cylinder with a radius of 99 μm and approximately 100,000 red blood cells. For weak scaling across nodes, we consider progressively larger cylinders which maintain the same number of red blood cells per node when densely packed.

3.2 Comparing Lagrangian and Eulerian Communication On-Node

In this section, we compare the efficiency of Lagrangian and Eulerian communication methods from Sect. 2.5 for performing the IB spreading operation. Accordingly, we focus on the three components of the simulation related to this task (the finite element model, IB spreading itself, and pertinent communication) and consider the runtime of these three components, rather than the runtime of the entire simulation.

An important difference between the two communication schemes is the amount of data to be transferred. For the Lagrangian scheme, communication size will be dependent by the number of cells located near to task boundaries. Assuming a non-pathological distribution of cells, this will vary with the density of cells in the flow or Hct. In contrast, the Eulerian scheme will have a uniform communication size independent of Hct. Further, the communication pattern for the Eulerian scheme is time-independent, while bookkeeping may be necessary to update the Lagrangian scheme as cells move and deform. In Fig. 5, we observe the intuitive result: runtime for Eulerian communication time is constant while the Lagrangian communication time varies directly with hematocrit.

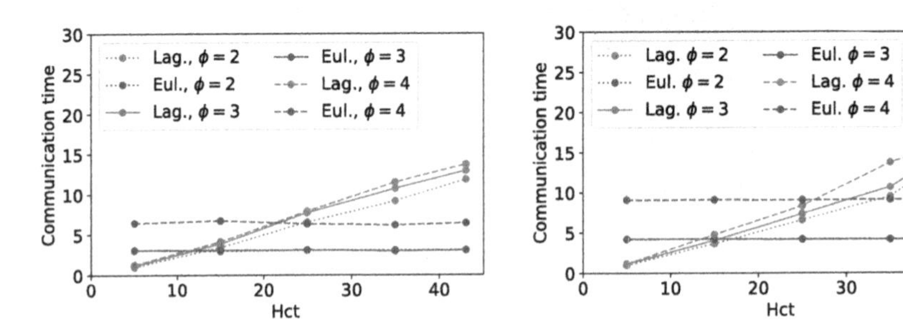

Fig. 5. Communication time for Lagrangian and Eulerian schemes for a DCC (Broadwell) node at left and a Vulcan (Blue Gene/Q) node at right. Communication times are measured in seconds and are normalized by the value of the Lagrangian scheme for $\phi = 2$ and Hct=5.

The size of the data to be transferred will also depend on the support of the delta function. For $\phi = 4$, the fluid halo increases to two grid points. This effectively doubles the amount of communicated data for the Eulerian scheme relative to the single grid point halo for $\phi = 2$ or 3. While the amount of Lagrangian data to be communicated is somewhat higher for $\phi = 4$, we observe in Fig. 5 that this increase is considerably more modest. Additionally, while hematocrit value at which Eulerian scheme begins to outperform is roughly similar between the two architectures, this cross-over value is consistently approximately 5% higher on Vulcan (Blue Gene/Q).

However, the merits of the two communication schemes also have to be judged in the context of the recomputation required and its impact on overall runtime. Figure 6 shows how the significance of recomputation varies not only with the

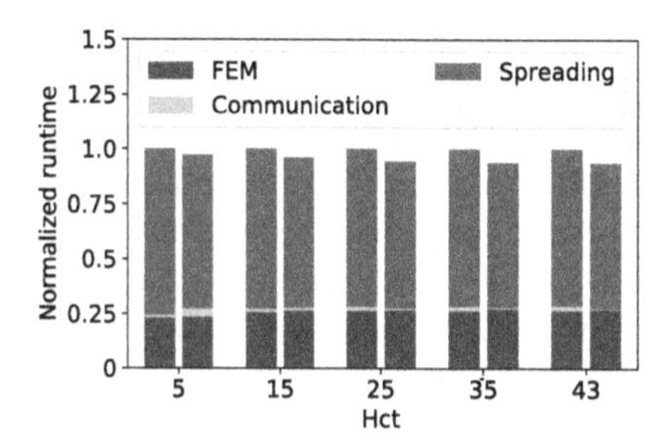

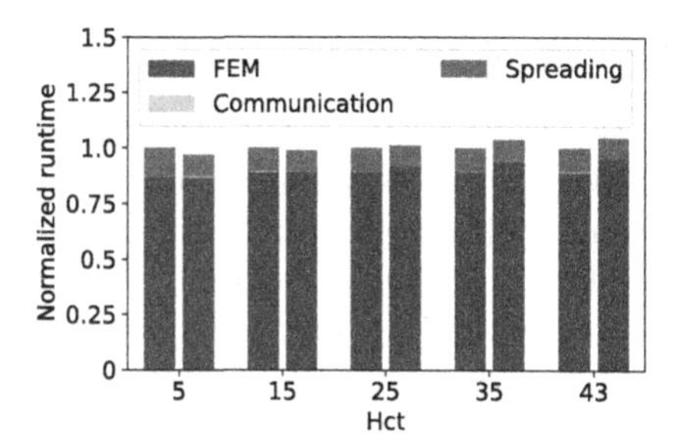

Fig. 6. For $\phi = 2$, we compare runtime on a DCC (Broadwell) node as a function of Hct for Lagrangian (left bar) and Eulerian (right bar) communication. Runtimes are measured in seconds and, for a given Hct, are normalized by the Lagrangian runtime. Left and right images show simple and subdivision finite element models, respectively.

communication scheme but also with the finite element model. As discussed above, the IB spreading operation performs expensive lookup operations when used with an indirectly addressed fluid grid. When paired with an inexpensive finite element model, we observe that the IB spreading recomputation performed by the Lagrangian scheme in the spreading operation becomes expensive relative to the finite element recomputation of the Eulerian scheme. As a result, the Eulerian scheme outperfoms in this framework, even at the low hematocrit values for which the communication cost exceeds than of the Lagrangian scheme.

Conversely, this situation is reversed for the subdivision finite element model. Due to the high computational expense of this model, the recomputation when computing forces on the cells exceeds that of the IB spreading operation. The Lagrangian scheme consequently proves more efficient for higher hematocrit values, with communication costs for either scheme being relatively inconsequential. This result is also relevant to other approaches for modeling cell mechanics, such as discrete element methods, which have reported that the force computation kernel is responsible for the majority of their runtime [18].

In summary, we observe that both communication and recomputation are associated with the relative performance of Lagrangian and Eulerian communication schemes. Looking solely at communication time, the Eulerian communication scheme clearly outperform its Lagrangian counterpart at the hematocrit values typical of blood flow. This advantage is most significant for smaller immersed boundary supports but remains even for $\phi = 4$. This result for a high density of immersed boundary vertices serves as a complement for the experience of [6], who found performing Eulerian communication was inefficient for a simulation with $\phi = 4$ and a density of immersed boundary vertices comparable to 10% Hct. However, we also find that the disparity between the cost of the finite element and spreading operations may render recomputation a more important factor than communication cost in determining the more efficient scheme, especially at higher cell densities.

3.3 Weak Scaling

As the purpose of a distributed memory parallelization scheme is to enable large simulations which require multiple nodes, the scalability of a communication scheme is also important. In contrast to the previous section, we now consider the scalabity of the full simulation, rather than the kernels and communication which differed in the Lagrangian and Eulerian communication schemes. Figure 7 shows weak scaling at 43% hematocrit for $\phi = 2$, 3, and 4 and using the simple finite element model. For weak scaling, we increase the problem size proportionately with the number of tasks, maintaining the same amount of work per task over successively larger task counts. To measure weak scalability, we normalize all runtimes by the runtime at the lowest task count.

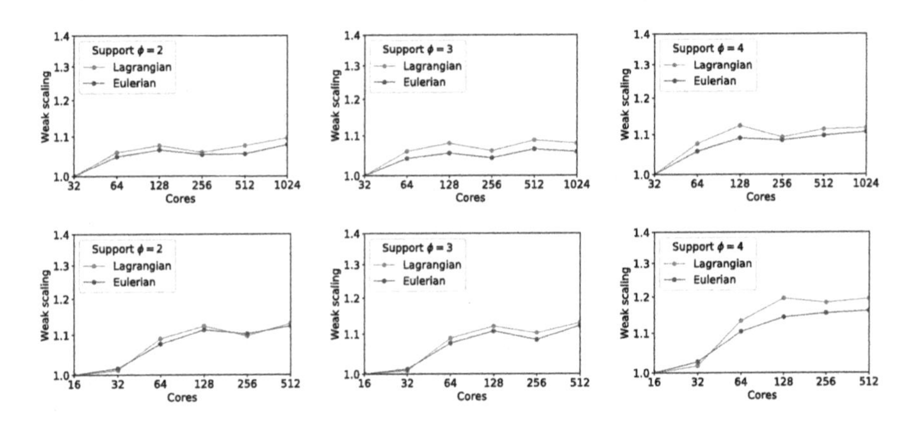

Fig. 7. Weak scaling for $\phi = 2$, 3, and 4 for DCC (Broadwell) in the top row and Vulcan (Blue Gene/Q) in bottom row

We observe broadly similar performance with the two architectures, although DCC (Broadwell) benefits from the much larger problem size per node. On both architectures, we observe a drop in performance between 32 and 64 tasks due to the maximum number of neighboring tasks being first encountered on the latter task count. A similarly marginal gain occurs with the Eulerian communication scheme for the IB spreading operation, which outperformed for this parameter set on a single node and maintains this modest advantage when the problem is weakly scaled across nodes. However, the primary influence on scalability comes from the delta function support, as performance with $\phi = 4$ is limited by larger communication and recomputation times due to the larger halo. In contrast, weak scaling remains around 89% parallel efficiency for $\phi = 2$ and 3.

3.4 Discussion

In this study, we investigate the factors influencing the performance of halo exchange for the immersed boundary method in the context of the hemodynamics application HARVEY. Focusing on the on-node performance of the IB spreading operation, we compare Lagrangian and Eulerian communication frameworks. In this comparison, the purpose is not to propose an optimal configuration based on our application, but to provide a starting point for evaluating IB method parallelization options for a given physical problem and model.

With respect to purely communication-related costs, we find that the intuitive cross-over for more efficient Eulerian than Lagrangian communication for the IB spreading operation occurred for a density of IB vertices relevant to many applications including blood flow. For physiological values of red blood cell hematocrit, Eulerian communication may provide an improvement, regardless of the delta function support. Conversely, for lower IB vertex densities and $\phi = 4$, we agree with the assertion of [6] that Eulerian communication may not be an efficient scheme. However, the exact cross-over point will nonetheless be variable: on systems with limited memory per node, like the Vulcan Blue Gene/Q, we see the hematocrit cross-over point, at which Eulerian scheme outperforms the Lagrangian, to be about 5% higher than the larger Broadwell nodes on the DCC cluster.

However, we find that communication costs must also be assessed in the context of the required recomputation. The support of the delta function and relative cost of the IB spreading and finite element model will influence the relative costs of communication and recomputation. We observe that while larger delta function support sizes necessarily increase compute costs, the additional communication costs may be modest relative to factors like recomputation. In future work, we plan to extend this study of immersed boundary halo exchange to simulations on heterogeneous CPU-GPU compute nodes, where the differences in recomputation and data motion cost are expected to become more significant.

Acknowledgments. We thank Thomas Fai and Charles Peskin for their comments and insight during the code development process. This work was performed under the auspices of the U.S. Department of Energy by LLNL under Contract DE-AC52-07NA27344. Computing support came from the LLNL Institutional Computing Grand Challenge program. Research reported in this publication was supported by the Office of the Director, National Institutes of Health under Award Number DP5OD019876. The content is solely the responsibility of the authors and does not necessarily represent the official views of the National Institutes of Health. Support was provided by the Big Data-Scientist Training Enhancement Program of the Department of Veterans Affairs and by the Hartwell Foundation.

References

1. Bailey, P., Myre, J., Walsh, S.D., Lilja, D.J., Saar, M.O.: Accelerating lattice boltzmann fluid flow simulations using graphics processors. In: 2009 International Conference on Parallel Processing, pp. 550–557. IEEE (2009)
2. Birgin, E., Lobato, R., Martínez, J.: A nonlinear programming model with implicit variables for packing ellipsoids. J. Global. Optim. **68**(3), 467–499 (2017)
3. Boedec, G., Leonetti, M., Jaeger, M.: Isogeometric FEM-BEM simulations of drop, capsule and vesicle dynamics in Stokes flow. J. Comput. Phys. **342**, 117–138 (2017)

4. Chen, S., Doolen, G.D.: Lattice Boltzmann method for fluid flows. Ann. Rev. Fluid Mech. **30**(1), 329–364 (1998)
5. Cirak, F., Ortiz, M., Schroder, P.: Subdivision surfaces: a new paradigm for thin-shell finite-element analysis. Int. J. Numer. Meth. Eng. **47**(12), 2039–2072 (2000)
6. Di, S., Xu, J., Chang, Q., Ge, W.: Numerical simulation of stirred tanks using a hybrid immersed-boundary method. China J. Chem. Eng. **24**(9), 1122–1134 (2016)
7. Fedosov, D., Caswell, B., Suresh, S., Karniadakis, G.: Quantifying the biophysical characteristics of plasmodium-falciparum-parasitized red blood cells in microcirculation. Proc. Nat. Acad. Sci. USA **108**(1), 35–39 (2011)
8. Givelberg, E., Yelick, K.: Distributed immersed boundary simulation in Titanium. SIAM J. Sci. Comput. **28**(4), 1361–1378 (2006)
9. Gounley, J., Draeger, E.W., Randles, A.: Numerical simulation of a compound capsule in a constricted microchannel. Procedia Comput. Sci. **108**, 175–184 (2017)
10. Guo, Z., Zheng, C., Shi, B.: Discrete lattice effects on the forcing term in the lattice Boltzmann method. Phys. Rev. E **65**(4), 046308 (2002)
11. Hochmuth, R., Waugh, R.: Erythrocyte membrane elasticity and viscosity. Ann. Rev. Physiol. **49**(1), 209–219 (1987)
12. Imai, Y., Omori, T., Shimogonya, Y., Yamaguchi, T., Ishikawa, T.: Numerical methods for simulating blood flow at macro, micro, and multi scales. J. Biomech. **49**(11), 2221–2228 (2016)
13. Krüger, H.: Computer simulation study of collective phenomena in dense suspensions of red blood cells under shear. Ph.D. thesis (2012)
14. Krüger, T., Varnik, F., Raabe, D.: Efficient and accurate simulations of deformable particles immersed in a fluid using a combined immersed boundary lattice Boltzmann finite element method. Comput. Math. Appl. **61**(12), 3485–3505 (2011)
15. Le, D.V.: Effect of bending stiffness on the deformation of liquid capsules enclosed by thin shells in shear flow. Phys. Rev. E **82**(1), 016318 (2010)
16. Li, X., et al.: Patient-specific blood rheology in sickle-cell anaemia. Interface Focus **6**(1), 20150065 (2016)
17. Mittal, R., Iaccarino, G.: Immersed boundary methods. Annu. Rev. Fluid Mech. **37**, 239–261 (2005)
18. Mountrakis, L., Lorenz, E., Malaspinas, O., Alowayyed, S., Chopard, B., Hoekstra, A.G.: Parallel performance of an IB-LBM suspension simulation framework. J. Comput. Sci. **9**, 45–50 (2015)
19. Ouro, P., Fraga, B., Lopez-Novoa, U., Stoesser, T.: Scalability of an Eulerian-Lagrangian large-eddy simulation solver with hybrid MPI/OpenMP parallelisation. Comput. Fluids **179**, 123–136 (2019)
20. Peskin, C.S.: The immersed boundary method. Acta Numer. **11**, 479–517 (2002)
21. Randles, A., Draeger, E.W., Oppelstrup, T., Krauss, L., Gunnels, J.A.: Massively parallel models of the human circulatory system. In: 2015 SC-International Conference for High Performance Computing, Networking, Storage and Analysis, pp. 1–11. IEEE (2015)
22. Randles, A.P., Kale, V., Hammond, J., Gropp, W., Kaxiras, E.: Performance analysis of the lattice Boltzmann model beyond Navier-Stokes. In: 2013 IEEE 27th International Symposium on Parallel & Distributed Processing (IPDPS), pp. 1063–1074. IEEE (2013)
23. Shrivastava, S., Tang, J.: Large deformation finite element analysis of non-linear viscoelastic membranes with reference to thermoforming. J. Strain Anal. **28**(1), 31–51 (1993)

24. Spandan, V., et al.: A parallel interaction potential approach coupled with the immersed boundary method for fully resolved simulations of deformable interfaces and membranes. J. Comput. Phys. **348**, 567–590 (2017)
25. Uhlmann, M.: Simulation of particulate flows on multi-processor machines with distributed memory. Technical report Centro de Investigaciones Energeticas Medioambientales y Tecnologicas (CIEMAT) (2004)
26. Wang, S., He, G., Zhang, X.: Parallel computing strategy for a flow solver based on immersed boundary method and discrete stream-function formulation. Comput. Fluids **88**, 210–224 (2013)
27. Wiens, J.K., Stockie, J.M.: An efficient parallel immersed boundary algorithm using a pseudo-compressible fluid solver. J. Comput. Phys. **281**, 917–941 (2015)
28. Wu, Z., Xu, Z., Kim, O., Alber, M.: Three-dimensional multi-scale model of deformable platelets adhesion to vessel wall in blood flow. Philos. Trans. R. Soc. A **372**(2021), 20130380 (2014)
29. Yu, Z., Lin, Z., Shao, X., Wang, L.P.: A parallel fictitious domain method for the interface-resolved simulation of particle-laden flows and its application to the turbulent channel flow. Eng. Appl. Comput. Fluid **10**(1), 160–170 (2016)
30. Závodszky, G., van Rooij, B., Azizi, V., Alowayyed, S., Hoekstra, A.: Hemocell: a high-performance microscopic cellular library. Procedia Comput. Sci. **108**, 159–165 (2017)

Future Ramifications of Age-Dependent Immunity Levels for Measles: Explorations in an Individual-Based Model

Elise Kuylen[1,2](✉), Lander Willem[2], Niel Hens[2,3], and Jan Broeckhove[1]

[1] IDLab, Department of Mathematics and Computer Science, University of Antwerp, Antwerp, Belgium
elise.kuylen@uantwerpen.be

[2] CHERMID, Vaccine and Infectious Disease Institute, University of Antwerp, Antwerp, Belgium

[3] Interuniversity Institute for Biostatistics and Statistical Bioinformatics, UHasselt, Hasselt, Belgium

Abstract. When a high population immunity already exists for a disease, heterogeneities, such as social contact behavior and preventive behavior, become more important to understand the spread of this disease. Individual-based models are suited to investigate the effects of these heterogeneities. Measles is a disease for which, in many regions, high population immunity exists. However, different levels of immunity are observed for different age groups. For example, the generation born between 1985 and 1995 in Flanders is incompletely vaccinated, and thus has a higher level of susceptibility. As time progresses, this peak in susceptibility will shift to an older age category. Simultaneously, susceptibility will increase due to the waning of vaccine-induced immunity. Older generations, with a high degree of natural immunity, will, on the other hand, eventually disappear from the population. Using an individual-based model, we investigate the impact of changing age-dependent immunity levels (projected for Flanders, for years 2013 to 2040) on the risk for measles outbreaks. We find that, as time progresses, the risk for measles outbreaks increases, and outbreaks tend to be larger. As such, it is important to not only consider infants when designing strategies for measles elimination, but to also take other age categories into account.

Keywords: Individual-based modeling · Epidemiology · Measles · Immunity

1 Introduction

Over the last decade, individual-based models have become increasingly popular within the domain of epidemiology [25]. In such a model, each individual is represented as a unique entity, thus offering the advantage of being able to

J. M. F. Rodrigues et al. (Eds.): ICCS 2019, LNCS 11536, pp. 456–467, 2019.
https://doi.org/10.1007/978-3-030-22734-0_33

model different levels of heterogeneity within a population. This makes it possible to incorporate system properties emerging from the unique behavior of several thousands or millions of individuals. Examples of heterogeneities that may be of importance for epidemiological models are social contact behavior, susceptibility, and preventive behavior - such as vaccination. Heterogeneity is especially important in the case of emerging diseases and diseases for which a high population immunity already exists [3].

This is true for measles, a vaccine-preventable disease that - at the moment - mainly affects young children and still caused approximately 110 000 deaths worldwide in 2017 [26]. Understandably, an important goal that the WHO has set is the global eradication of measles. To reach this goal, the WHO proposes a target vaccination coverage with two doses of the vaccine of 95% of children in a country, to ensure the entire population is protected from outbreaks through herd immunity [27]. As such, focus is often placed on the immunization of infants and young children.

However, older age groups are often ignored when estimating the risk for measles outbreaks and planning for their prevention [11]. In Belgium, the vaccine against measles was included in the basic vaccination scheme in 1985 [5]. Individuals born before this date have often been infected with measles in the past. As it is assumed that individuals who survive a measles infection acquire lifelong immunity, this age group is expected to have a high level of immunity. This was confirmed by a recent serological survey conducted in Belgium [20].

However, the generation born between 1985 and 1995 is incompletely vaccinated as a result of the introduction period of the vaccine [5]. In the future, their immunity level is expected to decrease, due to waning of vaccine-induced immunity. Recent multi-country surveys confirm that a growing fraction of susceptibles to measles are adolescents and young adults, indicating the need to focus on these age groups when planning for measles elimination [2,21,22]. This is also the conclusion of a study which aimed to determine age-dependent susceptibility to measles in Japan [14]. They conclude that supplementary vaccination for adults between 20 and 49 years old would be useful to prevent future outbreaks of measles.

The decades after the vaccine was introduced, uptake steadily increased - thus lowering the number of individuals susceptible to measles. This, in turn, led to a decrease in measles infections. Because of this, in combination with the aging of the generation born before the introduction of the vaccine, natural immunity has become less frequent. As of 2016, 96.2% of infants in Flanders, Belgium are vaccinated with the first dose of the MMR vaccine [4]. However, due to false reports on side-effects of the MMR (Measles - Mumps - Rubella) vaccine, vaccination levels have recently been declining in several countries - putting them at risk for dropping below the 95% immunization target [13].

The combination of these factors makes it necessary to look at the impact of different vaccination levels for different age groups on the risk of measles outbreaks, today as well as in the future. This is highlighted by a new analysis, which takes into account age-dependent social contact patterns to estimate herd immunity thresholds for different age categories [10].

A recent study [12] investigated how immunity levels for different age groups in Flanders could change in the future. They predict that, as a result of the factors described above, not only will age-dependent immunity levels shift in the future, but overall susceptibility to measles will also be higher. An individual-based model is particularly suited to investigate the impact of changing immunity levels in the future, because it is possible to model person-to-person interactions that differ based on age and on context of the interaction (e.g. schools, workplaces, daycare centers).

In this paper, we extend the individual-based model Stride [15] to include age-dependent immunity levels. We investigate the impact on outbreak occurrence and outbreak size of the shifting age-dependent immunity levels in Flanders for 6 different years from 2013 to 2040. We show that, as time progresses, the risk for a measles outbreak in Flanders increases, as does the predicted size of such an outbreak. Furthermore, in addition to infants, other age-groups experience a growing risk for measles infection.

2 Methods

We used Stride, an individual-based simulator for the transmission of infectious diseases [15]. Stride is a stochastic model: processes such as contacts between individuals and disease transmission have a probabilistic component. Furthermore, Stride is designed to be very versatile. By using different input files, it is possible to run simulations for a multitude of populations and diseases. The core logic of Stride is implemented in C++, making it highly portable and open to performance-optimization [23]. Finally, Stride is developed as an open-source project: its code can be found in a public Github repository [6]. More information about the structure and internal logic of Stride can be found in a previous publication [15].

We modified Stride to implement age-dependent immunity levels. Before the beginning of the actual simulation, the population is set up. During this process, each individual in the population is marked either susceptible or immune to the simulated disease. We use a target fraction of immune individuals for each age between 0 and 99 years. The unit of the age categories is one year. For each age category, we first calculate the target number of immune individuals, based on the immunity level and the total number of individuals in the age category. Next, we check for each randomly drawn person, whether the age bracket to which they belong contains enough immune individuals. If it does not, we mark the selected individual 'immune' and continue to randomly draw another individual from the population. We continue this process until all age brackets contain the target number of immune individuals.

We ran simulations for 6 different years from 2013 to 2040 (2013, 2020, 2025, 2030, 2035, 2040). The age-dependent immunity levels for each year that we used, were based on results from [12], and were obtained through personal communication with the authors. We received projections of immunity levels for the years mentioned above, for 500 municipalities in Belgium. To obtain the immunity levels that we used for our simulations, we took the average fraction of immune individuals for each age category over 500 bootstrap runs for all municipalities. All municipalities were treated the same: size differences between municipalities were not taken into account in calculating the mean immunity level per age category.

For comparison, we also ran simulations for each of the six calendar years using a constant immunity level which was identical over all age categories. The immunity levels used for these simulations were based on averages obtained from previous simulations that used age-dependent immunity levels. An overview of the average immunity level per simulated calendar year can be found in Table 1.

Immunity data for ages 0 to 85 years was available. Since Stride includes individuals up to 99 years of age, immunity levels for ages between 86 and 99 years were assumed to be the same as the immunity level for 85 year olds. Age-dependent susceptibility levels (1 - fraction of immune individuals in age category) that were used for each examined year can be seen in Fig. 1 (left).

The population we use consists of 600 000 individuals, representing a sample of the total population of Flanders - which consists of about 6 million people. The synthetic population is made up out of reference households, obtained from a survey in 2010–2011 [24]. Geographic distribution of these households is based on 2001 census data. Children are assigned to a daycare (0–2 years old), preschool (3–5 years old), primary (6–11 years old), secondary (12–17 years old) or tertiary (18–23 years old) school based on enrollment statistics acquired from Eurostat [8]. Adults (18–64 years old) are assigned to a workplace based on age-specific employment data and aggregated workplace size data from Eurostat and commuting data from the 2001 census. To account for general contacts, all individuals are also assigned to two 'communities'. One of these represents the general contacts made during the week, while the other represents those made during the weekend. We distinguish these two types of community for two reasons. Firstly, different contact rates apply for weekdays and weekend days. Secondly, by letting individuals make general contacts within two different communities, we are able to model the movement of an infectious disease between communities. Each community consists of 1000 individuals on average, which is in line with the size of communities used in a previous model [7].

The population is closed, meaning that no individuals are born or die over the course of the simulation. The contact rates that determine the probability of a contact occurring between two individuals of certain ages at a given location are based on data from a social contact survey in Flanders [24]. Separate social contact patterns are used for weekdays and weekend days. Holidays - both school and regular - are not taken into account.

We ran each simulation for 730 time-steps, which corresponds to 2 simulated years. This was done to allow each epidemic to run its course, until no more new infections were being recorded. At the beginning of each simulation, a single individual in the population is selected, and their health status is set to 'infected'. The natural history of the disease was modeled as follows. At the beginning of a simulation, it is determined for each individual how long it takes until they become infectious after becoming infected, how long it takes until they become symptomatic and finally, how long it takes for them to recover. Each of these durations is drawn from a distribution. Incubation periods are sampled from a lognormal distribution with median 12.5 and dispersion 1.25 [16]. After this period, individuals become infectious, which lasts 6 to 8 days [1]. After either 2 or 3 days of being infectious but asymptomatic, individuals begin to experience symptoms of their disease. This symptomatic period, during which persons stay at home and thus only make contacts within their household, lasts 6 to 8 days. In our model, we assume that each infected individual eventually becomes symptomatic.

When an infected individual has a contact with a susceptible individual, a chance for disease transmission exists. The probability for such a transmission event is calculated from the basic reproduction number R_0, which is supplied as an input parameter. For each calendar year that we simulated between 2013 and 2040, we tested input values of R_0 between 12 and 18, as the basic reproduction number for measles is commonly estimated to be in this range [9].

The relation between R_0 and the transmission probability was estimated by fitting a function to data resulting from a large number of simulations [7]. We tested 21 values for the transmission probability between 0 and 1, and started simulations on 7 different days of the week, to account for different contact rates during weekdays and weekend days. To obtain stable results, we ran 150 stochastic simulations for each of these scenarios, resulting in a total of 22050 runs.

At the beginning of each simulation, one infected individual was introduced into the population. All other individuals in the population were initialized as susceptible. Other parameters were identical to those used in the experiments described in this paper.

During each simulation, we tracked the number of secondary cases one infected individual caused in the completely susceptible population. We fitted the data-points we obtained to a quadratic function using the 'polyfit' method from the numpy Python package [17]. As such we derived the formula in Eq. (1).

$$\hat{R_0} = 0 + 32.862 * P(transmission) - 6.184 * P(transmission)^2 \tag{1}$$

Before the beginning of the simulation, after the population has been set up, we record for each individual their age in years and whether or not they are susceptible to the simulated disease. During each time-step of the simulation - which corresponds to one day - we keep track of how many individuals have been infected until then. This also includes individuals who have since recovered from the disease.

We executed 200 stochastic simulation for each combination of calendar year and R_0. Seeds to initiate our random number generator were generated from a non-deterministic, machine-specific source. To improve performance, several simulations were run in parallel, using the Python environment we created for Stride and the Python 'multiprocessing' package [19]. All simulations were run on a Linux machine, using 8 cores (16 threads).

3 Results

From the information on each individual's susceptibility, recorded at the beginning of every simulation, we calculated average age-dependent susceptibility levels. These are displayed in Fig. 1 (b), and are very close to the projected levels for each year, which are shown in Fig. 1(a). We observe that in 2013, there is a peak in susceptibility for individuals between - roughly - 18 and 28 years old, the generation born between 1985 and 1995. As time progresses, this peak shifts to an older age group. It also increases in height, as a result of the waning of vaccine-induced immunity. Vaccination coverage for infants and children remains stable.

We did the same for the simulations using uniform immunity levels for all age categories, the results of which can be seen in Fig. 1(c). The uniform immunity levels that were used for calendar years 2013–2040 can be found in Table 1. We see that, due to the number of individuals having natural immunity decreasing

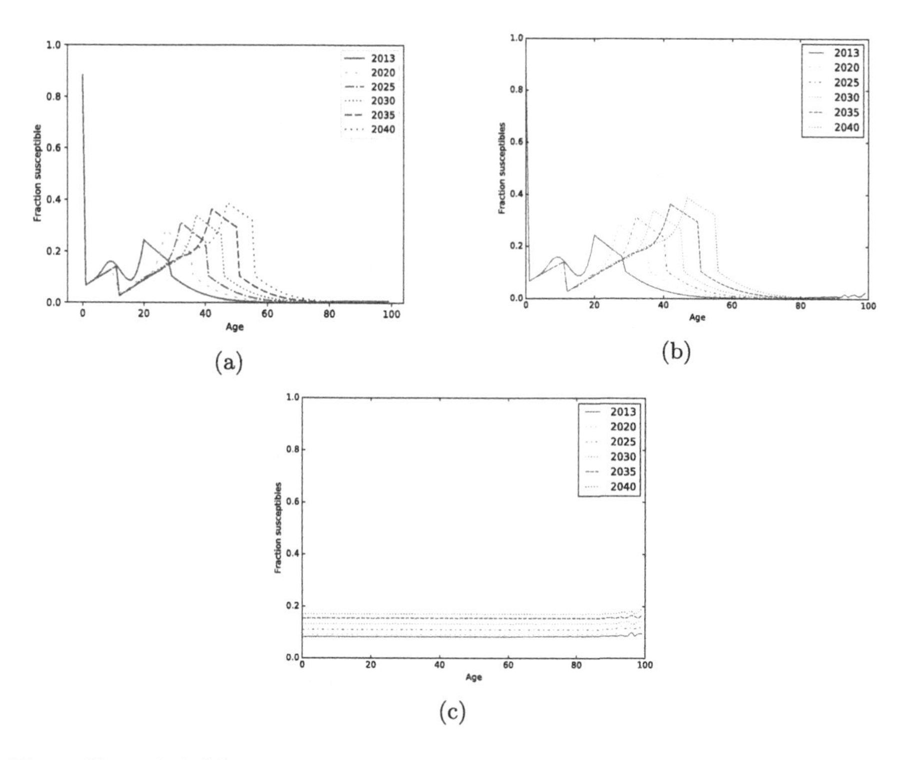

Fig. 1. Upper left (a): the projected fraction of susceptibles by age (in years) for years 2013–2040. Projections were based on a previous study [12]. Upper right (b): age-dependent susceptibility levels as observed in simulated populations for years 2013–2040, when using age-dependent immunity levels based on (a). Lower row, middle (c): age-dependent susceptibility levels as observed in simulated populations for years 2013–2040, when using uniform immunity levels for all age categories.

and the waning of vaccine-induced immunity, the overall immunity lowers as time progresses. Small variations in the oldest age categories can be attributed to a limited number of individuals of these ages being present in the simulated population.

Table 1. Percentage of immune individuals in the population for simulated years 2013, 2020, 2025, 2030, 2035 and 2040.

Year	Percentage immune
2013	91.61%
2020	90.81%
2025	88.91%
2030	86.67%
2035	84.57%
2040	82.90%

To get an idea of how often the introduction of an infected individual in the population leads to an actual outbreak, we first estimated an 'extinction threshold'. To do this, a histogram of the total number of infections at the end of each simulation was created. In Fig. 2, an overview of the results for the simulations with age-dependent immunity levels and R_0 12 can be seen. We created the same histogram for simulations that used a uniform immunity level for all age categories, and for simulations with R_0 values of 13, 14, 15, 16, 17 and 18. These yielded similar results, which are not shown here.

For each simulated calendar year, there is a certain fraction of runs where the epidemic dies out without causing many - or even any - secondary infections. We will refer to these as the extinction cases. To get a realistic idea of the size and frequency of outbreaks when no such extinction occurs, we set an 'extinction threshold'. When the total number of infected cases at the end of a simulation was higher than this threshold, we say that an outbreak occurred. Otherwise, the run was regarded as an extinction case. Using the results shown in Fig. 2 and similar histograms for other scenarios, we set the extinction threshold at 10 000 cases. A clear division can be seen: either the number of infected cases is close to zero, or there are at least 10 000 cases.

Using this threshold, we calculated the fraction of simulation runs for each combination of simulated year and R_0 value that resulted in an outbreak, both for simulations using age-dependent immunity levels and for simulations using uniform levels for all age categories. These results can be observed in Fig. 3. We can see that the fraction of outbreak occurrences increases as time progresses. This is the case when using age-dependent immunity levels as well as for simulations using uniform immunity levels. However, we can see that when age-dependent immunity levels are taken into account (Fig. 3(a)) the fraction of runs that leads to an outbreak is higher in the first simulated calendar years

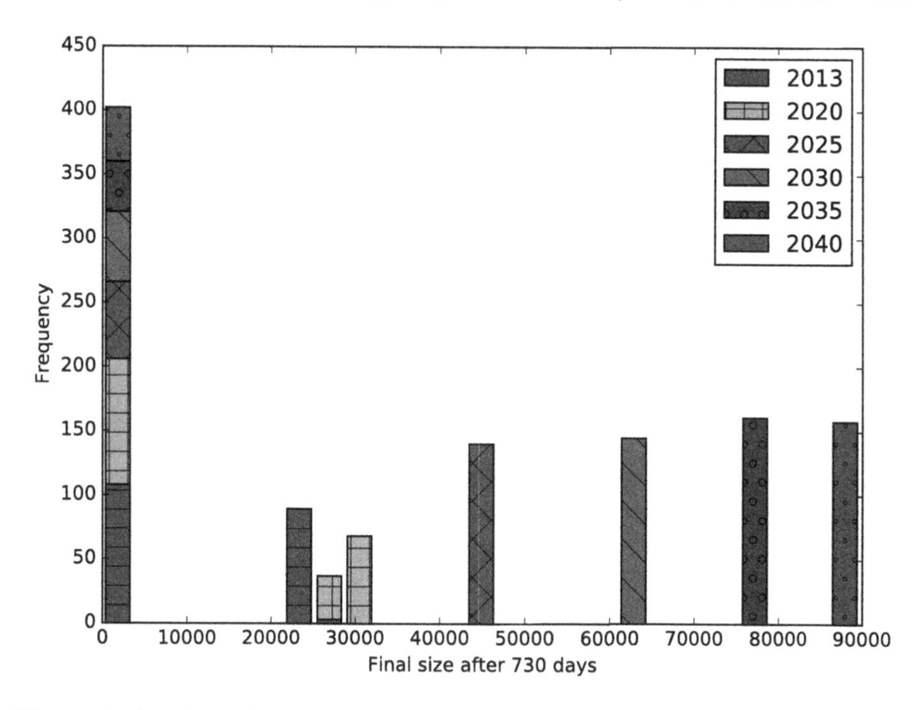

Fig. 2. Outbreak size frequencies, plotted by simulated calendar year (shown here for R_0 12 and for simulations using age-dependent immunity levels). For each simulated year shown here, 200 stochastic simulations were run.

compared to simulations using uniform immunity levels (Fig. 3(b)). We included results for simulations with values for R_0 of 12 (blue circles) and 18 (green triangles) (results for R_0 12, 14, 15, 16 and 17 not shown). The increasing trend can be observed for both R_0 values shown, but is less pronounced for higher values of R_0.

We also consider the final size of outbreaks when no extinction occurred. In Fig. 4, the distribution of outbreak sizes can be seen for simulations using age-dependent immunity levels (a) and uniform immunity levels for all age categories (b). In both cases, outbreak sizes increase as time progresses. However, when taking age-dependent immunity levels into account (a), the increase in size is slower in earlier years, and has a sharp increase from 2025 onward. On the other hand, for simulations using uniform immunity levels for all age categories (b), the increase in outbreak sizes is more linear. Furthermore, outbreak sizes in 2013 are already higher in the first case (a), compared to simulations in which immunity levels are uniform for all age categories (b). However, as time progresses the difference between (a) and (b) becomes less pronounced.

Results for R_0 values of both 12 and 18 can be observed in Fig. 4. The same general trends can be observed for runs with R_0 values of 13, 14, 15, 16 and 17, for which the results are not shown here.

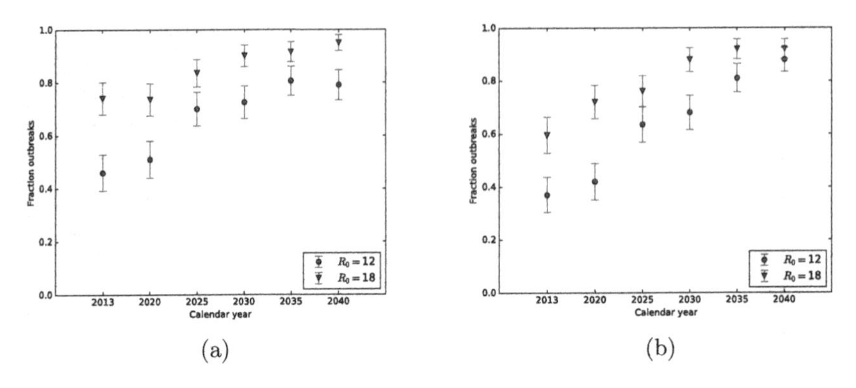

Fig. 3. Comparison of the fraction of simulation runs (out of 200) that result in an outbreak (extinction threshold = 10 000 cases) for simulations using age-dependent immunity levels (a) and uniform immunity levels for all age categories (b). Results shown here for R_0 12 and 18. Error bars indicate 95% percentile intervals. (Color figure online)

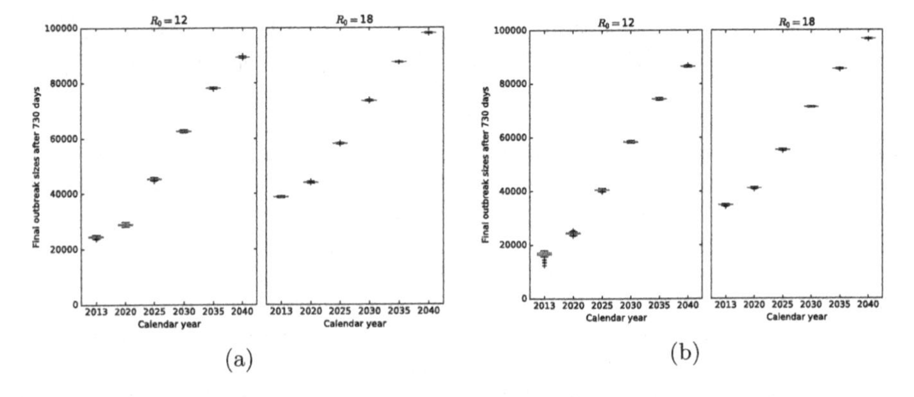

Fig. 4. Comparison of final outbreak sizes for simulations using age-dependent immunity levels (a) and uniform immunity levels for all age categories (b), when no extinction occurs (extinction threshold = 10 000 cases. Results here are shown for R_0 values 12 and 18.

We ran a total of 16800 simulations: 200 runs for 7 values of R_0 (12, 13, 14, 15, 16, 17 and 18) and 6 different simulated years (2013, 2020, 2025, 2030, 2035 and 2040), both using age-dependent immunity levels and uniform immunity levels for all age categories. It took about 26 h to run these, using 16 workers in the Python multiprocessing pool.

4 Discussion

We used an individual-based model to assess the impact of changing age-dependent immunity levels on the risk for measles outbreaks in Flanders. We were able to use detailed projections on the immunity levels of all ages from 0 to 85 years for calendar years 2013 to 2040. Since we used an individual-based model, we were also able to model contact rates that differed based on age and context. By doing this, instead of assuming homogeneous mixing, we could assess the impact of immunity levels differing by age more accurately.

We observed a difference between simulations that took into account age-dependent immunity levels and simulations where a uniform immunity level was assumed for all age categories. Even though average immunity levels were identical between these two scenarios for calendar years 2013 to 2040, we saw that when we took into account age-dependent immunity levels, the predicted risk for measles outbreaks was higher. This effect was the strongest in earlier calendar years, and became less pronounced as time progressed. When assuming uniform immunity levels for all age categories and an R_0 value of 12 the probability of an outbreak occurring in 2020 is about 0.3. By contrast, when taking into account age-dependent immunity levels, this probability increases to about 0.5. For higher values of R_0, the same trend can be observed, but the effect is less strong. This confirms that heterogeneities in the population become more important for estimating the risk for a measles outbreak as the overall immunity level of the population increases or the value of R_0 decreases. As we near the target of measles elimination, these will thus become more important when modeling the spread of this disease.

Furthermore, when outbreaks do occur, they are expected to become larger. This is because a larger part of the population will become susceptible to measles over time. The difference between the case in which age-dependent immunity levels are taken into account and when uniform immunity levels are assumed is observable in earlier calendar years, but fades as time progresses. This suggests that larger outbreak sizes are mainly due to a larger number of susceptibles available in the population. The older generation, which still has natural immunity, ages and dies out, while the fraction of infants and children immunized remains stable. The immunity level of the generation born between 1985 and 1995 was already lower, and is expected to further decrease due to the waning of vaccine-induced immunity. As such, even though the vaccination coverage of infants and young children remains high, the overall susceptibility in the population increases.

Regarding these results, some limitations need to be considered. First, the population in Stride is closed. For childhood diseases such as measles, a model in which new infants can be born into the population could yield more accurate results. In future work, we wish to add a demographic component to Stride, that would make this possible.

Secondly, there were a few factors that were not yet considered in the projected age-dependent immunity rates that we used. As the projections were based on serological data collected in Belgium in 2006, only humoral immunity is taken into account [20]. However, recent studies [18] suggest that, in cases where there is no humoral immunity protecting an individual, they can still be protected against a disease through cellular immunity. As such, the projections made here would overestimate the susceptibility for measles in the population.

Furthermore, we note that, for the projections we used, the assumption is made that no large measles outbreaks occur until 2040, which would again increase natural immunity. Finally, in the projections that we used, the recent increase in vaccine hesitancy was not taken into account. Due to, among others, false reports on side-effects of the MMR vaccine, vaccine uptake has been decreasing in some countries [13]. This could cause an additional increase in susceptibility, besides the one that was discussed in this paper.

When taking into account age-dependent immunity levels, we observed that, while the immunity level for infants and young children remains stable, the risk for a measles outbreak increases as time progresses. Moreover, the size of such outbreaks is expected to increase. As such, it is important to consider age-dependent immunity levels when planning for the elimination of measles. Other age groups - besides infants and young children - should be given attention as well when designing vaccination strategies. To increase the vaccination coverage for individuals born between 1985 and 1995 a catch-up campaign could, for example, be organized.

Those born between 1985 and 1995 have reached the age of becoming parents themselves today. An opportunity would be to immunize those that were left unvaccinated as infants when they visit a doctor for the immunization of their own children. In future work, it could be interesting to investigate the cost-effectiveness of such a campaign.

Finally, it would also be interesting to conduct the same research for other vaccine-preventable diseases and other populations. Using Stride, this should be feasible when input data on population immunity levels is available.

References

1. Anderson, R., May, R.: Infectious Diseases of Humans: Dynamics and Control. Oxford University Press, Oxford (1992)
2. Andrews, N., et al.: Towards elimination: measles susceptibility in Australia and 17 European countries. Bull. World Health Organ. **86**, 197–204 (2008)
3. Béraud, G.: Mathematical models and vaccination strategies. Vaccine **36**(36), 5366–5372 (2018)
4. Braeckman, T., et al.: Can Flanders resist the measles outbreak? Assessing vaccination coverage in different age groups among Flemish residents. Epidemiol. Infect. **146**(8), 1043–1047 (2018)
5. Braeye, T., Sabbe, M., Hutse, V., Flipse, W., Godderis, L., Top, G.: Obstacles in measles elimination: an in-depth description of a measles outbreak in Ghent, Belgium, spring 2011. Arch. Public Health **71**(1), 17 (2013)
6. Broeckhove, J., Kuylen, E., Willem, L.: Stride Github repository. https://github.com/broeckho/stride

7. Chao, D., Halloran, M., Obenchain, V., Longini, I.: FluTE, a publicly available stochastic influenza epidemic simulation model. PLoS Comput. Biol. **6**(1), e1000656 (2010)
8. European Commission: Eurostat. https://ec.europa.eu/eurostat/
9. Fine, P.: Herd immunity: history, theory, practice. Epidemiol. Rev. **15**(2), 265–302 (1993)
10. Funk, S., et al.: Target immunity levels for achieving and maintaining measles elimination. BioRxiv, p. 201574 (2018)
11. Hayman, D.: Measles vaccination in an increasingly immunized and developed world. Human Vaccines & Immunotherapeutics (2018)
12. Hens, N., et al.: Assessing the risk of measles resurgence in a highly vaccinated population: Belgium anno 2013. Eurosurveillance **20**(1), 20998 (2015)
13. Jansen, V., Stollenwerk, N., Jensen, H., Ramsay, M., Edmunds, W., Rhodes, C.: Measles outbreaks in a population with declining vaccine uptake. Science **301**(5634), 804 (2003)
14. Kinoshita, R., Nishiura, H.: Assessing age-dependent susceptibility to measles in Japan. Vaccine **35**(25), 3309–3317 (2017)
15. Kuylen, E., Stijven, S., Broeckhove, J., Willem, L.: Social contact patterns in an individual-based simulator for the tranmission of infectious diseases (Stride). Procedia Comput. Sci. **108**, 2438–2442 (2017)
16. Lessler, J., Reich, N., Brookmeyer, R., Perl, T., Nelson, K., Cummings, D.: Incubation periods of acute respiratory viral infections: a systematic review. Lancet. Infect. Dis. **9**, 291–300 (2009)
17. NumPy Developers: NumPy. http://www.numpy.org/. Accessed 04 Apr 2019
18. Plotkin, S.: Complex correlates of protection after vaccination. Clin. Infect. Dis. **56**(10), 1458–1465 (2013)
19. Python Software Foundation: Multiprocessing - Process-based parallelism. https://docs.python.org/3.4/library/multiprocessing.html. Accessed 14 Jan 2019
20. Theeten, H., et al.: Are we hitting immunity targets? The 2006 age-specific seroprevalence of measles, mumps, rubella, diphtheria and tetanus in Belgium. Epidemiol. Infect. **139**(4), 494–504 (2011)
21. Thompson, K.: What will it take to end human suffering from measles? Lancet. Infect. Dis. **17**(10), 1013–1014 (2017)
22. Trentini, F., Poletti, P., Merler, S., Melegaro, A.: Measles immunity gaps and the progress towards elimination: a multi-country modelling analysis. Lancet. Infect. Dis. **17**(10), 1089–1097 (2017)
23. Willem, L., Stijven, S., Tijskens, E., Beutels, P., Hens, N., Broeckhove, J.: Optimizing agent-based transmission models for infectious diseases. BMC Bioinf. **16**(1), 183 (2015)
24. Willem, L., Van Kerckhove, K., Chao, D., Hens, N., Beutels, P.: A nice day for an infection? Weather conditions and social contact patterns relevant to inlfuenza transmission. PLoS ONE **7**(11), e48695 (2012)
25. Willem, L., Verelst, F., Bilcke, J., Hens, N., Beutels, P.: Lessons from a decade of individual-based models for infectious disease transmission: a systematic review (2006–2015). BMC Infect. Dis. **17**(1), 612 (2017)
26. World Health Organisation: Measles. https://www.who.int/news-room/fact-sheets/detail/measles. Accessed 09 Jan 2019
27. World Health Organisation: Measles vaccines: WHO position paper - April 2017. Weekly Epidemiol. Rec. **92**(17), 205–228 (2017)

Evolution of Hierarchical Structure and Reuse in iGEM Synthetic DNA Sequences

Payam Siyari[1], Bistra Dilkina[2], and Constantine Dovrolis[1(✉)]

[1] Georgia Institute of Technology, Atlanta, GA 30332, USA
payamsiyari@gmail.com, constantine@gatech.edu
[2] University of Southern California, Los Angeles, CA 90007, USA
dilkina@usc.edu

Abstract. Many complex systems, both in technology and nature, exhibit hierarchical modularity: smaller modules, each of them providing a certain function, are used within larger modules that perform more complex functions. Previously, we have proposed a modeling framework, referred to as Evo-Lexis [21], that provides insight to some fundamental questions about evolving hierarchical systems.

The predictions of the Evo-Lexis model should be tested using real data from evolving systems in which the outputs can be well represented by sequences. In this paper, we investigate the time series of iGEM synthetic DNA dataset sequences, and whether the resulting iGEM hierarchies exhibit the qualitative properties predicted by the Evo-Lexis framework. Contrary to Evo-Lexis, in iGEM the amount of reuse decreases during the timeline of the dataset. Although this results in development of less cost-efficient and less deep Lexis-DAGs, the dataset exhibits a bias in reusing specific nodes more often than others. This results in the Lexis-DAGs to take the shape of an hourglass with relatively high H-score values and stable set of core nodes. Despite the reuse bias and stability of the core set, the dataset presents a high amount of diversity among the targets which is in line with modeling of Evo-Lexis.

Keywords: Complex systems · Hierarchical structure · Optimization · Hourglass effect · iGEM

1 Introduction

Hierarchically modular designs enhance evolvability in natural systems [15,16, 19], make the maintenance easier in technological systems, and provide agility and better abstraction of the system design [9,18].

This research is supported by DARPA's Lifelong Learning Machines (L2M) program, under Cooperative Agreement HR0011-18-2-0019, and by the National Science Foundation under Grant No. 1319549.

J. M. F. Rodrigues et al. (Eds.): ICCS 2019, LNCS 11536, pp. 468–482, 2019.
https://doi.org/10.1007/978-3-030-22734-0_34

In prior work in [21], we present *Evo-Lexis*, a modeling framework for the emergence and evolution of hierarchical structure in complex modular systems. There are many hypotheses in the literature regarding the factors that contribute to either the hierarchy or modularity properties. Local resource constraints in social networks and ecosystems [17], modularly varying goals [7,13,14], selection for more robust phenotypes [4,24], and selection for lower connection costs in a network [15] are some of the mechanisms that have been previously explored and shown to lead to hierarchically modular systems. The main hypothesis that Evo-Lexis follows is along the lines of [15], which assumes that systems in both nature and technology care to minimize the cost of their interconnections or dependencies between modules. We also studied the hourglass effect via Evo-Lexis. Informally, an hourglass architecture means that the system of interest produces many outputs from many inputs through a relatively small number of highly central intermediate modules, referred to as the "waist" of the hourglass. It has been observed that hierarchically modular systems often exhibit the architecture of an hourglass; for reference, in fields like computer networking [2], neural networks [10,11], embryogenesis [5], metabolism [8,23], and many others [19,22], this phenomena is observed. A comprehensive survey of the literature on hierarchical systems evolution, and the hourglass effect is presented in [19].

The motivation for this paper is that the Evo-Lexis model is quite general and abstract, and it does not attempt to capture any domain-specific aspects of biological or technological evolution. As such, it makes several assumptions that can be criticized for being unrealistic, such as the fact that all targets have the same length, or their length stays constant, or the fitness of a sequence is strictly based on its hierarchical cost. We believe that such abstract modeling is still valuable because it can provide insights into the qualitative properties of the resulting hierarchies under different target generation models. However, we also believe that the predictions of the Evo-Lexis model should be tested using real data from evolving systems in which the outputs can be well represented by sequences. One such system is the iGEM synthetic DNA dataset [1]. The target DNA sequences in the iGEM dataset are built from standard "BioBrick parts" (more elementary DNA sequences) that collectively form a library of synthetic DNA sequences. These sequences are submitted to the registry of standard biological parts in the annual iGEM competition. Previous research in [3,20] has provided some evidence that these synthetic DNA sequences are designed by reusing existing components, and as such, it has a hierarchical organization. In this paper, we investigate how to apply the Evo-Lexis framework in the time series of iGEM sequences, and whether the resulting iGEM hierarchies exhibit the same qualitative properties we observed in [21] which was solely based on abstract target generation models. We ask the following questions in this paper:

1. How can we analyze the iGEM dataset using the evolutionary framework of Evo-Lexis? How are the batches of targets formed? What properties of the iGEM batches are different than Evo-Lexis's setting?
2. When formed incrementally over the iGEM dataset, which are the architectural properties of Lexis-DAGs, and why?

2 Preliminaries

To develop *Evo-Lexis*, we extend the previously proposed optimization framework *Lexis* in [20]. Lexis models the most elementary modules of the system as symbols ("sources") and the modules at the highest level of the hierarchy as sequences of those symbols ("targets"). *Evo-Lexis* is a dynamic or evolving version of Lexis, in the sense that the set of targets changes over time through additions (births) and removals (deaths) of targets. *Evo-Lexis* computes an (approximate) minimum-cost adjustment of a given hierarchy when the set of targets changes over time (a process we refer to as "incremental design").

2.1 Lexis Optimization

Given an alphabet S and a set of "target" strings T over the alphabet S, we need to construct a Lexis-DAG. A Lexis-DAG D is a directed acyclic graph $D(V, E)$, where V is the set of nodes and E the set of edges, that satisfies the following three constraints:[1] **(a)** Each node $v \in V$ in a Lexis-DAG represents a string $\mathcal{S}(v)$ of characters from the alphabet S. The nodes V_S that represent characters of S are referred to as *sources*, and they have zero in-degree. The nodes V_T that represent target strings $T = \{t_1, t_2, \ldots, t_m\}$ are referred to as *targets*, and they have zero out-degree. V also includes a set of *intermediate nodes* V_M, which represent substrings that appear in the targets T. So, $V = V_S \cup V_M \cup V_T$. **(b)** Each node in $V_M \cup V_T$ of a Lexis-DAG represents a string that is the concatenation of two or more substrings, specified by the incoming edges from other nodes to

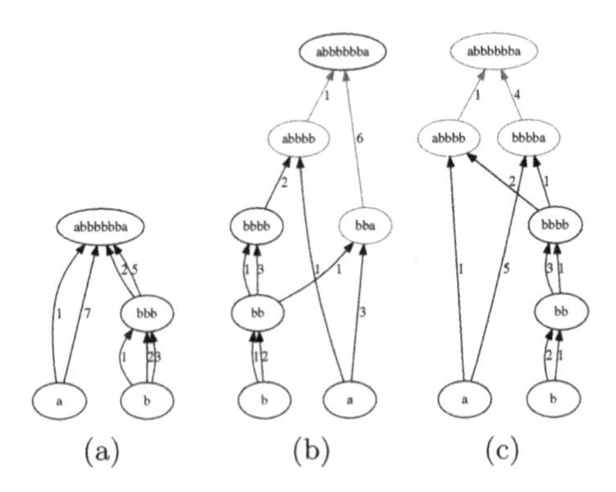

Fig. 1. Illustration of the Lexis-DAG for a single target $T = \{abbbbbba\}$ and sources $S = \{a, b\}$. Edge-labels indicate the occurrence indices: **(a)** A valid Lexis-DAG having both minimum number of concatenations and edges. **(b)** An invalid Lexis-DAG: two intermediate nodes are re-used only once. **(c)** An invalid Lexis-DAG: the top-layer string is not equal to the concatenation of its two in-neighbors (best viewed in color). (Color figure online)

[1] To simplify the notation, even though D is a function of S and T, we do not denote it as such.

that node. Note that there may be more than one edge from node u to node v. **(c)** A Lexis-DAG should only include intermediate nodes that have an out-degree of at least two, $\forall v \in V_M, d_{out}(v) \geq 2$ for a more parsimonious hierarchical representation. Figure 1 illustrates the concepts introduced here.

The Lexis Optimization Problem. The *Lexis* optimization problem is to construct a minimum-cost Lexis-DAG for the given alphabet S and target strings T. In other words, the problem is to determine the set of intermediate nodes V_M and all required edges E so that the corresponding Lexis-DAG D is optimal in terms of a given cost function $C(D)$. This problem can be formulated as follows:

$$\begin{aligned} & min_{(E,V_M)}\ C(D) \\ & s.t.\ D = (V, E) \text{ is a Lexis-DAG for } S \text{ and } T \\ & \text{where } C(D) = \mathcal{E}(D) = \sum_{v \in V} d_{in}(v) = |E| \end{aligned} \tag{1}$$

A natural cost function, as investigated in previous work [20], is the number of edges in the Lexis-DAG. The *edge cost* to construct a node $v \in V$ is defined as the number of incoming edges required to construct $\mathcal{S}(v)$ from its in-neighbors, which is equal to $d_{in}(v)$. The edge cost of source nodes is obviously zero. The edge cost $\mathcal{E}(D)$ of Lexis-DAG D is defined as the edge cost of all nodes, which is equal to the number of edges in D. With edge cost, the problem in Eq. (1) is NP-Hard [20]. This problem is similar to the *Smallest Grammar Problem* (SGP) [6] and in fact its NP-Hardness is shown by a reduction from SGP [20].

We solve the Lexis optimization problem in Eq. (1) with a greedy heuristic, called G-Lexis [20]. G-Lexis starts with the trivial flat Lexis-DAG, and at each iteration it chooses the substring ξ that maximally reduces the edge cost, when it is added as a new intermediate node to the Lexis-DAG and the corresponding edges are rewired by its addition.

Path-Centrality and the Core of a Lexis-DAG. After constructing a Lexis-DAG, an important question is to rank the constructed intermediate nodes in terms of significance or *centrality*. More formally, let $P_D(v)$ be the number of source-to-target paths that traverse node $v \in V_M$; we refer to $P_D(v)$ as the *path centrality* of intermediate node v. Path centrality can be computed as: $P(v) = P_S(v)\, P_T(v)$ where $P_S(v)$ is the number of paths from any source to v, and $P_T(v)$ is the number of paths from v to any target.[2]

An important follow-up question is to identify the *core* of a Lexis-DAG, i.e., a set of intermediate nodes that represent, as a whole, the most important substrings in that Lexis-DAG. Intuitively, we expect that the core should include nodes of high path centrality, and that almost all source-to-target dependency chains of the Lexis-DAG should traverse at least one of these core nodes. More formally, suppose K is a set of intermediate nodes and $\mathcal{P}^-(K)$ is the set of

[2] A similar metric, called *stress centrality* of a vertex, is studied in [12].

source-to-target paths after we remove the nodes in K from D. The core of D is defined as the minimum-cardinality set of intermediate nodes $Core(\tau) = \hat{K}$ such that the fraction of remaining source-to-target paths after the removal of $\hat{K}$ is at most τ:[3]

$$\begin{aligned} \hat{K} = argmin_{K \subseteq V_M} \quad & |K| \\ & s.t. \, |\mathcal{P}^{-}(K)| \leq \tau \, |\mathcal{P}^{-}(\varnothing)| \end{aligned} \tag{2}$$

where $|\mathcal{P}^{-}(\varnothing)|$ is the number of source-to-target paths in the original Lexis-DAG, without removing any nodes. We solve the core identification problem with a greedy algorithm referred to as G-CORE [20]. This algorithm adds in each iteration the node with the highest path-centrality value to the core set, updates the Lexis-DAG by removing that node and its edges, and recomputes the path centralities of the remaining nodes before the next iteration.

Hourglass Score. Intuitively, a Lexis-DAG exhibits the hourglass effect if it has a small core. We use a metric, named as Hourglass Score, or *H-Score*, in our study for measuring the "hourglass-ness" of a network. This metric was originally presented in [19]. To calculate the H-score, we create a flat Lexis-DAG D_f containing the same targets as the original Lexis-DAG D. Note that D_f preserves the source-target dependencies of D: each target in D_f is constructed based on the same set of sources as in D. However, the dependency paths in D_f are direct, without forming any intermediate modules that could be reused across different targets. So, by construction, the flat Lexis-DAG D_f cannot have a non-trivial core since it does not have any intermediate nodes. We define the H-score as follows: $H_D(\tau) = 1 - \frac{|Core(\tau)|}{|Core_f(\tau)|}$ where $Core(\tau)$ and $Core_f(\tau)$ are the core sets of D and D_f for a given threshold τ, respectively. Since that $Core_f$ can include a combination of sources and targets, it would never be larger than either the set of sources or targets, i.e., $|Core_f(\tau)| \leq min\{|S|, |T|\}$. Thus, $0 \leq H(\tau) \leq 1$. The H-score of D is approximately one if the core size of the original Lexis-DAG is negligible compared to the core size of the corresponding flat Lexis-DAG.

2.2 Evo-Lexis Framework and Key Results

The Evo-Lexis framework includes a number of components that are described below. A general illustration of the framework is shown in Fig. 2. In every iteration, the following steps are performed: **(1)** A batch of new targets is generated via a target generation model. **(2)** In the "expansion phase", the new targets are added incrementally to the current Lexis-DAG by minimizing the marginal cost of adding every new target to the existing hierarchy. We refer to this *incremental design* algorithm as INC-LEXIS, and it is described in detail [21]. **(3)** If the number of targets that are present in the system has reached a steady-state threshold, we also remove the batch of oldest targets from the Lexis-DAG.

[3] To simplify notation, we do not denote the core set as function of D.

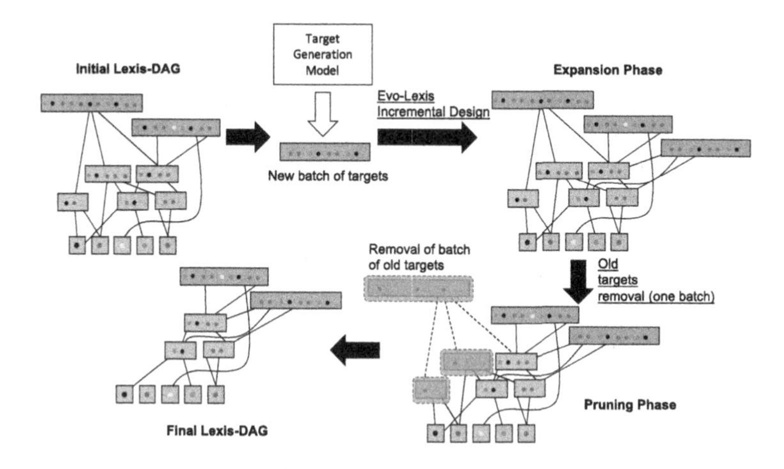

Fig. 2. A diagram of the Evo-Lexis framework.

In general, a system interacts with its environment in a bidirectional manner: the environment imposes various constraints on the system and the system also affects its environment. To capture this co-evolutionary setting in *Evo-Lexis*, we study how changes in the set of targets affect the resulting hierarchy but also how the current hierarchy affects the selection of new targets (i.e. whether a new candidate target is selected or not depends on its fitness or cost – and that depends on how easily that target can be supported by the given hierarchy). By incorporating well-known evolutionary mechanisms, such as tinkering (mutation), recombination, and selection, *Evo-Lexis* can capture such co-evolutionary dynamics between the generation of new targets and the hierarchy that supports them. Figure 3 is an overview of the following key results from the Evo-Lexis

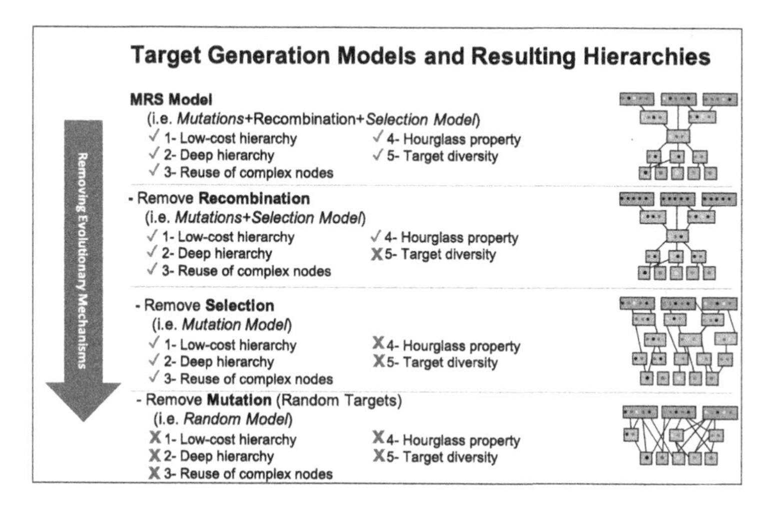

Fig. 3. Overview of results from Evo-Lexis.

model: **(i)** *Tinkering/mutation* in the target generation process is found to be a strong initial force for the emergence of low-cost and deep hierarchies. **(ii)** *Selection* is found to enhance the emergence of more complex intermediate modules in optimized hierarchies. The bias towards reuse of complex modules results in an hourglass architecture in which almost all source-to-target dependency paths traverse a small set of intermediate modules. **(iii)** The addition of *recombination* in the target generation process is essential in providing target diversity in optimized hierarchies.

3 iGEM Dataset

3.1 Preliminaries

The International Genetically Engineered Machine (iGEM) is an annual worldwide synthetic biology competition. The competition is between students from diverse backgrounds including biology, chemistry, physics, engineering, and computer science to construct synthetic DNA structures with novel functionalities.

Every year at the beginning of the summer, there is a "Distribution Kit" handed to teams which includes interchangeable parts (so called "BioBricks") from the Registry of Standard Biological Parts comprising various genetic components such as promoters, terminators, reporter elements, and plasmid backbones. Then, the teams try to use these parts and the new standardized parts of their own in order to build biological systems. The teams can build on previous projects or create completely new parts. At the end of the summer, all teams add their new BioBricks to the registry for further possible reuse in next years.

The iGEM Registry (i.e., the dataset we are working with) includes a set of standard biological parts. A [biological] part is a DNA sequence which encodes a biological function, e.g., a promoter or protein coding sequence. These biological parts are standardized to be easily assembled together and reused with other standardized parts in the registry. A "basic part" is a functional unit of a synthesized DNA that cannot be subdivided into smaller component parts. BBa_R0051 is an example of a promoter basic part. Basic parts have the role of sources in the Lexis setting. A "composite part" is a functional unit of DNA consisting of two or more basic parts assembled together. BBa_I13507 is an example of a composite part, consisting of four basic parts "BBa_B0034 BBa_E1010 BBa_B0010 BBa_B0012". The dataset we analyze is the set of all composite parts submitted to the registry from 2003 to 2017. In this dataset, the composite parts are represented by the string of their basic parts (i.e., a non-dividing representation). The sequence of iGEM composite parts can be considered as a sequence of target strings over a set of sources (i.e., basic parts). We have acquired the iGEM

Table 1. Basic statistics on iGEM dataset during 15 years (2003–2017)

# Sources	# Targets	Total length	Min/Max target length
7,889	18,394	107,022	2/100

data from https://github.com/biohubx/igem-data. All the BioBrick parts were crawled until Dec 28th 2017. In Table 1, the preliminary statistics about the dataset are listed. The dataset mostly presents targets of small length. The top 5 categories having the highest fraction of the targets belongs to those of length 5, 2, 3, 4 and 6, accounting for more than 70% of the dataset. Less than 10% of the targets have a length of more than 10.

3.2 Considering Annual Batches of Targets

The iGEM competition is conducted annually. Hence, it is reasonable to consider the sequences of targets as annual batches of targets arriving each year. This consideration is in line with the incremental design process in Evo-Lexis.

To show some differences between iGEM and Evo-Lexis, in Fig. 4, we can see how the number of sources, the number of targets, length statistics and source reuse statistics change over time. We can make the following observations from these figures:

1. The number of sources increases, where it was constant in Evo-Lexis.
2. In the first four years, the number of targets per year is noticeably small. Later on, the number of targets increases up to 2,000 and then fluctuates around 1,000 to 1,300 targets per year. In Evo-Lexis, the number of targets per batch is constant and they all have the same length.

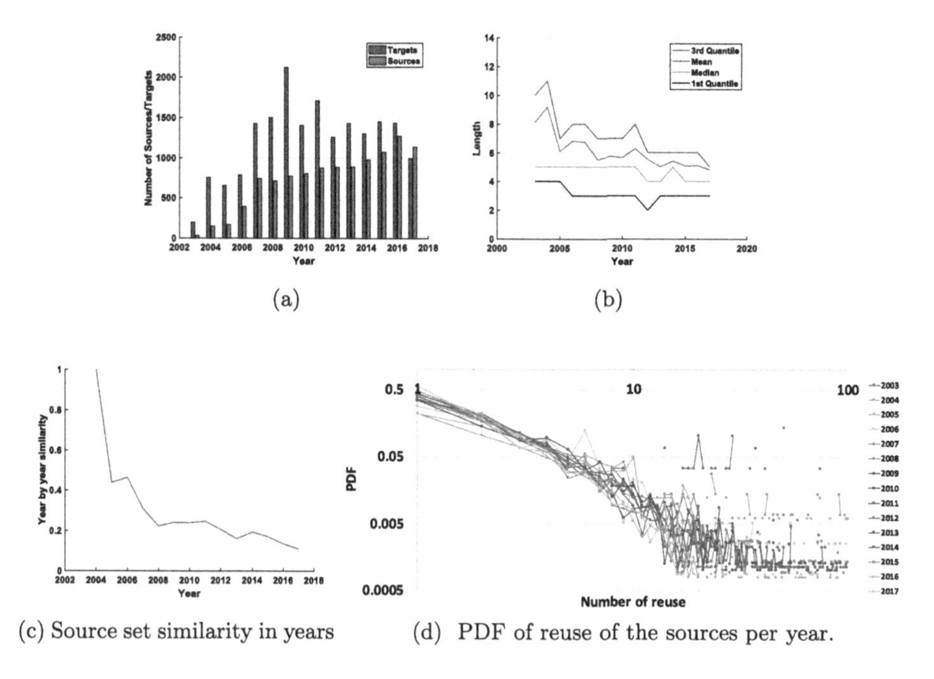

(a) (b)

(c) Source set similarity in years (d) PDF of reuse of the sources per year.

Fig. 4. Statistics of iGEM dataset when considered as yearly batches. Number of reuse is the number of times a source appear in a target in each year.

3. The mean and median of target lengths stay in the same range ($\in [5, 7]$) during all 15 years.
4. The reuse of sources (except for the beginning years) is extremely skewed in all years: few sources are used much more often than most of the sources (Fig. 4d). In Evo-Lexis, all sources are equally likely.

In the following sections, we show that how these differences between iGEM dataset and Evo-Lexis cause differences between the resulting Lexis-DAGs.

4 Analysis of iGEM Dataset in Evo-Lexis Framework

From this section on, we compare the results over iGEM with the results gathered from Evo-Lexis in [21]. We refer the reader to [21] for details of the model and parameter settings.

4.1 Lexis-DAG Cost Analysis

In this section, we observe how cost efficient the Lexis-DAGs over the iGEM dataset are. We consider an incremental setting similar to Evo-Lexis: In the first year, a clean-slate Lexis-DAG is constructed over the targets of that year. For the targets of the subsequent years, an incremental Lexis-DAG is constructed. Figure 5 shows how the normalized cost of the Lexis-DAGs varies over the years on iGEM. We observe major differences with Evo-Lexis; in Evo-Lexis the normalized cost remains almost constant.

To investigate the reasons for the above observations, in the same Fig. 5, we also track the cost reduction performance of the two stages of INC-LEXIS for each batch (as a reminder, in stage-1, we reuse intermediate nodes from previous Lexis-DAG and in stage-2, we further optimize the hierarchy using G-LEXIS). This experiment is done due to our interest in seeing how much stage-1 of INC-LEXIS contributes to the cost reduction on iGEM. There are two observations that we can make:

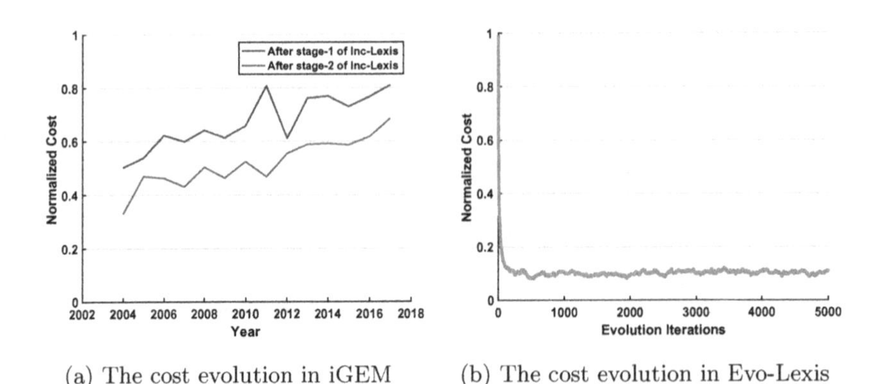

(a) The cost evolution in iGEM (b) The cost evolution in Evo-Lexis

Fig. 5. Comparison of cost evolution in iGEM and Evo-Lexis (from [21])

1. In most batches, more than 50% of the cost reduction is achieved by the stage-1, i.e., reuse stage. The contribution of stage-2 of Inc-Lexis is roughly constant throughout years. This suggests that iGEM targets reuse a significant amount of sequences from previous years in their own submissions.
2. There is an increasing trend in the normalized cost after stage-1. This observation means that the contribution of the reuse stage in Inc-Lexis decreases over the years. As mentioned, the contribution of stage-2 stays mostly constant. Hence, we can relate the increasing trend of the normalized cost to the fact that the amount of reuse reduces from year to year.

We can find the root-cause of the decrease of reuse over time on iGEM to the increase of the size of the set of sources. We have observed in Fig. 4a that there are many new sources that get introduced over the years. One of the requirements for reuse from one batch to another in Evo-Lexis is the fact that the set of sources does not drastically change (in fact it is constant in the Evo-Lexis framework). To investigate whether this is true in iGEM, we check the ratio of the sources from one year to the next that remain the same. Specifically, if we have $y_2 = y_1 + 1$, and if S_{y_1} & S_{y_2} are the set of sources in year y_1 & y_2 respectively, we check the ratio: $\frac{|S_{y_1} \cap S_{y_2}|}{|S_{y_1}|}$. This ratio, i.e., *year-by-year similarity*, is the fraction of sources that remain from the previous year. Figure 4c shows how this ratio changes from year to year. By year 2008, the ratio drops significantly to a value around 0.2 which means around 80% of the sources from the previous year are not reused. This reduces the amount of reuse that is possible in the iGEM dataset. The introduction of new sources is also propagated in individual targets. As time progresses, there is a higher probability to use more than X number of new sources per target. This observation is a further obstacle for reuse, especially given that the targets in iGEM are often short (5–7 subparts). Following the increase of the normalized cost, Fig. 6 shows that the DAGs get less deep and have lower average node length as time progresses. Overall, the results of this section show a number of differences between iGEM and Evo-Lexis:

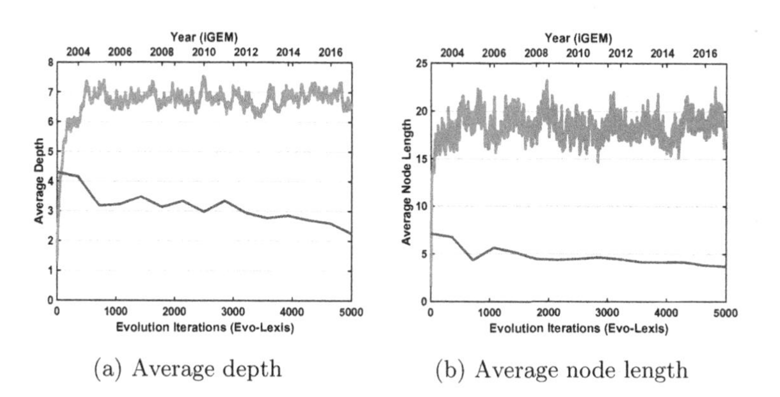

(a) Average depth (b) Average node length

Fig. 6. Average depth and node length in iGEM and Evo-Lexis (in green, [21])

1. In iGEM, the set of sources in each year has low similarity to the previous years, while in Evo-Lexis the source set is constant. The high amount of churn in the set of sources is the primary reason for the lower reuse in iGEM data compared to Evo-Lexis. The fact that the targets are shorter is another factor for iGEM's lower potential for reuse of longer intermediate nodes.
2. The normalized cost, depth and average node length are all lower in iGEM due to the reduced reuse potential as discussed above.

4.2 Hourglass Effect in iGEM

The following results in this section show that in all years, there is a small number of core nodes in the iGEM Lexis-DAGs. Figure 7 shows that such small cores make the topology of iGEM Lexis-DAGs consistent with an hourglass organization (high H-score values - more than 0.6 in Fig. 7c). In Evo-Lexis, we observe similar values of H-score for DAGs constructed using synthetic data. As observed, although the core size increases in iGEM over time, we see a steeper increase in the size of the flat DAG's core mostly due to the increase in set of sources. In Evo-Lexis, the core size shows a decreasing trend while the size of the core of the flat DAG does not significantly change, reflecting similarly high H-score values as in iGEM. Overall, we can see that the topology of the Lexis-DAGs in iGEM data is in line with the Evo-Lexis model, although the bias in selection of cost-saving nodes is not sufficiently large to cause a non-increasing normalized cost.

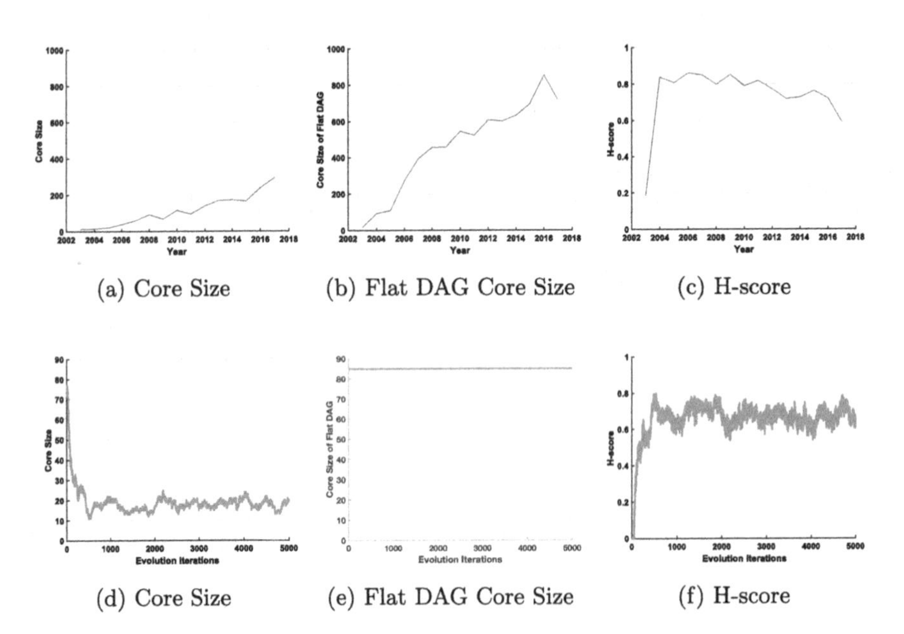

(a) Core Size (b) Flat DAG Core Size (c) H-score

(d) Core Size (e) Flat DAG Core Size (f) H-score

Fig. 7. Cores in iGEM and Evo-Lexis (bottom, [21]) ($\tau = 0.85$).

4.3 Diversity Among iGEM Targets

Another question is the degree of diversity among the targets of iGEM over time. We define the concept of *Normalized Diversity* as follows: Suppose we have a set of strings $T = \{t_1, t_2, ..., t_n\}$. The goal is to provide a single number that quantifies how dissimilar these elements are to each other.

- We first identify the *medoid* $\mathcal{M}_T$ of the set T, i.e., the element that has the lowest average distance from all other elements. We use Levenshtein distance as a measure of distance between targets: $\mathcal{M}_T = arg\ min_{m \in T} \sum_{t \in T} LD(t, m)$.
- To compute how diverse the elements are with respect to each other, we average the normalized distance of all elements from the medoid (distance is normalized by the maximum length of the two sequences in question). We call this measure σ_T, the *Normalized Diversity* of set T. The bigger the metric, the more diverse a set of strings is: $\sigma_T = \frac{\sum_{t \in T} \frac{LD[t, \mathcal{M}_T]}{max(|t|, |\mathcal{M}_T|)}}{|T|}$.

Figure 8 shows that the normalized diversity metric has a value of more than 0.5 throughout time and reaches up to 0.8 (this means that on average 50% to 80% of a target should be changed so that a target is converted to another in the set of targets in each year). Although such values of diversity are in line with Evo-Lexis, it is understandable that the diversity in iGEM is also partially impacted (towards higher values) by the introduction of new sources discussed before. Because of this reason, and the fact that the diversity is measured in a slightly different way in [21], we do not show a direct comparison in Fig. 8.

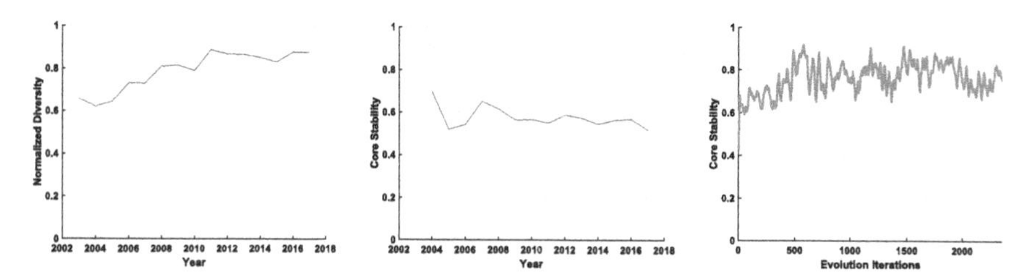

(a) Target diversity in iGEM (b) Core stability in iGEM (c) Core stability in Evo-Lexis

Fig. 8. Target diversity and core stability in iGEM over time.

4.4 Core Stability in iGEM Lexis-DAGs

We have already defined the core size and the H-score. Here we define an additional metric, related to the stability of the core across time.

We track the stability of the core set by comparing two core sets at two different times. A direct comparison of the core sets via the Jaccard index leads to poor results. The reason is that often the strings of the two sets are similar to each other but not completely identical.

Thus, we define a generalized version of Jaccard similarity that we call *Levenshtein-Jaccard Similarity*:

- Suppose we aim to compute the similarity of two sets A and B of strings. We define the mapping $A \to B$ where every element $a \in A$ is mapped to the most similar element $b \in B$. We also define the mapping $B \to A$ from every element $b \in B$ to the most similar element $a \in A$:

$$\begin{cases} A \to B = \{(a,b) \ s.t. \ a \in A \ \& \ b \in B \ \& \ b = arg \ max_{x \in B} Sim(a,x)\} \\ B \to A = \{(b,a) \ s.t. \ a \in A \ \& \ b \in B \ \& \ a = arg \ max_{x \in A} Sim(b,x)\} \end{cases} \quad (3)$$

  where $Sim(a,b)$ is the similarity of a to b and is calculated as: $Sim(a,b) = 1 - \frac{LD(a,b)}{max(|a|,|b|)}$. Notice that $max(|a|,|b|)$ is the maximum value of Levenshtein distance between a and b. This consideration ensures that if $a = b$ then $Sim(a,b) = 1$, and if a and b have the maximum distance then $Sim(a,b) = 0$.
- Considering both $A \to B$ and $B \to A$, we get the union of the two mappings and define the Levenshtein-Jaccard similarity as follows:

$$LevJac(A,B) = \frac{\sum_{(a,b) \in A \to B} Sim(a,b) + \sum_{(b,a) \in B \to A} Sim(b,a)}{(|A| + |B|)} \quad (4)$$

  We can see that if $A = B$ (all weights are equal to one) then $LevJac(A,B) = 1$. Also if none of the elements in A are similar to B (all the element pairs take zero similarity value), then $LevJac(A,B) = 0$.

As the results in Fig. 8c show, the core set in iGEM DAGs have relatively high values of the core stability measure (Eq. (4)), close to the values we observed in Evo-Lexis. This means that the core nodes stay similar across time, and there are no sudden changes in the content of the core set. One reason for this stability is that the set of core nodes includes several sources, and many of core sources get transferred to the next year.

Additionally, every year the focus of the iGEM designers is on specific parts, most of which are of high path centrality. For example, "BBa_B0010 BBa_B0012" (the most widely used "terminator" part) and "BBa_B0034" are almost always the top-2 central nodes (with the exception of year 2011). Also, some sources such as "BBa_R0011", always appear in the top-20 nodes in the core set. Remember that Fig. 4d shows that the reuse distribution of sources is highly skewed. In summary, the stability of the core set in iGEM is caused by the same reason with Evo-Lexis, which is the bias and selectivity towards using a specific set of nodes in consecutive years.

5 Conclusions

iGEM is a dataset that satisfies the basic assumption of Evo-Lexis framework: a sequence of target strings with potential temporal reuse of previously introduced substrings. Because of this compatibility, we chose to use this dataset in a case-study and contrast its qualitative properties with Evo-Lexis. We can summarize the answers to the questions posed in the abstract of this paper as follows:

- We observe that although incremental design can build efficient hierarchies over the iGEM targets, the normalized cost increases over time. This is due to the fact that the amount of reuse from previous years decreases mainly due to the frequent introduction of new sources over time. The small length of the targets in iGEM is also an additional factor for lowering the potential of reuse of the previously constructed parts in iGEM.
- The increasing normalized cost causes the Lexis-DAGs to become less deep and to contain shorter nodes on average as time progresses. This is different than Evo-Lexis. In addition, there is a high fraction of very short targets in each year in comparison to Evo-Lexis.
- The iGEM Lexis-DAGs present a bias in reusing specific nodes more often than the other nodes. This biased reuse results in the Lexis-DAGs to take the shape of an hourglass with relatively high H-score values and a stable set of core nodes over time. This observation is consistent with Evo-Lexis.
- The core sets over the years remain stable and similar to previous years in iGEM data despite the fact that the set of sources changes significantly and the target sets are diverse each year. Most of the stability is contributed by a small set of central sources and central intermediate nodes that are heavily reused in iGEM registry over time.

References

1. igem.org/Main_Page
2. Akhshabi, S., Dovrolis, C.: The evolution of layered protocol stacks leads to an hourglass-shaped architecture. In: SIGCOMM 2011, pp. 206–217. ACM (2011)
3. Blakes, J., et al.: Heuristic for maximizing DNA re-use in synthetic DNA library assembly. ACS Synth. Biol. **3**(8), 529–542 (2014)
4. Callebaut, W., Rasskin-Gutman, D.: Modularity: Understanding the Development and Evolution of Natural Complex Systems. Vienna Series in Theoretical Biology. MIT Press, London (2005)
5. Casci, T.: Hourglass theory gets molecular approval. Nat. Rev. Genet. **12**, 76 EP (2010)
6. Charikar, M., et al.: The smallest grammar problem. IEEE Trans. Inf. Theory **51**(7), 2554–2576 (2005)
7. Clune, J., Mouret, J.B., Lipson, H.: The evolutionary origins of modularity. Proc. Roy. Soc. London B: Biol. Sci. **280**(1755), 2012–2863 (2013)
8. Csete, M., Doyle, J.C.: Bow ties, metabolism and disease. Trends Biotechnol. **22**(9), 446–50 (2004)
9. Fortuna, M.A., Bonachela, J.A., Levin, S.A.: Evolution of a modular software network. PNAS **108**(50), 19985–19989 (2011)
10. Friedlander, T., Mayo, A.E., Tlusty, T., Alon, U.: Evolution of bow-tie architectures in biology. PLOS Comput. Biol. **11**(3), 1–19 (2015)
11. Hinton, G.E., Salakhutdinov, R.R.: Reducing the dimensionality of data with neural networks. Science **313**(5786), 504–507 (2006)

12. Ishakian, V., Erdös, D., Terzi, E., Bestavros, A.: A framework for the evaluation and management of network centrality. In: SDM, pp. 427–438. SIAM (2012)
13. Kashtan, N., Noor, E., Alon, U.: Varying environments can speed up evolution. PNAS **104**(34), 13711–13716 (2007)
14. Kashtan, N., Alon, U.: Spontaneous evolution of modularity and network motifs. PNAS **102**(39), 13773–13778 (2005)
15. Mengistu, H., Huizinga, J., Mouret, J.B., Clune, J.: The evolutionary origins of hierarchy. PLOS Comput. Biol. **12**(6), 1–23 (2016)
16. Meunier, D., Lambiotte, R., Bullmore, E.: Modular and hierarchically modular organization of brain networks. Frontiers in Neuroscience **4**, 200 (2010)
17. Miller, W.: The hierarchical structure of ecosystems: connections to evolution. Evol. Educ. Outreach **1**(1), 16–24 (2008)
18. Myers, C.R.: Software systems as complex networks: structure, function, and evolvability of software collaboration graphs. Phys. Rev. E **68**, 046116 (2003)
19. Sabrin, K.M., Dovrolis, C.: The hourglass effect in hierarchical dependency networks. Network Sci. **5**(4), 490–528 (2017)
20. Siyari, P., Dilkina, B., Dovrolis, C.: Lexis: An optimization framework for discovering the hierarchical structure of sequential data. In: SIGKDD 2016, pp. 1185–1194. ACM (2016)
21. Siyari, P., Dilkina, B., Dovrolis, C.: Emergence and evolution of hierarchical structure in complex systems. In: To Appear in Dynamics On and Of Complex Networks III: Machine Learning and Statistical Physics Approaches (2019)
22. Supper, J., Spangenberg, L., Planatscher, H., Dräger, A., Schröder, A., Zell, A.: Bowtiebuilder: modeling signal transduction pathways. BMC Syst. Biol. **3**(1), 67 (2009)
23. Tanaka, R., Csete, M., Doyle, J.: Highly optimised global organisation of metabolic networks. IEE Proc. Syst. Biol. **2**(4), 179–184 (2005)
24. Wagner, G.P., Pavlicev, M., Cheverud, J.M.: The road to modularity. Nat. Rev. Genet. **8**, 921 EP (2007). Review Article

Computational Design of Superhelices by Local Change of the Intrinsic Curvature

Pedro E. S. Silva[1], Maria Helena Godinho[1], and Fernão Vístulo de Abreu[2(✉)]

[1] CENIMAT/I3N, Departamento de Ciência dos Materiais, Faculdade de Ciências e Tecnologia, Universidade Nova de Lisboa, 2829-516 Caparica, Portugal
{pess,mhg}@fct.unl.pt

[2] Departmento de Física/I3N, Departamento de Física, Universidade de Aveiro, 3810-193 Aveiro, Portugal
fva@ua.pt

Abstract. Helices appear in nature at many scales, ranging from molecules to tendrils in plants. Organisms take advantage of the helical shape to fold, propel and assemble. For this reason, several applications in micro and nanorobotics, drug delivery and soft-electronics have been suggested. On the other hand, biomolecules can form complex tertiary structures made with helices to accomplish many different functions. A particular well-known case takes place during cell division when DNA, a double helix, is packaged into a super-helix—i.e., a helix made of helices—to prevent DNA entanglement. DNA super-helix formation requires auxiliary histone molecules, around which DNA is wrapped, in a "beads on a string" structure. The idea of creating superstructures from simple elastic filaments served as the inspiration to this work. Here we report a method to produce filaments with complex shapes by periodically creating strains along the ribbons. Filaments can gain helical shapes, and their helicity is ruled by the asymmetric contraction along the main axis. If the direction of the intrinsic curvature is locally changed, then a tertiary structure can result, similar to the DNA wrapped structure. In this process, auxiliary structures are not required and therefore new methodologies to shape filaments, of interest to nanotechnology and biomolecular science, are proposed.

Keywords: Tendril perversions · DNA folding · Design · Synthesis and processing

1 Introduction

In nature, a variety of ingenious mechanisms have been developed to fold, propel and assemble. Many of them use the ability to shape their structure in helical configurations [17]. For instance, *Erodium*, a flowering plant, uses a particular mechanism for disseminating its seeds. First, after flowering, the plant stores

J. M. F. Rodrigues et al. (Eds.): ICCS 2019, LNCS 11536, pp. 483–491, 2019.
https://doi.org/10.1007/978-3-030-22734-0_35

enough elastic energy in the fruits by creating a tension between awns and a surrounding tissue. After a given threshold, the tissue snaps and seeds are flung away from the plant. After reaching the ground, awns start a cycle of winding or unwinding depending on humidity, coiling when drying and straightening when wetting. These movements help seeds to move on the surface until finding a position to open a hole into the soil [6,21].

In biochemistry, the helical shape is very common and can be arranged in very different structures to accomplish different functions. For instance, DNA is made of two intertwined helices, and collagen, of three intertwined helices [14,16]. Furthermore, single, double and triple helices can also gain different shapes, creating super-helices—helices built with helices [13]. Their different shapes are critical to healthy functioning. For instance, chromosome segregation during cellular division requires considerable packaging to avoid DNA entanglement, while during transcription a more stretched fold must give access to RNA polymerase enzymes.

These are certainly only two simple examples that show how controlling the helical shape of filaments can be of great importance at very different length scales and for different purposes. Controlling the shape of artificial elastic filaments should also be highly desirable given the broad potential applications in soft-electronics [5,10], micro and nanorobotics [9,11], healthcare [1,23], and explains surging recent interest and technical developments in this field [4].

Two main mechanisms have been proposed for producing helices. Snir and Kamien suggested that entropic forces could be responsible for the helical folding of molecular chains [20]. The main idea is that the helix shape creates an excluded volume to solute molecules. In crowded environments, this creates an asymmetry which renders the helical configuration more stable. This type of mechanism can play an important role in the cellular packed environments explaining aggregation, orientation and organisation of co-linear similar molecules or structures.

The former mechanism of helical formation requires the existence of a crowded environment. This is not always available, as happens when helices appear in the macroscopic world, as in the curling behaviour observed in tendrils. In this case, helices are formed when changes in the material produce asymmetric internal stresses which create a variety of deformations, as combinations of stretching, bending and torsion [7]. This phenomenon can be easily illustrated when one side of a ribbon is run over with a blade, stretching one side of the ribbon relatively to the other. This modification creates an intrinsic curvature, whose intensity is related to the final number of loops. If instead of ribbons, this type of asymmetric stresses occur in linear filaments, then buckling instabilities can generate a spontaneous instability that turns the filament into a helical shape [8,12].

The way the asymmetric stresses are created in a filament can vary. For instance, a stretched elastic band can be glued upon a relaxed band of the same material [12]. In micro and nanotechnology, the strategy consists in modulating the concentration and the crosslinking density of temperature-responsive polymers [18,19,24]. In nature, stresses originate from asymmetric cellular organisations. These can even evolve in time, depending on factors such as cellular and water concentration or cellular composition [2].

In this work, we were inspired by the complex tertiary structures observed in biomolecules. In particular, we were attracted by the creation of helices made of helices—called superhelices—and by the peculiar shape of chromatin that takes place when DNA is packed during cell division.

We will show how it is possible to use computational modelling and simulation to design structures with shapes similar to those found in DNA. It will also be shown that these computational approaches can serve multiple purposes, since they can provide insights for experimental exploration and discovery, and they can also work as validation for the explanation of experimental results.

This paper is organised as follows. In the next section, we will discuss the main mechanisms that will be used to shape structures. Afterwards, we will describe two computational experiments. In the first case, the idea was to explain how superhelices could be generated. In the second case, we used the same strategy to experimentally shape a polymeric fibre. In practice, many factors can contribute to the final shape of the filament. Through computational/theoretical modelling it is possible to explain how the final shape is actually obtained.

2 Theory and Computational Methodology

Coiling linear filaments or ribbons has been achieved by several groups by creating an asymmetry along the main axis (longitudinal direction) of the filament [18,22]. For instance, setting up a bi-layered strip with different initial strains produces a mismatch between the two layers [12]. Upon release, one layer contracts more than the other, creating an intrinsic curvature and forming an arc. The curvature can increase by increasing the layers mismatch and forming a ring. The higher the curvature the higher the number of loops and smaller the radius. These helical structures are twistless helices and would collapse in a ring upon release and only if they are held a distance apart, they can have a pitch.

Helical curves can be described by using the Frenet-Serret (FS) frame, $Q = [\mathbf{T}, \mathbf{N}, \mathbf{B}]$, where $\mathbf{T}$ and $\mathbf{N}$ are tangent and normal vectors and $\mathbf{B}$ is the binormal vector given by the cross product of $\mathbf{T}$ and $\mathbf{N}$. The evolution of the FS frame can be written in terms of the Darboux vector, $\mathbf{\Omega}$, by the set of continuous differential equations:

$$Q_i' = \mathbf{\Omega} \times Q_i, \tag{1}$$

where $\mathbf{\Omega} = \kappa\mathbf{B} \pm \tau\mathbf{T}$, κ is the curvature and τ is the torsion. Here the curvature to torsion ratio is constant, a necessary and sufficient condition to define a helix, according to Lancrets theorem. Left- (L) and right-handed (R) helices differ by having opposite signs in the torsion factor.

The evolution of the FS frame is only discontinuous at inversion points (perversions). From a previous analysis [18], the equation for the evolution of the FS frame can be modified to hold the changes introduced by different types of perversions. With the application of three transformations at the perversion point,

Eq. 1 can be rewritten as:

$$Q_i' = (S\mathbf{\Omega}) \times (R_{\mathbf{\Omega},\alpha} R_{\mathbf{N},-2\theta} Q_i). \tag{2}$$

where S changes the handedness of the helix, $S\mathbf{\Omega} = S(\kappa\mathbf{B} + \tau\mathbf{T}) = \kappa\mathbf{B} - \tau\mathbf{T}$, $R_{\mathbf{N},-2\theta}$ expresses a rotation of the FS frame around the normal, θ is the angle between the tangent and the twisting vector, and $R_{\mathbf{\Omega},\alpha}$ adds further twisting to the perversion. $\alpha = 0$ corresponds to the description of symmetric perversions as commonly found in plant tendrils and gift ribbons. Both helices have the same centre line. By contrast, when $\alpha = \pi$ the antisymmetric perversion causes centre lines to be apart by twice of the helical radius.

In this work, we will use only $\alpha = \pi$ perversions to shape fibres. This is because these perversions can occur at well defined (engineered) positions, whereas symmetric perversions occur spontaneously at positions that can depend on the boundary conditions, but also on the unwinding process, being more difficult to control.

The way we will produce structures with different shapes (our tertiary structures) consists on connecting helices with opposite handedness through antisymmetric perversions. Therefore, left (L) and right (R) helical segments—our building blocks—are juxtaposed. It will be assumed that all L (or R) segments have the same length, but the length of L segments is different from the length of R segments. Different shapes can be obtained because antisymmetric perversions introduce a turning angle, which relates to the helix length helix by, $L = \Theta\sqrt{(a^2 + b^2)}$. Here Θ is the total turning angle, a the helical radius and b the height of the helix. In this work, for simplicity, a and b for all helical segments and, therefore, all segments have the same κ and τ. Different shapes can be obtained by changing the length of the two types of helical segments, and in this way by changing Θ.

In the next section we will show results obtained by running simulations using the molecular dynamics simulator LAMMPS (Large-scale Atomic/Molecular Massively Parallel Simulator) [3, 15]. Filaments are modelled by a set of beads arranged in a simple cubic lattice and connected to first and second neighbours by harmonic potentials, $V_{x_1,x_2} = k_h/2(l - l_{0,n})$, where k_h is the elastic constant, l is the distance between beads x_1 and x_2 and $l_{0,n}$ is the equilibrium bond distances ($l_{0,1} = \sigma$, for firsts neighbors and $l_{0,2} = \sqrt{2}\,\sigma$, for second neighbors). An intrinsic curvature is created by changing the equilibrium bond distances in one side of the rod. Then the pre-strain of the rod is becomes $\chi = (l_{0,1}' - l_{0,1})/l_{0,1}'$, where $l_{0,1}'$ is the modified equilibrium bond distance. Our simulations used a deterministic integration of the equations of motion using a NVE integrator (Verlet/Leap-frog method) to update beads positions and velocities on each time step (step size of $1 \times 10^{-3}\,\tau$).

3 Computer Simulations and Experimental Validation

To analyse how antisymmetric perversions can be used to modulate the shape of linear structures, two studies were performed. In a first study, the analysis uses a theoretical description of curves with perversions using Eqs. (1) and (2). Then computer simulations with LAAMPS were performed, which allow us to suggest that filaments with these types of shapes should be capable of being produced in practice. This is confirmed in our second study. There the same strategy to shape filaments is used on a polymeric fibre experimentally engineered to acquire a three-dimensional shape similar to those obtained in the previous study. In practice, the fibre obtained has a more complex shape, given the difficulties in controlling all experimental parameters, such as constant fibre thickness or equal building block lengths. Therefore, in a second stage of the study, we use the theoretical model to match the intrinsic curvature observed in the real fibre. This allows us to compare with the equivalent computer simulation.

3.1 Superhelices from Helical Blocks $\Theta_{\mathrm{R}} = 2\pi + \delta$ and $\Theta_{\mathrm{L}} = 2\pi$

The simplest (trivial) case considers R and L helical segments with the same length, i.e., with the same total turning angle, $\Theta_{\mathrm{R}} = \Theta_{\mathrm{L}}$. This case is shown in Fig. 1(a) $n = 0$, for $\Theta_{\mathrm{R}} = \Theta_{\mathrm{L}} = 2\pi$. Then each segment completes a full turn (2π), but as R and L helices - represented in red and black in Fig. 1(a) - have opposite handedness, every time one helix turns for one side, the following turns in the opposite direction. As a result, the full segment appears as it was made of two tied helices, united at the perversion points. The top view has the appearance ∞.

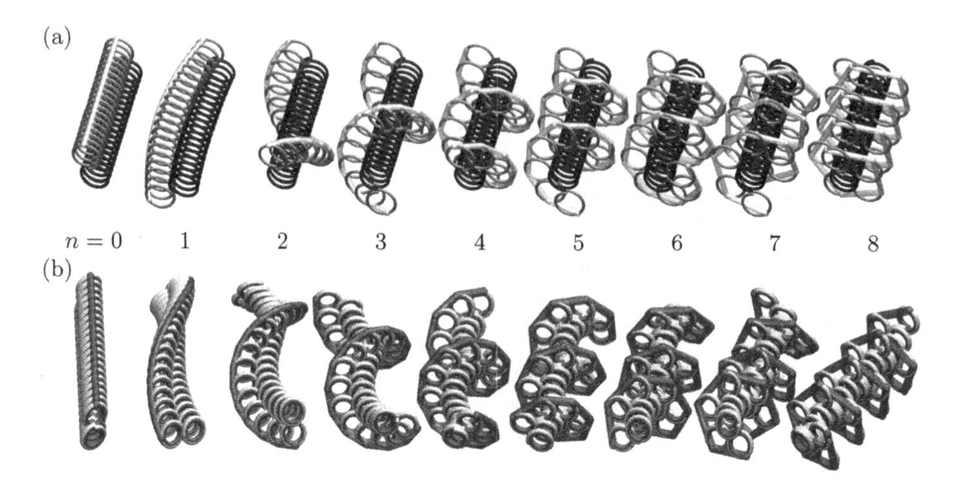

Fig. 1. Filaments obtained from the juxtaposition of 20 R and 20 L helix segments (a) Shapes predicted with the theoretical model describing the evolution of the centre line according to Eqs.(1) and (2) for helical segments with $\Theta_{\mathrm{R}} = 2\pi + \delta$ (represented in black) and $\Theta_{\mathrm{L}} = 2\pi$ (in red), and $\delta = 2\pi n/40$. (b) Computer simulations of filaments where the compression stresses (represented in red) act alternately on either side of the filament creating helices with antisymmetric perversions of $L_{\mathrm{R}} = (40+n)\,\sigma$ and with $L_{\mathrm{L}} = 40\,\sigma$. (Color figure online)

Keeping L helix blocks with exactly one loop, $\Theta_{\rm L} = 2\pi$, and slightly increasing the length of R helix blocks by δ, with $\delta = 2\pi n/40$ and $n = 1$ to 8, it can be observed that the red (L) helix blocks start to revolve around of the black (R) helix blocks. Also, in these particular cases, where one block (red) has an integer number of loops, all other blocks (black) are aligned. The alignment occurs because only helix segment always completes a 2π turn, while the other revolves slightly more. Increasing n, increases the number of loops of the superhelix until $\delta = \pi$. Then, the black helix block have two side-by-side red helix blocks (top view:). For $\delta > \pi$ the number of loops of the superhelix decreases until reaching an integer number of loops.

All these constructions were simulated for realistic elastic filaments by adjusting the curvature of one block with $L = 40\sigma$ to match one complete turn. Then, twenty R helix blocks with $L_{\rm R} = (40+n)\ \sigma$ alternated with twenty L helix blocks with $L_{\rm L} = 40\ \sigma$. When $n = 0$, the configuration matches the pattern predicted in Fig. 1(a) ($n = 0$).

For increasing n, rods develop a superhelix structure with an increasing number of turns. However, R helix blocks, which were aligned in Fig. 1(a), are now misaligned. This can be due to the fact that Eq. (2) describes the effect of the perversion in a simplified way, reducing its extent to a single point. In practice, perversions have an extension in which they deform the filament in a non-trivial way.

In any case, there is a good agreement in both approaches and, most importantly, in both approaches show that filaments with complex tertiary structures can be constructed using this simple strategy.

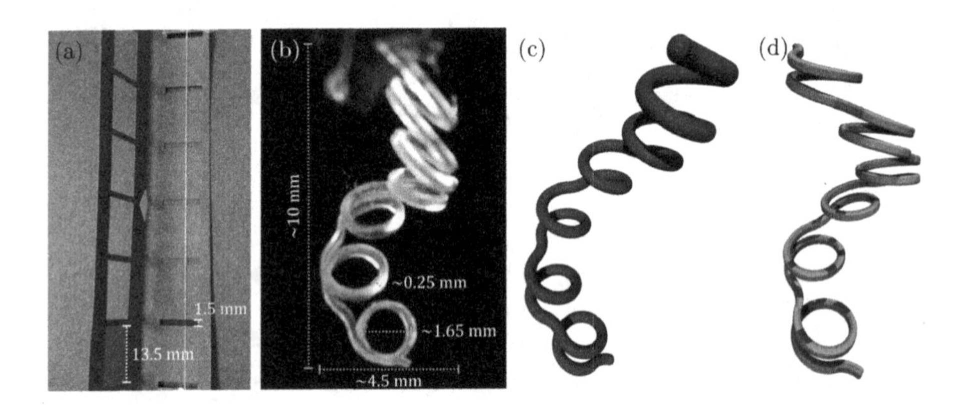

Fig. 2. Filaments with seven L and six R helix blocks with 1:9 ratio. (a) Pre-strained polymeric fibre in a black cardboard mask before irradiation with UV light, in which different sides alternately cover the fibre with different region lengths. (b) Upon release, the polymeric fibre displays a superhelix shape. (c) Matching theoretical filament by adjusting with the experimental result. (d) Rod obtained by computer simulations of the theoretical model quantities.

3.2 Designing Polymeric Superhelices

The former analysis instigated us to produce experimentally a superhelix with a real polymeric fibre. A 4 cm long fibre was pre-strained to 9.15 cm and, then, irradiated for 5 h, on both sides with ultraviolet (UV) and using the mask shown in Fig. 2(a). The fibre obtained upon release is shown in Fig. 2(b). The fibre displayed loops in a clear helical disposition and with different handedness.

Afterwards, we used the theoretical model to match the polymeric fibre, Fig. 2(c). Gravity forces altered the total height of the polymeric fibre. The torsion of the theoretical helix was adjusted to match with the experimental fibre, despite the later having no intrinsic torsion.

Then, using the adjusted quantities of the theoretical model, a rod with seven L helix blocks of length $L_{\mathrm{L}} = 15\sigma$ alternated with six R helix blocks of length $L_{\mathrm{L}} = 135\sigma$, Fig. 2(d) and pre-strain $\chi = 0.23$. One end of the rod was kept fix and an additional gravity-like force was used in all atoms, in such a way that all are under the influence of the same force. Overall, by visual comparison of Fig. 2(b) and (d), rod and polymeric fibre have a similar design.

4 Conclusions

A new method for designing filaments with complex tertiary structures resembling those of DNA, was presented. Interestingly, the creation of these structures does not require auxiliary structures to build up, as happens with histones in DNA. In this work, we also showed how a combination of analysis, in which computational and theoretical descriptions take an important part, help to build a thesis on how to design strategy to shape filament in effective ways. In particular, computational approaches offer a powerful tool to rapidly preview a given design.

Acknowledgement. This work was funded by FEDER funds through the COMPETE 2020 Program and National Funds through FCT—Portuguese Foundation for Science and Technology under projects numbers POCI-01-0145-FEDER-007688 (Reference UID/CTM/50025) and M-ERA-NET2/0007/2016 (CellColor).

References

1. An, B.W., et al.: Smart sensor systems for wearable electronic devices. Polymers **9**(12), 303 (2017). https://doi.org/10.3390/polym9080303
2. Armon, S., Efrati, E., Kupferman, R., Sharon, E.: Geometry and mechanics in the opening of chiral seed pods. Science **333**(6050), 1726–1730 (2011). https://doi.org/10.1126/science.1203874
3. Brown, W.M., Wang, P., Plimpton, S.J., Tharrington, A.N.: Implementing molecular dynamics on hybrid high performance computers – short range forces. Comput. Phys. Commun. **182**(4), 898–911 (2011). https://doi.org/10.1016/j.cpc.2010.12.021

4. Chen, Z., Huang, G., Trase, I., Han, X., Mei, Y.: Mechanical self-assembly of a strain-engineered flexible layer: wrinkling, rolling, and twisting. Phys. Rev. Appl. **5**(1) (2016). https://doi.org/10.1103/PhysRevApplied.5.017001
5. Cheng, Y., Wang, R., Chan, K.H., Lu, X., Sun, J., Ho, G.W.: A biomimetic conductive tendril for ultrastretchable and integratable electronics, muscles, and sensors. ACS Nano **12**(4), 3898–3907 (2018). https://doi.org/10.1021/acsnano.8b01372
6. Evangelista, D., Hotton, S., Dumais, J.: The mechanics of explosive dispersal and self-burial in the seeds of the filaree, Erodium cicutarium (Geraniaceae). J. Exp. Biol. **214**(4), 521–529 (2011). https://doi.org/10.1242/jeb.050567
7. Gerbode, S.J., Puzey, J.R., McCormick, A.G., Mahadevan, L.: How the cucumber tendril coils and overwinds. Science **337**(6098), 1087–1091 (2012). https://doi.org/10.1126/science.1223304
8. Goriely, A., Tabor, M.: Spontaneous helix hand reversal and tendril perversion in climbing plants. Phys. Rev. Lett. **80**(7), 1564–1567 (1998). https://doi.org/10.1103/PhysRevLett.80.1564, goriely_spontaneous_1998
9. Hu, C., Pané, S., Nelson, B.J.: Soft micro- and nanorobotics. Annu. Rev. Control Robot. Auton. Syst. **1**(1), 53–75 (2018). https://doi.org/10.1146/annurev-control-060117-104947
10. Jang, K.I., et al.: Self-assembled three dimensional network designs for soft electronics. Nat. Commun. **8**, 15894 (2017). https://doi.org/10.1038/ncomms15894
11. Lahikainen, M., Zeng, H., Priimagi, A.: Reconfigurable photoactuator through synergistic use of photochemical and photothermal effects. Nat. Commun. **9**(1) (2018). https://doi.org/10.1038/s41467-018-06647-7
12. Liu, J., Huang, J., Su, T., Bertoldi, K., Clarke, D.R.: Structural transition from helices to hemihelices. PLoS ONE **9**(4), e93183 (2014). https://doi.org/10.1371/journal.pone.0093183
13. Main, E.R., Jackson, S.E., Regan, L.: The folding and design of repeat proteins: reaching a consensus. Curr. Opin. Struct. Biol. **13**(4), 482–489 (2003). https://doi.org/10.1016/S0959-440X(03)00105-2
14. Motooka, D., et al.: The triple helical structure and stability of collagen model peptide with 4(s)-hydroxyprolyl-pro-gly units. Pept. Sci. **98**(2), 111–121 (2012). https://doi.org/10.1002/bip.21730
15. Plimpton, S.: Fast parallel algorithms for short-range molecular dynamics. J. Comput. Phys. **117**(1), 1–19 (1995). https://doi.org/10.1006/jcph.1995.1039
16. Pusarla, R.H., Bhargava, P.: Histones in functional diversification. FEBS J. **272**(20), 5149–5168 (2005). https://doi.org/10.1111/j.1742-4658.2005.04930.x
17. Silva, P.E.S., Vistulo de Abreu, F., Godinho, M.H.: Shaping helical electrospun filaments: a review. Soft Matter (2017). https://doi.org/10.1039/C7SM01280B
18. Silva, P.E.S., et al.: Perversions with a twist. Sci. Rep. **6**, 23413 (2016). https://doi.org/10.1038/srep23413
19. Silva, P.E.S., Godinho, M.H.: Helical microfilaments with alternating imprinted intrinsic curvatures. Macromol. Rapid Commun. **38**(5), 1600700 (2017). https://doi.org/10.1002/marc.201600700
20. Snir, Y., Kamien, R.D.: Entropically driven helix formation. Science **307**(5712), 1067–1067 (2005). https://doi.org/10.1126/science.1106243
21. Stamp, N.E.: Self-burial behaviour of erodium cicutarium seeds. J. Ecol. **72**(2), 611–620 (1984). https://doi.org/10.2307/2260070
22. Trindade, A.C., Canejo, J.A.P., Teixeira, P.I.C., Patricio, P., Godinho, M.H.: First curl, then wrinkle. Macromol. Rapid Commun. **34**(20), 1618–1622 (2013). https://doi.org/10.1002/marc.201300436

23. Wolinsky, J.B., Colson, Y.L., Grinstaff, M.W.: Local drug delivery strategies for cancer treatment: gels, nanoparticles, polymeric films, rods, and wafers. J. Controlled Release **159**(1), 14–26 (2012). https://doi.org/10.1016/j.jconrel.2011.11.031
24. Wu, Z.L., et al.: Three-dimensional shape transformations of hydrogel sheets induced by small-scale modulation of internal stresses. Nat. Commun. **4**, 1586 (2013). https://doi.org/10.1038/ncomms2549

Spatial Modeling of Influenza Outbreaks in Saint Petersburg Using Synthetic Populations

Vasiliy Leonenko[1(✉)], Alexander Lobachev[1], and Georgiy Bobashev[2]

[1] ITMO University, 49 Kronverksky Pr., Saint Petersburg 197101, Russia
vnleonenko@yandex.ru

[2] RTI International, 3040 Cornwallis Rd., P.O. Box 12194, Research Triangle Park, NC 27709, USA

Abstract. In this paper, we model influenza propagation in the Russian setting using a spatially explicit model and a detailed human agent database as its input. The aim of the research is to assess the applicability of this modeling method using influenza incidence data for 2010–2011 epidemic outbreak in Saint Petersburg and to compare the simulation results with the output of the compartmental SEIR model for the same outbreak. For this purpose, a synthetic population of Saint Petersburg was built and used for the simulation via FRED open source modeling framework. The parameters related to the outbreak (background immunity level and effective contact rate) are assessed by calibrating the compartmental model to incidence data. We show that the current version of synthetic population allows the agent-based model to reproduce real disease incidence.

Keywords: Epidemiology · Synthetic populations · Seasonal influenza · Agent–based modeling · FRED

1 Introduction

Today 55% of world's population lives in cities, and this number, according to UN predictions, is expected to reach 68% by 2050 [31]. Thus, the importance of cities for human societies is increasing over time. Due to their intricate structure, modern cities constitute a perfect example of complex systems, and our ability to understand them scientifically is limited [5]. So is the situation with the social and economic processes within them.

One of the processes intrinsically connected with urban structure is influenza epidemics. Seasonal influenza causes repetitive epidemic outbreaks resulting in

This work is financially supported by Ministry of Education and Science of the Russian Federation, Agreement #14.575.21.0161 (26/09/2017). Unique Identification RFMEFI57517X0161.

J. M. F. Rodrigues et al. (Eds.): ICCS 2019, LNCS 11536, pp. 492–505, 2019.
https://doi.org/10.1007/978-3-030-22734-0_36

high worker/school absenteeism, productivity losses and death cases due to disease complications. According to WHO [33], the corresponding annual number of mortality cases reaches whopping 500 thousand. To anticipate the incoming outbreaks and prepare healthcare infrastructure to fight with their detrimental ramifications, statistical and mathematical models are widely used. The predictive force of these models is limited by the fact that the influenza outbreak dynamics in urban settings is driven by a lot of factors, some of which are hard to be quantified. Along with the weather factors [20,27,29], the important role is played by the behavior of the human population, such as daily migration patterns resulting in different effective contact probabilities [6,30]. In fact, the bigger a city is, the milder is the response of the disease incidence to climate forcing and the more important the human-related factors become [8]. These factors also deserve attention due to occurrence of background immunity against influenza, which is a result of repetitive flu outbreaks caused by similar flu strains [17]. It might be assumed that peculiarities of commuting patterns of citizens and geographical distribution of their dwellings subsequently cause different distribution of effective contacts between the susceptibles and the infectives, leading to changes in immunity levels of the individuals in different cities. For instance, a highly connected city, where mass action law assumption [34] generally holds, might have an epidemic dynamics and consequently a distribution of the immune quite different from the city with apparent geographical clustering. The accumulation of these differences due to faster circulation of flu virus around the globe might be the reason of the failure of the approach which was earlier used to predict flu epidemics in Soviet Union [15]. The mentioned approach was based on the assumption that a forthcoming influenza outbreak dynamics could be predicted using the data from the cities already affected by the epidemic during the season under consideration, which worked in 1970s, but is not true anymore [22]. Modeling flu propagation using detailed population structures which (somewhat) accurately reflect the peculiarities of urban contact patterns might allow us to quantify the role of contact patterns on the formation of background immunity and to assess how the differences in city structures lead to different flu epidemic dynamics. This paper is considered to be the first step in the stated direction.

The aim of this work is to create a detailed description of urban population in the Russian setting and couple it with agent-based modeling framework to perform a simulation of flu dynamics. Using our previous results obtained in the field of flu outbreak modeling in Russia [19,21], we want to compare the output of a spatially explicit model with the one of standard SEIR compartmental model of Kermack–McKendrick type and to demonstrate the ability of the former to produce more plausible results than the latter. For this purpose, we regard influenza outbreak in Saint Petersburg in 2010–2011 as a case study. The city was chosen due to large populace (it's the second largest city in Russia), economic and cultural importance, and abundance of detailed data on influenza incidence (the records are available from 1935).

2 Synthetic Population

"Synthetic population" is a synthesized, spatially explicit human agent database (essentially, a simulated census) representing the population of a city, region or country. By its cumulative characteristics, this database is equivalent to the real population but its records does not correspond to real people – this fact helps avoid privacy issues. In our research, we employed the approach for synthetic population generation developed by RTI International [32], which was used by various research groups to create populations for 50 US states, along with another regions and countries. Statistical and mechanistic models built on top of the synthetic populations helped tackle a variety of research problems, including those connected with public health. Statistical analysis of opioid–related overdoses in Cincinnatti [4] can be named as an example.

According to the standard of RTI International, a synthetic population consists of several txt-files, each of them containing a table with every row being a single record corresponding to some entity – an individual, a household, a workplace, a school, etc. The full list of files with their short description is presented in Table 1. Sticking to the same standard, we generated the files corresponding to the population of Saint Petersburg. Since the data available for Saint Petersburg was not complete, we altered or omitted some of the methods, resulting in a simplified population, which, however, seems to satisfy our demands related to influenza modeling. The details of input data we used and the algorithms we employed to generate the population follow.

2.1 Household Data

The principal data source for our synthetic population is 2010 data from "Edinaya sistema ucheta naseleniya Sankt Peterburga" ("Unified population accounting system of Saint Petersburg") [11]. The data is represented in a form of Excel spreadsheets containing records with house addresses and the corresponding number of dwellers of certain age and gender (see Table 2).

To match the household addresses with the geographical coordinates and assess the plausibility of the obtained geographical data, a computational algorithm was developed and implemented using Python programming language. The details of the algorithm implementation follow.

Adding Object Coordinates

- For each record:
  - Form the address string using the information from the address fields of the record in the format "city" + "street" + "house".
  - Feed the address string to Yandex.Geocoder online service [36] which returns the latitude and the longitude of the object by this address.
  - Add coordinates to a record.

Table 1. File structure of a synthetic population for Saint Petersburg.

File	Contents
Files used in the current population version	
`households.txt`	Contains the location and descriptive attributes for each household. Household records in the `households.txt` file link to individual person records in the `people.txt` table
`people.txt`	Contains a record for each person, along with his or her age and sex. These synthetic person records link to the `households.txt` file (via the `sp_hh_id` field)
`schools.txt`	Contains a record for each school, along with its zip code, maximum capacity and coordinates
`workplaces.txt`	Contains a record for each workplace, along with its coordinates and size
Empty or omitted files	
`hospitals.txt`	Contains a record for each hospital, along with its coordinates, number of physicians and beds. Contains zero records in this version of the synthetic population
`gq.txt`	Contains a record for each general quarters, along with their type (prisons, student dorms, etc), coordinates and capacity Contains zero records in this version of the synthetic population
`gq_people.txt`	Contains a record for each person, which lives in general quarters, along with his or her age and sex. These synthetic person records link to the `gq.txt` file (via the `sp_gq_id` field). Contains zero records in this version
`pums_p.txt`	Contains personal records from the public use microdata series. Links to the `people.txt` file the `serialno` field. Absent in this version of the synthetic population
`pums_h.txt`	Contains household records from the public use microdata series. Links to the `households.txt` file the `serialno` field. Absent in this version of the synthetic population

Removing Implausible Data. Since the record addresses were apparently derived from handwritten data or manually typed, in some cases they are incomplete or contain typos. The geocoder we used always returns two coordinates as an output, no matter whether he processed the input successfully or not. If an address is not interpreted correctly, Yandex.Geocoder makes guesses on what the correct address should be, which often results in semi-random coordinates. We use an empirical algorithm to filter out those obviously senseless results. For this procedure, we rely on a number of empirical assumptions related to matches between the location (coordinate) and the text address. E.g., if there are multiple addresses to which only one pair of coordinates is assigned, we remove all such records from the database except the first one, summing the corresponding numbers of dwellers.

Table 2. Data source format.

Column names	Contents
`city, street, house`	Dwelling address
`g0, g1 ... g100`	Number of female dwellers of ages $0, 1, \ldots, 100$
`m0, m1 ... m100`	Number of male dwellers of ages $0, 1, \ldots, 100$

2.2 School Data

The list of schools and their addresses was formed manually using the data from the official web–site of the Government of Saint Petersburg [12]. The coordinates of schools were found using Yandex.Geocoder in the same fashion, as it was done for dwellings.

2.3 Workplace Data

The distribution of working places for adults and their coordinates were derived from the data obtained with the help of Yandex.Auditorii API [35]. Initially the data was available in a form of a .geojson file which consisted of relative workplace size assessments for each of the cells in a hexagonal grid (see Fig. 1).

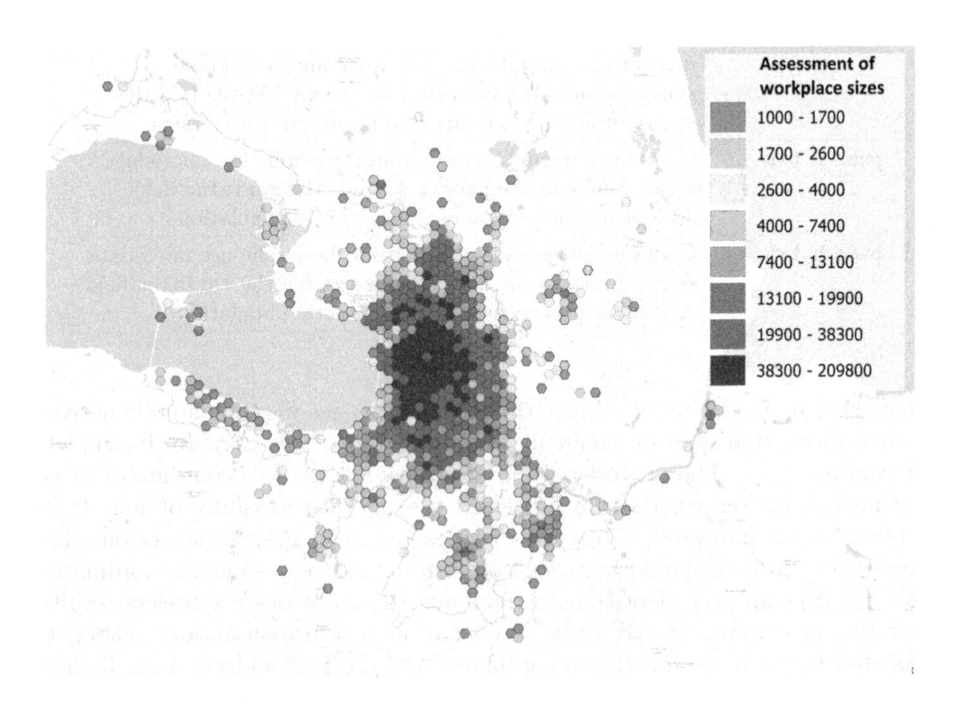

Fig. 1. The distribution of working places in St Petersburg. The numbers are given before the normalization.

This data was normalized using the official cumulative employment numbers [10]. Synthetic workplace records were created by assigning the calculated number of employees in each hexagonal cell to imaginary geographical location coinciding with the center of this cell.

2.4 Assigning People to Schools and Workplaces

We assumed that young people aged 7 to 17 attend schools, and the adults of working age (18 to 55 for males and 18 to 60 for females) might be working. Iterating through the list of records in `people.txt`, we were assigning each person to a closest school or working place, until they are filled to capacity or there is no more people to be assigned.

3 Agent-Based Modeling

An open-source framework FRED [13,26] was used for the simulations. The framework has discrete time, with the modeling step equal to one day. The epidemic process is initiated by assigning randomly an infectious status to some individuals in the population at the beginning of the simulation. In addition to that, the infection can be seeded according to a user-specified schedule, reflecting the external infection process.

The contacts among the individuals that lead to new infection cases are modeled in the following way.

- Each agent in the population potentially interacts with other agents with whom he shares activity locations. These locations include schools, workplaces, households and home neighborhoods (defined as 1 km square cells around the agent's household).
- During the weekends, schools are considered to be closed and most workers equally do not attend their working places. At the same time, the number of neighborhood contacts increases by 50%.
- The rate of effective contacts in a particular activity location depends on the expected number of contacts per infectious person per day and the infection transmission probability. The expected number of contacts is considered not to be dependent on the place size.

The output of the framework in a form of csv-files contains quantities and spatial distributions of individuals of four groups (susceptible, exposed, infected, recovered) at every time step. We used Python programming language and QGIS open source software to process the results and create maps and incidence graphs. An example of the map of influenza propagation is shown in Fig. 2.

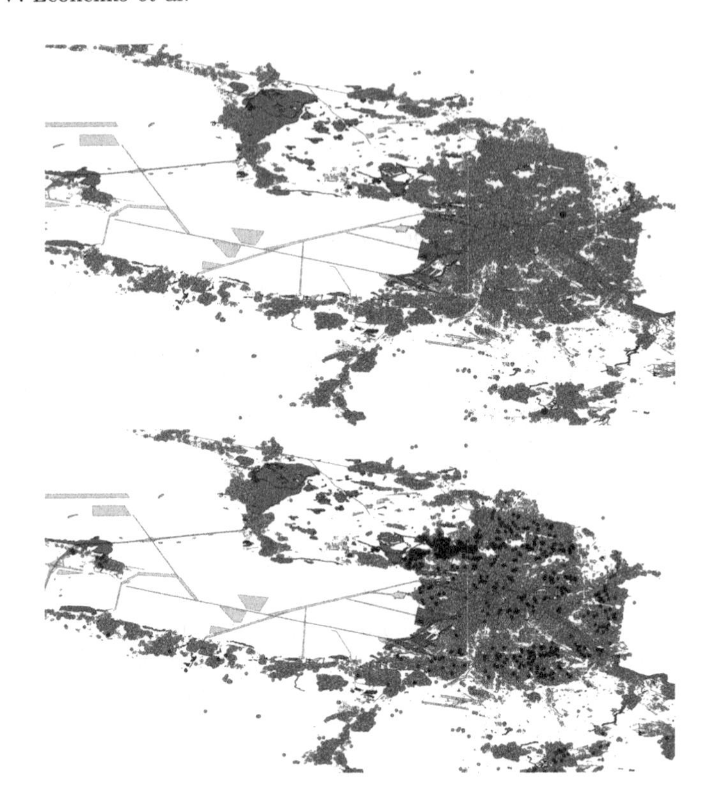

Fig. 2. The distribution of disease–free (green) and infected (red) households in Saint Petersburg: (a) Day 1, (b) Day 8. (Color figure online)

3.1 Influenza Incidence Data

The original dataset provided by the Research Institute of Influenza [1] contains cumulative weekly incidence, i.e., the number of new acute respiratory infection (ARI) cases per day in Russian cities, which includes influenza and other respiratory infections. Before the model fitting, we had to refine the incidence data by restoring the missed values and correcting the under–reporting biases. We also needed to extract flu incidence from the cumulative ARI incidence data. Corresponding algorithms are described in detail in [20], here we introduce briefly the sequence of operations.

- **Under-reporting correction.** Since infected people avoid visiting healthcare facilities during holidays, the corresponding weekly prevalence is lower than the actual number of newly infected. This under-reporting bias can be corrected by means of cubic interpolation [3] using the incidence registered in the adjacent weeks.

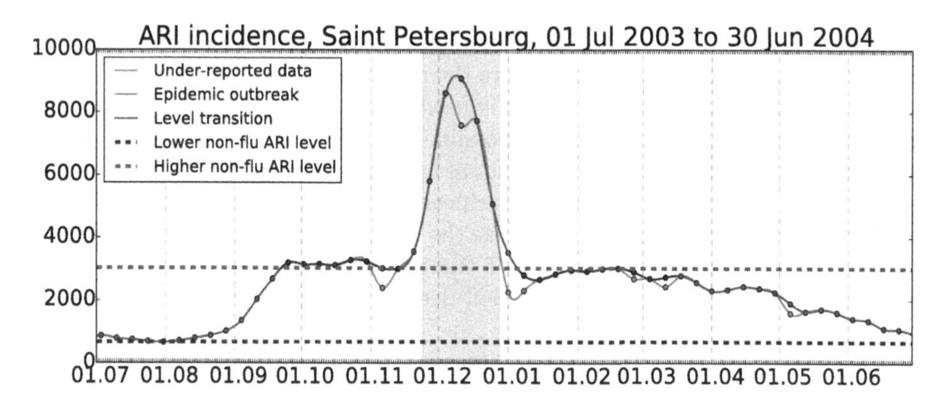

Fig. 3. Influenza outbreak data extraction from the interpolated ARI incidence. (Color figure online)

- **Bringing the incidence data to daily format.** Daily incidence is obtained with the help of cubic interpolation of weekly incidence. We assume that $n_{inf}^{Thu} = n_{inf}^{W}/7$, where n_{inf}^{W} is the weekly incidence taken from the database and n_{inf}^{Thu} is the daily incidence for Thursday of the corresponding week.
- **Extracting incidence data related to influenza outbreak.** At first, the algorithm finds higher non–influenza ARI incidence level, which corresponds to average daily number of newly infected during the months when influenza might occur in temperate regions (Fig. 3, red horizontal dashed line). The part of the graph, which is attributed to a flu outbreak (Fig. 3, red solid line), should have its peak well above the higher ARI level. It should also comply with the time period during which the ARI prevalence exceeds the non–epidemic ARI threshold assessed in the Flu Research Institute (Fig. 3, red rectangle). The beginning and ending of the extracted curve is chosen to match the higher ARI incidence level. The first incidence point of the curve is considered to be the first day of the epidemic outbreak.

3.2 Fitting a SEIR Model to Data

The model we use for fitting is a standard SEIR compartmental model in a form of a system of ordinary differential equations (see [19] for the details). Let $Z^{(dat)}$ be the set of incidence data points loaded from the input file and corresponding to one particular outbreak. Assume that the number of points is t_1, which equals the observed duration of the outbreak. The fitting algorithm selects the values of model parameters corresponding to the model output which minimizes the distance between the modeled and real incidence points:

$$F(Z^{(mod)}, Z^{(dat)}) = \sum_{i=0}^{t_1} (z_i^{(mod)} - z_i^{(dat)})^2,$$

Here $z_i^{(dat)}$ and $z_i^{(mod)}$ are the absolute incidence numbers for the i-th day taken from the input dataset and derived from the model correspondingly. The limited-memory BFGS optimization method is used to find the best fit [24]. Since the existence of several local minima is possible, the algorithm has to be launched several times with different initial values of input variables. The best fit is chosen as a minimum among the distances achieved from all the algorithm runs. To characterize the goodness of fit we utilize the coefficient of determination $R^2 \leq 1$. This coefficient shows the fraction of the response variable variation that is explained by a model [25]. The detailed description of the fitting procedure is available in [19,21].

3.3 Simulation

By fitting the SEIR model to ARI incidence for influenza outbreak in St Petersburg in 2010–2011 (see Fig. 5a), we assess two parameters: the background immunity level $1-\alpha$ and the effective contact intensity λ. The detailed data on the distribution of dwellers we used gave slightly less people in total than it was claimed by official statistics, so we normalized the susceptible ratio α to account for this. The obtained parameter values were used in FRED simulation along with the default values for influenza epidemics provided with the framework (see Table 3). The overall scheme of the described process is presented in Fig. 4.

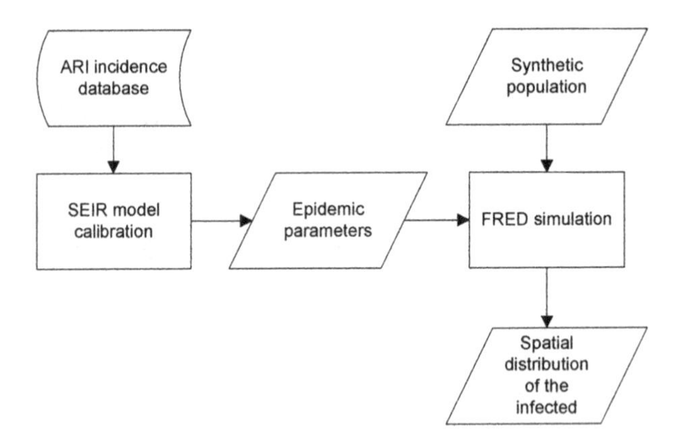

Fig. 4. FRED simulation scheme

The simulation was run 100 times with different seed values for the random number generator. Influenza incidence data obtained as a result of each simulation run was used to find confidence intervals for the daily influenza incidence—see Fig. 5b.

As Fig. 5 shows, despite the imminent uncertainty in synthetic population data, the output of the agent-based model demonstrates a satisfactory agreement with actual ARI incidence. Also, apart from the compartmental SEIR

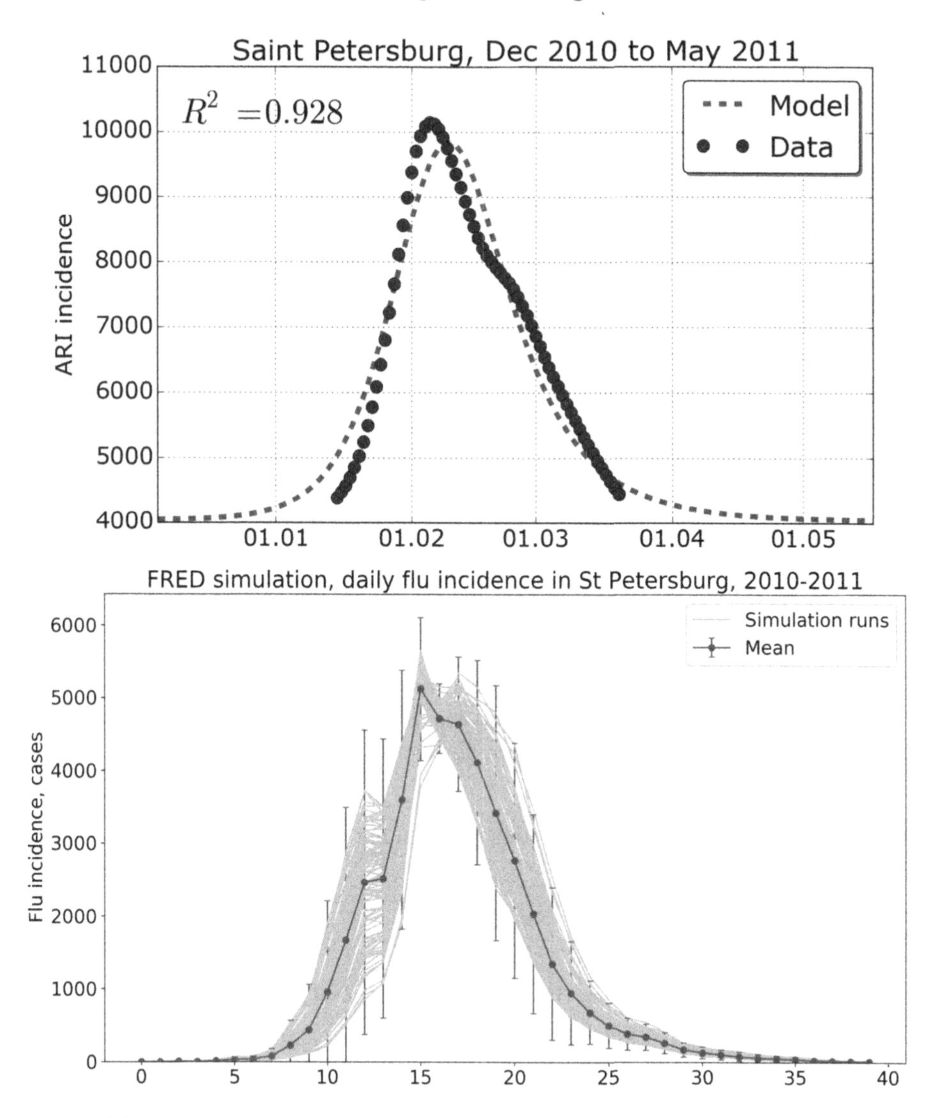

Fig. 5. (a) SEIR model calibration; (b) FRED output generated using the obtained parameter values. Note that the first graph demonstrates the cumulative ARI incidence (influenza and other acute respiratory diseases), whereas the second one shows disease incidence attributed solely to influenza, thus the baselines on the graphs correspond to a seasonal ARI level in the first case (around 4000 newly infected per week), and zero in the second case.

model, spatially explicit simulation demonstrates the right–skewed incidence curve, which better conforms to ARI data. This form of the curve might reflect different number of social connections between the individuals. The so called "superspreaders" are infected first and cause numerous infection cases, whereas

Table 3. Model parameters ('INF' stands for 'influenza', 'Is' is 'infectious symptomatic')

Parameter name	Description	Value	Source
`INF.transmissibility`	A coefficient that modulates the transmissibility of a condition	6.55	Estimated
`INF.S.susceptibility`	A ratio of susceptible individuals in the population	0.46	Estimated
`INF.E.duration_distribution`	Distribution of exposed state duration	lognormal	Default
`INF.E.duration_median`	Median of exposed state duration	1.9	Default
`INF.E.duration_dispersion`	Dispersion of exposed state duration	1.23	Default
`INF.Is.duration_distribution`	Distribution of infected state duration	lognormal	Default
`INF.Is.duration_median`	Median of infected state duration	5.0	Default
`INF.Is.duration_dispersion`	Dispersion of infected state duration	1.5	Default

Table 4. The incidence curve parameters obtained in the simulation compared to those of the real outbreak

Parameter name	Data	Simulation
Outbreak duration, days	65	51
Maximum incidence day (from the outbreak onset)	22	15
Maximum incidence height, cases per day	5756	5126

the less socialized persons got reached by the flu later and contribute somewhat to infection process, making the decline of disease incidence slower. The value $1 - \alpha$ of background immunity level used in both models leads to almost the same peak heights (see Table 4).

4 Discussion and Future Work

As we showed in this research, coupling of synthetic populations with agent–based models is a feasible approach which allows to perform spatially explicit influenza propagation modeling in Russian settings, even when a limited number of data is used to reconstruct the urban population. The apparent drawback of the current synthetic population is the procedure of assigning people to schools and workplaces, described in Sect. 2.4. The idea of picking the closest available spot in a school/workplace was favored for its simplicity, but it is obviously unrealistic. We plan to create several populations corresponding to different assigning algorithms (from picking a workplace/school at random within the whole city to seeking it in a limited radius from the person's home location) and to compare the results of the simulation runs. By doing that, we expect to quantify the variance in outbreak duration, peak day and peak height related to contact networks of different topological structures.

Another drawback of the performed simulation lies in the approximate nature of deriving disease–related parameters for FRED. Particularly, we cannot exactly match the value of `INF.transmissibility` to value of λ from the compartmental model, because they are not equivalent, although correlated. To obtain more realistic disease incidence generated by the model, global optimization techniques will be used by the authors in the same fashion as it was done earlier for the compartmental SEIR models [28]. Since repetitive simulations with different input parameter sets are computationally expensive, we consider applying methods for assessing and reducing the uncertainty in disease–related input parameters [2] which will decrease the state space of the model, and implementing an agent–based model using general–purpose computing on graphics processing units [23] to achieve a speedup.

In the current research we did not consider contacts in public transport. Although the research on the role of New York subway in disseminating influenza showed that its effect is slight [7], we still find it necessary to question this conclusion, because Russian commute patterns and the nomenclature of social groups which use metro may differ from the one of New York.

We assume that, in the long run, the influenza propagation modeling using synthetic populations will allow us to:

- classify the cities into several groups, depending on their geographical structure, contact pattern types and, subsequently, peculiarities of influenza dynamics and use models calibrated to one cities to predict epidemics in the others within the same group;
- reconstruct the pattern of background immunity formation as a function of urban and epidemic factors.

As a byproduct, synthetic populations created for the cities under consideration will be freely available and might facilitate solving urban issues with the help of modeling. The immediate gain we expect from the created synthetic population for Saint Petersburg is that research groups within our department, which work with problems such as transport planning [16,18], ambulance dispatching [9] and crime rate assessment [14], might want to switch from gathering data from scratch for every particular statistical or agent–based model to using a unified population. Since the synthetic population conforms to a certain defined standard, it might be easily reused in different projects and elaborated further by mutual efforts to everyone's benefit.

References

1. Research Institute of Influenza website. http://influenza.spb.ru/en/
2. Artzrouni, M., Leonenko, V.N., Mara, T.A.: A syringe-sharing model for the spread of HIV: application to Omsk, Western Siberia. Math. Med. Biol. J. IMA **34**(1), 15–37 (2015)
3. Baroyan, O., Basilevsky, U., Ermakov, V., Frank, K., Rvachev, L., Shashkov, V.: Computer modelling of influenza epidemics for large-scale systems of cities and territories. In: Proceedings of WHO Symposium on Quantitative Epidemiology, Moscow (1970)
4. Bates, S., Leonenko, V., Rineer, J., Bobashev, G.: Using synthetic populations to understand geospatial patterns in opioid related overdose and predicted opioid misuse. Comput. Math. Organ. Theory **25**(1), 36–47 (2019)
5. Bettencourt, L.M.: The origins of scaling in cities. Science **340**(6139), 1438–1441 (2013)
6. Cauchemez, S., Valleron, A.J., Boelle, P.Y., Flahault, A., Ferguson, N.M.: Estimating the impact of school closure on influenza transmission from sentinel data. Nature **452**(7188), 750 (2008)
7. Cooley, P., et al.: The role of subway travel in an influenza epidemic: a New York city simulation. J. Urban Health **88**(5), 982 (2011)
8. Dalziel, B.D., et al.: Urbanization and humidity shape the intensity of influenza epidemics in US cities. Science **362**(6410), 75–79 (2018)
9. Fu, X., Presbitero, A., Kovalchuk, S.V., Krzhizhanovskaya, V.V.: Coupling game theory and discrete-event simulation for model-based ambulance dispatching. Procedia Comput. Sci. **136**, 398–407 (2018)
10. Government of Saint Petersburg: Edinaya sistema ucheta naseleniya Sankt Peterburga (Unified population accounting system of Saint Petersburg). (in Russian) https://reestr-gis.spb.ru
11. Government of Saint Petersburg: Labor and employment committee. information on economical and social progress. (in Russian) http://rspb.ru/analiticheskaya-informaciya/razvitie-ekonomiki-i-socialnoj-sfery-sankt-peterburga/
12. Government of Saint Petersburg: Official web-site. https://www.gov.spb.ru/
13. Grefenstette, J.J., et al.: FRED (a framework for reconstructing epidemic dynamics): an open-source software system for modeling infectious diseases and control strategies using census-based populations. BMC Public Health **13**(1), 940 (2013)
14. Ingilevich, V., Ivanov, S.: Crime rate prediction in the urban environment using social factors. Procedia Comput. Sci. **136**, 472–478 (2018)
15. Ivannikov, Y., Ogarkov, P.: An experience of mathematical computing forecasting of the influenza epidemics for big territory. J. Infectol. **4**(3), 101–106 (2012). In Russian
16. Khodnenko, I., Kudinov, S., Smirnov, E.: Walking distance estimation using multi-agent simulation of pedestrian flows. Procedia Comput. Sci. **136**, 489–498 (2018)
17. Konshina, O., Sominina, A., Smorodintseva, E., Stolyarov, K., Nikonorov, I.: Population immunity to influenza virus a(h1n1)pdm09, a(h3n2) and b in the adult population of the Russian federation long-term research results. Russ. J. Infect. Immun. **7**(1), 27–33 (2017). https://doi.org/10.15789/2220-7619-2017-1-27-33. in Russian
18. Lantseva, A.A., Ivanov, S.V.: Assessment of pedestrian flow volumes through public transport modelling. Procedia Comput. Sci. **136**, 463–471 (2018)

19. Leonenko, V.N., Ivanov, S.V.: Fitting the SEIR model of seasonal influenza outbreak to the incidence data for Russian cities. Russ. J. Numer. Anal. Math. Modell. **31**(5), 267–279 (2016)
20. Leonenko, V.N., Ivanov, S.V., Novoselova, Y.K.: A computational approach to investigate patterns of acute respiratory illness dynamics in the regions with distinct seasonal climate transitions. Procedia Comput. Sci. **80**, 2402–2412 (2016)
21. Leonenko, V.N., Ivanov, S.V.: Influenza peaks prediction in Russian cities: comparing the accuracy of two SEIR models. Math. Biosci. and Eng. **15**(1), 209–232 (2018). https://doi.org/10.3934/mbe.2018009
22. Leonenko, V.N., Novoselova, Y.K., Ong, K.M.: Influenza outbreaks forecasting in Russian cities: Is Baroyan-Rvachev approach still applicable? Procedia Comput. Sci. **101**, 282–291 (2016)
23. Leonenko, V.N., Pertsev, N.V., Artzrouni, M.: Using high performance algorithms for the hybrid simulation of disease dynamics on CPU and GPU. Procedia Comput. Sci. **51**, 150–159 (2015)
24. Liu, D.C., Nocedal, J.: On the limited memory BFGS method for large scale optimization. Math. Program. **45**(1–3), 503–528 (1989)
25. van Noort, S.P., Águas, R., Ballesteros, S., Gomes, M.G.M.: The role of weather on the relation between influenza and influenza-like illness. J. Theor. Biol. **298**, 131–137 (2012)
26. Public Health Dynamics Lab: FRED wiki. https://github.com/PublicHealthDynamicsLab/FRED/wiki/
27. Seleznev, N.E., Leonenko, V.N.: Absolute humidity anomalies and the influenza onsets in Russia: a computational study. Procedia Comput. Sci. **119**, 224–233 (2017)
28. Seleznev, N.E., Leonenko, V.N.: Boosting performance of influenza outbreak prediction framework. In: Alexandrov, D.A., Boukhanovsky, A.V., Chugunov, A.V., Kabanov, Y., Koltsova, O. (eds.) DTGS 2017. CCIS, vol. 745, pp. 374–384. Springer, Cham (2017). https://doi.org/10.1007/978-3-319-69784-0_32
29. Shaman, J., Pitzer, V.E., Viboud, C., Grenfell, B.T., Lipsitch, M.: Absolute humidity and the seasonal onset of influenza in the continental United States. PLoS Biol **8**(2), e1000316 (2010)
30. Tamerius, J., Nelson, M.I., Zhou, S.Z., Viboud, C., Miller, M.A., Alonso, W.J.: Global influenza seasonality: reconciling patterns across temperate and tropical regions. Environ. Health Perspect. **119**(4), 439 (2011)
31. UN: 68% of the world population projected to live in urban areas by 2050. https://www.un.org/development/desa/en/news/population/2018-revision-of-world-urbanization-prospects.html
32. Wheaton, W.D., et al.: Synthesized population databases: a US geospatial database for agent-based models. Meth. Report (RTI Press) **2009**(10), 905 (2009)
33. WHO: Influenza (seasonal). Fact sheet No. 211, March 2014. http://www.who.int/mediacentre/factsheets/fs211/en/
34. Wilson, E.B., Worcester, J.: The law of mass action in epidemiology. Proc. Nat. Acad. Sci. **31**(1), 24–34 (1945)
35. Yandex: Auditorii. https://audience.yandex.ru/
36. Yandex: Geocoder. https://tech.yandex.com/maps/geocoder/

Six Degrees of Freedom Numerical Simulation of Tilt-Rotor Plane

Ayato Takii[1(✉)], Masashi Yamakawa[1], Shinichi Asao[2], and K. Tajiri[1]

[1] Kyoto Institute of Technology, Matsugasaki, Sakyo-ku, Kyoto 606-8585, Japan
kpp_fsl_ta@yahoo.co.jp

[2] College of Industrial Technology, 1-27-1, Amagasaki, Hyogo 661-0047, Japan

Abstract. Six degrees of freedom coupled simulation is presented for a tilt-rotor plane represented by V-22 Osprey. The Moving Computational Domain (MCD) method is used to compute a flow field around aircraft and the movement of the body with high accuracy. This method enables to move a plane through space without restriction of computational ranges. Therefore it is different from computation of such the flows by using conventional methods that calculate a flow field around a static body placing it in a uniform flow like a wind tunnel. To calculate with high accuracy, no simplification for simulating propeller was used. Fluid flows are created only by moving boundaries of an object. A tilt-rotor plane has a hovering function like a helicopter by turning axes of rotor toward the sky during takeoff or landing. On the other hand in flight, it behaves as a reciprocating aircraft by turning axes of rotor forward. To perform such two flight modes in the simulation, multi-axis sliding mesh approach was proposed which is a computational technique to enable us to deal with multiple axes of different direction. Moreover, using in combination with the MCD method, the approach has been able to be applied to the simulation which has more complicated motions of boundaries.

Keywords: Computational fluid dynamics · Unstructured grid · Flight simulation · Moving grid

1 Introduction

Recently, a tilt-rotor plane represented by V-22 Osprey has become to gain attention, which has features of both a helicopter and a plane. A tilt-rotor plane has a hovering function like a helicopter by turning axes of rotor to vertical during takeoff or landing. On the other hand in flight, it behaves as a reciprocating aircraft by turning axes of rotor to horizontal. These two flight modes are converted by rotating engine nacelles which are mounted to the tip of fixed wings. Although many numerical simulations have been conducted for tilt-rotor plane, most of them are simulations using overset grids because of its complicated shape [1–5]. Thus the geometric conservation law is not satisfied. Moreover, no coupled simulations have been performed between the flow field and the motion of a tilt-rotor plane with operating its flight control surfaces. To achieve a coupled simulation in two flight modes of tilt-rotor plane with high accuracy, we have

J. M. F. Rodrigues et al. (Eds.): ICCS 2019, LNCS 11536, pp. 506–519, 2019.
https://doi.org/10.1007/978-3-030-22734-0_37

already proposed a computational technique combining multi-axis sliding mesh approach and the Moving Computational Domain (MCD) method [6]. The MCD method is used to compute movement of a plane and unsteady flows occurred by the motion. The feature of this method is that fluid flows are created by movement itself of a moving object. Specifically, moving wall boundary condition generates flows. Therefore, the method is different from conventional methods that calculate a flow field around a static body placing it in a uniform flow like a wind tunnel. Applying this method to rotation of rotors, flows around them can be computed directly without a decrease in the accuracy caused by simplified computation model. In multi-axis sliding approach, the whole computational domain is divided into multiple domains. The domains are adjacent to each other and one can be embedded in another. Rotation of a body like a rotor and an engine nacelle is achieved by rotating whole computational domain which include them. Some domains can be also embedded in another one. A surface of a domain slides on a boundary between it and its adjacent domain. Physical variables are interpolated on the boundary between two domains.

As previous work, the technique has been adapted to flows around a tilt-rotor plane with three degrees of freedom using a bilaterally symmetric mesh and a symmetric boundary condition. In this paper, six degrees of freedom coupled simulation by using multi-axis sliding mesh approach is presented toward the practical simulation of a tilt-rotor plane. Next, the approach which has already proposed is intended for sliding surface of a flat plane or circular cylinder. In six degrees of freedom computation, it is more useful to be able to handle a spherical sliding surface. Therefore, the approach is improved to be able to deal with free-form surfaces. After some inspections are shown, its validity of the approach is also shown by applying to a flow around a tilt-rotor plane.

2 Numerical Approach

2.1 Governing Equation

As a governing equation, three-dimensional Euler equation is written in the conservation form as follows:

$$\frac{\partial \boldsymbol{q}}{\partial t} + \frac{\partial \boldsymbol{E}}{\partial x} + \frac{\partial \boldsymbol{F}}{\partial y} + \frac{\partial \boldsymbol{G}}{\partial z} = 0, \tag{1}$$

$$\boldsymbol{q} = \begin{bmatrix} \rho \\ \rho u \\ \rho v \\ \rho w \\ e \end{bmatrix}, \boldsymbol{E} = \begin{bmatrix} \rho u \\ \rho u^2 + p \\ \rho uv \\ \rho uw \\ u(e+p) \end{bmatrix}, \boldsymbol{F} = \begin{bmatrix} \rho v \\ \rho uv \\ \rho v^2 + p \\ \rho vw \\ v(e+p) \end{bmatrix}, \boldsymbol{G} = \begin{bmatrix} \rho w \\ \rho uw \\ \rho vw \\ \rho w^2 + p \\ w(e+p) \end{bmatrix}, \tag{2}$$

where $\boldsymbol{q}$ is the conserved quantity vector, $\boldsymbol{E}, \boldsymbol{F}, \boldsymbol{G}$ are the inviscid flux vectors. As unknowns, ρ is the density, u, v, w are the x, y, z components of the velocity vector and

e is the total energy per unit volume. The working fluid assumed to be perfect gas, the pressure p is defined as follows:

$$p = (\gamma - 1)\left[e - \frac{1}{2}\rho\left(u^2 + v^2 + w^2\right)\right], \tag{3}$$

where γ is the specific heat ratio and taken as 1.4 in this paper.

2.2 Moving Computational Domain Method

In this paper, we tried to simulate a continuous motion of a tilt-rotor plane from takeoff to hovering. To compute such free movement of the body, the Moving Computational Domain (MCD) method [7–10] was adapted. In the MCD method, a whole computational domain moves together with an object which exists inside the domain shown in Fig. 1. Using the method, that is, it becomes possible to simulate the movement of a tilt-rotor plane in infinite space. This method is one of the moving grid methods because vertices constructing computational domain also move, which is based on the unstructured moving grid finite volume method [11, 12]. In the method, fluxes are evaluated on a control volume in a space-time unified domain (x, y, z, t) to satisfy the geometric conservation law (GCL) [13]. The control volume is four-dimensional volume for three-dimensional flows. Flow variables are defined at the center of the cell in unstructured mesh. The flux vectors are evaluated using the Roe flux difference splitting scheme [14] with MUSCL approach and the Venkatakrishnan limiter [15]. To solve implicit algorithm, the two-stage Runge-Kutta method is adopted.

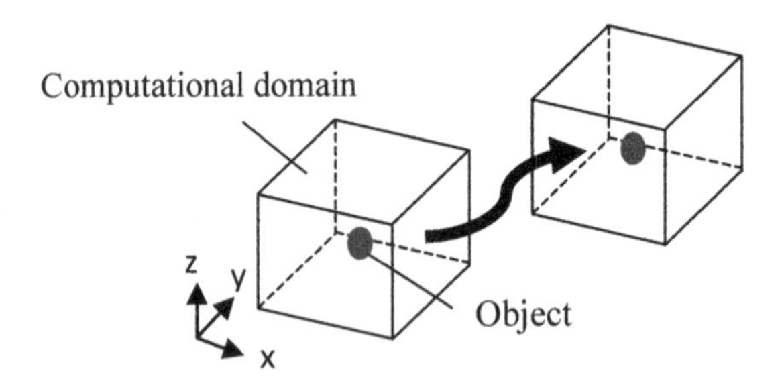

Fig. 1. Conceptual figure of the MCD

2.3 Multi-axis Sliding Mesh Approach

The multi-axis sliding mesh approach is moving grid method to represent local movement of part of an object by dividing the whole computational domain into multiple domains and sliding them. For example, the rotation of propeller is performed by rotating a whole domain containing the propeller. Although the shape of divided sub-domains can be arbitrary unless they interfere with each other, sphere and cylinder is suitable for a rotating body from the aspect of computational cost. Moreover, sub-domains can be nested (see Fig. 2). In this figure, a computational domain 3 is

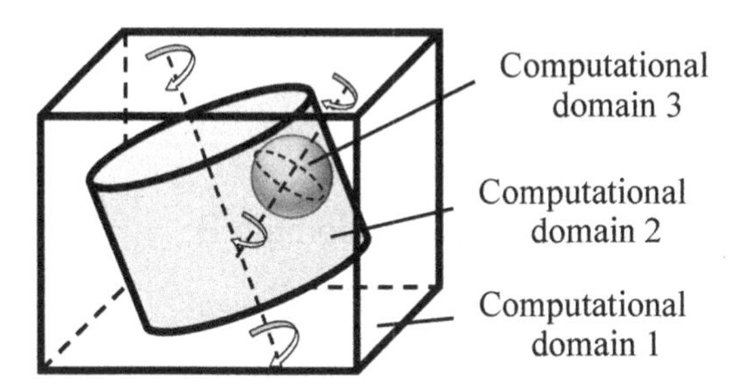

Fig. 2. Multi-axis sliding mesh approach

embedded in a computational domain 2. Also, the computational domain 2 is embedded in a computational domain 1. These domains have different axes of rotation respectively which can rotates independently. With combination of movement of subdomains, new motion is created. For example, as shown in Fig. 3, the vertical motion of an object is achieved by the combination of rotations of two circular domains and a horizontal movement of semicircular domain. The advantage of this approach is that mesh cells don't strain because all vertices of the mesh move as one. To transform domains, the MCD method is used. The multi-axis sliding mesh approach has a possibility to perform very flexible motion of objects without distorting computational meshes by combining with the MCD.

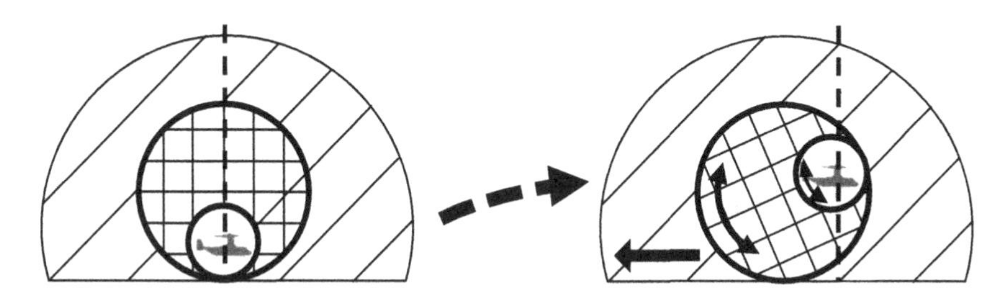

Fig. 3. Vertical motion by combining circles

In the approach, the surface of a computational domain is dealt with a boundary like the wall boundary and outer boundary although there is no physical boundary on the surface, because computational domains have each independent topology. In this paper, the conserved quantity $\boldsymbol{q}_b$ is determined from adjacency relationship between computational domains so as to satisfy the conservation law and interpolated into ghost cells (hereinafter called "sliding boundary condition").

Let us consider adjacent computational domains A and B to illustrate specific computational step of the sliding boundary condition, shown in Fig. 4. Focusing on the cell *i* facing the surface within computational domain A, the cell *j* within computational domain B exists which is adjacent to the cell *i* across the sliding surface. For the sliding boundary condition, physical values are interpolated based on the area of adjacent face between these cells. To calculate the overlapping area between the faces which may be

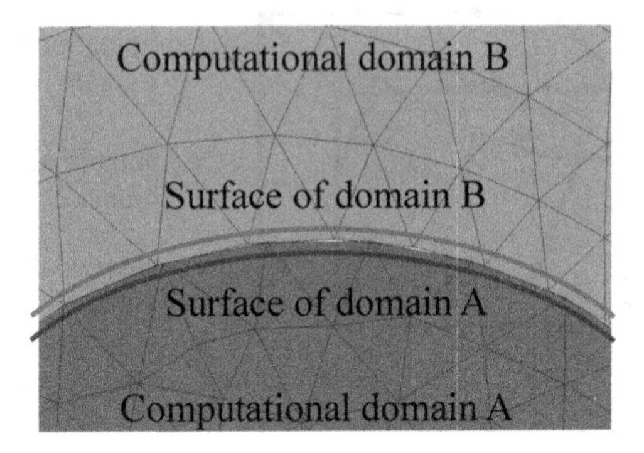

Fig. 4. Adjacent computational domains

non-coplanar, the boundary face of cell i is projected onto the boundary face of cell j. The area S_{ij} of adjacent face between cell i and cell j is defined by the overlapping area between the boundary face of cell i and the projected face. The cell i is generally adjacent to multiple cells within domain B. Thus, as the sliding boundary condition, the interpolated value $\boldsymbol{q}_{bi}$ of the cell i is defined by following equation:

$$\boldsymbol{q}_{bi} = \frac{\sum_{j \in i} \boldsymbol{q}_j S_{ij}}{\sum_{j \in i} S_{ij}}, \tag{4}$$

where $\boldsymbol{q}_j$ is the conserved quantity at the cell j and $\sum_{j \in i}$ means the summation over cells j which is adjacent to the cell i (Fig. 5).

2.4 Coupled Computation

In this paper, the position and rotation of a tilt-rotor plane is automatically determined by the fluid flow around it. Also, the flight control surfaces like the elevator and the pitch of rotor blades affect to the flow around the body. Then it determines the new position and rotation of the body. This series of mechanism is performed by coupled computation between a body and flows using following equations:

$$m \frac{d^2 \boldsymbol{r}}{dt^2} = \boldsymbol{F}, \tag{5}$$

$$I \frac{d\boldsymbol{\omega}}{dt} + \boldsymbol{\omega} \times I\boldsymbol{\omega} = \boldsymbol{T}, \tag{6}$$

where m is the mass, $\boldsymbol{r}$ is the position vector, $\boldsymbol{F}$ is the force vector, I is the inertia tensor (written in matrix form), $\boldsymbol{\omega}$ is the angular velocity vector and $\boldsymbol{T}$ is the torque vector.

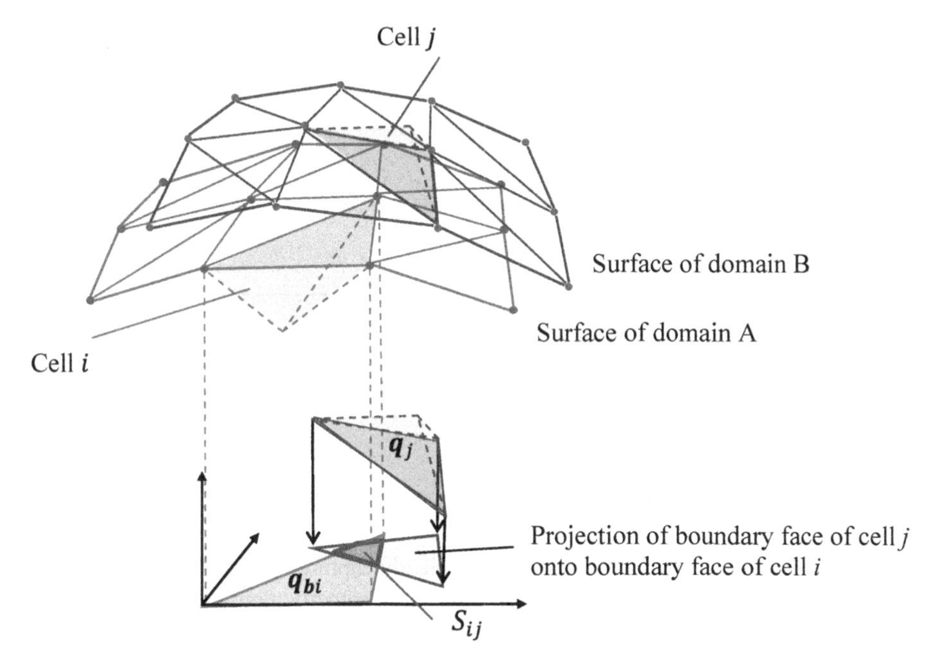

Fig. 5. Cells facing each other across sliding surface and overlapping area of them used to calculate interpolated value

3 Test Problems

To simulate a flow around a tilt-rotor plane, the multi-axis sliding mesh approach is inspected by applying it to test problems. In this section, the uniform flow and the shock tube problem are computed to verify the code.

3.1 Inspection for Geometric Conservation Law

The uniform flow passing through multi-axis sliding surfaces was computed to check whether the geometric conservation law is satisfied. The computational domain is consists of three sub-domains. The domain 1 is rotating spherical one which is embedded in the domain 2. The domain 2 is rotating cylindrical one which is embedded in the domain 3. The initial conditions of density, pressure, velocity components in the x, y and z directions are given by: $\rho = 1.0$, $p = 1.0/\gamma(\gamma = 1.4)$, $u = 1.0$, $v = 1.0$, $w = 1.0$. These values are given as the boundary condition as well.

Figure 6 illustrates the history of the error of density defined by Eq. (4). The order of error is under 10^{-12}. Moreover, we got the same result on the velocity and the pressure.

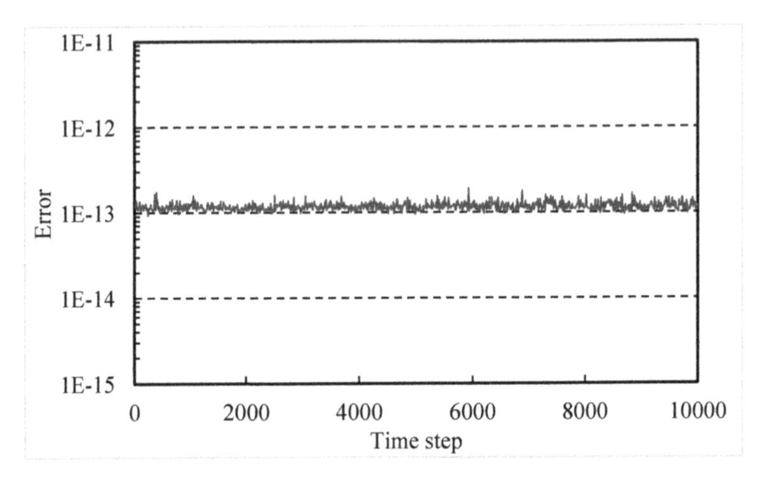

Fig. 6. History of error of density

Therefore, it was proven that this approach perfectly captured the uniform flow passing through multi-axis sliding surface and satisfied the geometric conservation law.

$$Error_i = \max\left(\frac{|\rho_i - \rho|}{\rho}\right). \tag{7}$$

3.2 Shock Tube Problem

The shock tube problem was applied to evaluate the approach compared with its exact solution. The computational domain is consists of three sub-domains. The domain 1 is spherical one which is embedded in the domain 2. The domain 2 is cylindrical one which is embedded in the domain 3 whose shape is rectangular. The domain 1 and domain 2 rotate as shown in Fig. 7. These rotating bodies have boundaries where a flow can path through and there is no shear force. Therefore the rotation of those doesn't affect the flow. In other words, the interpolation on sliding surface being accurate, the result of the computation using the approach must be equal to that of the

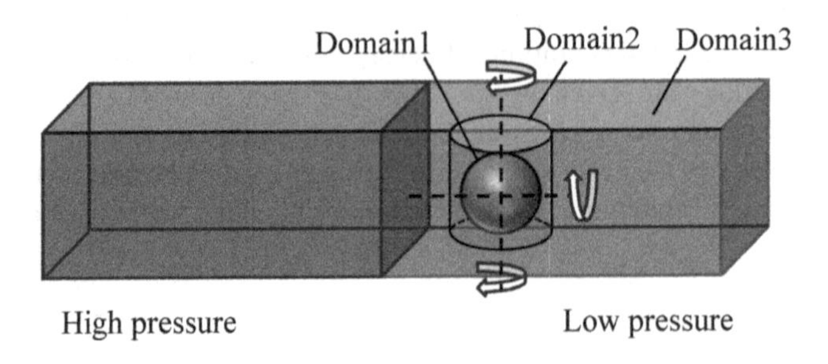

Fig. 7. Shock tube problem

computation using just a rectangular domain. Thus, the validity of the approach is shown by being applied to the shock tube problem.

The initial conditions of density, pressure, velocity components in the x, y and z directions are given by: $\rho_L = 1.0, p_L = 1.0/\gamma(\gamma = 1.4), u = 0.0, v = 0.0, w = 0.0$ at the left of the diaphragm, and $\rho_R = 0.1, p_R = 0.1/\gamma$ at the right of the diaphragm. The reflected condition is applied as the outer boundary condition. The number of cells is 1,005,115.

Figure 8 illustrates the result of the density distributions at t = 0.5 and t = 2.0. The result shows that the shock wave keeps the sharpness after passing through the sliding surfaces on the rotating bodies. Moreover, Fig. 9 shows the result of the density on the center line of the rectangular at t = 0.5, where that is equal to the exact solution. Thus, the validity of the approach is confirmed.

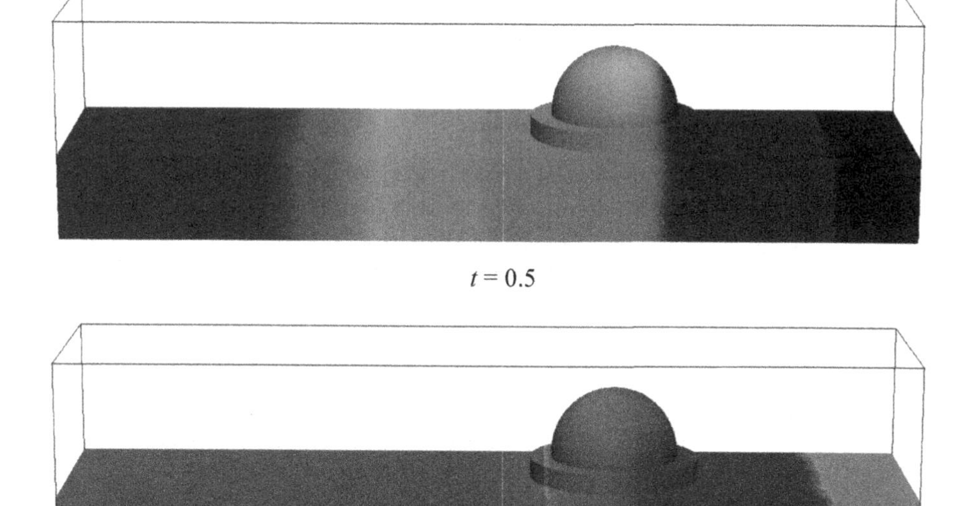

$t = 0.5$

$t = 2.0$

Fig. 8. Density contours

4 Application to Tilt-Rotor Plane

To achieve six degrees of freedom coupled simulation of a V-22 Osprey as a tilt-rotor plane, a combination approach with the multi-axis sliding mesh and the MCD method is used. In the simulation, hovering and yawing are performed.

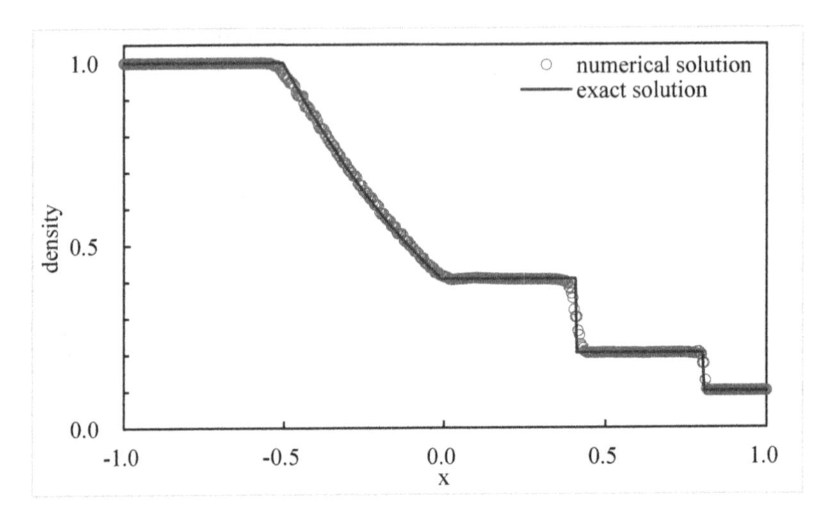

Fig. 9. Distribution of density

4.1 Computational Mesh and Conditions

In this computation, the computational domain is divided into 9 domains (see Fig. 10). Computational meshes of each domain are shown in from Figs. 11, 12, 13, 14 and 15. These meshes are generated by using MEGG3D [16, 17]. Here, the domains 1–6 are used to perform the rotations of two blade rotors and two engine nacelles. The domains 7–9 are used to perform the vertical motion for takeoff of the plane. Total number of

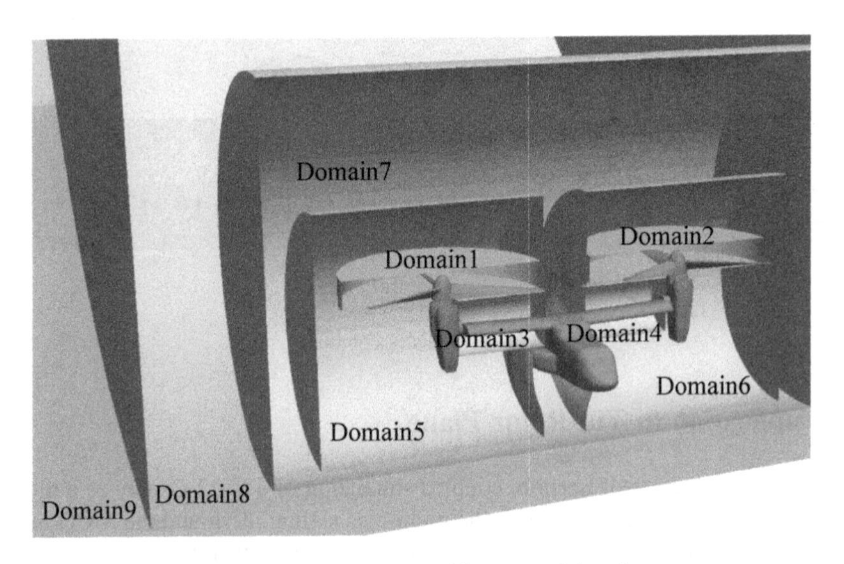

Fig. 10. Domain decomposition around the plane

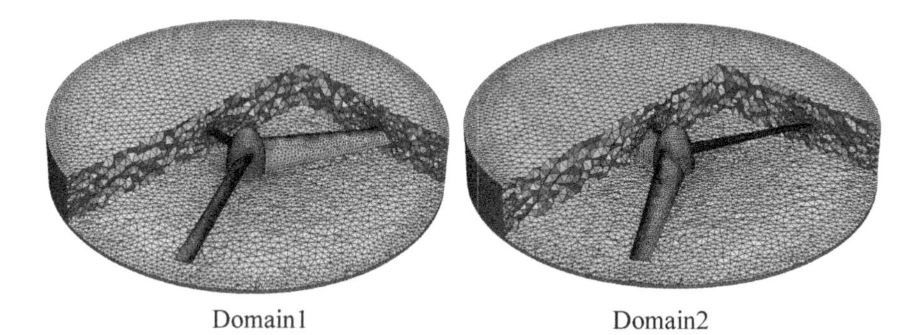

Fig. 11. Mesh around rotors

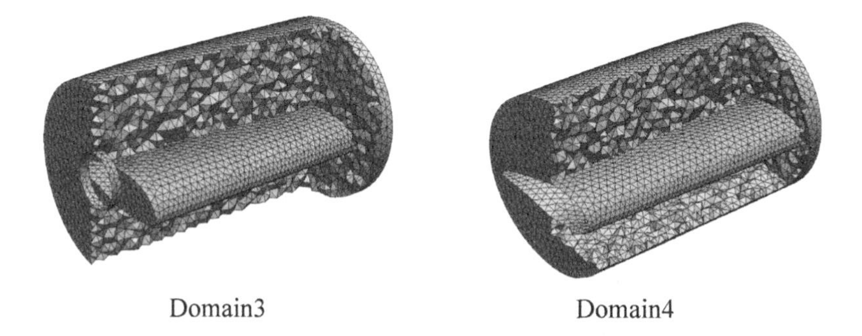

Fig. 12. Mesh around wings

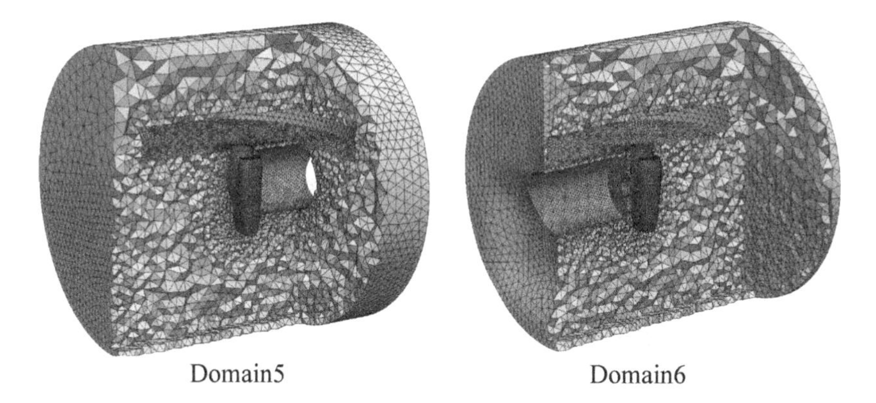

Fig. 13. Mesh around nacelles

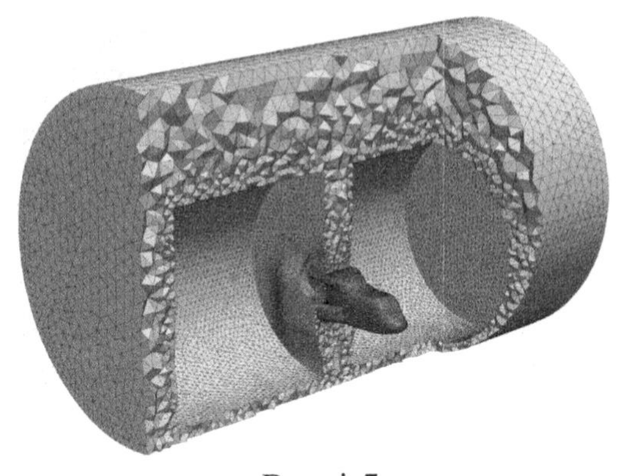

Fig. 14. Mesh around a body

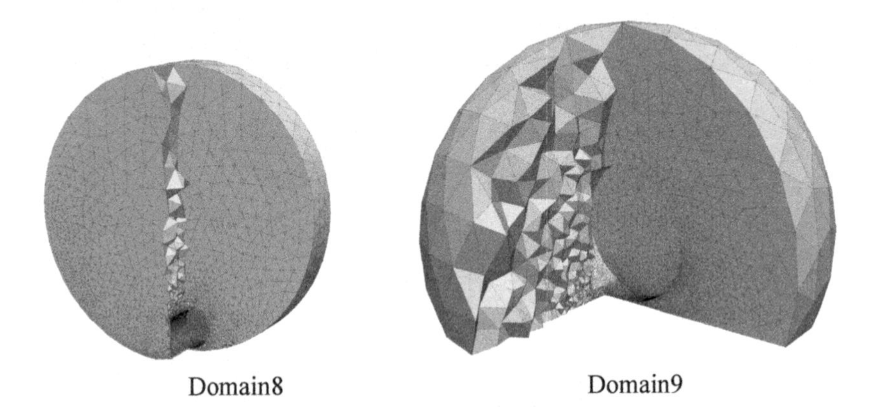

Fig. 15. Outer meshes

cells is 3,333,068. The initial conditions of density, pressure, velocity components in the x, y and z directions are given by: $\rho = 1.0$, $p = 1.0/\gamma(\gamma = 1.4)$, $u = 0.0$, $v = 0.0$, $w = 0.0$. The simulation is performed under the following conditions.

(1) Takeoff (The engine starts around the ground. It takes off in helicopter mode until it arrives 10.0 m high.)
(2) Hovering and yawing (The planes of rotors tilt in the opposite direction to yaw by up to 7°. The rotational speed of rotors is constant. The collective pitch is used to keep altitude and prevent from rolling.)

4.2 Results

Figure 16 shows that an isosurface of the velocity in hovering. It found that the flow around the tilt-rotor plane is bilaterally asymmetric although the body is almost

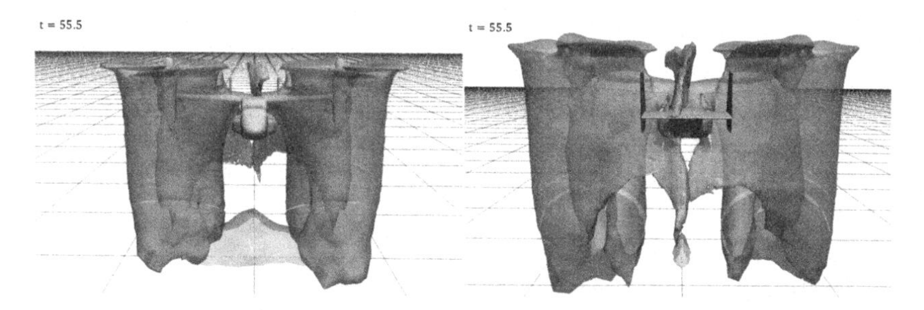

Fig. 16. Isosurface of velocity in hovering

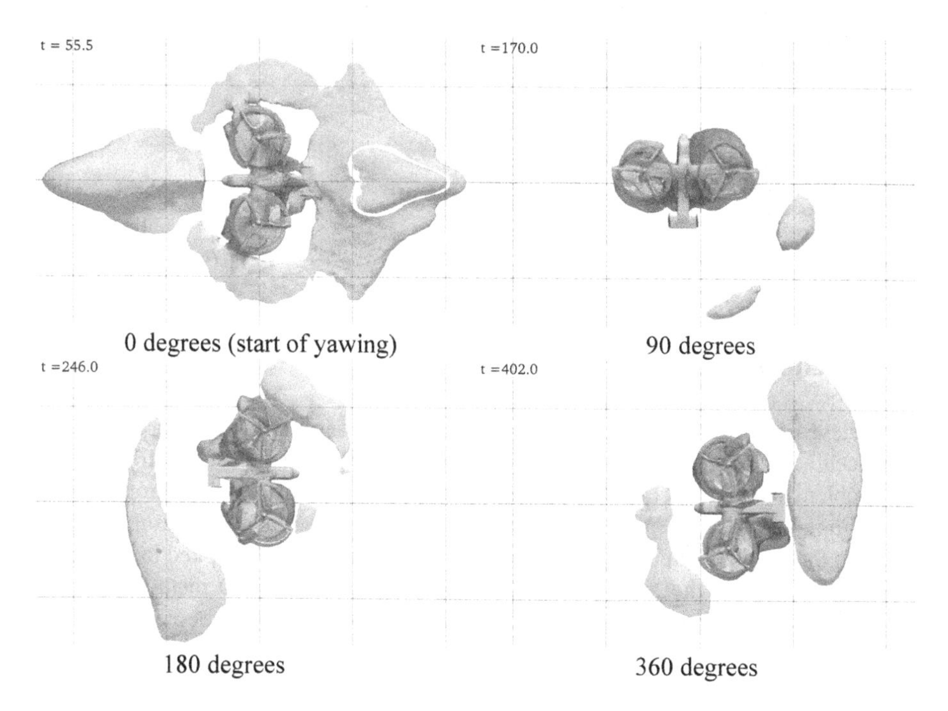

Fig. 17. Isosurface of velocity in yawing

bilaterally symmetric orientation. Figure 17 shows that an isosurface of the velocity in yawing. In the figure, a history from the start of yawing to one revolution is illustrated (0, 90, 180 and 360°). The flow around the plane is completely asymmetry. For the motion of the plane, yawing is achieved by tilting the planes of rotors in the opposite direction. This occurred by the fluid force due to a coupled simulation. Moreover, the plane moved in a rough circular motion viewed from the top by not only rotating but also translating (see Fig. 18). Using the approach, we could see the complicated phenomena occurred by direct computation of the motion of rotor blade. Furthermore, the flow was simulated continuously.

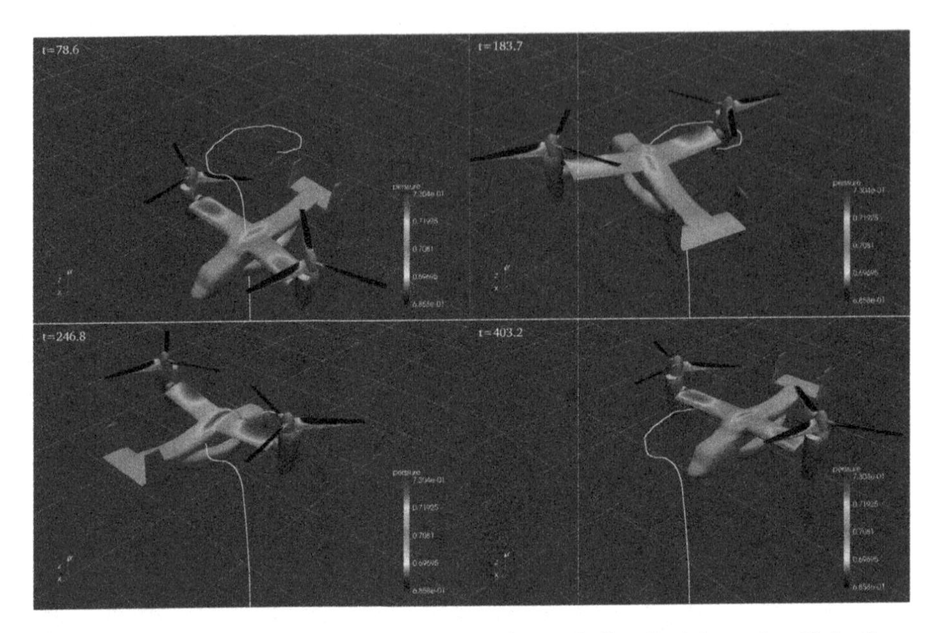

Fig. 18. Pressure contour and trajectory of center of mass (yellow line) in yawing (Color figure online)

5 Conclusions

To achieve six degrees of freedom coupled simulation around a tilt-rotor plane with high accuracy, the multi-axis sliding mesh approach was applied. The result of the first test problem showed that this method captured the uniform flow passing through multi-axis sliding surface and satisfied the geometric conservation law. The result of application to the shock tube problem showed that the approach can be applied to such flow. The combination of the multi-axis sliding approach and the MCD method, the flow occurred by a tilt-rotor plane was computed. Six degrees of freedom coupled simulation of hovering and yawing was simulated.

Acknowledgements. This publication was subsidized by JKA through its promotion funds from KEIRIN RACE and by JSPS KAKENHI Grant Number 16K06079.

References

1. Potsdam, M.A., Strawn, R.C.: CFD simulation of tiltrotor configurations in hover. In: American Helicopter Society 58th Annual Forum, Montreal, Canada, 11–13 June 2002
2. Lee-Rausch, E.M., Biedron, R.T.: Simulation of an isolated tiltrotor in hover with an unstructured overset-grid RANS solver. In: American Helicopter Society 65th Annual Forum, Grapevine, TX, 27–29 May 2009, pp. 1–19 (2009)
3. Gupta, V.: Quad Tilt Rotor Simulations in Helicopter Mode Using Computational Fluid Dynamics. University of Maryland, Maryland (2005)

4. Ye, L., Zhang, Y., Yang, S., Zhu, X., Dong, J.: Numerical simulation of aerodynamic interaction for a tilt rotor aircraft in helicopter mode. Chin. J. Aeronaut. **29**(4), 843–854 (2016)
5. Jimenez Garcia, A., Barakos, G.N.: Numerical simulations on the ERICA tiltrotor. Aerosp. Sci. Technol. **64**, 171–191 (2017)
6. Yamakawa, M., Chikaguchi, S., Asao, S.: Numerical simulation of tilt-rotor plane using multi axes sliding mesh approach. In: The 27th International Symposium on Transport Phenomena, Honolulu (2016)
7. Watanabe, K., Matsuno, K.: Moving computational domain method and its application to flow around a high-speed car passing through a hairpin curve. J. Comput. Sci. Technol. **3**(2), 449–459 (2009)
8. Yamakawa, M., Mitsunari, N., Asao, S.: Numerical simulation of rotation of intermeshing rotors using added and eliminated mesh method. Proc. Comput. Sci. **108**, 1883–1892 (2017)
9. Asao, S., et al.: Simulations of a falling sphere with concentration in an infinite long pipe using a new moving mesh system. Appl. Therm. Eng. **72**, 29–33 (2014)
10. Asao, S., et al.: Parallel computations of incompressible flow around falling spheres in a long pipe using moving computational domain method. Comput. Fluids **88**, 850–856 (2013)
11. Yamakawa, M., Matusno, K.: Unstructured moving-grid finite-volume method for unsteady shocked flows. J. Comput. Fluids Eng. **10–1**, 24–30 (2005)
12. Yamakawa, M., Takekawa, D., Matsuno, K., Asao, S.: Numerical simulation for a flow around body ejection using an axisymmetric unstructured moving grid method. Comput. Therm. Sci. **4**(3), 217–223 (2012)
13. Obayashi, S.: Freestream capturing for moving coordinates in three dimensions. AIAA J. **30**, 1125–1128 (1992)
14. Roe, P.L.: Approximate Riemann solvers parameter vectors and difference schemes. J. Comput. Phys. **43**, 357–372 (1981)
15. Venkatakrishnan, V.: On the accuracy of limiters and convergence to steady state solutions. AIAA Paper, 93-0880 (1993)
16. Ito, Y., Nakahashi, K.: Surface triangulation for polygonal models based on CAD data. Intern. J. Numer. Methods Fluids **39**(1), 75–96 (2002)
17. Ito, Y.: Challenges in unstructured mesh generation for practical and efficient computational fluid dynamics simulations. Comput. Fluids **85**, 47–52 (2013)

A Macroscopic Study on Dedicated Highway Lanes for Autonomous Vehicles

Jordan Ivanchev[1,2(✉)], Alois Knoll[2,3], Daniel Zehe[1,2], Suraj Nair[1,2], and David Eckhoff[1,2]

[1] TUMCREATE, 1 Create Way, Singapore 138602, Singapore
{jordan.ivanchev,daniel.zehe,suraj.nair,david.eckhoff}@tum-create.edu.sg
[2] Technical University of Munich, 3 Boltzmannstr., 85747 Munich, Germany
knoll@in.tum.de
[3] Nanyang Technological University, 50 Nanyang Ave, Singapore 639798, Singapore

Abstract. The introduction of Autonomous Vehicles (AVs) will have far-reaching effects on road traffic in cities and on highways. The implementation of Automated Highway Systems (AHS), possibly with a dedicated lane only for AVs, is believed to be a requirement to maximise the benefit from the advantages of AVs. We study the ramifications of an increasing percentage of AVs on the whole traffic system with and without the introduction of a dedicated highway AV lane. We conduct a macroscopic simulation of the city of Singapore under user equilibrium conditions with realistic traffic demand. We present findings regarding average travel time, throughput, road usage, and lane-access control. Our results show a reduction of average travel time as a result of increasing the portion of AVs in the system. We show that the introduction of an AV lane is not beneficial in terms of average commute time. Furthermore a notable shift of travel demand away from the highways towards major and small roads is noticed in early stages of AV penetration of the system. Finally, our findings show that after a certain threshold percentage of AVs the differences between AV and no AV lane scenarios become negligible.

Keywords: Autonomous vehicles · AV lane · AHS · Macroscopic simulation

1 Introduction

Autonomous vehicles have ceased to be only a vision but are rapidly becoming a reality as cities around the world such as Pittsburgh, San Francisco, and Singapore have begun investigating and testing autonomous mobility concepts [1]. The planned introduction of thousands of autonomous taxis, as currently planned in Singapore [1], poses a challenge not only to the in-car systems of the AV, but also to the entire traffic system itself.

Besides the deployment of more efficient ride sharing systems and the reduction of the total number of vehicles on the road, AVs can traverse a road faster

J. M. F. Rodrigues et al. (Eds.): ICCS 2019, LNCS 11536, pp. 520–533, 2019.
https://doi.org/10.1007/978-3-030-22734-0_38

while using less space. With the goal of achieving the maximum benefit in terms of traffic speeds and congestion reduction, the mixing of AVs and Conventional Vehicles (CVs) is a challenging problem [2]. One method to achieve this is the introduction of dedicated AV lanes on highways to allow AVs to operate more efficiently due to the absence of unpredictable random behaviour introduced by humans and the use of communication capabilities to coordinate local traffic, e.g. platoon organization [3].

Depending on the portion of AVs in the system, blocking certain lanes for CVs, and thereby limiting the overall road capacity for human drivers is certainly a step that can have considerable ramifications. While at the early stages, where only few AVs are on the road, it would constitute an incentive to obtain an AV, it could also possibly generate traffic congestion and increase travel times for other vehicles [4]. As the level of AV penetration in the road transportation system increases, the total congestion level would likely drop, however, the advantage of using AVs over CVs would gradually be diminished as well.

Lastly, converting highways into AHS could also affect the rest of the road network, as drivers of CVs may then choose a different route, caused by the changed capacity of the highway, which can lead to a mismatch between road network and traffic demand [5].

In this paper, we take a closer look at the benefits and drawbacks of introducing a dedicated AV lane on major highways Singapore. With a focus on the effect of an increasing percentage of AVs in the system, we study the impact of dedicated AV lanes in terms of capacity, travel time, and effect on other roads. We compare this scenario to a setting where no dedicated AV lanes are assigned. In short, our main contributions are:

- Using a macroscopic traffic simulation of the city-state of Singapore and using realistic travel demand, we show the impact of vehicle automation with and without AV lanes.
- We evaluate the overall impact of AV penetration on the system travel time.
- We discuss the city-scale effects, in particular how the demand and the throughput of other roads is affected.
- We quantify the effects caused by lane-access control.

2 Related Work

Automated Highway Systems (AHS) and their implications have received wide attention from both researchers and industry around the world. Investigations include general AHS policies and concepts [2,6], effects on travel times and capacity [6–13], traffic safety [6,8,13,14], and interactions between conventional human-driven vehicles and autonomous vehicles [15].

Several AHS studies and field trials were conducted as part of the California Partners for Advanced Transit and Highways (PATH) program: Tsao et al. discuss the relationship between lane changing manoeuvres and the overall throughput of the highway [9]. Their analytical and simulation results indicate a direct trade-off between the two. Lateral movement decreases the traffic flow, and higher traffic flow leads to longer lane change times. In this work, we ignore decreased throughput caused by lane changing and focus on a best-case city-wide benefits of dedicated AV lanes.

Godbole and Lygeros evaluate the increased capacity by the introduction of fully automated highways by treating the highway as a single-lane AHS pipe [14]. Similar to the work of Harwood and Reed [10], they study different platoon sizes and speeds but do not consider separated AV and CV lanes.

The pipeline capacity of AHSs was also studied by Michael et al. [11]. Their results show that longer AV platoons are favourable as they increase the capacity of the road due to lower inter-platoon distances. A high mixture of different vehicle classes, however, leads to a lower capacity. Lastly, the presented analytical model shows that as highway speed increases, the capacity reaches a saturation point after which the speed decreases again. This is inline with the findings presented in this paper when studying the throughput on only the AV lane.

Capacity analysis for managed lanes on highways were studied by Fakhariam Qom et al. using a combined macroscopic and mesoscopic analysis approach [16]. The authors analysed the capacity increase based on a mix of different headway time as well as the percentage of vehicles preferring the managed lanes over the general purpose lanes. Their study included one corridor with a fixed number of lanes, so they could not show effects on other general purpose links in the network. Ghiasi et al. also presented results on highway capacity in mixed traffic scenarios, analyzing the effect of different headway settings [17]. They show that when assumed that the headway an AV maintains to another AV is different from the headway it maintains to a human driven car, (and vice versa), the capacity does not necessarily increase steadily. They also show that the capacity strongly depends on the level of platooning. Assuming static headways for AVs and CVs, their findings are aligned with the ones presented in this paper.

In summary, it appears that while AHS seem to be a well-studied subject, a general evaluation of the introduction of AV lanes and their system-wide implications is still missing. Automated highways are mostly investigated in an isolated manner (and often also simplified to a 1-lane road) without taking into considering the rest of the road network. Using the city-state of Singapore as a case study, we show how travel times of both AVs and CVs are affected and that the introduction of dedicated AV lanes has a considerable effect on the entire road network by changing the distribution of traffic demand within the transportation system.

3 System Model

The goal of this study is to evaluate the allocation of one lane on every highway road to be used exclusively by autonomous vehicles. We investigate an increasing percentage of AVs in the system under two scenarios: with and without dedicated AV lanes on highways. In the former scenario, all roads that are not highways will exhibit normal traffic conditions as there will be a mixture of human drivers and AVs. All lanes on the highway roads will be accessible to AVs, while CVs will be able to utilize all lanes except one lane, which will be allocated exclusively for AVs usage. In the second scenario, all vehicles will share all lanes on all roads.

We model the different behaviour of AVs and CVs by means of smaller headway time, that is the time gap to the vehicle in front. In our scenario, we assume fully automated vehicles with no necessary human driver interaction (SAE Level 4+). We therefore further assume that AVs can afford a much smaller headway than normal vehicles since their reaction time is orders of magnitude smaller than that of humans [18]. A direct consequence of this is that effectively the capacity of the road is increased, as AVs need less space. The capacity (in cars per hour per lane) can then be calculated as follows:

$$C = \frac{3600}{h_{av} p_{av} + h_{cv} \left(1 - p_{av}\right)} \tag{1}$$

where h_{av} is the headway time for AVs (values may vary between 0.5 and 1 s, depending on level of comfort [13,17]), p_{av} is the percentage of AVs on the road segment, h_{cv} is the headway time for conventional vehicles set to 1.8 s. Equation 1 is based on [19], where it was derived from collected data on highway roads in Japan for varying percentages of vehicles with AHS.

If it is assumed that AVs maintain a different headway to CVs than they do to other AVs (and vice versa for CVs), then Eq. (1) can be extended according to Hussain et al. [20] where the denominator becomes: $p_{av}^2 h_{aa} + p_{av}(1 - p_{av})(h_{ac} + h_{ca}) + (1 - p_{av})^2 h_{cc}$, with h_{ac} being the headway of autonomous vehicles to conventional vehicles and so on. For the remainder of the paper we will assume a conservative AV headway of $h_{av} = 1s$ as well as $h_{aa} = h_{ac} = h_{av}$ and $h_{cc} = h_{ca} = h_{cv}$.

The primary measure we use to evaluate the impact on traffic caused by the introduction of an AV lane is the travel time of cars. The travel time $T = \sum_i^n t_i$ of a vehicle is determined by the traverse times of all n segments (or links) included in its route. The traverse time t_i of a segment i can be computed using the Bureau of Public Roads (BPR) function [21]:

$$t_i = \frac{l_i}{v_i} \left(1 + \alpha_i \left(\frac{F_i}{C_i w_i t}\right)^{\beta_i}\right) \tag{2}$$

where l_i is length of the road segment, v_i is the free flow velocity of the segment, F_i is flow, w_i is the number of lanes, t is time duration of the simulated period, C_i is the capacity of road segment i, α_i and β_i are parameters from the BPR function.

Free flow velocities $\hat{v}$ are extracted from historical GPS tracking data [22]. Parameters α_i and β_i are calibrated for different classes of roads depending on their speed limits using both GPS tracking data and a travel time distribution of the population for certain periods of the day. For a more detailed description of the calibration and validation procedures we refer the reader to [23].

4 City-Wide Simulation Study

We conduct a macroscopic city-wide simulation study of the city-state of Singapore to better understand the effects to be expected in a complex environment. We take a look at the traffic conditions on the highway, but also closely investigate the effect dedicated AV lanes have on the rest of the road network.

4.1 Methodology

In order to evaluate the scenarios of AV introduction in a road transportation system, we make use of an agent-based macroscopic simulation approach. The simulation consists of three steps: (1) agent generation, (2) route computation and (3) travel time estimation.

The underlying road network on which the traffic assignment is performed is modelled by means of a uni-directional graph, where each edge represents a road segment and nodes represent decision points at which a road may split or merge.

To introduce dedicated AV lanes, we alter the original graph by duplicating start and stop nodes of highway segments and creating a new edge to represent the AV lane, while removing one of the normal lanes from the original segment. In a way, we model the lane traffic dynamics by making the AV lane a separate road parallel to the original one. Connections are added between the start and end nodes and their respective copies so that vehicles can change lanes at the start/end of highway road segments. It must be noted, however, that the effect of traffic congestion due to the lane changing manoeuvres for joining or leaving the AV lane is not taken into consideration. Therefore, we have to assume that AVs perform lane changes in a non-obtrusive manner, meaning that their politeness on the road is very high, and they try to disrupt traffic flow as little as possible. This is assumption is not very far fetched given the great amount of safety requirements AVs need to satisfy.

Every agent is generated with an origin and destination sampled from an origin-destination (OD) matrix representing the travel demand of the system. Realistic traffic volume is modelled by synthesizing a sufficiently large vehicle population according to the available survey data.

The routes of the agents are computed using an incremental user equilibrium approach [24] aiming at representing reality in the sense that every driver is satisfied with their route and would not choose a different one given the current traffic situation. We assume a driver is satisfied when they are on the shortest route from origin to destination in terms of travel time. Routing is performed on the road network graph, where each edge has an attached weight representing the current traverse time of this road segment. Weights are updated after every batch route computation. To disallow conventional vehicles from using the AV lane, the weight of the edges representing these lanes are set to infinity when the route of a CV is computed and set back to their traverse time values for AV route computation.

Further assumptions regarding to the macroscopic simulation include:

- Agents want to minimize travel time and all perceive the traffic situation in the same way (there is no noise in the observations of the traffic states).
- Traffic is spread homogeneously in time during the simulation. In order for the BPR function (Eq. 2) to give a reasonable estimation of the traffic conditions, the flow F during the simulation time t needs to be spread homogeneously in time.
- Agents do not reroute while on their trip. In reality, drivers may change their trip plan dynamically according to unexpected events or observed traffic conditions ahead. As our traffic assignment provides information to the drivers about expected traffic conditions for the whole network, we assume that such events will not occur as drivers are given all the information they need prior to their trip.
- AVs use the same headway regardless of the car in front of them. This assumption is acceptable on this level of detail, in our opinion, as it accounts for simply removing the reaction time of the human driver.
- Lane changing manoeuvres performed by AVs do not have a considerable effect on traffic flow relative to the effect of lane changes in general due to the high level of politeness of AVs on the road dictated by the high level of imposed safety requirements. To quantify the real impact of lane changes based on the number of lanes, the traffic density and the speed, a microscopic model would be required.

4.2 Data and Scenario Description

We examine the city-state of Singapore with a total population of 5.4 million and around 1 million registered vehicles including taxis, delivery vans and public transportation vehicles. The fact that Singapore is situated on an island simplifies our scenario as the examined system is relatively closed with only two expressways leading out of the country. There are 3495 km or roads of which 652 km are major arterial roads and 161 km are expressways spreading over 715 squared km of land area. We have used publicly available data to acquire a unidirectional graph of the road network of Singapore, compromising of 240, 000 links and 160, 000 nodes. The number of lanes, speed limit and length of every link is available allowing us to extract information about its capacity.

For the purposes of our model we make use of two separate data sets. The first data set consists of GPS trajectories of a 20, 000 vehicle fleet for the duration of one month, providing information about recorded velocities on the road network during different times of the day [22]. The second one is the Household Interview Travel Survey (HITS) conducted in 2012 in the city of Singapore, which studies the traffic habits of the population. Information about the origin destination pairs, their temporal nature, and commuting time distribution during rush hour periods is extracted from it.

In order to achieve realistic traffic conditions, we estimated the number of agents based on the Singapore HITS data. We extracted the ratio of people who

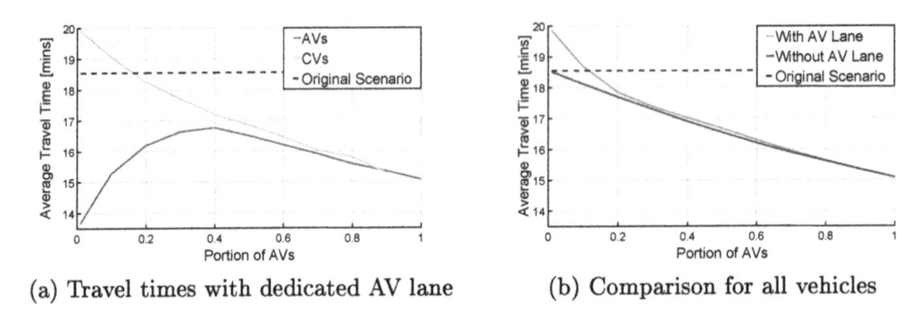

(a) Travel times with dedicated AV lane

(b) Comparison for all vehicles

Fig. 1. Average travel time of whole population, AVs and CVs as a function of AV percentage

actively create traffic on the streets (cab drivers, personal vehicle drivers, lorry drivers) and the total number of people interviewed.

We chose the morning commute hours (7:30 am–8:30 am) as the period which seems to be the most stable in terms of traffic volume and estimated the traffic demand consisting of 309, 000 agents.

Similar to the simplistic example the vehicle population is split into two parts: AV and CV. The percentage of AVs is varied in order to observe the effects in the initial stages of AV introduction, as well as the possible traffic situation if all vehicles were autonomous. We choose at random which vehicles will be autonomous.

In order to have a benchmark for measuring the efficiency of the suggested policy, we also evaluate the scenario where no AV lane is introduced and all vehicles can access all lanes on all roads. The analytical solution acquired for the simplistic example in the previous section indicates that the scenario with no AV lane will produce better or at least the same system performance as the introduction of the AV lane. Our case study will test our analytical results for a more realistic transportation system environment.

4.3 Effects on Average Travel Time

We evaluated the change in average commute time based on the percentage of AVs in the system. We compare two different settings: with and without a dedicated AV lane. As a baseline we use the average commute time without the existence of AVs (and thus no exclusive lanes). This time was found to be approx. 18.5 min and can be seen as the status quo.

We simulated both settings with the exact same travel demand, that is, identical origin-destination pairs for all vehicles. We then gradually increased the percentage of AVs in the system. The number of vehicles in the simulation was invariant; a higher percentage of AVs means that CVs were replaced with autonomous vehicles.

Figure 1 shows our result. The black dashed line serves as an illustration of the status quo. It can be observed that when introducing a lane exclusively for

AVs and thereby taking away capacity from conventional vehicles (Fig. 1a), the travel time for CVs increases, whereas the low percentage of AVs (driving on almost empty lane) travel considerably faster. This could be used as an incentive by policy-makers to increase the share of AVs in the transportation system. With an increasing percentage of AVs we observe that travel times for CVs decrease due to the fact that effectively vehicles move away from the common lane to the AV lane, which in turn increases the travel time on the AV lane. This behaviour can be observed until the AV lane is saturated (somewhere between 40% and 50% AV percentage).

From this point on, the choice of lane makes no difference for newly added AVs as they experience equal travel times regardless whether they choose to take the AV lane or the mixed lane. Therefore, the difference between average travel time of AV and CV decreases. Travel time of both AVs and CVs still decreases with the increase of the percentage of AVs since the capacity of the road network is effectively increased.

Figure 1b shows the comparison between the average travel times for the entire vehicle population with and without the introduction of an AV lane. It can be observed that the setting without an AV lane is always performing better than the AV lane one. After saturation of the AV lane the difference between the two curves becomes marginal. Before saturation of the AV lane, the capacity of the highway is not fully utilized and that, on average, the smaller travel times of the AVs cannot make up for the introduced delays for the CVs. We conclude that adding an AV lane, while initially being an incentive for early adopters, will noticeably penalize drivers of conventional vehicles at the early stages of AV introduction. Additional delays will be introduced by lane changing manoeuvres and other microscopic effects not considered in this macroscopic study [7,9].

4.4 Analysis of Effect of Headway Time

The headway time of vehicles is a crucial parameter for the computation of the road capacity (see Eq. 1) and therefore an important input for the travel time computation using the BPR function (see Eq. 2).

Although, AVs can afford to have smaller headways, this might negatively affect the comfort of the passengers, as small distances at high velocities can induce anxiety [13]. To this end, we examine the effect of varying the headway time (from 0.5s to 1s) for the city-scale simulation scenario with a dedicated AV lane on highways. It is therefore useful to evaluate the amount of overall travel time decrease as a function of the headway time.

In Fig. 2 we observe that depending on the headway, the improvement of overall travel time can vary between 20% and 26%. The difference between the improvement in the case of all AVs seems is bigger between headway 0.5 s and 0.75 s than between 1 s and 0.75 s.

This is due to the non-linear nature of the BPR function. As vehicles approach free flow velocities due to the increased capacity, any further improvement in capacity plays a smaller role and therefore the gains become less significant.

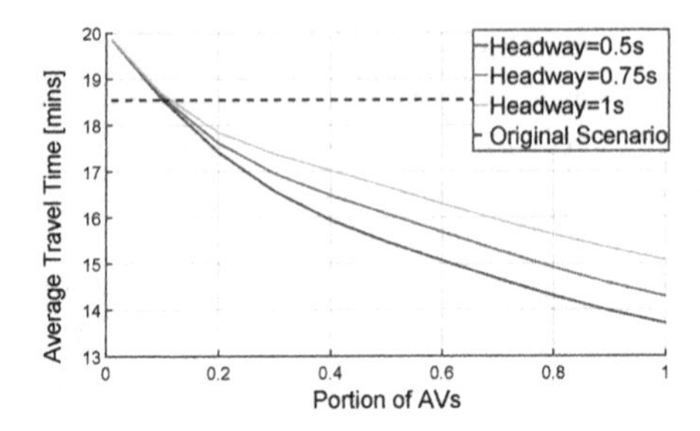

Fig. 2. Average travel time curves for varying headway time

It must be noted, that this happens only when traffic congestion is such that vehicles travel at free flow velocities.

If the system was more congested, we would not be able to observe the decrease of travel time improvement for smaller headway time, since the BPR function will still be in its non-linear part thus providing significant changes of travel time as the capacity is altered.

(a) Change of road throughput at 50% AVs (with AV lane)

(b) Change of road throughput at 50% AVs (without AV lane)

Fig. 3. Road throughput change caused by 50% AVs compared to 0% AVs. Blue colours represent higher throughput; red colours represent lower throughput (intensity indicates strength) (Color figure online)

4.5 Effects on Road Network Throughput and Traffic Distribution

We examine the traffic distribution on the entire road network to gain a better understanding of the changes occurring in traffic conditions as a results of the introduction of dedicated AV lanes. To the best our knowledge, existing literature focuses on the traffic changes on the highways and ramps only. In our work AV lanes are added only to the highways, while the rest of the network remained unchanged, however, we study the changes that occur to the throughput and demand for the whole system.

As a first step, we provide a qualitative measurement by visualizing the impact an introduction of 50% AVs has on the road network (compared to 0% AVs). To this end, we draw roads experiencing higher throughput in blue colours

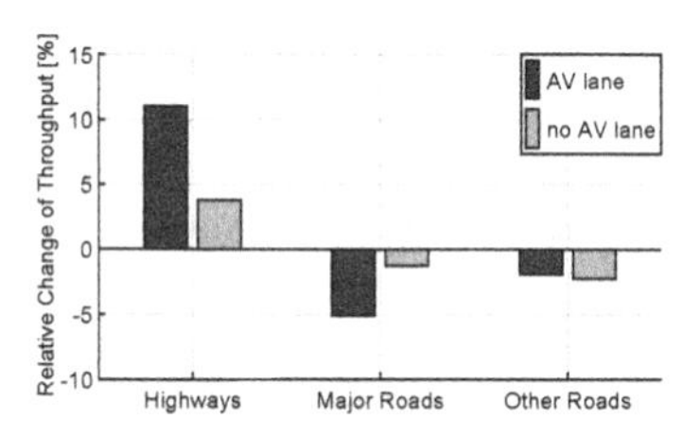

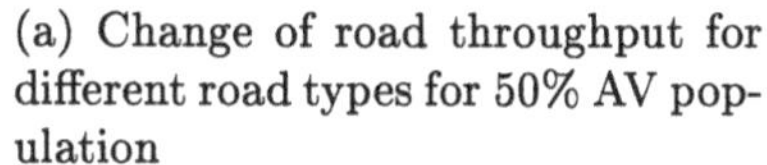
(a) Change of road throughput for different road types for 50% AV population

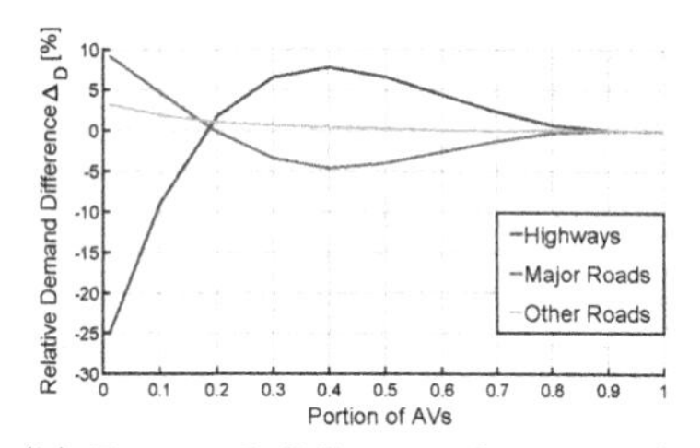

(b) Demand difference between AV lane setting and no AV lane setting for varying percentage of AVs

Fig. 4. AV introduction (with and without AV lane) implications on throughput Fig. 4a and demand Fig. 4b

and roads experiencing lower throughput in red colours. We show our results for both settings, i.e., with AV lane (Fig. 3a) and without AV lane (Fig. 3b).

It can be observed that in both cases the highways exhibit a considerably higher throughput of vehicles, which is expected since the capacity of the highway is technically increased by the introduction of autonomous vehicles. Other roads exhibit a slight decrease of throughput, which may be due to the changes in routing that are triggered by the AVs, which have a strong preference towards the highways. In other words, the AVs are, in a way, attracted to the highways, since they can traverse fast there and the capacity is sufficient. This, however, makes them willing to pass through more congested roads in order to get to the highway, which creates additional time losses for the regular vehicles. This argument is further strengthened by the fact that the roads leading to the highways do not exhibit increased throughput, although they exhibit higher demand since more vehicles want to use the highways. This means that the level of congestion on those roads is too high to allow an increase of throughput.

Comparing the throughput changes in the two scenarios, it can be observed that the increase of throughput on the highways for the case with no AV lane is smaller. Therefore, the change of routing triggered by the introduction of AVs is qualitatively the same but quantitatively different for the two examined scenarios. This can be observed in Fig. 4a. The relative increase of throughput for highways is almost three times higher for the dedicated AV lane case. Higher level of throughput increase on highways leads to higher level of throughput decrease on the major roads, which represent alternative routes. As mentioned before, if highways become too attractive for AVs, there can be negative effects on traffic conditions stemming from overly utilized roads which lead to the highways. The more balanced distribution of traffic in the no AV lane scenario could be the reason for its slight, however, consistent superiority over the dedicated AV lane case in terms of average travel time observed earlier in Fig. 1b.

Finally, we investigate the change in demand for different types of roads. We define demand for a road as the number of vehicles that have this particular road in their route under user equilibrium traffic assignment. Taking a closer look at the difference in travel demand between the scenario with AV lane and the scenario without the introduction of an AV lane, we measure the relative difference of travel demand between the scenarios. Formally, let the demand for road i for a given AV percentage k be $D^a_{i,k}$ for the AV lane scenario and $D^b_{i,k}$ for the benchmark scenario without an AV lane. Then, for the different classes of road (highways, major roads, other roads), we compute the difference in demand between the two examined scenarios relative to the AV lane scenario (a negative value therefore means that this particular type of road experiences less demand when AV lanes are introduced compared to not introducing AV lanes):

$$\Delta_{D,k} = \frac{\sum_i D^a_{i,k} - \sum_i D^b_{i,k}}{D^a_{i,k}} \tag{3}$$

Figure 4b shows our results for highways, major roads, and other roads. For highways, we observe that initially there is lower demand. This is due to the lower road capacity resulting from the dedicated lane, causing more CVs to avoid the highways. With more AVs in the system, the demand for the highways increases until the saturation point of the AV lane is reached. From this point on, the difference between the both settings becomes smaller, eventually converging to zero as the addition of more AVs cause the common lanes to achieve capacity values close to the one of the AV lane.

For the major roads we observe increased demand as CVs will favour these roads as an alternative to the more congested highways caused by the dedicated AV lane. This effect decreases until the saturation point of the AV lane.

At this moment the highways reach their maximum demand difference and therefore a smaller portion of the population takes the alternative routes utilizing major roads. Following the negative peak, the demand difference for major roads also converges to zero as the two scenarios become identical. The difference for smaller roads is less pronounced and can only be observed at a low percentage of AVs in the system. The higher utilization of major roads will also increase traverse time of these roads. Some vehicles will therefore choose to use minor roads as a third alternative.

It is interesting to note that after the saturation point of the AV lane the difference in travel times between the two scenarios is almost negligible, however, the actual assignment of traffic, as can be observed in Fig. 4b, is qualitatively different. This finding indicates that the introduction of an AV lane would not just affect the travel time of the population but also shift the route preferences of commuters.

4.6 Lane Access Control

We showed that AV lanes can provide faster traversal times if the number of cars on this lane is low enough. Once the number of AVs in the system exceed a certain threshold, this benefit diminishes. We therefore look at the benefits and trade-offs when only a subset of all autonomous vehicles (e.g. AV taxis, service vehicles, etc.) have access to the specialized lane.

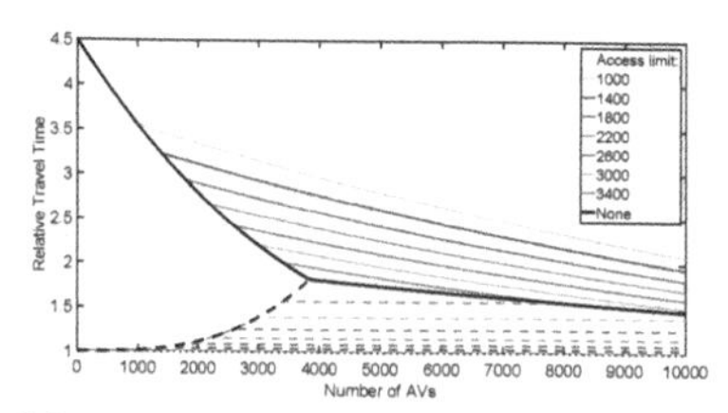

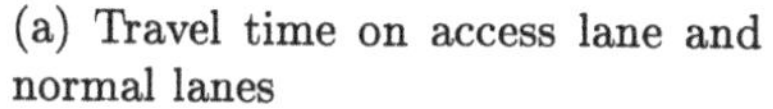
(a) Travel time on access lane and normal lanes

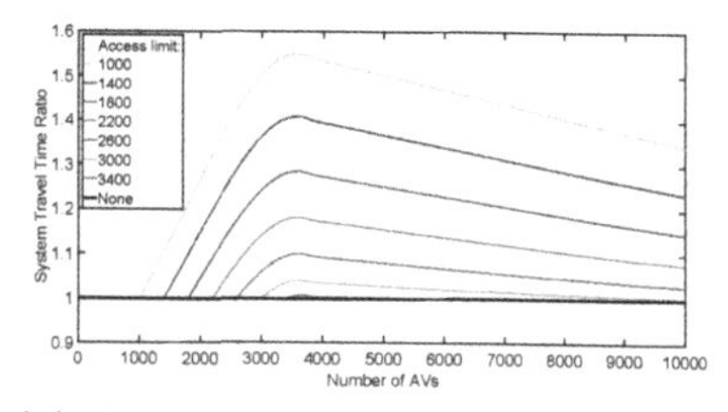

(b) System travel time comparison between access control and baseline

Fig. 5. Access control analysis. Figure 5a shows the travel time, relative to free flow conditions, on the normal lanes (solid lines) and access controlled AV lane (dashed line). Figure 5b shows the performance of the whole vehicle population relative to the no-access control scenario.

We analyse a simple example of a three-lane highway stretch with one AV lane and two normal lanes. We assume that 10,000 vehicles would like to traverse the road and, as before, we vary the percentage of AVs among those vehicles. Our results are shown in Fig. 5a. It can be seen that the set limit marks the point where the access control scenario diverges from the original one (solid black line). The result is a steady and fast commute on the AV lane for the vehicles with access (dashed lines) and a slower commute for the rest of the population (non-dashed lines). In the no-access control scenario (black lines), the traverse times along the AV and normal lanes is equal after saturation, i.e. where the black lines meet. When a limit is introduced, this is no longer the case and a smaller limit leads to a bigger difference between the two sets of vehicles on the road in terms of traverse time.

In order to better understand the trade-off between the fast commute of specialised vehicles and system travel time, Fig. 5b demonstrates the change of overall average system travel time relative to the scenario where there is no access control on the AV lane. This graph can be seen as the cost the system has to pay for having access control; higher values indicate longer traverse times for the vehicle population compared to the scenario where all AVs can use the dedicated lane. The difference grows as the access limit increases and peaks at the saturation point of the original scenario. This is because the difference of traverse times between AV and normal lanes in the base-line scenario is practically non-existent at (and after) this point. Furthermore, the difference in traverse times on AV lane and normal lanes decreases after this point. For these two reasons, a maximum ratio of differences between lane-access control and non-lane access control appears at the saturation point.

5 Conclusion and Future Work

In this article we demonstrated the effect of assigning one lane on highways exclusively for AVs. We showed that for lower percentages of AVs, or more precisely, before the dedicated lane is saturated, travel times for AVs can be significantly shorter, while at the same time CVs are delayed due to the reduced capacity of the highway.

Looking at the entire road network, we observe that also non-highways are affected as CVs will effectively be drawn away from the highways onto the major roads. This effect is especially pronounced at early stages of AV adaptation where the AV lane will remain mostly empty. Regardless of an introduction of the AV lane, we confirmed earlier findings that a larger number of AVs will have a positive impact on travel times for all vehicles. We further compared the scenario with AV lane introduction to a baseline scenario where no changes to traffic regulations are made. The latter scenario outperforms the former one over the whole range of AV percentages, however, the difference is of considerable amount only before the saturation point is reached. We also showed to which extent lane access control can still provide faster travel times to specialized AVs and quantified the effect on vehicles without access to the dedicated lane.

Future work includes micro (and submicroscopic) studies to better understand the impact of smart platooning strategies but also turbulences caused by lateral vehicle movement. Another interesting research direction would be replacing the User Equilibrium (UE) traffic assignment with an algorithm that looks for system optimum assignment (e.g. BISOS [25]), in order to check whether an AV lane would be more beneficial in such cases.

References

1. Henderson, J., Spencer, J.: Autonomous vehicles and commercial real estate. Cornell R. Estate Rev. **141** (2016)
2. Litman, T.: Autonomous vehicle implementation predictions – implications for transport planning. Victoria Transport Policy Institute, Technical report, November 2016
3. Segata, M., Bloessl, B., Joerer, S., Dressler, F., Lo Cigno, R.: Supporting platooning maneuvers through IVC: an initial protocol analysis for the join maneuver. In: Conference on Wireless On demand Network Systems and Services (WONS 2014), Obergurgl, Austria, pp. 130–137. IEEE, April 2014
4. Ivanchev, J., Zehe, D., Nair, S., Knoll, A.: Fast identification of critical roads by neural networks using system optimum assignment information. In: International Conference on Intelligent Transportation Systems (ITSC), Yokohama, Japan, pp. 1–6. IEEE (2017)
5. Ivanchev, J., Aydt, H., Knoll, A.: Spatial and temporal analysis of mismatch between planned road infrastructure and traffic demand in large cities. In: International Conference on Intelligent Transportation Systems, Las Palmas, Spain, pp. 1463–1470. IEEE (2015)
6. Kanaris, A., Ioannou, P., Ho, F.S.: Spacing and capacity evaluations for different AHS concepts. In: Ioannou, P.A. (ed.) Automated Highway Systems, pp. 125–171. Springer, Boston (1997). https://doi.org/10.1007/978-1-4757-4573-3_8
7. Cohen, S., Princeton, J.: Impact of a dedicated lane on the capacity and the level of service of an urban motorway. Procedia Soc. Behav. Sci. **16**, 196–206 (2011)
8. Carbaugh, J., Godbole, D.N., Sengupta, R.: Safety and capacity analysis of automated and manual highway systems. Transp. Res. Part C Emerg. Technol. **6**(1), 69–99 (1998)

9. Tsao, H., Hall, R., Hongola, B.: Capacity of automated highway systems: effect of platooning and barriers. Institute of Transportation Studies, University of California, Berkeley, Technical report UCB-ITS-PRR-93-26, February 1994
10. Harwood, N., Reed, N.: Modelling the impact of platooning on motorway capacity. In: Road Transport Information and Control Conference 2014 (RTIC 2014), London, UK. IET (2014)
11. Michael, J.B., Godbole, D.N., Lygeros, J., Sengupta, R.: Capacity analysis of traffic flow over a single-lane automated highway system. J. Intell. Transp. Syst. **4**(1–2), 49–80 (1998)
12. Hall, R., Chin, C.: Vehicle sorting for platoon formation: impacts on highway entry and throughput. Transp. Res. Part C Emerg. Technol. **13**(5), 405–420 (2005)
13. Van Arem, B., Van Driel, C.J., Visser, R.: The impact of cooperative adaptive cruise control on traffic-flow characteristics. IEEE Trans. Intell. Transp. Syst. **7**(4), 429–436 (2006)
14. Godbole, D.N., Lygeros, J.: Safety and throughput analysis of automated highway systems. Institute of Transportation Studies, University of California, Berkeley, Technical report UCB-ITS-PRR-2000-1, January 2000
15. Dresner, K.M., Stone, P.: Sharing the road: autonomous vehicles meet human drivers. In: Twentieth International Joint Conference on Artificial Intelligence (IJCAI 2007), vol. 7, Hyderabad, India, pp. 1263–1268, January 2007
16. Fakharian Qom, S., Xiao, Y., Hadi, M.: Evaluation of cooperative adaptive cruise control (CACC) vehicles on managed lanes utilizing macroscopic and mesoscopic simulation. In: 95th Annual Transportation Research Board Meeting, no. 16-6384, January 2016
17. Ghiasi, A., Hussain, O., Qian, Z.S., Li, X.: A mixed traffic capacity analysis and lane management model for connected automated vehicles: a Markov chain method. Transp. Res. Part B Methodol. **106**, 266–292 (2017)
18. Ioannou, P.A., Chien, C.-C.: Autonomous intelligent cruise control. IEEE Trans. Veh. Technol. **42**(4), 657–672 (1993)
19. Yokota, T., Ueda, S., Murata, S.: Evaluation of ahs effect on mean speed by static method. In: Fifth World Congress on Intelligent Transport Systems, Seoul, South Korea (1998)
20. Hussain, O., Ghiasi, A., Li, X., Qian, Z.: Freeway lane management approach in mixed traffic environment with connected autonomous vehicles. CoRR, vol. abs/1609.02946 (2016). http://arxiv.org/abs/1609.02946
21. Dafermos, S.C., Sparrow, F.T.: The traffic assignment problem for a general network. J. Res. Natl. Bur. Stand. Ser. B **73**(2), 91–118 (1969)
22. Sindhwani, M., Xin, Q.K.: Singapore traffic information platform: enabling traffic-aware applications & systems. In: 17th ITS World Congress, Busan, South Korea, October 2010
23. Ivanchev, J., Litescu, S., Zehe, D., Lees, M., Aydt, H., Knoll, A.: Determining the most harmful roads in search for system optimal routing. TU Munich, Technical report TUM-I1632, February 2016
24. Fisk, C.: Some sevelopments in equilibrium traffic assignment. Transp. Res. Part B Methodol. **14**(3), 243–255 (1980)
25. Ivanchev, J., Zehe, D., Viswanathan, V., Nair, S., Knoll, A.: BISOS: backwards incremental system optimum search algorithm for fast socially optimal traffic assignment. In: International Conference on Intelligent Transportation Systems (ITSC), Rio de Janeiro, Brazil, pp. 2137–2142. IEEE (2016)

An Agent-Based Model for Evaluating the Boarding and Alighting Efficiency of Autonomous Public Transport Vehicles

Boyi Su[1,2], Philipp Andelfinger[1,2](✉), David Eckhoff[1,3], Henriette Cornet[1], Goran Marinkovic[1], Wentong Cai[2], and Alois Knoll[2,3]

[1] TUMCREATE Ltd., Singapore, Singapore
{david.eckhoff,henriette.cornet,goran.marinkovic}@tum-create.edu.sg
[2] Nanyang Technological University, Singapore, Singapore
{bsu,pandelfinger,ASWTCAI}@ntu.edu.sg
[3] Technische Universität München, Munich, Germany
knoll@in.tum.de

Abstract. A key metric in the design of interior layouts of public transport vehicles is the dwell time required to allow passengers to board and alight. Real-world experimentation using physical vehicle mock-ups and involving human participants can be performed to compare dwell times among vehicle designs. However, the associated costs limit such experiments to small numbers of trials. In this paper, we propose an agent-based simulation model of the behavior of passengers during boarding and alighting. High-level strategical behavior is modeled according to the Recognition-Primed Decision paradigm, while the low-level collision-avoidance behavior relies on an extended Social Force Model tailored to our scenario. To enable successful navigation within the confined space of the vehicle, we propose a mechanism to emulate passenger turning while avoiding complex geometric computations. We validate our model against real-world experiments from the literature, demonstrating deviations of less than 11%. In a case study, we evaluate the boarding and alighting times required by three autonomous vehicle interior layouts proposed by industrial designers.

1 Introduction

Public transport systems based on autonomous mobility are currently in the focus of research groups around the world. Alongside components such as the powertrain and the facilities for reacting to the current driving situation, the interior layout of an autonomous vehicle (AV) has important implications on the efficiency of the overall transport system. The time required for passengers to board and alight determines the minimum dwell time required at each stop.

This work was financially supported by the Singapore National Research Foundation under its Campus for Research Excellence And Technological Enterprise (CREATE) program.

J. M. F. Rodrigues et al. (Eds.): ICCS 2019, LNCS 11536, pp. 534–547, 2019.
https://doi.org/10.1007/978-3-030-22734-0_39

Thus, the boarding and alighting time is a key metric when defining static or dynamic AV schedules.

In practice, once industrial designers have defined a set of possible AV interior layouts comprised of aspects such as the overall geometry and spacing as well as the size, number, placement, and orientation of seats, a quantitative comparison with respect to the required dwell time should be carried out. While real-world experimentation would provide accurate results, the costs for the creation of mock-ups and the acquisition of sufficient numbers of participants limits such experiments to small numbers of trials for the most promising AV designs.

We thus propose a simulation model of AV boarding and alighting processes, permitting large numbers of trials under various scenario conditions. We present an agent-based model of the passenger behavior. A number of challenges emerge due to the confined space for movement in the considered scenario that are not present in many common agent-based simulation scenarios (e.g., [6,9]).

To address these challenges, we adapt the collision avoidance behavior provided by the well-known Social Force Model (SFM) [18] with right-of-way extensions [10]. A novel size adaption method is proposed to allow agent to navigate through narrow corridors successfully without the need for complex geometric computations. The passenger behavior on the strategical level is carefully modeled according to the Recognition-Primed Decision [15] framework as a set of so-called experiences [17]. We validate our model in comparison to results from the literature based on experiments relying on human participants.

Our prototypical simulation tool accepts color-coded 2D representations of AV layouts from which executable scenarios are generated. Thus, industrial designers without experience in the simulation domain can rely on the simulation to evaluate vehicle designs. Our main contributions can be summarized as follows:

- We propose an agent-based model to evaluate the boarding and alighting times with respect to given vehicle interior layouts. We propose adaptations of well-known pedestrian simulation models to support the navigation in spatially confined scenarios.
- We present validation results showing less than 11% deviation compared to real-world experiments from the literature.
- The practical use of the model is demonstrated by evaluating the dwell times of three AV layouts defined by industrial designers.

The remainder of this paper is organized as follows: Sect. 2 introduces existing works fundamental to our model and related work in passenger simulation for public transport scenarios. Section 3 describes our basic simulation scenario. Section 4 describes our proposed passenger model. Section 5 provides validation and performance evaluation results. Section 6 provides a summary of our results.

2 Background and Related Work

In this section, we outline the models used to represent the high-level decision making and the low-level collision-avoidance behavior of agents. Further, we discuss related work on agent-based simulation of public transport scenarios.

2.1 Recognition-Primed Decision

Recognition-Primed Decision (RPD) is a conceptual framework that arose from the naturalistic decision-making movement, which seeks to understand human decision making. RPD is motivated by the insight that people frequently make decisions based on estimation and guesswork rather than purely rational processes. Each agent maintains a repertoire of patterns representing its previous experiences. When encountering a problem, the agent selects a matching situation from its repertoire and predicts the expected outcome based on the previous experience. A comprehensive description of RPD is given in [15]. In Sect. 4, we describe how the concepts of RPD are applied to our specific modeling problem.

Our agent-based model relies on the implementation of RPD in the CrowdTools simulation framework, which encapsulates the decision-making logic into experiences composed of one or more stages [17]. The decision-making process consists of three iterative steps: During *situation assessment*, the agent perceives situational cues, which are used to update its emotional state. During *experience matching*, the agent selects from its repertoire the experience that is the most similar to the current situation. During *experience execution*, the actions associated with the selected experience are carried out.

Four cognitive components implement the following tasks involved in the above process: The *perception system* detects the constraints imposed by the virtual environment and filters the relevant information for the decision making. The *working memory* stores predefined information and situational states. The *decision system* carries out the decision making based on the information from the perception system and the working memory. Finally, the *action system* executes the actions determined by the decision system. In Sect. 4, we present our mapping of the passenger behavior to the RPD concepts.

2.2 Collision Avoidance

Microscopic crowd simulations represent humans as autonomous agents that sense their surroundings, make decisions and carry out corresponding actions. The Social Force Model (SFM) [18] is a popular and well-studied model of the distance-keeping behavior of pedestrians. The SFM models the intention of a pedestrian as a driving force and the resistance between a pedestrian and its surrounding objects, i.e., neighboring pedestrians and stationary obstacles, as repulsive forces. The force acting on an agent i at time t is defined as follows:

$$m_i \frac{\mathrm{d}v_i}{\mathrm{d}t} = m_i \frac{v_i^0(t) - v_i(t)}{\tau_i} + F_{nb} + F_{obs}$$

The first summand represents the driving force, where m_i is the mass of the agent, $v_i^0(t)$ is the preferred velocity and $v_i(t)$ is the actual velocity; F_{nb} is the net force from neighboring agents, and F_{obs} is the force from the surrounding obstacles. The force received from neighbor j is defined as $f_{ij} = A\exp(-\frac{D_{ij}}{B})n_{ij}$. The time step size is denoted by τ_i. The parameters A, B determine the strength and range of the force. D is the distance between two agents. n_{ij} is the direction vector pointing from r_i to r_j. The force from an obstacle line l is determined in the same fashion based on the distance of agent i to the closest point on l.

Following the above definitions, for a pair of agents, the SFM defines a pair of opposite forces, which can lead to undesirable symmetric interactions [13]. As an example, suppose two agents are heading exactly towards each other in an open space. The agents experience forces opposite to their velocity, and decelerate without changing direction, leading to a standstill or collision. To solve this issue, a variant of the SFM introduces a right-of-way mechanism based on agent priorities [10]. The parameter B in the original SFM is replaced by $B_{ij} = B + R_{ji}r_i$, with the right-of-way of agent i over agent j defined as $R_{ij} = \max(1, p_i - p_j)$, if $p_i \geq p_j$, and 0 otherwise; p_i and p_j are the agents' priorities. The direction of the repulsive force is adapted according to R_{ij}. If agent i has higher priority than j, the repulsive force on j is assigned an angle between 0° and 90° to the preferred velocity of i.

2.3 Simulation of Passengers in Public Transport Scenarios

Several existing works have used agent-based simulations to determine the required dwell time of public transport vehicles. Perkins et al. [19] employ SFM to assess the dwell time of a train under various parameter combinations for the door width and placement as well as the crowd mix. Their focus is on the train-platform interface. Thus, in contrast to our work, the geometry and seat placement inside the vehicle as well as the passenger behavior after boarding are not considered.

Fletcher et al. [12] propose an automated procedure to optimize the geometry of a train's interior layout based on evolutionary algorithms and evaluation using agent-based simulation. While their approach provides interesting avenues for future research, the considered parameters of aisle width, seat width, and door width limit the range of possible layouts. The aim of our work is to support industrial designers when evaluating different layouts. Thus, we rely on color-coded floor plans as input so that designers are free in their decisions on aspects such as the seat placement and orientation and the placement of walls.

For parameterization and validation of passenger models, the existing works in the literature have relied on video footage [5,12,19,22], smart-card payment records [21], and experiments using mock-ups [14,20]. In our work, we validate against real-world experiments based on mock-ups from the literature. We are investigating data collection from virtual reality experiments [3] and real-world observations to achieve an accurate parameterization with respect to aspects such as the passengers' seat selection preferences.

3 Scenario

In the following, we describe the scenario considered in our simulations. The AV layout is provided as a color-coded 2D floor plan. As shown in Fig. 1, each layout represents a 6 m × 6 m station comprised of the vehicle of size 6 m × 2.7 m, and the platform (6 m × 3.3 m). The interior region includes regular seats and leaning seats (where passengers can lean on the wall), seat areas, standing areas, and a door. Outside the vehicle, queue areas and an alighting area are specified. Alighting passengers are assumed to exit the scenario through the alighting area. To simplify the boarding process, a vehicle entrance is defined as the position at which boarding agents choose their preferred destination.

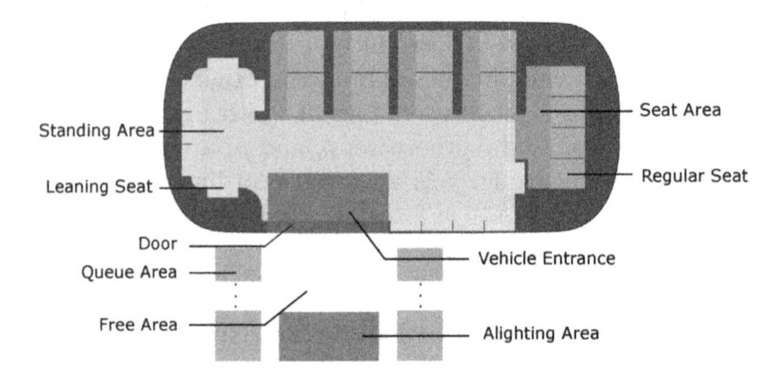

Fig. 1. Color-coded autonomous vehicle layout used as input to the simulation. (Color figure online)

We assume that when the simulation starts, the AV has stopped at the station and passengers are queuing at the platform. A configurable number of passengers then exits the vehicle, while some are assumed to stay inside to continue their trip. Once all alighting passengers have exited, the queuing passengers board the vehicle and move to their preferred seats or a location in the standing area. We are interested in the times taken for all alighting passengers to exit the vehicle, for the boarding passengers to enter the vehicle and to reach their desired locations.

4 Passenger Model

In this section, we describe the high-level decision-making and low-level distance-keeping behavior of our agent-based passenger model. Due to the highly confined environment within the vehicle, our model must address two main challenges: firstly, passengers must be able to pass through corridors narrower than their shoulder width. Secondly, the model must be able to cope with passenger passing each other in confined areas. We developed the model in the CrowdTools [7] simulation framework, which is an implementation of the RPD paradigm (cf. Sect. 2).

4.1 High-Level Decision Making

There are three types of agents in our model:

- **Alighting Passengers (AP):** APs are created in a seat or in the standing area and cross the vehicle center to reach the door. After exiting the vehicle, APs enter the alighting area to disappear from the scene.
- **Boarding Passengers (BP):** BPs are generated within the queue areas and remain idle until all APs have alighted. After subsequently entering the vehicle, each BP chooses a standing point or a seat. Once the desired location has been reached, the BPs become passive.
- **Passive Passengers (PP):** PPs are generated uniformly at random inside the vehicle, occupying a seat or space for standing. Larger numbers of PPs increase the interior density and prolong the boarding and alighting process.

Since the behavior of APs and PPs is relatively simple, our description focuses on the BP behavior. Table 1 lists the action to be executed in each stage in the experience set of BPs: *Navigate* represents long-distance navigation based on the approximated best-first search A^* algorithm [7]; *StraightWalk* represents direct translation only used on unobstructed paths; *WaitAtPoint* represents idling at the waiting point and returning to the position when being pushed away by neighbors; *Still* represents standing at the current position.

Figure 2 shows an excerpt of the decision-making when a BP arrives at the vehicle center, restricted to one stage for each experience. The BP first determines the experience that best matches the cue "Priority to Seats", which is "Go to seat". While the agent carries out the associated action of navigating to the seat, its perception system continuously evaluates the remaining distance and the occupation to the target seat. A violation of the agent's expectations is detected if the target is no longer available. The agent's current goal is considered to be achieved once the agent enters the seat. When an agent exits the current stage after having achieved the associated goal, it enters the subsequent stage of the current experience. In contrast, when the agent exits the stage through a violation of its expectations, a new round of experience matching is triggered.

4.2 Low-Level Collision Avoidance

We rely on the Social Force Model (SFM) with right-of-way extensions (cf. Sect. 2) for the fundamental collision-avoidance behavior among agents. To achieve plausible agent behavior in the confined environment of our boarding and alighting scenario, we propose three modifications to the model.

Size Adaption: The original SFM relies on a circular representation of agents, which overestimates the actual space occupied by a human. As a consequence, the model fails to support scenarios in which pedestrians travel through corridors narrower than their shoulder width. Some existing works propose elliptical [4] or even irregular polygonal representations [2,16] to solve this issue by introducing

Table 1. The experience set of a boarding passenger.

Experience	Stage	Action
QueueExp	Queue	WaitAtPoint
GetOnVehicleExp	GoToDoor	Navigate
	GoToVehicleCenter	StraightWalk
GoToSeatExp	GoToSeatFront	Navigate
	ArrivedSeat	Still
GoToStandingAreaExp	GoToStandingPoint	Navigate
	IdleInStandingPoint	WaitAtPoint

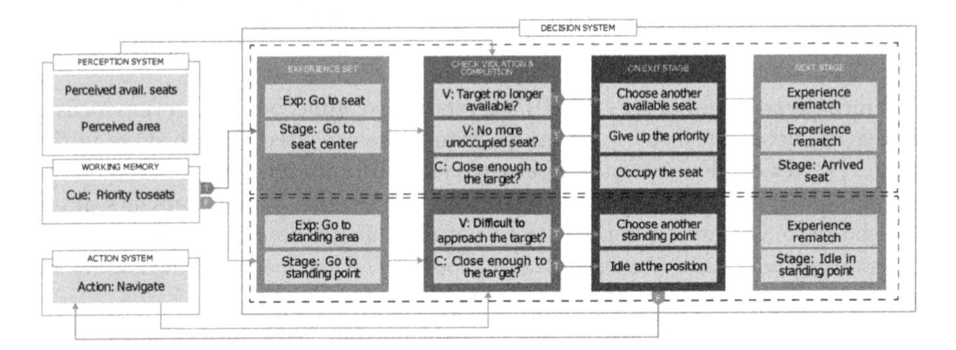

Fig. 2. An excerpt of the decision making of a boarding passenger when determining the next action after entering the vehicle.

lateral motion. However, these approaches substantially increase the complexity of the collision avoidance, both computationally and in terms of the need to address special cases, e.g., when there is insufficient space for an agent to rotate.

We propose a simple solution for the navigation through narrow corridors based on adapting the size of the circles representing agents. Our rationale is that 1. in real-world situations, the rotation performed by passengers to travel through corridors reduces their effective size in perpendicular direction to their target, and 2. if the agent is not directly facing an obstacle, its size in walking direction has only a minor effect on other agents. Thus, temporarily reducing the size of circular agents achieves a similar effect as a rotation of agents represented by more complex shapes, while avoiding the associated complexity.

As shown in Fig. 3a, the agent first perceives the surroundings to determine the closest object (obstacle line or agent). A side sensor S_{side} is generated by mirroring the vector v_{col} from the agent to the closest object across the front sensor S_{front}. A potential corridor is detected if S_{side} is obstructed, whereas S_{front} is unobstructed. When a corridor is detected, the agent reduces its radius for a configurable amount of time, after which the corridor detection is repeated. If no corridor is detected, S_{side} is rotated by 45° over the course of the next few time steps to achieve a sufficient coverage of the area ahead (cf. Fig. 3b).

Force Reduction. The magnitudes of the forces generated by the original SFM are excessively large when considering confined scenarios, where passengers may stand nearby each other or pass other passengers at close distances, while still aiming to avoid potential collision. This is contrary to the common use cases of SFM in large-scale scenarios with open spaces, where pedestrians tend to maintain relatively large distances from each other. Further, in our scenario, strong repulsive forces from stationary obstacles make it impossible for agents to traverse the seat edge areas and to enter seats.

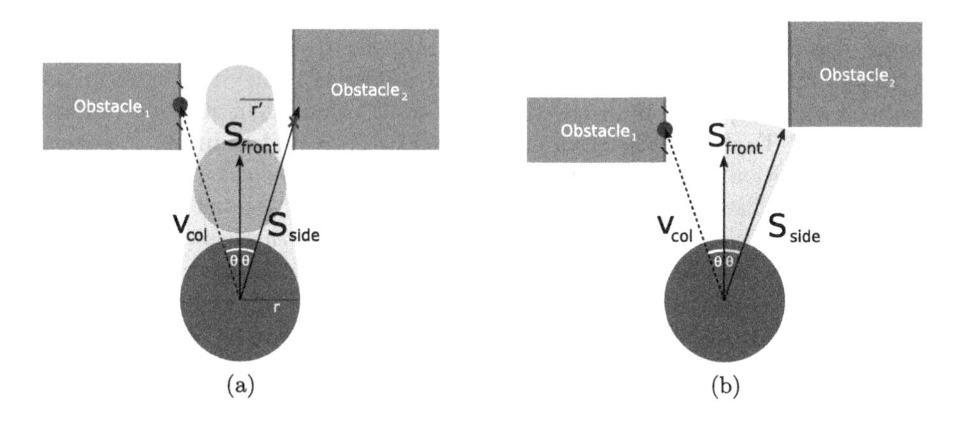

Fig. 3. Size adaption when traversing a narrow corridor.

We assume that passengers tend to maintain smaller distances to objects nearby within the vehicle than in an open space, and that they keep larger distance from neighboring passengers than from stationary obstacles. Although reducing the agent size makes it possible for agents to traverse paths that are narrower than their shoulder width, they may still fail to enter a corridor because of the strong repulsive force from the corridor walls. We address this issue by reducing the forces using two new scaling factors λ_{agent} and λ_{obs}:

$$m_i \frac{\mathrm{d}v_i}{\mathrm{d}t} = m_i \frac{v_i^0(t) - v_i(t)}{\tau_i} + \lambda_{agent} F_{nb} + \lambda_{obs} F_{obs}$$

We assign λ_{agent} and λ_{obs} values between 0 and 1. Reducing these forces also enables agents to stand both closer to each other and closer to walls, which is necessary to achieve high passenger densities commonly observed in public transport situations. In our simulation experiments, the specific values for the scaling factors were hand-tuned based on visual inspection of trial simulations, with values adapted according to the area in which each agent currently resides. The specific values are given in Sect. 5.

Modification of Right-of-Way Behavior: We employ an existing extension to the SFM that relies on agent priorities to model right-of-way behavior.

Three levels of the agent priority p are defined in our model. All types of agents start with an initial p value of 0. During the simulation, APs and BPs increase their priority to 1 when they are navigating inside the vehicle, and reset the value when they exit the vehicle or reach their destination inside the vehicle. An agent i with priority equal to 0 yields to another agent j with higher priority by moving in the direction perpendicular to the walking direction of j. For all agent types, p is set to -1 when the agent is seated. An agent with negative priority does not receive any social force, ensuring that the agent retains its position.

In certain situations, the above behavior fails to allow agents to pass each other successfully. When the repulsive force on an agent from a higher-priority neighbor is pointing to an obstacle, there is no space to step backward, as shown in Fig. 4a. Simply reversing the repulsive force does not solve this issue in our confined scenario, since both the back and front of an agent may be blocked by obstacles. To detect this situation, an extra step is added during the agent force computation: The agent first perceives the distance to the closest obstacle line *col*. If the distance only exceeds the agent's radius by a small positive ϵ, we assume that the agent cannot give way to its neighbor. In this case, the repulsive force is reoriented to be parallel to *col*, so that it is possible for the agent to avoid the approaching neighbor. The direction of the force from the higher-priority agent j on agent i is adjusted depending on the external distance D_{ij} as follows:

$$= \begin{cases} g(d_{avoid}, -v_j^0(t)), & D_{ij} > 2D_{min} \\ g(d_{avoid}, v_j^0(t)), & \text{otherwise} \end{cases}$$

where d_{avoid} is the direction from agent i to the closest obstacle line and D_{min} is a small positive value. The function $g(v_1, v_2)$ determines a unit vector that is perpendicular to v_1, with an orientation difference of less than 90° from v_2. In the more common cases of $D_{ij} > 2D_{min}$ as shown in Fig. 4b, where two agents are still positioned far from the point of a projected collision, agent i is driven by the adjusted force and walks towards agent j. Agent j triggers the size adaption of both agents, freeing up sufficient space for navigation. When two agents are close enough to each other so that both of their front sensors S_{front} are obstructed (cf. Fig. 4c), the size adaption algorithm is not activated. In this case, the agent i moves in the direction of the preferred velocity of agent j.

5 Evaluation

In this section, we present the validation of our model against results from real-world experiments from the literature. Further, we demonstrate the practical use of the model by comparing the boarding and alighting times required by different layouts of an autonomous vehicle created by industrial designers.

5.1 Parameter Setup

In Sect. 4, we described our adaptations of the Social Force Model to allow passengers to navigate the confined space inside the vehicle. To reduce the intensity of the forces from other agents and obstacles inside the vehicle, the scaling

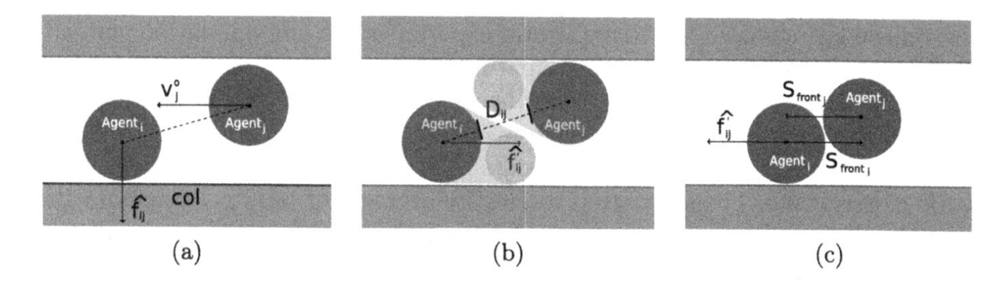

Fig. 4. Illustration of a situation that cannot be resolved by the original right-of-way SFM (a), and two solutions (b) and (c) depending on the distance between agents.

parameters are configured as follows for agents outside the vehicle, inside the vehicle, and on the seat edge: $\lambda_{agent} = 0.8, 0.6, 0.2$; $\lambda_{obs} = 0.2, 0.1, 0.01$.

To retain the distance kept by passengers even in confined spaces, the force generated by obstacles is reduced more sharply than that of agents. While our model allows modelers to specify passenger personas defined by preferred walking speeds and seat selection preferences, here we assume all passengers to be working adults, with a preferred speed of 1.4 m/s outside the vehicle, which is reduced to 0.56 m/s inside the vehicle, and to 0.28 m/s in the seat edge area. Seat selection is assumed to be uniformly at random. We are currently investigating approaches for data collection from virtual reality experiments [3] and real-world experiments to support the representation of different passenger types.

5.2 Validation Experiment

In 2010, Fernández et al. carried out experiments on the influence of the platform and vehicle design on boarding and alighting times as well as passenger saturation flows of a public transport vehicle [11,14]. The participants of the experiment repeatedly boarded or alighted a full-scale mock-up. We recreated

Table 2. Validation results.

Scenario and metric	Experiment	Simulation	Deviation	SD
Narrow door, Average alighting time	1.22 s/pass	1.23 s/pass	0.01 (0.82%)	0.080
Narrow door, Alighting saturation flow	0.85 pass/s	0.83 pass/s	0.02 (2.35%)	-
Wide door, Average alighting time	0.73 s/pass	0.81 s/pass	0.08 (10.96%)	0.050
Narrow door, Average boarding time	1.54 s/pass	1.56 s/pass	0.02 (1.30%)	0.169
Wide door, Average boarding time	1.18 s/pass	1.22 s/pass	0.04 (3.39%)	0.097

the vehicle layouts used in the real-world experiments in our simulation and validate our agent-based models by comparing the results from these experiments to our simulation results. In the experiments, 50 and 25 passengers took part in the alighting and boarding process, respectively.

Table 2 compares our simulation results with the measurements from the real-world experiments. The simulation results are averages of 100 iterations for each scenario. We can observe that the boarding and alighting times in our simulation are slightly longer than in the real-world experiments. Overall, given the observed deviations between 0.8% and 11%, we consider the model sufficiently accurate to provide estimates of real-world boarding and alighting times.

5.3 Case Study

We demonstrate the practical use of our model by evaluating the times taken for the boarding and alighting process when considering three different autonomous vehicle layouts. The following metrics are considered:

- **Alighting time:** The time until alighting passengers have exited the vehicle.
- **Boarding time:** The time until all alighting passengers have exited and all boarding passengers have entered the vehicle.
- **Settling time:** The time at until all boarding passengers have settled on a seat or in the standing area.

Figure 5 shows the interior layouts to be evaluated. Two of the layouts represent extreme cases with respect to the seat numbers and thus passenger comfort: "standing only" does not provide any regular seats, leaving a large amount of space for navigation inside the vehicle, whereas "maximum seating" provides 16 seats in total. The layout "balanced" aims to provide a sufficient number of seats, while still allowing passengers to navigate comfortably. Further, the layout leaves space near the door to place a wheelchair.

We populated each layout with 6 passive passengers, varying both the number of boarding and alighting passengers between 1 and 6. Given the 100 iterations for each parameter combination, 3600 iterations were executed for each layout.

The results shown in Fig. 6 indicate that, as expected, the spacious area for navigation in the layout "standing only" allowed for the shortest times for the

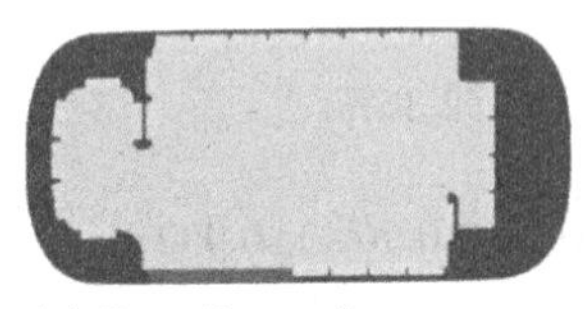

(a) Standing only: no regular seats, maximizes space for movement.

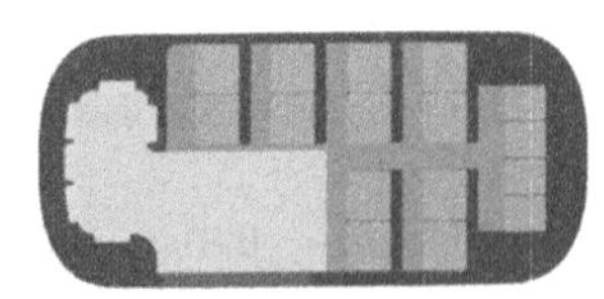

(b) Maximum seating: maximizes the number of regular seats.

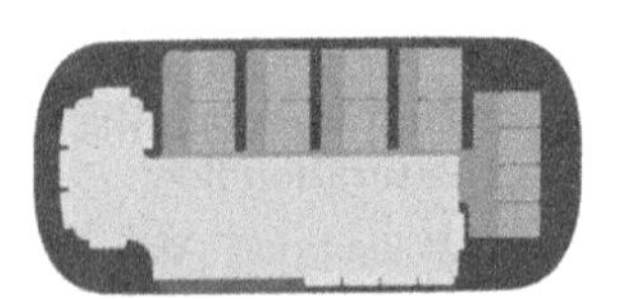

(c) Balanced: provides regular seats while preserving space for movement.

Fig. 5. The autonomous vehicle interior layouts evaluated in the case study.

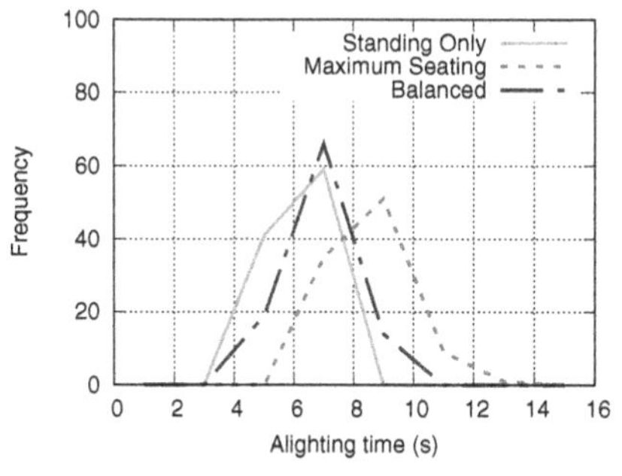

(a) Time until alighting passengers have exited the vehicle.

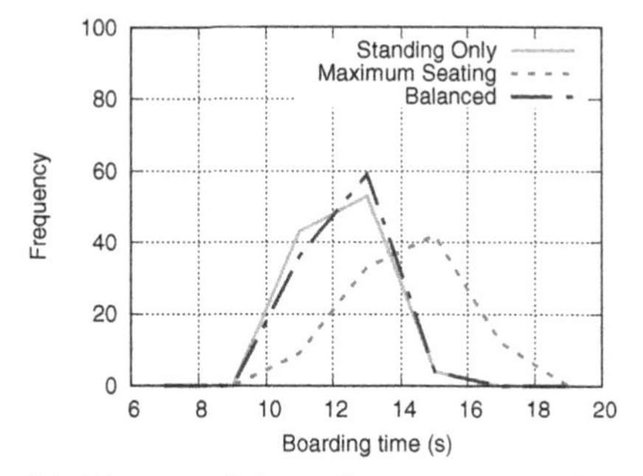

(b) Time until boarding passengers have entered the vehicle.

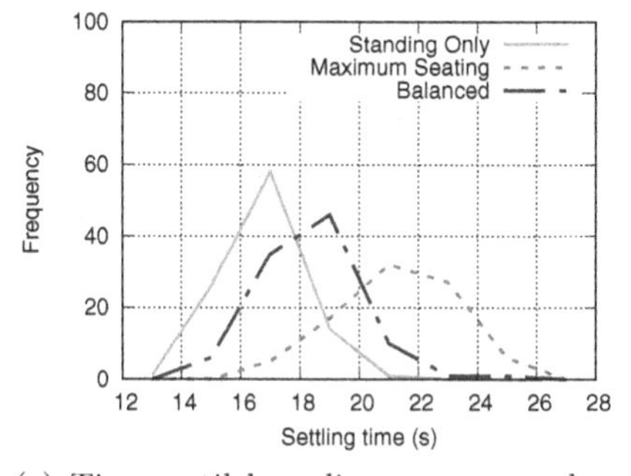

(c) Time until boarding passengers have reached their seats or standing positions.

Fig. 6. Simulation results for the three layouts in the case study.

alighting, boarding and settling processes, with averages of 6.0, 12.0, and 16.8 s. The layout "maximum seating" generated the largest values for all three metrics, with averages of 8.1, 13.6, and 20.8 s. The boarding and alighting times achieved with the layout "balanced" are only slightly longer than those for the "standing only" layout. Due to the time taken for passengers to enter their seats, the settling time is significantly larger than with the "standing only" layout. The effect of the settling time on the required dwell time depends on the point in time when the vehicle starts its trip. While the vehicle may start its trip as soon as all passengers have entered the vehicle, considerations of passenger comfort may suggest a delay to allow passengers to take their seats.

In the layout "maximum seating", passengers that have already taken their seats may make it cumbersome for others to get seated. Thus, the variability in settling times is particularly large with this layout.

According to our simulation-based evaluation, the lowest boarding, alighting, and settling times are achieved with the layout "standing only". While the purely quantitative evaluation permits comparisons in terms of efficiency, it must also be ensured that the selected layouts satisfy other requirements such as those given by Universal Design principles [8], as well as service quality indicators for public transportation as defined by standards such as EN 13186 [1].

6 Conclusion and Future Work

We presented a novel passenger model for evaluating public transport vehicle layouts in terms of the required dwell time. The low-level model is based on the traditional Social Force Model with additional right-of-way features. A dynamic adaption of the agent size emulates turning behavior. The RPD framework is used to define the passenger behavior, allowing for an adaptive decision-making based on the current situation faced by agents. We validated our model by replicating an existing mock-up experiment in our simulation. The observed deviations in boarding and alighting times were below 11%. We applied our simulation to autonomous vehicle layouts created by industrial designers, demonstrating substantial differences in the dwell times depending on the seat number and positioning. By relying on color-coded floor plans as the input to the simulation, different vehicle layouts can be evaluated easily. Of course, apart from efficiency metrics such as the dwell time, viable vehicle designs must also be evaluated in terms of the provided service quality through features such as step-free access, space for wheelchair users, and the presence of handrails.

In future work, we aim to extend our experiments to passengers with different characteristics. To this end, we are exploring data collection from virtual reality experiments [3] and real world observations. Potential future refinements include more complex passenger interactions in confined environments (e.g., exiting the vehicle to allow passengers to alight), joint decision-making (e.g., assisting elderly people), and crowd mixes at specific times of day (e.g., during peak hours).

References

1. Transportation – Logistics and Services – Public Passenger Transport – Service Quality Definition, Targeting and Measurement. Standard EN 13816:2002, European Committee for Standardization (2002)
2. Alonso-Marroquin, F., Busch, J., Chiew, C., Lozano, C., Ramírez-Gómez, A.: Simulation of counterflow pedestrian dynamics using spheropolygons. Phys. Rev. E **90**, 063305 (2014)
3. Andelfinger, P., et al.: Incremental calibration of seat selection preferences in agent-based simulations of public transport scenarios. In: Proceedings of the Winter Simulation Conference (WSC), Gothenburg, Sweden, pp. 833–844, December 2018
4. Best, A., Narang, S., Manocha, D.: Real-time reciprocal collision avoidance with elliptical agents. In: 2016 IEEE International Conference on Robotics and Automation (ICRA), pp. 298–305, May 2016
5. Bian, B., Zhu, N., Ling, S., Ma, S.: Bus service time estimation model for a curbside bus stop. Transp. Res. Part C Emerg. Technol. **57**, 103–121 (2015)
6. Bode, N.W.F., Wagoum, A.U.K., Codling, E.A.: Human responses to multiple sources of directional information in virtual crowd evacuations. J. R. Soc. Interface **11**(91), 20130904 (2014)

7. Cai, W.,et al.: COSMOS: CrOwd simulation for military OperationS. Technical report, School of Computer Eng., Nanyang Technological University, Singapore, July 2010
8. Connell, B.R., et al.: The Principles of Universal Design (1997). https://projects.ncsu.edu/www/ncsu/design/sod5/cud/about_ud/udprinciplestext.htm
9. Curtis, S., Guy, S.J., Zafar, B., Manocha, D.: Virtual Tawaf: a case study in simulating the behavior of dense, heterogeneous crowds. In: International Conference on Computer Vision Workshops, Barcelona, Spain, pp. 128–135, November 2011D
10. Curtis, S., Zafar, B., Gutub, A., Manocha, D.: Right of way: asymmetric agent interactions in crowds. Vis. Comput. **29**(12), 1277–1292 (2013)
11. Fernández, R., Zegers, P., Weber, G., Tyler, N.: Influence of platform height, door width, and fare collection on bus dwell time: laboratory evidence for Santiago de Chile. Transp. Res. Rec. J. Transp. Res. Board **2143**, 59–66 (2010)
12. Fletcher, D., Harrison, R., Karmakharm, T., Nallaperuma, S., Richmond, P.: Rate-Setter: roadmap for faster, safer, and better platform train interface design and operation using evolutionary optimisation. In: Proceedings of the Genetic and Evolutionary Computation Conference, Kyoto, Japan, pp. 1230–1237. ACM, July 2018
13. Gao, Y., Luh, P.B., Zhang, H., Chen, T.: A modified social force model considering relative velocity of pedestrians. In: International Conference on Automation Science and Engineering (CASE), Wisconsin, USA, pp. 747–751, August 2013
14. Helbing, D., Molnar, P.: Social force model for pedestrian dynamics. Phys. Rev. E **51**(5), 4282 (1995)
15. Klein, G.: The recognition-primed decision (RPD) model: looking back, looking forward. Nat. Decis. Mak., 285–292 (1997)
16. Langston, P.A., Masling, R., Asmar, B.N.: Crowd dynamics discrete element multi-circle model. Saf. Sci. **44**(5), 395–417 (2006)
17. Luo, L., Zhou, S., Cai, W., Lees, M., Low, M.Y.H., Sornum, K.: HumDPM: a decision process model for modeling human-like behaviors in time-critical and uncertain situations. In: Gavrilova, M.L., Tan, C.J.K., Sourin, A., Sourina, O. (eds.) Transactions on Computational Science XII. LNCS, vol. 6670, pp. 206–230. Springer, Heidelberg (2011). https://doi.org/10.1007/978-3-642-22336-5_11
18. Narang, S., Best, A., Manocha, D.: Interactive simulation of local interactions in dense crowds using elliptical agents. J. Stat. Mech. Theory Exp. **2017**(3) (2017). Article number: 033403
19. Perkins, A., Ryan, B., Siebers, P.O.: Modelling and simulation of rail passengers to evaluate methods to reduce dwell times. In: 14th International Conference on Modeling and Applied Simulation, MAS 2015, Rende, Italy, pp. 132–141 (2015)
20. Seriani, S., Fernandez, R.: Pedestrian traffic management of boarding and alighting in metro stations. Transp. Res. Part C Emerg. Technol. **53**, 76–92 (2015)
21. Sun, L., Tirachini, A., Axhausen, K.W., Erath, A., Lee, D.H.: Models of bus boarding and alighting dynamics. Transp. Res. Part A Policy Pract. **69**, 447–460 (2014)
22. Zhang, Q., Han, B., Li, D.: Modeling and simulation of passenger alighting and boarding movement in Beijing metro stations. Transp. Res. Part C Emerg. Technol. **16**(5), 635–649 (2008)

MLP-IA: Multi-label User Profile Based on Implicit Association Labels

Lingwei Wei[1,2], Wei Zhou[1,2(✉)], Jie Wen[1,2], Meng Lin[1,2], Jizhong Han[1,2], and Songlin Hu[1,2]

[1] School of Cyber Security, University of Chinese Academy of Sciences, Beijing, China
[2] Institute of Information Engineering, Chinese Academy of Sciences, Beijing, China
zhouwei@iie.ac.cn

Abstract. Multi-Label user profile is widely used and have made great contributions in the field of recommendation systems, personalized searches, etc. Current researches on multi-label user profile either ignore the associations among labels or only consider the explicit associations among them, which are not sufficient to take full advantage of the internal associations. In this paper, a new insight is presented to mine the internal correlation among implicit association labels. To take advantage of this insight, a multi-label propagation method with implicit associations (MLP-IA) is proposed to get user profile. A probability matrix is first designed to record the implicit associations and then combine the multi-label propagation method with this probability matrix to get more accurate user profile. Finally, this method proves to be convergent and faster than traditional label propagation algorithm. Experiments on six real-world datasets in Weibo show that, compared with state-of-the-art methods, our approach can accelerate the convergence and its performance is significantly better than the previous ones.

Keywords: Implicit association labels · User profile · Multi-label propagation

1 Introduction

As the consumer economy enters an era of “information overload” [1], a large number of personalized service platforms, including recommendation systems, begin to emerge to satisfy human’s more individualized demands. User profile, the fully understanding of users, is the basis of recommendation system [19] and exact-marketing [2, 3]. As a result, it is essential to provide an accurate and effective method to get user profiles.

Recently, user profile is widely studied by label propagation methods inferring user interests [4–6], user authority [21] and social attributes [20]. In order to get more accurate and abundant user profile, many researches prefer to apply multiple labels to analyze user profile. Some researches assumed labels were independent [5], ignoring the associations among them. However, it is not consistent with reality and cannot mine hidden label features very well. To overcome this limitation, Glenn et al. [1] considered the associations among labels to get user profile and obtained better performance.

J. M. F. Rodrigues et al. (Eds.): ICCS 2019, LNCS 11536, pp. 548–561, 2019.
https://doi.org/10.1007/978-3-030-22734-0_40

The associations in these works are explicit, indicating that there are some clear and definable connections among labels. For example, "Photography" and "Camera" are related, so they are called explicit association. As a result, a large number of methods based on explicit association were proposed and have achieved good performance [2, 22]. However, the existing methods rarely consider the implicit association of labels, where there are internal but not direct connections among them.

In many real-world applications, the associations among labels are complex [25]. With our observation, in addition to the explicit associations, there still exists some associations among implicit labels due to various reasons, such as uncertainty [23] or privacy issues [24]. For example, "Travel" and "health" are not related in any cases. However, they have many inherent connections, because people who like travelling always pay more attention to their health. Utilizing the implicit associations among labels, it is beneficial to make user profile more accurate and comprehensive.

To take advantage of this insight, a multi-label propagation method with implicit label associations (MLP-IA) is proposed to get user profile. We first design a probability matrix to record the implicit associations and then combine this probability matrix to multi-label propagation method to get more accurate user profile. Finally, we prove that the method is convergent and faster than traditional label propagation algorithms. To sum up, the main contributions of this study can be summarized as follows:

- **Insight.** We present a novel insight about associations among implicit association labels. In social platforms, due to users' social and living habits, there are still certain implicit associations among labels. At the same time, mining the associations is useful for the construction of user profile.
- **Method.** A multi-label propagation method with implicit label associations is proposed to get user profile. We first design a probability matrix to record the implicit associations and then combine the multi-label propagation method with this probability matrix to get more accurate user profile. Finally, we prove that the method is convergent and faster than traditional label propagation algorithm.
- **Evaluation.** We conduct experiments to evaluate our method on six real Weibo data sets of different sizes. The comparative experiments evaluate the accuracy and effectiveness of the proposed method. The results show our method can accelerate the convergence and the performance is significantly better than the previous methods.

The following chapters are organized as follows: In Sect. 2, related works are elaborated in details. The Sect. 3 explores our insights about the implicit association labels, and Sect. 4 describes the details of the proposed method and its efficiency. In Sect. 5, experiments and results are described. Finally, conclusions and future work are drawn in Sect. 6.

2 Related Works

The existing researches in user profile can be divided into two parts. One is to infer user's unknown attributes based on the user's own data by text-mining methods, and the other is to propagate labels by social-network structure.

2.1 Text-Mining Methods

There are many text-mining methods to extract user profile. The user's own data generally contains rich semantic information, so the user profile problem can be regarded as a text analysis problem [7, 8]. For user's interest profile, most researchers apply the topic model (LDA) to complete the keyword extraction on the blog, and then use TFIDF algorithm to select features [9–11]. However, text mining often has high complexity, and the extracted profiles are unstable because of the richness of semantics.

2.2 Social-Network Structure

The method is to label the unknown users based on the known users' labels by propagation, and multi-label algorithms are widely applied. Zhang et al. used multi-label propagation algorithm to mine user interests, and discovered potential interests of users through social relationships [6]. Xie et al. proposed the speaker-listener mechanism to update the label [12]. Dong et al. considered inference of gender and age using a social network, which is feasible only when the set of attribute values is extremely restricted [13]. To approach this, Chakrabarti et al. [14] proposed a method called EDGE-EXPLAIN and Besel et al. built interest profile by propagating activation functions, and proved the method was more suitable than the most advanced methods [15]. Ma et al. innovatively introduced label propagation to improve the accuracy of the semi-supervised learning algorithm [16].

Some scholars found there were links between labels. Recently, the explicit associations among labels have been taken into consideration, Glenn et al. [1] introduced the explicit association labels and the results showed that their method performed well. To the best of our knowledge, there are few researches on the associations among implicit association labels.

3 Priori Knowledge with Implicit Association Labels

As is observed, in many social platforms, because of hot spot events or other special reasons, there are certain associations among implicit association labels. For example, as is shown in Fig. 1, the node represents a user in Weibo and a directed edge indicates the user's social relationship. As highlighted with orange labels, we find the majority of users who like Entertainment in Weibo like Health as well.

We analyze the statistical characteristics of user interest labels in Weibo by correlation analysis [26] and a higher value indicates that there are certain associations among implicit association labels. We have show the top five label pairs in Fig. 2 and the value is the correlation score of interest labels.

The statistical results are explicable. For example, Fig. 2 shows that there are some associations between Health and Tourism. In the real world, users who like healthy lifestyle tend to pay more attention to tourism information. Obviously, they can enrich their lives through tourism and develop a healthy life with a relaxing lifestyle.

Based on our observation, implicit associations exists among labels and the features can be fully utilized to build a better user profile. Therefore, we will introduce the priori

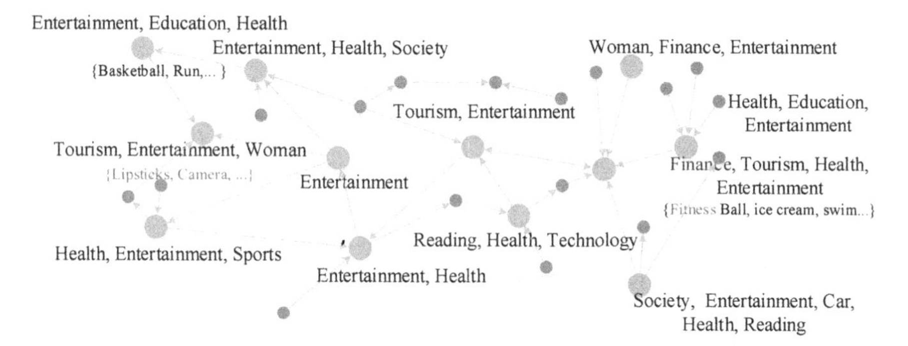

Fig. 1. An Example of interest labels propagation in Weibo. Note: nodes with labels indicate that users in Weibo where labels are the topic users are interested in. Green nodes are users with high influence such as V-plus users and orange nodes are some ordinary users. A directed edge indicates the user's social relationship. Some explicit labels are highlighted with blue color. (Color figure online)

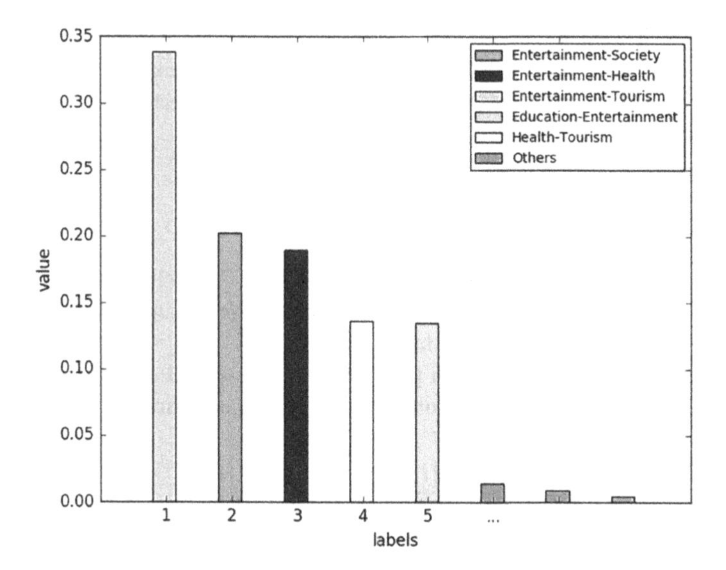

Fig. 2. Results of labels association analysis.

probability among implicit association labels to improve user profile model. The details will be illustrated in next section.

4 Our Model

This section mainly focuses on the improved multi-label propagation. Firstly, we will construct the priori knowledge to introduce the associations among implicit labels. And then two major matrixes in multi-label propagation algorithm will be initiation for propagation. Next, the model will be trained via labeled users and we will get the unlabeled users' label after a finite number of iterations.

The symbols mentioned in the paper are shown in Table 1.

Table 1. Symbols of our paper

Symbol	Description
I	A set of a pair of labels
P	A matrix of priori probability of associations among implicit labels
p_{ij}	An element of P, represents priori probability of implicit association between label i and label j
$F^{(t)}$	The t-th iteration of *LABEL*. And the initialization is recorded as $F^{(0)}$
T	A matrix of probability transferred among users
R	A set of users' social relationships and the element (u_i, u_j) represents u_i follows u_j. $\mathrm{R} = \{(u_i, u_j) \vert u_i, u_j \in U\}$
FANS	A set of users' fans and the element $\mathrm{FANS_i}$ represents the fans set of user i
FOLLOW	A set of users' followers and the element $\mathrm{FOLLOW_i}$ represents the user set of user i follows
C	A matrix that represents the connections between users
c_{ij}	An element of C, represents whether there is a connection between user i and user j

4.1 Priori Knowledge of Implicit Association Labels

From Sect. 3, a new insight about associations among implicit association labels has been found. We analyze the interest labels of users in Weibo and find there is a certain connection among different interest labels.

Specifically, we define the priori probability matrix P as Eq. 1 shows. The higher the value of $\mathrm{p_{ij}}$ is, the higher the probability of propagation among labels becomes.

$$\mathrm{p_{ij}} = \frac{\left|\left\{t \vert t \in I \wedge (l_i, l_j) \subseteq t\right\}\right|}{Z} \tag{1}$$

where $Z = \sum_{i=0}^{m} \sum_{j=0}^{m} \left|\left\{t \vert t \in I \wedge (l_i, l_j) \subseteq t\right\}\right|$. Some scholars have proved that the associations in social network are complex due to various reasons, such as uncertainty [23] or special events [24]. Therefore, we define that elements of I by co-occurrence, cultural associations, event associations or custom associations and so on. The detail is shown in Eq. 2.

$$\mathrm{I} = \mathrm{I_1} \cup \mathrm{I_2} \cup \mathrm{I_3} \cup \ldots \tag{2}$$

where $\mathrm{I_i}(\mathrm{i} = 1, 2, 3, \ldots)$ represent respectively a set of each user's interest label set, label sets sampled by cultural associations, event associations and custom associations.

4.2 Introduction of the Priori Knowledge

As our observation, in addition to other users' influence, the label will spread according to both other users and other labels, that is $F = U \cdot L \cdot P$. The labels will be propagated among users and each iteration is given by Eq. 3, where λ is a hyper parameter and it controls the influence of initialization.

$$\mathrm{F}^{(t+1)} = \lambda T \cdot F^{(t)} \cdot P + (1 - \lambda)F^{(0)} \tag{3}$$

The hyper parameter λ controls the influence of initialization. In each iteration, the users' labels will be updated by the neighbor node's labels and the implicitly associated labels. It's noted that after each iteration, the matrix F will be corrected by F_a for next correct propagation, which shows that our model is a semi-supervised learning method.

The loss function in the model uses the squared distance, as is shown in Eq. 4.

$$\mathrm{loss} = \left|\mathrm{F}^{(t+1)} - \mathrm{F}^{(t)}\right|^2 + \frac{1}{2}\|1 - \zeta\|^2 \tag{4}$$

In real world network, it is difficult to construct complete structure of networks because of privacy security. Therefore, we consider the influence of the integrity of social networks. In the model, a hyper parameter ζ is introduced.

We define $\zeta = \frac{\textit{the number of relationships in dataset.}}{\textit{the number of relationships in the real world.}}$, which represents the sparsity of social networks and the value indicates the integrity of a social network. When more relationships are added for constructing graph, ζ will tend to 1 and the model will get smaller loss value accordingly to make a better user profile.

4.3 Multi-label User Profile Based on Implicit Association Labels

Traditionally, given labels set $L = \{l_1, \ldots, l_m\}$, $U = \{u_1, \ldots, u_a, \ldots, u_{a+b}\}$, which contains a users with labels and b users without labels and their labels matrix $F = [F_a; F_b]$.

Firstly, if nodes are in a graph, multi-label propagation algorithm infers the labels via the aggregate labels of their neighbors until labels for all the nodes do not change [26]. The key to inference is the probability between nodes. In our model, two major matrixes, the user's initial interest vector matrix and label transfer matrix, are constructed. The details of two matrixes are as follows.

Initial Interest Vector Matrix. The initial interest vector matrix contains two parts. The one-hot method is used to build the labeled users' initial vector of interest labels. As is shown in Eq. 5, if the user is with the label, the value will be *1*. Otherwise, it will be *0*.

$$f_{ij} = \begin{cases} 1 & \text{if the user i is with the label j} \\ 0 & \textit{otherwise} \end{cases} \tag{5}$$

The unlabeled users' initial vector is zero vector. From this, we transform the multiple interest labels into the multi-label vector for propagation. It is worth noting that after each update iteration of label propagation algorithm, it is necessary to correct the labeled users' interest label matrix so as to obtain more accurate results of propagation.

Probability Transfer Matrix. The interest labels will spread among users. The label transfer matrix *T* is constructed via social relationships shown in Eq. 6. Here we introduce two matrixes D and W for better convergence, which had proved in [17]. Elements of D are computed by Eq. 7 and elements of W are computed by Eq. 8.

$$\mathrm{T} = \mathrm{D}^{-\frac{1}{2}} W \mathrm{D}^{-\frac{1}{2}} \tag{6}$$

$$\mathrm{d_{ii}} = \sum\nolimits_{j=1,\ldots,n} w_{ij} \tag{7}$$

$$\mathrm{w_{ij}} = \begin{cases} c_{ij} \times \mathrm{Sim(i,j)} & i \neq j \\ 0 & i = j \end{cases} \tag{8}$$

where *Sim*(*i*,*j*) indicates the similarity between user *i* and user *j* is expressed as is shown in the Eq. 9. That is, the less the ratio of the value is, the closer the distance is.

$$\mathrm{Sim(i,j)} = \frac{1}{\left| \log\left(\frac{|FANS_i|}{|FOLLOW_i|}\right) - \log\left(\frac{|FANS_j|}{|FOLLOW_j|}\right) \right| + 1} \tag{9}$$

Next, we will train the model via labeled users and the loss function is defined in Eq. 4. After each iteration in Eq. 3, the model will stop until loss is less than threshold that we set. The method is proved to be convergent in Sect. 4.4. The specific algorithm flow is shown in Algorithm 1.

Algorithm 1 MLP-IA

	Require: $\mathrm{L} = \{\mathrm{l}_1, \ldots, \mathrm{l}_m\}$, $\mathrm{U} = \{\mathrm{u}_1, \ldots, \mathrm{u}_a, \ldots, \mathrm{u}_{a+b}\}$, $\mathrm{I_i}(\mathrm{i} = 1,2,3,\ldots)$
1	Construct user interest label matrix F via Eq.5
2	Estimate the interity of social networks ζ
3	Construct set $\{\, t \mid t \in I \wedge (l_i, l_j) \subseteq t \,\}$
4	Construct priori probability matrix *P* via Eq.1
5	Estimate *Sim(i,j)* based on data sets.
6	Construct the matrix D and W via Eq.7 and Eq.8
7	Learn the probability transfer matrix T via Eq.6
8	**while** *loss >THRESHOLD* **do**
9	$\mathrm{F}^{(t+1)} = \lambda \mathrm{T} \cdot \mathrm{F}^{(t)} \cdot P + (1-\lambda)\mathrm{F}^{(0)}$
10	Fixed Source User Interest Label $\mathrm{F_a}$
11	Update loss via Eq.4
12	**end while**
13	Output $\mathrm{F_b}$

4.4 Analysis of Algorithms

Convergence Analysis. The convergence of our method is shown as follows. Let the user's label matrix be F. According to the labeled and unlabeled users, T can be divided into sub-matrices as is shown in Eq. 10, where subscript "a" indicates the user's label is known and subscript "b" indicates the user's label is unknown.

$$\mathrm{T} = \begin{bmatrix} T_{aa} & T_{ab} \\ T_{ba} & T_{bb} \end{bmatrix} \tag{10}$$

From the section above, we can see that the core formula of the label propagation algorithm proposed in this paper is Eq. 3. P matrix represents the co-occurrence relationship of labels, and $0 \leq \mathrm{p_{ij}} \leq 1$. F_a is the interest label of the source user, which is fixed and invariant. Therefore, our method is simplified as $\mathrm{F_b} \leftarrow T_{bb}\mathrm{F_b}P + T_{ba}\mathrm{F_a}P$, which leads to Eq. 11.

$$\mathrm{F_b} = \lim_{\mathrm{n}\to\infty} (\mathrm{T_{bb}})^{\mathrm{n}} F_b^{(0)} P^n + \left(\sum\nolimits_{i=1}^{n} (T_{bb})^{(i-1)}\right) T_{ba} F_a P^n \tag{11}$$

where $F_b^{(0)}$ is the initial value of F_b. Because T matrix is row-regular, T_{bb} is a submatrix of T, so it follows Eq. 12. Therefore, we can get Eq. 13.

$$\exists \gamma < 1, \sum_{j=1}^{u} (T_{bb})_{ij} \leq \gamma, \forall i = 1, \ldots, u \tag{12}$$

$$\begin{aligned} \sum\nolimits_j (T_{bb})_{ij}^n &= \sum_j \sum_k (T_{bb})_{ik}^{(n-1)} (T_{bb})_{kj} = \sum_k (T_{bb})_{ik}^{(n-1)} \sum_j (T_{bbb})_{kj} \\ &\leq \sum_k (T_{bb})_{ik}^{(n-1)} \gamma \leq \gamma^n \end{aligned} \tag{13}$$

P is priori knowledge and $0 \leq \mathrm{p_{ij}} \leq 1$, which can accelerate convergence. And $(\mathrm{T_{bb}})^{\mathrm{n}}$ indicates the sum of each line converges to 0, from which we can conclude in $(\mathrm{T_{bb}})^{\mathrm{n}} F_b^{(0)} P^n \to 0$. Thus the initial value of $F_b^{(0)}$ is inconsequential. Obviously, $\mathrm{F_b} = (\mathrm{I} - \mathrm{T_{bb}})^{-1} T_{ba} F_a$ is a fixed point.

Time Complexity Analysis. In the initialization of the label propagation algorithm, we need to establish an initial label for each Weibo user, and the complexity of the process is *O(n)*. In the propagation of the interest label, due to the convergence of our method, the iterations are fixed. And the time complexity of each iteration is *O(m)*. Therefore, the entire algorithm is nearly linear in time complexity.

5 Experiments

5.1 Dataset

Like Twitter, Weibo is the largest social network platform in China. To prove the universality and effectiveness of our method, we evaluate our method in different scale data sets in Weibo.

Firstly, we randomly get six different sets of users and their social datas such as followers, fans and blogs etc. The scale of the data sets are illustrated in Table 2. And different data sets are collected at different times. Due to the limit of Weibo, we just get part of their following users and obtain sparse social relationships of users.

Table 2. Dataset of our paper

Dataset	Number of labeled users	Number of unlabeled users	Size
1#	1129	4001	5130
2#	1355	4801	6156
3#	1581	5601	7182
4#	1807	6401	8208
5#	2033	7201	9234
6#	2200	8700	10900

Table 3. Baselines of our paper

Notation	Description
MLP	Multi-label propagation with any association among labels
MLP-EA	Introduce associations among explicit association labels
MLP-EIA	Introduce associations among both explicit and implicit association labels

Next, according to the characteristics of Weibo, we artificially labeled users' interest with interest labels based on their blogs and social relationships. The labeled users are selected if the user is marked with a "V" which means his identity had been verified by Sina. Analyzed by Jing et al. [18], these users were very critical in the propagation.

5.2 Comparisons and Evaluation Setting

To evaluate the performance of our method (MLP-IA), we compare it with other methods. Table 3 lists some compared baselines. We first compare with traditional Multi-Label Propagation (MLP) to evaluate the effectiveness of priori knowledge. Then we select Multi-Label Propagation Based on Explicit Association Labels (MLP-EA) to evaluate whether implicit association labels can perform better to mine the relationship than explicit association labels learning from [1]. In baseline three: Multi-Label Propagation Based on Explicit and Implicit Association Labels (MLP-EIA), we introduce associations among both explicit and implicit association labels to experiment, fully exploring the relationships among labels. Finally, to explore the impact of social relationship integrity on results, we make some new experiments in Sect. 5.4.

In the experiments, we will analyze the precision ratio and recall ratio of method which respectively represent the accuracy and comprehensiveness of user profile. And F1-Measure is a harmonic average of precision ratio and recall ratio, and it reviews the performance of the method.

5.3 Results and Analysis

The experiment results are shown in Fig. 3. From Fig. 3, we can see MLP has the highest precision ratio, and MLP, MLP-IA and MLP-EIA have stable result. However, as the size of data set increases, the precision ratio of MLP-EA decreases continuously. In the recall ratio, MLP-IA and MLP-EIA perform better than others and MLP have the worst result. Furthermore, in the F1-Measure, MLP-IA and MLP-EIA, introducing priori knowledge of association labels, achieve the best results.

Compared with MLP, the results show that although the precision ratio of our method is slightly reduced due to the introduction of priori knowledge, the recall ratio has been greatly improved, and the F1-Measure has also been improved. Therefore, the results prove that the association among labels is effectively mined based on implicit association labels and the user's interests can be well mined. But the recall ratio is less than MLP-IA. It indicates that although MLP can predict the interest label accurately, user profile is not complete and the convergence speed of model is rather slow.

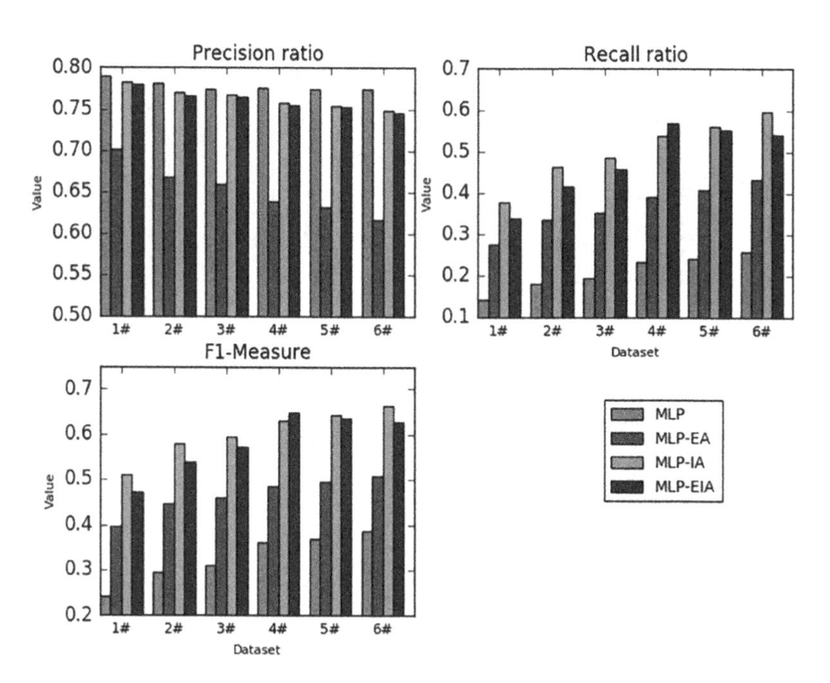

Fig. 3. Results of the precision ratio, the recall ratio and the F1-Measure.

In the MLP-EA, the associations among explicit labels was considered. The associations were calculated by word2vec before the application of label propagation. Results show that with the introduction of explicit associations among labels, the precision ratio basically remains unchanged and the recall ratio is improved especially in larger data sets, which proves the effectiveness of the explicit mining.

However, as the results show, our method performs better in recall ratio and F1-Measure. It can indicate association among implicit labels can perform better to mine the relationships among labels than explicit labels. In Weibo, posts are more arbitrary and it provides more features to make user profile. Nevertheless, there is much "noise" disturbing the results. Considering priori knowledge of association among implicit labels can avoid introducing textual "noise". On the other hand, the semantics of posts are too diverse to mine the associations among implicit association labels well. Instead of considering too many explicit details, it is more beneficial to explore the associations among implicit association labels.

Furthermore, we consider both the explicit labels and implicit association labels in MLP-EIA. The results show that our method has similar performance with MLP-EIA. It can prove that our method MLP-IA, introducing the implicit association labels, can capture the feature of users deeply and make user profile well. As is mentioned above, the associations among explicit labels includes too many features which are positive or negative for model training. So the model could not identify them well. Therefore, user profile model can be well constructed only by introducing implicit associations among labels.

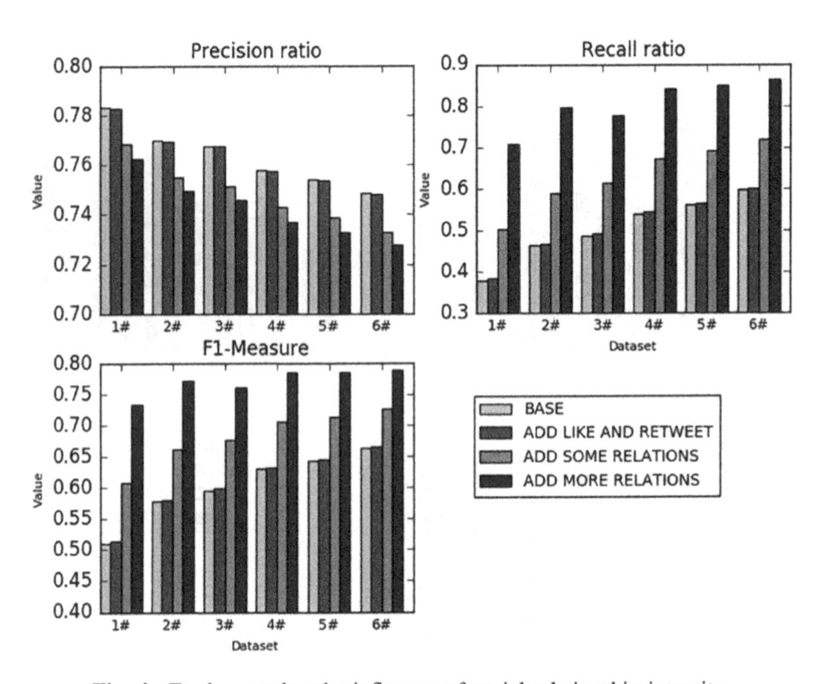

Fig. 4. Further results: the influence of social relationship integrity.

5.4 The Influence of the Social Relationship Integrity

To explore the impact of social relationship integrity on results, we conduct new experiments by adding more social relationships based on the same users' sets. In addition, we consider to add relationships of "LIKE" and "RETWEET", which also represents the interaction among users. The results are shown in Fig. 4. We can see that with the increase of known social relationship data, the performance of our method is gradually improving, especially in recall ratio. It proves that the model can identify more interest labels according to social relationship data.

The hyper parameter ζ in Eq. 7 which represents the sparsity of social networks. When adding more social relationships, the model will work on the more complete social network graph. The param will be equal to 1 and lead to a smaller loss. On the other hand, the matrix T in Eq. 6 will be dense after adding more social relationships, and the model can capture more features among users after each iteration.

Therefore, based on more relationships among users, we can build a more complete social network graph accordingly and explore more information about these interaction. And the richer the social relationships is, the higher the recall ratio of interest profile is.

5.5 Convergence

To evaluate the performances between the introduction of explicit labels and implicit association labels, we experiment with the convergence of the iterative times and time consuming. And the results are shown in Fig. 5.

The results intuitively show that our method converges faster with the introduction of the priori knowledge. And as the data scale increases, the time complexity is still low and time consumption does not increase exponentially. In particular, by comparison, our method takes the least iteration time. Therefore, the introduction of associations among implicit association labels can accelerate the result convergence, and accordingly it can be used in more real-time and large-scale recommendation systems.

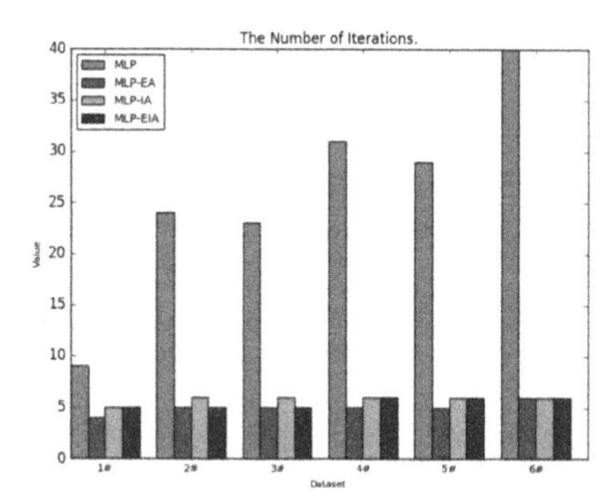

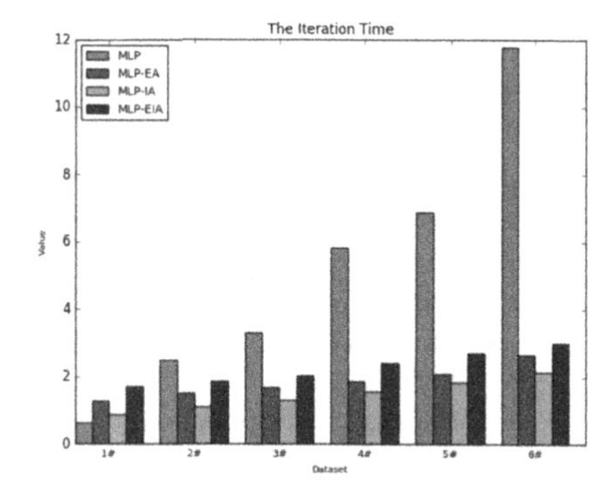

Fig. 5. The number of iterations and the iteration time. The left is the number of iterations and the right is the iteration time.

5.6 Summary

We compare our method with three baselines. The results show that our method can accelerate the convergence of propagation and make a significant increase in F1-Measure. However, baseline MLP-EA does not perform better in the experiments. The reason is, in the real world, explicit associations among labels include too many features which may be positive or negative for model training. The model could not identify them well. Instead of considering too many explicit details, it is more beneficial to explore the associations among implicit association labels.

Furthermore, our method achieves a similar performance with MLP-EIA. However, MLP-EIA takes much more time than our method. It proves that, although explicit associations among labels contains more features, it did not work in our model. Therefore, user profile could be well constructed by only introducing implicit associations among labels.

In addition, we explore the impact of social relationship integrity on results. The results show that, with the increase of known social relationship data, the performance of our method is gradually improving and it can identify more interest labels according to social relationship data.

6 Conclusion and Future Work

In this paper, we have studied the user profile by multi-label propagation. We proposed an improved multi-label propagation algorithm to utilize implicit association among labels and the implicit association labels can demonstrate more relationships among users. The experiments based on six real-world Weibo datasets have shown that our method accelerates the convergence and gets better performance than the previous methods.

Future work will pay more attention to improve the recall ratio of our method by extending the social relationships.

References

1. Boudaer, G., Loeckx, J.: Enriching topic modelling with users' histories for improving tag prediction in Q&A Systems. In: 25th International Conference Companion on World Wide Web, pp. 669–672. ACM, USA (2016)
2. de Souza, P.R., Durão, F.A.: RecTwitter: a semantic-based recommender system for Twitter users. In: Proceedings of the 24th Brazilian Symposium on Multimedia and the Web (WebMedia 2018), pp. 371–378. ACM, USA (2018)
3. Nurbakova, D.: Recommendation of activity sequences during distributed events. In: Conference, pp. 261–264 (2018)
4. Chang, P.S., Ting, I.H., Wang, S.L.: Towards social recommendation system based on the data from microblogs. In: International Conference on Advances in Social Networks Analysis and Mining, pp. 672–677. IEEE, USA (2011)
5. Li, R., Wang, C., Chang, C.C.: User profiling in an ego network: co-profiling attributes and relationships. In: International Conference on World Wide Web, pp. 819–830. ACM, USA (2014)

6. Zhang, J.L.: Application of tag propagation algorithm in the interest map of Weibo users. Programmer (5), 102–105 (2014)
7. Yao, Y., Yao, T.: Gender classification of Chinese Weibo users. In: Proceedings of the 2017 International Conference on E-commerce, E-Business and E-Government, pp. 5–8. ACM, USA (2017)
8. Xu, Y.N.: Modeling and application of social network user portrait based on text mining. Beijing University of Posts and Telecommunications, China (2016)
9. Chen, Z.M., Hu, Z.Y.: User profile study on UGC website. Comput. Syst. Appl. (2017)
10. Qin, F., Chen, Z., Zheng, X.: User behavior prediction based on emotion and interest. Comput. Syst. Appl. **27**(1), 28–34 (2018)
11. Tu, K., Ribeiro, B., Jensen, D., et al.: Online dating recommendations: matching markets and learning preferences. In: Proceedings of the 23rd International Conference on World Wide Web, pp. 787–792. ACM, USA (2014)
12. Xie, J., Kelley, S., Szymanski, B.K.: Overlapping community detection in networks: the state-of-the-art and comparative study. ACM Comput. Surv. **45**(4), 1–35 (2011)
13. Dong, Y., Yang, Y., Tang, J., et al.: Inferring user demographics and social strategies in mobile social networks. In: Proceedings of the 20th ACM SIGKDD International Conference on Knowledge Discovery and Data Mining, pp. 15–24. ACM, USA (2014)
14. Chakrabarti, D., Funiak, S., et al.: Joint label inference in networks. In: Companion of the the Web Conference, pp. 483–487 (2017)
15. Besel, C., Granitzer, M.: On the quality of semantic interest profiles for onine social network consumers. ACM, USA (2016)
16. Ma, C.: A method of social network user portrait analysis based on topic model. Anhui, China University of Science and Technology, China (2017)
17. Zoidi, O., Fotiadou, E., Nikolaidis, N., et al.: Graph-based label propagation in digital media: a review. ACM Comput. Surv. **47**(3), 48 (2015)
18. Jing, M., Yang, X.X.: The Characterization and composition analysis of Weibo "V". News and Writing **2**, 36–39 (2014)
19. Fernando, A., Ashok, C., Tony, J., et al.: Artwork personalization at Netflix. In: Proceedings of the 12th ACM Conference on Recommender Systems, pp. 487–488. ACM, USA (2018)
20. Chakrabarti, D., Funiak, S., Chang, J., et al.: Joint inference of multiple label types in large networks. Eprint arXiv, pp. 874–882 (2014)
21. Pal, A., Herdagdelen, A., Chatterji, S., et al.: Discovery of topical authorities in Instagram. In: International Conference on World Wide Web International World Wide Web Conferences Steering Committee (2016)
22. Liang, S.S., Zhang, X.L., et al.: Dynamic embeddings for user profiling in Twitter. In: Proceedings of the 24th SIGKDD Conference on Knowledge Discovery and Data Mining (KDD 2018), UK, pp. 1764–1773 (2018)
23. Peng, P., Wong, R.C.-W., Yu, P.S.: Learning on probabilistic labels. In: Proceedings of the 2014 SIAM International Conference on Data Mining, pp. 307–315. SIAM (2014)
24. Iyer, A.S., Saketha Nath, J., Sarawagi, S.: Privacy-preserving class ratio estimation. In: Proceedings of the 22nd ACM SIGKDD International Conference on Knowledge Discovery and Data Mining, pp. 925–934. ACM (2016)
25. Huai, M., Miao, C., Li, Y., et al.: Metric learning from probabilistic labels. In: Proceedings of the 24th ACM SIGKDD International Conference on Knowledge Discovery and Data Mining (KDD 2018), pp. 1541–1550. ACM, USA (2018)
26. Zhu, X., Ghanramani Z.: Learning from labeled and unlabeled data with label propagation. Carnegie Mellon University, Pittsburghers (2002)

Estimating Agriculture NIR Images from Aerial RGB Data

Daniel Caio de Lima[1(✉)], Diego Saqui[1,2], Steve Ataky[4], Lúcio A. de C. Jorge[3], Ednaldo José Ferreira[3], and José Hiroki Saito[1,5]

[1] UFSCar – Federal University of São Carlos, São Carlos, Brazil
daniel.lima@dc.ufscar.br
[2] IFMS – Federal Institute of Mato Grosso do Sul, Corumbá, Brazil
[3] Embrapa Instrumentation, São Carlos, Brazil
[4] UQUÀM – Université du Québec à Montréal, Montréal, Canada
[5] UNIFACCAMP – University Center of Campo Limpo Paulista, Campo Limpo Paulista, Brazil

Abstract. Remote Sensing in agriculture makes possible the acquisition of large amount of data without physical contact, providing diagnostic tools with important impacts on costs and quality of production. Hyperspectral imaging sensors attached to airplanes or unmanned aerial vehicles (UAVs) can obtain spectral signatures, that makes viable assessing vegetation indices and other characteristics of crops and soils. However, some of these imaging technologies are expensive and therefore less attractive to familiar and/or small producers. In this work a method for estimating Near Infrared (NIR) bands from a low-cost and well-known RGB camera is presented. The method is based on a weighted sum of NIR previously acquired from pre-classified uniform areas, using hyperspectral images. Weights (belonging degrees) for NIR spectra were obtained from outputs of K-nearest neighbor classification algorithm. The results showed that presented method has potential to estimate near infrared band for agricultural areas by using only RGB images with error less than 9%.

Keywords: Remote sensing · NIR image estimation · KNN · Spectral signature

1 Introduction

Remote Sensing (RS) has become an important system to obtain a huge amount of data, especially in precision agriculture processes. The association between information technology and agricultural procedures has been useful for obtaining and processing farm data, resulting in tools for productivity estimation, nutritional evaluation, water management, pests and diseases detection. Among several devices, multispectral and hyperspectral sensors attached to airplanes or Unmanned Aerial Vehicles (UAVs) can obtain images with detailed spectral information that helps identifying and distinguishing among materials spectrally similar [1].

J. M. F. Rodrigues et al. (Eds.): ICCS 2019, LNCS 11536, pp. 562–574, 2019.
https://doi.org/10.1007/978-3-030-22734-0_41

Properties extracted from reflectances in some ranges of the electromagnetic spectrum can be better evaluated by arithmetical combinations of different spectral bands [2]. These combinations usually use ranges from visible to Near Infrared (NIR) frequencies and are measures of vegetation activities. These measures are called Vegetation Indices (VIs) [3].

Especially for Brazilian agriculture, costs for acquiring data represent one of the most important barrier to the improvements provided by multispectral and hyperspectral sensors. Both sensors and their respective analytical platform are high-cost systems and they can not be offered as a set-off for familiar and/or small producers. For instance, the price of a hyperspectral camera is above tens of thousands of pounds. Furthermore, vehicles that transport these sensors (Drones and UAVs) are susceptible to mechanical and also human failures, leading to crashes during flight. Besides damaging the drone, the high cost spectral cameras could be damaged too. Common Red-Green-Blue (RGB) cameras are currently low cost sensors with potential to estimate some kind of VIs from visible spectral bands, such as Modified Photochemical Reflectance Index (MPRI) [4], that is applied to light use efficiency and water-stress. Nevertheless, the VI most used and accepted by agronomists is the Normalized Difference Vegetation Index (NDVI). This index uses NIR band and R band from RGB.

There are some papers that describe methods to obtain NIR images from ordinary cameras. Hardware alterations on cameras are described in [5, 6], removing NIR blocking filters from them. In [5], a new CFA (Color Filter Array) was developed to obtain a color image and a NIR image at the same time. A method to obtain NDVI images directly from a common camera is proposed in [6]. The first step to obtain this kind of images was removing the NIR blocking filter from one of the RGB channels of the camera and adding a low pass filter, allowing to obtain only NIR information. In this work, B channel from RGB was replaced for NIR. In 2016, a method to estimate NIR images from RGB images was proposed [7]. RGB images were captured by a camera attached to an UAV on different days and hours as well as NIR images obtained by a special camera, but these images were not from an aerial vehicle. After a regression analysis using RGB and NIR images, experiments showed that G channel from RGB is highly correlated to NIR, thus, they concluded that it is possible to estimate NIR images from G channel.

This paper introduces a method for estimating near infrared spectral information from RGB images, using R, G and B values and material endmembers. The purpose is making viable development of tools based on cheaper RGB cameras capable of estimating accurately NIR bands for VIs and other agricultural applications, making UAVs technology more attractive and accessible to familiar and small producers.

The rest of the text refers to the following: Sect. 2 – describing the hyperspectral image data used in the development, and the experiments for endmembers extraction; Sect. 3 – presenting the proposed method; Sect. 4 – describing the experimental results; and Sect. 5 – about conclusions and future works.

2 Spectral Data and Endmembers Extraction

2.1 Hyperspectral Images

Hyperspectral images were used as source of spectra to create a database of endmembers [8] for this experiment. Pure spectral signatures plays an important role on classification of materials on hyperspectral images. Because of the lack of a huge free hyperspectral image dataset, some well known images from literature were used to collect spectral information. These images have a ground truth image, allowing to assign each pixel spectral signature to different kinds of ground cover like vegetation, bare soil, minerals and water. Three public available hyperspectral scenes indicated in the literature were chosen to make spectral analysis: Indian Pines, Salinas and Pavia Centre. Information about these hyperspectral images can be obtained in [9]. RGB images were obtained using the three equivalent wavebands from the original hyperspectral images.

In addition to literature images, an image mosaic from a citrus cultivation in Lampa, Santiago, Chile, was used to do spectral analysis too. It was captured on January 17, 2011 by using a HySpex VNIR-1600 sensor attached to an airplane. The image was acquired between 12:49 and 12:54 p.m, with altitude of 2,500 m above sea level. Each image pixel corresponds to 0.5 m of spatial resolution, with 160 spectral bands, ranging from visible wavelengths to NIR (411.2 nm–988.9 nm). To generate a RGB image for this image (Fig. 1), bands 55 (611 nm), 41 (560.2 nm) and 12 (453.8 nm) were chosen to obtain respectively R, G and B channels.

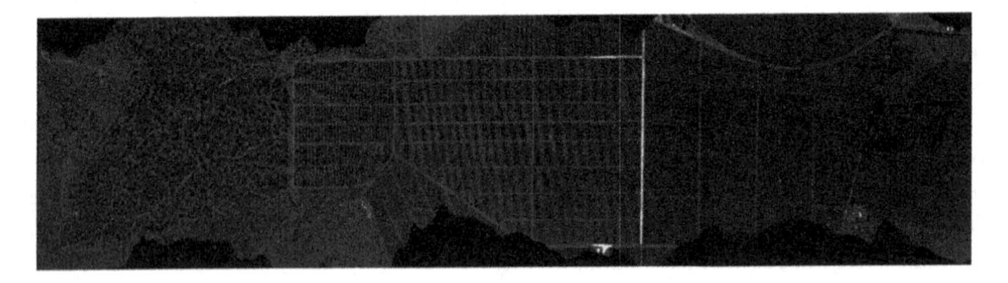

Fig. 1. RGB image from citrus cultivation in Lampa, Chile.

2.2 Endmembers Extraction Experiments

Pixels from selected images were taken to represent classes from their respective ground truths, resulting in 400 pixels per class, but for Indian Pines image was possible to get only 15 pixels per class, because some classes have less than 50 pixels in its ground truth. These selected pixels were used to find spectras that best represent each class (endmember). Automatic Target Generation Process (ATGP) [10], Pixel Purity Index (PPI) [11], N-FINDR [12] and Fast Iterative Pixel Purity Index (FIPPI) [13] were chosen to find these endmembers and verify the best algorithm to use on proposed method.

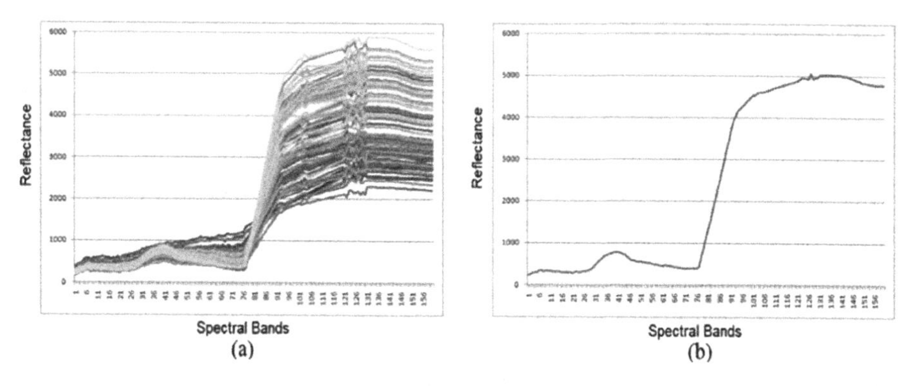

Fig. 2. Example of mean spectral signature. (a) 400 spectral signatures. (b) Mean spectral values from (a).

In addition to endmembers extraction results, an average of spectra was assessed for each class. An example of mean spectral signature can be seen at Fig. 2.

For the Lampa image, sixteen class labels were categorized for representing areas of soils (different conditions), lakes and crops. Each class from this image has an endmember associated with it. The same procedure is done to all images selected.

A total of 12,848 pixels were sampled to create a training database. In order to validate the method, 4,190 pixels spectra were sampled from other image areas. In Table 1 is shown the class label with their respective number of pixels.

Table 1. Training database samples distribution.

Class label	Number of samples (Pixels)
Dirt road 1	415
Dirt road 2	640
Lake 1	672
Lake 2	663
Plantation 1	861
Plantation 2	846
Plantation 3	660
Plantation 4	1,045
Plantation 5	976
Soil 1	1,024
Soil 2	900
Soil 3	820
Soil 4	877
Native soil 1	772
Native soil 2	647
Native vegetation	1,030
Total	12,848

3 Proposed Method

Figure 3 shows a block diagram that summarizes the proposed method. At the top, it is seen the hyperspectral image block, since the whole development is based on pixels sampled from this type of images. Then at the right side it is shown a block of endmembers extraction. These endmembers are hyperspectral pixels chosen to represent classes, as described in the previous section. At the left side it is seen the RGB image block, which represents the RGB pixel values extracted from the hyperspectral image data; and at the middle it is seen the ground truth pixels representing the classes. Using the RGB image input and the ground truth, the KNN classification is applied, to obtain the spectral signatures estimate of the RGB image pixels. It is important to explain that this block diagram is referred to the development diagram, since after this development, the NIR band estimation is based on RGB image and ground truth, without the use of the hyperspectral image data. The method will be described in detail in the following paragraphs, starting from KNN classification, because Hyperspectral images, RGB images acquisition, ground truth and Endmembers extraction have already been described in the previous section.

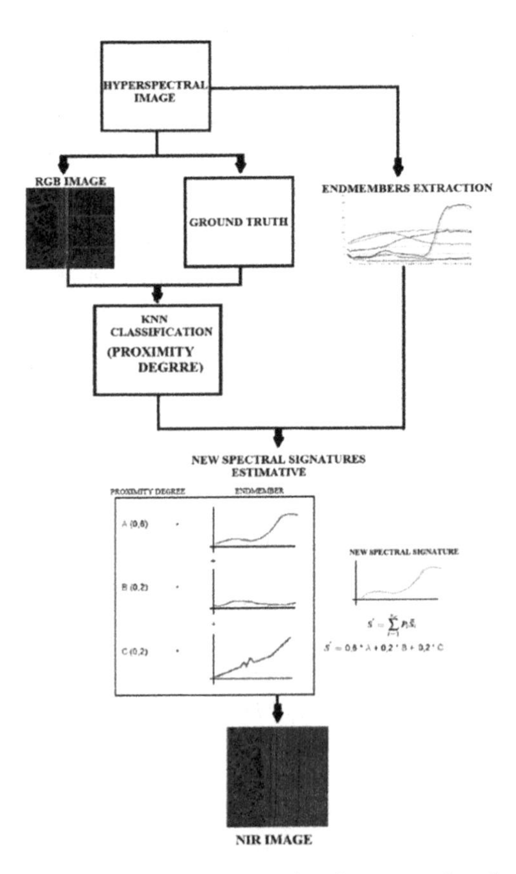

Fig. 3. Block diagram representing the proposed method.

The WEKA tool [14] was used for the experiments. The K-Nearest Neighbours (KNN) algorithm was chosen as the instance classifier. Input attributes were R, G, B and MPRI values.

MPRI is a VI based on normalized difference between two spectral bands in visible wavelength, specifically, red and green [5]. The MPRI equation is expressed as following:

$$MPRI = \frac{(R_{Green} - R_{Red})}{(R_{Green} + R_{Red})} \tag{1}$$

where R_{Green} is the green reflectance value and R_{Red} is the red reflectance value.

WEKA KNN returns a vector with a belonging degree to each class for the classified instance (pixel). At first, the algorithm creates an n dimension array called *dist*, which n is the number of classes from classification problem. Each *dist* element has an initial value, called *classifier correction*, defined by Eq. (2), where N is the number of instances.

$$correction = \frac{1}{MAX(1, N)} \tag{2}$$

For each k nearest neighbors the algorithm calculates a weight W using the distance between them and the sample to be classified, using the number of attributes (x) from input data, Eq. 3.

$$W_i = \frac{\sqrt{x}}{dist_i^2}, \ 1 \leq i \leq k \tag{3}$$

The *dist* array is updated on positions corresponding to each of k nearest neighbors, as can be seen on Eq. (4). After updating *dist*, the array is normalized, generating the proximity degree, or probability distribution P_i, Eq. (5).

$$dist_i = correction + W_i \tag{4}$$

$$P_i = Norm_i = \frac{dist_i}{\sum_{j=1}^{k} W_j}, \ 1 \leq i \leq k \tag{5}$$

P_i is used to calculate a weighted sum that origins new spectral signature, such as:

$$S' = \sum_{i=1}^{n_c} P_i \overline{S}_i \tag{6}$$

where n_c is the number of class labels, P_i is the probability of pixel belong to class i and $\overline{S}_i$ is the endmember array of class i. This new spectral signature calculated by Eq. (6) is based on the principle that hyperspectral pixels are composed by a mixture of endmembers from different targets [8]. Figure 4 illustrates how this step of obtaining new spectral signatures works, so that, at the left side it is shown the k = 3 nearest neighbors, represented by their endmember signatures, and their respective proximity degree, P_i, from the RGB pixel; and at the right side, the resulting estimative of spectral signature for the RGB pixel. Using this estimated spectral signature for all the pixels of the RGB image, it is possible to estimate its NIR band.

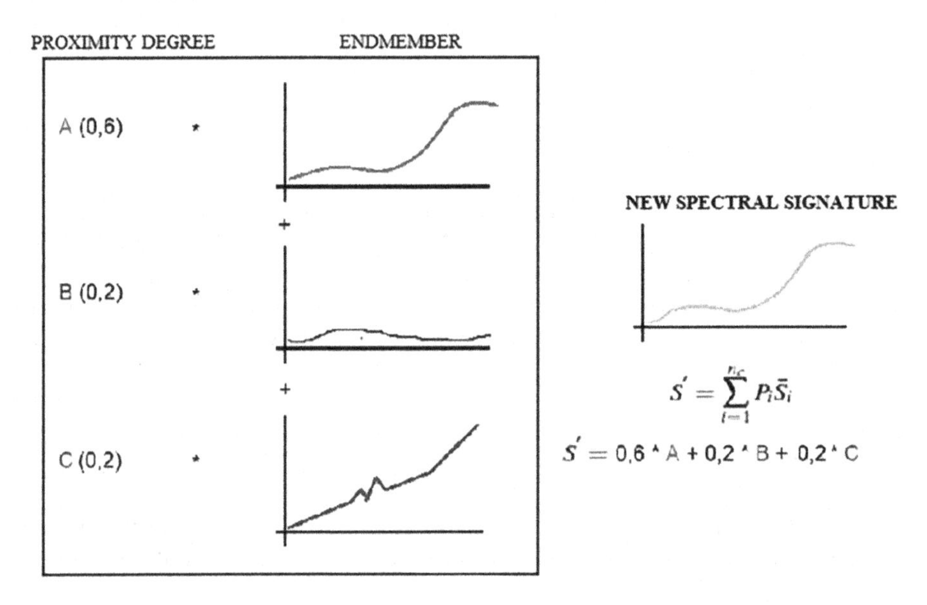

Fig. 4. Obtention of new spectral signatures from weighted sum using endmembers and KNN proximity degree.

In order to evaluate this experiment results, Root Relative Squared Error (RRSE) was calculated between classes' endmembers and validation data, wavelength by wavelength, using their respective KNN classification label. RRSE is given by:

$$RRSE(j) = \sqrt{\frac{\sum_{i=1}^{n}\left(\overline{S}_{j,i} - V_{j,i}\right)^2}{\sum_{i=1}^{n}\left(\overline{V}_j - V_{j,i}\right)^2}} \quad (7)$$

where n is the number of samples from validation dataset, $\overline{S}_{j,i}$ refers to endmember reflectance for wavelength j, $V_{j,i}$ is the validation dataset reflectance value for wavelength j. $\overline{V}_j$ is a mean value of reflectance values from validation dataset pixels for wavelength j.

4 Experimental Results

Endmembers were extracted with ATGP, PPI, N-FINDR, FIPPI and also mean spectral signature. Image segmentation with Spectral Angle Mapper (SAM) [15] and Spectral Information Divergence (SID) [16], spectral similarity measures, were used to verify which algorithm could be used to feed the estimation method with the best endmembers. SAM and SID algorithms use endmembers as a class reference pattern to classify spectral signatures of pixels from hyperspectral images, analyzing how far or near these pixels are from endmembers. Table 2 shows accuracy of segmentation performed after endmembers being extracted by these methods.

Table 2. Image segmentation accuracy (%) with SAM and SID.

INDIAN PINES					
	ATGP	PPI	N-FINDR	FIPPI	Mean Spectral Signature
SAM	25.66	22.64	23.68	31.02	33.33
SID	24.28	25.29	32.31	33.90	40.08
SALINAS					
	ATGP	PPI	N-FINDR	FIPPI	Mean Spectral Signature
SAM	29.59	44.45	52.10	54.72	62.14
SID	34.61	42.15	48.89	53.43	59.49
PAVIA CENTRE					
	ATGP	PPI	N-FINDR	FIPPI	Mean Spectral Signature
SAM	48.10	81.79	79.34	82.94	88.71
SID	69.14	58.96	73.06	57.61	84.27

According to Table 2, Mean Spectral signature outperformed classical algorithms of endmembers extraction, presenting the best segmentation results using SAM and SID classifiers. For the literature methods of endmembers extraction, FIPPI spectral signatures presented best results in segmentation task. Therefore, Mean Spectral signatures and spectral signatures selected by FIPPI were used on NIR estimative experiments.

The best performance on training datasets indicated 5 neighbors and weighted distance inverse (Euclidean distance) for KNN. Table 3 shows KNN classification accuracy using RGB and MPRI data from data set (Sect. 2.2) as input. As can be seen, Salinas image data showed best accuracy result, so it was chosen to make NIR estimative and also Lampa image, to verify estimative results on an image that doesn't have ground truth defined by remote sensing specialists.

The KNN classification accuracy to the Lampa image was 64.46%. After KNN classification, the RRSE was calculated between each sample spectra from validation

Table 3. KNN classification accuracy to the data sets.

Image	Accuracy (%)
Indian Pines	56.61
Salinas	93.76
Pavia Centre	69.39

data and the mean spectral signature which has the same label given by classifier. RRSE mean value between validation and average spectra is 58.88% to the Salinas data set and 52.86% to the Lampa data set (blue curve on Fig. 5). For experiments done with spectral signatures extracted with FIPPI, RRSE mean value between validation data and FIPPI endmembers is 67.68% to the Salinas data set and 70.22% to the Lampa data set (blue curve on Fig. 6).

Estimated spectral signatures were obtained by applying Eq. 6 for each sample of the validation dataset and RRSE was calculated for the two selected images using Mean Spectral Signatures (green curve on Fig. 5) and FIPPI spectras (green curve on Fig. 6). In average, RRSE for all wavelengths was 8.12% to the Salinas estimated image and 8.48% to the Lampa estimated image. RRSE results calculated between estimated

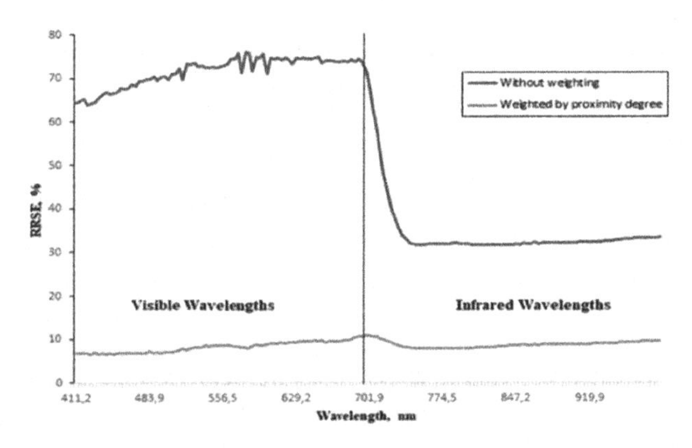

Fig. 5. RRSE between real validation data and average spectral signature (blue); and RRSE between new spectral signature and average spectral signature (green) (Lampa Image). (Color figure online)

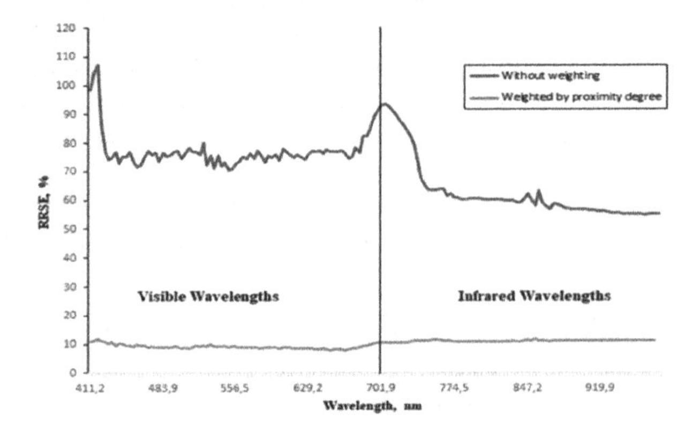

Fig. 6. RRSE between real validation data and spectral signatures extracted with FIPPI (blue); and RRSE between new spectral signature and endmembers extracted with FIPPI (Lampa image). (Color figure online)

spectras and FIPPI endmembers were 9.04% to the Salinas data and 10.14% for Lampa data in average. It is possible to see how RRSE get lower error values per band for estimated spectral signatures in comparison to the RRSE calculated in relation to the spectral signature from hyperspectral images for both Mean Spectral Signatures and FIPPI endmembers. Although new spectral signatures being a kind of endmembers mixture, they are still closer to the endmembers than spectral signatures from validation data for each classified pixel with KNN.

Entire Salinas image was used to show the power that this method has to estimate NIR spectral information, creating full RGB and NIR images for visual comparison. Since Lampa image has high dimension, an image region (400 $\times$ 400 pixels) was extracted from Fig. 1, also for a visual comparison between real images and those estimated by the proposed method. The estimative were performed using Mean Spectral Signatures and the results to the Salinas image are shown in Fig. 7. Results of estimation to the Lampa region image are shown in Fig. 8. Note that estimated images preserve a lot of transitions (high frequencies) among image components (classes), with good similarity with original images.

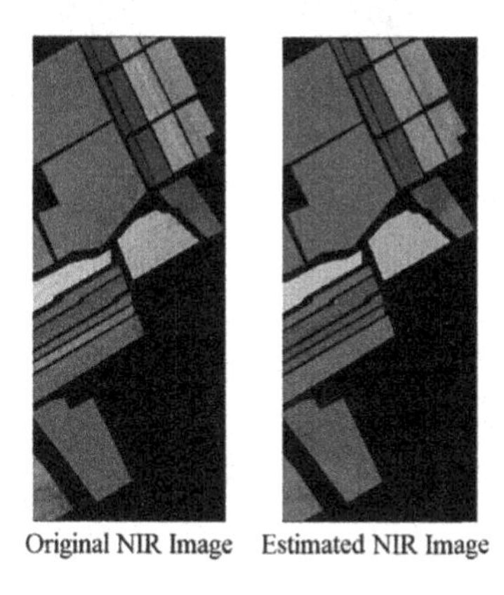

Fig. 7. Estimated image to Salinas Data Set.

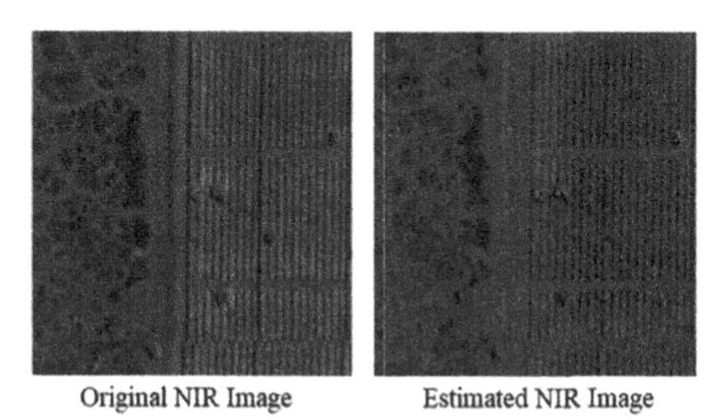

Fig. 8. Estimated image to Lampa image

At Figs. 9 and 10, NDVI pseudo color images are showed both NDVI calculated with original NIR data and NDVI calculated with estimated NIR data. NDVI values between 0.4 and 0.8 were considered to generate color ranges, because this NDVI value range shows the state of vegetation health. Red areas show uncovered soil and unhealthy vegetation. Colored areas (orange to green) shows different health state of vegetation, being green areas related to healthiest vegetation. Note that NDVI maps assessed using original NIR data and those based on NIR estimatives are very similar.

Original Data NDVI Estimated Data NDVI

Fig. 9. NDVI maps for Salinas image. (Color figure online)

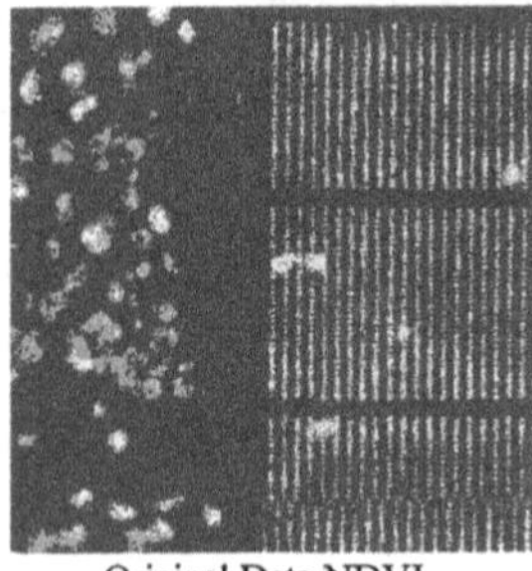
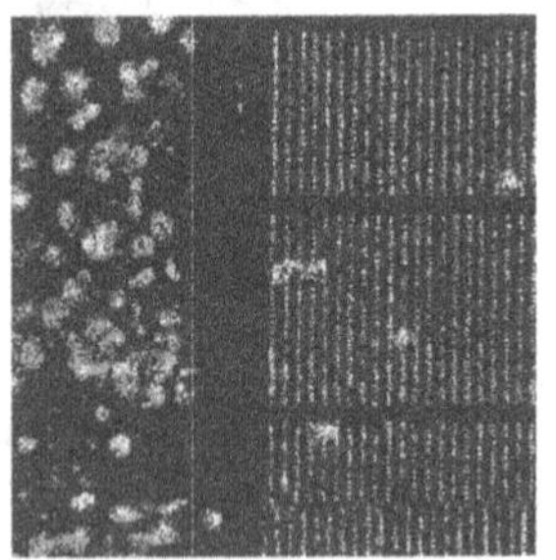

Estimated Data NDVI

Fig. 10. NDVI maps for Lampa image. (Color figure online)

5 Conclusion

In this paper, a method for estimating spectral signatures from RGB images using KNN was proposed. The use of a weighted sum using belonging degrees to classes has shown high potential for estimating spectral signatures for each image pixel, with an error smaller than 9% for NIR bands using Mean Spectral Signatures as endmembers, also resulting in quite similar NDVI maps. This kind of method makes feasible the use of accessible technologies to familiar and small producers. Some applications have high correlation between visible band and other bands, such as NIR. Thus, low-cost RGB cameras can be applied for obtaining adequate estimations in agriculture. Despite its good results, the method has limitations such as dependence on having 'pure' spectral signatures (endmembers) to estimating NIR bands and knowledge of the flight region to create a precise ground truth for KNN classification step. Ongoing work has shown many possibilities for improvements, including local correlations, measurements with a hand-held spectroradiometer and use of images with few bands (multispectral images).

Acknowledgements. This work was supported by the following Brazilian research agencies: Coordenação de Aperfeiçoamento de Pessoal de Nível Superior (CAPES) – Finance Code 001, and Conselho Nacional de Desenvolvimento Científico e Tecnológico (CNPq), project number 310310/2013-0; and Embrapa Instrumentation.

References

1. Mulla, D.J.: Twenty five years of remote sensing in precision agriculture: key advances and remaining knowledge gaps. Biosyst. Eng. **114**(4), 358–371 (2013)
2. Atzberg, C.: Advances in remote sensing of agriculture: context description, existing operational monitoring systems and major information needs. Remote Sens. **5**(2), 949–981 (2013)
3. Viña, A., Gitelson, A.A., Nguy-Robertson, A.L., Peng, Y.: Comparison of different vegetation indices for the remote assessment of green leaf area index of crops. Remote Sens. Environ. **115**(12), 3468–3478 (2011)
4. Yang, Z., Willis, P., Mueller, R.: Impact of band-ratio enhanced AWIFS image to crop classification accuracy. In: Proceedings of Pecora 17 (2008)
5. Lu, Y.M., Fredembach, C., Vetterli, M., Süsstrunk, S.: Designing Color Filter Arrays for the joint capture of visible and near-infrared images. In: 16th IEEE International Conference on Image Processing (ICIP), Cairo, Egypt, pp. 3797–3800 (2009)
6. Rabatel, G., Gorretta, N., Labbé, S.: Getting NDVI spectral bands from a single standard RGB digital camera: a methodological approach. In: Lozano, J.A., Gámez, J.A., Moreno, J.A. (eds.) CAEPIA 2011. LNCS (LNAI), vol. 7023, pp. 333–342. Springer, Heidelberg (2011). https://doi.org/10.1007/978-3-642-25274-7_34
7. Arai, K., Gondoh, K., Shigetomi, O., Miura, Y.: Method for NIR reflectance estimation with visible camera data based on regression for NDVI estimation and its application for insect damage detection of rice paddy fields. Int. J. Adv. Res. Artif. Intell. **5**(11), 17–22 (2016)
8. Plaza, J., Hendrix, E.M.T., García, I., Martín, G., Plaza, A.: On endmember identification in hyperspectral images without pure pixels: a comparison of algorithms. J. Math. Imaging Vis. **42**(2), 163–175 (2012)
9. HRSS: Hyperspectral Remote Sensing Scenes. http://www.ehu.eus/ccwintco/index.php?title=Hyperspectral_Remote_Sensing_Scenes. Accessed July 2017
10. Ren, H., Chang, C.-I.: Automatic spectral target recognition in hyperspectral imagery. IEEE Trans. Aerosp. Electron. Syst. **39**(4), 1232–1249 (2003)
11. H. G. Solutions: How does ENVI's pixel purity index work? (1999). http://www.harrisgeospatial.com/Company/PressRoom/TabId/190/ArtMID/786/ArticleID/1631/1631.aspx

12. Winter, M.E.: N-FINDR: an algorithm for fast autonomous spectral end-member determination in hyperspectral data. Int. Soc. Opt. Photonics Imaging Spectrosc. V **3753**, 226–276 (1999)
13. Chang, C.-I., Plaza, A.A.: A fast iterative algorithm for implementation of pixel purity index. IEEE Geosci. Remote Sens. Lett. **3**(1), 63–67 (2006)
14. Hall, M., Frank, E., Holmes, G., Pfahringer, B., Reutemann, P., Witten, I.H.: The WEKA data mining software: an update. SIGKDD Explor. **11**(1), 10–18 (2009)
15. Kruse, F.A., et al.: The spectral image processing system (SIPS) – interactive visualization and analysis of imaging spectrometer data. In: AIP Conference Proceedings, vol. 283, no. 1, pp. 192–201 (1993)
16. Chang, C.-I.: An information-theoretic approach to spectral variability, similarity, and discrimination for hyperspectral image analysis. IEEE Trans. Inf. Theory **46**(5), 1927–1932 (2000)

Simulation of Fluid Flow in Induced Fractures in Shale by the Lattice Boltzmann Method

Rahman Mustafayev and Randy Hazlett(✉)

University of Tulsa, Tulsa, OK 74133, USA
Randy-Hazlett@UTulsa.edu

Abstract. With increasing interest in unconventional resources, understanding the flow in fractures, the gathering system for fluid production in these reservoirs, becomes an essential building block for developing effective stimulation treatment designs. Accurate determination of stress-dependent permeability of fractures requires time-intensive physical experiments on fractured core samples. Unlike previous attempts to estimate permeability through experiments, we utilize 3D Lattice Boltzmann Method simulations for increased understanding of how rock properties and generated fracture geometries influence the flow. Here, both real induced shale rock fractures and synthetic fractures are studied. Digital representations are characterized for descriptive topological parameters, then duplicated, with the upper plane translated to yield an aperture and variable degree of throw. We present several results for steady LBM flow in characterized, unpropped fractures, demonstrating our methodology. Results with aperture variation in these complex, rough-walled geometries are described with a modification to the theoretical cubic law relation for flow in a smooth slit. Moreover, a series of simulations mimicking simple variation in proppant concentration, both in full and partial monolayers, are run to better understand their effects on the permeability of propped fractured systems.

Keywords: Fractures · Shale · CFD

1 Background

The structure of fractures and their complexity will generally depend on two main conditions: (1) the type of material undergoing the breakage, determined by its physical and chemical properties and (2) total mechanical stress acting on the material. The general definition of a fracture for the purpose of this work is any discontinuity in a rock volume created during rupturing of a rock mass, generating surface with annihilated cohesion.

In its most simple representation, we can imagine a fracture as flow in a smooth slit. Laminar flow between smooth parallel plates can be posed and solved analytically [1], yielding a cubic law relationship between permeability (k) or transmissivity (T) and geometrical parameters: aperture (h), cross sectional area (A), and width (W).

$$T = kA = \frac{Wh^3}{12} \tag{1}$$

J. M. F. Rodrigues et al. (Eds.): ICCS 2019, LNCS 11536, pp. 575–589, 2019.
https://doi.org/10.1007/978-3-030-22734-0_42

Flow in real fractures, however, is impacted by wall roughness, aperture, and shear displacement. According to Hakami [2], there are three main aspects to fluid flow prediction in single fractures (see Fig. 1): fluid properties, fracture void geometry, and imposed boundary conditions. We focus on the elements in boldface.

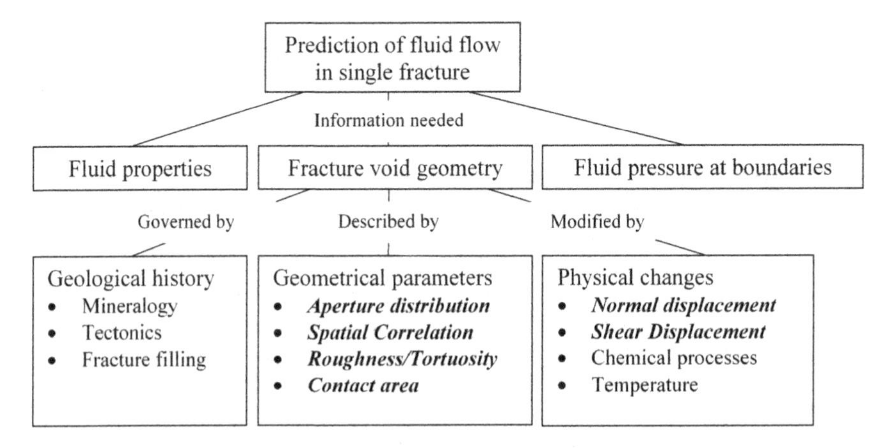

Fig. 1. Main factors altering the flow in a single fracture. Context of this work is in bold italic. Figure modified from [2].

Although once considered of no commercial hydrocarbon potential, with recent advancement in completion and stimulation practices in long horizontal wells, oil and gas production from shale formations is made economically feasible and is an increasingly important contributor in the fossil fuels portfolio with global potential [3]. We focus on the fracture void geometry of fractured shale and the subtopics of geometrical parameterization of the system and physical changes produced by the fracturing process, namely, normal and shear displacement.

1.1 Surface Characterization

Surface roughness can be regarded as any irregularity or deviation of the surface structure from the mean smooth plane. The larger these deviations, the rougher the surface, and in case they are relatively small, a surface is considered smooth. Fracture roughness is controlled by stress conditions which affect the crack propagation pattern, lithology of the rock matrix, including all types of heterogeneities, and finally, secondary physical and chemical processes, such as weathering, erosion, and mineral precipitation.

Standards from the field of tribology, studying effect of surface roughness on lubrication, wear and friction [4], and some amplitude parameters proposed by Stout and Blount [5] were considered. The most straight-forward parameter to describe roughness, representing the mean of absolute profile height deviations from the mean plane, is average roughness (S_a).

$$S_a = \frac{1}{N_x N_y} \sum_{j=1}^{N_y} \sum_{i=1}^{N_x} \left| z(x_i, y_j) - m \right| \tag{2}$$

where z is the vertical height at any point, N_x and N_y are total number of data points in x and y direction respectively, and m is the mean of all data points.

$$m = \frac{1}{N_x N_y} \sum_{j=1}^{N_y} \sum_{i=1}^{N_x} z(x_i, y_j). \tag{3}$$

However, the most widely used parameter, known in statistics as standard deviation, is root mean square (RMS) roughness (S_q) given by

$$S_q = \sqrt{\frac{1}{N_x N_y} \sum_{j=1}^{N_y} \sum_{i=1}^{N_x} \left(z(x_i, y_j) - m \right)^2} \tag{4}$$

We also have the difference between extrema, S_t, skewness, S_{sk}, and kurtosis, S_{ku}.

$$S_t = \max\left(z(x_i, y_j)\right) - \min\left(z(x_i, y_j)\right) \tag{5}$$

$$S_{sk} = \frac{1}{N_x N_y S_q^3} \sum_{j=1}^{N_y} \sum_{i=1}^{N_x} \left(z(x_i, y_j) - m \right)^3 \tag{6}$$

$$S_{ku} = \frac{1}{N_x N_y S_q^4} \sum_{j=1}^{N_y} \sum_{i=1}^{N_x} \left(z(x_i, y_j) - m \right)^4 \tag{7}$$

Other statistical characterizations of note are the joint roughness coefficient (JRC) [6], semi-variogram [7], and fractal dimension [8].

1.2 Fracture Characterization

While there are many reasonable ways to characterize roughness of a single surface, flow in fractures is between two similar surfaces. Many of the above concepts can be generalized to properties of the plane pair or the space they create, such as a semi-variogram on point aperture. The surfaces are similar due to the fracture generation process from an intact material. However, the fracturing process generates debris, giving dislocations in one or both surfaces, and any shear displacement creates aperture distributions with potential impingement "pillars" at points of contact and potential elimination of spatial correlation regarding flow, despite similarities only a short distance away. Since shale has significant clay and organic content, the resulting fracture is also sensitive to further imposed stress with plastic flow or creep, yielding apertures changing with time.

When considering flow in the fracture, we also have the notion of tortuosity ($\tau = L_a/L$), as the arclengths for streamlines, L_a, are longer than the sample length, L. Yet another measure of tortuosity, T_s, was defined by Belem et al. [9] as a normal component of the average roughness ratio, R_s, which could also be extended to fractures as the average between top and bottom surfaces. The average roughness ratio compares the actual surface area to that of a nominal plane spanning the same region.

1.3 Flow Characterization

Early attempts were experimental in nature [10–12] and explored extensions of flow in rough conduits with a relative roughness parameter, $\varepsilon/2h$, where ε is absolute roughness height and $2h$ is equal to hydraulic diameter. Witherspoon et al. [13] incorporated a friction factor to model experiments on granite, basalt, and marble fractures. Other authors introduced a relative roughness ratio [14–16], the JRC coefficient [17], and a contact area fraction [18], c, to account for periodic collapse of the flow area by upper and lower surface contact. Many of these cubic law corrective models strongly depend upon the nature of the average aperture used.

2 Procedure

2.1 Experimental

We consider real rough-walled profiles with both uniform and variable aperture field distributions, including taking into account asperities at points of contact between the two surfaces. All of the models described herein are based on 3D surface profiles acquired from longitudinal fractures created by Brazilian tests performed on four 1 in. diameter, 1.5 in. long shale core samples that did not contain visible macroscopic fractures (see Fig. 2).

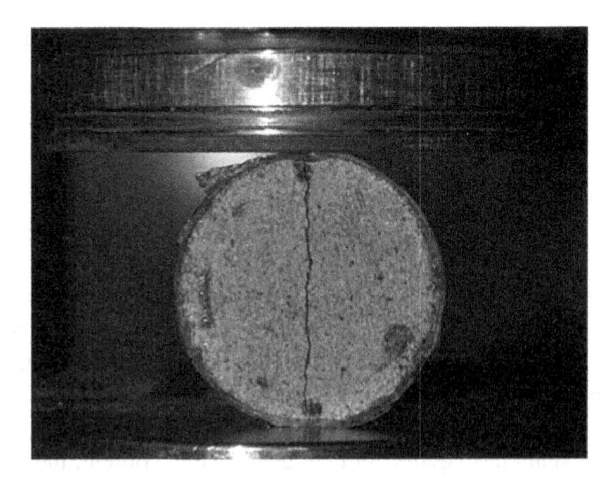

Fig. 2. Fracture creation using the Brazilian test procedure.

The Brazilian test is commonly used for material tensile strength determination and can essentially be described as uniaxial normal stress compression process leading to failure.

Resulting fracture faces are digitized for surface topography using a commercial optical profilometer manufactured by Nanovea (see Fig. 3). The main characteristics of any profilometer are its maximum measurement range and resolution in x and y directions, determined by the stepping of the sample stage holder, and resolution, measurement range and accuracy in the z direction, dictated by the particular optical pen installed. Our pen had a measurement range of 27 mm with a vertical resolution of 600 nm and vertical accuracy of 3000 nm. We used steps of 100 μm in both x and y directions in the surface characterization of all samples. The data required preprocessing to convert the cloud point data to an STL mesh that was importable to Exa's PowerDelta tool to ensure a dense compatible boundary suite that would serve as no-slip boundary conditions in PowerFlow, Exa's Lattice Boltzmann Method flow simulator.

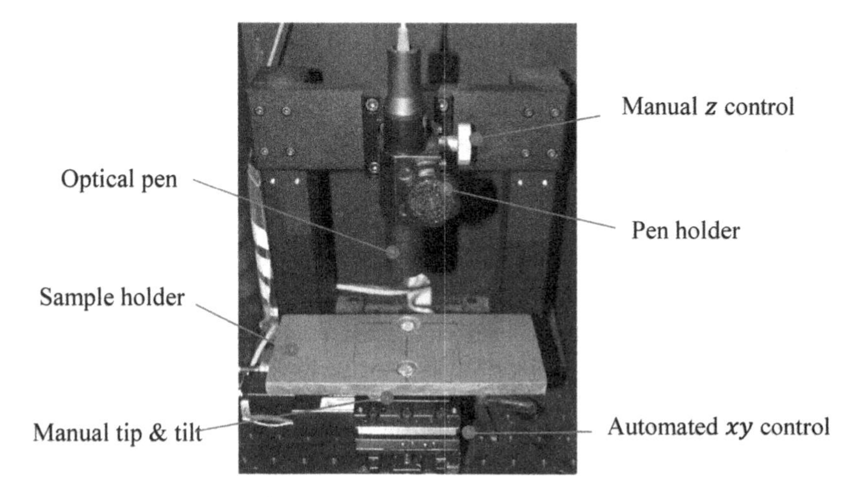

Fig. 3. Surface profilometer used for data acquisition.

Figure 4 indicates the two types of simulations that dictate the level of geometrical characterization needed to correlate with flow behavior. We could entertain exclusively normal translation of the upper surface to achieve a nearly constant aperture, or we could allow normal and shear displacement to yield two unmatched surfaces and variable aperture.

Fig. 4. Flow geometries without and with displacement in the direction of fracturing.

Due to the debris created in the fracturing process, however, we did not have ideal mating surfaces. Rather than work with both surfaces with dislocations and a possible debris field, we instead duplicated and translated a digital representation of the lower surface to create an upper surface and flow volume. This is illustrated in Fig. 5 where we had to construct the two surfaces, define the flow field, and encapsulate the open fracture into a regular solid geometry.

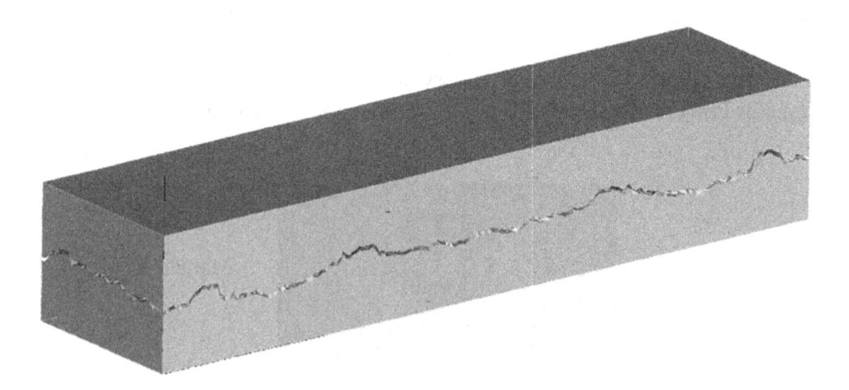

Fig. 5. Dual surfaces to fracture to 3D volume transformation for simulation input.

In field stimulation operations, the fractures are created by exerting enough pressure transmitted by a fluid to crack the rock. Release of the pressure would collapse the fracture, which could, in some cases, heal. To maintain an open conduit, suspended sand or other proppant material is included in the pressurizing fluid. The proppant, often coated to produce a bond with the formation to avoid return of solids, maintains an aperture consistent with the sand grain diameter. We simulate the permeability of a proppant-ladden fracture as spherical elements placed between the fracture faces. Rather than simulate the delivery process, we investigate the impact of proppant density as deviations from monolayer coverage. That is, we start with a closed packing arrangement and create lower coverage with random removal of particles.

2.2 Computational

Method. We employ a commercial, three dimensional, Lattice Boltzmann simulator (PowerFlow) provided by Exa Corporation. Application of Lattice Boltzmann Method (LBM) has several advantages compared to other numerical simulation schemes [19]:

1. Mesoscopic level of operations – simplified microscopic description allows more natural description of small scale effects and detailed modelling of highly complex geometric boundaries.
2. Simple and automatic volume discretization – simulation volumes can easily be meshed into a lattice of cubic voxels without any need for solid boundary adaptations.

3. Inherent parallelism – simulations can be run on multiple processors due to the localized nature of the performed operations.

Even though LBM is highly parallel, computational demand required for simulations to converge is quite high – larger domains of simulation can require somewhere from a couple of days up to a week of run time on 100 cores to obtain reasonable results.

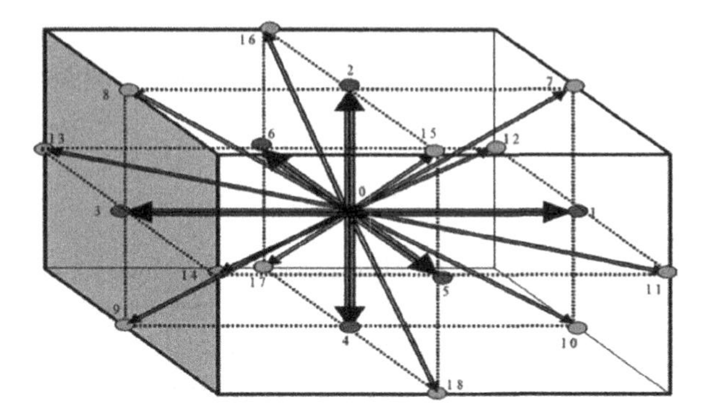

Fig. 6. D3Q19 lattice arrangements for 3D problems [21].

Unlike molecular dynamics, instead of analyzing fluid as a collection of individual particles, LBM treats fluid volume as "collection of particles represented by a particle velocity distribution function at each grid point" [20]. Another critical part of the LBM development is introduction of a simplified collision operator introduced by Bhatnagar, Gross and Krook [21], which considers a stencil with 19 possible nodes for discrete particle velocities (see Fig. 6).

$$u_{lat} = \frac{v_{lat} Re}{Resolution} \tag{8}$$

$$u_{max} = \frac{h^2}{8\mu} \frac{dP}{dL} = \frac{3}{2} u_{lat} \tag{9}$$

$$g = \frac{dP}{dL} / \rho_{lat} \tag{10}$$

where u_{lat} in all of the equations represents average lattice velocity, v_{lat} is kinematic lattice viscosity, u_{max} is maximum lattice velocity, ρ_{lat} is lattice density and g is gravity or acceleration applied to induce fluid flow. Notice that the main two parameters used to control other simulation parameters are Reynold's Number, *Re,* and resolution, λ.

$$k_{lat_{sim}} = \frac{v_{lat} u_{lat_{sim}}}{g} \tag{11}$$

$$k_{sim} = k_{lat_{sim}} \left(\frac{l_{char}}{\lambda} \right)^2 \tag{12}$$

After reaching a steady state, we extract average simulated lattice velocity $u_{lat_{sim}}$, which we further use to calculate simulated permeability. Conversion from permeability in squared lattice length units to square meters is done using the voxel size which essentially is determined by the ratio of characteristic length l_{char} and λ. Characteristic length in all of our simulations is chosen to be equal to the aperture height separating surfaces of a fracture.

An important task before the start of simulations with real rock surfaces is to perform verification exercises. This is achieved by running a benchmark case with parallel plate geometry and comparing simulation results with analytical solution given by Eq. 13.

$$u(z) = 4u_{max}\frac{z}{h}\left(1 - \frac{z}{h}\right) \tag{13}$$

where z is elevation above the bottom plate surface in the range [0, h]. Comparison of the velocity profiles and computed permeability with varying aperture shows a good match (see Fig. 7) between simulation and theory.

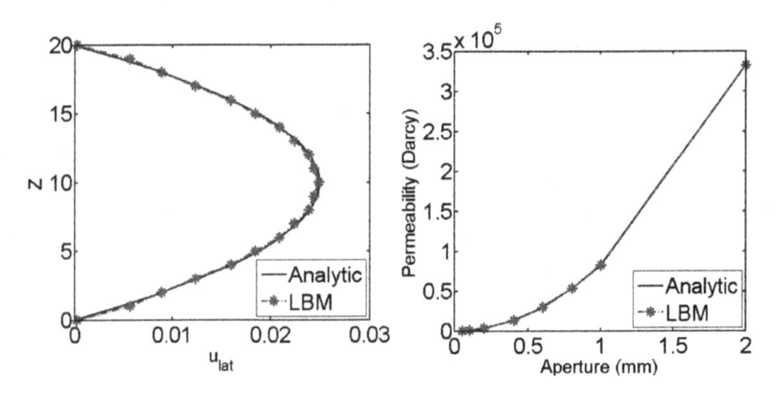

Fig. 7. Comparison of analytic and LBM simulation velocity profiles and permeability values for steady state flow between parallel plates.

Based on these results and on the measure of error between analytic and LBM permeability always lower than 0.7% for all the steady state Poiseuille flow cases in Fig. 7, we assume the correctness of the simulator.

Validation. To determine proper values for *Re* and λ, we once again run a series of test simulations. First runs were made using parallel plate geometry with h = 0.8 mm. Error percent was calculated using the analytic cubic law solution.

Resolution. Based on Fig. 8, we can easily conclude that the most appropriate Resolution to run the simulations is ten lattice units. If we go above this number, we increase accuracy, but the improvement will not be significant and will incur a higher cost in both discretization and simulation time.

Reynolds Number. The next critical parameter affecting both accuracy and speed control is Re. Flow in smooth parallel plates indicated an abrupt change in flow behavior for Re $\geq$ 20, although errors compared to parallel plate analytical solution were less than 1%. Flow between two complementary rough surfaces separated by a 0.2 mm vertical aperture gave the results presented in Fig. 9. In the absence of reference solution, we assumed the result from the simulation with the smallest Re number to be the most correct one and a basis for computed error. Based on the rough fracture simulations, we concluded that the maximum Re number value we should utilize to run our simulations without sacrificing accuracy is ten.

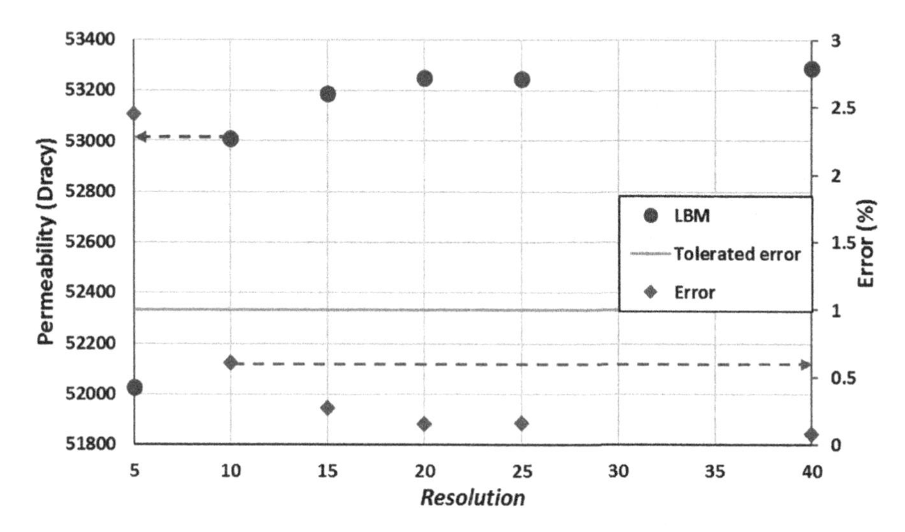

Fig. 8. Permeability change with *Resolution* variation for parallel plate geometry

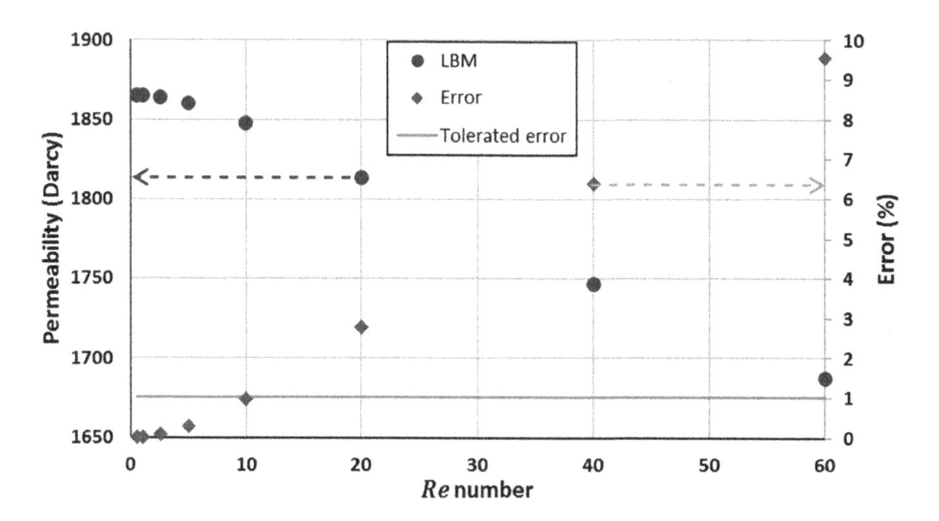

Fig. 9. Permeability change with ***Re*** number variation for rough mated surfaces (h = 0.2 mm).

3 Results

Summary statistics on four fractured shale samples are given in Table 1. LBM simulation results are provided in Table 2.

3.1 Aperture

Based on a series of simulations with apertures varying form 0.02 mm up to 0.4 mm, we propose the correlation

$$k = \frac{h^2}{12\tau_x \tau_y T_s^2} \tag{14}$$

which is compared with the simulation and smooth slit cubic law results in Fig. 10. Since these simulations only translate the duplicate fracture face normal to the direction of flow, the aperture is everywhere constant. The correlation includes measures of tortuosity in both x and y directions, since flow is actually 3D in nature, and a third measure of tortuosity, T_s, based upon area roughness ratio.

Table 1. Summary statistics of four core samples

Property		Name	Defined in	Sample			
				A1	B1	C1	D1
Size (μm)	x	Length	-	4200	4200	4200	10000
	y	Width	-	2100	2100	2100	5000
m (μm)		Mean vertical height	Eq. 3	225	314	254	330
S_t (μm)			Eq. 5	351	625	473	649
S_a (μm)		Average roughness	Eq. 2	44	91	58	82
S_q (μm)		RMS roughness	Eq. 4	55	113	73	103
S_{sk}			Eq. 6	−0.4567	−0.0371	−0.0303	0.2171
S_{ku}			Eq. 7	3.4323	2.569	2.916	2.915
JRC [24]	x	JRC coefficient	[24]	12.59	23.08	20.20	22.61
	y	JRC coefficient	[24]	13.93	20.55	18.76	20.68
JRC [23]	x	JRC coefficient	[23]	12.65	49.55	35.39	45.71
	y	JRC coefficient	[23]	15.40	36.84	29.49	35.91
τ_x		x tortuosity	1.2	1.031	1.121	1.086	1.111
τ_y		y tortuosity	1.2	1.037	1.090	1.072	1.087
R_s		Average roughness ratio	1.2	1.066	1.202	1.152	1.190
T_s		Normal component of R_s	[9]	1.066	1.20	1.15	1.19

Table 2. LBM simulation results for mated surfaces

Aperture, mm	Permeability, Darcy		
	Sample A1	Sample B1	Sample D1
0.02	25.86	20.08	22.94
0.04	103.9	76.70	87.74
0.08	234.5	299.0	341.6
0.10	420.5	654.1	749.4
0.15	647.0	1163	1331
0.20	1470	1842	2128
0.30	2663	4327	5559
0.40	6164	8058	8900

3.2 Shear

When considering lateral or shear translation, aperture is no longer constant, as illustrated in Fig. 11, and Eq. 14 can no longer adequately capture the added complexity.

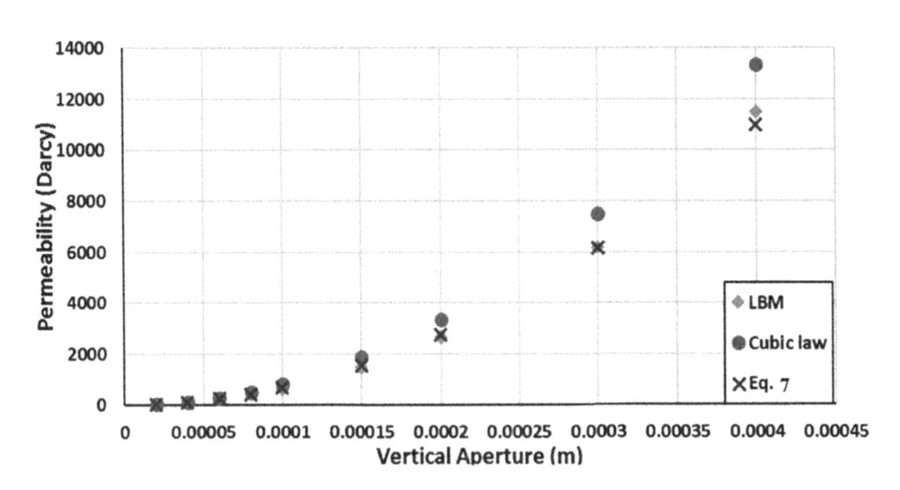

Fig. 10. Comparison of simulation results with cubic law and proposed equation for surface A1.

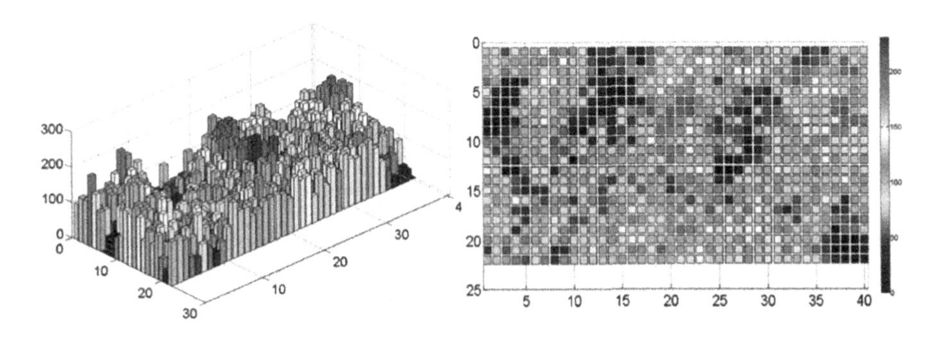

Fig. 11. Typical aperture distribution created with vertical displacement plus lateral shear.

We propose modifications to arrive at a new correlation shown in Eq. 8.

$$k = \frac{h_G^2}{12\tau_x\tau_y T_s^2\left[1 + \frac{S_q}{2h_G}\right]}\left(\frac{1-c}{1+c}\right) \tag{15}$$

where the RMS roughness, S_q, and contact area fraction, c, are also introduced. A variety of mean apertures were tested: arithmetic mean aperture, geometric average aperture, arithmetic mean excluding zero values (NNZ), and the mechanical aperture (distance between mean planes) with the geometric mean aperture, h_G, providing the closest match in Eq. 15 to the simulated permeability, as shown in Fig. 12.

3.3 Proppant Density

When using a relatively large dispersed solid to bridge surfaces and maintain an aperture, it is believed that the impact of the surface asperities and roughness play a much less dominant role. The roughness and shape distribution of the proppant could possibly become the focal point affecting resulting flow behavior, but these elements are controllable. The preliminary work in this area, therefore, uses smooth parallel plates with variable concentration of proppant to examine first order effects. Starting with closed packing and monolayer filling, lower concentrations ($f < 1$) were obtained by random removal of spheres. The observed relation for concentration dependent permeability is displayed in Fig. 13 and captured by the curve fit

$$k = 7.78 \times 10^4\, e^{-4.14f} \tag{16}$$

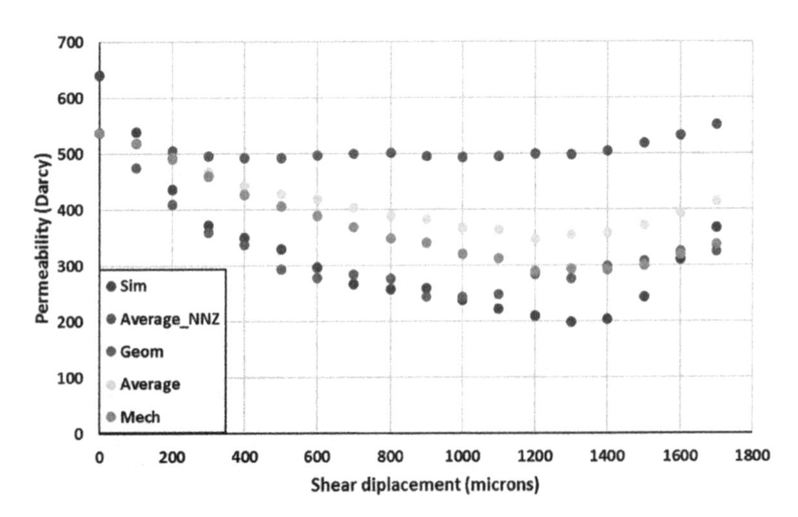

Fig. 12. Results generated using Eq. 4 for different average apertures vs simulation results (surface A1). Legend: Geom – h_G, Average – h_A, Average_NNZ – h_A excluding zero apertures, Mech - h_m.

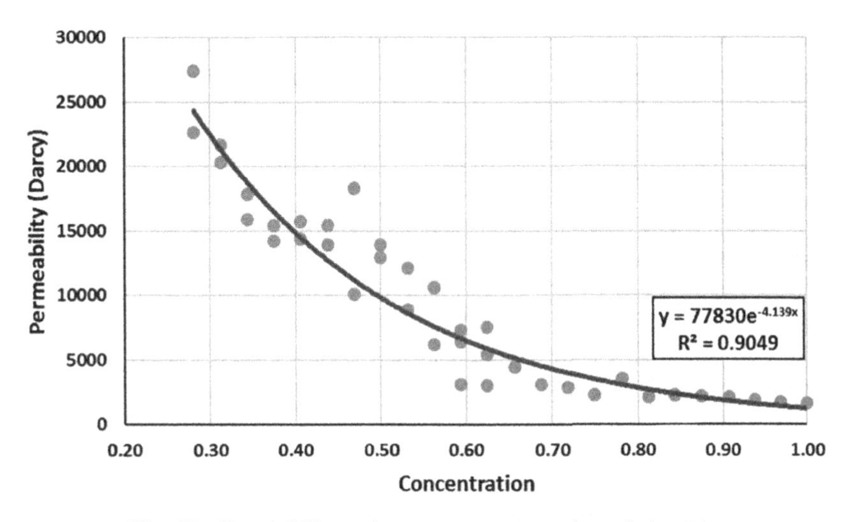

Fig. 13. Permeability and proppant concentration relationship.

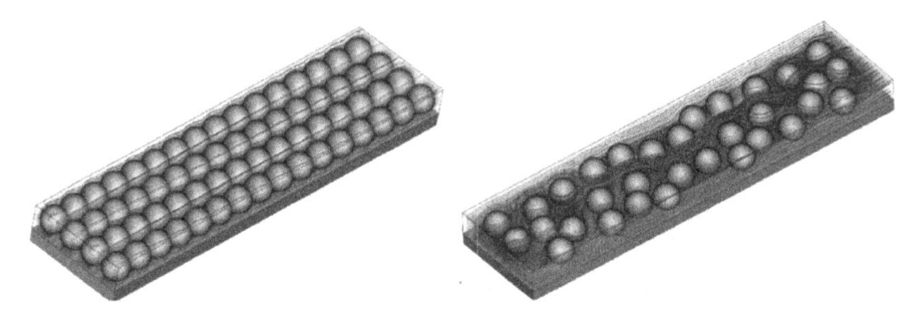

Fig. 14. Velocity streamlines in full monolayer (top) and partial monolayer (bottom). Uniform flow through all channels in full monolayer while in partial monolayer flow mostly occurs through the biggest channel.

with a relatively high correlation coefficient (R^2 = 0.905). The scatter seen in this figure at low values of concentration is due to multiple placement patterns at each of these concentrations. While field efforts target placement of increasing proppant volume [22], LBM simulations clearly demonstrate the value in only partial coverage, provided proppant is delivered throughout the rock failure zone.

4 Conclusions

A procedure was assembled for systematic characterization of induced fractures, with regard to impact on fluid flow, and their re-assembly to create digital fractures of uniform aperture and those with aperture distributions created through shear displacement of one face. In both cases, the simple flow in a smooth slit model required empirical modification using phenomenological parameters related to either frictional drag or pathlength extension to successfully represent the pressure drop relationship to flow rate. Elemental work with propped fractures supplements flow characterization with size and concentration of artificially indicated the value in partial fracture filling in sustainable fracture conductivity (Fig. 14).

5 Future Work

It is highly desirable to extend this work to multiphase flow due to the known bimodal wettability in such rocks with organic and inorganic porosity. The rock matrix was treated as impermeable, though real systems actually feed the fracture. Additionally, the rock matrix can be treated as a system under stress and undergoing plastic flow, resulting in partial embedment and loss of permeability. The impact of fracture roughness in proppant studies was ignored and should be investigated, as well as characterization of shape, distribution, and surface roughness of proppant. A test matrix that includes all the major commercial shale plays would be advantageous, as the surface properties examined should be functions of brittleness, related to mineralogy and organic content.

References

1. Bird, B., Stewart, W., Lightfoot, E.: Transport Phenomena. Wiley, New York (1960)
2. Hakami, E.: Aperture Distribution of rock fractures. Royal Institute of Technology, Stockholm (1995)
3. U.S. Energy Information Administration (EIA). https://www.eia.gov/analysis/studies/worldshalegas/. Accessed 12 Feb 2019
4. Thomas, T.: Rough Surfaces. Imperial College, London (1999)
5. Stout, K., Blount, L.: Three-Dimensional Surface Topography, 2nd edn. Penton Press, London (2000)
6. Barton, N.: Review of a new shear-strength criterion for rock joints. Eng. Geol. **7**(4), 287–332 (1973)
7. Kelkar, M., Perez, G.: Applied Geostatistics for Reservoir Characterization. Society of Petroleum Engineers, Richardson (2002)
8. Brown, S., Scholz, C.: Broad bandwidth study of the topography of natural rock surfaces. J. Geophys. Res. Solid Earth **90**(B14), 12575–12582 (1985)
9. Belem, T., Homand-Etienne, F., Souley, M.: Quantitative parameters for rock joint surface roughness. Rock Mech. Rock Eng. **33**(4), 217–242 (2000)
10. Lomize, G.: Flow in fractured rocks. Gosenergoizdat Moscow **127**, 197 (1951)
11. Louis, C.: A study of groundwater flow in jointed rock and its influence on the stability of rock masses. Imperial College of Science and Technology (1969)
12. Iwai, K.: Fundamental studies of fluid flow through a single fracture. University of California, Berkeley (1976)
13. Witherspoon, P., Wang, J., Iwai, K., Gale, J.: Validity of cubic law for fluid flow in a deformable rock fracture. Water Resour. **16**(6), 1016–1024 (1980)
14. Elrod, H.: A general theory for laminar lubrication with Reynolds roughness. J. Tribol. **101**(1), 8–14 (1979)

15. Zimmerman, R., Kumar, S., Bodvarsson, G.: Lubrication theory analysis of the permeability of rough-walled fractures. Int. J. Rock Mech. Min. Sci. Geomech. Abstr. **28**(4), 325–331 (1991)
16. Renshaw, C.: On the relationship between mechanical and hydraulic apertures in rough-walled fractures. J. Geophys. Res. Solid Earth **100**(B12), 24629–24636 (1995)
17. Olsson, R., Barton, N.: An improved model for hydromechanical coupling during shearing of rock joints. Int. J. Rock Mech. Min. Sci. **38**(3), 317–329 (2001)
18. Walsh, J.: Effect of pore pressure and confining pressure on fracture permeability. Int. J. Rock Mech. Min. Sci. Geomech. Abstr. **18**(5), 429–435 (1981)
19. Otomo, H., et al.: Simulation of residual oil displacement in a sinusoidal channel with the lattice Boltzmann method. Comptes-Rendus de Mecanique de l'Academie des Sciences **343**, 559–570 (2015)
20. Eker, E., Akin, S.: Lattice Boltzmann simulation of fluid flow in synthetic fractures. Transp. Porous Media **65**(3), 363–384 (2006)
21. Bhatnagar, P., Gross, E., Krook, M.: A model for collision processes in gases: small amplitude processes in charged and neutral one-component systems. Phys. Rev. **94**, 511–525 (1954)
22. Carroll, J., Wethe, D.: Chesapeake energy declares 'Propageddon' with record frack. Bloomberg: Markets. https://www.bloomberg.com/news/articles/2016-10-20/chesapeake-declares-propageddon-with-record-frack-in-louisiana. Accessed 14 Feb 2019

Incentive Mechanism for Cooperative Intrusion Response: A Dynamic Game Approach

Yunchuan Guo[1,2], Xiao Wang[1,3], Liang Fang[1], Yongjun Li[1,2], Fenghua Li[1,2], and Kui Geng[1(✉)]

[1] Institute of Information Engineering, Chinese Academy of Sciences, Beijing, China
gengkui@iie.ac.cn

[2] School of Cyber Security, University of Chinese Academy of Sciences, Beijing, China

[3] School of Computer and Communication Engineering, University of Science and Technology Beijing, Beijing, China

Abstract. Multi-hop D2D (Device-to-Device) communication is often exposed to many intrusions for its inherent properties, such as openness and weak security protection. To mitigate the intrusions in time, one of significant approaches is to establish a Cooperative Intrusion Response System (CIRS) to respond to intrusion activities during data transmission. In CIRS, user equipments that act as relays (RUEs) are assumed to actively help destination nodes to respond to intrusion activities. However, this assumption is often invalid in multi-hop D2D communication because the RUEs are selfish and unwilling to spend extra resources on undertaking response tasks. To address this problem, a game approach is proposed to encourage RUEs to cooperate. In detail, we formulate an incentive mechanism for CIRS in multi-hop D2D communication as a dynamic game and achieve an optimal solution to help RUEs decide whether to participate in detection or not. Theoretical analysis shows that only one Nash equilibrium exists for the proposed game. Simulations demonstrate that our mechanism can efficiently motivate potential RUEs to participate in intrusion detection and response, and it can also block intrusion propagation in time.

Keywords: Device-to-device · Cooperative intrusion response · Incentive mechanism · Game theory

1 Introduction

Multi-hop device-to-device (D2D) communication [15], which enables direct data transmission between source user equipments (SUEs) and destination user equipments (DUEs) with the assistance of other user equipments (UEs) acting as

Supported by the National Key Research and Development Program of China (No. 2016YFB0801001) and the National Natural Science Foundation of China (No. U1836203).

J. M. F. Rodrigues et al. (Eds.): ICCS 2019, LNCS 11536, pp. 590–603, 2019.
https://doi.org/10.1007/978-3-030-22734-0_43

relays (RUEs), provides a promising solution for mobile network operator to meet the growing users demands (such as higher throughput, lower transfer delay, and better power efficiency). Now multi-hop D2D communication has been widely used in various application, e.g., vehicle-to-vehicle networks [27], unmanned aerial vehicle (UAV) assisted wireless communications [11], and near field communications (NFC) [6].

However, multi-hop D2D communication may be exposed to many intrusions because of its openness, weak security protection on mobile UEs and the direct data transmission without the fixed network infrastructures. For example, in D2D-based vehicle-to-vehicle (V2V) communication [27], adversaries may disguise as normal vehicle equipment and send malicious packets via D2D communication to invade and compromise the destination vehicle equipment [12], thus threatening the lives of the people at destination vehicles. To mitigate the intrusions in a timely way, one significant thing that should be done is to establish a Cooperative Intrusion Response System (CIRS) for D2D communication to detect and respond to intrusion activities during data transmission. Through this approach, intrusion activities are detected and responded in time, communication overhead (e.g., the total volume of data traffic) are drastically reduced, thus communication security is improved.

Motivation: However, the CIRS cannot efficiently work in multi-hop D2D networks, because the RUEs are selfish and unwilling to spend extra resources on undertaking the intrusion detection and response tasks. Hence, a selfish RUE would not participate in responding intrusion events unless a satisfying incentive is given to compensate its extra cost. Without adequate participation of RUEs, the performance of a CIRS will be drastically decreased. To address this problem, an incentive mechanism, which motivates RUEs to promptly respond to intrusions, is required.

Considering the incentive resources being used to incentivize participation, existing incentive mechanism can be roughly divided into two categories: social-aware and financial-aware [7]. In the social-aware incentive mechanism [20], two social phenomena (i.e., social trust and social reciprocity) are used to find the social relationships among UEs and identify the best relays. However, privacy leakage is a serious challenge in social-aware incentive mechanism [22], because the process of identifying social relationships among UEs is usually accompanied by extra private information leakage. Compared with the social-aware incentive, the financial-aware incentive mechanism, which allocates financial resources to cooperators to incentivize participation, is a more desirable incentive paradigm in a practical application. However, existing CIRS and financial-aware incentive mechanisms are suffer from two problems, respectively: (1) **Low response accuracy.** In existing CIRS, response activities are operated based on aggregated monitored data from different sensors, thus most CIRS suffers from the loss of accuracy and the response accuracy is low. (2) **False-reporting attack.** When a packet is normal, malicious RUEs without carrying out detections might claim that they have detected the packet and have not found any abnormal data in this packet. Through this approach, they expects to win more rewards from DUEs.

Contribution: To address the above problems, in this paper, a dynamic game approach is utilized to establish a decentralized incentive mechanism for CIRS, which promote the response accuracy and mitigates the potential false-reporting attack. Our main contributions are as follows.

(1) In this paper, we formulate an incentive mechanism for CIRS in multi-hop D2D communication as a dynamic game and achieve the only one Nash equilibrium for RUEs to decide whether to participate in detection or not.
(2) We evaluate the benefit and cost of DUE and RUEs to analyze the proposed game. A reputation-based spot-check mechanism is also proposed to mitigate the potential false-reporting attack.
(3) Simulations demonstrate that our mechanism can efficiently motivate potential RUEs to participate in intrusion detection and response, and can also block intrusion propagation in time.

The remainder of this paper is organized as follows. In Sect. 2, we discuss the related work. We introduce the system model and discuss the spot-check mechanism in Sect. 3. Payoffs of RUEs and DUE are evaluated in Sect. 4. In Sect. 5, we formulate the incentive mechanism as a dynamic game and analyze its Nash equilibrium in Sect. 5. Simulations are provided in Sect. 6 to demonstrate the validity of proposed results. Section 7 draws the conclusion.

2 Related Work

2.1 D2D Communication

D2D communication [10,19] has received considerable attention in recent years and can be divided into two categories: standalone D2D and network-assisted D2D. UEs in standalone D2D organize communications by themselves and transfer messages directly without fixed network infrastructures (e.g., base stations) [3]. However, it is a big challenge for standalone D2D UEs to establish, maintain and control the communication only by themselves, which requires high complexity of the UEs. As a solution to this challenge, network-assisted D2D communication, which utilizes fixed network infrastructures for communication organization and resources allocation, has been widely studied. Zhou *et al.* [26] proposed a bargaining game to promote security and efficiency in network-assisted D2D with the presence of malicious eavesdroppers. Though network-assisted D2D works better than standalone D2D in practical applications, those two D2D paradigms could be failed due to long distance for their one-hop structure.

To solve the above problems, multi-hop D2D communication problems have been widely studied and applicated in various fields [6,11,27]. Zhou *et al.* [27] addressed the dependable D2D content distribution problem using a coalition formation game approach to optimize peer discovery, route selection, and spectrum allocation jointly. The spectrum trading contract was designed in [11] for D2D-based UAV-assisted cellular networks to better serve local mobile users.

Liu *et al.* [13] designed multi-hop D2D communication protocol and algorithm to address resource allocation problem for the general multi-hop D2D communication underlying cellular networks. Liu *et al.* [14] proposed three wireless power transfer policies in the power transfer model to analyzed the physical layer security in energy constrained D2D communication. Xu *et al.* [12] investigated the interplay between incentives and interdependent security risks in D2D offloading, and designed security-aware incentive mechanisms.

Multi-hop D2D communication provides an efficient D2D communication scheme with a variety of advantages such as improved spectral efficiency, and increased network capacity. Unfortunately, due to the weak security protection on ordinary mobile UEs, D2D communication may be exposed to many intrusions. In the past few decades, researchers are mainly focused on the security issues in single-hop D2D communication. So far, however, there has been little discussion about the security problems in multi-hop D2D communication.

2.2 Incentive Mechanism

Existing work investigates the incentive mechanism in wireless networks can be roughly divided into two categories: social-aware and financial-aware. Social-aware incentive mechanism for wireless networks is studied in [1,5,18]. Chen *et al.* [20] proposed a social-trust and social-reciprocity-based framework to promote cooperation among devices for multi-hop D2D communication. A cooperative video multicast system was developed in [4] to provide incentive for clients to share video packets with each other based on social ties in D2D communication. Gao *et al.* [8] formulates the dynamic social-aware peer selection problem as a dynamic optimization problem and proposes the drift-plus-penalty ratio algorithm to solve it. However, privacy leakage is a serious and inevitable problem in social-aware incentive mechanism.

Considering the privacy leakage issue, financial-aware incentive mechanism [12] is a more desirable incentive mechanism paradigm in practical application. Yang *et al.* [21] designed and analyzed platform-centric and user-centric financial-aware incentive mechanisms for mobile phone sensing. Guo *et al.* [9] formulated the incentive mechanism for CIDS as an evolutionary game to maximize nodes utility and motivate nodes to cooperate. However, financial-aware incentive mechanism may suffer serval attacks, and among them false-reporting attack is common and inevitable [25]. Zhang *et al.* [23,24] studied the free-riding and false-reporting problem in crowdsourcing and designed an incentive mechanism to motivating providers to complete their assigned tasks.

3 Basic Idea and System Model

3.1 Communication-Response Model

Figure 1 gives the communication process of multi-hop D2D and the mechanism of CIRS. Its detail processes are as follows. First, DUE requests a file with size

D from SUE. After SUE receives the request from DUE, it computes the routing path uses via routing algorithms (e.g., interference aware touting algorithms [16]) and obtains a set RP of RUEs where $RP = \{\mathrm{RUE}_0, \mathrm{RUE}_1, ..., RUE_{N-1}\}$ and N is the number of RUEs. Then, SUE sends a packet containing the file to its nearby RUE. When a neighboring RUE receives the packet and detects its potential threat (e.g., virus, Trojan) with detection rate α via intrusion detection technology (e.g. pattern matching). If without any threat included in this packet, RUE will relay the packet to its next nearby RUE in the routing path. Otherwise, it will pre-undertake corresponding countermeasures (e.g., interrupting the current communication and isolating the packet) and upload the threaten evidence to a trusted third party (TTP). In our work, we assume that the probability of malicious packets is ρ and no collusion between UEs exists.

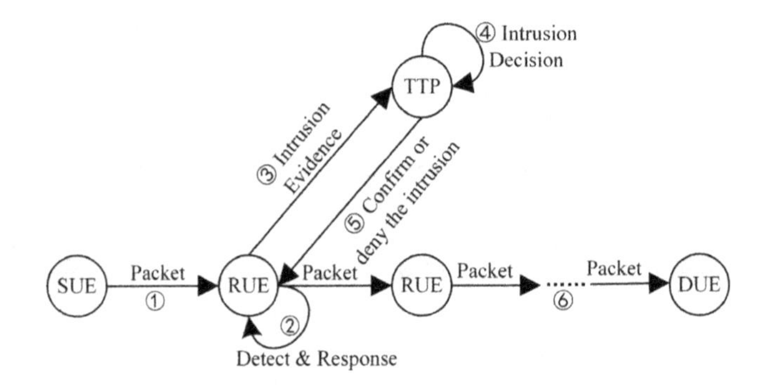

Fig. 1. Communication-response model.

3.2 Incentive Mechanism

As described in Sect. 1, RUEs are uninterested in participating in intrusion detection and response without sufficient incentive. To address this problem, a dynamic game-theoretic approach is proposed to stimulate a selfish RUE to detect and respond to an intrusion event. In our approach, RUE first evaluates the benefit and cost for detecting and responding to the potential intrusion, and then takes action based on its decision. After the multi-hop D2D communication is completed, DUE will decide to pay a reward only to the RUE who has worked correctly. Furthermore, due to the reward for RUE is paid after the communication is completed, some RUEs (called false-reporters) may lie to DUE that they have detected the packet in order to get rewards without detecting the packet if no intrusion is found by the RUEs before them. To address this problem, we design a reputation-based spot-check mechanism.

3.3 Spot-Check Mechanism

To mitigate false-reporting attack, we design a reputation-based spot-check mechanism. We assume that all UEs have a reputation score *rep*. If a RUE's reputation is less than a preset threshold rep_{th}, it will receive less reward gained from intrusion detection than normal RUEs. The process of spot-check mechanism is described as follows.

First, DUE notifies the SUE to start the spot-check activity. After receiving the packet, SUE intentionally sends a malicious packet mp with no to its nearby RUE with the destination of DUE. If the RUE claims that it has detected and then relays mp to the next RUE, we can regard this RUE as a false-reporter and reduce the RUE's reputation. Finally, DUE rewards the RUEs which correctly responds to mp. The spot-check mechanism runs and repeats irregularly when the multi-hop D2D routing path is idle, and the running status is only known to SUE, DUE. In the following section, We assume that all UEs are normal and have a reputation score above the threshold rep_{th}.

4 Benefit and Cost

4.1 RUE (User Equipment as Relay)

To establish the multi-hop D2D communication, RUE selection algorithm, which can find the optimal RUE and generate routing path for each UE, is significant. As the issue of RUE selection algorithm has been fully discussed in [13], in this paper, we will not investigate this problem and will assume the optimal routing path for multi-hop D2D communication has been selected. Here we consider the $\mathrm{RUE_i} \in \mathrm{RP}$ as the $(i+1)$th RUE that receives packets from SUE, where $i = 0, 1, 2, ..., N-1$. Hence, we evaluate the detection cost and response cost of $\mathrm{RUE_i}$ as follows.

Detection Cost: Each kind of intrusions has a unique attack pattern that can be recognized by attack pattern matching algorithms (e.g. Aho-Corasick algorithm [2]). In this paper, $\mathrm{RUE_i}$ matches the packet to existing attack patterns to detect the potential intrusion activity. The number of attack patterns for match is m_i and the computational complexity of the pattern matching algorithm that $\mathrm{RUE_i}$ selected is $cpma(m_i, D)$, where D is the size of the packet. Hence, the detection cost of $\mathrm{RUE_i}$ can be defined as follows.

$$C_{detection_i} = \lambda_{dc} \cdot cpma(m_i, D), \tag{1}$$

where the λ_{dc} is the cost unit for computational complexity of the pattern matching algorithm selected by $\mathrm{RUE_i}$.

Response Cost: If the packet is detected to be malicious, $\mathrm{RUE_i}$ will consume its resources to undertake countermeasures. Here we consider two types of resources: the memory space and the energy of $\mathrm{RUE_i}$. If intrusion response requires too many resources or RUE's idle resources are limited, the response cost will be expensive for RUEs. Here we donate the memory and energy utilization for undertaking countermeasures as M_{u_i} and E_{u_i}, respectively. Furthermore, the idle memory and energy are described as M_{f_i} and E_{f_i}, respectively. Hence, the response cost of $\mathrm{RUE_i}$ can be defined as follows.

$$C_{response_i} = \lambda_{rc} \cdot (\frac{M_{u_i}}{M_{f_i}} + \frac{E_{u_i}}{E_{f_i}}), \quad (2)$$

where the λ_{rc} is the cost unit for resource utilization in undertaking countermeasures.

4.2 DUE (User Equipment as Destination)

After the multi-hop D2D communication is completed, DUE will be in one of three states: (1) **State 1.** DUE has received the normal packet; (2) **State 2.** DUE is invaded by the malicious packet; (3) **State 3.** No packet reaches DUE because the packet from SUE is detected to be malicious by $RUE_i \in RP$. Under above states, DUE gains three different benefits as follows.

State 1: In this state, DUE receives normal packet and no intrusion activity happens. The total benefit of DUE is gained from the packet received, and can be divided into two parts. One part is the fixed benefit F that gains from receiving the packet successfully. The other gains from DUE's interest in the content of the packet, where the interest factor per unit of packet size is θ. If DUE is interested in the received packet, the larger the packet size D is, the higher benefit DUE can gain from it. Hence, the benefit of DUE can be defined as follows.

$$B_{s_1} = \lambda_{s_1} \cdot (F + \theta D), \quad (3)$$

where λ_{s_1} is the cost unit for DUE's interest.

However, B_{s_1} donates the sum of N benefits gained from RUEs in RP. As a result of this, the benefit of DUE that gains from RUE_i can be given as follows.

$$B_{s_1 i} = \lambda_{s_1} \cdot \frac{F + \theta D}{N}, \quad (4)$$

State 2: In this state, DUE receives malicious packet and is invaded. The benefit of DUE is negative and depends on the risk of the exploited vulnerability. Here we consider the factors proposed in [17] to evaluate the exploited vulnerability, and the risk can be calculated by weighting all those factors. The risk from exploiting vulnerability v_j can be defined as r_j with $0 < r_j < 1$. Hence, the benefit of DUE under the state 2 can be defined as follows.

$$B_{s_2} = -\lambda_{s_2} \cdot r_j, \quad (5)$$

where λ_{s_2} is the cost unit for the risk of exploiting a vulnerability.

Under this state, DUE is invaded and all the RUEs that have detected the packet should be responsible for the intrusion. Moreover, the RUEs who receives and detects the packet early should have greater responsibilities than RUEs after them. As a result of this, the negative benefit of DUE gained from RUE_i can be given as follows.

$$B_{s_2 i} = -\lambda_{s_2} \cdot \frac{2 \cdot r_j \cdot (N - i)}{N(N+1)}, \quad (6)$$

State 3: No packet reaches DUE because the packet is detected to be malicious and an intrusion is responded by $\mathrm{RUE_i} \in \mathrm{RP}$. If the intrusion packet is detected by $\mathrm{RUE_i}$ with a small serial number i, which means the intrusion is blocked in time, the benefit of DUE will be high. Hence, the benefit gained from $\mathrm{RUE_i}$ can be given as follows.

$$B_{s_3 i} = \lambda_{s_2} \cdot \frac{(N-i) \cdot r_j}{N}. \tag{7}$$

5 Dynamic Game and Its Analysis

We define the game as a triplet $G = \{\{\mathrm{DUE}\} \cup RP, S, U\}$, where RP is the set which consists of the RUEs in routing path, S donates the strategy space, and U is the set of players' utilities. Here we assume that the probability that the packet is malicious is ρ. The game tree can be seen in Fig. 2, where the leaf nodes present the players' utilities with the tuple $U = (U_{DUE}, U_{RUE})$. We define the strategy combination as a tuple $S = (S_{RUE}, S_{DUE})$, where $S_{RUE} = (detect, no\_detect)$ and $S_{DUE} = (pay, no\_pay)$. We analyze the game in two levels.

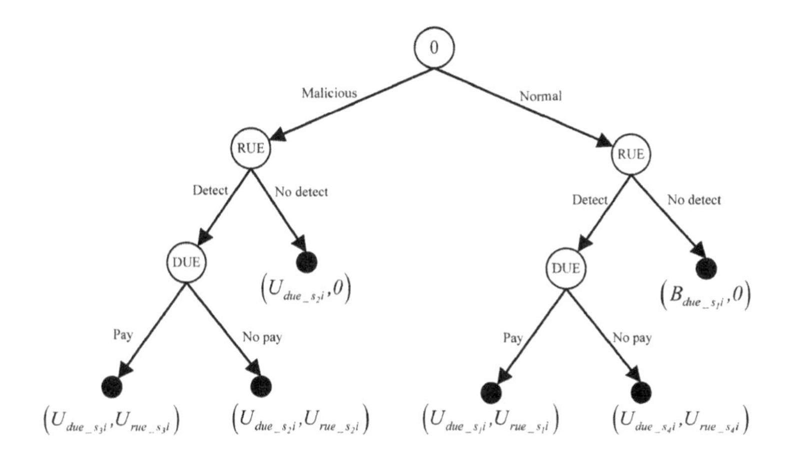

Fig. 2. The dynamic game tree.

5.1 DUE Level

The benefits of DUE gained from $\mathrm{RUE_i}$ in this game are described in (4), (6), and (7). After the communication is completed, there are four possible states of DUE and RUE. Therefore, the rewards for RUEs are defined as P_{s_1}, P_{s_2}, P_{s_3}, P_{s_4}. As the benefits of DUE are described in Sect. 4.2, for each $\mathrm{RUE_i}$, DUE's four different utilities are as follows.

When the multi-hop D2D communication is completed, if DUE has received the normal packet and $\mathrm{RUE_i}$ detects correctly, the DUE's utility gained from $\mathrm{RUE_i}$ can be defined as follows.

$$U_{due\_s_1i} = B_{s_1i} - P_{s_1i} = \lambda_{s_1} \cdot \frac{F + \theta D}{N} - \frac{P_{s_1}}{N}, \tag{8}$$

where P_{s_1i} is the reward for $\mathrm{RUE_i}$.

When the multi-hop D2D communication is completed, DUE is invaded by the malicious packet, the DUE's utility gained from $\mathrm{RUE_i}$ can be defined as follows.

$$U_{due\_s_2i} = B_{s_2i} - P_{s_2i} = -\lambda_{s_2} \cdot \frac{2 \cdot r_j \cdot (N - i)}{N(N+1)}, \tag{9}$$

where $P_{s_2i} = 0$ because DUE is invaded but no intrusion is detected by $\mathrm{RUE_i}$.

When the multi-hop D2D communication is completed, no packet reaches DUE because the packet is detected to be malicious by $\mathrm{RUE_i} \in \mathrm{RP}$, the DUE's utility gained from $\mathrm{RUE_i}$ can be defined as follows.

$$U_{due\_s_3i} = B_{s_3i} - P_{s_3i} = \frac{(N-i)}{N} \cdot (\lambda_{s_2} \cdot r_j - P_{s_3}) \tag{10}$$

where P_{s_3i} is the reward for $\mathrm{RUE_i}$ and it is high for the $\mathrm{RUE_i}$ with a small serial number i because they have blocked the propagation of malicious packet in a timely way.

When the multi-hop D2D communication is completed, if DUE has received the normal packet and $\mathrm{RUE_i}$ has detected incorrectly and responded to it, the DUE's utility gained from $\mathrm{RUE_i}$ can be defined as follows.

$$U_{due\_s_4i} = B_{s_1i} - 0 = \lambda_{s_1} \cdot \frac{F + \theta D}{N}. \tag{11}$$

5.2 RUE Level

The costs of $\mathrm{RUE_i}$ in this game are described in (1) and (2), and the rewards that DUE can pay are given in (8)–(11). Therefore, the utilities of $\mathrm{RUE_i}$ can be expressed as follows.

When the multi-hop D2D communication is completed and no intrusion happens, if $\mathrm{RUE_i}$ has detected the packet correctly, the utility of $\mathrm{RUE_i}$ can be defined as follows.

$$U_{rue\_s_1i} = P_{s_1i} - C_{detection_i} = \frac{P_{s_1}}{N} - \lambda_{dc} \cdot cpma(m_i, D). \tag{12}$$

When the multi-hop D2D communication is completed and intrusion happens, if $\mathrm{RUE_i}$ hasn't found the intrusion after detecting the packet, the utility of $\mathrm{RUE_i}$ can be defined as follows.

$$U_{rue\_s_2i} = P_{s_2i} - C_{detection_i} = -\lambda_{dc} \cdot cpma(m_i, D). \tag{13}$$

When the multi-hop D2D communication is completed and intrusion is detected and responded by $\mathrm{RUE_i}$, the utility of $\mathrm{RUE_i}$ can be defined as follows.

$$\begin{aligned} U_{rue\_s_3i} &= P_{s_3i} - C_{detection_i} - C_{response_i} \\ &= \frac{(N-i) \cdot P_{s_3}}{N} - \lambda_{dc} \cdot cpma(m_i, D) - \lambda_{rc} \cdot (\frac{M_{u_i}}{M_{f_i}} + \frac{E_{u_i}}{E_{f_i}}), \end{aligned} \tag{14}$$

When the multi-hop D2D communication is completed and no intrusion happens, if RUE$_i$ has detected the packet incorrectly and responded to it, the utility of RUE$_i$ can be defined as follows.

$$\begin{aligned} U_{rue\_s_4 i} &= 0 - C_{detection_i} - C_{response_i} \\ &= -\lambda_{dc} \cdot cpma(m_i, D) - \lambda_{rc} \cdot (\frac{M_{u_i}}{M_{f_i}} + \frac{E_{u_i}}{E_{f_i}}), \end{aligned} \tag{15}$$

5.3 Equilibrium Analysis

Theorem 1. *The strategy $s = (detect, pay)$ is the only Nash equilibrium of the dynamic game, if DUE's rewards for RUEs satisfy the conditions as follows.*

$$\begin{cases} \lambda_{rc}\frac{1-\alpha}{\alpha}(\frac{M_{u_i}}{M_{f_i}} + \frac{E_{u_i}}{E_{f_i}}) \cdot N < P_{s_1} < \lambda_{s_1} \cdot (F + \theta D), \\ P_{s_2} = 0, \\ (\lambda_{dc} cpma(m_i, D) + \lambda_{rc}\frac{1}{\alpha}(\frac{M_{u_i}}{M_{f_i}} + \frac{E_{u_i}}{E_{f_i}})) \cdot N < P_{s_3} < \lambda_{s_2} \cdot r_j, \\ \rho \cdot P_{s_3} + (1-\rho) \cdot \frac{P_{s_1}}{N} < \alpha\rho \cdot \lambda_{s_2} + \lambda_{s_2}\frac{\rho \cdot (1-\alpha)}{\alpha} \cdot \frac{2 \cdot r_j}{N+1} \\ \qquad + \lambda_{s_1}\frac{(2\alpha - 1)(1-\rho)}{\alpha}(F + \theta D). \end{cases} \tag{16}$$

Proof. Here the method of inverse analysis is used. We can divide the problem into three aspects.

(1) P_{s_2} is supposed to be 0 as described in Sect. 5.1. As DUE's utility should be positive when it decides to rewards the RUEs detected the packet correctly. Therefore, according to (3) and (7), we can get the first condition as follows.

$$\begin{cases} P_{s_1} < B_{s_1} = \lambda_{s_1} \cdot (F + \theta D), \\ P_{s_2} = 0, \\ P_{s_3} < B_{s_3} = \lambda_{s_2} \cdot r_j. \end{cases} \tag{17}$$

(2) According to the Fig. 2 and the utilities of RUE$_i$ defined in (12)–(15), the condition below should be satisfied to ensure the utility of "detect" is higher than "no_detect".

$$\begin{cases} \alpha U_{rue\_s_3 i} + (1-\alpha) U_{rue\_s_2 i} > 0, \\ \alpha U_{rue\_s_1 i} + (1-\alpha) U_{rue\_s_4 i} > 0. \end{cases} \tag{18}$$

Now using the utilities in (12)–(15), we can get condition two as follows.

$$\begin{cases} P_{s_1} > \lambda_{rc}\frac{1-\alpha}{\alpha}(\frac{M_{u_i}}{M_{f_i}} + \frac{E_{u_i}}{E_{f_i}}) \cdot N, \\ P_{s_3} > (\lambda_{dc} cpma(m_i, D) + \lambda_{rc}\frac{1}{\alpha}(\frac{M_{u_i}}{M_{f_i}} + \frac{E_{u_i}}{E_{f_i}})) \cdot N. \end{cases} \tag{19}$$

(3) According to the Fig. 2 and the utilities of DUE defined in (8), (9) (10) and (11), the condition below should be satisfied to ensure the utility of "pay" is higher than "no_pay".

$$\alpha(\rho \cdot U_{due\_s_3 i} + (1-\rho) \cdot U_{due\_s_1 i}) > (1-\alpha)(\rho \cdot U_{due\_s_1 i} + (1-\rho) \cdot U_{due\_s_4 i}). \tag{20}$$

Now using the utilities in (8)–(11), we can get condition 3 as follows.

$$\begin{aligned} \rho \cdot P_{s_3 i} + (1-\rho) \cdot P_{s_1 i} < \alpha\rho \cdot B_{s_3 i} + \frac{(2\alpha-1)(1-\rho)}{\alpha} B_{s_1 i} \\ - \frac{\rho \cdot (1-\alpha)}{\alpha} B_{s_2 i}. \end{aligned} \tag{21}$$

As described in Sect. 5.1, we have

$$\begin{aligned} \rho \cdot \frac{(N-i)}{N} P_{s_3} + (1-\rho) \cdot \frac{P_{s_1}}{N} < \alpha\rho \cdot \lambda_{s_2} \frac{(N-i)r_j}{N} \\ + \lambda_{s_1} \frac{(2\alpha-1)(1-\rho)}{\alpha} (F + \theta D) \\ + \lambda_{s_2} \frac{\rho \cdot (1-\alpha)}{\alpha} \cdot \frac{2 \cdot r_j \cdot (N-i)}{N(N+1)}. \end{aligned} \tag{22}$$

The final result can be calculated as follows.

$$\begin{aligned} \rho \cdot P_{s_3} + (1-\rho) \cdot \frac{P_{s_1}}{N} < \alpha\rho \cdot \lambda_{s_2} + \lambda_{s_2} \frac{\rho \cdot (1-\alpha)}{\alpha} \cdot \frac{2 \cdot r_j}{N+1} \\ + \lambda_{s_1} \frac{(2\alpha-1)(1-\rho)}{\alpha} (F + \theta D). \end{aligned} \tag{23}$$

Combining the three conditions (17), (19) and (23), we can get the final condition (16) to ensure the Nash equilibrium $s = (detect, pay)$ of the dynamic game. □

6 Experiment Evaluation

In the experiment, we adopt a taxi scenario to simulate multi-hop D2D communication where each mobile device in a taxi is a D2D UE. Data was gathered from 8:00:00 a.m. to 8:59:59 a.m. on August 13, 2015 including 10088 GPS records of 442 taxis in the Changping area in Beijing, China. During this period, we assume that: (1) Each taxi has one mobile device in it and it communicates with others via multi-hop D2D networks; (2) Distance of D2D communication is 100 m, and taxis within the scope of the communication can transmit packets with each other; (3) Energy consumption is only considered in intrusion detection and response; (4) Size of the packet is $D = 2000$. Probability of a malicious packet is $\rho = 0.5$. The potential number of RUEs $\in RP$ is $N = 10$.

As described in Sect. 4.2, we define the values of parameters as follows. For the aspect of RUE, we assume that the multi-pattern matching algorithm is Aho-Corasick algorithm [2], thus $cpma(m_i, D)$ can be calculated as $m_i \cdot D$.

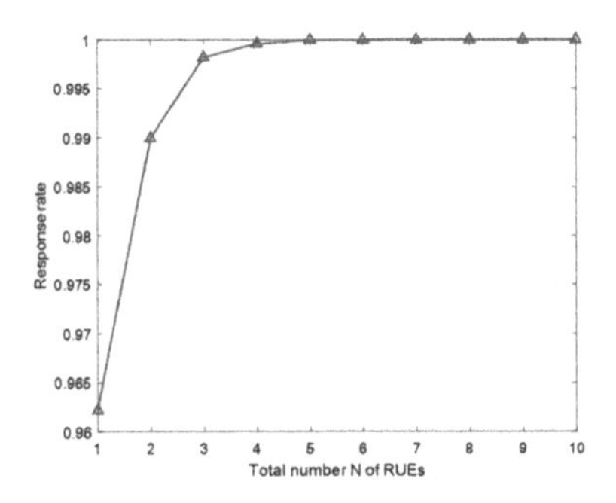

Fig. 3. Response rate.

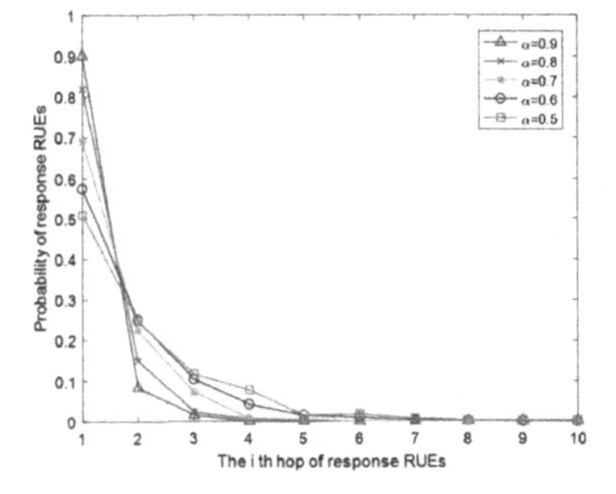

Fig. 4. Timeliness of response.

The remaining parameters are as follows. λ_{dc} and λ_{rc} are (5×10^{-6}) and 25, respectively. The detection rate is $\alpha = 0.9$. The number of attack patterns for match is $m_i = 1000$. The memory and energy utilization for undertaking countermeasures are $M_{u_i} = 20$ and $E_{u_i} = 10$, respectively. The free memory and energy are $M_{f_i} = 100$ and $E_{f_i} = 100$, respectively. For the aspect of DUE, the parameters are as follows. λ_{s1} and λ_{s2} are 10 and (2×10^3), respectively. The risk is a constant $r_j = 0.9$. Fixed benefit F is 25, and interest factor θ is 2.5×10^{-3}. Therefore, according to Theorem 1, the value of P_{s_1} and P_{s_3} can be set as 150 and 1000, respectively. Without the special statement, we set the parameters value described above as default.

Response Rate: We pick different number N in order to show the proportion of total response number in the number of malicious packet in D2D communication. Figure 3 shows the change of response rate over the total number N, respectively. From Fig. 3, we can see that, given the detection rate $\alpha = 0.9$, the rate of response increases with the growth of RUEs' number N if the intrusion happens. Figure 3 shows the compensation for the single detection node.

Timeliness of Response: We pick different detection rate α and the result is presented in Fig. 4. The abscissa in Fig. 4 is the ith hop of the response RUE and the ordinate is the probability of response RUEs at specific hop i. If the probability that RUE responds to the malicious packet is high, intrusion activity could be blocked in time. From Fig. 4, we can see that, with the growth of the detection rate α, more malicious packet is responded by the RUE with smaller hops. This means that intrusion will be blocked in a timely way before the malicious packets arrive DUE.

7 Conclusion

Multi-hop D2D communication may be exposed to many intrusions for its inherent properties, such as openness and weak security protection. To mitigate the intrusions in time, in this paper, we formulate an incentive mechanism for CIRS in multi-hop D2D communication as a dynamic game and achieve an optimal solution to help RUEs decide whether to participate in detection or not. Theoretical analysis shows that the only Nash equilibrium exists for the proposed game. To mitigate the false-reporting attack, we proposed a spot-check mechanism on the basis of binary reputation score. Simulations demonstrate that our mechanism can efficiently motivate potential RUEs to participate in intrusion detection and response, and can also block intrusion propagation in time.

References

1. Ahmed, M., Li, Y., Waqas, M., Sheraz, M., Jin, D., Han, Z.: A survey on socially aware device-to-device communications. IEEE Commun. Surv. Tutor. **20**(3), 2169–2197 (2018)
2. Aho, V.A., Corasick, J.M.: Efficient string matching: an aid to bibliographic search. Commun. ACM **18**(6), 333–340 (1975)
3. Asadi, A., Wang, Q., Mancuso, V.: A survey on device-to-device communication in cellular networks. IEEE Commun. Surv. Tutor. **16**(4), 1801–1819 (2014)
4. Cao, Y., Jiang, T., Chen, X., Zhang, J.: Social-aware video multicast based on device-to-device communications. IEEE Trans. Mob. Comput. **15**(6), 1528–1539 (2016)
5. Chen, G., Tang, J., Coon, J.P.: Optimal routing for multi-hop social-based D2D communications in the internet of things. IEEE Internet Things J. **5**(3), 1880–1889 (2018)
6. Coskun, V., Ozdenizci, B., Ok, K.: The survey on near field communication. Sensors **15**(6), 13348–13405 (2015)
7. Feng, W., Yan, Z., Zhang, H., Zeng, K., Xiao, Y., Hou, Y.T.: A survey on security, privacy, and trust in mobile crowdsourcing. IEEE Internet Things J. **5**(4), 2971–2992 (2018)
8. Gao, Y., Xiao, Y., Wu, M., Xiao, M., Shao, J.: Dynamic social-aware peer selection for cooperative relay management with D2D communications. IEEE Trans. Commun. **67**(5), 3124–3139 (2019)
9. Guo, Y., Zhang, H., Zhang, L., Fang, L., Li, F.: Incentive mechanism for cooperative intrusion detection: an evolutionary game approach. In: Shi, Y., Fu, H., Tian, Y., Krzhizhanovskaya, V.V., Lees, M.H., Dongarra, J., Sloot, P.M.A. (eds.) ICCS 2018. LNCS, vol. 10860, pp. 83–97. Springer, Cham (2018). https://doi.org/10.1007/978-3-319-93698-7_7
10. Haus, M., Waqas, M., Ding, A.Y., Li, Y., Tarkoma, S., Ott, J.: Security and privacy in device-to-device (D2D) communication: a review. IEEE Commun. Surv. Tutor. **19**(2), 1054–1079 (2017)
11. Hu, Z., Zheng, Z., Song, L., Tao, W., Li, X.: UAV offloading: spectrum trading contract design for UAV assisted cellular networks. IEEE Trans. Wirel. Commun. **17**(9), 6093–6107 (2018)
12. Xu, J., Chen, L., Liu, K., Shen, C.: Designing security-aware incentives for computation offloading via device-to-device communication. IEEE Trans. Wirel. Commun. **17**(9), 6053–6066 (2018)
13. Liu, T., Lui, J.C.S., Ma, X., Jiang, H.: Enabling relay-assisted D2D communication for cellular networks: algorithm and protocols. IEEE Internet Things J. **5**(4), 3136–3150 (2018)
14. Liu, Y., Wang, L., Zaidi, S.A.R., Elkashlan, M., Duong, T.Q.: Secure D2D communication in large-scale cognitive cellular networks: a wireless power transfer model. IEEE Trans. Commun. **64**(1), 329–342 (2016)

15. Shamganth, K., Sibley, M.J.: A survey on relay selection in cooperative device-to-device (D2D) communication for 5G cellular networks. In: 2017 International Conference on Energy, Communication, Data Analytics and Soft Computing (ICECDS), pp. 42–46. IEEE (2017)
16. Tang, J., Xue, G., Zhang, W.: Interference-aware topology control and QoS routing in multi-channel wireless mesh networks. In: ACM Interational Symposium on Mobile Ad Hoc Networking and Computing (2005)
17. Ten, C.W., Liu, C.C., Manimaran, G.: Vulnerability assessment of cybersecurity for scada systems. IEEE Trans. Power Syst. **23**(4), 1836–1846 (2008)
18. Wang, H.M., Xu, Y., Huang, K.W., Han, Z., Tsiftsis, T.A.: Cooperative secure transmission by exploiting social ties in random networks. IEEE Trans. Commun. **66**(8), 3610–3622 (2018)
19. Wang, M., Yan, Z.: A survey on security in D2D communications. Mob. Netw. Appl. **22**(2), 195–208 (2017)
20. Xu, C., Proulx, B., Gong, X., Zhang, J.: Exploiting social ties for cooperative D2D communications: a mobile social networking case. IEEE/ACM Trans. Netw. **23**(5), 1471–1484 (2015)
21. Yang, D., Xue, G., Fang, X., Tang, J.: Crowdsourcing to smartphones: incentive mechanism design for mobile phone sensing. In: Proceedings of the 18th Annual International Conference on Mobile Computing and Networking, pp. 173–184. ACM (2012)
22. Zhang, C., Sun, J., Zhu, X., Fang, Y.: Privacy and security for online social networks: challenges and opportunities. IEEE Netw. **24**(4), 13–18 (2010)
23. Zhang, X., Xue, G., Yu, R., Yang, D., Tang, J.: You better be honest: discouraging free-riding and false-reporting in mobile crowdsourcing. In: 2014 IEEE Global Communications Conference, pp. 4971–4976. IEEE (2014)
24. Zhang, X., Xue, G., Yu, R., Yang, D., Tang, J.: Keep your promise: mechanism design against free-riding and false-reporting in crowdsourcing. IEEE Internet Things J. **2**(6), 562–572 (2015)
25. Zhang, Y., van der Schaar, M.: Reputation-based incentive protocols in crowdsourcing applications. In: 2012 Proceedings of IEEE INFOCOM, pp. 2140–2148. IEEE (2012)
26. Zhou, Q., Lu, W., Chen, S., Yang, L., Wang, K.: Promoting security and efficiency in D2D underlay communication: a bargaining game approach. In: GLOBECOM 2017–2017 IEEE Global Communications Conference, pp. 1–6, December 2017. https://doi.org/10.1109/GLOCOM.2017.8254089
27. Zhou, Z., Yu, H., Chen, X., Yan, Z., Mumtaz, S., Rodriguez, J.: Dependable content distribution in D2D-based cooperative vehicular networks: a big data-integrated coalition game approach. IEEE Trans. Intell. Transp. Syst. **19**(3), 953–964 (2018)

A k-Cover Model for Reliability-Aware Controller Placement in Software-Defined Networks

Gabriela Schütz[1,2](✉)

[1] Institute of Engineering of the University of Algarve, Campus da Penha, Faro, Portugal
gschutz@ualg.pt
[2] Center for Electronic, Optoelectronic and Telecommunications (CEOT), Campus de Gambelas, Faro, Portugal

Abstract. The main characteristics of Software-Defined Networks are the separation of the control and data planes, as well as a logically centralized control plane. This emerging network architecture simplifies the data forwarding and allows managing the network in a flexible way. Controllers play a key role in SDNs since they manage the whole network. It is crucial to determine the minimum number of controllers and where they should be placed to provide low latencies between switches and their assigned controller. It is worth to underline that, if there are long propagation delays between controllers and switches, their ability of reacting to network events quickly is affected, degrading reliability. Thus, the Reliability-Aware Controller Placement (RCP) problem in Software-Defined Networks (SDNs) is a critical issue. In this work we propose a k-cover based model for the RCP problem in SDNs. It simultaneously optimizes the number and placement of controllers, as well as latencies of primary and backup paths between switches and controllers, providing reliable networks against link, switch and controller failures. Although RCP problem is NP-hard, the simulation results show that reliabilities greater than 97%, satisfying low latencies, were obtained and the model can be used to find the optimum solution for different network topologies, in negligible time.

Keywords: Software-defined network · Controller placement · Reliability · k-cover problem

1 Introduction

Software-Defined Network (SDN) decouples control and data planes simplifying the data forwarding and allowing the network management in a flexible way. The SDN control plane is crucial to the network performance [9]. It handles

Supported by the UID/MULTI/00631/2013 project of the Portuguese Science and Technology Foundation (FCT).

J. M. F. Rodrigues et al. (Eds.): ICCS 2019, LNCS 11536, pp. 604–613, 2019.
https://doi.org/10.1007/978-3-030-22734-0_44

state distribution, control applications and network connectivity for propagating events to switches and also between multiple controllers. Network failures that disconnect the control and data planes could block requesting instructions from switches to controllers and may cause packet loss and network unsatisfactory performance [4]. The optimal number and placement of controllers, as well as the assignment of controllers to switches, play a very important role towards performance and reliability of SDN [2].

Although a switch can detect a control path failure, it has no capacity to establish a new route and connection will be lost until a backup control path is found. The distance between switches and their assigned controller affects propagation latency and restoration time, so low latency paths are required. Providing in advance backup control paths with acceptable latency allows quick restoration of the control plane against path failure, since a switch can initiate its backup path as soon as it detects a control path failure [7]. Similarly, planning in advance, for each switch, low latency connections to two different controllers over two disjoint paths allows quick restoration of the control plane against controller failure or congestion.

The above reasoning grounds our approach. In this work we propose a mathematical model based on the k-cover problem to plan a reliable SDN, enhancing the protection of the control plane against link, switch, and controller failures. It determines the optimum number of controllers and their placements, constrained to: (i) every switch must be connected to two different controllers, a primary and a backup controller, over two disjoint control paths; (ii) every switch must be connected to its assigned primary controller over two disjoint paths; (iii) control paths (primary and backup) latencies must be bellow a given threshold.

The remainder of this article is organised as follows: Sect. 2 is a short review of related works. The proposed model is described in Sect. 3, which includes the mathematical formalization. The experimental simulation is described in Sect. 4, while Sect. 5 presents and analyses the results. Section 6 draws some conclusions from the obtained results.

2 Related Work

Reliability-Aware Controller Placement (RCP) is a particular case of the Control Placement Problem (CPP). Several works have already addressed different issues related to the CPP. This section briefly overviews some works on fault tolerant and reliable controller placement towards the improvement of network resilience. Network elements failures may cause the disconnection between controllers and switches. Zhang et al. [8] call lost nodes to these switches that are unable to connect the controller due to failures. They minimize the number of lost nodes using a min-cut based controller placement algorithm to obtain a partition of the network, such that inside each partition switches and the respective controller are well connected. Hock et al. [3] define performance and resilience metrics in the controller placement problem and implement a framework to evaluate the entire solution space. In [4] the expected percentage of control path loss, defined as the number of broken control paths due to network failures, is used to characterize the SDN reliability and for a given number of controllers they maximize the SDN reliability through a binary integer programming. Muller et al. [5] formulate the problem as a binary integer programming to maximize the average number of disjoint paths between switches and controllers. They also propose heuristics for defining lists of backup controllers to deal with controller failure. Ros and Ruiz [6] develop a heuristic for the fault tolerant controller placement problem where reliability thresholds must be satisfied. Their results show that if each node connects to two or three controllers, it can provide more than five nines reliability and also that, generally, ten controllers are enough, being its number more related to the network topology than to the network size. Vizarreta et al. [7] present two controller placement strategies for a resilient control plane. One strategy considers that switches have to be connected to a controller over two disjoint paths and the other considers that switches have to be connected to two different controllers over two disjoint paths. They evaluate their two approaches in comparison to the unprotected case.

In this work the planning of primary and backup control paths in advance, as in [7], is also considered, but this approach is different because it determines the minimum number of controllers that ensure, simultaneously, disjoint primary and backup control paths providing required latencies and also primary and backup controllers for each switch, as mentioned in Sect. 1, modelled and formalized as a k-Cover Problem.

3 Problem Formalization

3.1 Problem Overview

When deploying multiple controllers, the reliability and resilience of SDN reside on a controller placement highly fault tolerant. Clearly, more controllers can increase the control network reliability, but also imply on more communications to exchange information, harder network management and overall cost increase [6]. It is advisable to place as few controllers as possible, taking into consideration that too few controllers would increase latency and decrease reliability. Thus the main goal is to find the appropriate number and locations of controllers to ensure control plane reliability and satisfy low propagation delay between switches and their assigned controllers.

This approach achieves the above goal. It minimizes the number of controllers ensuring the existence of at least two disjoint paths, primary and backup, between each switch and one controller and the assignment of two controllers, primary and backup, to each switch. All paths between switches and controllers satisfy the required latency. It is assumed that each switch communicates with the primary controller over the primary control path. If the primary controller fails than communication is quickly restored to the backup controller over a disjoint path, meaning that disconnection is avoided. If the primary control path to the primary controller fails (due to a link or switch failure) then the disjoint backup path to that controller is promptly initiated. Thus, for each switch, we compute its control paths reliability, denoted by (R_s), as the probability of no communication disconnection between switch and controller, as follows:

$$R_s = r_s \cdot [r_{C_p} \cdot (1 - (1 - r_p)(1 - r_b)) + (1 - r_{C_p}) \cdot r_{C_b} \cdot r_{dp}] \tag{1}$$

where, for simplicity, failure probability of a path, a link, a switch or a controller is denoted by f_*, where $*$ is equal to l for link; s for switch; C_p and C_b for primary and backup controllers; p and b for primary and backup disjoint paths to primary controller; dp for disjoint backup path to backup controller, being the reliability of a component $r_* = 1 - f_*$. The failure probability of a path from i to j is computed as

$$f_{p(i,j)} = 1 - \Pi_{l \in p(i,j)}(1 - f_l)\Pi_{s \in p(i,j)-\{i,j\}}(1 - f_s). \tag{2}$$

Average network reliability, denoted by R, is calculated as the average of the switches control paths reliabilities,

$$R = \frac{\sum_s R_s}{\text{number of switches}} \tag{3}$$

Obviously, there are two disjoint paths between switches and a controller if the degree of every node is equal or greater than two and the network has no articulation points. Next, we introduce some cover definitions applied to this problem.

Definition 1: A switch is covered by a controller if the path between them provides the required latency.

Definition 2: A switch is k-covered if it is covered by at least k different controllers.

Definition 3: A network is k-covered if every switch is k-covered and k is the degree of the coverage.

It is clear that the number of controllers to achieve a k coverage degree increases directly with k. Hence, the bigger is k the more controllers are needed. The network must be at least 2-covered to assure connectivity in the presence of a controller failure.

3.2 Mathematical Formalization

In the following mathematical formalization, the network is represented as an undirected graph $G(V, E)$, where $V = \{1, 2, ..., N\}$ is the set of nodes (switches) and E is the set of edges (bidirectional links) connecting nodes. $V_c \subseteq V$ denotes the subset of switches $(v \in V)$ hosting a controller. We assume a uniform demand and equal amount of traffic forwarded between switches and controllers. Since the propagation latency is the largest part of latency and the length of a communication link is proportional to the propagation delay it introduces, we assume that path length is equivalent to path latency. The latency of a primary control path between controller i and switch j is the length of the shortest path between them and is denoted by d^p_{ij}, the respective disjoint backup control path latency, to the primary controller or to the backup controller, denoted by d_{ij}^{b} or d_{ij}^{dp} is the length of the shortest path between controller i and switch j in the sub-graph obtained by removing the links and intermediate nodes of the primary control path. Δ_p and Δ_b are, respectively, primary and backup control paths latency threshold.

We define, below, constants $a_{ij}, \forall i, j \in V$ used to ensure that j can be covered by a controller placed in i only if the two disjoint shortest paths between i and j satisfy the respective required latencies.

$$\forall i, j \in V, a_{ij} = \begin{cases} 1, & \text{if } d_{ij}^{p} \leq \Delta_p \wedge d_{ij}^{b} \leq \Delta_b \\ 0, & \text{otherwise} \end{cases} \tag{4}$$

The coefficients of the objective function are equal to the average of the weighted sum of the distances of primary and backup paths, calculated as follows:

$$f_i = \frac{\sum_{j \in V} (\alpha d_{ij}^{p} + \beta d_{ij}^{b}) a_{ij}}{\sum_{j \in V} a_{ij}}, \forall i \in V \tag{5}$$

The binary decision variables are:

$$x_i = \begin{cases} 1, & \text{if the location of switch } i \in V \text{ is choosen to place a controller} \\ 0, & \text{otherwise} \end{cases} \tag{6}$$

The RCP is formalized as a 2-Cover Problem, as follows:

$$\min \sum_{i \in V} f_i x_i. \tag{7}$$

subject to:

$$\sum_{i \in V} a_{ij} x_i \geq 2, \forall j \in V \tag{8}$$

$$x_i \in \{0, 1\}, \forall i \in V. \tag{9}$$

The objective function (7) minimizes the number of controllers weighted by the average of primary and backup path lengths. Constraints (8) ensure that every switch is, at least, covered by 2 different controllers, using disjoint paths, both providing feasible latencies. Constraints (9) define variables as binary.

The optimum solution of this formalization obtains the minimum number of controllers and their locations, such that each switch is connected, at least, to two controllers by two disjoint paths towards each controller, complying with the required latencies. The assignment of controllers to switches is implicit, since for each switch the primary controller is the nearest one; the backup controller is the nearest controller that can be connected over a disjoint path. Primary and backup paths between a switch and a controller will be disjoint by construction of $a_{ij}, \forall i, j \in V$, given in (4). Therefore, considering the obtained set of controllers $V_c = \{i \in V : x_i = 1\}$, we define the binary assignment variables of primary and backup controllers to each switch, as follows: $\forall j \in V$, $y_{ij}^{Cp} = 1$, if $arg(min_{i \in V_c} d_{ij}^{p}) = i$ and $y_{ij}^{Cb} = 1$, if $arg(min_{i \in V_c} d_{ij}^{dp}) = i$.

4 Experimental Setup

4.1 Network Topologies

This approach was tested in all topologies with lower node degree greater than or equal to 2, available online in SNDlib database [1] (so, abilene, brain, ta2 and zib54 topologies were excluded). Euclidean distances were computed and associated to the links. We define, as usually, the *Diameter* of a network as the maximum shortest path between any two nodes in the network. Table 1 summarizes networks parameters. Five small networks (dfn-bwin, dfn-gwin, di-yuan, geant and nobel-us) were not included in Table 1 because no feasible solution exists for the 2-cover problem given in (7)–(9). In fact, for at least one switch, j, $\sum_{i \in V} a_{ij} < 2$ thus constraints (8) can not be satisfied.

Table 1. Network parameters

Ref.	Topology	N	$\|E\|$	Average degree	Diameter	
					Edges	Length (km)
1	pdh	11	34	6.18	3	670.36
2	polska	12	18	3.00	4	746.15
3	atlanta	15	22	2.93	5	612.79
4	newyork	16	49	6.12	3	522.75
5	nobel-germany	17	26	3.06	6	1893.81
6	ta1	24	55	4.58	4	1012.42
7	france	25	45	3.60	5	1215
8	janos-us	26	84	6.46	8	2172.61
9	norway	27	51	3.78	7	1255.56
10	sun	27	102	7.56	7	1068.92
11	nobel-eu	28	41	2.93	8	1807.7
12	india35	35	80	4.57	7	1582.08
13	cost266	37	57	3.08	8	2039.76
14	janos-us-ca	39	122	6.26	10	2776.39
15	giul39	39	172	8.82	6	1079.67
16	pioro40	40	89	4.45	7	1774.95
17	germany50	50	88	3.52	9	2583.56

4.2 Parameters

Propagation latency depends on the shortest paths and network topology, so we have considered for latency limits, in Eq. (4), $\Delta_p = 0.5 \times$ diameter and $\Delta_b = 0.6 \times$ diameter. Without loss of generality, we assumed equal weight for primary and backup paths, so we used $\alpha = \beta = 0.5$ in Eq. (7).

5 Results

The optimum solution was obtained by exactly solving the 2-Cover Problem formalized above, (7)–(9), and afterwards, as described in Sect. 3.2, a procedure to assign primary and backup controllers to switches was applied. The algorithm was implemented in Matlab, taking an average execution time of 1.019 s, and varying between 0.217 s and 2.69 s, on an Intel i5-3210M CPU. The results obtained are presented and analysed below, concerning the optimum number of controllers, reliability and latency.

5.1 Number of Controllers

The optimum number of controllers obtained by the 2-Cover Problem, is shown in Table 2. Four controllers are enough in 58.8% of the networks. It is worth to underline that this number of controllers ensure, for each switch, two disjoint paths for two different controllers and also two disjoint paths for the primary controller, being all paths within the required latency. It is noted that the number of controllers depends greatly on the topology, mainly to cope with the requirement of connect two controllers to each switch over two disjoint control paths. The main difficulty occurs with topologies where two adjacent switches have degree 2 and both are adjacent to another switch, forming a triangle, because the latter is an articulation point and therefore creates a bottleneck, implying that is necessary to place a controller in one of the two-degree switches, to cover only these two switches. For instance, networks 6 to 11 have a similar number of switches but only network 7, which presents two bottlenecks (three switches forming a triangle twice), needed 6 controllers. Results also show that there is not a linear relation between the minimum number of controllers and the number of switches or the link density or the diameter. However, there is a tendency for networks with more switches, smaller link density and larger diameter to require fewer controllers (networks 13 to 17) and vice versa (networks 1 to 4).

Table 2. Number of controllers

Network Ref.	5, 6, 8, 9, 11, 13, 14, 15, 16, 17	1, 4, 10, 12	2, 3, 7
$\|V_c\|$	4	5	6

5.2 Reliability

For each network, we compute the average network reliability, R, given in (3). It was considered, as in [7], the same failure probability for all the switches (including those hosting a controller) and the same failure rate per length of the link. We defined switch failure probabilities of 0.5%, 1% and 2% and link failure probabilities of 0.1% and 0.5% per 100 km. Figure 1 plots R for these 6 scenarios.

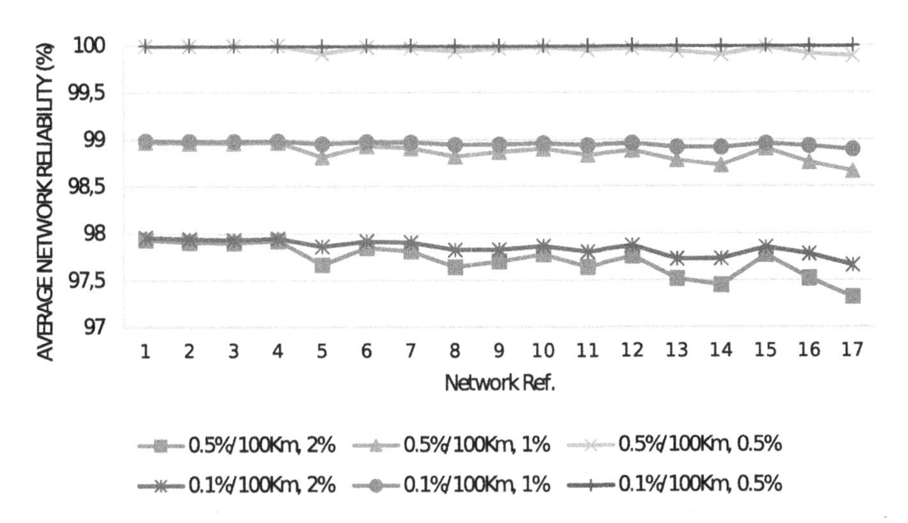

Fig. 1. Average network reliability, considering 6 scenarios combining 2 probabilities for link failure per 100 km with 3 probabilities for switch failure (f_l, f_s).

We have obtained R values ranging from 97.32% to 99.99% being all networks average equal to 98.89%. Therefore our approach ensures high average reliability. Figure 1 shows that, for each network, R presents almost the same values when only link failure probability varies. For the same link failure probability, R decreases around 1% with each switch failure probability increment. So, we conclude that switches (including controller) failure probabilities have greater impact on the network reliability than link failure probabilities.

The Average Control Path Availability used in [7] is the equivalent to R for theirs strategies. They only plot results for network 13, considering link failure probability of 0.1% per 100 km. Our results for that network outperformed theirs since we obtained reliabilities equal to 97.73%, 98.92% and 99.998% for node failure probabilities of 2%, 1% and 0.5%, respectively.

5.3 Propagation Latencies

As stated in Sect. 3.2 we have considered that the propagation latency is measured by the control path length. The average control path length to: primary controller over primary path ($L_p^{C_p}$) and over backup path $((L_b^{C_p}))$; to backup controller $((L_{dp}^{C_b}))$, over a disjoint backup path with length d^{dp}, are computed as follows:

$$L_p^{C_p} = \frac{1}{N} \sum_{i \in V_c} \sum_{j \in V} d_{ij}^p y_{ij}^{C_p} \tag{10}$$

$$L_b^{C_p} = \frac{1}{N} \sum_{i \in V_c} \sum_{j \in V} d_{ij}^b y_{ij}^{C_p} \tag{11}$$

$$L_{dp}^{C_b} = \frac{1}{N} \sum_{i \in V_c} \sum_{j \in V} d_{ij}^{dp} y_{ij}^{C_b} \tag{12}$$

Figure 2 plots these average control path lengths and the diameter.

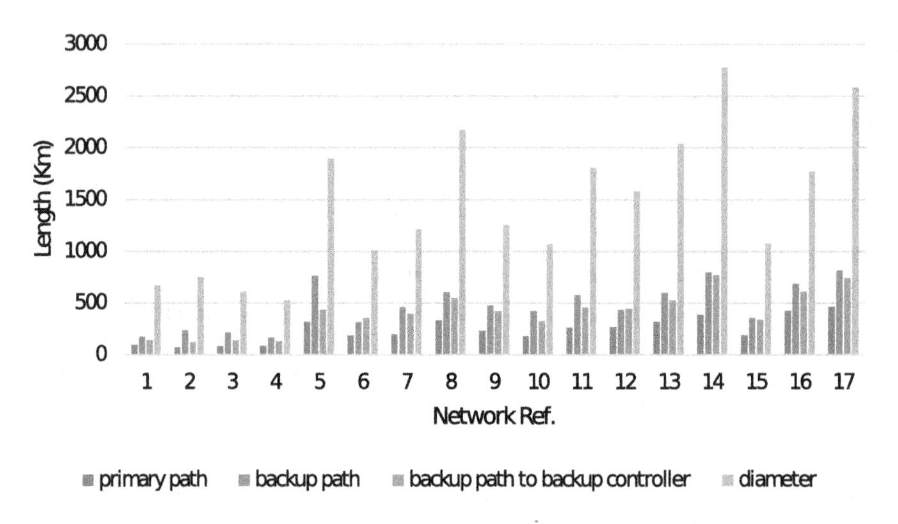

Fig. 2. Average control path length

As expected the primary path has the smallest average length. Only Networks 1 and 12 present a backup path to the primary control slightly lower than the backup path to the backup controller. For each network the average control path length ratios to diameter were computed and for the 3 control paths (p, b and dp) they range from 9.77% to 24.12%, 26.01% to 40.22% and 16.45% to 35.19%, respectively. Therefore we can state that planning in advance paths concerning the protection against failures can be obtained with low latencies for disjoint primary and backup paths.

Vizarreta et al. [7] plot average latency results considering 2 and 4 controllers for five selected SNDlib topologies, 2, 12, 13, 14, and 17, considering their 2 strategies as mentioned in Sect. 2. Our approach needed more than 4 controllers for networks 2 and 12 thus path length results are not comparable. Comparing average control paths lengths, considering 4 controllers, we can see that our results present lower values for networks 13 (321, 599 and 528 km) and 14 (388, 801 and 773 km) and higher values for network 17 (462, 818 and 741).

6 Conclusion

In this article we have presented a 2-cover based approach for the RCP problem in SDN. It was able to find the minimum number of controllers, their placement and the assignment of controllers to switches, satisfying low propagation latencies, below defined limits, while ensuring for each switch the assignment of two controllers (primary and backup) and the existence of at least two disjoint paths between each switch and its assigned controllers. Thus, it is foreseen the quickly restoration of communication to the backup controller over a disjoint path, when primary controller fails and also a promptly initialization of a disjoint backup path when the primary path fails. Results show that the proposed approach is able to determine, in all tested topologies, a highly reliable controller placement with low latencies. The approach proved also to be computationally efficient and scalable, as its performance is independent of network dimension. Therefore, it can be used to efficiently solve the considered RCP problem, under the assumptions discussed in this article and it can be easily extended to consider different amounts of traffic between switches and controllers, capacity constraints and switches with different protection levels.

References

1. Snd-lib. http://sndlib.zib.de
2. Heller, B., Sherwood, R., McKeown, N.: The controller placement problem. In: Proceedings of the First Workshop on Hot Topics in Software Defined Networks, HotSDN 2012, pp. 7–12. ACM, New York (2012). https://doi.org/10.1145/2342441.2342444. http://doi.acm.org/10.1145/2342441.2342444
3. Hock, D., Hartmann, M., Gebert, S., Jarschel, M., Zinner, T., Tran-Gia, P.: Pareto-optimal resilient controller placement in SDN-based core networks. In: Proceedings of the 2013 25th International Teletraffic Congress (ITC), pp. 1–9, September 2013. https://doi.org/10.1109/ITC.2013.6662939
4. Hu, Y., Wang, W., Gong, X., Que, X., Cheng, S.: On reliability-optimized controller placement for software-defined networks. China Commun. **11**(2), 38–54 (2014). https://doi.org/10.1109/CC.2014.6821736
5. Müller, L.F., Oliveira, R.R., Luizelli, M.C., Gaspary, L.P., Barcellos, M.P.: Survivor: an enhanced controller placement strategy for improving SDN survivability. In: 2014 IEEE Global Communications Conference, pp. 1909–1915, December 2014. https://doi.org/10.1109/GLOCOM.2014.7037087
6. Ros, F.J., Ruiz, P.M.: Five nines of southbound reliability in software-defined networks. In: Proceedings of the Third Workshop on Hot Topics in Software Defined Networking, HotSDN 2014, pp. 31–36. ACM, New York (2014). https://doi.org/10.1145/2620728.2620752. http://doi.acm.org/10.1145/2620728.2620752
7. Vizarreta, P., Machuca, C.M., Kellerer, W.: Controller placement strategies for a resilient SDN control plane. In: 2016 8th International Workshop on Resilient Networks Design and Modeling (RNDM), pp. 253–259, September 2016. https://doi.org/10.1109/RNDM.2016.7608295
8. Zhang, Y., Beheshti, N., Tatipamula, M.: On resilience of split-architecture networks. In: 2011 IEEE Global Telecommunications Conference - GLOBECOM 2011, pp. 1–6, December 2011. https://doi.org/10.1109/GLOCOM.2011.6134496
9. Zhang, Y., Cui, L., Wang, W., Zhang, Y.: A survey on software defined networking with multiple controllers. J. Netw. Comput. Appl. **103**, 101–118 (2018). https://doi.org/10.1016/j.jnca.2017.11.015. http://www.sciencedirect.com/science/article/pii/S1084804517303934

Robust Ensemble-Based Evolutionary Calibration of the Numerical Wind Wave Model

Pavel Vychuzhanin(✉), Nikolay O. Nikitin(✉), and Anna V. Kalyuzhnaya

ITMO University, 49 Kronverksky Pr., St. Petersburg 197101, Russian Federation
pavel.vychuzhanin@gmail.com, nikolay.o.nikitin@gmail.com

Abstract. The adaptation of numerical wind wave models to the local time-spatial conditions is a problem that can be solved by using various calibration techniques. However, the obtained sets of physical parameters become over-tuned to specific events if there is a lack of observations. In this paper, we propose a robust evolutionary calibration approach that allows to build the stochastic ensemble of perturbed models and use it to achieve the trade-off between quality and robustness of the target model. The implemented robust ensemble-based evolutionary calibration (REBEC) approach was compared to the baseline SPEA2 algorithm in a set of experiments with the SWAN wind wave model configuration for the Kara Sea domain. Provided metrics for the set of scenarios confirm the effectiveness of the REBEC approach for the majority of calibration scenarios.

Keywords: Evolutionary algorithm · SWAN wind wave model · Ensemble modelling · Robust optimisation · Model calibration

1 Introduction

The various tasks of offshore development and coastal shipping make it necessary to use the regional configurations of the numerical wind wave models to reproduce historical extreme events and predict potential hazards. To obtain the forecasts and hindcasts of desired quality, the suitable physical parameters of models should be identified for the specific simulation conditions.

The numerical model calibration of ocean wind wave model involves the fitting of simulation results with the in-situ and satellite wave measurements. The purpose of calibration is the identification of the physical parameters set that allows minimising the discrepancy between the model and observations.

However, it is a sophisticated task to calibrate the model manually even with the metocean experts' involvement. The modern wind wave models are computationally intensive, and each simulation run can take hours to compute. Also, a dramatically low time-spatial coverage of the available historical wave measurements and low quality of atmospheric reanalyses in some regions (like

J. M. F. Rodrigues et al. (Eds.): ICCS 2019, LNCS 11536, pp. 614–627, 2019.
https://doi.org/10.1007/978-3-030-22734-0_45

the Arctic seas, in particular the Kara Sea region described in the paper) makes it hard to validate the parameter set reliably. The obtained parameters with minimal discrepancy can be very specialised in case of over-fitting to the low number of observed data points and can actually decrease the quality of long-term simulation results in non-observed locations or time ranges [3].

There are many well-known optimisation approaches that can be applied to automate the parameters' tuning for environmental models as well as [8]. Despite this, in the paper a task-specific robust evolutionary algorithm is proposed. It allows to make reliable calibration decisions in situations with high environmental uncertainty and tries to ensure a tolerable solution identification.

At the moment, the modern atmospheric reanalysis still has quality issues in the Arctic region [7]. We proposed an algorithm that establishes artificial diversity for wind velocity fields. It was used to generate the probabilistic ensemble of input wind fields to take the impact of the surface forcing uncertainty into account. Then, the multi-objective fitness function was used to achieve the trade-off between robustness and performance of the optimised model.

We conducted a set of experiments to verify the effectiveness of the proposed approach against the baseline SPEA2 algorithm using the Kara Sea domain and the SWAN (Simulating WAves Nearshore) [2] model as the case study. The nine spatially distributed points were chosen to analyse the performance and robustness of the model's configurations obtained after calibration in one-month training runs of the model. The several configurations with different subsets of calibration and validation points were compared to estimate the statistical metrics of optimisation effectiveness for both algorithms.

This paper is structured as follows. Section 2 describes the problem statement and mathematical formalisation of the robust optimisation task. Section 3 provides an overview of various calibration approaches and their applicability for the problem. Section 4 contains a detailed description of the baseline SPEA2 algorithm and the proposed robust algorithm. Section 5 is dedicated to the experimental studies (model configuration, datasets, results and metrics). Section 6 summarises the obtained results and highlights of the key findings.

2 Problem Statement

As it was noted in the introduction, coverage of observed met-ocean data (especially oceanic observations) is extremely sparse. Although, reliable information about met-ocean characteristics is needed in many regions (e.g. Arctic seas). That's why during last decades it became a common practice to obtain the information about met-ocean events and processes from forecasting or hindcasting (retrospective) simulation results from numerical hydrodynamic models. Nevertheless, for solving such task the numerical models should be fitted (through model parameters) to the certain water area. Taking into account few spatial points and small sizes of datasets with observations, there is a serious risk of model overfitting when model fits to specific features of observed data instead of fitting to common features of the target region. Description of the solution to this problem is the main goal of this article.

Hydrodynamic model fitting through the tuning of model parameters (or model calibration) can be formulated as an optimisation task. For this purpose, it is reasonable to present the simulation process in a general mathematical notation (1).

$$Y = \{Y_1, Y_2, ..., Y_k\} = M(\xi \mid \theta), \tag{1}$$

where $Y = \{Y_1, Y_2, ..., Y_k\}$ denotes multivariate output data (simulated fields, e.g. wave heights), $M(\bullet)$ is the model operator, ξ is the input data (boundary and initial conditions), θ is the set of model parameters.

With that, the tuning of model parameters (or model calibration) can be formalized in terms of multi-objective optimisation in the model parameter space and written as:

$$\begin{aligned} \theta_{opt} &= \arg\min_\theta F(\theta), \\ F(\theta) &= \mathcal{G}(f_i(\theta, Y, \{x, y\})), \end{aligned} \tag{2}$$

where $\mathcal{G}(\bullet)$ is an operator for multiobjective transformation to F, f_i is the objective function, i = 1 ... n , $\{x, y\}$ are spatial coordinates of a point-of-interest.

In a case of wind waves hindcasting, the poor time and spatial coverage of observations make the model optimisation much harder. The over-fitting of the solution to the specific events represented in small data samples can cause a non-optimal model configuration with lower quality under different external conditions. One of the ways to improve the robustness of optimisation results is to enlarge training dataset with new instances with relatively small artificial disturbances. This issue makes it necessary to take the simulation uncertainty factors into account.

The uncertainty in the wind wave model can be represented not only by disturbances in design variables [18]. There are deviations in the environment variables that can be represented through input data sets diversity (for the SWAN model the wind forcing obtained from atmospheric reanalysis is most important). In this case input data ξ should be transformed to ensemble realisation $\{\xi\}_n = \{\xi_1, ..., \xi_n\}$ by addition of artificial disturbance (or noise) and Eq. (2) transforms into Eq. (3). A detailed description of the ensemble procedure is given in Sect. 4.3.

$$\begin{aligned} \theta_{rob} &= \arg\min_\theta \tilde{F}(\theta \mid \{\xi\}_n), \\ \tilde{F}(\theta \mid \{\xi\}_n) &= \mathcal{G}(\tilde{f}_i(\theta \mid \{\xi\}_n, Y, \{x, y\})). \end{aligned} \tag{3}$$

An ensemble objective function $\tilde{f}_i$ defines landscape of objective function over the space of parameters considering ensemble of input states $\{\xi\}_n$. As an example, ensemble fitness function can be represented by the expected function for the ensemble of runs with small disturbances in input data (shown in Eq. (4)). This approach can be used to produce better solutions for the set of diverse environmental scenarios and increase the expected performance.

$$\tilde{f}(\theta \mid \{\xi\}_n) = \int_{-\infty}^{\infty} f(\mathbf{x}, \xi + \boldsymbol{\delta}) \cdot p(\boldsymbol{\delta}) d\boldsymbol{\delta} \tag{4}$$

As an example of the hydrodynamic model for experimental studies, third-generation wind wave model SWAN [2] was chosen. The wind waves are surface waves in the oceans and seas that caused by the interaction between water masses and sea-level wind. Wind waves models of third-generation (e.g. SWAN) allow to simulate the wave spectra and to reconstruct characteristics of waves (e.g. heights, periods, directions). The SWAN model can be described with the action balance equation (5).

$$\frac{\partial}{\partial t}N + \frac{\partial}{\partial x}c_x N + \frac{\partial}{\partial y}c_y N + \frac{\partial}{\partial \sigma}c_\sigma N + \frac{\partial}{\partial \theta}c_\theta N = \frac{S}{\sigma}, \tag{5}$$

where on the left-hand side $N = \frac{E}{\sigma}$ denotes the wave action density and E is an energy of wave spectrum, σ is the relative frequency, θ is the group wave direction, c is the group velocity in corresponding space. The right-hand side represents the source and sink term in a form Eq. (6).

$$S = S_{in} + S_{ds} + S_{nl}, \tag{6}$$

where S_{in} is the input energy obtained by wind, S_{ds} is the energy of dissipation and S_{nl} denotes the energy of wave-wave nonlinear interaction.

These three terms represent the genesis of wave energy sources/sinks and are a powerful handle for wave model fitting. From this point of view, it is convenient to express energy sources through model parameters. Wind energy is characterised by the drag function (DRG), wave dissipation—by the wave breaking (STMP) and bottom friction (CFW) functions. Energy flow from non-linear interactions is relatively small and wasn't taken into account in the current paper.

In the frame of this article, the experimental study (Sect. 5) was provided to assess the practical effectiveness of the proposed robust calibration method in comparison with the general-purpose calibration algorithms. The SWAN model configuration for the Kara Sea was chosen as a case study because of the importance of this region for offshore industrial development and extremely low density of sensors in areas of interest.

3 Related Work

Model calibration or tuning is a subject with extensive literature [8,25]. The conservative approach is to estimate the parameters in an expert way [10,16]. It includes the development of several candidate sets of parameters based on previous simulation experience and manual individual adjustment of each parameter. The quality metrics for the model quality assessment are calculated with the comparison of model time series and historical values obtained from the reanalyses and observations.

Since the manual "trials-and-errors" method is time-consuming and gives solution only for particular model setup, the automatic calibration of models is widely used for different aspects of environmental simulations like atmospheric [6] and ocean [26] forecasting tasks. As a basic approach, the space-filling design for the parameter space can be used [24] for model calibration. However, the high-resolution configurations of wind wave models of 3rd generation are computationally-intensive and require a lot of time to process the appropriate date range and spatial domain. This problem makes it necessary to reduce the number of runs required for calibration.

There are many well-known optimisation methods applied to environmental models like derivative-free optimisation [22], various Bayesian optimisation methods [5] and surrogate-assisted methods [9].

However, the evolutionary (genetic) algorithms are efficient enough to perform a robust solution search [14] in complex parameter space with a lack of historical data for quality assessment [20]. The applicability of evolutionary algorithms for SWAN wave model calibration is demonstrated in [12].

The robust optimal design approaches have a lot of applications in many fields [27]. They are often based on Monte Carlo methods that allow representing the uncertainty from different sources [4]. The perturbation-based ensemble allows sampling the modelling uncertainty in a more systematic way [19]. A set of simulation with small differences induced by stochastic modifications allows to increase the variability of the calibration dataset and improve the quality of models [17].

Nevertheless, the discrepancy usually simulated as additional noise in model output and observations [1] without taking the actual sources of external uncertainty (e.g. wind forcing for reanalysis) into account. The task of a reliable calibration of a wave model for a specific domain with poor observational coverage makes it necessary to implement the approach that combines the ensemble-based diversity of environmental variables with multi-objective evolutionary optimisation.

4 Evolutionary Algorithms for Models Fitting

We compared the robust wave model calibration with a baseline solution—the multi-objective evolutionary algorithm that estimates the most suitable solution without taking uncertainty into account. The other approach is based on the same algorithm with modified fitness functions—it estimates the performance and robustness of the solution with the ensemble of forecasts obtained from several model's runs with noised inputs. The source code of both algorithms was implemented in Python and available in [23].

4.1 Baseline Approach

The commonly used SPEA2 multi-objective optimisation algorithm [28] was chosen as a baseline solution for the calibration task. In terms of evolutionary algorithms, in our case, each individual corresponds to a genotype represented by a certain set of model parameter and the phenotype (values of the objective function) are the errors of the model predictions, corresponding to these parameters. At each iteration of evolution, the Pareto-optimal set of individuals is selected according to the values of the fitness function, and all non-dominated solutions are saved in the archive. Then the mating pool is filled with a binary tournament selection and recombination and a mutation operator are applying for each individual. The resulting mating pool becomes a new population at the next iteration of the algorithm.

Despite the fact that some modern evolutionary algorithms outperform SPEA2 in some synthetic tasks [13], we decided to base the experiments on a well-studied [12] algorithm to separate the impact from the proposed ensemble-based modifications from other features' influence.

4.2 Robust Ensemble-Based Evolutionary Calibration (REBEC) Approach

The main disadvantage of the baseline algorithm is that the model variables optimise exactly for the specific conditions that were used for the fitness function evaluation. It allows to maximise the performance for the observed case, but the solution found can be unstable even after small changes in external conditions. The lack of the time-spatial coverage of observational data for wave parameters in target regions makes it complicated to take the different external uncertainties (e.g. forcing-induced, resolution-induced, etc.) into account.

The more robust approach to model parameters optimisation can be implemented using the ensemble of wave models configured using different input data sets. We can form the stochastic ensemble of wave models with the perturbed wind forcings and search for more robust model parameters using this ensemble instead of a single model with certain forcing.

For this purpose, we can adapt the baseline SPEA2 algorithm (that was introduced above) by changing the fitness assignment strategy: for a given genotype, the set of phenotypes corresponding to the elements of the ensemble is estimated and based on its values the robust metric is calculated. The flowchart of the proposed algorithm is presented in Fig. 2.

It is important to find a compromise between performance and robustness of the obtained solution [11], so the fitness function for the algorithm is based on the composite estimation robustness and performance metrics. The performance can be calculated as a vector of root-mean-square errors (RMSE) against observations for a set of target points, and the robustness can be simulated in various ways [15]. Figure 1 depicts the set of the ensemble error surfaces that are used for metric calculation.

We tried to use the mean-variance as a robust metric, but it causes the domination of the solutions with low wind drag and, consequently, near-zero wind-induced variability. So, the ensemble mean was chosen as a trade-off metric.

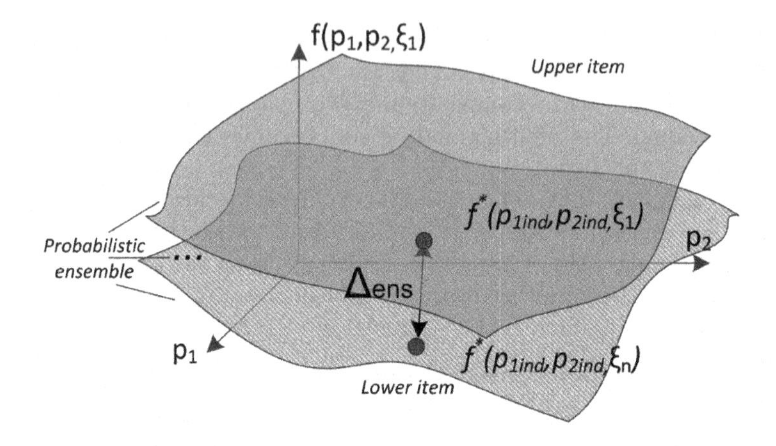

Fig. 1. The landscape of an objective function for a probabilistic ensemble

The pseudocode of the final implementation of the robust algorithm is presented in Algorithm 1.

```
Data: Initialised ensemble, populationSize,
archiveSize, crossoverRate, mutationRate
Result: best individual from archive
pop ← InitPopulation(populationSize)
archive ← ∅
while not ConvergenceCriterion() do
    for individ in pop do
        ensObjectives ← ∅
        for model in ensemble do
            ensObjectives [model ] ← CalculateObj(individ, model)
        end
        bestByObjectives ← TakeBestByMean(ensObjectives, ensAmount)
        individ.objectives ← Mean(bestByObjectives)
    end
    union ← archive + pop
    for individ in union do
        individ.fitness ← CalculateFitness(individ)
    end
    archive ← TakeNonDominated(union, archiveSize)
    matingPool ← BinaryTournamentSelection(archive, populationSize)
    pop ← CrossoverAndMutation(matingPool, crossoverRate,
    mutationRate)
end
```

Algorithm 1. The pseudocode of the implemented REBEC algorithm

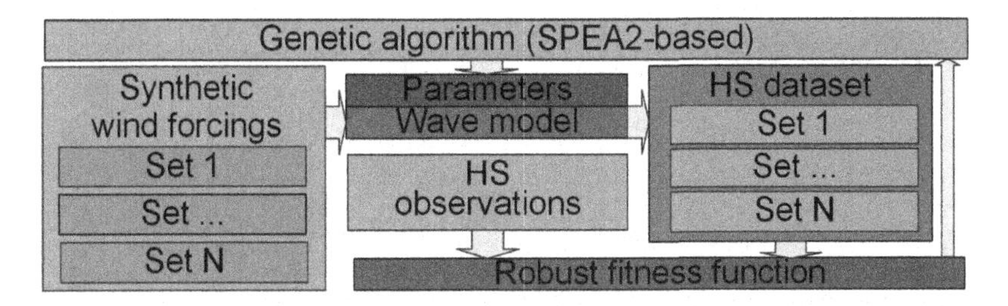

Fig. 2. The main logical blocks and interconnections of proposed robust evolutionary algorithm

4.3 Synthetic Input Data Generation with Artificial Noise

To implement the proposed probabilistic optimisation method, we developed the supplementary algorithm that allows to add specific noise to wind velocity variables—U (eastward) and V (northward) vector components from atmospheric reanalysis data that is used by the wave model as an external forcing.

The algorithm starts from uniform scattering of the randomly-located sources of artificial noise in the gridded data. To obtain the realistic wind field after the application of noise, the time-spatial correlation terms are added to control the noise spreading from the source.

The noise function for the wind vector component U produced by one noise point can be written as:

$$f^*(j,t) = N(0,\sigma) \cdot corr(U_j, V_j) \cdot corr(U_t, U_{t-1}) \tag{7}$$

where j is the spatial index of source points, t is the time step index, σ is the standard deviation parameter of the Gaussian distribution, U is the matrix of wind U-components.

Then, the aggregated noise from N source points for specific data point induced by all source points can be obtained as:

$$f(i,t) = \sum_{j=1}^{N} f^*(j,t) \cdot corr(U_i, U_j) \tag{8}$$

where i is the spatial index of the data point, j is the spatial index of the noise point, t is the time step index, N is the number of noise points, U is the matrix of wind U-components.

The example of the wind field augmented with noise by the described method (with σ equal to 25% of basic value magnitude) is presented in Fig. 3.

It can be seen that the common wind patterns are similar but some wind speed variability exists. The additional post-processing procedure was applied to the perturbed model runs output to suppress the non-realistic wind height peaks in the observed calm periods. However, the near-peaks variability was preserved.

In this way, the ten wind data sets augmented with artificial noise were generated and used in an experimental study.

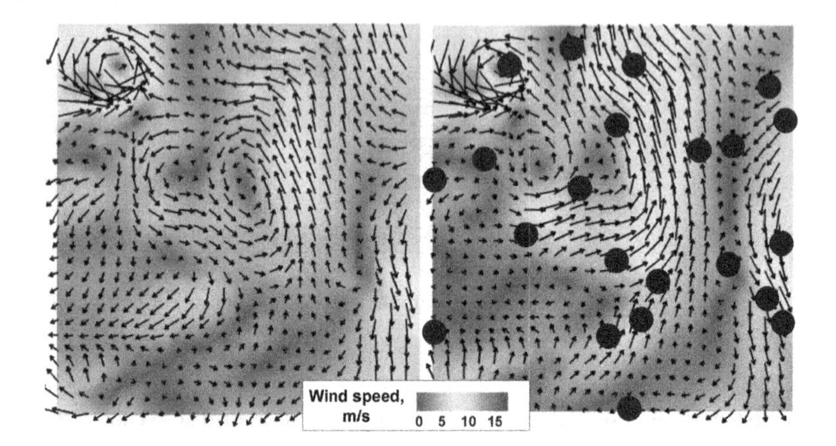

Fig. 3. The example of comparison of (a) basic ERA-Interim for Kara Sea region and (b) wind field augmented with noise for the same region. The blue marks depicts the noise source points locations. (Color figure online)

5 Experimental Study

The case study for the calibration task is based on the SWAN model configuration for the Kara Sea region. The significant wave height (Hsig) variable was chosen as a target variable. Moreover, the results in nine representative points were analyzed (P1–P9 presented in Fig. 4) to take into account possible spatial variability of the optimal solution.

5.1 Synthetic Data for Wave Observations

The wave observations data are required for the validation of the model quality and calibration algorithm effectiveness. However, such data often cannot be obtained from open data sources. To perform a reproducible experiment with Kara Sea configuration, we used the simulation results from the high-resolution WaveWatch III model [21] configuration. The systematic biases of synthetic observations against model were removed. Then we analysed the error metrics for the significant wave height variable against real observations in points 1–3 (RMSE is 0.29 m and MAE is 0.21 m). We accept the quality of the WaveWatch III output as sufficient to be used as the reference dataset for the optimisation algorithms' evaluation.

To maintain the variability of experimental scenarios, we prepared 18 subsets of synthetic observational points to be used for calibration. They consist of observation points located in various spatial areas with different depths and distances to the coast. To reproduce various scenarios, the calibration subsets were initialised with random point groups of a certain size: from a single-point situation to the all-points-instead-one case. For each subset, the points that were not used during the calibration were assigned to validation sets. It allows to compare the effectiveness of the robust algorithm with baseline SPEA2 and analyse the dependency of results from selected observation points set.

5.2 Model Configuration

The SWAN model was configured with the regular curvilinear grid in cartesian coordinates. The initial conditions were obtained for a preliminary monthly spin-up run. The boundary conditions were not set (since the control points were distanced from the grid boundaries, see Fig. 4). The simulation dates range was set from 20140814.120000 to 20140915.000000. The time step for integration was defined as 120 min and output time step is 3 h. The parameterisations GEN3, COLLINS, QUADRUPL, TRIAD and DIFFRACtion were enabled. The output was configured to obtain the significant wave height (HS) values in 9 spatial points. Their locations are specified in Fig. 4.

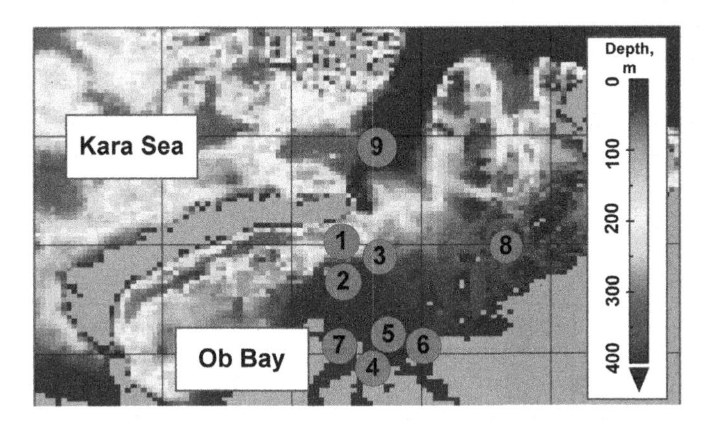

Fig. 4. The part of the bathymetry of the simulation domain: the Kara Sea and Ob bay. The land cells are shaded with a grey mask. The locations of observation points and their indices are specified with green marks. (Color figure online)

5.3 Sensitivity Analysis

The sensitivity analysis of model parameters described in Sect. 2 was performed to estimate their significance. We ran the set of experiments with every parameter independently modified by additive noise with Gaussian distribution with $\sigma/\mu = 0.25$ assumption. The results are obtained from 50 experiments with sequential noising of each variable from the chosen set.

The wind drag was identified as the most sensitive parameter with a high relative output and input variance ratios, the wave steepness is the second one and the sensitivity bottom friction coefficient is quite low for most of the comparison points. In can be concluded that the SWAN model error function has a wide "plateau" with similar error values and many local minimums in the "valley" area that can affect the algorithm's convergence and robustness.

5.4 Validation of REBEC Approach

A set of experiments was conducted to compare the results of optimisation experiments. The initial population for both calibration approaches was produced using Latin hypercube sampling (LHS) in the parameter space. The parameters for both calibration algorithms was chosen as: population size is set to 20 individuals, the number of generations—60, the archive size—5 individuals, the probability of mutation and crossover—0.2.

Two objective functions were chosen for model results quality assessment: the mean absolute error (MAE) and root mean square error (RMSE).

The calibration for every scenario was repeated 100 times to obtain the distribution of the relative improvement of RMSE and MAE against the model configuration with default parameter values (DRF = 1.0, CFW = 0.015, STPM = 0.00302). Also, the mean relative standard deviation of the calibrated parameters set is provided. The boxplots for the results with different scenarios are presented in Fig. 5.

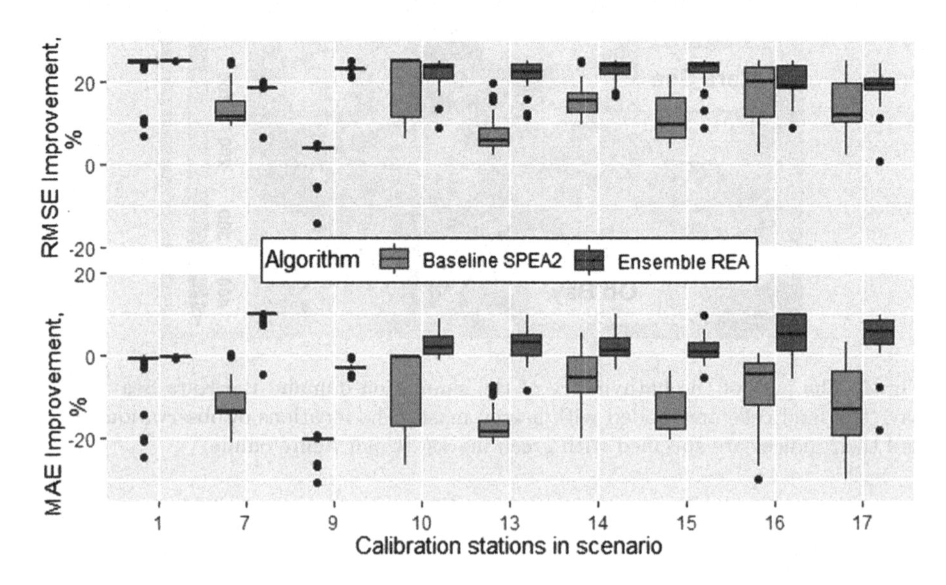

Fig. 5. The comparison of the baseline and robust algorithms' performance on the validation set of stations in all scenarios. The RMSE and MAE metrics are presented as an improvement against the corresponding values for the default configuration.

It can be seen that the variance of metrics for the robust algorithm is lower and the quality is better. The detailed metrics for all scenarios and stations sets are provided in Table 1.

As can be seen, the robust approach provides a better or equivalent improvement of model performance for the validation points in all groups of scenarios. The standard deviation for both model parameters and relative improvement

Table 1. Error metrics for the baseline and robust algorithms. The "test" block contains the metrics for the verification points. The "train" block contains the metrics for the calibration points. The boldface numbers indicate the best metrics for all station sets (the higher improvement and lower standard deviation is better)

Scenario	Algorithm	Validation points							Calibration points						
		Improvement, %						Par. SD	Improvement, %						Par. SD
		RMSE			MAE				RMSE			MAE			
		Mean	Max	SD	Mean	Max	SD		Mean	Max	SD	Mean	Max	SD	
1–9	BL	**11**	**22.1**	7.8	−9.2	1.8	7.1	3.3	6.6	**26.4**	13.7	2.1	13.6	6.8	3.3
	RB	11	15	**2.7**	**4**	**6.9**	**2**	**2.7**	**11.3**	16.4	**3**	**2.5**	**6.5**	**3**	**2.7**
10–14	BL	16.6	**23.6**	4.9	−7.2	0.7	5.2	2.7	**24.4**	**29.1**	4.1	7.9	12.8	2.7	2.7
	RB	**18.1**	21	**2.8**	**4.1**	**9.4**	**2.5**	**2.4**	22.9	26.9	**3.9**	**8.7**	**14.4**	**2.2**	**2.4**
15–18	BL	17.6	**25.1**	5.7	−6.6	2.8	6.5	3.1	**27.2**	**34.2**	6.2	**11.2**	**17.1**	**2.8**	3.1
	RB	**18.7**	24.7	**4.1**	**5.4**	**10**	**4.1**	**3**	23.4	33.4	**5.1**	11.2	17.9	3.2	**3**
All	BL	14.0	**23**	6.6	−8.4	1.7	6.4	3.2	14.5	**27.9**	9.7	4.7	**13.5**	4.9	3.2
	RB	**14.6**	18.3	**3.1**	**4.6**	**8.7**	**2.7**	**2.8**	**15.8**	21.6	**3.9**	**5.2**	10.5	**2.9**	**2.8**

values are also lower than the baseline. In can be concluded that the optimal algorithm choice for validation points varies in different scenarios. The scenarios 1–9 operate with a single-point calibration set. The performance of the robust algorithm for this group of validation points is similar to baseline RMSE (but outperforms it for the MAE metric and calibration points metrics). For the other scenarios, the gain is near 1–2% RMSE and 10% MAE against the baseline.

Also, the calibration set quality averaged for all scenarios for the robust approach also outperforms the baseline. The standard deviation of the obtained metrics is smaller for all scenarios, as well as the mean standard deviation for model parameters. We can claim that a robust approach is effective for the cases with several spatially scattered points that can be applied for calibration. It is important to notice that the calibration points' quality is not affected in a negative way.

6 Conclusion

In the paper, the practical approach to the calibration of numerical wave models under data quality and availability constraints was proposed. The algorithm for the simulation of artificial data diversity was implemented and applied to the ERA-Interim reanalysis wind data. The regional configuration of the SWAN model was used as a case study for the parameters tuning algorithm effectiveness evaluation.

The proposed REBEC approach was compared with the baseline SPEA2 algorithm in a set of experiments. The lower variability and better performance metrics for the spatially distributed calibration and verification points were obtained. It confirms the effectiveness of the robust calibration approach for the simulation domains with a small number and poor coverage of real observations. However, the negative impact of the proposed approach for computational performance (several simulations should be performed for each candidate parameters set) makes the robust optimisation potentially non-preferable for the model configurations with the sufficient spatial coverage of observations and high-quality atmospheric reanalyses.

The source code of the algorithms for calibration, pre- and post- processing as well as the configuration files for SWAN are available in an open repository [23].

Acknowledgements. This work is financially supported by National Center for Cognitive Research of ITMO University.

References

1. Bhat, K.S., Haran, M., Goes, M., Chen, M.: Computer model calibration with multivariate spatial output: a case study. In: Chen, M.-H., Müller, P., Sun, D., Ye, K., Dey, D. (eds.) Frontiers of Statistical Decision Making and Bayesian Analysis, pp. 168–184. Springer, New York (2010). https://doi.org/10.1007/978-1-4419-6944-6
2. Booij, N., Ris, R.C., Holthuijsen, L.H.: A third-generation wave model for coastal regions: 1. Model description and validation. J. Geophys. Res. Ocean. **104**(C4), 7649–7666 (1999)
3. Brynjarsdóttir, J., OHagan, A.: Learning about physical parameters: the importance of model discrepancy. Inverse Probl. **30**(11), 114007 (2014)
4. Che, J., Wang, J., Li, K.: A Monte Carlo based robustness optimization method in new product design process: a case study. Am. J. Ind. Bus. Manag. **4**(07), 360 (2014)
5. Cornejo-Bueno, L., Garrido-Merchán, E.C., Hernández-Lobato, D., Salcedo-Sanz, S.: Bayesian optimization of a hybrid system for robust ocean wave features prediction. Neurocomputing **275**, 818–828 (2018)
6. Duan, Q., et al.: Automatic model calibration: a new way to improve numerical weather forecasting. Bull. Am. Meteorol. Soc. **98**(5), 959–970 (2017)
7. Fredriksen, L.E.: An evaluation of the reanalyses ERA-Interim and ERA5 in the Arctic using N-ICE2015 data. Master's thesis, UiT The Arctic University of Norway (2018)
8. Hourdin, F., et al.: The art and science of climate model tuning. Bull. Am. Meteorol. Soc. **98**(3), 589–602 (2017)
9. James, S.C., Zhang, Y., O'Donncha, F.: A machine learning framework to forecast wave conditions. Coast. Eng. **137**, 1–10 (2018)
10. Jin, K.R., Ji, Z.G.: Calibration and verification of a spectral wind-wave model for Lake Okeechobee. Ocean Eng. **28**(5), 571–584 (2001)
11. Jin, Y., Sendhoff, B.: Trade-off between performance and robustness: an evolutionary multiobjective approach. In: Fonseca, C.M., Fleming, P.J., Zitzler, E., Thiele, L., Deb, K. (eds.) EMO 2003. LNCS, vol. 2632, pp. 237–251. Springer, Heidelberg (2003). https://doi.org/10.1007/3-540-36970-8_17
12. Kovalchuk, S.V., et al.: A conceptual approach to complex model management with generalized modelling patterns and evolutionary identification. Complexity **2018**, 15 pages (2018)

13. Li, M., Yao, X.: An empirical investigation of the optimality and monotonicity properties of multiobjective archiving methods. In: Deb, K., Goodman, E., Coello Coello, C.A., Klamroth, K., Miettinen, K., Mostaghim, S., Reed, P. (eds.) EMO 2019. LNCS, vol. 11411, pp. 15–26. Springer, Cham (2019). https://doi.org/10.1007/978-3-030-12598-1_2
14. Liu, Y., Khu, S.T.: Automatic calibration of numerical models using fast optimisation by fitness approximation. In: International Joint Conference on Neural Networks, IJCNN 2007, pp. 1073–1078. IEEE (2007)
15. McPhail, C., Maier, H., Kwakkel, J., Giuliani, M., Castelletti, A., Westra, S.: Robustness metrics: how are they calculated, when should they be used and why do they give different results? Earth's Future **6**(2), 169–191 (2018)
16. Mortlock, T.R., Goodwin, I.D., Turner, I.L.: Calibration and sensitivities of a nearshore SWAN model to measured and modelled wave forcing at Wamberal, New South Wales, Australia (2014). https://doi.org/10.7158/C14-016.2014.12.1
17. O'Donncha, F., Zhang, Y., Chen, B., et al.: Ensemble model aggregation using a computationally lightweight machine-learning model to forecast ocean waves. arXiv preprint arXiv:1812.00511 (2018)
18. Paenke, I., Branke, J., Jin, Y.: Efficient search for robust solutions by means of evolutionary algorithms and fitness approximation. IEEE Trans. Evol. Comput. **10**(4), 405–420 (2006)
19. Rougier, J., Sexton, D.M., Murphy, J.M., Stainforth, D.: Analyzing the climate sensitivity of the HadSM3 climate model using ensembles from different but related experiments. J. Clim. **22**(13), 3540–3557 (2009)
20. Schmitt, C., Rey-Coyrehourcq, S., Reuillon, R., Pumain, D.: Half a billion simulations: evolutionary algorithms and distributed computing for calibrating the simpoplocal geographical model. Environ. Plan. B Plan. Des. **42**(2), 300–315 (2015)
21. Tolman, H.L., et al.: User manual and system documentation of WaveWatch III TM version 3.14. Technical note, MMAB Contribution 276, 220 (2009)
22. van Vledder, G.P.: Calibration of SWAN 40.20 for field cases Petten, Slotermeer and Westerschelde (2003)
23. Vychuzhanin, P., Nikitin, N., Deeva, I.: The source code of the robust evolutionary algorithm for SWAN model calibration (2019). https://github.com/ITMO-NSS-team/SwanEvolution
24. Wainwright, J., Mulligan, M.: Environmental Modelling: Finding Simplicity in Complexity. Wiley, Chichester (2005)
25. Williams, J.J., Esteves, L.S.: Guidance on setup, calibration, and validation of hydrodynamic, wave, and sediment models for shelf seas and estuaries. Adv. Civil Eng. **2017**, 25 pages (2017)
26. Williamson, D.B., Blaker, A.T., Sinha, B.: Tuning without over-tuning: parametric uncertainty quantification for the NEMO ocean model. Geosci. Model Dev. **10**(4), 1789–1816 (2017)
27. Zang, C., Friswell, M., Mottershead, J.: A review of robust optimal design and its application in dynamics. Comput. Struct. **83**(4–5), 315–326 (2005)
28. Zitzler, E., Laumanns, M., Thiele, L.: SPEA 2: Improving the strength Pareto evolutionary algorithm. TIK-Report 103 (2001)

Approximate Repeated Administration Models for Pharmacometrics

Balazs Nemeth[1(✉)], Tom Haber[1,2], Jori Liesenborgs[1], and Wim Lamotte[1]

[1] Hasselt University - tUL - Expertise Center for Digital Media, Wetenschapspark 2, 3590 Diepenbeek, Belgium
{balazs.nemeth,tom.haber,jori.liesenborgs,wim.lamotte}@uhasselt.be
[2] Exascience Lab, Imec, Kapeldreef 75, 3001 Leuven, Belgium

Abstract. Improving performance through parallelization, while a common approach to reduce running-times in high-performance computing applications, is only part of the story. At some point, all available parallelism is exploited and performance improvements need to be sought elsewhere. As part of drug development trials, a compound is periodically administered, and the interactions between it and the human body are modeled through pharmacokinetics and pharmacodynamics by a set of ordinary differential equations. Numerical integration of these equations is the most computationally intensive part of the fitting process. For this task, parallelism brings little benefit. This paper describes how to exploit the nearly periodic nature of repeated administration models by numerical application of the method of averaging on the one hand and reusing previous computational effort on the other hand. The presented method can be applied on top of any existing integrator while requiring only a single tunable threshold parameter. Performance improvements and approximation error are studied on two pharmacometrics models. In addition, automated tuning of the threshold parameter is demonstrated in two scenarios. Up to 1.7-fold and 70-fold improvements are measured with the presented method for the two models respectively.

Keywords: Pharmacometrics · Monte Carlo sampling · Hamiltonian Monte Carlo · High-performance computing · Hierarchical models · Approximation · Importance sampling

1 Introduction

One of the key questions of drug development, which pharmacometrics is concerned with, is what dosage regimen is safe and effective for individuals within a population. In this field, models from pharmacokinetics (PK) and pharmacodynamics (PD) characterize the interactions between a drug and an organism. Here, PK describes how a drug is affected by the organism, and PD describes the effect of the compound on the organism. The use of tools in this field requires both theoretical knowledge of biological systems and statistical expertise [14].

J. M. F. Rodrigues et al. (Eds.): ICCS 2019, LNCS 11536, pp. 628–641, 2019.
https://doi.org/10.1007/978-3-030-22734-0_46

Therefore, methods that are easy to use, like the one described in this paper, are of great interest.

Due to the complexity of these models, sufficient data is required to derive meaningful conclusions, but clinical data is typically sparse. Therefore, the common approach is to pool data from multiple drug trails and subjects within those trials. In this context, it is imprecise to merely consider the data as an unstructured collection of observations. Rather, with each observation, additional valuable information is available. This includes from which subject an observation is taken, his or her weight and height.

To incorporate this information, mixed effect models are used. Since PK and PD models typically rely on ordinary differential equations (ODEs), simulation requires computationally intensive numerical methods. An integrator is configured to ensure some level of accuracy in the result. Depending on the ODEs, the size of the steps that are taken is limited. More importantly, models with repeated administration hamper performance further. In these models, the simulation of dosing events causes the integrator to invalidate any gathered knowledge about the ODEs and take small steps. In addition, after a dosing event, computational time is spent on determining what step size to use.

Estimating parameters for these models in a reasonable amount of time requires not only the right mathematical tools, but also techniques from computer science. For example, within a drug trial, a compound is tested on multiple subjects and to determine the model parameter quality, each subject can be simulated in parallel. After parallelization, the most computationally intensive part is the numerical integration. Although parallel numerical integration has been studied [11], only limited improvements are possible [13].

Instead, the method outlined in this paper exploits the periodic behavior of models in pharmacometrics by reusing previous computations and employing the method of averaging to form an approximation of the model. It is applicable on top of any numerical integrator and besides a single parameter, no additional input from the user is required. To de-emphasize the existence of the parameter, it is important to note that it can be tuned automatically in a use-case dependent manner. Two examples are discussed to demonstrate this.

The remainder of this paper is structured as follows. Section 2 lists related work. Two examples of repeated administration models are discussed in Sect. 3. Section 4 discusses how these are used when data is sparse. The approximation method is presented in Sect. 5. Next, experimental results are shown in Sect. 6, and the paper is concluded and directions for future work are provided in Sect. 7.

2 Related Work

Dunne et al. [5] studied the application of the method of averaging in pharmacometrics, but their approach consisted of transforming the model by hand followed by solving it symbolically. The automated method presented in Sect. 5 partially relies on the same observations but differs in two ways. First, it does not require the user to manually alter the model. Second, for models that combine both PK and PD, all portions of the model are handled while the approach outlined by Dunne et al. focuses mainly on dealing with the PD portion where no periodicity is observed.

Conrad et al. [3] tackle computationally expensive models by constructing and gradually refining approximations of the posterior for Bayesian inference during Markov Chain Monte Carlo (MCMC) sampling. Their approximation method uses previous evaluations in a shrinking region to interpolate the posterior function. Similarly, Gong et al. [6] propose an adaptive refinement strategy that builds a surrogate model to explore a target distribution. Compared to these approaches where no knowledge of the underlying model is used, the approximation described in this paper works at the level of the model itself. As such, the two approaches are complementary.

Rasmussen [15] considers Hybrid Monte Carlo (HMC) on Bayesian integrals. In his work, gradients of the posterior are approximated using a Gaussian Process. He notes that to guarantee that the samples generated by HMC are unbiased, accurate posterior evaluations are only required at the end of a set of leapfrog iterations. Similarly, in Sect. 6, gradients are computed from the approximation and the final accept-reject step relies on the real model.

3 Repeated Administration Models

This paper considers two models to exemplify what is seen in drug development when patients are administered a compound periodically. While the details of the models are less important for the work presented in this paper, they are listed here to describe their structure. Each model in this paper, denoted by f, is built using a set of ODEs parametrized by a vector ϕ. The set of q equations in f is denoted by $S = \{S_i(t)\}_1^q$.

Data to which these models are fit consists of a dosage regimen D and a sequence of observations (y_j, x_j). Each dosing event (a, c, t) in D adds some amount a of a compound to any state identified by c in model f at time t. Without loss of generality, the first dose is administered at $t = 0$, and all observations and dosing events are sorted by increasing time t. To fit ϕ, prediction $\hat{y}_j$ need only be made at x_j and Algorithm 1 outlines how to obtain predictions. It relies on a subroutine that implements an integrator of which the state is stored in $\mathcal{I}$.

The execution time of the integrator is mainly determined by the range spanned by x_j and the number of dosing events falling in that range since. Repeatedly stopping the integrator to simulate dosing events is the main cause for slowdown; as noted in Sect. 1, the integrator cannot take large steps when the internal state is changed. The method presented in Sect. 5 avoids this.

3.1 Nimotuzumab Model

The first model characterizes PK behavior of Nimotuzumab, a humanized monoclonal antibody mAb, in patients with advanced breast cancer [16]. The system of coupled differential equations in Eq. 1 describes the dynamics of this model.

Algorithm 1. Using an integrator to collect predictions $\hat{y}_j$.

```
Input: x_1, ..., x_n, D, and S
Result: ŷ_1, ..., ŷ_n
k = 1; I = InitializeIntegrator(S)
(a, c, t) = GetDose(D, k)
for j = 1, ..., n do
    while t ≤ x_j do
        IntegrateTo(I, t)
        AddToState(I, c, a)
        k = k + 1; (a, c, t) = GetDose(D, k)
    end
    IntegrateTo(I, x_j)
    ŷ_j = GetState(I)
end
```

$$\begin{cases} \frac{dC_{\text{tot}}(t)}{dt} & = -(k_{\text{e}} + k_{\text{pt}}) \cdot C(t) + k_{\text{tp}} \cdot A_{\text{t}}(t) - \left(\frac{k_{\text{int}} \cdot R_{\text{tot}} \cdot C(t)}{k_{\text{ss}} + C(t)}\right) \\ \frac{dA_{\text{t}}(t)}{dt} & = k_{\text{pt}} \cdot C(t) \cdot v_1 - k_{\text{tp}} \cdot A_{\text{t}}(t) \\ \frac{dR_{\text{tot}}(t)}{dt} & = k_{\text{syn}} - k_{\text{deg}} \cdot R_{\text{tot}}(t) - \left(\frac{(k_{\text{int}} - k_{\text{deg}}) \cdot C(t) \cdot R_{\text{tot}}(t)}{k_{\text{ss}} + C(t)}\right) \\ C(t) & = 0.5 \cdot \Big[C_{\text{tot}}(t) - R_{\text{tot}}(t) - k_{\text{ss}} \\ & \quad + \sqrt{(C_{\text{tot}}(t) - R_{\text{tot}}(t) - k_{\text{ss}})^2 + 4 \cdot k_{\text{ss}} \cdot C_{\text{tot}}(t)}\Big] \end{cases} \quad (1)$$

Observations to which this model is fit consist of measured free concentrations of the mAb compound $C(t)$, at a particular time t, determined by the total mAb concentrations $C_{\text{tot}}(t)$, the total target concentration $R_{\text{tot}}(t)$ and the steady state rate constant k_{ss}. The change in the amount of free mAb in tissue compartments $A(t)$ depends on $C(t)$ and k_{pt} and k_{tp} which denote tissue-serum and serum-tissue rate constants respectively. The other constants that need to be estimated are the elimination rate k_{el}, the degradation rate k_{deg}, zero-order kinetic synthesis k_{syn} and irreversible internalization rate k_{int}. Note that there is a bidirectional influence between the compartments and $C(t)$ since it also appears on the right hand side. The model parameter vector ϕ is $[cl, v_1, Q, v_2, k_{\text{ss}}, k_{\text{int}}, k_{\text{syn}}, k_{\text{deg}}]$, where $k_{\text{e}} = cl/v_1$, $k_{\text{pt}} = Q/v_1$ and $k_{\text{tp}} = Q/v_2$.

Figure 1 shows an example of the evolution of ODE states in time for the Nimotuzumab model from Eq. 1 with parameters $cl = 9.93 \times 10^{-4}$, $v_1 = 1.38$, $Q = 4.00 \times 10^{-3}$, $v_2 = 44$, $k_{\text{ss}} = 12.71$, $k_{\text{int}} = 3$, $k_{\text{syn}} = 1$ and $k_{\text{deg}} = 7$. There are ten dosing events, each adding 50 milliliters intravenously. Programmatically, this is done by adding the same amount to $C_{\text{tot}}(t)$ at each dosing event. During the first few dosing intervals, the concentration of the compound increases until the rate at which it is eliminated balances the rate at which the compound is added to the system. While $A_{\text{t}}(t)$ increases perpetually due to the bidirectional interplay between it and the compartments, *nearly* periodic behavior is observed in $C_{\text{tot}}(t)$ and $R_{\text{tot}}(t)$. Note that measurements are also taken after the final dosing event as $C(t)$ drops.

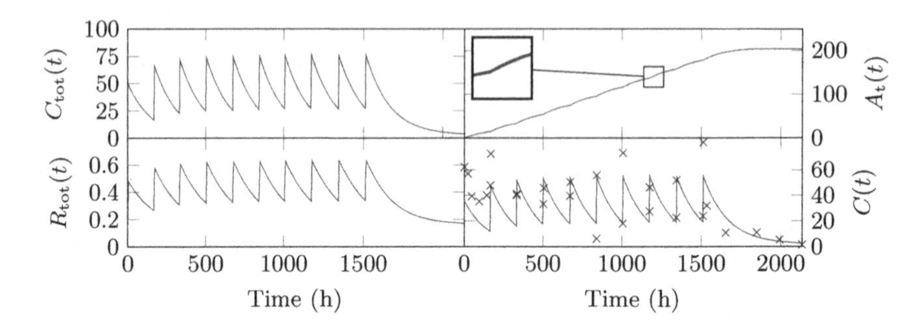

Fig. 1. The ODE states from the Nimotuzumab three-compartment model with ten dosing events. The state for $C_{\rm tot}(t)$, $A_{\rm t}(t)$ and $R_{\rm tot}(t)$ in function of time is shown on the left and top right, and the projected value $C(t)$ with observations shown as red crosses on the bottom right. After the first few dosing events, $C_{\rm tot}(t)$ and $R_{\rm tot}(t)$ exhibit close to periodic behavior. The plots were created by supplying a dense sequence of time points for x_j to Algorithm 1. The inset on $A_{\rm t}(t)$ is discussed in Sect. 5.

3.2 Canagliflozin Model

Canagliflozin is a drug for type-2 diabetes treatment. The model in Eq. 2 for this drug consists of both a PK and a PD portion. The former is modelled by a two-compartment model [9] denoted by the gut compartment $A_G(t)$, the central compartment $A_C(t)$ and the peripheral $A_P(t)$. Following Dunne et al. [4], the latter is captured by glycated haemoglobin (HbA1c) denoted by $H(t)$.

$$\begin{cases} \frac{\mathrm{d}A_{\rm G}(t)}{\mathrm{d}t} &= -k_{\rm a} \cdot A_{\rm G}(t) \\ \frac{\mathrm{d}A_{\rm C}(t)}{\mathrm{d}t} &= k_{\rm a} \cdot A_{\rm G}(t) - k_{23} \cdot A_{\rm C}(t) + k_{32} \cdot A_{\rm P}(t) - k_{\rm e} \cdot A_{\rm C}(t) \\ \frac{\mathrm{d}A_{\rm P}(t)}{\mathrm{d}t} &= k_{23} \cdot A_{\rm C}(t) - k_{32} \cdot A_{\rm P}(t) \\ \frac{\mathrm{d}H(t)}{\mathrm{d}t} &= k_{\rm in} + Ef - k_{\rm out} \cdot H(t) \\ C(t) &= A_{\rm C}(t)/v \\ Ef &= (Ef_{\rm c} + Ef_{\rm p})\frac{H(0)-5}{8-5} \\ Ef_{\rm c}(t) &= E_{\rm max}\frac{C(t)}{EC_{50}+C(t)} \end{cases} \tag{2}$$

For this model, $\phi = [k_{\rm out}, H(0), Ef_{\rm p}, EC_{50}, E_{\rm max}]$, where $Ef_{\rm p}$ represents the placebo effect, $k_{\rm in} = H(0) \cdot k_{\rm out}$, EC_{50} is the exposure that gives half-maximal effect and $E_{\rm max}$ is the maximal effect of the drug. The remaining parameters are fixed. A simulation with $k_{\rm out} = 10.24 \times 10^{-4}$, $H(0) = 7.72$, $Ef_{\rm p} = -0.482$, $EC_{50} = 60.34$ and $E_{\rm max} = -0.736$ is shown in Fig. 2. The remaining parameters are $k_{\rm a} = 3.86$, $k_{23} = 0.101$, $k_{32} = 0.0928$, $k_{\rm e} = 0.174$ and $v = 92.2260$. Similarly to Nimotuzumab, periodic behavior is observed for the PK portion.

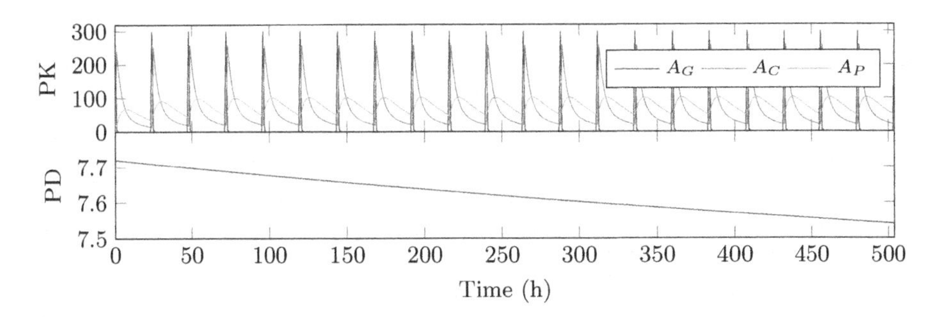

Fig. 2. Canagliflozin PK/PD model for the first 21 dosing events. Periodic behavior is observed after a few dosing events for the PK portion of the model shown at the top. The PD portion, shown at the bottom, does not stabilize.

4 Hierarchical Models

Pharmacometrics deals with models where the amount of available data is limited. Therefore, mixed effects models are used where data is grouped and structured into a hierarchy according to some classification [2]. The data considered in this paper is structured as shown in Eq. 3.

$$y_{ij} = f(x_{ij}, \phi_i) + \epsilon_{ij}, \quad i = 1, \ldots, M, \quad j = 1, \ldots, n_i \tag{3}$$

A one-way classification is used resulting in a hierarchy with two layers. The first layer represents the population as a whole, and the second layer consists of individuals. The number of individuals is denoted by M, each of which has n_i observations. The function f, parameterized by ϕ, describes the structural model exemplified by those from Sect. 3. As these models capture PK or PD behavior or both, x_{ij} will be the jth time point at which an observation was taken for the ith individual. The residuals $\epsilon_{ij} \sim \mathcal{N}(0, \sigma)$ account for the intra-individual variance. With a slight abuse of notation, the individual parameters ϕ_i are given by $\mu + \eta_i$ where $\eta_i \sim \mathcal{N}(0, \Omega)$ and Ω captures inter-individual variance. Here, η_i and μ are called the random and fixed effect respectively. While Eq. 3 only allows for additive error, its purpose is to be illustrative. It is important to note that the framework is general enough for other likelihood models as well. The goal is to estimate μ, Ω, and σ.

5 Approximating Models

In a model, states are classified either as periodic or non-periodic. Typically, the PK portion is periodic and the PD portion is non-periodic, but this need not be the case. In the integrated states, three phases are distinguished. The first phase spans over all dosing events for which the system has not yet entered periodicity. The second phase is the periodic phase typically taking up the majority of time in repeated dosing models as noted in Sect. 3. The start of this phase is detected based on a threshold τ that defines when a state is classified as periodic. The final phase starts at the last dosing event and ends at the last observation. In Fig. 1, depending on τ, the second phase could start at 500 h.

The goal is to avoid stopping and altering the state of the integrator to simulate dosing events since this increases execution time substantially. During the first interval of the second phase, all periodic states for the remaining observations are collected. The value of all non-periodic states is collected during the full length of the second phase by applying the method of averaging numerically.

In clinical trials, it is common to have dosage regimens where all dosing events add the same amount of a compound in the same way, i.e. $a_i = a_j$ and $c_i = c_j$ for any pair of dosing events i and j in Algorithm 1. However, it is possible to generalize the presented method where multiple runs of periodic behavior are observed. Since the models targeted in this paper only use dosage regimens with a fixed dosing amount, such extensions are left as future work. As will be shown in Sect. 6, the efficacy of the presented method depends on the time spent in periodic phases.

In reality, doses will never be spaced *exactly* uniformly throughout time. For example, one of the individuals in the Nimotuzumab data set with 10 dosing events, has the last dose administered at 1512.2 h after the start of the trial. The average dosing interval is thus approximately 168.02, but the dosing intervals for this individual are between 167.33 and 170.07. In case varying intervals are captured by the model, noise is added complicating periodicity detection. Therefore, a preprocessing step ensures that the events are spaced equally at the cost of potentially introducing some error in the final approximation.

If the mean time between doses is $\Delta t = t_{|D|}/(|D|-1)$, then the time for dosing event k is set to $t'_k = (k-1) \cdot \Delta t$. Next, each observation j is shifted according to the offset to the dosing event before it. Concisely, x_j is shifted to $x'_j = t'_k + z_j$, where z_j is computed as follows. If t_k denotes the time of the dosing event before it, then $z_j = \min(x_j - t_k, \Delta t - \epsilon)$. Here, capping the offset at $\Delta t - \epsilon$ ensures that the observation is not shifted to the next interval when it is close to the end since doing so introduces a large error due to the rapid rise in compound concentration after a dose. Figure 3 illustrates this process for an exaggerated example; as for the Nimotuzumab example shown above, the variance in dosing intervals for real use cases is typically much smaller. For models in which the dosing intervals are fixed, like for the Canagliflozin model, data need not be preprocessed.

After preprocessing, integration can start. For any model f, three different sets of equations S, S' and $\tilde{S}$ are used. Here, S is the original unaltered set of equations used during the first and third phase. During the first interval of the second phase, S' is used and the method of averaging is applied numerically during the remaining intervals in the second phase using $\tilde{S}$. The details of these sets of equations will be introduced next.

Integration commences on the set of equations $S = \{S_i(t)\}_1^q$ in f. At each dosing event k, all states in S are partitioned into r periodic states $P = \{P_i(t)\}_1^r$ and $q-r$ non-periodic states $N = \{N_i(t)\}_{r+1}^q$ by using some threshold τ and the

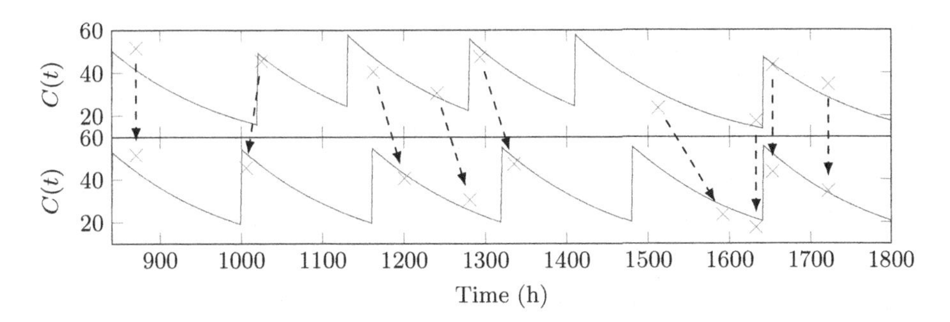

Fig. 3. Dosing events are shifted to ensure that each dosing interval is the same. All observations, shown as red crosses, associated with each dosing interval are shifted accordingly. The 7th observation is an example of an observation that, without capping, would be shifted to the next interval.

criteria $|(S_i(t'_k) - S_i(t'_{k-1}))/S_i(t'_k)| < \tau$. If $|P| > 0$, the state of the integrator $\mathcal{I}_{\text{real}}$ is copied to $\mathcal{I}_{\text{approx}}$. At this time, denoted by t_α below, the second phase is entered and integration continues using $\mathcal{I}_{\text{approx}}$.

During the first interval of the second phase, integration continues with S', a set of equations constructed by adding the equations $\frac{\mathrm{d}*}{\mathrm{d}P'_i(t)}t = P_i(t)$ to those in S for a total of $2|P|+|N|$ equations. The value of $P'_i(t_\alpha)$ is set to 0. These additional equations will be used to compute the average for use in the remaining intervals of the second phase. After one dosing interval, integration continues using $\tilde{S}$, constructed by taking the equations $\tilde{P} = \{\mathrm{d}\tilde{P}_i(t)/\mathrm{d}t = 0\}_1^r$ together with the states in N. The initial value for the states in $\tilde{P}$ is $P'_i(t_\alpha + \Delta t)/\Delta t$. In other words, the states in P are replaced by a constant equal to the mean value during a dosing interval. This is how the method of averaging is applied numerically. The values of the states in N are then collected during the second phase at each x'_j. Finally, at the last dose, integration continues using S restoring the state of the states in P to those saved in $\mathcal{I}_{\text{real}}$. The top left of Fig. 4 demonstrates when each of these sets is used.

The states of P during the second phase are collected at times $t_\alpha + z_j$ for all observations j for which $x'_j > t_\alpha$. Note that if integration can only continue forward in time, all z_j need to be sorted. This can be seen as moving observations to the first interval of the second phase. Figure 4 shows the output for the Nimotuzumab model from Fig. 1. Note that except for a different value of the integrated states, preprocessing and shifting of observations and events is not reflected in the output.

Let $c(t_0, t_1, S)$ denote the computational cost of using an integrator between time t_0 and t_1 on a set of equations S. The total cost of integration can be broken down into $c(0, t_\alpha, S)$, $c(t_\alpha, t_{\alpha+\Delta t}, S')$, $c(t_{\alpha+\Delta t}, t_{|D|}, \tilde{S})$ and $c(t_{|D|}, t_{n_i}, S)$. Since doses need not be simulated in $\tilde{S}$, $c(t_{\alpha+\Delta t}, t_{|D|}, \tilde{S}) \ll c(t_{\alpha+\Delta t}, t_{|D|}, S)$. Some overhead is introduced by preprocessing the data and using S' for one interval,

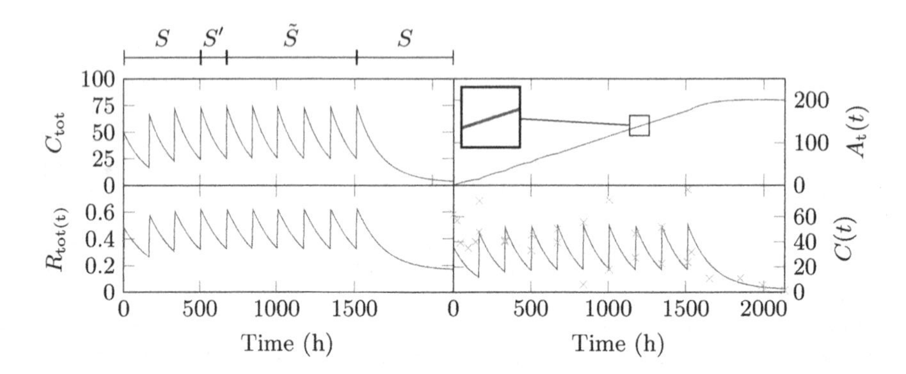

Fig. 4. Approximation of the Nimotuzumab three-compartment model with ten dosing events. Different sets of equations are used at different times. The sets are $S = P \cup N$, $S' = P' \cup N$ and $\tilde{S} = \tilde{P} \cup N$. These are only shown in the top left, but the change in equations effects all states. The choice for τ defines the phases. Here, the first phase spans $[0, 504]$, the second phase spans $[504, 1512.2]$ of which the first interval is $[504, 672]$ and the third phase starts at 1512.2. Compare all results with Fig. 1 and note how $A_t(t)$ is smoothed out due to applying the method of averaging numerically. However, preprocessing and event shifting happens transparently. The effect of approximation on the other states is barely visible.

but this is typically much smaller than the reduction in execution time obtained by avoiding simulation of doses between $t_{\alpha+\Delta t}$ and $t_{|D|}$.

Note that states in P are distinguished from those in N by τ. If τ is set too low, all states remain non-periodic and there is no second and third phase. In this case, no cost reduction will be made while some error will still be introduced by the preprocessing step. On the other hand, if all states are marked as periodic, then $c(t_\alpha + \Delta t, t_{|D|}, \tilde{S}) = 0$ since it can be skipped completely and larger cost reductions are expected. Note also that if all measurements after the last dose fall within a span of Δt, integration does not need to switch back to S from $\tilde{S}$.

A useful aspect of the outlined approach is that S' and $\tilde{S}$ can be constructed from S without symbolic manipulation. Integrator implementations require the user to provide a function that, given $S_i(t)$, returns a vector of which the ith component represents $\frac{\mathrm{d}*}{\mathrm{d}S_i(t)}t$. Multiplying this vector with the bit vector where all the components corresponding to states in N are set to 1 is a straightforward way to transform S into $\tilde{S}$.

6 Performance Evaluation

Test data is taken from an online resource [17] for the Nimotuzumab model and is generated synthetically for the Canagliflozin model using the parameter estimates from Dunne et al. [4]. The Stochastic Approximation Expectation Maximization (SAEM) algorithm from Kuhn et al. [10] is used to fit a complete hierarchical model, described in Sect. 4. It is difficult to obtain a clear understanding of how well the presented approximation performs by comparing SAEM directly. Instead, the SAEM algorithm is run on the real model and the parameters at which the likelihood is evaluated are logged. The CVODE solver from the SUNDIALS software package [8] is used as the integrator implementation.

The evaluation time together with the log-likelihood value of the classical approach from Algorithm 1 is measured for the collected parameters. The same is measured for the approximate model with different choices for τ. Figure 5 illustrate the influence of τ on both the relative error of the log-likelihood and the speedup between the real and the approximate model. For $\tau = 0$, no speedup is expected since no states will be classified as periodic. Since doses are shifted for the Nimotuzumab model, some error is still introduced. This is not the case for the Canagliflozin model as it does not take into account varying dosing intervals. In both models, the slowdown with $\tau = 0$ is due to computing and sorting z_j, and the additional bookkeeping that is needed to compare the value of each state with τ. Note the difference in speedup between the two models. The Canagliflozin data contains individuals with a much larger number of dosing events than those in the data for the Nimotuzumab model.

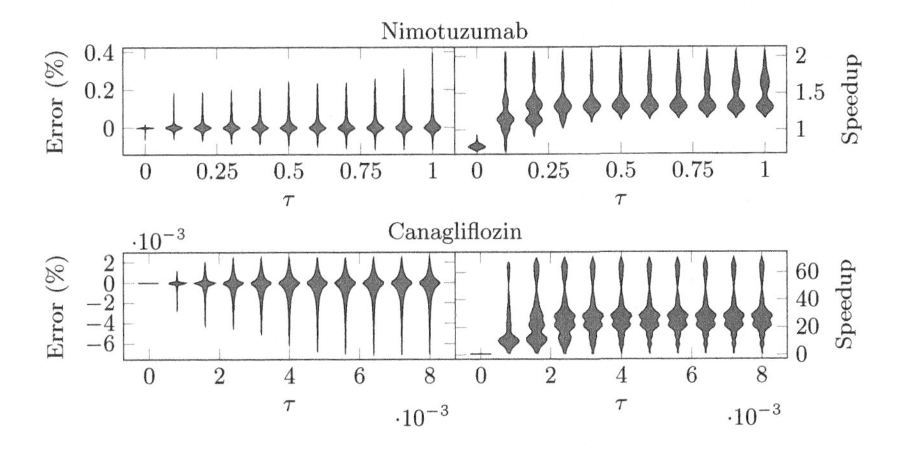

Fig. 5. Violin plots showing relative error and speedup as the threshold τ increases for the Nimotuzumab model at the top and for the Canagliflozin model at the bottom. A larger τ increases the probability of introducing a larger error. At the same time, a higher speedup factor is obtained. While both models show the same behavior as τ increases, there is a difference in scale of the error and τ due to a different number of dosing events in the data and structural differences between the models.

Next, data is generated synthetically with an increasing number of doses to show that the total time spent by the integrator in the second phase determines the improvements that can be obtained by using the approximate model. In Fig. 6, τ increases from 0 to 0.008, showing that with more dosing events, and hence more periodic behavior, a larger increase in performance is observed.

Recall from Sect. 4 that $\phi_i = \mu + \eta_i$. In algorithms like SAEM, one of the steps involves integrating out random effects η_i for a given individual. Due to

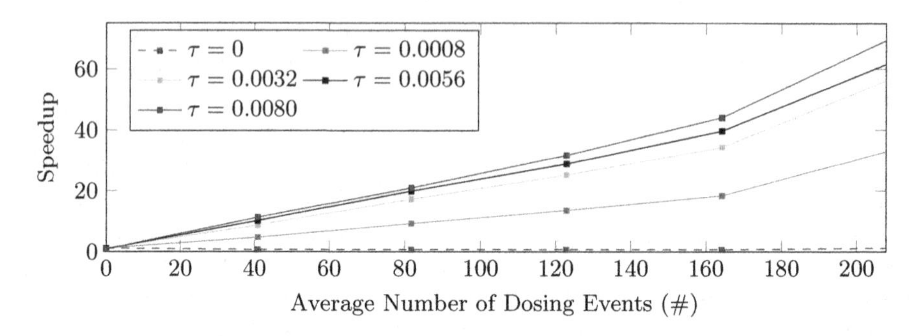

Fig. 6. Speedup for varying τ and varying number of observations for the Canagliflozin model. With more observations, the second phase makes up a larger fraction of the total execution time. Hence, there is a more opportunity to reduce execution time. Although not clearly visible, with $\tau = 0$, a slowdown of up to 25% is seen.

the complexity of the models, MCMC samplers are used. Using the approximate model directly in this step results in biased estimates as the introduced errors change the distribution of random effects. As shown above, through the choice of τ, accuracy is sacrificed for performance. Two ways are discussed to use the approximation without introducing bias. A function that weights both the accuracy and the performance aspects is given for each. The same function can then be used to tune τ automatically. While tuning brings with it some computational costs, estimating parameters of hierarchical models takes orders of magnitude longer so it is worth spending some time on the tuning process. The objective is to find a sufficiently good value for τ and not necessarily the optimum. Therefore, tuning can be done on a subset of individuals.

One way to use the approximation is with HMC. Here, new positions are proposed by following the gradient L times and performing an accept-reject step at the final position. If gradients are computed from the approximate model and the accept-reject relies on the real model, the samples obtained remain unbiased [15]. Note that in scenarios where L is large, larger reductions in execution time are possible. Since the gradients are only approximate, proposals will be of lower quality. For example, if the real and the approximate gradients differ too much, the proposed positions will have low mass and many points will be rejected. In turn, this lowers the effective sample size (ESS), a metric used to evaluate the information content of dependent samples. Tuning τ is accomplished by maximizing ESS per unit time. Figure 7 shows this metric for Canagliflozin using $L = 4$ while varying τ. Clearly, the optimal value for τ depends on the choice of L.

As noted above, generating samples directly with any MCMC sampler from the random effects distribution built with the approximate model will introduce bias due to the errors. Another way to use the approximation is through importance sampling, where bias is corrected by weighting each sample [12]. These weights, obtained by taking the ratio between the density of the real and the approximate model, can be computed in parallel. If there is too much difference

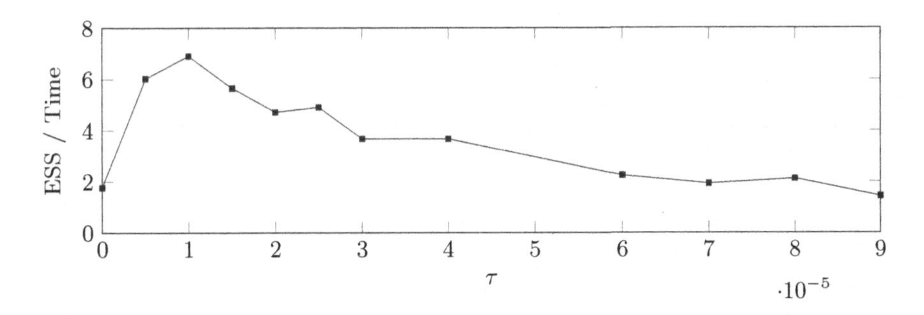

Fig. 7. Effective sample size per unit time while varying τ for the Canagliflozin model. This metric can be used to tune τ automatically.

between the importance distribution and the target distribution, expectations computed from samples will exhibit more variance, denoted by $\sigma_\mathcal{T}$. An estimator $\hat{\sigma}_\mathcal{T}$ is built by repeated sampling. A value for τ that trades off between computational efficiency and quality is chosen by minimizing $\hat{\sigma}_\tau$ while keeping time fixed. With multiple random effects, the covariance estimator $\hat{\Sigma}_\tau$ is used instead. Figure 8 shows this for the Nimotuzumab example. In this case, τ is tuned by minimizing $|\hat{\Sigma}_\tau|$, the determinant of the covariance matrix.

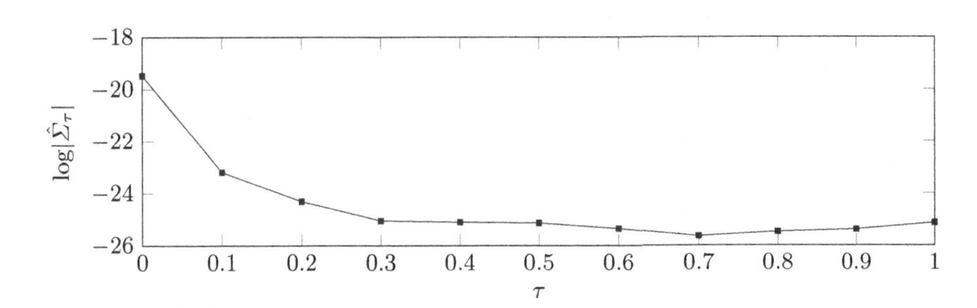

Fig. 8. The value of $\log|\hat{\Sigma}_\tau|$ in function of τ for the importance sampling estimator. By setting τ to 0.7, an appropriate trade-off between approximation accuracy and computational cost is made.

7 Conclusion and Future Work

This paper introduces an approximation of repeated administration models that exploits past computation efforts and employs the method of averaging numerically. In case of models with varying dosing intervals, a preprocessing step allows for detection of periodic behavior at the cost of adding some error to the approximation. The actual improvements vary depending on the model and the parameters of the model. On one of the test models, up to 70-fold reductions in run-time were measured while introducing only on the order of 10^{-3} relative error. Since fitting a hierarchical model can take up to hours or even days depending on the configuration parameters of algorithms like SAEM, these improvements have a tremendous impact on the end-users.

The approximation relies on setting the threshold τ to detect repetitive behavior in ODE states. It determines both the error and speedup of using the approximation instead of the real model. Incorporating a self-adjusting mechanism to automatically set τ for an MCMC sampler was discussed. Different objective functions can be devised depending on the use-case to tune τ, some of which will be studied in future work.

Speculative parallelism is a method to parallelize sequentially dependent tasks [7]. It has previously been applied to the classic Metropolis-Hastings MCMC sampler [1] where the sequence of accept-reject choices are guessed to predict the chain positions. Verification of these predictions then proceeds in parallel. A benefit of the speculative approach is that the collected samples are unaffected. Similarly, the approximation method presented in this paper can be applied to predict the chain, after which verification can occur in parallel. As in Sect. 6, it is again possible to tune τ. Here, τ trades off between the prediction accuracy and the time spent creating the prediction.

The choice of τ does not bound the error in the approximation. Tolerance bounds are typically already provided as parameters for numerical integration methods. Therefore, a promising direction of future work is to consider the change in integration results by entering the second phase one interval later.

Acknowledgments. The work presented in this paper was funded by Johnson & Johnson. The authors would like to thank Pieter Robyns and Nick Michiels for their valuable feedback on this work.

References

1. Angelino, E., Kohler, E., Waterland, A., Seltzer, M., Adams, R.P.: Accelerating MCMC via parallel predictive prefetching. arXiv preprint arXiv:1403.7265 (2014)
2. Carey, V.J., Wang, Y.-G.: Mixed-effects models in S and S-Plus (2001)
3. Conrad, P.R., Davis, A.D., Marzouk, Y.M., Pillai, N.S., Smith, A.: Parallel local approximation MCMC for expensive models. SIAM/ASA J. Uncertain. Quantif. **6**(1), 339–373 (2018)
4. de Winter, W., et al.: Dynamic population pharmacokinetic–pharmacodynamic modelling and simulation supports similar efficacy in glycosylated haemoglobin response with once or twice-daily dosing of canagliflozin. Br. J. Clin. Pharmacol. **83**(5), 1072–1081 (2017)
5. Dunne, A., et al.: The method of averaging applied to pharmacokinetic/pharmacodynamic indirect response models. J. Pharmacokinet. Pharmacodyn. **42**(4), 417–426 (2015)
6. Gong, W., Duan, Q.: An adaptive surrogate modeling-based sampling strategy for parameter optimization and distribution estimation (ASMO-PODE). Environ. Model. Softw. **95**, 61–75 (2017)

7. Grama, A., Karypis, G., Kumar, V., Gupta, A.: Introduction to Parallel Computing, 2nd edn. (2003)
8. Hindmarsh, A.C., et al.: SUNDIALS: suite of nonlinear and differential/algebraic equation solvers. ACM Trans. Math. Softw. (TOMS) **31**(3), 363–396 (2005)
9. Hoeben, E., De Winter, W., Neyens, M., Devineni, D., Vermeulen, A., Dunne, A.: Population pharmacokinetic modeling of canagliflozin in healthy volunteers and patients with type 2 diabetes mellitus. Clin. Pharmacokinet. **55**(2), 209–223 (2016)
10. Kuhn, E., Lavielle, M.: Coupling a stochastic approximation version of EM with an MCMC procedure. ESAIM Probab. Stat. **8**, 115–131 (2004)
11. Lions, J.-L., Maday, Y., Turinici, G.: A "parareal" in time discretization of pde's. Comptes Rendus de l' Académie des Sciences. Série I. Mathématique, p. 332, January 2001
12. MacKay, D.J.C., Mac Kay, D.J.C.: Information Theory, Inference and Learning Algorithms. Cambridge University Press, Cambridge (2003)
13. Minion, M.: A hybrid parareal spectral deferred corrections method. Commun. Appl. Math. Comput. Sci. **5**(2), 265–301 (2011)
14. Owen, J.S., Fiedler-Kelly, J.: Introduction to Population Pharmacokinetic/pharmacodynamic Analysis with Nonlinear Mixed Effects Models. Wiley, Hoboken (2014)
15. Rasmussen, C.E.: Gaussian processes to speed up hybrid Monte Carlo for expensive bayesian integrals. Bayesian Stat. **7**, 651–659 (2003)
16. Rodríguez-Vera, L., et al.: Semimechanistic model to characterize nonlinear pharmacokinetics of Nimotuzumab in patients with advanced breast cancer. J. Clin. Pharmacol. **55**(8), 888–898 (2015)
17. Trame, M.N.: Page 2018 nlmixr workshop materials (2018). https://github.com/nlmixrdevelopment/PAGE-2018

Evolutionary Optimization of Intruder Interception Plans for Mobile Robot Groups

Wojciech Turek, Agata Kubiczek, and Aleksander Byrski(✉)

AGH University of Science and Technology, Krakow, Poland
{wojciech.turek,olekb}@agh.edu.pl, agat.kubiczek@gmail.com

Abstract. The task of automated intruder detection and interception is often considered as a suitable application for groups of mobile robots. Realistic versions of the problem include representing uncertainty, which turns it into NP-hard optimization tasks. In this paper we define the problem of indoor intruder interception with probabilistic intruder motion model and uncertainty of intruder detection. We define a model for representing the problem and propose an algorithm for optimizing plans for groups of mobile robots patrolling the building. The proposed evolutionary multi-agent algorithm uses a novel representation of solutions. The algorithm has been evaluated using different problem sizes and compared with other methods.

Keywords: Intruder detection · Mobile robot motion planning · Evolutionary computing

1 Introduction

The problem of securing buildings from unauthorized intrusion is often considered a very suitable application for groups of mobile robots. Repetitive monitoring of a given area, systematic observation and possible interaction with hostile intruders fits perfectly the popular Dull-Dirty-Dangerous rule of robotization. This is one of few tasks where robots can be more efficient than humans, even when leaving financial expense aside.

These reasons have motivated many researchers to consider various variants of the problem over the last few decades. The comprehensive survey presented in [7] locates the background of the problem in operations research, graph theory, classical search theory and differential games. Existing approaches are grouped into two main categories: *pursuit-evasion* games and *probabilistic search*.

The *pursuit-evasion* games, including the well-recognized *cops and robbers* game [1], focus on finding the worst-case guarantee of finding an intruder in the specified environment. The assumptions typically include faultless motion and detection and the aim is to minimize the number of pursuers, which provide the guarantee.

J. M. F. Rodrigues et al. (Eds.): ICCS 2019, LNCS 11536, pp. 642–655, 2019.
https://doi.org/10.1007/978-3-030-22734-0_47

More realistic assumptions of detecting an intruder require using different methods and lead to probabilistic results rather than guarantees. The problems described as *probabilistic search* focus on modeling robots' sensors inaccuracy and uncertainty of intruder's decisions.

The variant of the intruder interception problem, presented in this paper, is considered an open problem in the domain. A group of mobile robots operates in a known, indoor environment. Each robot is able of detecting an intruder with limited probability. Intruder's decisions are uncertain, there is no guarantee of particular decision in specific circumstances. The environment can be equipped with static sensors, also imperfect. The aim is to reduce the probability of intruder's presence in the secured area.

In this paper we present the definition of the intruder interception problem, which represents realistic assumptions of intruders' unpredictability and inaccuracy of detection. We propose a novel approach to modeling of the problem, which represents crucial features of the problem. The model of intruder(s) was constructed based on a quasi-probabilistic measure spreading across the graph nodes (parts of building's model). The robots were dispatched based on redirection lists present in the nodes. Such redirection lists were used for defining an evolutionary algorithm (encoded in a dedicated genotype and evaluated by a predefined fitness function) for finding and optimizing plans for mobile robots groups. We evaluate the solution in different scenarios and compare the results with a reference, greedy algorithm. We also compare two variants of evolutionary algorithm, showing significant superiority of the Evolutionary Multi-Agent approach.

In the next section the excerpt from state-of-the-art regarding the intruder interception is provided. Next the problem is discussed and its novel model is presented. In the subsequent section the planning algorithm is shown and later the experimental evaluation is presented. Finally the paper is concluded, showing other applications of the proposed method.

2 Existing Approaches to the Intruder Interception Problem

The considered problem of indoor intruder interception using mobile robots is located within the wider area, often referred to as target detection and tracking. A comprehensive review of related work has been presented by Chung et al. in [7] and recently further extended by Robin et al. [14]. The area includes problems like target identification, tracking and following, however the most challenging open problems are related to finding moving targets [9].

The target detection and tracking taxonomy (fragment presented in Fig. 1) divides the mobile search problem into three categories. The firs one, *Capture*, is also often referred to as pursuit-evasion. Particular problems within this category share common aim: to ensure that no intruder was left in the secured area. Providing such guarantees forces strong assumptions concerning motion and detection model, moving the solutions far from reality.

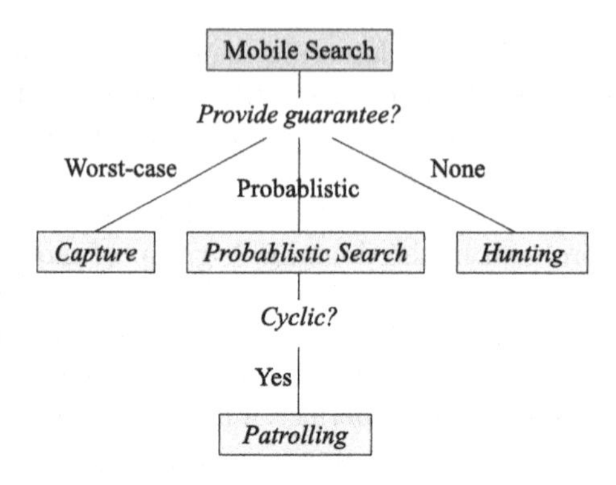

Fig. 1. Taxonomy of mobile search problems [14]

Typical approach to pursuit-evasion problem modelling is using graphs for representing environments. With rooms represented by vertexes and accessibility relation modelled by edges, the problem becomes a discrete planning task, often called the *graph-clear* [2]. In its basic form, each pursuer clears a node in which it is located. Pursuers must form a formation, which splits the graph into clear and potentially contaminated parts. Among various existing variants, the work presented in [11] is important in the context of our work. Besides the new variant of the model, which reflects additional features of reality, the authors show, that the problem of finding the minimal number of pursuers is NP-hard for general graphs.

Two other categories of mobile search problems, that is probabilistic search and hunting, do not provide guarantees and therefore focus on finding minimal number of robots needed. The task is defined as a planning problem, which aims at minimization of intruder presence probability over time, with many variants, of course. Typically more realistic models of the environment, the intruders and perception are considered. In [8] the environment is represented as a grid, and the intruder is modelled using the partially observable Markov decision process. In [13] a particle filter is used to represent possible locations of intruder; the aim is to maximize of number of visible particles while potential fields are used to control the pursuers. An example of optimization is presented in [15], where local optimization of trajectories is used. The common feature of the considered problems is the exponential complexity which prevents from finding global optimum.

Analysis of the existing approaches shows that among the three major problem definition dimensions (environment model, target motion model, pursuer sensing model) the environment model implies most consequences. The dependency between the used environment model and problem complexity has been discussed in [10], where the need for simplifying the model is underlined. Most of existing approaches use metric maps or grid representations, although graph-based models are typically far more efficient. However, automated transformations of metric data into a meaningful topological model is not straightforward.

The solution proposed in this paper uses our previously proposed method [16]. It generates a list of spaces from a provided list of walls, identifies passages between spaces and generates a graph. The graph lacks semantic information, however it can be automatically enriched with various metric data derived from the features of the spaces.

According to [14], further investigation is needed in the area of realistic models and their efficient processing. Also focus on sub-optimal solutions is expected due to the nature of the problem. The variant of the intruder interception problem, presented in this paper, is located within these areas.

3 Problem Definition and Model

In order to define precisely the problem considered in this paper, let us assume that there is a known building. The topological map of the building is known, it is composed of rooms and passages between rooms. It can be created manually or by using methods presented in [16]. In the normal situation nobody incidental is present (no human security), only the robots and potential intruder(s). In the case of intruder(s) trespassing, he should be quickly intercepted – and this is the task of the robot-security group.

Certain knowledge regarding the presence of intruders in a particular rooms can become available at any time – e.g. derived from a set of static sensors, present in strategic places of the building. Alternatively, the information about possible intrusion can be deduced from the structure of a building, assuming rooms with windows on the ground floor more prone.

Certain knowledge about possible movements of the intruder can also be available. The intruder's velocity, aims and decisions can be considered: unknown, random, deliberative or adversarial. The knowledge about intruder's decisions have probabilistic character – only the probability of certain actions can be predicted.

The group of robots can be located in one place or spread around the building. Robots can be heterogeneous, differing in velocity and perception capabilities. Each robot is able to detect an intruder located in the same room, with a certain probability, as the intruder can be overlooked.

The planning task for the group of robots is centralized. The goals of the planning is to:

- plan the movement of the robots in such way, that the probability of intruders' presence is minimized in the shortest possible time,
- adopt to the detected changes and new information when one is available – the system should work continuously.

With these assumptions, let us model the building as a graph, $G = (V, E)$, where V is a set of labels of the nodes. $V = \{v_1, \ldots, v_k\}, k \in \mathbb{N}$ (k is the maximum number of vertices). Each node is associated with one room, while one room is represented by d nodes, where d is the number of passages of the room. So a node can be interpreted as one passage (door) connecting a room with other

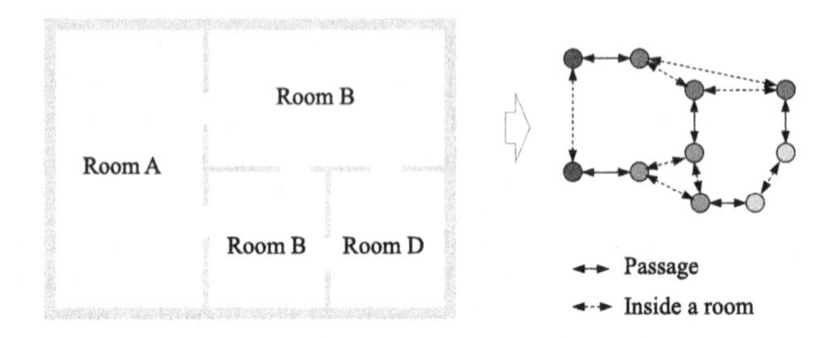

Fig. 2. Example of the proposed graph model of a building

rooms (see Fig. 2). Each passage is represented by two nodes, one in each of the connected rooms.

E is the set of edges $E = \{e_{1,2}, e_{1,3}, \ldots, e_{j,k}\}, i, j \in \mathbb{N}, \{i, j\} \in \{1, \ldots, k\}^2$, $i \neq j$. Typically there are edges fully connecting all nodes associated with one room. Additional edges represent passages – these can be directional. Each edge has a weight, which is proportional to the distance (or travel time) between passages represented by the nodes.

By using this modelling method, a complex building can be represented as a weighted graph, representing the topology of rooms and passages, enriched with distances between passages within the rooms. It allows finding paths and storing additional information about model state (information about robots and intruders).

Let us define a probabilistic model of an intruder, which can be located in a certain room, by defining a quasi-probabilistic measure of intruder's presence in a certain room $\mathbb{R} \ni x_n \geq 0$. This value can exceed 1 (so it is not a probability per se), and can be treated as a certain estimate of a number of intruders present at a certain node of the graph.

To define the dynamics of intruders and robots, let us assume the existence of discrete time units. In each time unit an intruder can stay at the current graph node n or start moving to an adjacent node. The decisions are driven by intruder's model, a certain probability p_i. The presence measure changes in each of the nodes according to the following equation:

$$x'_n \leftarrow x_n - \sum_{i=1}^{m} p_i x_n + \sum_{i=1}^{m} r_i \tag{1}$$

where x'_n is the presence measure in the next time unit (short for $x_n(t+1)$ as the next observed value of $x_n(t)$) and p_i is the probability of moving of an intruder from n-th to i-th node. $\mathbb{R}^+ \ni r_i \geq 0$ is the presence metric value of an intruder approaching from the i-th, which has been subtracted from the x_i value k time steps earlier (k depends on the edge length between i and n and the intruder's velocity).

The values of intruder presence measures are computed for each of the nodes in each observable moment. Thus overall we get a certain quasi-probabilistic measure of the presence of intruders in the graph, that gets "distributed" through its nodes and edges according to the Eq. 1.

When the robot comes to visit certain room (any of the room's nodes), the intruder presence measure in the room (nodes and edges) is decreased according to the following equation:

$$x'_n \leftarrow x_n \cdot (1 - p_d) \quad (2)$$

where p_d is the probability of detecting of the intruder by the robot.

4 Planning Algorithm

Based on the graph depicting the building, defined in the previous section, the planning algorithm is constructed. Note, that the graph represents presence measure of one or more intruders, and this presence distribution changes over time. At each observable time moment, the planning algorithm can assess the place of the most probable spotting of the intruder. Here two algorithms will be presented, the former is very easy though efficient, and will be used as a reference one, while the latter is one of tangible results of the research presented.

4.1 Greedy Reference Algorithm

Assuming $R = \{r_1, \ldots, r_k\}$ the set of robot labels, a simple greedy algorithm, assigning the robots, present in the node v_x, to the next nodes may be proposed. The algorithm takes into consideration the number of already realized robot visits in the neighboring nodes:

```
function AssignRobotsToNextNodes(v_x, map)
    neighbors ← getNeighborsOfNode(v_x, map)
    sortAccordingToAscendingVisitCount(neighbors)
    nextNode ← getFirst(neighbors)
    incrementVisitCount(nextNode)
    return nextNode
end function
```

thus the robots present in the node v_x will be assigned with the next target – one of the neighboring nodes, while the less visited nodes will be prioritized.

Note that the proposed algorithm uses the local knowledge present in the graph to dispatch the robots, and although it is centralized, it works very efficiently, even for relatively large graphs. Such algorithm must be used instead of deterministic, "brute-force" planning as the considered problem apparently belongs to NP class (it is quite similar to VRP problems).

Locally-focused algorithms are prone to stucking in local extrema, therefore being aware that such algorithm can deliver certain (and easy to attain), useful, sub-optimal result, we might turn towards using state-of-the-art global optimization algorithm to make these results better.

4.2 Plan Optimization

Finding satisfactory solutions to the considered, *NP*-hard planning problem justifies using general-purpose global-optimization metaheuristics, like the well-known evolutionary algorithms. It is to note that some of such algorithms were subjected to detailed formal analysis, showing the proof that they will be able to reach the solution, wherever it is in the search space [3,17]. Every possible evolutionary algorithm (and many other metaheuristics), in order to solve a problem, require a proper definition (and encoding) of a problem, construction of the fitness function in the problem domain and construction of the variation operators. The main novelty of the proposed approach is the solution definition and encoding, which does not focus on paths of particular robots, but rather represents routes to cover.

Let us introduce the concept of a dispatch list $s \in S$, which is a list of node labels of a certain length dl. A valid dispatch list for node v can contain only labels of nodes adjacent to v and the label of v. A valid dispatch list can contain duplicates of node labels. The dispatch list encodes consecutive redirections of robots entering the node; once the list is finished, dispatching starts from the first item again.

The search space S includes all the possible dispatch lists of a certain length dl for a given graph (of n nodes):

$$S = V^{dl \cdot n} \tag{3}$$

a vector of dispatch lists, belonging to such space, can be depicted as:

$$S \ni s = (d_1, d_2, \dots, d_n) \tag{4}$$

where

$$d_i = (v_r, \dots, v_s) \tag{5}$$

where $||d_i|| = dl$ and $(v_r, \dots, v_s)$ are all adjacent to v_i or equal v_i.

As an example, imagine we have simple, full graph consisting of four nodes: V = {A, B, C, D}. Considering dispatch lists of the length equal to 3, one of possible solutions of the problem can look as follows:

$$((A, B, C), (D, C, B), (B, C, A), (D, C, A)) \tag{6}$$

This sample solution will be interpreted as a sequence of orders: "When the first robot arrives to node A, send it again to node A, then send the second one to B. When the first robot arrives to node B, send it to node D, then send the second one to C, the third one to B and the fourth one again to D", and so on.

The goal is to find such dispatch list, that the intruder will be localized in the shortest possible time. Coming back to the definition of the presence measure, we can construct the fitness function as a simple sum of the presence measure over the whole graph after a fixed time:

$$f = \sum_{i=1}^{n} x_i \tag{7}$$

and check this value after certain, predefined time of dispatching the robots according to the current schedule. Thus the evaluation function is defined. Computing the fitness function value requires performing the simulation of the process: intruder spreading and robots movements.

The remaining steps are crossover and mutation. The representation treated as dispatch lists for all the nodes may be very easily subjected to any crossover working with combinatorial problems, e.g. discrete crossover. So the offspring will randomly inherit subsequent elements of the dispatch list either from one or another parent, e.g. let us focus only on the dispatch list for two parents and offspring:

$$crossover((\mathbf{B}, \mathbf{D}, B), (D, C, \mathbf{B})) \rightarrow (B, D, B) \tag{8}$$

assuming that these two dispatch lists belong to the same node (A), and this node is connected to all other nodes.

The mutation is also very simple. Focusing on the dispatch list for a certain node, its elements can be randomly altered:

$$mutation(D, \mathbf{C}, B) \rightarrow (D, B, B) \tag{9}$$

again the considered node A is connected with all other ones.

The presented approach has been implemented and tested using two evolutionary metaheuristics: an Evolutionary Algorithm and an Evolutionary Multi-Agent System.

4.3 Evolutionary Algorithm

The first type of evolutionary algorithms, namely genetic algorithm was proposed by Holland and updated by Rechenberg, Schwefel, Fogel, Michalewicz and many others [18]. This optimization metaheuristic requires using a dedicated encoding of a problem as a genotype vector, a random construction of a population of individuals (a set of such vectors) and submitting this population to a cycle of operations, which would process the population according to evolutionary inspiration by Darwin's theory of evolution. Holland's original algorithm used binary encoding and proportional selection. There do exist many other popular encodings, e.g. real-valued, discrete, tree-based and many more (relevant to particular problems to be solved).

In the presented solution we are using an Evolutionary Algorithm (similar to the one proposed by Michalewicz [12]). It consists in realization of the following sequence of operations: (1) random initialization of a population (minding the constraints of the problem), (2) evaluation of the individuals by running the simulation of intruders and robots, (3) selection of a mating pool, (4) generation of the next population by means of crossover i.e. creation of new individuals based on randomly selected parents, (5) introduction of random aberrations to the solutions of the new population (mutation). At this point the algorithm returns to (2). The stopping condition of the loop is specified by the limit of real computations time.

4.4 Evolutionary Multi-agent System

Another evolutionary-type algorithm used in the experiments presented in this paper is the Evolutionary Multi Agent-System [6]. This metaheuristic has a firm formal background proving its correctness [3]. It may be treated as an approach (proposed first by Cetnarowicz in 1996 and developed significantly since then) to put together autonomy of agent-based systems with the optimization capabilities of evolutionary-inspired metaheuristics.

Agents in EMAS represent solutions to a given optimization problem. They are located on islands representing distributed structure of computation. The islands constitute local environments, where direct interactions among agents may take place. In addition, agents are able to change their location, which makes it possible to exchange information and resources all over the system.

In EMAS, phenomena of inheritance and selection – the main components of evolutionary processes – are modeled via agent actions of *death* and *reproduction*. As in the case of classical evolutionary algorithms, inheritance is accomplished by an appropriate definition of reproduction. Core properties of the agent are encoded in its genotype and inherited from its parent(s) with the use of variation operators (mutation and recombination). Moreover, an agent may possess some knowledge acquired during its life, which is not inherited. Both inherited and acquired information (phenotype) determines the behavior of an agent. It is noteworthy that it is easy to add mechanisms of diversity enhancement, such as allotropic speciation (cf. [5]) to EMAS. It consists in introducing population decomposition and a new action of the agent based on moving from one evolutionary island to another.

Many optimization tasks, which have already been solved with EMAS and its modifications, have yielded better results than certain classical approaches. They include, among others, optimization of neural network architecture, multi-objective optimization, multimodal optimization and financial optimization. EMAS has thus been proved to be a versatile optimization mechanism in practical situations. A summary of EMAS-related review is given in [4].

5 Experiments and Results Evaluation

In order to evaluate the two algorithms implementing the presented plan optimization method, a series of experiments was carried out. Three different building models have been used in the tests. The first one was a ring of 8 rooms, each having 2 doors. The second (Fig. 3) was far more complex, with four stories and multiple rooms. The last one, presented in Fig. 4, is a model of a real, existing building.

The experiments were performed on a single computer with 4 core CPU and 12 GB of RAM. Each test case has been run 30 times. The diagrams present aggregated results of all launches. Each case is represented by the series of data: the best result obtained in any launch for the given fitness evaluations count, the worst result for the given evaluations count, and the average value for all launches.

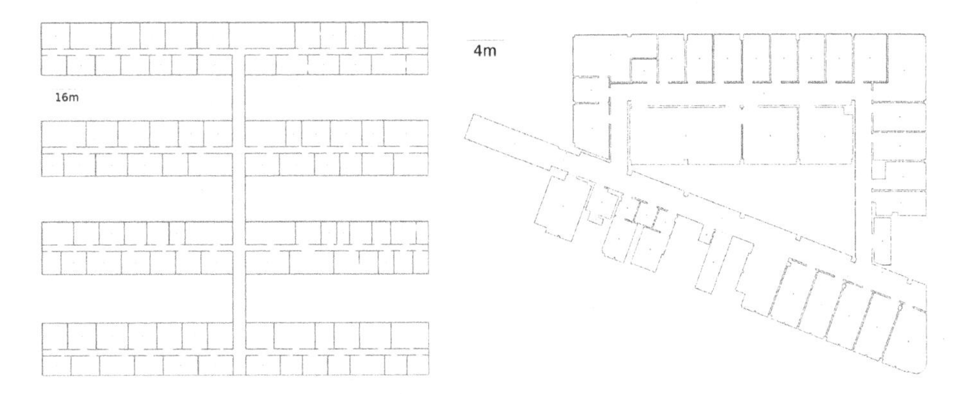

Fig. 3. Large, multi-storey building

Fig. 4. Real-life building

5.1 Test Cases

The purpose of the experiments was to evaluate the performance of the algorithm depending on parameters configuration and the test building structure. Four test cases have been proposed. The first one includes simple building scheme and its goal is to compare the results of two different evolutionary algorithms and a greedy one for the same parameters configurations. Another two test cases involve the real-life building and were created to examine the effect configuration has on evolutionary algorithm performance. Two parameters have been analyzed - dispatch list's length and population size. The last test case involves complex multi-storey building. It examines how number of robots exploring the building affects the results of evolutionary and greedy algorithm.

5.2 Results

The results of the simple building experiment are presented in Fig. 5. Separate series illustrate the fitness values for EMAS and for conventional evolutionary algorithms. Constant value represents the greedy algorithm result.

Both evolutionary algorithms produce much better results than the greedy approach and is this case the fitness reduces to value close to zero. That means the solution close to the optimum could be achieved. At first the conventional EA produced better results but eventually the best results come from EMAS. Moreover, the EMAS offers much better results for worst-case scenario (comparing worst results achieved in any launch for the given fitness evaluations count).

The experiments involving real-life building were designed to examine the influence of configuration on the results. In the first case the dispatch list's length was analyzed. The results are presented in Fig. 6, for better readability only average series were included. It can be observed that there is a dispatch list length which yield the best solutions. For the analyzed building list of 10–12 elements seems optimal. Increasing the length of the list improves the results to

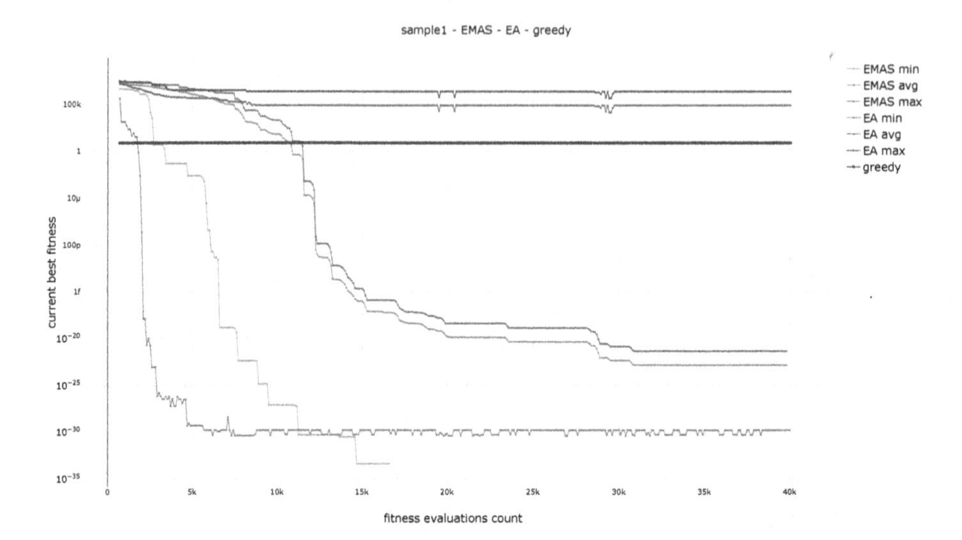

Fig. 5. Results of both evolutionary and greedy algorithms for the small building case

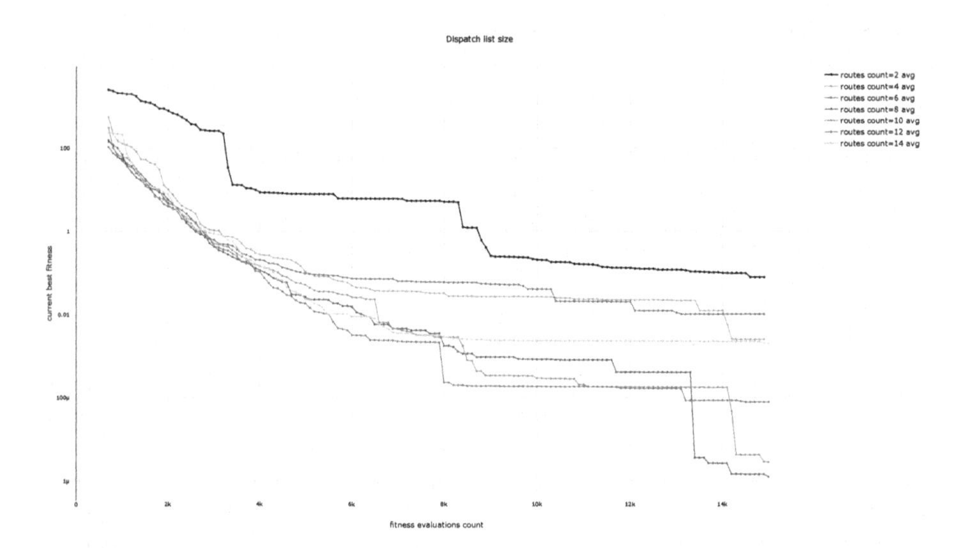

Fig. 6. Evolutionary algorithm results for real-life building calculated for various dispatch list lengths

some extent, as the solution space extends. However, when the list is too long the last elements of solution are never used and they impede the computation.

The second real-life scenario involved examining the population size's impact on the performance. In order to obtain best results, the dispatch list size has been set to the value determined in previous experiment. The results are presented in Fig. 7 - as before, only average series are included. Similar to the dispatch

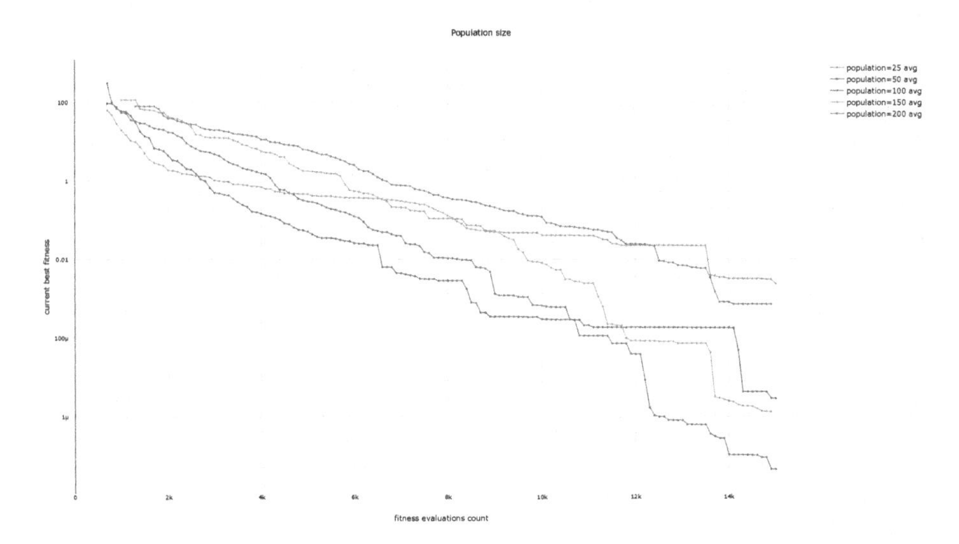

Fig. 7. Evolutionary algorithm results for real-life building calculated for various population sizes

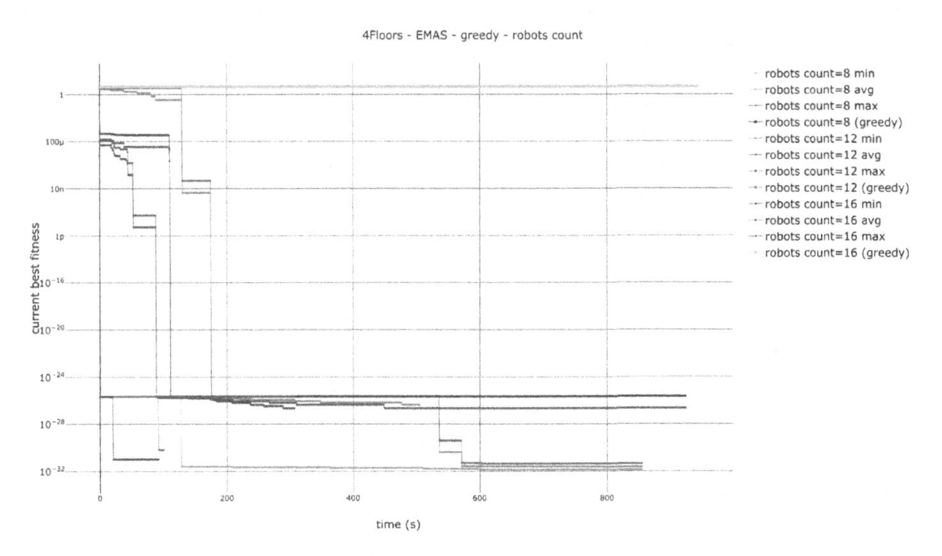

Fig. 8. Results of EMAS and greedy algorithms calculated for multi-storey building and various robots count

list size, an optimum population size can be determined and for this case equals 100 individuals. If the population is too small it becomes less diverse and this can cause stopping the algorithm in local optimum. However, too big population implies more fitness evaluations and therefore fewer epochs of evolution.

The last experiment examined the influence of number of robots exploring the building on the algorithms' performance. The results are presented in Fig. 8.

This test case involved complex multi-storey building and comparison of EMAS and greedy approach results. As can be expected, for both methods performance improved with the increasing robots' count, however the improvement in case of the greedy algorithm is hardly visible, also due to logarithmic scale on the vertical axis. The EMAS algorithm found far better solutions and made use of more numerous groups of robots. Larger number of robots allowed finding good solutions faster, which is visible in the first 200 s of optimization.

6 Conclusions and Further Work

In this paper a novel approach to planning of intruder interception by a group of mobile robots is proposed. A dedicated intruder model has been constructed, based on a quasi-probabilistic presence measure. Actual plan of the mobile robots movement was constructed based on redirection lists distributed among the nodes of the graph. These lists became parts of directly encoded genotype, subjected to a process of evolutionary optimization.

The evolutionary approach to solving this problem (based on EA and EMAS algorithms) were compared with a dedicated (very natural) greedy algorithm and it turned out that the evolutionary approach (in particular EMAS) prevailed significantly, being able to lower the presence measure in the whole graph to the greatest extent.

The proposed method of intruder localization by the means of multiple robots turns out to be quite similar to VRP or MTSP problems. In future we are planning to test existing VRP/MTSP benchmarks in order to prove the applicability of the proposed model to those very important hard computation problems, showing the generality of the proposed approach.

Acknowledgments. The research presented in this paper was partially supported by the funds of Polish Ministry of Science and Higher Education assigned to AGH University of Science and Technology.

References

1. Aigner, M., Fromme, M.: A game of cops and robbers. Discrete Appl. Math. **8**(1), 1–12 (1984)
2. Alspach, B.: Searching and sweeping graphs: a brief survey. Le matematiche **59**(1–2), 5–37 (2004)
3. Byrski, A., Schaefer, R., Smołka, M.: Asymptotic guarantee of success for multi-agent memetic systems. Bull. Pol. Acad. Sci. Tech. Sci. **61**(1) (2013)
4. Byrski, A., Drezewski, R., Siwik, L., Kisiel-Dorohinicki, M.: Evolutionary multi-agent systems. Know. Eng. Rev. **30**(2), 171–186 (2015)
5. Cantú-Paz, E.: A survey of parallel genetic algorithms. Calculateurs Paralleles, Reseaux et Systems Repartis **10**(2), 141–171 (1998)
6. Cetnarowicz, K., Kisiel-Dorohinicki, M., Nawarecki, E.: The application of evolution process in multi-agent world (MAW) to the prediction system. In: Proceedings of the 2nd International Conference on Multi-Agent Systems (ICMAS 1996). AAAI Press (1996)

7. Chung, T.H., Hollinger, G.A., Isler, V.: Search and pursuit-evasion in mobile robotics. Auton. Rob. **31**(4), 299 (2011)
8. Ferrari, S., Fierro, R., Tolic, D.: A geometric optimization approach to tracking maneuvering targets using a heterogeneous mobile sensor network. In: Proceedings of the 48th IEEE Conference on Decision and Control (CDC), pp. 1080–1087 (2009)
9. Garg, V., Tiwari, R.: A chronological review of the approaches used for multi robot navigation. In: International Conference on Recent Trends in Engineering, Science Technology - (ICRTEST 2016), pp. 1–8, October 2016
10. Hollinger, G., Singh, S., Djugash, J., Kehagias, A.: Efficient multi-robot search for a moving target. Int. J. Rob. Res. **28**(2), 201–219 (2009)
11. Kolling, A., Carpin, S.: Pursuit-evasion on trees by robot teams. IEEE Trans. Rob. **26**(1), 32–47 (2010)
12. Michalewicz, Z.: Genetic Algorithms + Data Structures = Evolution Programs. Springer, Heidelberg (1996). https://doi.org/10.1007/978-3-662-03315-9
13. Mottaghi, R., Vaughan, R.: An integrated particle filter and potential field method applied to cooperative multi-robot target tracking. Auton. Rob. **23**(1), 19–35 (2007)
14. Robin, C., Lacroix, S.: Multi-robot target detection and tracking: taxonomy and survey. Auton. Rob. **40**(4), 729–760 (2016)
15. Sarmiento, A., Murrieta-Cid, R., Hutchinson, S.: An efficient motion strategy to compute expected-time locally optimal continuous search paths in known environments. Adv. Rob. **23**(12–13), 1533–1560 (2009)
16. Turek, W., Cetnarowicz, K., Multan, M., Sośnicki, T., Borkowski, A.: Modeling buildings in the context of mobile robotics. Research Reports of Warsaw University of Technology. Electronics (194), 165–174 (2014)
17. Vose, M.D.: The Simple Genetic Algorithm: Foundations and Theory. MIT Press, Cambridge (1998)
18. Whitley, D., Sutton, A.M.: Genetic algorithms - a survey of models and methods. In: Rozenberg, G., Bäck, T., Kok, J.N. (eds.) Handbook of Natural Computing, pp. 637–671. Springer, Heidelberg (2012). https://doi.org/10.1007/978-3-540-92910-9_21

Author Index